The University of Wisconsin

PUBLICATIONS IN MEDIEVAL SCIENCE

PUBLICATIONS IN MEDIEVAL SCIENCE

1

The Medieval Science of Weights (Scientia de ponderibus): Treatises Ascribed to Euclid, Archimedes, Thabit ibn Qurra, Jordanus de Nemore, and Blasius of Parma.
Edited by Ernest A. Moody and Marshall Clagett.

2

Thomas of Bradwardine: His "Tractatus de proportionibus." Its Significance for the Development of Mathematical Physics.
Edited and translated by H. Lamar Crosby, Jr.

3

William Heytesbury: Medieval Logic and the Rise of Mathematical Physics.
By Curtis Wilson.

4

The Science of Mechanics in the Middle Ages.
By Marshall Clagett.

The Science of Mechanics in the Middle Ages

THE
Science of Mechanics in the Middle Ages

MARSHALL CLAGETT

MADISON, 1959

The University of Wisconsin Press

LONDON — OXFORD UNIVERSITY PRESS

Published by The University of Wisconsin Press
430 Sterling Court, Madison 6, Wisconsin

Published in The United Kingdom and Commonwealth
(excluding Canada) by Oxford University Press

Copyright © 1959 by Marshall Clagett

Distributed in Canada by Burns and MacEachern, Toronto

Printed in The Netherlands by N. V. Drukkerij G. J. Thieme, Nijmegen

Library of Congress Catalog Card Number 59-5309

To
Alexandre Koyré,
Anneliese Maier,
AND Ernest Moody
Historians of Early Mechanics

Preface

THIS study constitutes the fourth volume in the University of Wisconsin's *Publications in Medieval Science*. In a sense it complements and extends the subject matter of the first three volumes, while still keeping the same major purpose of presenting documentary material on which to base the study of medieval science. Its objective is somewhat wider, since it attempts to make available material useful for the evaluation of a larger area of medieval mechanics than does any of the former volumes. It is hoped that it will fill a lacuna in the literature of early mechanics—particularly in English. It hardly need be said that its preparation would have been impossible without the assistance of many prior studies in mechanics and much encouragement and financial aid from institutions and individuals alike.

It is difficult to know where to begin in acknowledging the assistance I have received over the past years in the preparation of this volume. Something from each of my various European trips has gone into it—even when the principal objective of the trip was the preparation of some other study in medieval science, since on each trip I enriched my collection of films of medieval physical and mathematical manuscripts. Of the many institutions who have so generously supported my research I must mention as prior in point of time Columbia University, for a fellowship grant in 1939 that allowed me to come in contact for the first time with the manuscript sources on which so much of this work is based. Following the war, I received assistance from the College of Letters and Science of the University of Wisconsin and the Research Committee of the Graduate School of the University of Wisconsin. And accompanying this assistance has been the continuing encouragement of the Dean of the College of Letters and Science, Mark Ingraham, whose general interest in the history of science and particular interest in my own work has never flagged. Similar encouragement has come from President Conrad Elvehjem–formerly Dean of the Graduate School–whose paramount interest in modern science has never blinded him to the importance of uncovering the early backgrounds to that science.

In addition to this University support, support has also come from private foundations. I can only inadequately thank the John Simon Guggenheim Memorial Foundation for two fellowships that allowed me time and money to pursue my studies. Most recently I owe thanks to the National Science Foundation of the United States for a very generous grant which allowed me a profitable year of manuscript study in 1955–56. The completion of this volume would have been considerably delayed without that grant. Nor should I omit acknowledging with thanks the travel assistance given me by the American Philosophical Society that made one of my visits to Europe possible. Furthermore a year's leave at the Institute for Advanced Study at Princeton made possible by a grant from the Institute and one from the Research Committee of the University of Wisconsin has allowed me freedom to guide the volume through the last stages of publication.

Turning from institutions to individuals, I must initially express my debt to three eminent historians of mechanics—Professor Alexandre Koyré, of the Institute for Advanced Study at Princeton and the École Pratique des Hautes Études of Paris, with whom I have in recent years had such a warm and intellectually profitable relationship, and to whom I put the final test of friendship—the reading of this manuscript. To Dr. Anneliese Maier of the Vatican Library, whose careful and penetrating investigations must be the point of departure for any study of medieval mechanics, as they have been for mine. And to my friend and erstwhile collaborator Professor Ernest Moody of the University of California at Los Angeles, from whom I have learned so much over the past years and who also read this work in typescript. Furthermore, Professor Moody allowed me to include one of his own translations in Chapters 4 and 5 of this study. I have also greatly profited from studying the works of my master, Lynn Thorndike, Professor Emeritus, Columbia University, and those of my friends Professor E. J. Dijksterhuis of the Universities of Leiden and Utrecht and Dr. Alistair Crombie of Oxford University.

My work was also read in typescript by Professor I. B. Cohen of Harvard University, Professor Curtis Wilson of St. John's College, and Dr. John Murdoch of Harvard University. It was read in both typescript and galley proof by Father James A. Weisheipl, O.P., the House of Dominican

[Preface xi]

Studies, River Forest, Illinois. All of these friends took time away from heavy academic duties to read this volume. Many of their suggestions have been incorporated into the finished product, although—needless to say—on none of these kind people is to be put blame for any of the defects that still remain. I must also thank my colleague Professor Aaron Ihde of the Departments of Chemistry and the History of Science for reading Chapter 2, Dr. Michael Hoskin of Leicester University for reading parts of Chapters 4 and 5, and Mr. Stillman Drake for reading parts of Chapters 1 and 2. Parts of Chapter 6 benefited from the work of my former students, Drs. Edward Grant of Harvard University and Tom Smith of the University of Oklahoma.

I cannot leave this preface without one final acknowledgment, to our departmental secretary Mrs. Loretta Freiling, who has so patiently typed and retyped this long and complex book.

ERRATA

Page 72: note 6, line 2. For Vol. *1* read Vol. 2.

Page 232: line 10. For 22 read 24.

Page 250: second line of variant readings for line 40. For B read *B*.

Page 319: in variant readings for line 54. For BD:D *BG* read D:BD *BG*.

Page 439: line 4. For (F/R' read (F/R).

Page 513: note 14, line 2. For 456 read 426.

Page 521: note 33, line 16, insert comma after *movet*. Line 17, delete comma after *potest*. Line 20, for *eum* read *cum*.

Page 525: line 6. After "impetus" insert "(see passage 6)."

Page 538: lines 2 and 7. For chord read cord.

Page 590: line 11. For position read positions.

Page 608: line 26. After "philosophy" insert "was serious."

Page 609: column 1, line two from bottom. After *nisi* insert *velocius* and delete [amplius].

Page 648: note 55. For Vol. *1* read Vol. 2.

Page 651: line 13. For 1400 read 1401.

Page 680: line 9. For 633 read 663.

Page 682: line 1. For 581 read 587.

Acknowledgments

I WISH to thank the following publishers for permission to make use of portions of the books indicated. Full bibliographical citations are given in the notes.

Abelard-Schuman, Ltd., for permission to quote from *Greek Science in Antiquity*, by Marshall Clagett.

Harcourt, Brace and Company, Inc., for permission to adapt freely a passage of the translation by Edward MacCurdy of the *Notebooks of Leonardo da Vinci*.

Harvard University Press, for permission to make use of translations in *A Source Book in Greek Science*, by M. R. Cohen and I. E. Drabkin, and of the translation by P. H. Wicksteed and F. M. Cornford of Aristotle, *The Physics*, in the Loeb Classical Library.

Northwestern University Press, for permission to use, slightly altered, portions of the translation by H. Crew and A. de Salvio of Galileo, *Dialogues Concerning Two New Sciences*.

Oxford University Press, for permission to quote from translations in *Mathematics in Aristotle*, by Sir Thomas L. Heath.

University of Chicago Press, for permission to quote from Galileo, *Dialogue on the Great World Systems*, translated by T. Salusbury and revised by G. de Santillana.

I am grateful also to the following libraries for permission to reproduce pages from manuscripts in their possession: Biblioteca Nazionale Centrale, Florence; Biblioteca Vaticana; Bibliothèque Nationale, Paris; Bodleian Library, Oxford; and Columbia University Libraries.

Contents

Preface ix
Acknowledgments xii
List of Plates xvi
Note on Texts, Translations, and Abbreviations xvii
Introduction xix

Part I: Medieval Statics

Chapter 1. The Greek and Arabic Forerunners of Medieval Statics . . 3

   1.1 *The Book on the Balance*, attributed to Euclid.—24
   1.2 Archimedes, *On the Equilibrium of Planes* or *On Centers of Gravity*. Book I, Postulates, Propositions 1–7.—31.
   1.3 Hero of Alexandria, *Mechanics*. Book I, Chapters 2–8, 23–24, 33–34; Book II, Chapters 7–8, 22.—38.
   1.4 Archimedes, *On Heaviness and Lightness* (A Fragment of the *Floating Bodies*). —52.
   1.5 al-Khāzinī, *Book of the Balance of Wisdom*. Introduction, Sections 1–5; Chapter 1, Sections 1–9; Chapter 5, Sections 1–3.—56.

Chapter 2. Jordanus de Nemore and Medieval Latin Statics 69

   2.1 *The Theory of Weight*, attributed to Jordanus de Nemore. Book I, Propositions 6, 8, and 10 (R1.06, R1.08, R1.10).—104.
   2.2 *An Anonymous Commentary on the Elements of Jordanus on Weights*. Proposition 8.—109.
   2.3 Johannes de Muris, *The Four-Parted Work on Numbers*. Book IV, Second Tract.—113.
   2.4 Albert of Saxony, *Questions on the [Four] Books on the Heavens and the World of Aristotle*. Book III, Questions 2–3.—136.
   2.5 Marcus Trivisano of Venice, *On the Macrocosm or Greater World*. Book I.—146
   2.6 Galileo Galilei, *Mechanics* (on the lever and the inclined plane).—150.

Part II: Medieval Kinematics

Chapter 3. Gerard of Brussels and the Origins of Kinematics in the West   163

    3.1 Gerard of Brussels, *On Motion*. Book I, Suppositions 1–8, Proposition 1; Book II, Proposition 1.—187.

Chapter 4. The Emergence of Kinematics at Merton College   .   .   .   199

    4.1 Thomas Bradwardine, *Treatise on the Proportions of Velocities in Movements*. Chapter IV, Part 2.—220.
    4.2 Albert of Saxony, *Questions on the Eight Books of the Physics of Aristotle*. Book VI, Question 5.—223.
    4.3 Thomas Bradwardine, *On the Continuum*. Definitions 1–3, 7–14, 23–24; Suppositions 6–9; Conclusions 22–24, 26.—230.
    4.4 William Heytesbury, *Rules for Solving Sophisms*. Part VI, Local Motion.—235.
    4.5 *On Motion* (A Fragment), attributed to Richard Swineshead.—243.
    4.6 John of Holland, *On Motion*. Definitions.—247.
    4.7 Galileo Galilei, *The Two New Sciences*. Third Day, definitions of uniform motion and uniform acceleration.—251.

Chapter 5. The Merton Theorem of Uniform Acceleration   255

    5.1 William Heytesbury, *Rules for Solving Sophisms*. Part VI (continued).—270.
    5.2 *Proofs of Propositions Posited in the Rules for Solving Sophisms*, attributed to William Heytesbury.—284.
    5.3 Richard Swineshead, *The Book of Calculations: Rules on Local Motion*. Second Supposition to Conclusion 38, and other Conclusions.—290.
    5.4 John Dumbleton, *The Summa of Logical and Natural Things*. Part III, Chapters 10, 11.—305.
    5.5 *De motu incerti auctoris*.—326.

Chapter 6. The Application of Two-Dimensional Geometry to Kinematics   331

    6.1 Nicole Oresme, *On the Configurations of Qualities*. Part I, Chapters 0–4, 8, 10, 11, 13; Part II, Chapters 1, 3, 4, 5, 8; Part III, Chapters 1, 7, 8.—347.
    6.2 Giovanni di Casali, *On the Velocity of the Motion of Alteration*. An extract on uniformly difform quality.—382.
    6.3 Jacobus de Sancto Martino, *On the Latitudes of Forms* (According to the Doctrine of Nicole Oresme). Introduction; Propositions 19, 21–24, 26–27, 30, and Notanda.—392.
    6.4 Blasius of Parma, *Questions on the Treatise on the Latitudes of Forms*. Question III.—402.
    6.5 Galileo Galilei, *The Two New Sciences*. Third Day, Theorems I and II, Corollary 1.—409.
    6.6 *The Journal of Isaac Beeckman*, "On a Stone Falling in a Vacuum."—417.

Part III: Medieval Dynamics

Chapter 7. Aristotelian Mechanics and Bradwardine's Dynamic Law of Movement 421
- 7.1 *An Anonymous Treatment of Peripatetic Dynamics.*—445.
- 7.2 Nicole Oresme, *On the Book of the Heavens and the World of Aristotle.* Book I, Question 12.—463.
- 7.3 *A Brief Tract on Proportions Abridged from the Book on Proportions of Thomas Bradwardine, the Englishman.*—465.
- 7.4 Francischus de Ferraria, *On the Proportions of Motions* (extracts).—495.

Chapter 8. John Buridan and the Impetus Theory of Projectile Motion . . 505
- 8.1 Franciscus de Marchia, *On the Sentences of Peter Lombard* (A *Reportacio* of the Fourth Book).—526.
- 8.2 John Buridan, *Questions on the Eight Books of the Physics of Aristotle.* Book VIII, Question 12.—532.

Chapter 9. The Free Fall of Bodies 541
- 9.1 John Buridan, *Questions on the Four Books on the Heavens and the World of Aristotle.* Book II, Question 12.—557.
- 9.2 Albert of Saxony, *Questions on the [Four] Books on the Heavens and the World of Aristotle.* Book II, Question 4.—565.
- 9.3 Nicole Oresme, *On the Book of the Heavens and the World of Aristotle.* Book I, Chapter 17.—570.
- 9.4 *The Manuscripts of Leonardo da Vinci.* M 44r–v, M 45r, M 47r–v, M 48r, M 49r.—572.
- 9.5 Galileo Galilei, *The Two New Sciences.* Third Day, experiment of rolling balls on inclined plane.—576.

Chapter 10. Mechanics and Cosmology 583
- 10.1 John Buridan, *Questions on the Four Books on the Heavens and the World of Aristotle.* Book II, Question 22.—594.
- 10.2 Nicole Oresme, *On the Book of the Heavens and the World of Aristotle.* Book II, Chapter 25.—600.
- 10.3 Nicolas Copernicus, *On the Revolutions of the Celestial Orbs.* Book I, Chapters 7–8.—610.
- 10.4 *Questions on the Eight Books of the Physics* [*in the Nominalist Manner*], attributed to Marsilius of Inghen. Book III, Question 7.—615.

Part IV: The Fate and Scope of Medieval Mechanics

Chapter 11. The Reception and Spread of the English and French Physics, 1350–1600 629
Chapter 12. Medieval Mechanics in Retrospect 673
Bibliography 683
Index 699

List of Plates

Facing Page

1. A page from a thirteenth-century manuscript of the *De ratione ponderis*, illustrating the bent lever proposition — 72

2. A page from the holograph copy of William Moerbeke's translation of the works of Archimedes — 176

3. A page from a thirteenth-century manuscript of the *Liber de motu* of Gerard of Brussels — 177

4. A page from a fourteenth-century manuscript of Swineshead's *Calculationes*, showing the Oresme configuration system — 272

5. A page with the statement of the Merton uniform acceleration theorem. From William Heytesbury's *Regule solvendi sophismata* — 273

6. A page from a fourteenth-century manuscript of Swineshead's *Calculationes*, showing the Oresme configuration system — 344

7. A page from a fifteenth-century manuscript of Oresme's *De configurationibus qualitatum* — 345

8. A page from a fourteenth-century manuscript of *De latitudinibus formarum* — 440

9. A page from a late thirteenth-century manuscript giving the text of the *Liber de ponderoso et levi* attributed to Euclid — 441

Note on Texts and Translations of Documents

I HAVE made all of the translations except in those few cases where another translator has been indicated. In the translations arbitrary paragraphs and passage numbers have often been added to facilitate coordination with the commentary following the document. I have in general not enclosed these numbers in parentheses or in brackets, although whenever numbers have been added to the Latin texts I have been careful to use brackets. Throughout the texts, translations, and footnotes I have used square brackets to supplement the text and parentheses to enclose parenthetical editorial comments. I have even followed this practice when giving an indirect paraphrase of a passage originating in a Greek or medieval work.

The Latin texts of the documents are given when the texts are currently available only in manuscript or rare early printed editions. The Latin, Greek, or Arabic texts of the other documents can be easily found by consulting the editions cited at the beginning of the translations.

In the Latin texts here included some critical apparatus has been given. Generally two or three (or sometimes more) manuscripts have been used, although on rare occasions I have been forced to use a single copy. In most cases critical editions of these works still remain to be done, and in some cases my students and I are working on the texts. Thus I do not pretend to have used all the manuscripts or to have noted all variant readings. I have here satisfied myself that the passages used are cogent and make mathematical sense. It is doubtful whether the meaning or sense of our selections will be changed appreciably when the critical edition of these documents has been completed. But even if this should be the case, it seemed important to print these texts as they are, so that this material will be available to the reader interested in the history of science. From these texts he can make a first-hand evaluation of some important problems in the development of medieval mechanics.

As regards the texts, I should warn the reader that I have not attempted to give pure "diplomatic" transcriptions, which are often so puzzling to the modern reader. I have instead punctuated at will as I thought the meaning demanded. Similarly I have been free with capitalization, replacing many lower-case letters, particularly those that stand for the subjects of geometrical or mechanical discourse and where the passages include many such letters.*

In some cases I have given more in the Latin text of the document than in the translation. And in one case—Document 5.5—only the Latin has been given since its translation resembles closely other documents already presented.

Diagrams have been given in the translations and are not repeated in the Latin texts.

---

* In selections where only a few letters are used I have often left them to stand as they are in the manuscripts.

Introduction

THE key role of physics in general and mechanics in particular in the development of modern science has often been recognized, for it was in the mechanical area of early modern physics that the first thoroughgoing application was made of the mathematical and experimental techniques so crucial to the growth of modern science. But an admission of the important role of mechanics in the early modern period does not mean that we must search only the mature mechanical works of the seventeenth century for the beginnings of modern science. It is an obvious fact to the historian of science that the physical concepts of a Galileo or a Descartes, or even a Newton, radical as they may seem, were conditioned in many ways by the ancient and medieval learning that survived into the early modern period. And thus anyone who is honestly interested in the enormously complex *historical* process of the formation of modern science must examine in detail the germinal concepts of the preceding periods. Such an examination will reveal the elements of continuity (and thereby also of novelty) in the new science. This examination will give some insight into how a protoscientific theory was criticized and emended until it was no longer a cogent whole. It will also show how the very points of criticism of the older system became points of departure for the new. It will show, in short, how medieval mechanics—largely Aristotelian with some traces of Archimedean character—was continually modified to the point where it was seriously undermined, thus requiring a new mechanical system—and it was the Galilean-Newtonian system of the seventeenth century that fulfilled that requirement. This volume attempts a documentary analysis of some of the crucial criticism and modification of Aristotelian mechanics that took place from the thirteenth through the fifteenth century. In the course of this analysis both the ancient antecedents and the early modern consequences of medieval mechanics will also be examined briefly.

Medieval mechanics as a field of careful historical research is certainly not one with a long history. The delay in research in medieval mechanics was, I believe, due to the common view held in some quarters since the seventeenth century that science—and particularly mechanics—was an invention of the seventeenth century, and, if it had any significant antecedents, those antecedents lay not in the Middle Ages but rather in antiquity. And since there is an element of truth in this view, even those early students who gave some attention to medieval mechanics, such as Charles Thurot and Giovanni Vailati in the nineteenth century, treated medieval opinion merely as an offshoot of Greek mechanics and not as an object of independent research. Thurot's investigations of 1868–69 into the history of the principle of Archimedes (see the Bibliography), while they treated the medieval views of hydrostatics only incidentally, nevertheless had the distinct advantage of being based on the direct study of both manuscripts and early printed editions of the works of medieval authors. Vailati's important studies in the history of statics, originally published at the end of the nineteenth century (and later collected in his *Scritti* in 1911), concentrated on the antique period but were the first to recognize the importance of the statical texts attributed to the thirteenth-century mathematician Jordanus de Nemore in the growth of the concept of virtual velocities. However, unlike Thurot, Vailati did not investigate medieval manuscript sources.

Thus it was actually reserved to one individual to change the investigation of medieval mechanics from an incidental bypath to a field of investigation where the principal objective of research was to estimate the views of the medieval schoolmen on mechanical problems. This was the eminent French scientist Pierre Duhem, whose *Les Origines de la statique* (1905–6) brought to light the rich content of medieval statical treatises that had received little or no attention prior to his time. His *Études sur Léonard de Vinci* (1906–13), although very badly organized, attempted the same thing for kinematics and dynamics. It was in the latter work that Duhem discussed the fruitful medieval emendations of the Aristotelian mechanics of free fall and projectile motion and first outlined the kinematic description of uniform and accelerated motion that took place at Oxford and Paris in the fourteenth century. It was as the result of Duhem's investigation of manuscripts and early printed sources that John Buridan, Nicole Oresme, and other schoolmen of fourteenth-century Paris and Oxford emerged as key figures in the development of late medieval physics. It was Duhem who uncovered the medieval *impetus* theory—a kind of protomomentum concept—which under the stimulation of the scholastic discussion of Buridan and others was to exert some influence on late medieval and even early modern physical thought. So rich were Duhem's investigations—partially emended and corrected by numerous articles and his monumental *Le Système du monde* (1913–16, 1954–57; see the Bibliography for Duhem's writings)—that one can say that in a sense the succeeding study of medieval mechanics has been largely devoted to an extension or refutation of Duhem's work.

But in spite of their obvious importance, Duhem's investigations were not without serious defects. He made extravagant claims for the modernity of medieval concepts. Thus Buridan's impetus theory appears to emerge in Duhem's hands as a theory of inertia; and Oresme is considered as the inventor of analytic geometry because of his system of graphing qualities and movements, and as a precursor of Copernicus because of his discussion of the possibility of the earth's rotation. Furthermore, Duhem's procedure of presenting only parts of crucial passages—often out of context and then only in French translation without the equivalent Latin passages—made it almost impossible to evaluate Duhem's judgments without an extensive search of the manuscripts. E. J. Dijksterhuis' summary of Duhem's research in his *Val en Worp* (1924), while not substantially challenging the views of Duhem, presented some of the passages from the Latin texts. Other scholars like B. Jansen and E. Borchert—in studies on Peter John Olivi and Oresme respectively—accepted Duhem's conclusions for the most part and went on to extend them to the subjects of their investigation.

It was only with the publication of Anneliese Maier's studies that Duhem's works were given a thoroughgoing review and re-evaluation. Her general conclusions are found in one of the most important of her many publications, *Die Vorläufer Galileis im 14. Jahrhundert* (1949). Miss Maier's work has been based on detailed manuscript investigations, and her principal studies are cited in the Bibliography. The result of Maier's magnificent studies has been to place the mechanical ideas uncovered by Duhem in their proper medieval setting and to show their essential divergences from the later concepts of modern mechanics. Thus, to give only one example, Miss Maier shows that Buridan's *impetus* is at best to be considered a rejected analogue to momentum rather than identical to it. Furthermore, Miss Maier's studies opened up areas of medieval natural philosophy not studied by Duhem. The succeeding pages of this volume will demonstrate the great debt owed by me and all recent students of medieval mechanics to Miss Maier's work. A less detailed but important critique of Duhem's thesis has been made by that eminent historian of mechanics, A. Koyré. His *Études Galiléennes* (1939) emphasizes the point that it was only when Galileo abandoned some of the medieval views (like one version of the impetus theory) that he made essential progress.

It should be obvious from a study of Duhem and his critics that the one essential for an adequate review of medieval mechanical doctrines is the publication of the texts on which judgments have to be made. Maier has already done important service by complementing her critical comments with the texts of important passages. Similarly the distinguished student of medieval philosophy, Ernest Moody, has demonstrated the importance of making available the full texts of some of the crucial scholastic treatises in which are found the most interesting of the medieval mechanical ideas. He first gave us the *Quaestiones super libris quattuor de caelo* of John Buridan in 1942, where are found in their full context many of Buridan's important ideas, including a discussion of the impetus theory, of the possible rotation of the earth, and so on. Moody also initiated the project, in which I joined him as collaborator, to make available *all* of the principal medieval statical treatises. This project resulted in a volume entitled the *Medieval Science of Weights* (1952), the first volume in the University of Wisconsin's *Publications in Medieval Science*. Our purpose in that volume was to present the complete texts so that historian and scientist alike could judge the claims of Duhem for the importance of medieval statics. I forbear at this point to mention the many other editors and authors who have assisted in extending and clarifying medieval mechanics. Needless to say, the detailed studies in this volume have made use of a considerable number of monographs and editions, and their use has been acknowledged in the appropriate places.

Now it is my hope with this volume to treat some of the crucial problems of medieval mechanics in such a way that, while I offer my own judgment, the reader has before him sufficient textual material to evaluate for himself the significance of the medieval solutions for the growth of mechanics.

In my examination of medieval mechanics I have quite largely neglected (1) medieval discussions of methodology such as those which have been reviewed so excellently by A. C. Crombie in his *Robert Grosseteste and the Origins of Experimental Science*, 1100–1700, and (2) many of the important fringe areas between physics and philosophy so ably examined by A. Maier in the works we have already noted. I have, in short, concentrated on presenting the substantial content and objectives of a few of the mechanical doctrines of the medieval period which were framed in mathematical terms or which had important consequences for a mathematical mechanics. In treating the content of medieval mechanics I have adopted the convenient but somewhat anachronistic division of mechanics into statics, kinematics, and dynamics. Concepts and proofs important for all three of these divisions often appear during the Middle Ages in the same work and are intertwined one with another. On the other hand, we shall note a growing propensity to treat problems falling within the three categories separately; and we find, therefore, works almost entirely devoted to statics (even if from a dynamic viewpoint), as well as mechanical treatises with separate sections devoted to dynamic problems and to kinematic problems.

In the first part of this work I am concerned with statical problems. It is a somewhat shorter section, since, as I noted above, Ernest Moody and I have already published an extensive volume on medieval statics. But even in this section I hope to give some new material and enough selections and introductory critical remarks to make a connected history of the content of medieval statics.

It will be evident from the material I have presented in the first two chapters that medieval statics, like the other aspects of medieval mechanics, depends greatly on the mechanical concepts and their analysis given by Greek mechanicians: the Aristotelian author of the *Mechanica*, Archimedes, Hero, and others. The selections in the first chapter are from Greek and Arabic treatises attributed to Archimedes, Euclid, Hero, and al-Khāzinī. They illustrate Greek ideas that were to prove influential to medieval statics. It will be clear from my presentation of the medieval statical documents themselves in the second chapter that one of the most precious heritages from Greece was the employment of mathematical-deductive proofs in statics. The form of such proofs originated about the time of Euclid and Archimedes and was stamped on most succeeding treatments of statics. In addition to inheriting the Greek form of proof and analysis, the medieval statical authors inherited several important theorems that were to become the focal points of both medieval and modern statics. Among these theorems we can single out the general law of the lever as applied to both straight and bent levers. Furthermore, the problem of the inclined plane, which leads to a correct procedure of analyzing or resolving forces when properly solved (which it was not in antiquity), came into the Middle Ages and there received a brilliant and correct solution which in some respects surpasses both the solutions of Stevin and Galileo. Also inherited in primitive form was the principle of virtual velocities which is found in nascent form in the *Mechanica* attributed to Aristotle and more clearly in the *Mechanics* of Hero. We shall see that in the Middle Ages this principle was applied in the formal mathematical proofs of theorems relative to the law of the lever for straight and bent levers and to the equilibrium of weights on oppositely inclined planes. In essence, all of the medieval proofs show that unless the well-known lever and inclined plane theorems are accepted as true, the principle of virtual velocities is violated. In such proofs the medieval students were clearly foreshadowing the modern dynamic approach to statics that only became thoroughly established with the work of John Bernoulli and Lagrange, although Stevin, Galileo, and other scientists of the early modern period were not uninfluenced by this approach. As we shall see in Chapter 2, it was probably in the works attributed to the famous thirteenth-century mathematician Jordanus de Nemore that a new and important form of the principle of virtual velocities originated. The study of the subsequent history of medieval statics after the first half of the thirteenth century shows that, while further original contributions do not appear to have been made, there was at least some continuing treatment of statics and some improvement in the precise statement of the principles involved. The further interest in statics is demonstrated by the prevalence of numerous late medieval copies of the thirteenth-century treatises associated with the name of Jordanus, and by a number of commentaries on the earlier treatises. Perhaps the most important contribution of this late medieval period is found in a commentary on one of the earlier treatises. For in this commentary, dating, I believe, from the late fourteenth century, we find a discussion of the medieval form of the principle of virtual velocities and of its use by Jordanus. In this commentary, the principle emerges with explicit expression. As I have shown, the principal medieval statical tracts were available in printed form in the sixteenth century and, no doubt, exerted some influence on early modern scientists.

Associated closely with the medieval statical treatises was at least one important treatise in hydrostatics, the so-called *De insidentibus in humidum* or *De ponderibus Archimenidis*. This treatise of the thirteenth century, whose background and main theorems are studied in Chapter 2, depends in a fundamental way on the mathematical treatment of the density problem that originated in quite different form in the genuine work of Archimedes, *On Floating Bodies*. The medieval *De insidentibus* uses, apparently for the first time, the expression "specific weight" to distinguish density or the intensity of weight from gross weight. Furthermore, Archimedes' principle is there expressed in Latin for the first time. On the whole, hydrostatics did not particularly flourish in the late Middle Ages, although certain schoolmen, such as Johannes de Muris, Albert of Saxony, and Blasius of Parma, were not uninfluenced by the *De insidentibus*. Nicholas of Cusa in the fifteenth century was perhaps dependent in part on the hydrostatics of the *De insidentibus*, and later Galileo also seems to have known that treatise.

The medieval developments in statics owed something to Aristotelian dynamics but for the most part fell outside of the formal scholastic treatment of dynamics, although on occasion the schoolmen of the fourteenth century used the conclusions of the statical treatises in arguing the validity of dynamical theorems. On the other hand, the achievements of medieval kinematics were very much more an integral part of the scholastic discussions of Aristotelian statements regarding force and motion. This is evident from my detailed treatment of kinematics in the second part of the volume. Early kinematicists made simple proportionality statements comparing completed movements in terms of the space traversed in equal times or in terms of the times necessary to traverse some given space. Generally speaking, ancient authors did not assign to velocity a magnitude consisting of a ratio of space to time. However, from the period of Gerard of Brussels' *Liber de motu* in the thirteenth century, schoolmen began to think of velocity as a magnitude, although its definition was still not given as a ratio of the unlike magnitudes of space and time (see Chapter 3).

It was at Merton College, Oxford, between about 1328 and 1350, that real advances were made in kinematics. These contributions to kinematics appear chiefly in the works of Thomas Bradwardine, William Heytesbury, Richard Swineshead, and John Dumbleton; I have discussed and illustrated them in Chapters 4 and 5. Particularly important was the development of a concept of instantaneous velocity and consequently of an analysis of various kinds of acceleration. This analysis, as I have shown in Chapter 4, grew out of the discussion of the philosophical problem of the intension and remission of forms. It led to the distinction of "quality" or intensity of velocity from "quantity" of velocity. In cases of acceleration, the quality of velocity was thought to vary from instant to instant. And so instantaneous velocity was considered as the intensity or quality of velocity at an instant; it was measured, mathematically speaking, by the space which *would* be traversed if a body were moved for a given time at the velocity it had at the instant.

From their consideration of acceleration the Merton authors derived the so-called Merton theorem of uniform acceleration. This theorem equated (with respect to space traversed in a given time) a uniform acceleration and a movement uniform at the speed possessed by the uniformly accelerating body at the middle instant of the time of acceleration. In the course of the fourteenth century this theorem or rule was given many proofs, and the documents presented in Chapter 5 illustrate the most original of those proofs. The analysis of qualities made at Oxford and the consequential analysis of local motion passed to Paris and to other parts of Europe. At Paris not long after 1350 Nicole Oresme perfected (if not invented) a kind of graphing system which employed two-dimensional figures to represent variations in permanent qualities and in movements. In the cases of motion, the base line of such a figure represented time and the perpendiculars raised at points along that base line represented the velocities at instants referred to the points on the base line. The area of the whole figure, dimensionally equivalent to the distance traversed in the movement, represented for Oresme the quantity of the motion or its "total" velocity. This system has been described in detail in Chapter 6 and illustrated with extensive selections from the works of Oresme and his successors. Among the selections I have included are the geometric proofs of the Merton theorem given by Oresme and his contemporary, Giovanni di Casali. It was the geometric two-dimensional kind of proof that Galileo was to use in the course of developing the law of free fall, as has been clearly demonstrated in Chapter 6.

In certain respects, the development of the kinematic description of motion at Oxford represented an important modification of Aristotelian views regarding the categories of quantity and quality. Similarly, medieval discussions of dynamics produced significant changes in the Aristotelian discussion of the role of force and resistance in the production of motion, as the third part of my volume shows in detail. For Aristotle, local motion was a process demanding force and resistance in substantial contact, whether it was the natural motion of elements to their natural places or the unnatural (violent) motion of bodies forced from their natural places. In either kind of motion, the space traversed in a given time was apparently conceived as being directly proportional to the force producing the motion and inversely proportional to the resistance hindering (but necessary to) the motion. Needless to say, the Aristotelian definition of force differed fundamentally from that later adopted by Newton, since for Aristotle force was that which produced motion and for Newton it was that which produced acceleration. Now Aristotle's rules relating the distance traversed to force and resistance were submitted to criticism in late antiquity, particularly by John Philoponus in the sixth century. While Aristotle's view led to the opinion that velocity is proportional to the *ratio* of force to resistance, Philoponus' treatment resulted in the opinion that velocity follows the *arithmetic excess* of force beyond resistance, a view given precise mathematical expression by Thomas Bradwardine in his *De proportionibus motuum in velocitatibus* of 1328 but rejected by him in favor of another opinion. As has been shown in Chapter 7, Bradwardine adopted a peculiar opinion regarding the relation of velocity to force and resistance, to the effect that velocity is exponentially related to the ratio of force to resistance, i.e., that an arithmetic increase in velocity follows a geometric increase of some original ratio of force to resistance. This view was totally unrelated to experimental investigations, but it nevertheless held sway for over a century before being generally abandoned. One of its most important by-products was, however, that it related velocity to instantaneous changes. In doing so, it foreshadowed the differential type of equation and in fact helped to stimulate the kinematic developments at Merton College which we have already mentioned.

While Bradwardine's solution of the problem of describing motion in terms of force and resistance was almost universally accepted by the Parisian school of John Buridan and his successors in the fourteenth century, Buridan's most significant contribution to dynamics was in the elaboration of a special theory to explain two pivotal problems of mechanics: (1) the continuance of projectile motion after the cessation of the initial force of projection and (2) the acceleration of bodies as they fall. These problems had occupied a central position in dynamical discussions since the time of Aristotle. The difficulty for Aristotle of explaining the continuance of projectile motion is evident, for Aristotle's theory of violent motion demanded the continued substantial contact of the motor and the thing moved. Aristotle, we have shown in Chapter 8, apparently held that the initial motor communicated motive force to the medium (i.e., the air) which was particularly apt for motion; the air then acted as the continuing motor. Such a theory evoked severe criticism in antiquity from Philoponus—largely on empirical grounds. In place of the Aristotelian theory, Philoponus suggested that the initial mover or projector impressed an incorporeal kinetic force into the projectile and it was this force that kept the projectile in flight until the impressed force was destroyed by the weight of the projectile and perhaps also by the resistance of the air. This theory of impressed force was the starting point in a long history of efforts to explain momentum or inertial effects. In Chapter 8, I have presented a document showing how John Buridan in the fourteenth century developed a new form of the impressed force theory, a form that is known as the *impetus* theory. As in the older theory of Philoponus, impetus in Buridan's theory was considered as a motive force impressed into the projectile; it was the impetus which kept the projectile in motion. But contrary to the older theory, impetus was described by Buridan as being of permanent rather than self-expending nature. This made it a kind of analogue to inertia, particularly when Buridan suggested that the continuing motions of the heavenly bodies could be explained by the impression of impetuses by God. Furthermore, Buridan described the immediate quantitative measure of impetus in terms of the quantity of matter of the projectile and the velocity imparted to the projectile. Such terms are obviously similar to the quantities used later to measure momentum or quantity of motion in Newtonian mechanics. Although Buridan's description of impetus was continually available from the fourteenth century through the sixteenth, most later authors abandoned Buridan's "inertial" impetus for a "self-expending" impetus.

The impetus theory was also applied by Buridan and others to solve the second crucial problem of mechanics, the acceleration of falling bodies, as is demonstrated in Chapter 9. In this solution, it was supposed that the continuing presence of the source of motion, represented by the weight of the body, continually impressed more and more impetus into the falling body and thus greater and greater velocity. It is clear that when the ontology of impetus changed from some kind of force to an effect like quantity of motion or momentum in early modern mechanics—with the consequent abandonment of Aristotelian natural philosophy—Newton's second law of motion and thus his definition of force were not far off. For with such a changed meaning for impetus, we then have a case where a continuing force (gravity) is directly producing acceleration, i.e., a continuingly increasing velocity, rather than producing an intermediate but increasing force which in turn produces an acceleration.

It is a point of interest that late medieval schoolmen in their detailed study of kinematics did not apply their kinematic theorems regarding uniformly accelerated motion to the description of freely falling bodies. Still, Jordanus in the thirteenth century and Oresme in the fourteenth seem to have believed that the velocity of fall was directly proportional to the time of fall rather than to the distance of fall, as was commonly held since antiquity.

From the other medieval mechanical questions of interest to the historian of mechanics, I have singled out in Chapter 10 the questions as to whether (1) the earth rotates, (2) motion is an entity distinct from the moving body, and (3) whether there can exist a plurality of worlds. I have given these questions special attention because they reflect the spread of some basic mechanical ideas such as the relativity of the detection of motion, the assumption of a closed mechanical system, and the application of the concept of center of gravity to large bodies. It was at Paris in the writings of Buridan and Oresme that these mechanical concepts were most interestingly applied to the questions we singled out above. But these and the other mechanical concepts had widespread circulation throughout Europe in the course of the fourteenth, fifteenth, and sixteenth centuries. That circulation and the fate of these ideas I have traced in Chapter 11.

On the whole, I have attempted to show in this study how the medieval schoolmen treated certain critical problems involving the causal and descriptive aspects of equilibrium and of "natural" and "forced" motion. While the solutions of these problems as presented here and summarized in Chapter 12 will be seen to lie at least in a general way within the basic framework of Aristotle's natural philosophy, still these solutions reveal important aspects of the medieval logical and descriptive analysis that were to prove useful in early modern times when the Aristotelian framework was abandoned.

Part I

MEDIEVAL STATICS

Chapter 1

The Greek and Arabic Forerunners of Medieval Statics

ERNEST MOODY and I have already shown in our volume on medieval statics that it was the interpenetration of two Greek traditions, the Aristotelian and Euclidian-Archimedean traditions, which resulted in the most original medieval statical work.[1] We also showed that in the works attributed to the eminent mathematician of the early thirteenth century, Jordanus de Nemore, there can be found the following important statical ideas and procedures: (1) the use of the principle of virtual velocities in the *proofs* of statical theorems, thus foreshadowing the modern dynamic approach to statical problems; (2) a dynamic demonstration of the application of the law of the lever to bent levers, thus revealing a surer understanding of the principle of static moment; (3) the idea of resolving a force into components as exemplified by the medieval concept of "positional" gravity, which in its best applications is a rhetorical equivalent to the modern expression defining the component of force, $F = W \sin a$; and (4) the use of (1) and (3) to give an elegant and correct solution of the inclined plane problem.

Now it is fitting that I should preface my detailed examination of these medieval statical doctrines by some remarks on the contributions of their Greek and Arabic antecedents. I hope to present in this first chapter enough of this early material to permit us to assay properly the originality of the medieval contributions.

There is some evidence that we ought to start the history of statics in antiquity with Archytas of Taras (*fl. ca.* 400–365 B.C.). We are told by

[1] *The Medieval Science of Weights* (Madison, 1952), pp. 6 and 7.

Diogenes Laertius that Archytas was the first to expound mechanics by means of mathematical principles.[2] Such an exposition has not come down to us. In view of the content of the succeeding mechanical treatises, it seems reasonable to suppose that Archytas treated the same basic instruments or machines into which later authors resolved all machines: the lever, the wheel and axle, the pulley, the wedge, and the screw. It would be most surprising if Archytas did not give in his treatise the law of the lever found already in the earliest extant mechanical treatises. But without further evidence as to the specific content of Archytas' treatise, we are forced to pass on without delay to our earliest surviving treatises.

It appears likely that the earliest extant mechanical treatise is the work entitled *Mechanica*[3] and attributed to Aristotle. The attribution to Aristotle has been questioned mainly on the ground that the treatise's attention to practical problems is "quite un-Aristotelian,"[4] which is doubtful reasoning at best, considering the enormous range of the intellectual activity of Aristotle during his last years at the Lyceum. But ordinarily it is agreed by those who would question its attribution to Aristotle that the treatise was composed by an Aristotelian shortly after Aristotle's time. Some would suggest without evidence that its author was Strato the Physicist, who succeeded to the headship of the Lyceum in 287 B.C., but for the present I shall make no attempt to argue further the question of the identity of the author, whom in common with the usual practice I shall call Pseudo-Aristotle.

Whether or not the *Mechanica* is the earliest extant mechanical treatise, it proved of the greatest importance for the later history of statics, even though we know of no medieval translation of it either into Arabic or into Latin. It exercised its influence on other mechanical writings—like the *Mechanics* of Hero—which were in turn influential in the course of statical history. Let us then examine its contents in some detail.

The most important point for our study of the *Mechanica* is that, like the later medieval treatises and unlike the Archimedean treatises, it takes a dynamic approach to problems of statics. But at the same time, its approach is imprecise and nonmathematical (i.e., without formal mathematical proofs, for the most part). In the beginning,[5] the author indicates

---

[2] *Lives*, VIII, 79–83.

[3] Alternate Latin titles are *Problemata mechanica* and *Quaestiones mechanicae*. I have followed Apelt in using *Mechanica*.

[4] *Mechanica*, translation of E. S. Forster (Oxford, 1913), preface.

[5] *Mechanica*, edition of Otto Apelt (Leipzig, 1888), pp. 95–96, Bekker no. 847b. In my quotations I have generally followed the Forster translation cited in note 4, although occasionally changing it. Even while using Apelt and Forster I have employed Bekker numbers throughout.

that it seems strange that a small force (ἰσχύς) can move a large weight (βάρος) and that one can move with a lever a weight he cannot move without the lever. "The original cause (αἰτία) of all such phenomena is the circle . . . (847b 16–17). The phenomena observed in a balance (ζυγόν) can be referred to the circle, and those observed in the lever (μοχλός) to the balance; while practically all the other phenomena of mechanical motion are connected with the lever. Furthermore, since no two points on one and the same radius travel with the same rapidity (ἰσοταχῶς), but of two points that which is further from the fixed center is the quicker (ὂν θᾶττον), many marvelous phenomena occur in the motions of circles . . ." (848a 12–19). It is then to the peculiarities of circular motion that we owe many curious mechanical actions. It is, for example, because of the fact that the velocity of a weight on a radial arm increases as we move the weight along the arm away from the fulcrum, that actions of a balance and a lever can be explained, according to the author. Or as he puts it, "the radius which extends further from the center is displaced more quickly than the smaller radius when the near radius is moved by the same force" (848b 1–9). "More quickly" in this context means a larger arc in the same time.

The reason why the velocity increases as we go farther from the fulcrum is explained in the following manner. Fundamentally we ought to consider a radius in motion around a fulcrum as undergoing two essential displacements, a natural tangential movement and an unnatural movement toward the center due to the fact that it is a constrained system. "Now if the two displacements of a body are in a fixed proportion, the resulting displacement must necessarily be a straight line, and this line is the diagonal of the figure, made by the lines drawn in this proportion" (848b 9–13). The author then gives as an example the familiar figure of a parallelogram of velocities (see Fig. 1.1). AH is the diagonal resultant of the two displacements (i.e.

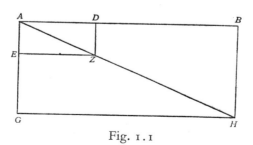

Fig. 1.1

velocities) represented by the magnitudes AB and AG. A similar parallelogram of velocities was also given by Hero of Alexandria in his *Mechanics* (see Doc. 1.3 below, passage I.8). Now, the author goes on to say, if a point "is moved in two displacements in no fixed ratio for any time, its [resulting] displacement cannot be in a straight line... [but in such a case] a curve is produced" (848b 25–35). Remember that the author considers the movement of a radius of a circle to consist in two displacements, one tangential and the other towards the center. If the proportion of these two movements were constant, the resultant velocity would be represented by the chord *BG* (see Fig. 1.2), but actually the displacement is along the

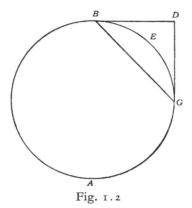

Fig. 1.2

arc *BEG*. The continual interference with the tangential motion by the varying motion of constraint toward the center produces the circular motion. He then goes on to show by geometrical example that the movement on the end of a shorter radius is less than that on the end of a longer radius because it is more interfered with.

And, if one of two displacements caused by the same forces is more interfered with and the other is less, it is reasonable to suppose that the motion more interfered with will be slower than the motion less interfered with; which seems to happen in the case of the greater and lesser of the radii of circles. For on account of the extremity of the lesser radius being nearer the stationary center than that of the greater, being, as it were, pulled in a contrary direction, towards the middle, the extremity of the lesser moves more slowly. This is the case with every radius, and it moves in a curve, naturally along the tangent and unnaturally towards the center, and the lesser radius is always moved more in respect of its unnatural motion; for being nearer to the retarding center it is more constrained (849a 8–19).

In the case of the lever, the author appears to be contending that a given weight on a lever arm has a greater effectiveness, i.e. greater force, on a longer arm because it has a greater velocity there, i.e., because it sweeps out a greater arc in the same time, being possessed with a larger proportion of natural motion without interference. In short, the effectiveness of any given weight depends not only on its free weight but on its velocity due to its position on a moving radius. If we make the absolute free weights inversely proportional to the velocities (i.e. to the arcs simultaneously traversed by the weights on the ends of the lever arms), then the total effectiveness of each weight should be the same and equilibrium should obtain. But it is obvious in this case that the arm lengths are directly proportional to the potential, simultaneous arcal displacements. Thus equilibrium obtains when the free weights are inversely proportional to the lever arm lengths. Such seems to be the reasoning of the author of the *Mechanica* when he states as follows the law of the lever:

Now since a longer radius moves more quickly than a shorter one under pressure of an equal weight; and since the lever requires three elements, viz. the fulcrum—corresponding to the cord of a balance and forming the center—and two weights, that exerted by the person using the lever and the weight which is moved; this being so, as the weight moved is to the weight moving it, so, inversely, is the length of the arm bearing the weight to the length of the arm nearer to the power. The further one is from the fulcrum, the more easily one will raise the weight; the reason being that which has already been stated, namely, that a longer radius describes a larger circle. So with the exertion of the same force the motive weight will change its position more than the weight which it moves, because it is farther from the fulcrum (850a 36–850b 6).

As Vailati,[6] Duhem,[7] and others have pointed out, this passage seems to contain a primitive application of the basic principle of virtual velocities, or work, to a problem in statics. Of course, one might say that this is a

---

[6] Giovanni Vailati, "Il principio dei lavori virtuali da Aristotele a Erone d'Alessandria," *R. Accademia delle scienze di Torino, Atti*, Vol. *32* (1897) 940–62. Reprinted in his *Scritti* (Leipzig, Florence, 1911), pp. 91–106.

[7] P. Duhem, *Les Origines de la statique* (Paris, 1905–6), Vol. *1*, 5–12; Vol. *2*, 291–302. Duhem attempts to connect the dynamic approach of Pseudo-Aristotle with the so-called Aristotelian law of movement. In Chapter 7 below we have summarized the dynamics of Aristotle. Furthermore we have shown later in this first chapter how the author of the *Liber karastonis* starts from Aristotle's law of movement, i.e., velocity is proportional to the motive force, and arrives at the law of the lever. This seems to substantiate Duhem's attempt to connect the two. Ernst Mach in his *The Science of Mechanics*, translated by T. J. McCormack, 5th ed. (LaSalle, Ill., 1942), pp. 12–13, 98–99, 105–6, while not able to accept Duhem's conclusion concerning the dependence of Pseudo-Aristotle on Aristotle's dynamic law, admits that Pseudo-Aristotle has used in germinal form the principle of virtual velocities.

far cry from the modern doctrine which John Bernoulli states in this way:[8] "I call *virtual velocities* those acquired by two or more forces in equilibrium when one imparts to them a small movement; or if the forces are already in motion.[9] The *virtual velocity* is the element of velocity which each body gains or loses, with respect to the velocity already acquired, in an infinitely small time, its direction being considered."

It is well known that Lagrange considered the principle of virtual velocities the most general and fundamental to the science of statics and that he applied it in manifold ways.[10]

Now the principle as used and applied in the Pseudo-Aristotelian treatise is different in at least two major respects from its modern counterpart: (1) the virtual or potential displacement is not posited as being infinitely small; and (2) the virtual displacement is arcal rather than rectilinear. It is the *vertical, rectilinear* potential displacement that the modern dynamic approach to statics assumes in computing the virtual work involved. We shall see shortly that a later Greek mechanical author, Hero of Alexandria, was aware of this, although he does not make a special point of it. It was the greater glory of the medieval mechanician to sharpen this concept of virtual velocities in terms of vertical displacements and above all to apply it in *formal, mathematical* proofs, which was the case in neither the Pseudo-Aristotelian nor the Heronian mechanics.

Before passing on to the Archimedean tradition in antiquity, we ought to note one or two other points concerning the statics of the Pseudo-Aristotle that were to reappear later in the medieval treatises. Chapter 2 of the *Mechanica* (850a) contains what is essentially a correct analysis of the stability of the balance beam when supported from above and the instability of such a beam supported from below. The author recognizes that a material beam when tilted has the greater portion of its weight in the elevated arm beyond the line of suspension. "In that case the side on which the greater part of the beam is must necessarily sink until the line which divides the beam into two equal parts reaches the actual perpendicular

---

[8] Johannes Bernoulli, *Discours sur les loix de la communication du mouvement qui a mérité les Eloges de l'Académie Royale des Sciences, aux années 1724 et 1726*, in his *Opera Omnia* Vol. *3* (Lausanne, 1742), 23. "J'Appelle *vitesses virtuelles*, celles que deux ou plusieurs forces mises en équilibre acquierent, quand on leur imprime un petit mouvement; ou si ces forces sont déja en mouvement. La *vitesse virtuelle* est l'élement de vitesse, que chaque corps gagne ou perd, d'une vitesse déja acquise, dans un tems infiniment petit, suivant sa direction."

[9] I am following the edition, but there should be a comma here and it should read probably: "Or if the forces are already in motion, the *virtual velocity*. . . ."

[10] J. L. Lagrange, *Mécanique analytique*, in *Oeuvres de Lagrange*, Vol. *11* (Paris, 1888), 22.

(i.e., the actual line of suspension), since the weight now presses on the side of the beam which is elevated." Hence the beam rights itself. But the contrary is true of the beam supported from below, since the greater portion of the beam is depressed and there is nothing to right it again. Problems involving the wheel and the axle (i.e., windlasses), the wedge, and the pulley he attempts to solve in terms of the lever (chaps. 13, 17, and 18). Thus there are mentioned in the *Mechanica* four of the five simple machines described later by Hero of Alexandria. Only the screw is missing.

If the *Mechanica* attributed to Aristotle was the foundation of the dynamic tradition, we have to look elsewhere for the first extant evidences of the mathematical-statical, nondynamic approach. It is not beyond possibility that the *Book on the Balance* attributed to Euclid is the earliest treatise embodying this approach. Unfortunately, we have only an Arabic text of this work, wherein it is assigned to Euclid.[11] But at least we are told in another treatise, going back ultimately to the Hellenistic period—the so-called *De canonio*, extant only in a medieval Latin translation—that the law of the lever was proved by Euclid, Archimedes, and others.[12] This may very well be a reference to the proof contained in this *Book on the Balance*, which circulated under the name of Euclid. That neither Hero nor Pappus mentions Euclid's proof shakes our confidence in the genuineness of the fragment but is by no means conclusive evidence that Euclid did not compose such a work. I have translated as Document 1.1 the Arabic text of the so-called Euclidian treatise. It seems to me to be important in the history of statics for the following reasons: (1) if by Euclid, it would be the earliest *mathematical* demonstration of the law of the lever; (2) like the proof of Archimedes, it is entirely statical in its approach—no dynamic considerations enter into the proof of, or conditions for, equilibrium.

As the reader examines the treatise, he will notice that it rests on the following axioms: (1) the assumption of equilibrium in the special case of the lever where equal weights are at equal distances from the fulcrum (cf. Doc. 1.2, Archimedes, *On the Equilibrium of Planes*, Bk. I, post. 1); (2) the assumption that a weight suspended anywhere along a line at right angles to a beam exerts the same force for the rotation of the beam; and (3) the assumption that a weight suspended in the line of the vertical passing through the fulcrum of a balance does not disturb the equilibrium

---

[11] For published text, see Document 1.1. For a discussion of its authorship, see the commentary to Document 1.1, passage 1.

[12] Moody and Clagett, *Medieval Science of Weights*, p. 66.

of a balance. Following these assumptions, the author of this treatise gives an ingenious proof of the proposition that if we move one weight on one arm of a balance a given distance away from the fulcrum while we are moving its equal weight placed on the same arm of the balance the same distance toward the fulcrum, the equilibrium is not disturbed. This proposition is then applied to the case where the lever arms are in inverse ratio to the weights, the ratio being 3:1. For he supposes first that the shorter arm is increased until it equals the longer arm and he supposes that a unit weight is left on the original longer arm, while a weight equal to it is placed on the other extremity (see Fig. 1.3). Since the equal weights

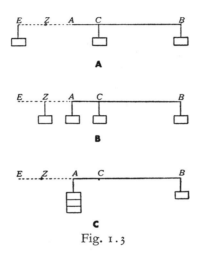

Fig. 1.3

are at equal distances, this beam is in equilibrium. Now supposing that another weight equal to each of the two suspended weights is hung in the line of the fulcrum, the equilibrium is undisturbed. If we move the weight at E to Z while we move its equal weight at C to A, the equilibrium is undisturbed, by the previously proved proposition, since $CA = ZE$. Now suppose that we hang another weight at C equal to each of the other weights. Again the equilibrium is undisturbed. If we move this new weight at C to A while we move the weight at Z to A, the equilibrium is again undisturbed, since $CA = AZ$. But now we have three weights at A, one at B, and $BC = 3CA$, and equilibrium remains. This is what we wished to prove. Presumably the author would conclude that the same reasoning holds for any case where the weights are in the inverse ratios of the lever arms, although he does not specifically tell us so.

Now Archimedes' famous proof as contained in the *Equilibrium of*

Planes, Book I, propositions 6 and 7[13] (see Doc. 1.2), is in the same tradition as this proof attributed to Euclid, but with these important additions: (1) the introduction of the concept of center of gravity, and (2) the complete generalization of the relationships of weights and distances to include irrationals. As a matter of fact, by the time Archimedes comes to prove the law he is actually no longer talking about physical weights but rather about homogeneous geometrical magnitudes possessing "weight" proportionate to the quantity of magnitude; and of course, as in the Euclidean proof, the material beam has become a weightless line. Archimedes, like Euclid, makes use of the special case of equilibrium of the balance of equal arm lengths supporting equal weights. In both proofs the special case is a postulate, which, though it may ultimately rest on experience, in the context of a mathematical proof appears to be a basic appeal to geometrical symmetry.

Of all the ancient proofs of the law of the lever, that of Archimedes is certainly the best known and the most discussed. Hence we shall try here only to assay its basic character. The purpose of Archimedes' proof is to demonstrate how the case of *any* lever wherein the arms are inversely proportional to the suspended weights reduces to or is identical with the special case of equal arms and equal weights. This is done (1) by converting the weightless beam of unequal arm lengths into a beam of equal arm lengths, and then (2) distributing the two unequal weights analyzed into rational component parts over the extended beam uniformly so that we have a case of equal weights at equal distances. Finally (3) the proof utilizes propositions concerning centers of gravity (proved previously in another treatise) to show that the case of the uniformly distributed parts of the unequal weights over the extended beam is in fact identical with the case of the composite weights concentrated on the arms at unequal lengths. Further it is shown separately in proposition 7 that if the law is true for rational magnitudes, it is true for irrational magnitudes as well.

The severest criticism of this proof is, of course, the classic discussion by E. Mach,[14] which stresses two points: (1) experience must have played a predominate role in the proof and its postulates in spite of its apparently mathematical-deductive form; and (2) any attempt in fact to go from the special case of the lever to the general case by replacement on a lever arm of the expanded weights by weights concentrated at the center of gravity

---

[13] See English paraphrase and translation by T. L. Heath, *The Works of Archimedes* (Cambridge, 1897), pp. 189–94.

[14] Mach, *Science of Mechanics*, edit. cit., pp. 13–17, 19–20, 24–28.

must assume that which has to be proved, namely, the principle of static moment. This is not the place to enter into the extensive literature that has grown up around this criticism by Mach, but we can say that quite successful defenses or clear explanations of Archimedes' procedure have been prepared by J. M. Child[15] and G. Vailati.[16]

The student of medieval mechanics must admit that the actual content of the proof of Archimedes had little direct influence on medieval Latin statics. Of course, the *Equilibrium of Planes* and its commentary by Eutocius were translated into Latin in 1269 by William Moerbeke.[17] However, little evidence of the use of this treatise by medieval mechanicians has been uncovered. But the indirect influence of the mathematical form of Archimedes' proofs is another matter. The concept of a tight, deductive, mathematical proof of the law of the lever certainly influenced statics in late antiquity and during the Islamic period and led to the general acceptance by medieval mechanicians of the necessity of mathematical demonstrations in mechanics. Furthermore, the concept of center of gravity introduced by Archimedes, so essential to his whole approach, was used extensively in late antiquity (e.g. by Hero and Pappus) and also by the Arabic authors.[18]

---

[15] J. M. Child, "Archimedes' Principle of the Balance and Some Criticisms Upon it," in C. Singer, editor, *Studies in the History and Method of Science*, Vol. 2 (Oxford, 1921), 490–520.

[16] Vailati, "Del concetto di centro di gravità nella statica di Archimede," *R. Accad. d. Scienze di Torino, Atti*, Vol. *32* (1896–97), 500 *et seq.* reprinted in his *Scritti*, pp. 79–90. By the same author, "La dimostrazione del principio della leva data da Archimede," *Atti del Congresso Internazionale di Scienze Storiche*, Vol. *12* (1904), reprinted in his *Scritti*, pp. 497–502. For other discussions, see W. Wundt, *Logik*, Vol. 2 (Stuttgart, 1907), 306–8; O. Hölder, *Die mathematische Methode* (Berlin, 1924), pp. 39–45; W. Stein, "Der Bergriff des Schwerpunktes bei Archimedes," in *Quellen und Studien zur Geschichte der Mathematik, Astronomie, und Physik*, Abt. B, Studien, Bd. *1* (1931), 229–30; V. Lenzen, "Archimedes' Theory of the Lever," *Isis*, Vol. *17* (1932), 288–89; V. Lenzen, "Reason in Science," *Reason*, Vol. *21* (1939) 81–83; Dora Reimann, "Historische Studie über Ernest Machs Darstellung der Entwicklung des Hebelsatzes," *Quellen und Studien zur Geschichte der Mathematik, Astronomie, und Physik*, Abt. B, Bd. *3* (1936), 554–92. See the recent and excellent summary of the various critiques in E. J. Dijksterhuis, *Archimedes* (Copenhagen, 1956).

[17] See Clagett, "The Use of the Moerbeke Translations of Archimedes in the Works of Johannes de Muris," *Isis*, Vol. *43* (1952), 236–42.

[18] See E. Wiedemann, "Beiträge zur Geschichte der Naturwissenschaften VII— Über arabische Auszüge aus der Schrift des Archimedes über die schwimmenden Körper," *Sitzungsberichte der Physikalisch-medizinischen Sozietät in Erlangen*, Vol. *38* (1906), 157, n. 3. Particular attention should be called to the ample remarks on center of gravity contained in the first lecture of al-Khāzinī's *Book of the Balance of Wisdom* partially edited and translated by N. Khanikoff in *Journal of the American Oriental Society*, Vol. *6* (1860), 25–38. Some of these remarks have been translated in Document 1.5.

And while as a specific doctrine it played little or no role in Latin statics before the sixteenth century, its influence is clearly seen in a number of medieval treatises which use the procedure of replacing a material section of a beam by a weight hung from the middle point of that section of the beam.

This idea is found, for example, in the widely popular treatise *De canonio*, which, although it exists only in medieval Latin translation, clearly is a translation from the Greek.[19] This treatise consists of four propositions which determine in the geometric manner problems relative to the steelyard or Roman balance, i.e., the material balance of unequal arm lengths. Thus the treatise finds out how one can determine the weight to hang on the shorter arm of a Roman balance in order to hold the balance in equilibrium with no weight on the longer arm. This treatise assumes as proven the general law of the lever. It also assumes that the weight of a symmetrical, homogeneous beam is proportional to the length of the beam. The material beam is reduced to a geometrical case by treating the problem in the following manner. Assume that equal segments of a beam similarly placed on each side of the fulcrum counterbalance each other and thus make their equal counterbalanced segments of the beam as if they were weightless. With such a reduction made, we can then say that the weight to be hung on the shorter arm (the weight z in Fig. 1.4) must be such that

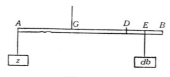

Fig. 1.4

it will balance the excess of the longer arm of the beam over the shorter arm (i.e., the material segment DB). But, as the author of the *De canonio* says, "it has been demonstrated in the books which speak of these matters, that it makes no difference whether the weight of DB is equally distributed along the whole line DB, or whether it is suspended from the mid-point of that segment."[20] It is here, of course, that we see the ultimate influence of the doctrine of center of gravity and the proof of Archimedes. With DB replaced by a weightless segment DB, we can instead hang a weight db equal to material segment DB from the midpoint E. The whole problem has now been reduced to a problem involving the general law of the lever as applied to weightless beams.

[19] For text, translation and discussion of this work, see Moody and Clagett, *Medieval Science of Weights*, pp. 57–75.
[20] Ibid., p. 65

This same "replacement" doctrine reflective of Archimedes' procedure is also evident in a treatise which like the *De canonio* goes back to a Greek original and which also like the *De canonio* takes up the problem of the Roman balance—the so-called *Liber karastonis*, existing in both Arabic and Latin editions.[21] As I have pointed out elsewhere in my text of this work, the work was composed in its present extant form by the Arabic mathematician Thābit ibn Qurra (d. 901), who was attempting to revise and clarify a Greek work which we may call for want of its Greek title by the Latin title *Cause karastonis*.[22] We shall concern ourselves later with the importance of this treatise for the continuance and further growth of the Pseudo-Aristotelian dynamic tradition; here it is important to note only its use of the center of gravity idea. For example, it is proved in Proposition VI that the portion of a material beam segment which is "a continuous expanded weight of equal thickness" may be replaced by a weight equal to the segment and hung at the middle point of the segment (now considered immaterial) without disturbing the prior equilibrium.[23] The point of great interest about this proof in the Arabic version is that in fact it reverses the Archimedean procedure. Archimedes went from propositions regarding center of gravity to the general law of the lever. With the *Liber karastonis*, however, we find the law of the lever proved first on dynamic grounds and then used to prove the identity of the force action on a lever arm of (1) a weight distributed continuously over a segment of a lever arm and (2) that same weight concentrated at the midpoint or center of gravity of the distributed weight. I have discussed this proof in detail elsewhere and so refrain from repeating its details here.[24] Suffice it to say, the procedure followed in the *Liber karastonis* is no longer open to the objection of circularity believed by Mach to be present in Archimedes' proof.

In point of time the latest of the more or less intact mechanical works is that of Hero of Alexandria, whom recent research now tends to date about A.D. 62.[25] Unfortunately, we possess only a fragment of the Greek text of Hero's work, but we do have virtually the complete text in Arabic under the name *On the Lifting of Heavy Things (Fī raf' al-'ashyā' al-thaqīlat)*.

---

[21] *Ibid.*, pp. 79–117 for the Latin text and English translation. A German translation of the Arabic text is given by E. Wiedemann, "Die Schrift über den Qarastûn," *Bibliotheca Mathematica*, 3 Folge, Vol. *12* (1911–12), 21–39.

[22] Moody and Clagett, *Medieval Science of Weights*, p. 79.

[23] *Ibid.*, pp. 102–4.

[24] *Ibid.*, pp. 368–71.

[25] O. Neugebauer, "Über eine Methode zur Distanzbestimmung Alexandria-Rom bei Heron," *Det Kgl. Danske Videnskabernes Selskab. Historisk-filologiske Meddelelser* 26:2 (1938), 23; 26:7, (1939). Cf. *Centaurus*, Vol. *1* (1950–51), 117–31.

The Arabic text has been published twice.[26] Although it is quite clear that Hero's work was not known in the Latin West in the Middle Ages, Hero's *Mechanics* is important to us for several reasons; and since no English translation exists, I have taken the opportunity here to include as Document 1.3 some pertinent selections from the *Mechanics* which I have translated from the Arabic text. (1) In the first place, the document reveals the influence of Archimedes on Hero, particularly in respect to the doctrine of centers of gravity and to the statement of the law of the lever (see passage 1.24). (2) Even more interesting to students of medieval mechanics is Hero's extension of the law of the lever to bent or irregular levers (see I.33 and I.34). He clearly reveals an understanding of the principle that regardless of the angle of the lever arms at the fulcrum, effective weight (i.e. static moment) is determined by the weight and by the horizontal distance to the line of the vertical running through the fulcrum. Of course, unlike the later medieval treatment of the problem, Hero has no formal mathematical proof. From Hero we get the impression that Archimedes has already proved this generalized law for irregular levers. (3) Hero's *Mechanics* also stands as a forerunner to the medieval Latin treatises in respect to another common problem of statics, the problem of the force exerted by a weight on an inclined plane. While Hero's treatment of this problem is ingenious (see I.23), it is incorrect and far inferior to the beautiful proof of the medieval author of the *De ratione ponderis* (see Chapter 2, pp. 106–7, and Doc. 2.1, prop. 10 with its commentary). Incorrect as Hero's proof is, yet it is superior to the later efforts of the Greek mathematician Pappus (*ca.* 300 A.D.?).[27]

[26] Hero's *Mechanics* was published first by B. Carra de Vaux, *Les Mécaniques ou L' Élévateur de Héron d'Alexandrie* (Paris, 1894) as an extract from the *Journal asiatique*, Ser. 9. Vol. *1* (1893), 386–472; Vol. 2 (1893) 152–269, 420–514; then it was published later by L. Nix in Vol. 2 of *Heronis Alexandrini opera quae supersunt omnia*, Vol. 2, Fasc. II (Leipzig, 1900).

[27] For the treatment of the inclined plane problem by Pappus see F. Hultsch, editor, *Pappi Alexandrini collectionis quae supersunt; e libris manuscriptis edidit, latina interpretatione et commentariis instruxit Fridericus Hultsch*, Vol. *3* (Berlin 1878), 1054–58 (Bk. VIII, prop. 10). For an English translation, see M. R. Cohen and I. E. Drabkin, *A Source Book in Greek Science* (New York, 1948), pp. 194–96. As these editors point out, Pappus' solution can be reduced in modern terms to the following formula: $F = C/(1 - \sin a)$ where F is the force required to just move a weight up an inclined plane of inclination a and C is the force needed to move that same weight in a horizontal plane. This is erroneous in two major respects: (1) It assumes that a force proportional to the weight is necessary to move a weight on a horizontal plane which is untenable by inertial physics. This fallacy was noted by Galileo (see Doc. 2.6 below). (2) As the angle a approaches 90° the force F increases without limit; whereas it should increase toward the force needed to lift the weight vertically.

(4) If Hero shows himself influenced by Archimedes in his use of centers of gravity, the influence on him of Pseudo-Aristotle is even more apparent, particularly in his treatment of the five simple machines by which "a known weight is moved by a known power *(al-qūwat)*." He tells us in Book II, chapter 1, that these machines rest on a single principle and, as in Pseudo-Aristotle's account, this principle is the principle of the circle: the greater the radius, the greater the effectiveness of a moving weight on the end of the radius. He had already, in Book I, chapter 2–4, 6, and 7, repeated conclusions about the movement of circles much like those given in the *Mechanica* of Pseudo-Aristotle, and the reader may examine these conclusions in Document 1.3 below. Hence it is the "principle of the circle" which is the cause of our being able to move a large weight with a small force.

The principle of the circle as outlined by Hero can be explained as follows. We have two circles with the same center A and with diameters BG and DE (see Fig. 1.5). The circles are "mobile" around point A and

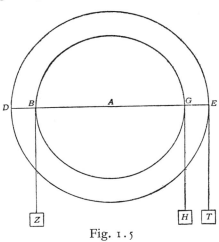
Fig. 1.5

are in a plane perpendicular to the plane of the horizon. If we hang equal weights Z and H from points B and G respectively, then equilibrium results, since both the weights and the distances are equal. If we make BG the beam of a balance suspended from A, and if we shift the weight H suspended from G, suspending it rather from E, the circle turns. But if we increase the weight Z, equilibrium will result when $T/Z = BA/AE$. And thus the line BE acts as a balance beam turning about the suspension point A. Hero says once more that Archimedes has already proved this in his *Equilibrium of Planes*.

(5) But while all machines are similar to the circle or lever, the fundamental principle that explains why one can lift a large weight with a small force is for Hero the principle of virtual velocities or virtual work. True, this principle was expressed by Hero in a primitive form: "the ratio of force to force is [inversely] as the ratio of time to time" (see Doc. 1.3, passage II.22). This principle is elsewhere called the "principle of slowing up." A careful study of the examples shows that what is really involved is distance, i.e., that the ratio of force to force is inversely as the distances through which the forces act. Thus, for example, a small force applied to the last wheel in a wheel-axle train has to be continually applied through several revolutions of that wheel to lift a weight through the distance of one revolution. The same "principle of slowing up" is applied to the lever (II.25) and to the wedge and screw (II.28). Quite clearly this "principle of slowing up" is one of the "causes" that Hero regards as scientific knowledge and by which we explain questions of mechanics. For example in a question (II.34, quest. n) which we have not translated

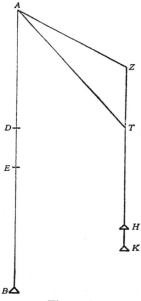

Fig. 1.6

but here paraphrase for the sake of economy, Hero falls back on the work principle to explain why it is the more difficult to move a weight suspended by a cord, the closer to the point of support the cord is pushed. The example he gives (see Fig. 1.6) is of weight B supported by a chord AB attached at A. Then if we take hold of the chord at point D and swing it over so that the chord is now in the position AZH, it being held at Z, it is clear, Hero tells us, that H is higher than point B. Further, if we hold the line at E and draw it as before, the total line will now be ATK, i.e., T will be lower than Z, and furthermore the weight will now be at a point K lower than H. "Thus when we draw the weight starting from point E, it arrives at K, and when we draw it, starting from D, it arrives at H. Thus *one lifts the weight further* when he starts from D than when he starts from E, and in order to carry the weight higher, a greater force is necessary than to carry the weight less high because to lift a weight higher demands a longer time." We are quite obviously here in the presence of the principle of work used to explain why a greater force has to be exerted.

One might suggest that for Hero the principle of work does not involve the "virtual" or "potential" work which would be applied to statical problems and theorems, but rather that the principle is applied by him only to machines which perform "actual" work, so that he is saying that when machines are in operation what is gained in force is lost in "time." But his association of the lever with the balance and the circle cuts across the boundary between "potentiality" and "actuality." In the late sixteenth and seventeenth century, as will be seen in the next chapter, writers like Stevin deliberately tried to keep a dynamic principle of work out of problems of statics where no movement takes place and, like Hero, to reserve the principle of work or its equivalent for the explanation of why machines develop mechanical advantage. Of course, as we shall show, Galileo completely accepted the principle as applying equally well to the balance in equilibrium and to the lever in motion.

We can mention briefly one final Greek author, namely, Pappus of Alexandria, whose eighth part of his *Mathematical Collection* was predominantly influenced by Hero and Archimedes. Pappus exercised no direct influence on medieval Latin statics. We have already mentioned his erroneous treatment of the inclined plane problem, where he introduces as a factor to be considered the force necessary to move the weight on a horizontal plane (see note 27 above). As Document 2.6, appended to the next chapter, demonstrates, Galileo was acquainted with and rejected Pappus' solution of the inclined plane problem.

So far as Arabic statics is concerned, it is important to notice that both of the Greek traditions which I have just outlined were well represented. This is true even though there is little evidence of an Arabic translation of either the *Mechanica* of Pseudo-Aristotle or the *Equilibrium of Planes* of

Archimedes[28]—the principal sources of the two traditions (although perhaps indirect evidence favors a translation of the *Equilibrium of Planes*). However, it will be recalled that our only extant text of Hero's *Mechanics* contains significant vestiges of both the dynamical-Aristotelian and statical-Archimedean traditions. Similarly, the Aristotelian tradition is even more pointedly represented in the *Liber karastonis* of Thābit ibn Qurra, which the reader will remember was a revision of a Greek work whose title we know only by its Latin translation as *Cause karastonis*. We have already noticed how the *Liber karastonis* indirectly reflects the Archimedean procedure using centers of gravity, although it reverses the procedure of Archimedes by deducing the idea of center of gravity from the principle of static moment. But more important, the author attempts to derive the law of the lever from Aristotle's dynamical rules of movement. Following the Latin text, we can reconstruct the reasoning in the following manner.[29]

Two fundamental axioms seem to underlie his proof. (1) In the case of the lever, the force or power of movement (in Latin, *virtus motus*) of a weight on the extremity of the lever arm is proportional to both the weight and a potential arc swept out by the point of extremity. In modern terms, $F \propto WA$. (2) Equilibrium results when the force of movement on one side of the lever is equal to the force of movement on the other side of the lever. Both of these assumptions were implicit in the earlier *Mechanica* of Pseudo-Aristotle, and no doubt the original Greek author of the *Cause karastonis* was influenced by that treatise.

The author appears to have derived the first assumption directly from the Aristotelian rules of movement, for he states as his first proposition the Aristotelian dynamic law:[30] "In the case of two spaces which two moving bodies describe in the same time, the proportion of the one space to the other is as the proportion of the power of the motion of that which

---

[28] Not only have I not been able to find any extant manuscripts of either of these works mentioned in the catalogues of the Arabic collections of at least the western European libraries, but the standard Arabic biographers do not seem to refer to them. See E. Wiedemann, "Beiträge zur Geschichte der Naturwissenschaften III," *Sitzungsberichte der Physikalisch-medizinischen Sozietät in Erlangen*, Vol. *37* (1905), 234, 247. However, Wiedemann (p. 249) following Carra de Vaux raises the possibility that the views on center of gravity extant in al-Khāzinī's *Balance of Wisdom* (see commentary to Document 1.3, passage I.24), which al-Khāzinī assigns to 'Abū Sahl and Alhazen, may have originated in the lost work of Archimedes *On Balances*, which Vailati has so penetratingly reconstructed. But it is conceivable that these Arabic authors drew their statements about centers of gravity from Hero and the *Equilibrium of Planes*.

[29] Cf. my comments in Moody and Clagett, *Medieval Science of Weights*, pp. 363–66.

[30] *Ibid.*, p. 91.

cuts the one space to the power of the motion of that which cuts the other space." If we assume that with a constant time the Aristotelian dynamic rule [31] holds that $F \propto WS$, i.e., that the force is proportional to the weight and the distance, and we notice that in the case of lever arms the distance is arcal—then the first assumption follows, $F \propto WA$, where A is the arcal distance swept out or potentially swept out. The second assumption is, of course, an appeal to the basic idea of virtual velocities. If we set the lever in motion, then $W_1 A_1 = W_2 A_2$.

The proof of the law of the lever of Thābit then proceeds as follows: [32] (1) Equal weights at equal distances are in equilibrium, "since the power of motion at the two points [of the extremities] is equal." The power of motion (WA) on each extremity would obviously be equal since equal weights are posited, and, being at equal distances, they would sweep out equal arcs. (2) Equal weights at unequal distances are no longer in equilibrium, since the power of motion at the more distant extremity is now greater. (3) If the weights are no longer kept equal, but if they are selected so that they are inversely proportional to their distances from the fulcrum, equilibrium prevails. This is because the equality of their forces of motion has been restored. He reasons that this is so as follows: If we have weights $W = W_1 \neq W_2$ and arm lengths $l_1 > l_2$, from (2), $W_1 A_1 > W A_2$. To produce equilibrium we have to add to W some weight r such that $(W + r)A_2 = W_1 A_1$. If $W_2 = W + r$, then $W_2 A_2 = W_1 A_1$. But by a previous proposition (2), it is clear that $A_2/A_1 = l_2/l_1$. Hence by substitution we conclude that when $W_2 l_2 = W_1 l_1$, then $W_2 A_2 = W_1 A_1$. But since $W_2 A_2 = W_1 A_1$ represents a condition of equilibrium, so does $W_1 l_1 = W_2 l_2$, and this is the law of the lever.

As we have pointed out in the discussion of Greek statics above, the author of the *Liber karastonis* goes on to prove with the law of the lever that, without disturbing equilibrium, we can replace a material section of a beam by a weight equal to that material section but suspended from the midpoint of a line replacing that section. [33] With this proved, he then arrives at the same formulation as the *De canonio* for the weight necessary to hang on a shorter arm of a Roman balance to equilibrate the additional weight of the longer arm. We here simply note without going into the details of the proof that he arrives at the following conclusion (see Fig.

---

[31] See Chapter 7 below for a discussion of the Aristotelian rules of movement.
[32] Moody and Clagett, *Medieval Science of Weights*, pp. 92–97.
[33] Ibid., pp. 102–9 (prop. 6).

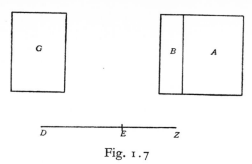

Fig. 1.7

1.7):[34] $e = (db \cdot AB)/2\,GA$ where e is the desired weight, AB is length of the beam, GA is the length of the shorter arm, db is the weight of the excess of the longer arm over the shorter, and the length of that excess is DB.

The dynamic influences so prominently present in the *Liber karastonis* were further demonstrated in a small work appearing in Arabic and Latin manuscripts, bearing the title in Latin of *De ponderoso et levi*, and attributed to Euclid. We have quoted from this work extensively in Chapter 7 below. This work was of importance to the hydrostatic developments among the Arabs and in the Latin West but played no significant role in the development of *statics* as such. But it is indicative of the great concern of the Arabs with problems of specific gravity. A number of treatises considering such problems of specific gravity have been singled out and studied—including a brief fragment of Archimedes' *Floating Bodies* which includes some of the axioms and enunciations of the original work without their demonstrations[35] (see Doc. 1.4). The particular interest of this fragment is that in it the postulate is introduced by a definition of specific weight. But, of course, a number of the treatises contain such a definition, and the Latin work *De insidentibus in humidum* attributed to Archimedes, which undoubtedly had as its principal source one or more Arabic works, contains such a definition (see Chapter 2 below).

It is of interest to note that one of the Arabic manuscripts of the fragment of the genuine *Floating Bodies* of Archimedes also contains a number of more general statements of concern to statics:[36]

---

[34] *Ibid.*, pp. 110–15 (prop. 8). See also pp. 371–72.

[35] Wiedemann, "Beiträge, VII," *Sitzungsberichte der Physikalisch-medizinischen Sozietät*, Vol. *38* (1906), 152–62 translates and discusses the Archimedean fragment. In "Beitrage, VI," *Sitzungsberichte*, Vol. *38* (1906), 10 n. 1, numerous Arabic works on specific gravity are mentioned. Cf. the summary, pp. 163–80.

[36] The Arabic text of this fragment was not available to me. I have accordingly rendered the German translation of Wiedemann (*Sitzungsberichte*, *38*, 157–59) into English.

[Title]. That which the geometer understands by the [following] expressions: "weight," "the heavy body," "center of gravity," "distance," and "the ratio of weight to weight is as the ratio of distance to distance inversely."

If we assume any point and about it a number of magnitudes, and if we assume a power in each of these magnitudes pressing towards the point initially assumed and given, then the given point is the middle point of the whole. All the given magnitudes which are about the point constitute the "heavy body." The force by which it [the heavy body, concentrated] in that point pushes toward the center of the universe is its "weight." Then we assume that this weight will push towards the middle point of the universe. The middle point of the universe must be contiguous to (i.e., identical with) a single point. Now that point of this body with which the middle point of the universe coincides, is the "center of gravity" of this body. The "distance" is the line which is drawn between the centers of gravities of two weights.

We [now] turn ourselves to the expression "the ratio of the weight to the weight is as the ratio of the distance to the distance inversely." If one draws the distance, i.e., the connecting line in the space between the centers of gravities of two weights in the plane which is parallel to the horizon, then that expression signifies that the ratio of the weight to the weight is as the ratio of the magnitudes of the distances which the common point—i.e., the place of suspension of the mechanism—divides one from the other, and indeed the ratio of one to the other is an inverse one . . . .

This treatment of centers of gravity is in conformity with the Aristotelian idea of the center of the world and is essentially dynamical—and, of course, is empirically without significance as a method to locate centers of gravity.

As a matter of fact, although centers of gravity play a minor role in Western medieval statics, they loom large in Arabic treatises—particularly in treatises concerned with specific weight. It is worth pointing out that the treatise on the *Balance of Wisdom* by al-Khāzinī (*fl.* 1115–22), whose main objective is the determination of specific gravity, is much concerned with the concept of center of gravity; and we have, consequently, included in our volume some of the sections on centers of gravity and weight (see Doc. 1.5). Since al-Khāzinī a little later in his work includes in separate chapters both the fragment of the *Floating Bodies* of Archimedes and the *Liber de ponderoso et levi* attributed to Euclid, it is not surprising to find the influence of both present in this discussion of centers of gravity and weight.

With our remarks on the highlights of Greek and Arabic statics now complete, we are prepared to consider in the next chapter the achievements in statics of the Western Latin mechanicians, and particularly the statical advances of the celebrated mathematician, Jordanus de Nemore.

Document 1.1

*The Book on the Balance**
Attributed to Euclid

1. [DEFINITION] Weight is the measure of the heaviness and lightness of one thing compared to another by means of a balance.

2. [Axiom I] When there is a straight beam of uniform thickness, and there are suspended on its extremities two equal weights, and the beam is suspended on an axis at the middle point between the two weights, then the beam will be parallel to the plane of the horizon.

3. [Axiom II] When two weights—either equal or unequal—are placed on the extremities of a beam, and the beam is suspended by an axis on some position of it such that the two weights keep the beam on the plane of the horizon, then if one of the two weights is left in its position on the extremity of the beam and from the other extremity of the beam a straight line is drawn at a right angle to the beam in any direction at all, and the other weight is suspended on any point at all of this line, then the beam will be parallel to the plane of the horizon as before.

This is the reason that the weight is not changed when the cord of one of the two sides of the balance is shortened and that of the other is lengthened.

[Propositions]

4. [Prop. 1] This being assumed, we posit straight line AB (see Fig. 1.8) as a beam of a balance whose axis is at point C, and we draw CE at right angles to line AB, and we extend it in a straight line to point D, and we make line CD equal to CE, and we complete the square CH by drawing parallels. Then we place equal weights at points A, H, and E.

And so I say that these three weights keep lines AB, ED parallel to the horizon (i.e., in equilibrium).

---

* Translated from the Arabic text of F. Woepcke in the *Journal asiatique*, Ser. 4, Vol. *18* (1851), 217–32.

[Euclid, *On the Balance* 25]

The proof of this: One weight has been placed on one of the extremities of line *AB* at point *A*. From the other extremity we have drawn a line at right angles, the line *BH*, and we have placed on it a weight equal to the

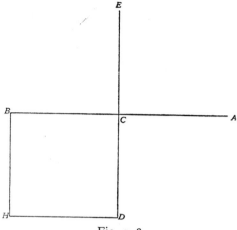

Fig. 1.8

weight which is at point *A*. And so the two weights maintain the line *AB* parallel to the horizon [by Axiom II]. For the same reason it is necessary that the two weights which are at points *E*, *H* keep line *ED* parallel to the horizon. Thus weights *A*, *E*, *H* will keep lines *AB*, *ED* parallel to the horizon.

It is clear that if the weight which is at point *H* is removed to point *B* from which the line *BH* was drawn at right angles, then with weight *A* it maintains line *AB* parallel to the horizon, just as it was necessary when the weight was at point *H*. The line *ED* will accordingly not be in equilibrium, since the weight *E* will make it incline on its side. But if weight *E* is moved to point *C*, or if weight *E* is left on its place and a weight equal to it is placed at point *D*, then the weight *E* balances the line *ED* and it will be parallel to the horizon. We conclude from this that the weight which is at point *H* was one weight which stood in place of two weights at points *B*, *D*, each of which was equal to it.

5. [Prop. 2] With this assumed, we posit line *TH* (see Fig. 1.9), and we divide it into two [equal] parts at point *C*, and we describe on lines *TC*, *CH* two circles *TEC*, *CBH* in the same plane. And let point *C* be the axis of the balance. Let us take two equal lines *CZ*, *TW*, and draw two lines *WA*, *ZE* at right angles to line *TH*. And we draw lines *TA*, *CE*. We draw line *ACB* in a straight line. We draw line *BH*. Then we place three equal weights at points *A*, *E*, *H*.

Then it is known from what we have already proved that these three weights maintain the two lines *AB*, *EC* parallel to the horizon, and also the plane of the two circles and all of the lines described therein. It is

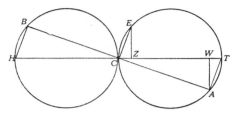

Fig. 1.9

clear that if the weight which is at *A* is moved to point *W* from which the line *WA* is drawn at right angles to line *TH*, the line *TH* remains parallel to the horizon, and the planes of the two circles will incline in the direction of the weight *E*. And so if weight *E* is moved to point *Z* from which was drawn *ZE* at right angles to line *TH*, then both line *TH* and the planes of the two circles remain parallel to the horizon. It is already evident that if line *TH* is divided into two parts at *C*, and point *C* is the axis of the balance, and a weight is placed on one of its extremities, namely point *H*, and the line *CZ* is taken as equal to the line *TW*, and two weights each equal to the weight *H* are placed at points *Z*, *W*, then the three weights will maintain the line *TH* parallel to the horizon.

In the same way we demonstrate that if the weight placed at point *Z* is moved away from the axis toward *W* a certain amount, and the weight which is at *W* is brought nearer to the axis by the same amount, then the line *TH* will remain parallel to the horizon, balancing the weights.

6. [Prop. 3] With this assumed, we posit line *AB* (see Fig. 1.10) as a beam of a balance, and we divide it into two [equal] parts at point *C*, and we make point *C* the axis. We divide line *AC* into as many equal parts as

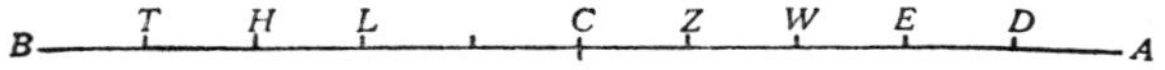

B———T—H—L———C—Z—W—E—D———A

Fig. 1.10

we wish. And so we make five divisions at points *D*, *E*, *W*, and *Z*. We divide the line *CB* into similar quantities at points *T*, *H*, *L* . . . . It is known from what we have deduced that when three equal weights are taken and placed at points *D*, *Z*, *B*, then the weights keep line *AB* parallel to the horizon. But the excess of line *CB* over line *DC* is that by virtue of which weight *B* outweighs weight *D*, and it is *TB*. And *TB* is equal to *ZC* in

[Euclid, *On the Balance*   27]

length, and it was already evident that it is equal to it in force of weight (*qūwat al-thiql*).

Then we move weight D to point E and weight B to point T, and we leave weight Z in its place. Then the three weights, according to what we proved in what went before, keep the line AB parallel to the horizon. And the excess of TC over EC is TH, and TH is equal to ZC in length, and ZC is equal to TH in force of weight, just as we proved. And ZC then is equal to TB in force of weight. Thus TB is equal to TH in force of weight.

In the same way we demonstrate that all of the quantities which are taken from the line CB and which are equal in length are equal in force of weight. It is then clear that the diminution of force of weight when the weight is moved from B to T is equal to the diminution that occurs when a weight is moved from T to H. The same reasoning applies to all the quantities of equal lengths taken from CB.

7. [Prop. 4] When a beam of a balance is taken and divided into unequal segments and its axis is at the point of division, and two weights are taken—the ratio of one to the other being like the ratio of the segments of the beam—and the lighter of the weights is suspended on the extremity of the longer of the segments and the heavier of the weights is suspended on the extremity of the shorter segment, then the beam is balanced in weight and parallel to the horizon.

Exemplification: The beam AB (see Fig. 1.11) is divided at point C

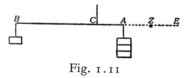

Fig. 1.11

into two unequal segments, and two weights are suspended at points A, B, and the ratio of weight A to weight B is as the ratio of distance CB to the distance CA. Then I say that the two weights A, B maintain the beam AB parallel to the horizon.

Demonstration: We increase CA in distance by the quantity AE so that EC is equal to CB. Let the distance EC be three times AC. Thus when the weight A is removed, and a weight equal to weight B is placed at point E, and another weight equal to weight B is placed at point C, the beam EB is balanced. It is known from what went before that the movement of weight E to Z and the movement of weight C to A balances the beam EB. And because, if one adds at point A another weight equal to the first, its force will be as the force of the first, it follows from this that if the weight Z which was at point E is moved to point A, and there is also placed at A another weight equal to each of the two equal weights moved from Z to A and from C to A, the beam AB will be balanced and the three equal weights which are at point A, and each of which is equal to the weight B, will, with weight B, keep the beam AB parallel to the horizon. But the distance EC is the same number of multiples of AC as all the weights at point A are of one of them, and each one of these weights is equal to the weight B, and the distance EC is equal to the distance CB. Thus CB is the same number of multiples of distance AC as the weights at A are of weight B. And the ratio of the weight A to weight B is as the ratio of distance CB to the distance CA. And so the two weights A, B maintain the beam AB parallel to the horizon, and this is what we wished to prove. The *Book of Euclid* is completed. I have found the book in another copy attributed (or belonging) to the Banū Mūsā [which?] I have collated *('āraḍtu)* with the copy of Abū 'l-Ḥusain al-Ṣūfī.

COMMENTARY

1. The question of whether this work is truly by Euclid is a difficult one. Woepcke thought not, since no Greek or Arabic author mentions such a work.[37] However this is by no means conclusive. It seems to be quite likely that the text was translated from the Greek and that in all probability there existed a Greek text bearing the name of Euclid. As Woepcke pointed out, the text of the medieval *De canonio* speaks of the law of the lever as "having been demonstrated by Euclid and by Archimedes (sicut demonstratum est ab Euclide et Archimede, et aliis)." Furthermore, it appears certain that the *De canonio* was translated from the Greek rather than the Arabic.[38] Hence, the author of the *De canonio*, writing in Greek, apparently knew of a demonstration of the law of the lever attributed to Euclid, and in Greek. Woepcke seems to feel that the explicit of the treatise speaks of a copy attributed to the Banū Mūsā (lī Banū Mūsā). I suggest this may merely mean that the copy belonged to the Banū Mūsā. Note that the scribe speaks of a copy which belongs to one Abū 'l-Ḥusain al-Ṣūfī. Woepcke identifies him with the astronomer

---

[37] F. Woepcke, "Notice sur des traductions arabes de deux ouvrages perdus d'Euclide," *Journal asiatique*, Ser. 4, Vol. 18 (1851), 217.

[38] Moody and Clagett, *Medieval Science of Weights*, pp. 58–59.

[Euclid, *On the Balance* 29]

Abū 'l-Ḥusain 'Abd al-Raḥmān ibn 'Umar al-Ṣūfī al-Rāzī (*fl.* 903–86). Presumably this roughly dates the text at hand.

2. Compare Euclid's first axiom with the first postulate of Archimedes' *On the Equilibrium of Planes* (Doc. 1.2).

3. Axiom II is crucial for Euclid's demonstration. It holds that so long as the weight is in a line perpendicular to the beam, it exerts the same force. He notes that this axiom explains why lengthening or shortening the cord of a balance pan does not disturb equilibrium. See propositions E.3 and P.3 of the medieval Latin *Elementa Jordani* and *Liber de ponderibus* (Moody and Clagett, *Medieval Science of Weights*, pp. 132–33, 156–57), both of which hold that inequality in the length of the pendants does not disturb the equilibrium.

4. The first proposition applies Axiom II to a particular case. The reader is reminded that the weights are thought of as acting perpendicular to the plane of the page. Thus, if ED and AB, BH, and HD are thought of as rigid but weightless rods, equal weights at H and E tend to produce the equilibrium of rod ED, while the equality of A and H tends to produce equilibrium in AB.

5. As in the previous proposition, the weights are thought of as acting in a direction perpendicular to the paper. By a slight extension of proposition 1 and by using Axiom II it is clear that BA (and EC) are in equilibrium if equal weights are at H, E, and A, since $BH = EC$ and $BC = AC$. If BA is in equilibrium about C, and EC is in equilibrium about C, then the plane determined by these lines is in equilibrium. Thus all the lines in the plane, including TH, are in equilibrium. One does not disturb the equilibrium if one moves E along EZ to Z, and A to W, since EZ and AW are both perpendicular to HT (and Axiom II is applied). This would be true for any points Z and W so long as $CZ = WT$, and hence it would be true that if we moved a weight Z (equal to H) the same distance from Z as we moved an equal weight W from T, the equilibrium of TH would be maintained.

6. In proposition 3 the author introduces a concept which he calls "force of weight." He concludes here that equal lengths of the beam are equal in "force of weight." And thus when an equal weight is moved a unit distance towards the fulcrum on one side of the balance, the same diminution is produced as when it is moved another unit distance toward the fulcrum.

7. Finally we are at last in the position to prove the law of the lever, which he proves only for the special case of weights and distances being

in a 3:1 ratio. But the understanding is clearly there that the same reasoning would apply to any ratio. It has been noted in the text of the chapter that this proof is Archimedean in character and has no trace of the Aristotelian-Heronian dynamic approach.

Document 1.2

Archimedes, *On the Equilibrium of Planes*
or
*On Centers of Gravity**

[Book I, Postulates]

WE POSIT:

1. Equal weights (βάρεα) suspended at equal distances are in equilibrium (ἰσορoπεῖν); equal weights suspended at unequal distances are not in equilibrium but incline toward the weight suspended at the greater distance.

2. When weights are in equilibrium at certain distances, if something is added to one of them, they will not be in equilibrium but will incline toward the weight to which something is added.

3. Similarly, if something is taken away from one of the weights, they are not in equilibrium but incline toward the weight from which nothing was taken.

4. In two equal, similar, and coinciding plane figures, the centers of gravity (κέντρα τῶν βαρέων) also mutually coincide.

5. The centers of gravity of unequal but similar weights will be similarly situated. We say that points are similarly situated in relation to similar figures when the straight lines drawn from these points to equal angles make equal angles with the corresponding sides.

6. If magnitudes (μεγέθεα) are in equilibrium when suspended at certain distances, then magnitudes equal to them, suspended at the same distances, will also be in equilibrium.

7. The center of gravity of every figure whose perimeter is concave in the same direction is necessarily inside of the figure.

---

* Translated from the Greek text of J. L. Heiberg (Leipzig, 1913). Compare the English paraphrase of T. L. Heath, *The Works of Archimedes* (Cambridge, 1897).

[Propositions]

With these suppositions, [we propose the following:]

1. Weights (βάρεα) in equilibrium at equal distances are equal.

For if they are unequal, when the excess of the greater has been removed the remaining weights will no longer be in equilibrium, since something will have been taken from one of the weights which are in equilibrium [postulate 3]. [But this is absurd by postulate 1.] Hence when weights suspended at equal distances are in equilibrium, these weights are equal.

2. Unequal weights suspended at equal distances are not in equilibrium but incline toward the greater weight.

For taking away the excess, they will be in equilibrium, because equal weights suspended at equal distances are in equilibrium [postulate 1]. Therefore, if we add that which has been removed, it will incline toward the greater weight, since one will have added something to one of the weights in equilibrium [postulate 2].

3. Unequal weights suspended at unequal distances can be in equilibrium, the greater weight being suspended at the lesser distance.

Let A, B be unequal weights (see Fig. 1.12), and let A be the greater.

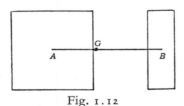

Fig. 1.12

Let them be in equilibrium when suspended at distances AG, GB. It is necessary to demonstrate that AG is less than GB.

For if it is not less, take away the excess by which A exceeds B. Since one has taken away something from one of [two] weights in equilibrium, it inclines at B [postulate 3]. But it does not so incline, for if GA is equal to GB, the weights will be in equilibrium [for the equal weights are at equal distances; postulate 1]; and if GA is greater than GB, then the inclination will be at A, since equal weights suspended at unequal distances are not in equilibrium but the inclination is at the [weight] suspended at the greater distance [postulate 1]. Therefore GA is less than GB.

Thus it is clear [conversely] if the unequal weights suspended at unequal distances are in equilibrium, then the greater weight is at the lesser distance.

4. If two equal magnitudes (μεγέθεα) do not have the same center of

[Archimedes, *Equilibrium of Planes*

gravity (κέντρον τοῦ βάρεος), the center of gravity of the magnitude composed of both these magnitudes is at the middle point of the straight line joining the centers of gravity of the magnitudes.

Let A be the center of gravity of weight A (see Fig. 1.13), and B be

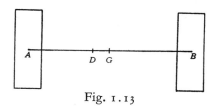

Fig. 1.13

that of B. Having drawn the line AB, let it be divided into two equal parts at G. I say that point G is the center [of gravity] of the magnitude composed out of the two magnitudes.

For if not, let the center of gravity [of the magnitude composed of both magnitudes A, B] be at D, if that is possible [for it has been demonstrated that it is on line AB].* Since, therefore, point D is the center of gravity of the magnitude composed of A, B, equilibrium results with point D supported. Therefore magnitudes A, B are in equilibrium at distances AD, DB. But this is impossible, [for equal weights at unequal distances are not in equilibrium; postulate 1]. Therefore, it is clear that G is the center of gravity of the magnitude composed of A and B.

5. If the centers of gravity of three magnitudes are situated on the same straight line, and the magnitudes are equal in weight, and the distances between the centers [of gravity] are equal, the center of gravity of the composite of all the magnitudes will be a point which is also the center of gravity of the middle [magnitude].

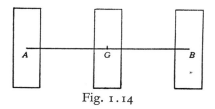

Fig. 1.14

Let there be the three magnitudes, A, B, G (see Fig. 1.14), and their centers of gravity, points A, B, G, situated on [the same] straight line. Let [the magnitudes] A, B, G be equal, and let AG and GB be equal lines. I

* Probably in the lost work *On Balances*, but certainly not here; hence this phrase is bracketed.

say that the center of gravity of the magnitude composed of all these magnitudes is the point G.

For since magnitudes A, B have the same weight, their center of gravity will be at point G, since AG and GB are equal. But the center of gravity of [magnitude] G is also the point G. Hence it is clear that the center of gravity of the magnitude composed of all [the magnitudes] will be the point which is also the center of gravity of the middle [magnitude].

Corollary I

Now from all this it is evident that if the centers of gravity of any odd number of magnitudes are situated on the [same] straight line, and if the magnitudes situated at an equal distance from the middle one have the same weight, and if the distances between their centers of gravity are equal—then the center of gravity of the magnitude composed of all the magnitudes will be the point which is also the center of gravity of the middle one of them.

Corollary II

Further, if the magnitudes are even in number, and their centers of gravity are situated on [the same] straight line, and if the middle ones, as well as those [on each side] which are equally distant from them, have equal weight, and the distances between their centers of gravities are equal, the center of gravity of the magnitude composed of all the magnitudes will be the middle point of the line joining all the centers of gravity of the magnitudes, as represented below (see Fig. 1.15).

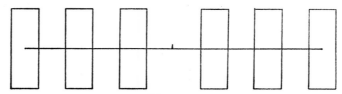

Fig. 1.15

6. [Two] commensurable magnitudes are in equilibrium when they are inversely proportional to the distances at which they are suspended.

Let the commensurable magnitudes be A, B (see Fig. 1.16), whose centers [of gravity] are A, B. Let there be a certain length ED, and let the magnitude A be to the magnitude B as the length DG to the length GE. It is necessary to demonstrate that the center of gravity of the magnitude composed of both A, B is [the point] G.

[Archimedes, *Equilibrium of Planes*

For since A is to B as DG is to GE, and A and B are commensurables, then DG and GE are commensurables, i.e., as one straight line to another. Hence there is a common measure of EG, GD. Let it be N. Let it be

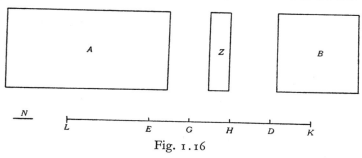

Fig. 1.16

supposed that EG is equal to each of the [lines] DH, DK, and that DG is equal to EL. And since DH is equal to GE, so also DG is equal to EH. Hence LE is equal to EH. Therefore LH is double DG, and HK is double GE. Hence N also measures each line LH, HK, since it measures their halves. But A is to B as DG is to GE, and DG is to GE as LH is to HK, since each [of the lines LH, HK] is twice as much as each [of the lines DG, GE]. Hence A is to B as LH is to HK. As many times as N is contained in LH, Z is also contained in A. Therefore [by Euclid, *Elements*, V, def. 5*] LH is to N as A is to Z. But also KH is to LH as B is to A [Euclid, *Elements*, V.7, corol.]. Hence by reason of equality [Euclid, *Elements*, V.22], KH is to N as B is to Z. And so as many times as N is contained in KH, Z is also contained in B. But it has been demonstrated that Z measures A. Thus Z is the common measure of both A, B. Therefore with LH divided into parts equal to N, and A into parts each equal to Z, the segments each equal to N which are in LH will be equal in number to the magnitudes each equal to Z which are in A. Then if on each of the segments in LH is placed a magnitude equal to Z so that it has its center of gravity on the midpoint of the segment, then all of the magnitudes are equal to A and the center of gravity of the composite of all [the magnitudes] will be [point] E; for all the parts are even in number, and on each side of point E have been placed parts equal in number, LE being equal to HE [prop. 5, corol. II]. Similarly it will be demonstrated that, if in the individual segments of KH [each equal to N] there is placed a magnitude equal to Z so that its center of gravity is in the midpoint of the segment, all the magnitudes would be equal to B, and the center of

---

* As Heiberg suggests; but perhaps preferable is VII, def. 20, interpreted as applying to commensurable magnitudes and their common measure.

gravity of the composite of all [the magnitudes] will be at point D. Therefore A will be [as] situated at E, and B [as] at D. Now the mutually equal magnitudes placed on the line segments are even in number, and their centers are equally distant apart. It is clear, therefore, that the center of gravity of the composite magnitude is the middle point of the line on which are situated the centers of gravity of the middle magnitudes [by prop. 5, corol. II]. And since LE is equal to GD, and EG is equal to DK, hence LG will also be equal GK. Thus the center of gravity of the magnitude composed of all [the magnitudes] is point G. Therefore with A placed at E, and B at D, they will be in equilibrium about G.

7. Now also, if the magnitudes are incommensurable, they will similarly be in equilibrium when suspended at distances inversely proportional to their magnitudes.

Let the incommensurable magnitudes be AB, G, (see Fig. 1.17) and the

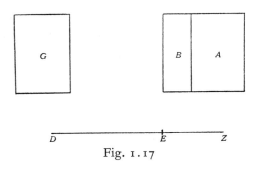

Fig. 1.17

distances DE, EZ, and let AB be to G as ED is to EZ. I say that the center of gravity of both AB, G [together] is E.

For if AB placed at Z, and G at D are not in equilibrium, then either AB is too great in relationship [to G] to be in equilibrium or it is not great enough. Let it be too great. Then take away from AB a quantity less than the excess by which AB is more than G [that is, less than the quantity whose removal] is necessary to re-establish equilibrium, but a quantity just large enough so that the remainder A is commensurable with G. Since, therefore, magnitudes A, G are commensurable, and the ratio A to G is less than the ratio DE to EZ, A and G suspended at distances DE and EZ so that A is placed at point Z, and G at D, will not be in equilibrium [prop. 6]. For the same reason it would never be [that they will be in equilibrium] if G is greater than that which is necessary to keep AB in equilibrium.

COMMENTARY

As I have indicated in the body of the text, Archimedes' proof of the law of the lever has been the occasion of extensive commentary both on the grounds of the nature of his postulates and the possible circularity of his proof when he employs the concept of centers of gravity to pass from the special case of equal weights at equal distances to the general law of the lever. I think it out of place in this treatment of medieval mechanics to review that commentary, but the reader can find the principal discussions in notes 14–16 of this chapter. He can also find a paraphrase of the proofs of propositions 6 and 7 in my recent volume *Greek Science in Antiquity* (New York, 1955), Appendix IV.

Document 1.3

Hero of Alexandria, *Mechanics*[*]

BOOK I, CHAPTER 2. ON CIRCLES. Circles fixed on the same axis will always move in the same direction, and this is the direction in which the axis is moved. But circles which are on two [different] axes but which are in contact with each other by teeth are moved in opposite directions; the one turns to the right and the other to the left. If the two circles [which are engaged] are equal, a single rotation of the one to the right corresponds to a single rotation of the other to the left. But if they are unequal so that one of them is larger than the other, the smaller one turns several times before the larger one makes a single revolution. [The number of turns] depends on the ratio of their sizes.

I.3. In accordance with this introductory statement, let us rotate two equal circles, the one is *HEKD* (see Fig. 1.18) and the other is *ZCTE*,

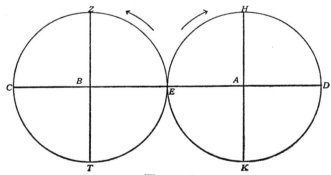

Fig. 1.18

around their centers, *A* and *B*. They touch at point *E*. When they move from point *E* in the same time through distances equal to half the circumferences, then in that time point *E* describes the arc *EHD* in traveling to point *D*, having moved through the same movement as point *C* over the arc *CTE*. There will be some points which move in the same direction and some which move in opposite directions. The points similarly placed

[*] Selections translated from the Arabic text of L. Nix (Leipzig, 1900).

[Hero of Alexandria, *Mechanics*

in the circles move in opposite directions, while those symmetrically placed move in the same direction.... [Then follows further directional analysis of the movement of various points where Hero speaks of upward and downward vertical movements and right and left lateral movements and he concludes:] It is for this reason that likeness and difference [in direction] are only relative, and so we must distinguish for each movement the likeness and oppositeness of movement....

I.4. *On different circles.* Let the circles be unequal with centers at points A and B (see Fig. 1.19). Let the larger circle be the circle whose

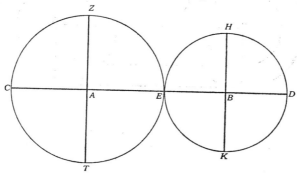

Fig. 1.19

center is at point A. In the case of these circles the relations will not be so perfect as in the case of equal circles. Let there be two points whose revolutions begin from E. And for example, let us make the diameter CE double the diameter ED. Then the arc EZC will be double the arc EHD, as Archimedes has already shown. And so in the time which point E in its movements describes the arc EZ in the direction of C the point E' will describe arc EHD in its oppositely directed movement.... [Then he discusses the directional changes taking place as the result of the difference of size between the two circles; chapter 5 makes an analysis of cases where there are more than two circles.]

I.6. Furthermore, large circles do not always have a movement more rapid (*'asra'*) than smaller circles. But it also happens that small circles are moved more quickly than large. For when the two circles are fixed on the same axis, the larger circles are moved more rapidly than the smaller. But if the circles are separated on different axes, but in the same apparatus, as in the case of wagons with many [but different-sized] wheels, the small circles move more swiftly than the large. Their forward displacements (literally, "movements") are the same, and they are each of them moved in the same time. Hence it is necessary for the smaller circle to revolve many times while the large wheel revolves once. And this is why the small wheel is moved more rapidly.

I.7. And it could also be that the movements of the large circle and the small circle are equally swift even when both are fixed in their motion around the same center. Let us think of two circles fixed on the same center, the center A (see Fig. 1.20). And let there be a tangent to the larger

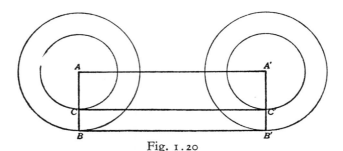

Fig. 1.20

circle, line BB'. And let us join the points A and B. Thus line AB will be perpendicular to the line BB'. And the line BB' is parallel to the line CC', and the line CC' is tangent to the smaller circle. Also let us draw from point A a line parallel to the other two lines, and this is the line AA'. And so if we imagine that the large circle rolls along line BB', the small circle passes along line CC'. And when the large circle has already made one revolution, it will appear to us that the small circle has made one revolution also. Thus the position of the circles will be the position of the circles whose center is at A', and line AB will be in the position of line $A'B'$. For this reason line BB' will be equal to line CC'. And line BB' is the line which the larger circle traverses when it makes a single revolution. And line CC' is the line along which the smaller circle is rolled in a single revolution. And so the movement of the smaller circle is equally as fast as the movement of the larger circle, because the line BB' is equal to the line CC', and things which traverse equal distances in equal times have movements of equal swiftness. In thinking about it, one might consider the conclusion absurd because it is not possible for the circumference of the larger circle to be equal to the circumference of the smaller circle. And so we say that the circumference of the smaller circle has not just turned upon the line CC', but it has traversed the path of the larger circle along with that circle. And so it happens that the movement of the smaller circle is equal in swiftness to the movement of the larger circle as a result of the two movements....

I.8. It is possible for a single point which is moved in two equal (i.e., uniform) movements to describe two unequal lines.

We shall now prove this. Let us suppose a rectangle $ABCD$ (see Fig. 1.21). AD is its diagonal. Let point A traverse the line AB uniformly.

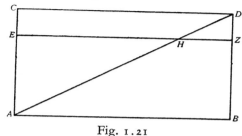

Fig. 1.21

And let line AB be moved over the lines AC and BD with a uniform movement so that it is always parallel to line CD. And let the time during which the point A moves to B be equal to the time in which line AB moves to CD. And so I say that point A is moved over two unequal lines in the same time. The proof of this: When line AB moves during a certain time such that it has arrived at the position EZ, point A moved on the line AB will be at the same time on the line EZ. And the ratio of the line AC to the line AB, i.e., to the line CD, will be as the ratio of the line AE to the line which is between point E and the point in movement on it. But the line AC is to the line CD as AE is to EH. And so then the point which is moved on line AB will have arrived at point H on the diagonal AD. And so we prove similarly that the point which traverses line AB never deviates from line AD. And so it moves on each of the lines AD and AB in the same time, and the lines AD and AB are unequal in length. Thus this point which is moved uniformly, simultaneously describes two unequal lines. But, as we said, the movement of the point on the line AB is a simple one, while its movement on the diagonal AD is a movement compounded of the movement of AB on the two lines AC and BD and of the movement of A on the line AB....

I.23.... Let us assume that we wish to move a weight on an inclined plane. And let its surface be smooth and even, and in the same way [we let] the part of the weight which it supports [be smooth]. It is necessary for us in this situation to apply to the other side some force or some weight strong enough first to balance the weight, i.e. to put it in equilibrium, with, in addition, an excess of force over the weight large enough to lift up the weight. In order to show our statement to be correct, we demonstrate it for a given cylinder (see Fig. 1.28 in commentary below). Since the cylinder does not touch the surface of the plane over a large area, it will naturally roll downward. Then let us imagine a plane passing through the line of contact of the cylinder with the [inclined] plane and perpendicular to that incline. It will be apparent to us that this [new] plane will pass through the axis of the cylinder and cut it into two equal parts—for when there is a circle with a line tangent to it and a line is erected from the point of tangency at right angles, then that line will go through the center of the circle. Let us also erect on that line [of contact], i.e. the line of the cylinder, another plane perpendicular to the horizon. It will not be the plane first erected but will divide the cylinder into two unequal parts, the smaller of which will be in the upper region and the larger of which will be in the lower region. The larger of the sections will overpower the smaller, since it is the larger; and so the cylinder will roll. If, on the other side of the plane passing perpendicular to the horizon, we take from the larger section the quantity by which it exceeds the smaller, then the divisions will be in equilibrium and their total weight will be immobile on the line of contact with the surface [of the incline], and so it will not tend in either direction, i.e. either upward or downward. Therefore we need a force equal to the difference [of the sections] for their equilibrium. And so when we add a little increase to this force, it will overpower the weight.

I.24. I now hold that it is necessarily obligatory to us who are treating of the mechanical arts to consider what is inclination *(al-mail)* and what is center of gravity, either in an actual body or in a noncorporeal one. For although inclination and tendency *('inhirāf)* are not truly spoken of except in bodies, yet no one objects if we say that in plane and solid geometrical figures the center of inclination and the center of gravity is a certain point. Archimedes has already explained the question sufficiently.... Posidonius (?), one of the companions of the Porch (i.e. Stoics?), has already defined the center of inclination and gravity physically. For he said that the center of gravity or of inclination is a point such that when the weight is suspended from that point, it (the weight) is divided into two equal parts. Accordingly Archimedes and the mechanicians who followed him particularized this definition and distinguished between the point of support and the center of gravity. As for the point of support, it is the point on a body or an incorporeal figure such that when the object is suspended from that point, its parts are in equilibrium. By this I mean, it is neither depressed nor elevated. Thus equilibrium occurs when one thing counterbalances another, as is seen in a balance as it swings to the

[Hero of Alexandria, *Mechanics*

plane of the horizon or to a plane parallel to that plane. So Archimedes says weights will not incline either upon a line or upon a point. As for "upon a line"—when the weight is supported from (literally, "upon") two points of that line so that the line will not "incline" and the plane including that line which is perpendicular to the horizon remains perpendicular to the horizon—however the line may be moved, the weight does not "incline." When we have said the weight is "inclining," we mean only its descent downward, i.e. its movement toward the earth. As for equilibrium which is "upon a point," it occurs when the weight is suspended from that point and the body is moved in any way, its parts, one to the other, remaining equal.

One weight counterbalances another if they are suspended from two points of a line (beam) divided into segments, which line is itself suspended from the point of division, and the line is parallel to the horizon, the ratio of weights one to the other being inversely as the ratio of their distances from the point of suspension. That weights suspended in this way are in equilibrium, Archimedes has already demonstrated in his book on the equilibria of figures, where levers are employed....

I.33. [As regards the irregular or bent lever] Archimedes has proved that in this case also the ratio of the weight to the weight is as the ratio of the distance to the distance inversely. As for the distances in the case of irregular and inclined beams, it is necessary that we imagine them as follows (see Fig. 1.22). We extend the cord issuing from point *C* up to

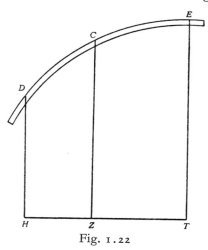

Fig. 1.22

point *Z*, and we draw a line which we imagine goes through *Z* and is equal to *HZ* and let it be fixed, i.e., let it be at right angles to the cord.

And so with the two cords issuing from the two points *D*, *E*, i.e. cords *DH*, *TE*, the distance which is between the line *CZ* and the weight which is at point *E* is marked by *ZT* [and the distance between *CZ* and weight *D* is marked by line *HZ*. Hence,] with the immobility of the balance, *ZH* is to *ZT* as the weight which is suspended from point *E* to the weight which is suspended at point *D*. This is what was proved formerly.

I.34. Let there be a wheel or movable pulley on an axis of center *A* (see Fig. 1.23). Let its diameter be line *BC* parallel to the horizon. Let

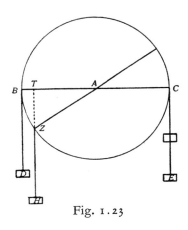

Fig. 1.23

two cords be suspended from the two points *B*, *C*; and let them be *BD* and *CE*; and let us suspend on them two equal weights. It is thus apparent to us that the pulley will not incline on either side, because the two weights are equal and the two distances from point *A* are equal. Then let the weight which is at *D* be greater than the weight which is at *E*. Thus it is apparent to us that the wheel inclines in the direction of *B*, and point *B* inclines with the weight. And so it is necessary for us to know at what position will the heavier weight at *B*, when it descends, come to rest. Let the point *B* descend, and let it reach the point *Z*, and the cord *BD* becomes the cord *ZH*. Then the weight will be at rest. It will be apparent to us then that the cord *CE* will wrap itself on the rim of the wheel; and the support of the weight will be from point *C*, since the part which was wrapped is not suspended. And so let us extend *ZH* to point *T*. Then, because the two weights are in equilibrium, the ratio of one weight to the other will be as the ratio of the distances between *A* and the cords. And so *AC* is to *AT* as the weight at *H* is to the weight at *E*. When we make the ratio *AC* to *AT* the same as the ratio of weight to weight and we draw on line *BC* the line *ZT* at right angles, it will be apparent to us that the wheel moves

[Hero of Alexandria, *Mechanics* 45]

from point B to point Z and [there] comes to rest. And this is the law (*al-qaul*) also for other weights....

II.7. We have already completed the description and explanation of the construction and use of the aforementioned five powers (i.e. machines). We will now speak as follows of the cause whereby each of these instruments is able to move large weights with a small power.

Let us posit two circles having the same center, namely point A (see Fig. 1.24). Let their diameters be lines BC, DE. Let the two circles rotate

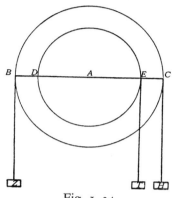

Fig. 1.24

about point A, which is their center; and let the circles be perpendicular to the horizon. Let us suspend two equal weights from points B and C, and they will be at points Z, H. Hence it is apparent to us that the circle will not incline in any direction, because the two weights Z, H are equal and the distances BA, AC are equal. And so BC will be a beam of a balance which is moved about a point of suspension, namely, the point A. If we transfer the weight which is at C, suspending it at E, the weight Z will incline downward and set the circles revolving. But when we increase the weight at T, it balances the weight at Z; and [then] the ratio of weight T to weight Z will be as the ratio of distance BA to distance AE. Then we imagine the line EB as a balance which turns about A as a point of suspension. Archimedes has already demonstrated this in his book on the equilibrium of inclination. And so it will be evident from this that it is possible to move a large weight by a small force; for, the circles having the same center and the large weight being on any arc of the smaller circle and the small force being on any arc of the large circle, the ratio of the radius of the larger circle to the radius of the smaller circle is greater than the ratio of the larger weight to the small force which moves it. And so verily the small force overpowers the large weight.

II.8. Since we have shown our example regarding the circle to be correct, we now wish to demonstrate this on (each of) the five powers (i.e. machines). After doing this, we shall make clear their demonstration *(burhān)*. The ancients *(al-qudamā')* advanced as a premise this principle *(muqaddimat)*. And so let us now demonstrate this first for the instrument which is called the lever.... Let there first be a lever parallel to the ground, and let it be line AB (see Fig. 1.25), and let the weight to be moved by

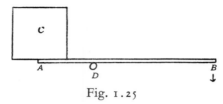

Fig. 1.25

the lever be at point A. It is weight C. Let the moving force be applied to B, and let the stone which is under the lever and on which it moves be at point D. Let BD be greater than line DA.... [When we lift up or push down on B, the weight C moves, thereby describing a smaller circle than B]. If the ratio of line BD to DA is equal to the ratio of the weight at C to the force at B, then the weight C balances the force B. If the ratio of BD to DA is greater than the ratio of the weight to the power, then the force overpowers the weight, because these are two concentric circles and the weight is on the arc of the smaller circle while the motive power is on the arc of the larger circle. And so it is clear to us that the same thing took place for the lever as for the two concentric circles. And so, then, the lever which moves the weights operates by the same cause as operates in the case of the two circles....

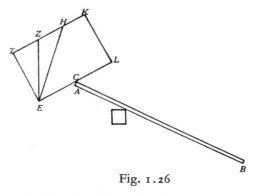

Fig. 1.26

[Hero then proceeds (II.9) to describe the case where the lever is used to upturn a heavy weight C (see Fig. 1.26). He shows that in this case the force *B* develops a larger counterforce at *A* which supports or lifts at any moment section *EHKL* of weight *C*. For since *ZE* is perpendicular to the horizon, section *EZT* balances its equal section *EZH* around the fulcrum *E*. This leaves *EHKL* to be lifted, and the force necessary to maintain it or lift it is computed by the law of the lever. Naturally the force at *B* has to be continually less since the excess section *EHKL* continually diminishes. Finally] the weights [of the sections of *C*] will be in a position of equilibrium, so that it is not necessary to have any force, when the plane passing through point *E* and perpendicular to the horizon divides the weight into two equal parts. This procedure with regard to the lever is related to the circle.... The balance is also related to the circle; which is evident because the circle is a kind of balance. [The other machines are explained in somewhat the same manner. As Hero puts it,] that the five powers (i.e., machines) which move the weight are similar to circles which are mounted on a single center has already been proved by the figures which we have drawn in the foregoing. But I hold that they are even more directly similar to the balance than to circles....

II.22. In this instrument (the wheel and axle) and all machines like it of great power, there occurs a slowing down, because by the amount that the moving power is weaker in relationship to the weight moved, by such an amount do we increase the time. Thus the ratio of force to force is as the ratio of time to time [inversely]. For example, when a power of two hundred talents is applied to wheel *B* (see Fig. 1.27), and it puts the weight

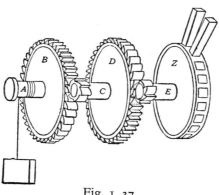

Fig. 1.27

[one thousand talents] in movement, a complete turn of *B* is necessary in order for the weight to be lifted the length of the circumference of the axle *A*. If the movement is to be accomplished by the movement of the wheel *D*, however, it is necessary that the axle *C* turn five times in order that the axis A complete a single turn, since the diameter of wheel B is five times that of axle C; and five turns of C are equivalent to one turn of B.... The wheel D makes the wheel B move, and the five turns necessary to move the wheel D take five times the time of a single turn, and two hundred talents is five times forty talents. And so the ratio of the motive force to the force of the thing moved is an inverse one, and it is so for a multitude of axles and wheels....

COMMENTARY

I.2–7. These sections belong more properly to our history of kinematics of Chapter 3, but we include them here for two reasons: (1) to keep all of the sections from the same document together, since the whole document represents the first time there has been any extended English translation from Hero's *Mechanics*; and (2) to show the dependence of Hero on the previous considerations of the Pseudo-Aristotle in his *Mechanica*. Sections I.2–4 are further developments of remarks made in the introductory section of Pseudo-Aristotle's *Mechanica* (848a), particularly where the latter says that "some people contrive so that as the result of a single movement a number of circles move simultaneously in contrary directions." I.6 grows out of considerations taken up in chapter 7 of the Pseudo-Aristotelian *Mechanica*, where the movements of larger and smaller circles are treated. Unfortunately, Hero does not mention or develop the anticipation of the inertial idea found in chapter 7 of the Aristotelian work, where it is said, "Some people further assert that the circumference of a circle keeps up a continual motion, just as bodies which are at rest remain so, owing to their resistance." I.7, takes up the problem of the so-called wheel of Aristotle and as a problem comes directly out of Pseudo-Aristotle's *Mechanica*, chapter 24. This problem has an interesting history which I. E. Drabkin has traced in his paper in *Osiris* (for full citation, see Bibliography). Since it has little interest for the problems we are treating, I forbear to discuss it here.

I.8. This is Hero's version of the parallelogram of velocities. We have already discussed briefly the Aristotelian version of the parallelogram. Neither Pseudo-Aristotle nor Hero explicitly go from a parallelogram of velocities or displacements to one of forces, although both hold that the velocity or displacement is directly proportional to the force producing the velocity (see Chapter 7 below). At any rate, the early efforts to *resolve* movements into vector components must certainly have been one of the key background developments to the idea of *resolving* forces. Hence these passages belong to a history of statics and dynamics as well as kinematics.

I.23. Although no drawing accompanies Hero's discussion of the inclined plane, the procedure is clear. As Cohen and Drabkin have shown,[39]

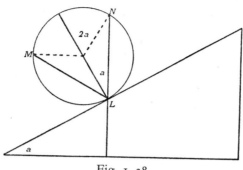

Fig. 1.28

if we take a drawing like the accompanying figure (see Fig. 1.28), then we can reduce the procedure to the following modern terminology

$$F = \frac{\text{area of segment } MLN}{\text{area of circle } O} \cdot W,$$

where F is the force necessary to keep the cylinder in equilibrium on an inclined plane of inclination a and W is the weight of the cylinder.

(2) Rewriting (1) we get

$$F = W \frac{\left[\frac{a}{90} \pi + 2 \cos a \sin a\right]}{\pi}.$$

It can be noted that while this solution is incorrect, it at least has the advantage over the only other ancient solution of the problem, that of Pappus, that when a is 90°, $F = W$, as it should. For Pappus' solution see note 27 above. As I have indicated earlier in the chapter, Hero's exposition of the inclined plane problem is inferior to the medieval treatment in the *De ratione ponderis* (see Doc. 2.1, prop. 10). Needless to say, it is also inferior to the treatments of Stevin and Galileo (see Doc. 2.6).

I.24. This section demonstrates the dependence of Hero on the prior work of Archimedes. This is particularly true for Archimedes' treatment of centers of gravity and the law of the lever. Hero's citation of Archimedes' definition and the distinction of center of gravity from point of support is especially precious historically, since the original work in which Archimedes gave such a definition is now lost. It is supposed that this definition appeared in the work *On Balances* (περὶ ζυγῶν) cited by Pappus.[40]

[39] Cohen and Drabkin, *Source Book*, pp. 199–200.

[40] J. L. Heiberg, *Archimedis Opera Omnia*, Vol. 2 (Leipsig, 1913), 548.

On the basis of the statements of Pappus together with this passage from Archimedes, Giovanni Vailati attempted a very subtle reconstruction of Archimedes' theory of centers of gravity, which he believed was fashioned by Archimedes to serve as a basis of the deductions contained in the *Equilibrium of Planes*,[41] and the reader interested in statics is strongly urged to consult Vailati's exciting reconstruction. One other point worth noting concerning this section is that Hero refers (in addition to Archimedes) to another mechanician that L. Nix reconstructs as Posidonius. But Hero would place him earlier than Archimedes (d. 212 B.C.). This would make identification with Poseidonius the Stoic (d. *ca.* 50 B.C.) impossible.

In addition to this passage from Hero's work, the Arabs also had available the important passages on centers of gravity presented by al-Khāzinī (*fl.* 1115–22) and drawn by him from Abū Sahl ibn Bishr of Qūhistān (d. by 850) and Alhazen (Ibn al-Haitham) of Baṣrah, (d. *ca.* 1039) and which ultimately show the influence of Archimedes. These passages are of such importance that I shall now present their substance in the translation of N. Khanikoff (see Doc. 1.5).

I.33 and 34. In these two chapters, Hero clearly demonstrates that he understands the principle of static moment in its general form, namely, that the force of a weight on an irregular lever arm depends not only on the weight but as well on the *horizontal* distance to the line running vertically through the fulcrum, a principle later used and understood by the medieval author of the *De ratione ponderis* and by Leonardo da Vinci. Notice that, as in the case of the inclined plane principle, there is no formal proof of the bent lever theorems. Again we find a formal mathematical presentation in the *De ratione ponderis;* but it should be noticed that the medieval theorem (given in Doc. 2.1, prop. 8) has regard only for the equilibrium of *equal* weights at equal horizontal distances, although the author of that treatise certainly understood the more general statement as regards any weights. It seems quite apparent from Hero's statement that Archimedes preceded him in the explanation of irregular levers, perhaps in the lost treatise *On Balances*.

II.7 and 8. These passages have been sufficiently commented upon in the text. They illustrate that Hero follows the Pseudo-Aristotelian *Mechanica* in his attribution of the mechanical advantage of the lever and similar machines to "the principle of the circle." However, in explaining the principle of static moment in II.7 he does not specifically put the radii in movement as had the Pseudo-Aristotle. This however is done in effect in

---

[41] See the articles by Vailati cited in note 16 above.

[Hero of Alexandria, *Mechanics*

II.8 where Hero describes his principle of "slowing up," which on careful reading emerges as the basic concept of work. Mechanical advantage is obtained by a small force in a machine lifting a large weight by the fact that it must act longer (i.e., in actuality, farther). The reader is directed to the text, where the whole matter has been adequately explained.

Document 1.4

Archimedes, *On Heaviness and Lightness* (A Fragment of the *Floating Bodies*)*

[Definition, Postulate]

1. [DEFINITION] Some bodies and liquids are heavier than others. We say that one body is heavier than another body, or one liquid is heavier than another liquid, or that a body is heavier than a liquid, when, after taking an equal volume of each of them and weighing them, one of them proves heavier than the other. But if their weights are equal, it is not said that one of them is heavier than the other. It is only said to be heavier when one has more weight.

2. [Postulate] We posit in regard to the nature of a liquid that its parts are equivalent in position [i.e., lie evenly] and are continuous, and that anything which thrusts more than they [the parts] drives along anything which thrusts less. And each of its individual parts will be pressed by that which is perpendicularly above it, so long as the liquid is not enclosed in something and something else does not press upon it.

[Propositions]

3. [Prop. I.2] The figure *(shakl)* of every liquid is that of a sphere.

4. [Prop. I.3] If a body equal in weight to some liquid is let down into the liquid, it sinks only until its surface is level with the surface [of the liquid].

5. [Prop. I.4] If any body lighter than a liquid is let down in the liquid, it is not submerged completely, but some part of it will project above the surface of the liquid.

6. [Prop. 1.5] If a body lighter than a liquid is let down into it, a certain amount of it is submerged. If one takes a quantity of liquid equal in volume to that amount of the body which is submerged, one finds that

* Translated from the Arabic text of H. Zotenberg in the *Journal asiatique*, Ser. 7, Vol. *13* (1879), pp. 509–15.

[Archimedes, *Heaviness and Lightness* 53]

the weight of this quantity of liquid is equal to the weight of the whole body.

7. [Prop. I.6] If any body lighter than a liquid is [forcibly] covered over by the liquid [i.e., submerged therein], then its ascent will take place with a force equal to the force of the excess in weight of the quantity of liquid equivalent in volume to the body beyond the weight of that body.

8. [Prop. I.7] If any body heavier than a liquid is let down into it, [it sinks below in it (Gotha MS), and] its weight when submerged is equal to the excess in weight of that body [in air] beyond the weight of a quantity of the liquid equivalent in volume to the body.

9. [Prop. I.8] If any body lighter than a liquid and shaped in the figure of a segment of a sphere is let down into the liquid and care is taken that the base of the inserted body is not in the liquid, then the figure will be upright *(qā'imā)* so that the axis of the segment coincides with the vertical. If afterwards it [the axis] is inclined so that the base enters the liquid, it will not remain inclined but will right itself perpendicularly to the horizon.

10. [Prop. II.1] If any body lighter than a liquid is let down in it, the ratio [in weight] of that body to the weight of liquid equivalent in volume to that body is as the ratio of the weight of that part which is submerged to the weight of the whole body. So ends the Treatise of Archimedes.

COMMENTARY

1. The paragraph numbers for this document are arbitrary numbers assigned by me, while the numbers in brackets refer to the numbers of the equivalent propositions in the Greek text. The Arabic text of this fragment is known from at least two manuscripts: Paris, BN Fonds suppl. arabe 952 bis; and Gotha, MS 1158, 2. As indicated under the title above, my translation has been made from the published text, which Zotenberg transcribed from the Paris manuscript. A German translation from the Gotha manuscript has been published by E. Wiedemann in the *Sitzungsberichte der Physikalisch-medizinischen Sozietät in Erlangen*, (Vol. *38* [1906], 152–62; for full title see the Bibliography). In addition to circulating independently, this fragment was included by al-Khāzinī in his *Book of the Balance of Wisdom* (first maqālat, chap. 2; unpublished as a whole, but for partial publication of other parts of this text, see Doc. 1.5). Perhaps a more extended translation of the *Floating Bodies* served as the basis for a treatise by al-Kindī in his ninth-century tract called the *Great Treatise on Bodies which are Immersed in Water* (see Wiedemann, *Ibid.*, p. 160). Al-Khāzinī also summarized the conclusions of the fragment once more in

chapter 6 of the first maqālat (translated by Wiedemann, *ibid.*, Vol. *40* [1908], 133–35). Compare also the passages from Fakhr al-Dīn al-Rāzī (d. 1210), translated by E. Wiedemann in the *Archiv für die Geschichte der Naturwissenschaften und der Technik* (Vol. *2* [1910], 394–98).

Note initially that this fragment contains a definition of specific weight not found in the Greek text now extant. It is assumed that this definition was added either in late antiquity or after the translation into Arabic. However, there is no way to know now when the *first* definition of specific weight was made; but it can hardly be doubted that it was Greek in origin. As Document 1.5 reveals, Menelaus did a treatise on the hydrostatic balance, and it is at least reasonable to suppose that it contained some definition of specific weight. Furthermore, as we shall note in the next chapter, the Latin poem *Carmen de ponderibus* that seems to date ultimately from late antiquity had such a definition of the specific weight of both solids and liquids, and this poem certainly reflects commonplace Greek and Roman metrological and hydrostatical knowledge. One last point remains to be noted in connection with the definition of specific weight in the Arabic fragment of the *Floating Bodies;* and this is that the Gotha manuscript omits the following: "or one liquid is heavier than another liquid, or that a body is heavier than a liquid."

2. In the last part of the first postulate beginning "so long as . . ." the Arabic text appears to be closer to the reading of the pristine Greek text than either Greek MS *B* (now lost, but reconstructed from the Latin Moerbeke translation of 1269) or MS *C* (discovered by Heiberg in 1906; see his edition of the *Opera* of Archimedes, Vol. *2*, 318). In the first place, like MS *C*, it contains the negative equivalent of μή which is left out by *B*; but more important, it substitutes the Arabic equivalent for καθειργμένον ("enclosed" or "contained") for καθιέμενον (*descendens*, Moerbeke; "sunk") in both *B* and *C*.

3–10. Proposition I.1 of the Moerbeke-Greek text is omitted here in the Arabic text—perhaps because of its strictly mathematical nature, since it is a locus definition of a spherical surface. Propositions I.2–8, and II.1 as given in the Arabic text are fairly close to the meaning if not wording of the Moerbeke and Greek texts. I.2 differs somewhat in the Moerbeke text—it is missing in *C*; in the Moerbeke text (MS, Vat. Ottob. lat. 1850, f. 55v) it runs: "Omnis humidi consistentis ita ut maneat inmotum superficies habebit figuram spere habentis centrum idem cum terra." (The surface of every liquid at rest has the figure of a sphere whose center is the same as that of the earth). Proposition I.6 has a different reading in the Gotha manuscript, as follows: "If any body heavier than a liquid is let down into it, it sinks below in it, and that part of it (the liquid) which [seeks to] lift it (the body) is equal in volume to this body." Also, after I.8 the Gotha manuscript includes the following comment: "The cause of the uprightness in the liquid is that which Thābit ibn Qurra has declared in his *On the Rolling and the Nonrolling of the Sphere*, namely, that the center of gravity *(markaz thiql)* of this segment is situated on its axis. Accordingly it (the segment) is stable only when the axis coincides with the vertical perpendicular to the surface of the liquid in which the segment of the sphere is immersed." Note also that the Arabic text omits the postulate before I.8 contained in the Greek and Moerbeke texts. Proposition I.9 is also omitted in the Arabic text. It also should be obvious to the reader that the proofs are omitted for all of the propositions given.

Document 1.5

al-Khāzinī, *Book of the Balance of Wisdom**

1. [INTRODUCTION] SECTION 1.... The balance of wisdom is something worked out by human intellect and perfected by experiment and trial, of great importance on account of its advantages and because it takes the place of ingenious mechanicians. Among these [advantages] are: 1. exactness in weighing: this balance shows variation to the extent of a *mithqāl*, or a grain, although the entire weight is a thousand *mathāqīl*, provided the maker has a delicate hand, attends to the minute details of the mechanism, and understands it; 2. that it distinguishes pure metal from its counterfeit, each being recognized by itself, without any refining; 3. that it leads to a knowledge of the constituents of a metallic body composed of any two metals, without separation of one from another....

2. Section 2. The Theory of the Balance of Wisdom. This just balance is founded upon geometrical demonstrations and deduced from physical causes in two points of view: 1. as it implies centers of gravity, which constitute the most elevated and noble department of the exact sciences, namely, the knowledge that the weights of heavy bodies vary according to the difference of distance from a common point—the foundation of the steelyard; 2. as it implies a knowledge that the weights of heavy bodies vary according to difference in rarity or density of the liquids in which the body weighed is immersed—the foundation of the balance of wisdom.

To these two principles the ancients directed attention in a vague way, after their manner, which was to bring out things abstruse, and to declare dark things in relation to the great philosophies and the precious sciences. We have, therefore, seen fit to bring together, on this subject, whatever useful suggestions their works, and the works of later philosophers, have afforded us, in connection with those discoveries which our own meditation, with the help of God and His aid, has yielded.

* Translation of N. Khanikoff—slightly altered—from his own Arabic text in the *Journal of the American Oriental Society*, Vol. 6 (1860), 1–128.

[al-Khāzinī, *Book of the Balance*

3. Section 3. Fundamental Principles of the Art of Constructing This Balance. Every art, we say, has its fundamental principles upon which it is based and its preliminaries to rest upon, which one who would discuss it must not be ignorant of. These fundamental principles and preliminaries class themselves under three heads: 1. those which rise up [in the mind] from early childhood and youth, after one sensation or several sensations, spontaneously; which are called first principles and common familiar perceptions; 2. demonstrated principles, belonging to other departments of knowledge; 3. those which are obtained by experiment and elaborate contrivance....

4. Section 4. Institution of the Water Balance.... It is said that the [Greek] philosophers were first led to think of setting up this balance and moved thereto by the book of Menelaus addressed to Domitian [reigned 81–96 A.D.], in which he says: "O King, there was brought one day to Hiero, King of Sicily, a crown of great price.... Now it occurred to Hiero that this crown was not of pure gold but alloyed with some silver.... He therefore wished to ascertain the proportion of each metal contained in it, while at the same time he was averse to breaking the crown, on account of its solid workmanship. So he questioned the geometricians and mechanicians on the subject. But no one sufficiently skillful was found among them, except Archimedes the geometrician, one of the courtiers of Hiero. Accordingly, he devised a piece of mechanism which, by delicate contrivance, enabled him to inform King Hiero how much gold and how much silver was in the crown, while yet it retained its form." That was before the time of Alexander.* Afterwards Menelaus [himself] thought about the water-balance and brought out certain universal arithmetical methods to be applied to it; and there exists a treatise by him on the subject.... Subsequently, in the days of al-Ma'mūn (reigned 813–33), the water balance was taken into consideration by the modern philosophers.... Some one of the philosophers who have been mentioned added to the balance a third bowl, connected with one of the two bowls, in order to ascertain the measure, in weight, of the rising of one of the two bowls in the water; and by that addition, somewhat facilitated operations....

5. Section 5. Forms and Shapes of the Water-Balance.... Balances used in water are of three varieties of shape: 1. one with two bowls arranged in the ordinary way, called "the general simple balance," to the beam of which are frequently added round-point numbers; 2. one with three bowls for the extreme ends, one of which is suspended below another and is the

---

* In fact almost a century later.

water-bowl, which is called "the satisfactory balance" or "the balance without movable bowl"; 3. one with five bowls, called "the comprehensive balance," the same as the balance of wisdom, three of the bowls of which are a water-bowl and two movable bowls.... [Upon such balances are fixed points indicating] the specific gravities of metals, relatively to a determined sort of water....

6. Chapter 1. Main Theorems Relative to Centers of Gravity, According to Abū Sahl [ibn Bishr] of Qūhistān (d. by 850) and Ibn al-Haitham (Alhazen, d. *ca.* 1039) of Baṣra.... Section 1. 1. Heaviness is the force with which a heavy body is moved towards the center of the world. 2. A heavy body is one which is moved by an inherent force, constantly, towards the center of the world. Suffice it to say, I mean that a heavy body is one which has a force moving it towards the central point and constantly in the direction of the center without being moved by that force in any different direction; and that the force referred to is inherent in the body, not derived from without, nor separated from it—the body not resting at any point out of the center, and being constantly moved by that force, so long as it is not impeded, until it reaches the center of the world.

7. Section 2. 1. Of heavy bodies differing in force some have a greater force, which are dense bodies. 2. Some of them have a lesser force, which are rare bodies. 3. Any body whatever exceeding in density has more force. 4. Any body whatever exceeding in rarity has less force. 5. Bodies alike in force are those of like density or rarity, of which the corresponding dimensions are similar, their forms being alike as to gravity. Such we call bodies of like force. 6. Bodies differing in force are those which are not such. These we call bodies differing in force.

8. Section 3. 1. When a heavy body moves in liquids, its motion therein is proportioned to their degrees of liquidness, so that its motion is most rapid in that which is most liquid. 2. When two bodies alike in volume, similar in shape, but differing in density, move in a liquid, the motion therein of the denser body is the more rapid. 3. When two bodies alike in volume, and alike in force, but differing in shape, move in a liquid, that which has a smaller surface touched by the liquid moves therein more rapidly. 4. When two bodies alike in force, but differing in volume, move in a liquid, the motion therein of the larger is the slower.

9. Section 4. 1. Heavy bodies may be alike in gravity, although differing in force and differing in shape. 2. Bodies alike in gravity are those which, when they move in a liquid from some single point, move alike—I mean, pass over equal spaces in equal times. 3. Bodies differing in gravity are

those which, when they move as just described, move differently; and that which has the most gravity is the most rapid in motion. 4. Bodies alike in force, volume, shape, and distance from the center of the world are like bodies. 5. Any heavy body at the center of the world has the world's center in the middle of it; and all parts of the body incline, with all its sides, equally, towards the center of the world; and every plane projected from the center of the world divides the body into two parts which balance each other in gravity, with reference to that plane. 6. Every plane which cuts the body without passing through the center of the world divides it into two parts which do not balance each other with reference to that plane. 7. That point in any heavy body which coincides with the center of the world, when the body is at rest at that center, is called the center of gravity of that body.

10. Section 5. 1. Two bodies balancing each other in gravity with reference to a determined point are such that, when they are joined together by any heavy body of which that point is the center of gravity, their two [separate] centers of gravity are on the two sides of that point, on a straight line terminating in that point—provided the position of that body [by which they are joined] is not varied; and that point becomes the center of gravity of the aggregate of the bodies. 2. Two bodies balancing each other in gravity with reference to a determined plane are such that, when they are joined together by any heavy body, their common center of gravity is on that plane—provided the position of that body [by which they are joined] is not varied; and the center of gravity common to all three bodies is on that plane. 3. Gravities balancing each other relatively to any one gravity, which secures a common center to the aggregate, are alike. 4. When addition is made to gravities balancing each other relatively to that center and the common center of the two is not varied, all three gravities are in equilibrium with reference to that center. 5. When addition is made to gravities balancing each other with reference to a determined plane, of gravities which are themselves in equilibrium with reference to that plane, all the gravities balance with reference to that plane. 6. When subtraction is made from gravities balancing each other, of gravities which are themselves in equilibrium and so the center of gravity of the aggregate is not varied, the remaining gravities balance each other. 7. Any heavy body in equilibrium with any heavy body does not balance a part of the latter with the whole of its own gravity nor with more than its own gravity, so long as the position of neither of the two is varied. 8. Bodies alike in force, alike in bulk, similar in shape, whose centers of gravity are equally distant

from a single point, balance each other in gravity with reference to that point and balance each other in gravity with reference to an even plane passing through that point; and such bodies are alike in position with reference to that plane. 9. The sum of the gravities of any two heavy bodies is greater than the gravity of each one of them. 10. Heavy bodies alike in distance from the center of the world are such that lines drawn out from the center of the world to their centers of gravity are equal.

11. Section 6. 1. A heavy body moving towards the center of the world does not deviate from the center; and when it reaches that point, its motion ceases. 2. When its motion ceases, all its parts incline equally towards the center. 3. When its motion ceases, the position of its center of gravity is not varied. 4. When several heavy bodies move towards the center and nothing interferes with them, they meet at the center; and the position of their common center of gravity is not varied. 5. Every heavy body has its center of gravity. 6. Any heavy body is divided by any even plane projected from its center of gravity into two parts balancing each other in gravity. 7. When such a plane divides a body into two parts balancing each other in gravity, the center of gravity of the body is on that plane. 8. Its center of gravity is a single point.

12. Section 7. 1. The aggregate of any two heavy bodies joined together with care as to the placing of one with reference to the other has a center of gravity which is a single point. 2. A heavy body which joins together any two heavy bodies has its center of gravity on the straight line connecting their two centers of gravity; so that the center of gravity of all three bodies is on that line. 3. Any heavy body which balances a heavy body is balanced by the gravity of any other body like to either in gravity, when there is no change of the centers of gravity. 4. One of any two bodies which balance each other being taken away and a heavier body being placed at its center of gravity, the latter does not balance the second body; it balances only a body of more gravity than that has.

13. Section 8. 1. The center of gravity of any body having like planes and similar parts is the center of the body—I mean, the point at which its diameters intersect. 2. Of any two bodies of parallel planes, alike in force and alike in altitude—their common altitude being at right angles with their bases—the relation of the gravity of one to the gravity of the other is as the relation of the bulk of one to the bulk of the other. 3. Any body of parallel planes which is cut by a plane parallel with two of its opposite planes is thereby divided into two bodies of parallel planes, and the two have [separate] centers of gravity which are connected by a straight line

[al-Khāzinī, *Book of the Balance*

between them, and the body as a whole has a center of gravity which is also on this line. So that the relation of the gravities of the two bodies, one to the other, is as the relation of the two portions of the line [connecting their separate centers of gravity and divided at the common center], one to the other, inversely. 4. Of any two bodies joined together, the relation of the gravity of one to the gravity of the other is as the relation of the two portions of the line on which are the three centers of gravity— namely, those pertaining to the two taken separately, and that pertaining to the aggregate of the two bodies—one to the other, inversely.

14. Section 9. 1. Of any two bodies balancing each other in gravity with reference to a determined point, the relation of the gravity of one to the gravity of the other is as the relation of the two portions of the line which passes through that point and also passes through their two centers of gravity, one to the other. 2. Of any two heavy bodies balancing a single heavy body relatively to a single point, the one nearer to that point has more gravity than that which is farther from it. 3. Any heavy body balancing another heavy body relatively to a certain point, and afterwards moved in the direction towards that other body while its center of gravity is still on the same straight line with the [common] center, has more gravity the farther it is from that point. 4. Of any two heavy bodies alike in volume, force, and shape, but differing in distance from the center of the world, that which is farther off has more gravity.

[In the second chapter al-Khāzinī includes the fragment of Archimedes' *Floating Bodies* already here translated as Document 1.4. In chapter 3 is given the text of the *De ponderoso et levi*, attributed to Euclid (and whose text and English translation have been published by Moody and Clagett in *The Medieval Science of Weights* and parts of which are quoted in Chapter 7 of this volume).]

15. Chapter 5. Theorems Recapitulated for the Sake of Explanation.... Section 1. Differences in the Weights of Heavy Bodies at the Same Distance from the Center of the World. 1. I say that elementary bodies—differing in this from the celestial spheres—are not without interference, one with another [as to motion], in the two directions of the center and the circumference of the world, [as it appears] when they are transferred from a denser to a rarer air, or the reverse. 2. When a heavy body, of whatever substance, is transferred from a rarer to a denser air, it becomes lighter in weight; from a denser to a rarer air, it becomes heavier. This is the case universally with all heavy bodies. 3. When one fixes upon two heavy bodies, if they are of one and the same substance, the larger of the two in

bulk is the weightier of them. 4. When they are of two different substances, and agree in weight, and are afterwards transferred to a denser air, both become lighter; only that the deficient one, that is, the smaller of the two in bulk, is the weightier of them, and the other is lighter. 5. If the two are transferred to a rarer air, both become heavier; only that the deficient one, that is, the smaller of the two in bulk, is the lighter of them in weight and the other is the heavier.

16. Section 2. When a heavy body moves in a liquid, one interferes with the other; and therefore water interferes with the body of anything heavy which is plunged into it and impairs its force and its gravity in proportion to its body. So that gravity is lightened in water in proportion to the weight of the water which is equal [in volume] to the body having that gravity; and the gravity of the body is so much diminished. As often as the body moving [in the water] is increased in bulk, the interference becomes greater. This interference, in the case of the balance of the wisdom, is called the rising up [of the beam]. 2. When a body is weighed in air and afterwards in the water-bowl, the beam of the balance rises in proportion to the weight of the water which is equal in volume to the body weighed; and therefore, when the counterpoises are proportionally lessened, the beam is brought to an equilibrium parallel with the plane of the horizon. 3. The cause of the differing force of the motion of bodies, in air and in water, is their difference of shape. 4. Yet when a body lies at rest in the water-bowl, the beam rises according to the measure of the volume of the body, not according to its shape. 5. The rapidity of the motion of the beam is in proportion to the force of the body, not to its volume. 6. The air interferes with heavy bodies, and they are essentially and really heavier than they are found to be in that medium. 7. When moved to a rarer air, they are heavier; and on the contrary, when moved to a denser air, they are lighter.

17. Section 3. 1. The weight of any heavy body of known weight at a particular distance from the center of the world varies according to the variation of its distance therefrom; so that as often as it is removed from the center, it becomes heavier, and when brought near to it, is lighter. On this account, the relation of gravity to gravity is as the relation of distance to distance from the center. 2. Any gravity inclines toward the center of the world; and the place where the stone having that gravity falls upon the surface of the earth is its station; and the stone, together with its station, is on a straight line drawn from the center of the world to the station mentioned. 3. Of any two like triangles, standing on one of the

[al-Khāzinī, *Book of the Balance*

great circles of the surface of the earth, the distance between the [two] apexes is greater than between the bases, because the two are on two straight lines drawn from the center of the world, making the two legs of a triangle of which the apex is the center of the world and the base [includes] the two apexes. When the stations of the two figures are connected [by a straight line], we get the shape of two similar triangles, the larger of which as to legs is the broader as to base. 4. The plane of incidence of a perpendicular line from the center of the world falling upon any even plane parallel with the horizon is the middle point of that plane, and the part of it which is nearest to the center of the world. Thus let the plane be *ab* (see Fig. 1.29), the center *h* and the perpendicular line upon the

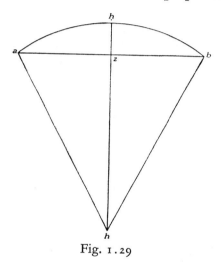

Fig. 1.29

[plane] *ab* from the center *hz*—that is the shortest line between the center and the plane. 5. Let any liquid be poured upon the plane *ab*, and let its gathering-point be *h*, within the spherical surface *ahb* [formed by attraction] from the center *h*; then, in case the volume of the liquid exceeds that limit, it overflows at the sides of *ab*. This is so only because any heavy body, liquid or not, inclines from above downwards and stops on reaching the center of the world; for which reason the surface of water is not flat, but on the contrary, convex, of a spherical shape....

COMMENTARY

1. It should be noted at the outset that al-Khāzinī flourished at the end of the eleventh and the beginning of the twelfth centuries. He was a Greek slave raised in the 'Persian" city of Merv, where he received a good

scientific education. Extant in addition to the *Book of the Balance of Wisdom* are some astronomical tables that give the positions of fixed stars for the year A.H. 509 (1115–16; see the article "al-Khāzinī," *Encyclopedia of Islam*, Vol. 2, 937–38).

The introductory section of the *Book of the Balance of Wisdom* is included simply to show the over-all objective of the treatise, the treatment of the so-called balance of wisdom. I am not interested here in the actual construction of the balance of wisdom or of any of its predecessors, although they are described in some detail in other sections of al-Khāzinī's treatise. I am interested here in the general and theoretical statements made by the author.

2. In section 2 of the introduction, the author points to the fundamental principles that underlie the hydrostatic balance. The first is purely statical and embraces the idea of center of gravity and the variation of effective weight on a lever arm depending on the distance from the fulcrum. The second is hydrostatic—the weights of immersed bodies vary according to the rarity or density of the liquids in which they are immersed.

3. Methodologically, the maker and user of this balance (and presumably the follower of any similar art) depends on (1) commonly accepted and familiar principles which by either single or repeated impression become immediately acknowledged, (2) demonstrations (presumably logical and mathematical?), and (3) experiment and active contrivance.

4–5. Of the ancient authors who treated the hydrostatic balance Menelaus is singled out—his tract being extant in Arabic (but not in Greek). From the tract of Menelaus, al-Khāzinī draws a version of the well-known crown story associated with Archimedes and King Hiero of Syracuse. It serves to introduce the alloy problem. In our next chapter we shall notice and describe the techniques of solving this problem deduced from Archimedes' *Floating Bodies* and the so-called principle of Archimedes contained therein and in addition from the method mentioned by Vitruvius based on volumetric comparisons. The other determinations in medieval Latin literature are also discussed in Chapter 2 below.

The Arabs were themselves much taken with the alloy problem. E. Wiedemann has discussed the many treatments of the problem among the Arabs (see the Bibliography below and particularly the article in the *Sitzungsberichte*, Vol. *38* [1906], 163–80). He cited in particular the following tracts: (1) al-Bīrūnī, *On the Relationship Which Exists in Volume between Metals and Precious Stones*, a treatise which includes a description of al-Bīrūnī's "conical vessel" for specific gravity determinations by means of finding the ratio of the weight of water displaced to the weight of a substance in air; (2) Abū Manṣūr al-Nairīzī, *On the Determination of the Quantitative Composition of Mixed Bodies;* (3) ʿUmar al-Khayyāmī, a tract on the determination of the contents of alloys; and (4) a treatise on specific weight attributed to Plato (and probably translated from the Greek). While Wiedemann gives modern formulations in detail for the methods employed in these various treatises, we give here only his general conclusions (*ibid.*, p. 180).

The [last] three treatments given above describe three different methods for the determination of the quantities of the components in the mixtures of two substances. In the case of Manṣūr the volumes of the components and mixtures are compared. It is the simplest method. In the tract ascribed to Plato, the weight loss in water as a fraction of the weight in water forms the basis of calculation. Most complicated is the procedure of Omar Khayyām which uses as a point of departure the relation of the weights in air and water.

Although we have not given any of the parts of the al-Khāzinī treatise which describe the specific gravity determination, we can note that his theoretical description of the problem reduces to the following formulation (*Journal of American Oriental Society*, Vol. 6 [1860], 104, 126):

$$X = W \frac{\frac{1}{d'} - \frac{1}{\text{Sp. gr.}}}{\frac{1}{d'} - \frac{1}{d''}}$$

where X is the weight of silver in the alloy, W is the absolute weight of the body examined, "Sp. gr." its specific gravity, d' that of gold, and d'' that of silver.

6. The first section of the first chapter consists mainly in an Aristotelian definition of heaviness. Heaviness is defined as the force with which a heavy body moves toward the center of the world. This force is inherent in the body and is not extrinsic. It acts so long as the heavy body is away from the center of the world.

7. The second section is also Aristotelian, but with some additions that owe themselves to the *Liber de ponderoso et levi* attributed to Euclid and quoted below in Chapter 7. Bodies of greater density are of greater force, assuming that the bodies are of equal volume and are of similar shapes.

8. The conclusion of section 2 is—according to our author—borne out, for if we put two bodies alike in volume in a liquid, the denser moves faster (cf. *De caelo*, IV.1, 308a 29-33). But if the forces (weights) are the

same, the body with greater volume (and therefore of greater surface) will move more slowly.

9. This section, reflecting the so-called Aristotelian law which related force (gravity) and speed directly (see Chapter 7 below), was undoubtedly drawn from the treatise already mentioned, the *De ponderoso et levi*. Bodies alike in gravity are defined as those which move over equal distances in equal times. Every heavy body at the center of the world would have the world's center as its center. Every plane passing through the center of the world would thus divide the body into two parts which balance each other in gravity. Then the center of gravity of a body is defined as that point which coincides with the center of the world when the body is at rest at the center of the world. This is, of course, entirely a theoretical rather than an operational definition.

10. This section may reflect something of the *Equilibrium of Planes* but perhaps also the lost Archimedean treatise *On Balances*. The conclusion of this whole section is that, regardless of the addition or subtraction made in gravity to a system in equilibrium, so long as the common center of gravity remains unchanged, the equilibrium is retained. It is rather interesting that some of these statements are of the same form as the geometric axioms of Euclid's *Elements* (Book I). The axiom "equals added to equals yield equals" is adapted to gravities. Gravities in equilibrium with reference to a plane added to gravities in equilibrium with respect to the same plane maintain the equilibrium. The same holds for subtracting gravities. "The whole is greater than its part" is reflected by statements 7 and 9 in this fifth section.

11. In the first statement we have clear indication that for al-Khāzinī there is no momentum concept for bodies in motion. A body going to the center of the world would simply cease to move on arrival there. Compare this with the "impetus" idea as revealed in Document 9.3 where Oresme indicates that a body falling to the center of the world would be carried past the center by the impressed force it would have acquired in fall, and thus it would ascend a certain distance before again falling towards the center, and it would continue to oscillate for a time about the center.

12. This section is somewhat repetitious so far as centers of gravity are concerned. Again the point is made that when there is no change in the centers of gravity of a system, however else it might be altered, the system remains in equilibrium.

13. This section makes the point that in a symmetrical and homogeneous body (or "one having like planes and similar parts") the center of gravity is

[al-Khāzinī, *Book of the Balance* 67]

the center of the body. What the author has in mind is a beam of uniform thickness and of homogeneous material. Such bodies are then thought of as being composed of a series of parallel equal planes so that any such body cut by a plane between, but parallel to, two of these opposite plane elements is divided into two bodies composed in turn of plane elements. Each has its center of gravity, and the center of gravity of the whole is in a line connecting the two separate centers of gravity. Then follows the first statement of the law of the lever. The weights are inversely as the segments of the line connecting their centers of gravity.

14. The lever law is here restated. Then it is declared that when two weights successively counterbalance a third heavy body, the one that maintains equilibrium when placed nearer the fulcrum has more [absolute] gravity. On the other hand a weight has greater [effective] gravity the farther it is from the fulcrum. Finally we are told that a given body has greater weight the farther it is from the center of the world. This is explained further in chapter 5, where it is pointed out that the air gets rarer the farther out we are from the center of the earth and thus has less buoyant effect. This is against general Aristotelian theory which held that bodies have more gravity the closer they are to the center of the world. This idea is ignored and the Archimedean analysis of the immersed bodies is used as a point of departure.

15. Section 1 of chapter 5 recapitulates the principles of Archimedes' *Floating Bodies*. We learn that of bodies of one and the same substance, the larger in bulk (i.e. in volume) is the heavier. When they are of different substances and are of equal weights, and further are transferred to a denser medium, both become lighter, but the one of lesser bulk is now the heavier. The situation is reversed if they are transferred to a rarer medium.

16. Section 2 continues the outline of Archimedean principles and we see the "principle of Archimedes" stated. But non-Archimedean considerations of the motion of the bodies in liquids are introduced. Here "shape" is an important factor according to this author. Differences in shape produce differences in interference and hence differences in velocity. But when a body is at rest in a medium, only the *volume* of liquid it has displaced is important for determining the reduction of its weight. Furthermore, the rapidity of the *movement of the beam*, he holds, is in proportion to the force of the body and not to its volume.

17. Since the rarity of the medium increases as we go out from the center of the world, a given body is continually heavier in that medium the farther out it is from the center of the world. "On this account the relation

of gravity is as the relation of distance to distance from the center [of the world]."⁴² Also included in this third section is a concept generally accepted by the Greek philosophers—including Aristotle and Archimedes—that the surface of liquids at rest is spherical with the center of the sphere being the center of the world.

⁴² This is somewhat strange, for al-Khāzinī seems to be saying that gravity varies directly as distance from the center of the world. Yet he also says that increase in gravity depends on the density of the medium in accordance with Archimedes' principle. Now if he believed that the density of the medium varied directly as distance from the center of the world, the gravity would not increase *directly* as the distance from the center. It would appear that we should not make too precise the proportionality statements of al-Khāzinī.

Chapter 2

Jordanus de Nemore and Medieval Latin Statics

AS in the case of medieval mathematics, the maturation of medieval statics awaited the extensive translating activity of the twelfth and thirteenth centuries. Of the various Greek and Arabic treatises mentioned in Chapter 1, the following were rendered into Latin during this period: (1) the *De canonio*, from the Greek; (2) the *De ponderoso et levi* attributed to Euclid, from the Arabic; (3) the *Liber karastonis*, from the Arabic by Gerard of Cremona; (4) the *Equilibrium of Planes* of Archimedes[1] with the

[1] The earliest connection of Archimedes with lever and balance problems in Latin is found in the *Epistola de proportione et proportionalitate* of Ametus filius Iosephi, translated by Gerard of Cremona from the Arabic. The pertinent passage reads (MS Paris, BN lat. 9335, f. 66r; cf. M. Curtze in the supplement volume to *Euclidis opera omnia* of J. L. Heiberg and H. Menge [Leipzig, 1899], pp. xxvii–xxix, who transcribes the passage from Vienna, Nat.-bibl. cod. 5277, f. 309v): "In carastone quoque, cum fuerit perpendicularis eius equidistans superficiei orizontis, erit proportio longioris duarum partium ipsius ad breviorem earum sicut proportio gravioris ponderis ad levius pondus. Et in corda divisa cum portante erit proportio longioris sectionis ad minorem sicut proportio vocis minoris ad vocem longioris, cum percutiuntur cum una re et una potentia. Iam ergo ostendimus diffinitionem proportionalitatis ei, qui huius declarationis ordinem assecutus est, et quid nomen eius significat ei, qui hanc non consequitur ordinem, ne qui hanc considerat epistolam ab hac scientia sit vacuus. Arsamides *(i.e. Archimedes)* quoque ponderum proportionalitatem diffinivit dicens: 'Pondera proportionalia diversa sunt que uno ponderantur angulo.' Per quod voluit intelligi, ut, cum primum ponderum ponitur in lance trutine et secundum eorum in altera lance, et suspenditur trutina suspensorio suo, erit angulus quem circumdat statera et suspensorium trutine unus ad tertium et quartum cum tertium fuerit positum in loco primi et quartum in loco secundi, et similiter si quintum in loco primi et tertii ponatur et sextum in loco secundi et quarti. Et cum primum etiam et secundum in duabus lancibus ponuntur, et tertium et quartum in duabus lancibus alterius trutine, et quintum et sextum in duabus lancibus trutine tertie, anguli qui sunt inter suspensoria trutinarum et sta-

commentary of Eutocius, from the Greek by William Moerbeke in 1269; (5) the *Floating Bodies* of Archimedes, also from the Greek by Moerbeke in 1269. There were in addition two works either translated from or dependent in part on Arabic originals: (6) the *Liber de insidentibus in humidum* or *De ponderibus* attributed to Archimedes and (7) a short fragment summarizing the propositions of the *Liber karastonis*, but without proofs.

All of these works except (4) and (5) circulated widely as part of the medieval statical corpus. The *Equilibrium of Planes* and *Floating Bodies*,

teras earum sunt etiam uni. Per hoc autem quod in verbis eius invenitur, scilicet ex diversis, voluit intelligi quod, cum primum ponderum fuerit equale secundo, statera trutine erit cum suspensorio ipsius coniuncta, neque erit inter ea angulus. Yrinus (*i.e.* Hero) autem diffinivit proportionalitatem dicens: 'Pondera proportionalia diversa sunt ea que, cum appensa fuerint, erunt linee ordinate unicuique antecendenti earum et consequenti super equales angulos superficiei orizontis.' Quod etiam in nullo separatur ab eo secundum quod Arsamides ponderum proportionalitatem diffinivit. Ostendam ergo illud, et ponam duas ex perpendicularibus trutinarum *ab*, *gd* dispositas quattuor ponderibus *a*, *b*, *g*, *d*, et sit proportio *a* ad *b* sicut proportio *g* ad *d*, et sit unaqueque earum in duo media divisa super duo puncta *u* et *h*, et due statere earum sint due linee erecte *eu*, *zh*, et duo suspensoria earum sint due linee *tu*, *kh*, quarum queque secundum rectitudinem usque ad orizontis superficiem ad duo puncta *l* et *m* producatur, et protrahantur due perpendiculares *ab* et *gd* secundum

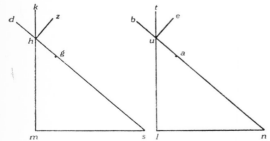

rectitudinem donec occurrant superficiei orizontis in duobus punctis *n* et *s*. Manifestum est igitur quod due linee *tl*, *km*

sunt due perpendiculares supra orizontis superficiem. Protraham etiam in superficie orizontis duas lineas *nl*, *sm*. Et quia, secundum quod dixit Yrinus, angulus *anl* est equalis angulo *gsm*, et duo anguli *uln*, *hms* sunt recti, erunt duo trianguli *unl*, *hsm* similes, et duo anguli *nul*, *shm* erunt equales; ergo duo anguli *tub*, *khd* sunt equales. Cum ergo minuerimus eos ex duobus rectis angulis *eub*, *zhd*, remanebunt duo anguli *eut*, *zhk* equales, qui sunt duo anguli ponderum, quod Arsamides diffinivit. Et illud est quod voluimus ostendere." The fragment of Archimedes here mentioned does not come from any extant work of Archimedes. It is perhaps from the lost περὶ ζυγῶν. The fragment of Hero is also unknown to me. The whole explanation of the definitions of Archimedes and Hero is rather odd. Notice that Gerard has rendered the beam of the balance as *perpendicularis trutine*, certainly an inappropriate translation, and the needle or vertical indicator as *statera*, a term often used for the whole balance and synonymous with *trutina*. The object of Ametus' "demonstration" is to show that Hero's definition is fundamentally the same as that of Archimedes. The substance of the definitions as interpreted by Ametus is that one set of weights is said to be proportional to another when the one can be substituted for the other in a balance without changing the angle between the vertical indicator and the beam (Archimedes) or without changing the angle that the beam makes with the plane of the horizon (Hero). Notice that earlier in the passage Ametus uses the term *caraston*, which refers to a balance of unequal arm lengths, but that

on the other hand, appear in only two manuscripts[2] and seem to have exerted little direct influence on medieval statics and hydrostatics. From the first three works the medieval mechanician learned what were the principal problems and concepts of Greek statics. Even more important, he learned the necessity of mathematical proofs in mechanics—of starting with geometrized physical postulates and proceeding therefrom by formal deductive steps to fundamental statical theorems.

To the seven works listed above, we can perhaps add another, (8) the *Mechanica* attributed to Aristotle. The question of whether or not a Latin translation of this work was made in the Middle Ages is difficult to answer. The principal fact weighing against such a translation is the absence of any extant manuscripts. Furthermore, the standard medieval library catalogues have no reference to such a translation. On the other hand, as Haskins points out, the celebrated emperor Frederick II in his *De arte venandi cum avibus* makes what appears to be a clear-cut reference to the work, calling it by the Latin title *Liber de ingeniis levandi pondera* and claiming that therein is contained the statement "that a larger circle is more effective in lifting a weight."[3] Furthermore, it is possible that the author of the Latin treatise *De ratione ponderis*, which we shall discuss shortly, had read certain of the problems in the *Mechanica* and had derived therefrom some of his propositions.[4] However, I know of no other citation or use of the *Mechanica* in the thirteenth or fourteenth centuries. But in the early fifteenth century, a license for the export of books from Bologna, dated August 18, 1413, includes the title *Repertorium super mechanica Aristotilis* (!).[5] The fact that a repertorium had been composed certainly suggests that a prior translation had been made. And it is not improbable that the *Liber mechanicorum* listed among a group of books at

---

the beam is considered as weightless, for he notes that the law of the lever applies to it; ordinarily when the term *caraston* is used a material beam possessing weight is understood. Obviously the word "inversely" must be added to Ametus' statement of the proportionality of weights and arm lengths.

[2] M. Clagett, "The Use of the Moerbeke Translations of Archimedes in the Works of Johannes de Muris," *Isis*, Vol. *43* (1952), 237, 240.

[3] C. H. Haskins, *Studies in Medieval Science*, 2d ed. (Cambridge, Mass., 1927), pp. 316–17; *De arte venandi cum avibus*, MS Vat. Pal. lat. 1071, 13c, ff. 23v–24r: "Portiones circuli quas faciunt singule penne sunt de circumferentiis equidistantibus, et illa que facit portionem maioris ambitus et magis distat a corpore avis iuvat magis sublevari aut impelli et deportari, quod dicit Aristotiles in libro de ingeniis levandi pondera dicens quod magis facit levari pondus maior circulus."

[4] E. A. Moody and M. Clagett, *Medieval Science of Weights* (Madison, 1952), pp. 205–11, and the discussions of propositions R3.02–R3.03, R3.06, pp. 402–4.

[5] *Chartularium Studii Bononiensis*, Vol. 2 (Bologna, 1913), 222, item 18.

Padua in 1401 is the translation we seek.[6] In view of these facts, we must keep an open mind on the question of a possible translation of the *Mechanica*.

The importance of medieval statics centers not only on the translation of Greek and Arabic statical treatises but also on the original composition of a series of works attributed to Jordanus de Nemore and composed in the early thirteenth century. Before listing these works and discussing their contributions, a word must be said about Jordanus. He has often been indentified with Jordanus of Saxony, who served as Master General of the Dominican Order from 1222 to 1237.[7] This identification was based largely on a statement made in the fourteenth century by Nicholas Trivet.[8] Against such an indentification is the fact that in the writings attributed to Jordanus of Saxony there is neither evidence of any mathematical interests nor any use of the place name "de Nemore" or of the appellation "Nemorarius." In addition, it appears that Jordanus taught at Toulouse.[9] But the University of Toulouse was not founded until 1229, at least so far as the arts faculty is concerned. Hence, if Jordanus was teaching at Toulouse as late as 1229, it is doubtful that he could have been the Dominican Master General who took up his duties in 1222, although, of course, this is not impossible. The only evidence from the mathematical works that connects our Jordanus with the Saxon is that at least one scientific manuscript bears the name Jordanus de Alemania,[10] i.e., Jordanus of Germany. Regardless of whether or not the mathematician and the Dominican are identical, we can conclude that his statical (and at least part of his mathematical) activity was already completed by 1260 and probably a great deal earlier than that time. This we know from the fact that the two main statical works attributed to Jordanus, the *Elementa* and the *De ratione ponderis*, as well as several of his mathematical works, were listed in the

---

[6] A. Gloria, *Monumenti della Università di Padova (1318–1405)*, Vol. *1* (Padua, 1888), 385.

[7] Moody and Clagett, *Medieval Science of Weights*, pp. 121–22.

[8] *Ibid.*, pp. 122, 373. Most of the succeeding information concerning Jordanus' life is also taken from the summary in Moody and Clagett.

[9] M. Curtze, "Jordani Nemorarii de triangulis libri quatuor," *Mitteilungen des Coppernicusvereins für Wissenschaft und Kunst zu Thorn*, 6 Heft (Thorn, 1887), p. vi.

[10] *Ibid.* Curtze believed Jordanus the mathematician to be identical with the Dominican, although he cited a letter from H. Denifle supporting the contrary view. Curtze has noted one occasion in a scientific manuscript of the appellation "Jordanus de alemania." His attempt to explain "Nemorarius" as meaning "born of the wood" is controverted by the fact that the only early form of the name appears to be "de Nemore." Hence it looks as though this is a place name. But this place has not been identified with any surety.

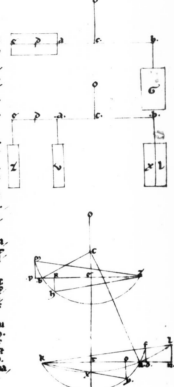

Plate 1: A page from a thirteenth-century manuscript of the *De ratione ponderis*, illustrating the bent lever proposition. MS Bodleian Library, Auct. F. 5.28, f. 128r.

Biblionomia of Richard of Fournival,[11] a catalogue composed sometime between 1246 when Richard was mentioned as being chancellor of the cathedral church at Amiens and 1260 when he was mentioned as being dead.[12]

There are three principal statical treatises attributed to Jordanus. Continuing our list of medieval statical texts available in Latin, we can list and number them as follows: (9) *Elementa Jordani super demonstrationem ponderum* (or *Elementa Jordani de ponderibus*), (10) *Liber de ponderibus* (called for convenience, Version *P*), and (11) *Liber de ratione ponderis*. To these can be added several variant versions and commentaries of the *Elementa* and the *Liber de ponderibus* belonging to the period from about 1275-1500. These I have numbered from (12) to (17), and I shall discuss them later in this chapter. I hope ultimately to publish them as a companion volume to the *Medieval Science of Weights*.

Because of the fact that all three of the works assigned to Jordanus share a number of common enunciations of propositions, E. A. Moody has suggested (without any evidence of an Arabic original, to be sure) that a skeleton set of propositions (i.e., enunciations without proofs) was translated from the Arabic, and then later to this set of propositions Latin scholars added other propositions and commentaries.[13] But it seems to me just as plausible that the most "primitive" of these texts, namely, the *Liber de ponderibus* (Version *P*), was translated from the Greek or Arabic and that Jordanus, being dissatisfied with the character and content of the proofs given in Version *P*, abandoned those proofs and supplied a new set. Still one further possibility is that the author of Version *P*, a thoroughly convinced "Aristotelian," came upon an entirely original *Elementa* of Jordanus and decided to give very much shorter proofs, changing their form to suit his Aristotelian ideas. Whatever was the origin of the *Elementa* and the *Liber de ponderibus*, it seems certain that Jordanus or some other author almost contemporary with him expanded and corrected the nine propositions of the *Elementa*, thereby producing the *De ratione ponderis* with its forty-five propositions. There is no sure evidence to

---

[11] L. Delisle, *Le Cabinet des manuscrits de la Bibliothèque Nationale*, Vol. 2 (Paris, 1874), 526. "43. Jordani de Nemore liber philothegny CCCCXVII propositiones continens. Item eiusdem liber de ratione ponderum et alius de ponderum proportione." Other codices in the *Biblionomia* contain other mathematical works of Jordanus.

[12] *Histoire littéraire de la France*, Vol. 23, 717.

[13] Moody and Clagett, *Medieval Science of Weights*, p. 15.

support Thurot's conjecture that the *De ratione ponderis* is a translation from a Greek original going back to Ptolemy's περὶ ῥοπῶν.[14]

Since all three of these texts have been published with English translations and commentaries,[15] we shall in this section merely summarize their important contributions to statics, supplying in Document 2.1 only the most interesting of the medieval proofs. The first of the three treatises, the *Elementa*, is the one most surely assigned to Jordanus. For in this work the author refers to a work entitled *Philotegni* in a manner that suggests he was referring to his own work, and indeed Jordanus composed a work entitled *Liber Philotegni de triangulis* where the reference of the *Elementa* can be found.[16]

Let us examine first some of the seven suppositions and nine propositions that make up the *Elementa*. In the first place in suppositions 4 and 5 we find emerging the concept of a component of force in a constraint system:[17] "[A weight] is heavier positionally, when, at a given position, its path of descent is less oblique. A more oblique descent is one in which, for a given distance, there is a smaller component of the vertical."

These suppositions are saying (1) that force along an "oblique" path is inversely proportional to the obliqueness, while (2) obliqueness is measured by the ratio of a given segment of the oblique path to the amount of the vertical intercepted by that path. In modern terms we would say that, so long as the oblique path is a straight line, then the force F along the incline or oblique path is computed as follows: $F = W \sin a$, where W is the free weight, and a is the angle of the inclination of the oblique path. While the oblique path used in the propositions of the *Elementa* turns out to be arcal rather than rectilinear—as we shall see in a moment—in the *De ratione ponderis*, the oblique paths are rectilinear as well as arcal. This will be evident by an examination of propositions 9 and 10 of the *De ratione ponderis*. In proposition 9 the author declares that a weight is of the same heaviness at any position along the inclined plane,[18] that is to say, its positional gravity or force along the incline is the same anywhere on the incline. The proof he gives for this statement is based on showing that, if we take equal segments of the inclined path anywhere along the path, these equal segments will intercept equal lengths of the vertical. Nothing could be clearer, then, assuming suppositions 4 and 5 and prop-

---

[14] C. Thurot, "Recherches historiques sur le Principe d'Archimède," *Revue archéologique*, Vol. *19* (1869), 117.

[15] Moody and Clagett, *Medieval Science of Weights*, texts V–VII.

[16] *Ibid.*, pp. 130, 379.

[17] *Ibid.*, pp. 128–29.

[18] *Ibid.*, pp. 188–89.

osition 9, than that the force of weight on the incline decreases as the ratio of a segment of the incline to its vertical intercept increases, which ratio is of course nothing but the secant of a. But if the force or weight is inversely as the secant a, it is directly as the sine a. Furthermore, proposition 10, which we have included in Document 2.1 below and which we shall discuss at somewhat greater length later in this chapter, makes clear what is meant by obliquity or declination, when it says: "The proportion of declinations [of two oppositely inclined planes] is not one of the angles but of the lengths of the [inclined] lines measured to a horizontal intersecting line by which they (the inclined lines) partake equally of the vertical (i.e., to a line which would cut off equal projections of these lines on the vertical)."

As is abundantly clear from the proof itself in proposition 10, the force along the inclined plane is inversely as the obliquity or declination. Hence we very clearly have the modern formulation present in this proposition of the *De ratione ponderis*. Now no one will deny that, as long as the author of the *De ratione ponderis* applies his idea of component force or positional gravity to oblique paths which are straight lines, he is entirely correct in his procedure. But it should also be remarked that he retains in propositions 2 to 5 an application of the principle to arcal paths, an application that constituted the exclusive use of the principle in the *Elementa*. For, as in the *Elementa*, we find in these propositions 2 to 5 an attempt to measure the positional gravity of a weight on a lever arm by taking an arbitrary arc of the path which the weight would describe when the lever is in motion, and seeing how much of the vertical *this arc* would intercept. For example, examine proposition 5 of Part I of the *De ratione ponderis* which states that "if the arms of the balance are unequal, then if equal weights are suspended from their extremities, the balance will be depressed on the side of the longer arm." The proof rests on the assumption that positional gravity or effective weight is measured by the intercept of a segment of the weight's arcal path on the vertical. Thus if A and B are equal weights and we take the quarters of circles through which A and B could move, then A is positionally heavier, since arc AG intercepts CG on the vertical which is greater than CF intercepted by arc BF. Actually this procedure is legitimate so long as the mechanician uses the intercepts of quadrants, since in effect he is comparing the total vertical distances through which the forces could act. And as we shall see shortly, Jordanus knew and applied the principle of virtual velocities. Hence one might say that this principle is implicit in the treatment of proposition 5.

The method of measuring effective weight by comparing the intercepts made by arcs on vertical lines was based ultimately on the analysis in the *Mechanica* of the Pseudo-Aristotle, which, as we saw in the first chapter, suggested that the nearer to the fulcrum the weight was placed, the less it possessed of natural, vertical movement.

The inherent difficulty of using the intercepts of arcs when those arcs are not quadrants of a circle is shown particularly in the analysis of the stability of a balance beam suspended from above which is found in proposition 2 of the *Elementa*. Let us suppose we have a balance beam *BC* suspended from above as illustrated in the accompanying diagram (see Fig. 2.1).[19] Jordanus held that *C* moves downward to a position of hori-

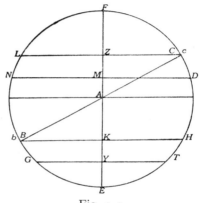

Fig. 2.1

zontal equilibrium because, he believed, a weight at *C* is positionally heavier in its raised position. He thought that this was so because if we take any pair of equal arcs *CD* and *BG*, which represent the potential movements downward of two equal weights at *B* and *C*, *CD*, although equal to *BG*, will intercept more of the vertical than will *BG*—that is, *ZM* is greater than *KY*. Thus a weight at *C* is positionally heavier. Ingenious reasoning, but wrong.

In the *De ratione ponderis*[20] the author raised and discussed an objection that could be made to this kind of analysis. The objection was this. If *C* is positionally heavier than *B*, why cannot we support a heavier weight at *B* by the weight at *C*, although in fact we cannot? The author of the *De ratione ponderis*, recognizing that we cannot, posits a subtle solution which we can paraphrase as follows: If *B* is taken to be heavier than *C*, it must be heavier by some definite amount, say *k*. Now if we take smaller and

[19] *Ibid.*, pp. 130–31.   [20] *Ibid.*, pp. 178–79.

smaller equal arcs, CD and BG, the difference in their obliquity (measured rather peculiarly by certain horn angles) approaches zero, or to put it in another way which is in accord with the spirit of his technique of taking vertical components as the measure of obliquity, the ratio of these vertical components, to wit, ZM to KY, approaches one, i.e., the excess of ZM beyond KY becomes smaller and smaller and approaches zero. Hence, according to our author, we can always take arcs small enough so that the excess in positional gravity of C over B [represented by the excess of ZM over KY] is less than any given excess of B over C, i.e., less than k. Hence C will never by its superior positional gravity be able to balance a heavier weight at B.

This argument, as Duhem realized,[21] contains the germ of the method which would in fact refute the idea that an elevated weight has greater positional gravity by reason of its elevation and which would also refute therefore the very proposition which held that a weightless beam suspended from above would seek horizontal equilibrium. For as we take smaller and smaller arcs, the limit of the ratio of ZM/KY is one. If we proceed to the limit, the pretended positional advantage enjoyed by C vanishes, and hence we must conclude that C has no advantage due to its elevated position.

So much then for our brief discussion of the concept of positional gravity or components of force as given in the *Elementa* and the *De ratione ponderis*. Perhaps an even more important contribution to statics by these two treatises was the use of the principle of virtual velocities or displacements to prove theorems of statics. In the *Elementa* we find a proof of the law of the lever which clearly uses the principle of virtual velocities as applied to vertical distances. This proof is repeated without alteration in the longer treatise *De ratione ponderis*, and we have given it in Document 2.1. If the reader will study this proof, he will see that it consists essentially of these steps: (1) Either a lever with weights inversely proportional to the lever arms is in equilibrium or it is not. (2) If it is assumed that it is not, then one weight or the other descends. (3) But if one of the weights descends, it would perform the same action (i.e., work) as if it lifted a weight equal to itself and placed at an equal distance from the fulcrum through a distance equal to the distance it descended. But equal weights at equal distances are in equilibrium. Thus a weight does not have the force sufficient to lift its equal weight the same distance it descends. It, therefore, does not have the force to perform the same action—

[21] P. Duhem, *Les Origines de la statique*, Vol. *1* (Paris, 1905), 140.

namely, to lift a proportionally smaller weight a proportionally longer distance.

In the course of this proof Jordanus does not explicitly name the principle of virtual velocity as a principle, but he does say that if weights *a* and *l* on the same side of a lever (see Fig. 2.6) are shown to be inversely proportional to their vertical distances *GD*, *ML*, when the lever moves, then "that which suffices to lift *a* to *D* would be sufficient to lift *l* through [the distance] *LM*." Such a statement certainly implies acceptance of the work principle. At least one commentator in the fourteenth century is somewhat more explicit concerning the principle and its justification in his treatise known as the *Aliud commentum*. This commentator believes that the justification of the principle lies in the first proposition of the *Elementa*, whose statement and proof by Jordanus is most ambiguous.[22] The first proposition states that "the proportion of the velocity of descent, among heavy bodies, is the same as that of weight, taken in the same order; but the proportion of descent to the contrary ascent is the inverse proportion."[23] While this appears at first glance to be an affirmation, regarding weights moving freely, that the velocity is proportional to the weight, the fourteenth century commentator whose commentary was published with the *Liber de ponderibus* (Version P) in 1533 interprets this proposition as refering to weights on a balance. This, then, involves interpreting the velocity of descent or ascent as the vertical rectilinear distances through which the weights on a balance or a lever arm move in the same time. Our commentator, whose remarks we shall quote fully in a moment, then interprets the first part of the theorem in the sense that the effective weight of a given weight on a lever arm depends on, or is directly proportional to, the velocity of descent as well as to the free weight. Moreover, as we shall see, the commentator in the course of his interpretation of the second part holds that as one decreases the proportion of actual weights by moving the second weight farther out on the other arm, the ratio of rectilinear ascent to descent increases and the increase is in the same proportion as the decrease of the ratio of the actual weights. The principle of work, then, as understood by the commentator is essentially this: What suffices to lift a weight W through a vertical distance H will suffice to lift a weight kW through a vertical distance H/k and it will also suffice

---

[22] G. Vailati, "Il principio dei lavori virtuali etc.," Chapter 3 in his *Scritti* (Leipzig, Florence, 1911), pp. 99–103; Moody and Clagett, *Medieval Science of Weights*, pp. 426–28.

[23] Moody and Clagett, *Medieval Science of Weights*, pp. 128–29.

[Medieval Latin Statics

to lift a weight W/k through a vertical distance kH. But here is the whole passage from the commentator's remarks:[24]

> ...the author does not understand the first part of the theorem to mean that if the weight a, left to its own nature, moves freely in a certain medium, or traverses it, that b—i.e., the lesser weight—, if left to its own nature moves more slowly, or traverses less of that distance of descent in the same time, according to the ratio which b has to a. For this author is not concerned to treat of the movement of the heavy body left to its own nature, but only of the motion of the heavy body when it is suspended on a balance, and is resisted by a weight placed on the other arm of the balance. This is made clear from the second part of the theorem, in which the author treats of the ascent of weights—when however no weight ascends naturally, in a medium in which it would naturally descend if left to its own nature. But a ascends on the arm of a balance, by reason of the force exerted by the weight of the other arm descending. Since therefore the proportion which the author seeks to establish, for the case of ascent, is proved by means of the first part of the theorem, this proof would not be valid unless, in the first part of the theorem, the descent were assumed to take place on a balance, just as in the second part of the theorem the ascent is assumed to take place on a balance. And if it is understood in this way, it must then have reference to the equality or inequality of the arms of the balance....
>
> ...it is necessary that the conclusion be understood in the following manner: (see Fig. 2.2) That in a balance $BACD$, whose center is A, a weight g, when

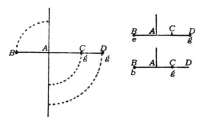

Fig. 2.2

suspended at C, is related to the same weight g, when suspended at D, according to the proportion of the whole descent which it can have at C to the whole descent which it can have at D. But by reason of the fact that it cannot descend to a lower position, except according to the length of the radius whose circum-

---

[24] *Ibid.*, pp. 299–303. I have made a correction to Moody's translation of this passage. In the sentence beginning the third paragraph, "To show the sufficiency..." Moody has "...another weight, such as would be equally able, if placed at D, to raise g to the vertical position." The Latin text correctly indicates that it is g which is to be placed at D, "...aliud pondus quod equefaciliter levaret g in D situ...." Capital letters designating points are here italicized, although this practice was not followed in Moody and Clagett.

ference it describes, it follows, on this interpretation, that the weight g, at position C, is related to the same weight g at position D, according to the ratio of CA to DA, so that the weight g, when placed at C, would be able to descend to the vertical position with equal ease, with a weight on the other arm of the balance which would be related to the weight first raised, in the proportion of CA to DA. And this takes care of the first part of the theorem.

To show the sufficiency of the second part of the theorem, I say that if the weight b would suffice to raise a weight g placed at C, to the vertical position, another weight, such as would be equally able to raise g to the vertical position if g were placed at D would be related to b according to the ratio of DA to CA. And so, if this interpretation is true for the one case, it seems that it will be true for all cases, since according to this exposition the gravity varies only by reason of the variation in positions. Therefore, if in one case the gravity of the same weight is varied according to the proportion of the lever arms, there is no reason why this should not be so in any case whatever. If therefore we interpret the argument in this manner, the author's theorem holds good; otherwise not. And if we understand it in this way, it has bearing on the eighth theorem (i.e., the proof of the law of lever; see in Doc. 2.1), in proof of which this theorem is invoked.

But it seems that this exposition is not adequate to the meaning of the theorem, because the theorem states that as weight is to weight, so is velocity to velocity—whereas in our exposition we did not discuss velocity. We can, therefore, argue in this manner: Let the weight e be placed in the same position as b, and let the weight g, placed at D, be related to it (i.e., to e at B), as g when placed at C is related to the weight b. It seems therefore that g at D exerts an equal force on e (at B), as that same weight g, placed at D, exerts on the weight b. Hence g, at D, is able to raise e to the vertical position just as quickly as it is able, when placed at C, to raise b to the vertical position. But b or e arrive just as quickly as g to the vertical position; therefore the velocity of g, when placed at D, is related to its velocity when placed at C, according to the ratio of DA to CA—by the fifth proposition of Archimedes' *De curvis superficiebus*, according to which the ratio of the radii, or of the diameters, or of the circumferences, is the same; therefore, etc.

If however this argument is not persuasive, it does not matter whether the velocity is proportional or not, as long as this holds: If g, placed at D, suffices to raise e, then g placed at C suffices to raise b. And on this account the first conclusion of the text of Jordanus has another wording, namely, that "between any heavy bodies the proportion of force and of weight is the same, in the same order." And this likewise suffices for proving the eighth theorem, in proof of which this theorem is assumed. And this is sufficient as explanation of this theorem.

While in the *Elementa* Jordanus uses the principle of virtual velocities for the proof of the law of the lever, the author of the *De ratione ponderis* (who, of course, may have been Jordanus himself) extends its use to the

[Medieval Latin Statics

proof of other statical theorems. In the first place, he employs it to prove the special case of the equilibrium of the bent lever on whose extremities hang equal weights at equal distances from the vertical line passing through the fulcrum. The principle is also used in the case of the equilibrium of two weights on oppositely inclined planes, which weights are directly proportional to the lengths of the inclines. I have thought the proofs of these theorems of such importance in the history of statics that I have included them in their entirety in Document 2.1 below, even though they have been previously translated and published.

Turning first to the case of the bent lever, one should initially realize that, when the author of the *De ratione ponderis* takes up this case, he appears to be doing so to correct an error that had appeared in the *Elementa*. In the sixth proposition of the *Elementa* we read:[25] "When equal weights are suspended at unequal distances from the axis, and if the longer arm is bent until its end is at the same distance from the vertical [through the axis] as the end of the shorter arm is, and if the latter remains unmoved, then the weight on the longer arm will become positionally lighter than the other weight." The attempted proof of this theorem makes use of an incorrect application of positional gravity to arcs. The reasoning is that, if we take a segment of the vertical *EM* (see Fig. 2.3),

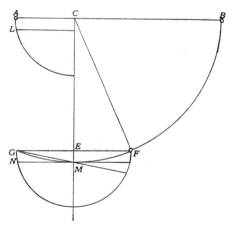

Fig. 2.3

the arc *AL*, or rather its equal *GN* which intercepts *EM*, is more nearly a vertical path than the arc *FM*. Hence the weight at *A* is positionally heavier than an equal weight at *F*.

Now the author of the *De ratione ponderis* leaves out this proposition

[25] *Ibid.*, pp. 136–37.

and substitutes for it an entirely correct proposition: "If the arms of a balance are unequal and they make an angle at the center of motion (i.e., the fulcrum), and if their termini are equally distant from the vertical line [passing through the fulcrum], in this disposition equal suspended weights will be of equal heaviness (see Doc. 2.1, prop. 8)." The reader will notice in examining the proof that it is an indirect proof. Either the weights are in equilibrium or they are not. If they are not, then one descends. But such a descent would involve a weight moving through a vertical distance in such a way as to lift a weight equal to it through a vertical distance greater than that of its descent. But this is impossible [since it violates the principle of virtual velocities].

In this proposition, it should be further noted, the principle of static moment is clearly anticipated in the sense that the author emphasizes that both the weight and the *horizontal distance* to the vertical are the factors determining the effective force of a weight on a bent lever arm. True, it involves only the special case of equal weights at equal horizontal distances. But that, like Hero, our author recognized the general lever law as applied to bent levers appears evident from his proof of proposition 1 (R3.01) in Part III:[26]

If the axis is on a perpendicular above the beam of the balance, then, however great a weight be suspended from either of the arms, it will not be possible for it to descend to a position directly below the axis.

Let the balance beam, for example, be *ABC* (see Fig. 2.4), the perpendicular

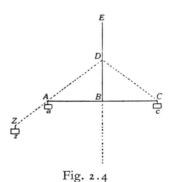

Fig. 2.4

BDE, with the axis at *D*, and a greater weight *a* (at *A*) than the weight *c* (at *C*). Then draw the lines *DA* and *DC*, and let *DA* be extended to *Z* in such a manner that *DAZ* is to *DA* as the weight *a* is to the weight *c*. Let there be weight *z* (at *Z*) equal to *c*. When, therefore, the three weights *a*, *c*, and *z* are in this manner attached to the beam *ABC*, their revolution is around the axis *D*, just as if they

[26] *Ibid.*, pp. 204–5, 401–2; cf. prop. R3.05, pp. 208–11.

[Medieval Latin Statics 83]

were hanging on the lines DAZ and DC. But as so placed, z will tend to be at the same distance from the vertical passing through D, as c will; *and [weight] a likewise will be at a proportional distance from the vertical (distabit quoque et [pondus] a proportionaliter a directo eiusdem).* Therefore a will not be able to descend as far as the vertical.

The section which I have italicized is a clear indication that the author understands the application of the lever law to the bent lever, both in the special case of equal weights which will be in equilibrium at equal horizontal distances from the vertical and in the more general case of a proportionally heavier weight which at a proportionally less horizontal distance balances the other weight.

Now let us turn to the proof of the theorem relative to inclined planes, which we have noticed makes use of the work principle. Like the bent lever proof, this is an indirect one. Either the weights which are directly proportional to the lengths of the inclines are in equilibrium or they are not. If not, then one weight descends, drawing the other up. But in drawing the weight up, it is acting in exactly the same way as if it pulled a weight equal to itself an equal vertical distance. But this is impossible, for the force along equally inclined planes is constant, and so a weight would not have the force to move a weight equal to itself and situated on an equally inclined plane. But the movement of a proportionally heavier weight a proportionally smaller vertical distance (i.e., the movement as posited) would be the same (i.e., would take precisely the same work) as the movement of an equal weight an equal distance. Hence this posited movement is also impossible. Q.E.D.

Such is the second remarkable application of the work principle to statics made by the author of the *De ratione ponderis*. I shall not attempt here to describe in detail the rest of the contents of the *De ratione ponderis*. I should like to remind the reader that while the *Elementa* contains only nine propositions, to which the four propositions of the *De canonio* were often appended, the *De ratione ponderis* contains forty-five propositions arranged in four parts. The first part ends with the elegant inclined plane proof which we have just discussed. The second part is an elaboration of the *De canonio* and thus gives propositions relative to the Roman balance. It does not, however, contain any significantly original propositions. The third part contains numerous special lever problems which show the influence of the *Mechanica* attributed to Aristotle. It analyzes correctly problems regarding the stability and instability of the balance. As we have indicated, it reveals a correct understanding of the principle of static moment, making use of the principle as already revealed in the special bent lever case of proposition 8 of Part I. Furthermore, it should be noticed that among these theorems in Part III is the first really correct analysis of a problem which had already appeared in the *Mechanica* attributed to Aristotle, the *Mechanics* of Hero, and the *On Architecture* of Vitruvius. This is the problem of the computation of the forces exerted by a weight on the supports of a beam from which the weight hangs. The solution is offered in proposition 6 of Part III of the *De ratione ponderis*, which reads as follows:[27] "A weight not suspended at the middle [of a beam supported at each end] makes the shorter part heavier according to the ratio of the longer part to the shorter part." The proof involves assuming that the weight e (see Fig. 2.5) is divided into two weights d and

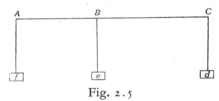

Fig. 2.5

f, such that $d/f = AB/BC$. We imagine that d and f are suspended at C and A. Then each of them suspended at these points "will be of the same heaviness as the weight e, (i.e., they will exert the same force) so long as we imagine in each case that the opposite end [of the beam] is the fulcrum. Hence for those who at A and C are carrying the weight e suspended at B, the heaviness at A will be to the heaviness at C as the length CB is to the length BA."

The fourth part of the *De ratione ponderis* contains propositions relative to dynamics and we shall have occasion in a later chapter to discuss certain of these propositions.

The third treatise attributed to Jordanus—and I would say falsely so—is the so-called *Liber de ponderibus*, known as Version P. It was much less fruitful in producing original statical arguments. It does have an interesting introduction which demonstrates that its author is thoroughly influenced by Aristotelian dynamics. It has been treated at some length by E. A. Moody in our statics volume, and I refer the reader to that discussion without further comment.[28]

Of the various texts that formed the corpus of the medieval science of

[27] *Ibid.*, pp. 210–11, 404–5. The drawing on page 210 is incorrect; weights d and f should be interchanged.

[28] *Ibid.*, pp. 145–65, 384–88.

weights, only one has not yet been treated in detail: the thirteenth-century *De ponderibus Archimenidis* or *De insidentibus in humidum*, a treatise on hydrostatics attributed to Archimedes. This treatise, however, was not the first consideration of hydrostatics in the Latin West. Before its appearance, hydrostatic knowledge, or rather written expression of hydrostatic knowledge, was meager. Vitruvius, in the ninth book of his *De architectura*, mentions the famous story of Archimedes and the determination of the component weights of gold and silver in Hiero's crown (or sacred wreath).[29] The method described by Vitruvius[30] is one comparing the volume $[V]$ of a crown [of weight W] with the volume $[V_1]$ of a weight $[W]$ of gold and with the volume $[V_2]$ of the same weight $[W]$ of silver. Then the component weights $[w_1$ and $w_2]$ of gold and silver are readily related by the following proportionality (and thus individually determined since their sum W is known):

$$\frac{w_1}{w_2} = \frac{V_2 - V}{V - V_1}.$$

There is also a section on the crown problem in the *Carmen de ponderibus*, a work on weights and measures that appears in several codices of Priscian's grammatical work.[31] It possibly dates from about A.D. 500. Two methods

---

[29] Vitruvius, *On Architecture*, Edition and translation of F. Granger, Vol. 2 (London, 1934), 204–7. "Tum vero ex eo inventionis ingressu duas fecisse dicitur massas aequo pondere, quo etiam fuerat coronoa, unam ex auro et alteram ex argento. Cum ita fecisset, vas amplum ad summa labra implevit aquae, in quo demisit argenteam massam. Cuius quanta magnitudo in vasum depressa est, tantum aquae effluxit. Ita exempta massa quanto minus factum fuerat, refudit sextario mensus, ut eodem modo, quo prius fuerat, ad labra aequaretur. Ita ex eo invenit quantum ad certum pondus certam aquae mensuram responderet. Cum id expertus esset, tum auream massam similiter pleno vaso demisit et ea exempta, eadem ratione mensura addita invenit ex aquae numero non tantum esse: minore quanto minus magno corpore eodem pondere auri massa esset quam argenti. Postea vero repleto vaso in eadem aqua ipsa corona demissa invenit plus aquae defluxisse in coronam quam in auream eodem pondere massam, et ita ex eo, quod fuerit plus aquae in corona quam in massa, ratiocinatus reprehendit argenti in auro mixtionem et manifestum furtum redemptoris." For the distinction between "crown" and "wreath," see E. J. Dijksterhuis, *Archimedes* (Copenhagen, 1956), p. 19.

[30] Cf. T. L. Heath, *The Works of Archimedes* (Cambridge, 1897), pp. 260–61.

[31] F. Hultsch, *Metrologicorum scriptorum reliquae*, Vol. 2 (Leipzig, 1866), 95–98:

> 125 "argentum fulvo si quis permisceat auro,
> quantum id sit quove hoc possis deprendere pacto,
> prima Syracusi mens prodidit alta magistri.
> regem namque ferunt Siculum quam voverat olim
> caelicolum regi ex auro statuisse coronam,
> 130 conpertoque dehinc furto—nam parte retenta
> argenti tantundem opifex immiscuit auro—

of solving the crown problem are given in this section. The first follows (note 31, lines 125–62):

The first, profound mind of the Syracusan Master has disclosed how—if one should mix silver with tawny gold—you can determine how much of it [the silver] there is in this mixture. For they say that the Sicilian king once ordered a crown of gold to be fashioned, one suitable for a heavenly king. But he learned that he

---

    orasse ingenium civis, qui mente sagaci,
    quis modus argenti fulvo latitaret in auro,
    repperit inlaeso quod dis erat ante dicatum.
135 quod te, quale siet, paucis, adverte, docebo.
    lancibus aequatis quibus haec perpendere mos est
    argenti atque auri quod edax purgaverit ignis
    impones libras, neutra ut praeponderet, hasque
    summittes in aquam: quas pura ut ceperit unda,
140 protinus inclinat *pars* haec quae sustinet aurum;
    densius hoc namque est, simul aëre crassior unda.
    at tu siste iugum mediique a cardine centri
    intervalla nota, quantum discesserit illinc
    quotque notis distet suspenso pondere filum.
145 fac dragmis distare tribus. cognoscimus ergo
    argenti atque auri discrimina: denique libram
    libra tribus dragmis superat, cum mergitur unda.
    suma dehinc aurum cui pars argentea mixta est
    argentique meri par pondus, itemque sub unda
150 lancibus impositum specta: propensior auri
    materies sub aquis fiet furtumque docebit.
    nam si ter senis superabitur altera dragmis
    sex solas libras auri dicemus inesse,
    argenti reliquum, quia nil in pondere differt
155 argentum argento, liquidis cum mergitur undis.
    haec eadem puro deprendere possumus auro,
    si par corrupti pondus pars altera gestet.
    nam quotiens ternis pars inlibata gravarit
    corruptam dragmis sub aqua, tot inesse notabis
160 argenti libras, quas fraus permiscuit auro.
    pars etiam quaevis librae, si forte supersit,
    haec quoque dragmarum simili tibi parte notetur.
    Nec non et sine aquis eadem deprendere furtum
    ars docuit, quam tu mecum experiare licebit.
165 ex auro fingis librili pondere formam,
    parque ex argento moles siet; ergo duobus
    dispar erit pondus paribus, quia densius aurum est.
    post haec ad lancem rediges pondusque requires
    argenti, nam iam notum est quod diximus auri,
170 fac et id argento gravius sextante repertum.
    tunc auro cuius vitium furtumque requiris
    finge parem argenti formam pondusque notato:

was defrauded. For the artisan kept out a certain part of the gold and [instead] mixed in the same amount of silver. The king beseeched the genius of the city to find, with his sagacious mind, how much silver was hidden in the tawny gold, without altering this object dedicated to the gods. Listen, I shall teach you in brief how it is done. In each of the pans of a balance customarily used for weighing these [metals] place equal weights of silver and gold purified by gnawing fire, so that neither of them is in excess (i.e., they are in equilibrium). And then plunge them into water. Since they are immersed in water, the pan bearing the gold immediately inclines, for it is heavier due to the fact that the water is denser *(crassior)* than the air. Re-establish equilibrium and note the intervals from the central point, how far it will have withdrawn thence and by how many points the chord is distant from the suspended weight. Suppose it is distant by three drachmas. We know then the difference between the gold and silver: when immersed in water one pound of the one exceeds the other by three drachmas.

Before going on with the rest of this passage, we should note that if we

```
           altera quo praestat leviorque est altera moles
           sit semissis onus: potes ex hoc dicere quantum
     175   argenti fulvo mixtum celetur in auro.
           nam quia semissem triplum sextantis habemus,
           tres inerunt auri librae, quodque amplius hoc est,
           quantumcumque siet, fraus id permiscuit auro....

           quare diversis argenti aurique metallis,
           quis forma ac moles eadem est par addito pondus,
           argento solum id crescit, nihil additur auro.
           sextantes igitur quos tu superesse videbis,
     190   in totidem dices aurum consistere libris,
           parsque itidem librae sextantis parte notetur.
           quod si forte parem corrupto fingere formam
           argento nequeas, at mollem sumito ceram.
           atque brevis facilisque tibi formetur imago
     195   sive cybi seu semiglobi teretisve cylindri,
           parque ex argento simuletur forma nitenti,
           quarum pondus item nosces. fac denique dragmas
           bis sex argenti, cerae tres esse repertas:
           ergo in ponderibus cerae argentique liquebit,
     200   si par forma siet, quadrupli discrimen inesse.
           tum par effigies cera simuletur eadem
           corruptae, cuius fraudem cognoscere curas.
           sic iustum pondus quod lance inveneris aequa
           in quadruplum duces; quadupli nam ponderis esset,
     205   si foret argenti moles quae cerea nunc sit.
           cetera iam puto nota tibi—nam diximus ante—
           quo pacto furtum sine aquis deprendere possis.
           haec eadem in reliquis poteris spectare metallis."
```

assume the difference between the loss of weight of gold and silver at three drachmas (and if we use, as the poem tells us earlier, 1 Attic libra equal to 75 drachmas), then the difference of weight in water between gold and silver is 1/25 of the original weight, which compares well with the figure of 1/23.3 computed below with modern values.

Then take the gold in which the silver is mixed, as well as an equal weight of pure silver. And also immerse them in water after they have been placed on a balance. The golden material becomes heavier under water and will [thus] indicate the deception. For if [starting with an air weight of six pounds] one exceeds the other by eighteen drachmas, we say that there are in it only six pounds of gold; the rest of the silver [has no effect,] because silver does not differ in weight from silver when immersed in liquids. We are able to learn of these mixtures with pure gold, if some other part [than the eighteen additional drachmas] supports the equivalent weight of the corrupted specimen. For as many times as the unblemished part outweighs the corrupted under water by three drachmas, you will know that there will be present the same number of pounds of silver which the cheat has mixed with the gold. Furthermore, if by chance there is some fraction of a pound in excess, this would also be known to you by a similar fraction of drachmas.

This first method, then, does not compare volumes in the manner of the account of Vitruvius but rather uses a hydrostatic balance; and thus it basically rests on Archimedes' principle. Hence this determination involves the comparison of the losses of weights in water. The procedure described is first to weigh any equal air weights of gold and silver in water. This gives the relationship of the losses of weight of gold and silver in water. The crown and an equal air weight of silver are then weighed in water. We now have enough information to find the desired component weights. We can show this technique symbolically as follows: Archimedes' principle (prop. 7 of the *Floating Bodies*; see Doc. 1.4, prop. I.7) leads to this formula:

$$\frac{w_1}{w_2} = \frac{F_2 - F}{F - F_1},$$

where w_1 and w_2 are, as before, the partial weights of gold and silver in a crown of weight W; F_2, F, F_1 are the losses of weight in water of equal weights W of the silver, the crown, and the gold. Now by the first step in this poem, the ratio determined is $\frac{F_1}{F_2} = a$. By the next step F and F_2 are determined. Thus we can find the following relationship:

$$\frac{w_1}{w_2} = \frac{F_2 - F}{F - F_2 a}.$$

Since W is the sum of w_1 and w_2 and is known, each of these weights can now be found. A final observation concerning the first method of treating the alloy problem in the *Carmen de ponderibus* is that it speaks of submerging the whole balance instead of just a water bowl in water when one is conducting determinations of losses of weight in water.

Immediately following the first method of solving the problem, the author of this poem describes briefly a second method in which the hydrostatic balance is not involved (note 31, lines 163–79, 186–91):

> This same art has taught how to recognize the deception even without the use of water, and you can try it out *(experiare)* with me. You fashion a gold form of one pound weight, and let there be an equal amount *(moles, i.e., volume)* of silver. Hence the weight of these two equal [volumes] will be unequal, because gold is the denser. After this, put them in a balance, and you will require a weight of silver, because of what we have said about the gold. Do this and the gold will be found to be heavier by a sixth [weight] *(sextans)*. Then make a form of silver equivalent [in volume] to the gold whose corruption and fraud you seek, and note the weight. The amount by which the one mass is heavier and the other is lighter is a one-half weight *(semissis onus)*. You can from this tell how much silver is hidden in the tawny gold. For because we have a one-half [weight] which is triple a sixth [weight], there will be three pounds of gold in it; and by the amount that this is more—however much it is—so much the cheat has mixed with the gold.... Wherefore in different metals of gold and silver where the form and mass (i.e. volume) is the same, only the silver is increased in weight [to bring about equilibrium]; nothing is added to the gold. Hence you say that there are just as many pounds of gold as there are sixth [weights] which you see to be in excess, and in like manner the fraction of a pound is known from the fraction of a sixth [weight].

The second method of solving the crown problem is "volumetric" but not like the method described by Vitruvius. Initially the author declares that if we take equal volumes of gold and silver, the gold weighs more, because the gold is denser *(densius)*. Here then we have implied a definition of specific weight (perhaps the first—and at any rate the clearest—of the early descriptions of density available in early Latin works). Having related gold and silver by their weights for an equal volume, the author then suggests that we make a volume of silver equivalent to that of the suspected alloy. We then note how much the silver weighs. Then, placing both the alloy and the silver in a balance, we add to the silver enough to bring about equilibrium. This addition is then the key to how much silver has been used to corrupt the pure gold. If it is the proper multiple, computed on the basis of the known difference of weight of a pound of gold and of an equivalent volume of silver, the specimen is pure gold. If it is less than the computed multiple, some silver has been introduced, and the amount of silver depends on how much less the addition is than it ought to be. This method, rather vaguely described in this passage, is further developed by the author of the *De insidentibus in humidum* in the thirteenth century and by Johannes de Muris in the fourteenth century. I have shown in detail in the commentary to Document 2.4 the arithmetical procedures of such a method. It should be pointed out further that the figures given by the poet for this second method are confused and inaccurate—particularly the basic fact cited to the effect that a pound of gold is heavier by a *sextans* than an equivalent volume of silver. A *sextans* here and in the earlier part of the poem is 1/6 of a *libra*. This does not appear to fit at all with the relation of the specific gravities of gold and silver implied in the first method, where the gold is 1/25 heavier in water than an equivalent air weight of silver.

Just following the second method in the *Carmen de ponderibus*, a further passage (note 31, lines 192–208) tells how to make volumes of wax and silver equivalent to the suspected object by making a pattern of wax and a casting of silver. Although the figures here seem to me to be completely confused, there are further evidences of this process in the Middle Ages. One manuscript of the tenth century[32] gives figures for the comparative weights of wax and several metals replacing it in fusion that lead to fair approximations of specific gravities, e.g., a specific gravity of 10.2 for silver, 11.8 for lead, and 18.6 for gold.

Before leaving the *Carmen de ponderibus*, note should also be made of the section just preceding the treatment of the crown problem. There we find described a hydrometer of the type we would call an aerometer,[33]

[32] The manuscript is Paris, BN lat. 12292; it is discussed by M. Berthelot, *La Chimie au moyen âge*, Vol. *1* (Paris, 1893), 175–76.

[33] Hultsch, *Metrologicorum scriptorum reliquae*, Vol. 2, 94–95:

"ducitur argento tenuive ex aere cylindrus,
quantum inter nodos fragilis producit harundo,
105 cui cono interius modico pars ima gravatur,
ne totus sedeat totusve supernatet undis.
lineaque a summo tenuis descendat ad imam
ducta superficiem, tot *quae* aequa in frusta secatur
quot scriplis gravis est argenti aerisve cylindrus.
110 hoc cuiusque potes pondus spectare liquoris.

[Medieval Latin Statics

which is similar to one ascribed to Pappus by al-Khāzinī in his *Book of the Balance of Wisdom*[34] and also to that described by the patristic bishop Synesius of Cyrene in a letter to the famous Hypatia.[35] This hydrometer of the *Carmen de ponderibus* is probably the ultimate source of the crude hydrometer later described in the fourteenth century both by Albert of Saxony (see Doc. 2.4) and by Blasius of Parma.[36] There is implied here, as in the other passages, a definition of specific weight, without however using the term "specific": "If just as much liquid is taken in each of two cases, the 'heavier' will be greater in weight. If equal weights are taken, the 'rarer' fluid is that which is greater [in volume]." The Pseudo-Archimedean *Liber*

---

      nam *si* tenuis erit, maior pars mergitur unda;
      si gravior, plures modulos superesse notabis.
      quod si tantumdem laticis sumatur utrimque,
      pondere praestabit gravior; si pondera secum
115 convenient, tunc maior erit quae tenuior unda est;
      ac si ter septem numeros texisse cylindri
      hos videas latices, illos cepisse ter octo,
      his dragma gravius fatearis pondus inesse.
      sed refert aequi tantum conferre liquoris,
120 ut dragma superet *gravior*, quantum expulit undae
      illius aut huius teretis pars mersa cylindri."

---

[34] al-Khāzinī, *Book of the Balance of Wisdom*, edition and translation of Khanikoff (see Bibliography), pp. 40–53, and particularly pp. 40–41. "It is evident, from the theorems already stated, and from what is to be presented respecting the relations between the gravities of bodies, that the relation of any volume of a heavy body to any volume of another heavy body, in direct ratio, when the two weigh alike in air, is as the relation of gravity to gravity inversely, [when the two are weighed] in water. The force of this fundamental principle, once conceded, leads to the construction of an instrument which shows us the exact relations in weight of all liquids, one to another, with the least labor, provided their volumes are of the same volume, definitely determining the lightness of one relatively to another. ...We shall, therefore, speak of the construction of this instrument, the marking of lines upon it, and the development of a rule for the putting upon it of arithmetical calculation and letters...."

[35] Cf. Synesius of Cyrene, *Epistolae*, Epistola XV, Migne, Patr. lat., Vol. *66*, cc. 1351–52 (cf. edition of Hercher, [Paris, 1873]). And see A. Fitzgerald, *The Letters of Synesius of Cyrene, Translated into English etc.* (London, 1926), Letter 15, p. 99: "*To the Philosopher (Hypatia)* I am in such evil fortune that I need a hydroscope (i.e., hydrometer). See that one is cast in brass for me and put together. The instrument in question is a cylindrical tube, which has the shape of a flute and is about the same size. It has notches in a perpendicular line, by means of which we are able to test the weight of the waters. A cone forms a lid at one of the extremities, closely fitted to the tube. The cone and the tube have one base only. This is called the baryllium. Whenever you place the tube in water, it remains erect. You can then count the notches at your ease, and in this way ascertain the weight of the water."

[36] Moody and Clagett, *Medieval Science of Weights*, pp. 274–79, 421.

de insidentibus in humidum was to generalize this and the preceding definition and to introduce the term "specific" to distinguish the weight due to density from gross or numerical weight, as we shall see shortly.

One more short hydrostatic notice prior to the period of translations is worth observing. Like the *Carmen de ponderibus* this piece concerns the alloy problem. It is a brief paragraph that appears independently in a manuscript now at Paris (BN lat. 12292, f. 1r), which is judged to be of the tenth century.[37] It was alo included as an item in the well-known book of recipes entitled the *Mappae clavicula*.[38] In the independent version it reads as follows:[39]

[37] Thurot, *op. cit.* in note 14, p. 112, gives the following transcription of this passage (another version is found in the *Mappae clavicula*, see note 38): "*De probatione auri et argenti*. Omne aurum purum cuiuslibet ponderis omni argento similiter puro eiusdem tamen ponderis densius est parte vicesima. Quod ita probari potest. Si purissimi auri libra cum aeque puri argenti simili pondere sub aqua conferatur in statera xii denariis, id est vicesima sui parte, aurum gravius argento vel argentum levius auro invenietur. Quapropter si inveneris opus aliquod auro formatum cui argentum per mixtionem inesse videatur scireque volueris quantum in eo contineatur argenti, sume argentum sive aurum, et examinato suspecti operis pondere, non minus pesantem massam de utrovis metallo fabricato, atque utrumque, et opus scilicet et massam, staterae lancibus imponito, aquisque immergito. Si argentea fuerit massa quam fecisti, opus ponderabit; si aurea fuerit, allevato opere, aurum inclinabitur. Hoc tamen ita fiet ut quot partibus inclinabitur aurum totidem partibus sublevet argentum, quia quicquid in ipso opere fuerit sub aqua praeter solitum ponderis ad aurum propter densitatem pertinet, quicquid autem levitatis ad argentum propter raritatem est referendum. Et ut hoc facilius possit adverti, con[sidera]re debes tam in gravitate auri quam in levitate argenti xii denarios significare libram, sicut prima lectionis huius fronte praefixum est.*"

[38] Thomas Phillipps, "The *Mappae clavicula*; a Treatise on the Preparation of Pigments during the Middle Ages," *Archeologia*, Vol. *32* (1847), 225, item 194: "*(De auri pondere.)* Omne aurum purum cujuslibet ponderis omni argento similiter puro eiusdem tamen ponderis densius est parte sui xxiiii. et insuper ccxl. quod ita probari potest. Si purissimi auri libra cum eque puri argenti simili pondere sub aqua conferatur xi. denariis, id est xxiiii et ccxl sui parte, aurum gravius argento, vel argentum levius auro, invenietur. Quapropter, si opus aliquod inveneris formatum, cui argentum per commixtionem inesse videatur, scireque volueris quantum auri, quantumve in eo argenti, contineatur, sume argentum sive aurum, et examinato inspectione pondere non minus pensantem massam de utrovis metallo fabricato, atque utrumque opus, scilicet, et massam stantem lancibus imponito, aquisque inmergito. Si argentea fuerit, allevato opere, aurum inclinabitur: hoc tamen ita fiet, ut quot partibus inclinatur aurum, totidem partibus sullevetur argentum; quia, quicquid in ipso opere fuerit, sub aqua preter solitum ponderis ad aurum, propter densitatem, pertinet; quicquid autem levitatis ad argentum, propter raritatem, conferendum est. Et ut hoc facilius possit adverti, considerare debes tam in gravitate auri, quam in levitate argenti, denarios xi. signi[fi]care libram, sicut prima lectionis huius fronte prefixum est.*" Cf. the discussion of Berthelot, *La Chimie au moyen âge*, Vol. *1*, 169–71.

[39] See note 37.

[Medieval Latin Statics   93]

On the testing of gold and silver. Pure gold of any weight at all is denser *(densius)* by one-twentieth part than the same weight of silver, also pure. This can be proved as follows: If one weighs in a balance under water a pound of the purest gold and also an equal weight of pure silver, one will find the gold heavier than the silver, or the silver lighter than the gold, by 12 *denarii*, i.e. by a twentieth part of it. Accordingly if you find some object fashioned of gold in which silver seems to be mixed, and you wish to know how much silver is contained in it, take some silver or gold, and having weighed the suspected object, take the same weight of either metal and place both of them—the object and the weight of metal—in the pans of a balance. If the weight which you have employed is silver, the object will be heavier. If the weight is gold, the gold will incline (i.e. be heavier) and the object will be elevated. This, moreover, takes place in such a way that by however many parts the gold is heavier by that same number of parts will the silver be lighter. This is because [if we weigh it against the silver weight] any surplus of weight of the object under water is due to the gold being denser; while [if we weigh the object against a gold weight] any levity (i.e., deficiency in weight) is due to the silver being rarer. And so that this can be more readily observed, you must realize that per pound the gold is heavier [than the silver in water] and the silver lighter [than the gold in water] by 12 *denarii*, just as was affirmed in the beginning of this statement.

While somewhat incomplete, this short piece gives a method that is obviously of the same basic nature as the first method of the section of the *Carmen de ponderibus* analyzed above. The figure of 1/20 part heavier is very rough; and in fact the version found with *Mappae clavicula* gives a better figure of 1/24 + 1/240 part ("... est parte sui xxiiii et insuper ccxl"), which for a pound makes the gold 11 *denarii* heavier than the silver in water. Using specific gravities of 19.3 and 10.5 for gold and silver respectively, the actual difference in weight would be close to 1/23.3 part of the original weight. For these authors the pound is 12 ounces and each ounce is 20 *denarii*.

This brings us to Pseudo-Archimedes' *De insidentibus in humidum*, which, we suggested in our text of this work, is a work compounded out of early Latin and Arabic sources.[40] It was in all probability done in the thirteenth century, since there are a number of manuscripts but none earlier than the thirteenth century. The first thing to note is the definition of "specific" weight and its distinction from "numerical" (gross) weight.[41]

---

[40] Moody and Clagett, *Medieval Science of Weights*, pp. 37–38.

[41] *Ibid.*, pp. 41–43. Consult also the Latin text of Document 2.3 below, where the definitions and postulates form a part of Johannes de Muris' *Quadripartitum numerorum*, Book IV, tract 2. I have made changes in Moody's translation in the light of my distinction between specific weight and specific gravity.

(Definitions)

....7. The relation of one heavy body to another can be considered in two ways: in one way, according to species; in another way, according to number.

8. It is considered according to species, as when we seek to compare the specific weight of gold to that of silver; and this should be done on the basis of equal volumes of gold and silver.

9. The weight of one body is compared to that of another, according to number, when we wish to determine, by weighing, whether a mass of gold is heavier than a mass of silver, irrespective of the volumes of the given masses.

10. One body is said to be heavier than another, numerically, if when these bodies are suspended at the ends of the balance beam, its arm of the balance inclines downward; or, if its weight is equal to the weight of a greater number of *calculi*.

(Postulates)

1. No body is heavy in relation to itself—so that water is not heavy in water, nor oil in oil, nor air in air.

2. Every body is of greater weight in air than it is in water.

3. Of two bodies equal in volume, the one whose weight is equal to that of a greater number of *calculi* is of greater specific weight.

4. Of two bodies of the same kind, the proportion of volumes to weights is the same.

5. All weights are proportional to their *calculi*.

6. Bodies are said to be equal in specific weight, when the weight of equal volumes of them is equal.

The definition of specific weight here given is the first, so far as I know, to use the term "specific," i.e., the first to use a different term to distinguish specific from gross weight. Perhaps the terms "specific" and "numerical" were borrowed from their use in a different context in the *Topics* of Aristotle.[42] It should be observed that the expression 'specific weight" is not limited like the modern "specific gravity" to a comparison of the densities of substance and of water. For the schoolmen it is, in a sense, interchangeable with "density," although it should be recognized that in the scholastic circles of the fourteenth century density is defined as the quantity of *matter* in relationship to a given volume, while specific weight

---

[42] Thurot, *op. cit.* in note 14, p. 114, quotes the Boethius translation of *Topics* I.7 as follows: "Numero enim, aut specie aut genere, idem solemus appellare: numero quidem, quorum nomina plura, res autem una, ut indumentum et vestis; specie autem quae, cum sint plura, indifferentia secundum speciem, ut homo homini equus equo...."

was understood as the *weight* in relationship to a given volume. We shall have more to say about the scholastic discussions of density later. In these pages I have used the term "specific weight" always in its scholastic sense, while I reserve "specific gravity" for its modern meaning.

A second point worth noting in our brief characterization of the *De insidentibus in humidum* is that Archimedes' principle is found there—for the first time explicitly expressed in Latin. It is proposition 1 and runs:[43] "The weight of any body in air exceeds its weight in water by the weight of a volume of water equal to its own volume." Furthermore, all of the propositions, including those connected with the alloy problem, are in the mathematical form of the Archimedean tradition, although in places the proofs of the extant text are corrupt and certainly bear no direct relationship to proofs found in the genuine *Floating Bodies* of Archimedes; and indeed, except for the first proposition, neither do the enunciations. I forbear to discuss the contents of the treatise further, since a complete text, an English translation, and an analysis have already been given in the *Medieval Science of Weights*; and my commentary to Document 2.3 below goes over much the same ground.

The influence of the *De insidentibus* in the fourteenth century is evident. Johannes de Muris in 1343 repeats a major part of it in his *Quadripartitum numerorum* (see Doc. 2.3), but he replaces the mathematical demonstrations with numerical examples. His empirical data is very rough, as I shall point out in the commentary to Document 2.3. His main purpose is merely to illustrate the arithmetical rules involved in problems of specific weight, and accordingly he does not seem to mind that he uses fantastic data. Like the author of the *De insidentibus*, Johannes concerns himself with the alloy problem, i.e., the determination by weighing and simple arithmetical rules of the components parts of an alloy or mixture. His treatment is perhaps an extension of the second method found in the *Carmen de ponderibus*. He treats the alloy problem by (1) preparing volumes of the component substances equal to the volume of the alloy, (2) weighing the equal volumes, (3) finding the parts by volume of the components, and finally (4) finding the parts by weight of the components. This technique is described in detail in Document 2.3 and its commentary. It would appear to be an entirely theoretical rather than practical method, at least in the pages of Johannes' treatise. Furthermore, the modern reader must always keep in mind that a source of error for such volumetric determinations is that usually the volume of the mixture is not simply the sum

---

[43] Moody and Clagett, *Medieval Science of Weights*, pp. 42–43.

[96]   Jordanus de Nemore and]

of the volumes of the components. So far as I know, the medieval authors were unaware of this.

About a generation or less after Johannes' treatise, Albert of Saxony also takes up hydrostatic problems in his *Questions on the Four Books of the Heavens and the World of Aristotle*. He no doubt ultimately draws his use of the expression "specific weight" from the Pseudo-Archimedean treatise —perhaps through the intermediary of Johannes de Muris. However he passes over Archimedes' principle in the determination of specific weights in the case of bodies denser than the fluid in which they are immersed—no doubt because he is making comparisons of specific weights without actually using a balance. He uses instead the doubtful method of comparing (but not quantitatively) specific weights by means of the speed of fall through the liquid (see Doc. 2.4 and its commentary). This he probably draws from the "Peripatetic" *Liber de ponderoso et levi* attributed to Euclid (see Chapter 7 below). Blasius of Parma in the fourth book of his *De ponderibus*, written at the end of the fourteenth century, does the same thing.[44] He mixes the traditions of the *Carmen de ponderibus* (through Albert of Saxony—for his crude hydrometer), of the Pseudo-Archimedean *De insidentibus* (which he knew first hand and probably also in the form given by Johannes de Muris), and that of the *Liber de ponderoso et levi* (perhaps through Albert of Saxony).

Less directly connected with any of these traditions—although perhaps not uninfluenced by the *De insidentibus in humidum*—were the logical discussions before the middle of the century at Merton College. These discussions, arising in part out of the Aristotelian treatment of "heavy" and "light," are concerned with the arithmetical (but certainly not experimental) measure of changing density and rarity. Most subtle (but unproductive of any real quantitative treatment of mass), they concern themselves with "matter" *(materia)* and "volume" *(quantitas* or *magnitudo)* as the factors in determining the measure of density and rarity.[45] These discussions

---

[44] *Ibid.*, pp. 272–79.

[45] For example, read the discussion of Richard Swineshead in his *De raritate et densitate*, a part of his *Liber calculationum*, MS Cambridge, Gonv. and Caius 499/268, f. 176r: "Sequitur inquirere penes quid raritas et densitas attendantur. Due positiones rationales sunt invente, quarum una ponit quod raritas attenditur penes proportionem quantitatis subiecti ad eius materiam et densitas penes proportionem materie ad quantitatem. Secunda ponit quod raritas attenditur penes quantitatem non simpliciter sed in materia proportionata vel in comparatione ad materiam.... Pro istis est notandum quod plus de materia in pedali terre quam in pedali ignis...." For an interesting discussion of the scholastic treatment of the problem of the quantity of matter, see A. Maier, *Die Vorläufer Galileis im 14. Jahrhundert* (Rome 1949), 26–52. Her conclusion (p. 51) re-

[Medieval Latin Statics 97]

form a part of the growing interest in the distinction of intensity factors from capacity factors—a development which we shall discuss more fully in Chapter 4 below. From these discussions arose the concept of specific weight as "intensive" weight and gross weight as "extensive" weight. This is neatly illustrated in a section drawn from the *De macrocosmo* of Marcus Trivisano of Venice (see Doc. 2.5), who died in 1378 before completing this work.[46] Incidentally Trivisano in this same passage abandoned the Aristotelian concept of "heavy" and "light" in favor of the idea of relative heaviness; i.e. fire is not intrinsically light but merely the least heavy of the elements.

In the early fifteenth century, Nicholas of Cusa can be connected at least remotely with the tradition of the Pseudo-Archimedean treatise, although his treatment of the problem of specific gravity in his *De staticis experimentis* (Book IV of the *Idiota*) completely abandons the mathematical demonstrations characteristic of the thirteenth-century treatise;[47] even so, his vaunted experiments are little more than "thought" experiments.

garding these discussions is worth noting: "Aber trotz all dem ist es nie zu einer wirklichen begrifflichen Klärung der quantitas materiae gekommen, oder gar zu einer Definition, die tatsächlich eine quantitative Erfassung der Masse ermöglicht hätte." Of Swineshead's treatment she says (p. 52, note): "Eine exaktere Bestimmung hat nicht einmal der *Calculator* zu geben vermocht."

[46] *De macrocosmo*, Part VI (MS of Johns Hopkins; transcription of George Boas, f. 103r, p. 113): "Adverte quisque praesens opus inspexerit quod Dominus Marcus Trivisano de contrata Sancti Martialis Venetiarum huiusmet operis auctor quippe Aurelius idipsum non complevit, morte praeventus, anno Domini millesimotrigentesimoseptuagesimooctavo, cuius animam Iesu Christo feliciter commendare dignetur quicumque in eo legerit. Amen." Cf. G. Boas, "A Fourteenth Century Cosmology," *Proceedings of the American Philosophical Society*, Vol. 98 (1954), 50–59.

[47] Nicholas of Cusa, *Idiota de staticis experimentis*, in *Opera omnia iussu et auctoritate Academiae Litterarum Heidelbergensis ad codicum edita*, Edition of L. Bauer (Leipsig, 1937), pp. 123–25 (cf. *Opera* [Basel, 1565], p. 174). "ORATOR... Sed quaeso, si totius hominis pondus in comparatione ad aliud aliquod animal quaereres, quomodo procederes? IDIOTA. Hominem in libra ponerem, cui simile pondus appenderem in alia parte. Deinde hominem in aquam mitterem, et iterum extra aquam ab alia parte aequale appenderem, et diversitatem ponderum annotarem, faceremque itidem cum animali dato, et ex varia diversitate ponderum quaesitum annotarem. Post hoc attenderem ad ponderum hominis et animalis diversitatem extra aquam; et secundum hoc moderarem inventum et conscriberem. ORATOR. Hanc moderationem non capio. IDIOTA. Ostendam tibi, inquit. Et accepto ligno levi, cuius pondus ut tria, et aquae eiusdem magnitudinis ut quinque, ipsum in duas divisit inaequales partes, quarum una habuit duplam magnitudinem, alia simplam; ambas in cuppam altam posuit et cum fuste tenuit ac aquam superfudit; et fuste retracta ascenderunt ligna ad aquae superficiem, et maius lignum citius quam minus. Ecce, aiebat, tu vides diversitatem motus in identitate proportionis ex eo evenire, quia in levibus lignis in maiori est plus levitatis.... ORATOR... Sed dicito: quomodo resistit aqua, ne

It is a fair question to ask whether the interesting beginnings in statics of the first half of the thirteenth century were continued throughout the rest of the Middle Ages. We can also ask whether these beginnings had significant influence on the works of early modern mechanicians. Aside from the interest in hydrostatics and the density problem already mentioned in the previous section, we can detail in answer to the first question the following evidence of a continuing concern with statics during the later Middle Ages:

---

descendat lignum? IDIOTA. Ut maior gravedo minori. Quare, si lignum rotundum in ceram presseris et extraxeris locum aqua implendo, et huius aquae pondus similiter et ligni notaveris, comperies, si pondus ligni excedit pondus aquae, lignum descendere, si non, natare et super aquam partem proportionalem ligni manere secundum excessum ponderis aquae super pondus ligni.... IDIOTA... Nam certissimum est aliud esse pondus auri, aliud argenti et ceterorum in aequalitate magnitudinis; et aliud cuiuslibet pondus in aëre, aliud in aqua, aliud in oleo aut alio liquore. Unde, si quis pondera illa omnia signata teneret, ille profecto sciret, quantum unum metallum est gravius alteri in aëre et quantum in aqua. Hinc, data quacumque massa, per ponderum eius diversitatem in aëre et aqua scire posset, cuius metalli massa foret et cuius mixturae. Et sicut dictum est de aëre et aqua, ita etiam de oleo dici posset aut alio quocumque humore, in quo experientia facta fuisset. ORATOR. Sic absque massae fusione et metallorum separatione mixtura attingeretur, et ingenium istud in monetis utile foret ad sciendum, quantum cupri immixtum sit auro aut argento." Cf. the English translation of Henry Viets, "De staticis experimentis of Nicolaus Cusanus," *Annals of Medical Science*, Vol. *4* (1922), 115–35 (complete work, this passage pp. 128–29): "ORATOR... But I should like to know how you would proceed, if you were seeking the weight of an entire man, in comparison with any other animal. IDIOT. I should place the man on a scale and should hang a weight equal to him on another part of it; then I should put the man into water, and again outside the water I should hang a weight, equal to him, from another part; and I should note the difference of the weights. Also I should do the same with the given animal, and I should note down what I had learned from the various differences of the weights. After this I should direct my attention to the difference of the weights of man and animal out of the water, and then adjust and write down what I had found. ORATOR. I do not grasp this principle of adjusting. IDIOT. I will show you. *And having taken a light piece of wood, whose weight was as III, while that of water filling the same space was as V, he divided it into two unequal parts, one of which was double the size of the other. He put both in a tall cask and held them down with a stick, while he poured water over them, and when he withdrew the stick, the pieces of wood jumped to the surface of the water, and the larger piece quicker than the smaller.* Look, you see the difference of motion in exact proportion with its cause, for in light pieces of wood, there is more lightness in the larger one.... ORATOR... But tell me how the water hinders the wood sinking. IDIOT. Because the greater heaviness hinders the less. Wherefore if you have pressed a round piece of wood into wax and drawn it out, filling its place with water, and have noted the weight of this water and similarly of the wood, you will find that if the weight of the wood exceeds the weight of the water, the wood sinks; if not, it floats, and a part of the wood remains above the water in proportion to the excess of the weight of the water over the weight of the wood.... IDIOT... For it is very certain that gold has one weight,

[Medieval Latin Statics

1. In the first place there was some reproduction of the principal versions of the earlier treatises: The *Liber de ponderibus*, the *Elementa de ponderibus* of Jordanus, the *De ratione ponderis* attributed to Jordanus, the *Liber karastonis*, the *De canonio*, and so on. This we know from the existence today of a number of manuscripts of these various treatises dating from the late thirteenth, the fourteenth, and the fifteenth centuries.

2. At the same time, many of the schoolmen of the fourteenth century, while not primarily concerned with problems of statics, nevertheless had the occasion to cite the thirteenth-century treatises. As I have already pointed out in the *Medieval Science of Weights*, this group of schoolmen included Thomas Bradwardine, the anonymous author of the *Tractatus de sex inconvenientibus*, John Dumbleton, Roger Thomas, Francischus de Ferraria, Albert of Saxony, and Marsilius of Inghen.[48]

3. There were, in addition, a number of commentaries and reworkings of the earlier treatises made from the later days of the thirteenth century through the fourteenth and fifteenth centuries. The first of these—which, continuing the number system applied to statical treatises at the beginning of the chapter, we can list as number (12)—was a commentary dating from the second half of the thirteenth century. I have called it the "Corpus Christi Version" of the *Elementa* from its earliest manuscript, and from it I have included here as Document 2.2 the proof of proposition 8 (the law of the lever). It is of interest to note that this "commentator" confused the essential point of the lever proof of Jordanus, as I have pointed out in the commentary to this document. Another commentary (13) that dates from the thirteenth or early fourteenth century is the one which I call the "Pseudo-Euclid Version" of the *Elementa* of Jordanus. It is marked by the frequent citation of Euclid and by an elaboration of the geometrical principles involved; it was attributed to Euclid in some of the manuscripts.

silver another and other metals in relation to their quantity. And the weight of anything you like is one thing in air, another in water, another in oil, or in other liquid. Then, if anyone had all these weights registered, he would know forthwith how much heavier one metal is than another in air, or how much in water. Hence whatever mass were given, through the difference of its weights in air and water, he could tell to what metal or to what mixture, the mass belonged, and as it has been settled regarding air and water, so also it could be settled concerning oil, or any other fluid, in which an experiment has been made. ORATOR. So without melting the mass, or separating the metals, the mixture would be found out, and this system would be useful in mints to find out how much copper has been mixed with gold or silver." In the orator's second speech I have changed Dr. Viets' translation of "capio" from "approve of" to "grasp"; and in the very last speech I have changed his translation of "fusio" from "pouring out" to "melting."

[48] Moody and Clagett, *Medieval Science of Weights*, pp. 231, 413.

I have published elsewhere proposition 8 from this treatise.[49] It, too, confuses the lever law, but in a fashion different from that of the "Corpus Christi Version."

Most interesting of the commentaries was that from which we have already quoted a long extract above to illustrate the commentator's appreciation of Jordanus' use of the work principle in the first and the eighth propositions of the *Elementa*. This treatise (14) we have called the *Aliud commentum*; in all probability it dates from the second half of the fourteenth century, since the two earliest manuscripts of it date from that time.[50] In this group of new versions and commentaries we can also place two interesting but somewhat inferior works of Blasius of Parma, works which were composed at the end of the fourteenth century. The first is entitled (15) *Tractatus Blasii de ponderibus*. The other is called (16) *Questiones super tractatum de ponderibus*. Both of these treatises were discussed (and the first one published) in the *Medieval Science of Weights*, (pp. 231-79, 413-21). It is sufficient to say here that (1) they were inferior to the principal treatises attributed to Jordanus, and (2) they show continued influence of the Hellenistic-medieval corpus of statical works. A final commentary on the *Elementa* of Jordanus is that of Henry Anglegena, probably of the fifteenth century; we can number it (17). I hope to publish in the future the full texts of the versions and commentaries numbered from (12) to (17), and I shall at that time include a detailed analysis of their contents and objectives.

Our second and final question concerning the influence of the medieval statical corpus on early modern mechanics is more difficult. Duhem has shown in the first volume of his *Les Origines de la Statique* that Leonardo da Vinci was influenced by the statical works attributed to Jordanus.[51] It has often been pointed out that Leonardo anticipated to a certain extent the concept of static moment when he described the horizontal distance to the vertical line through the fulcrum as "the potential lever arm." But as we have shown, the same concept was held by both Hero and the author of the *De ratione ponderis*. Unlike the latter author, Leonardo gives no formal proof of the bent lever theorem.

It is clear as we pass into the sixteenth century that the main achievements of medieval statics became generally available as the result of two

---

[49] *Ibid.*, pp. 308-11.
[50] *Ibid.*, p. 295.
[51] P. Duhem, *Les Origines de la statique*, Vol. 1 (Paris, 1905), 13-33 deals with Leonardo's statics, and chapter 8, pp. 156-93 deals with the medieval influences on Leonardo. Cf. Duhem, *Études sur Léonard de Vinci*, Vol. 1 (Paris, 1906), 310-16 *et passim*.

publications: (1) the publication by Peter Apian in 1533 of the so-called *Liber de ponderibus*—Version *P*—together with the very interesting fourteenth-century commentary whose discussion of propositions 1 and 8 of the *Elementa de ponderibus* of Jordanus we have already mentioned; (2) the publication in 1565 of the Tartaglia copy of the text of the *De ratione ponderis*. The reader will find the full titles of both of these publications in the Bibliography appended to this work. It should be noticed in addition that Tartaglia in his Italian treatise *Quesiti et inventioni diversi* (Venice, 1546; pp. 81–97) reproduced various propositions of Part I of the *De ratione ponderis*, commenting on and paraphrasing the material. Incidentally, the famous forerunner of Galileo, G. Battista Benedetti, gave a brief but quite unsatisfactory refutation of the views of Jordanus and Tartaglia in his *Diversarum speculationum...liber*, since his most severe charge against Jordanus and Tartaglia is that they have assumed the direction of force of weights on balances to be at right angles to the beam and parallel with each other, while Benedetti assumes that, since they incline toward the center of the world, such inclinations cannot be in parallel directions[52]—a criticism which is obviously of no real moment.

It is not our intention here to inquire extensively into the history of statics in the sixteenth and seventeenth centuries. It will be sufficient for us to remark that even the briefest glance at the works of the most important mechanicians of this period reveals to us their indebtedness to the dynamic tradition originating with the *Mechanica* attributed to Aristotle and carried on by the medieval authors. Even Stevin, who speaks against the use of dynamic considerations in treating problems of equilibrium,[53]

---

[52] Giovanni Battista Benedetti, *Diversarum speculationum mathematicarum et physicarum liber* (Turin, 1585), pp. 148–51 for criticism of Jordanus. "De quibusdam erroribus Nicolai Tartaleae circa pondera corporum et eorum motus, quorum aliqui desumpti fuerunt a Jordano scriptore quodam antiquo....(p. 150)...Omnis autem error in quem Tartalea, Iordanusque lapsi fuerunt ab eo, quod lineas inclinationum pro parallelis vicissim sumpserunt, emanavit."

[53] E. J. Dijksterhuis, *The Principal Works of Simon Stevin*, Vol *1*, *General Introduction, Mechanics* (Amsterdam, 1955), 507, 509, "The reasons why equal gravities at equal arms (i.e. arm lengths) are of equal apparent weight are known by common knowledge, but not so the cause of the equality of apparent weight of unequal gravities at unequal arms [inversely] proportional thereto, which cause the Ancients, when they inquired into it, considered to reside in the circles (i.e. circumferences) described by the extremities of the arms, as appears in Aristotle's *In Mechanicis* and his successors. This we deny, and we give the following reason therefor: E. *That which hangs still does not describe a circle;* A. *Two gravities of equal apparent weight hang still;* E. *Therefore two gravities of equal apparent weight do not describe circles.* And consequently there is no circle. But where there is no circle, the circle cannot be that in which resides any cause, so that the circles are not here that

owes something to the dynamic tradition, and his use of the expression *staltwicht* for apparent weight seems to reflect directly the medieval *gravitas secundum situm*. His celebrated proof of the inclined plane law [54] rests on the absurdity of perpetual motion, which appears to have dynamical implications.

As a final example of the influence of the dynamic tradition on early modern statics, we can note briefly some considerations of Galileo. For in spite of the fact that Galileo was one of Archimedes' greatest admirers, he was very much dependent on the dynamic approach. We have given as Document 2.6 a selection from the *Mechanics* of Galileo, based on lectures delivered at the University of Padua in the 1590's and published first in French translation in 1634 and then later in Italian in 1649. In the first section he applied the principle of virtual displacements to the confirmation of the law of the lever. The inclined plane principle, however, he derived from the lever principle, after which he went on to analyze movement on an inclined plane in terms of the work principle, which he assumed in the following form: Force is to force inversely as the distances through which these forces pass, or, as he puts it in another way, by the amount there is a gain in force by the same amount is there loss in time and velocity. The similarity with Hero's *Mechanics* is obvious. As I have noted in the commentary to Document 2.6, Galileo continued to use the principle of virtual velocities throughout his whole career.

It would not be without interest to conclude this section on medieval statics with a general estimate of medieval statics. [55] We can see how the history of this one branch of physics in an early period illustrates some of the truisms regarding the general development of science that occasionally escape attention. First it illustrates the success which emerged when the ordinary fruits of human experience are analytically abstracted and generalized to form the first principles of a science. Thus from an analytic intuition of what is gained and what is lost in the use of a lever came the general principle of virtual work. In the second place, the study of medieval statics illustrates the significant achievements that could be and were

in which resides the cause of the equality of apparent weight."

[54] *Ibid.*, pp. 174–79. This proof involving the wreath of spheres about the inclines is so well known I forbear to repeat it here; however we might quote the heart of the proof, p. 179: "This descent on the one and ascent on the other side will continue forever, because the cause is always the same, and the spheres will automatically perform a perpetual motion, which is absurd."

[55] This concluding paragraph is substantially as in Moody and Clagett, *Medieval Science of Weights*, pp. 6–7.

made when the abstractions and generalizations which served as principles were given even the most elementary mathematical form, and further when the logical implications following from the first principles were themselves developed in the language of quantity. For example, from his initial concept of positional gravity mathematically expressed—a brilliant intuition of component forces—Jordanus proceeded by the use of the principle of virtual displacements and the theorems of plane geometry to deduce correctly a general proposition relating interconnected weights on oppositely inclined planes to the lengths of those planes. Similarly the neat geometrical extension of his first principles led him to his correct theorem regarding the bent lever. Lastly, our study of medieval statics reveals the great importance for scientific development of the fact that natural science was an integral and connected part of the general arts program. As we have said earlier, the originality and success of Jordanus' efforts in statics stemmed in part from the union of a philosophical approach (that of Aristotle and his successors) with a more rigorous mathematical tradition (that of Euclid and Archimedes). A student of the arts faculty of a medieval university would almost certainly come in contact with both of these traditions in the course of his study. The junction, then, of the philosophical and mathematical traditions in statics was but one illustration of the more general interplay between the two traditions. Most of us who have investigated the origins of Western science acknowledge this interplay by affirming that the principal forebears of modern science were in fact the twin traditions of Greek philosophy and mathematics.

Document 2.1

*The Theory of Weight**
Attributed to Jordanus de Nemore

1. PROPOSITION 6 (R1.06). If the arms of a balance are proportional to the weights of the suspended bodies in such a way that the heavier weight is suspended on the shorter arm, the suspended weights will be equally heavy according to position.

As before, let the rule (i. e. the beam) be ACB (see Fig. 2.6), the sus-

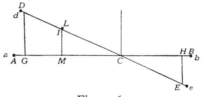

Fig. 2.6

pended weights a and b, and let the proportion of b to a be the same as that of AC to BC. I say that the balance will not move in any direction. For suppose the balance would descend from b and assume the oblique line DCE in place of ACB. And if the suspended weight d is the same as a while e is the same as b, and the line DG descends perpendicularly [to the line ACB] and EH ascends [to the same line ACB], it is clear that, since triangles DCG and ECH are similar, the proportion of DC to CE is that of DG to EH. But DC is to CE as b is to a. Hence DG is to EH as b is to a. Then let CL equal CB and also CE, and let l equal b in weight, and let us drop a perpendicular LM. Then because LM and EH clearly are equal, DG will be to LM as b is to a and as l is to a. But, as it was shown, a and l are inversely proportional to their contrary [upward] movements. Hence, that which suffices to lift a to D would be sufficient to lift l through

---

* For the Latin text, see E. A. Moody and M. Clagett, *Medieval Science of Weights* (Madison, 1952), pp. 182–90. The translation, however, is my own, but it agrees substantially with that of Ernest Moody.

[Jordanus, *Theory of Weight*   105]

LM. Since, therefore, *l* and *b* are equal and *LC* equals *CB*, *l* does not follow *b* in a contrary movement, nor will *a* follow *b* as is proposed....

2. Proposition 8 (R1.08). If the arms of a balance are unequal and they make an angle at the center of motion (i.e. the fulcrum), and if their termini are equally distant from the vertical line [passing through the fulcrum], in this disposition equal suspended weights will be of equal heaviness.

Let the center be *C* (see Fig. 2.7), the longer arm *AC* and the shorter

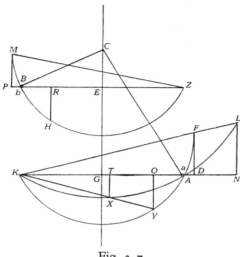

Fig. 2.7

one *BC*, and let *CEG* descend perpendicularly [to the horizontal]. Let the equal lines *AG* and *BE* fall on *CEG* perpendicularly. When, then, there are equal weights *a* and *b* suspended, they will not move from this position. For let *AG* and *BE* be projected equally to *K* and *Z* respectively, and on them draw portions of circles *MBHZ* and *KXAL* around the center *C*. Construct also the arc *KYAF* similar and equal to the arc *MBHZ*, and let the arc *AX* and *AL* be equal to each other and similar to arcs *MB* and *BH*, and *AY* and *AF*. Then if *a* is heavier than *b* in this position, let *a* descend to *X* with *b* ascending to *M*. Draw the lines *ZM*, *KXY*, *KFL* and construct the perpendicular *MP* on line *ZBP*, as well as the perpendiculars *XT* and *FD* on *KGT*. Because *MP* is equal to *FD* which is greater than *XT* by similar triangles, *MP* is greater than *XT*. Therefore, *b* ascends further vertically than *a* descends, which is impossible since they are equal [in weight]. Again, let *b* descend to *H*, drawing *a* to *L*, and let the perpendicular *HR* be constructed on *BZ*, and *LN* and *YO* on *KAN*.

Then LN will be greater than YO and thus greater than HR, from which an impossibility is similarly deduced....

3. Proposition 10 (R1.10). If two weights descend through paths of different obliquity, and the declinations are in the same proportion as the weights taken in the same order (i.e., if the declinations are directly proportional to the weights), the weights will have the same power for descent (i.e., they will be in equilibrium).

Let line ABC (see Fig. 2.8) be parallel to the horizon and let BD be

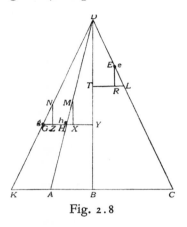

Fig. 2.8

erected orthogonally on it. From BD let lines DA and DC descend, and let DC be of greater obliquity. Then I say that the proportion of declinations is not one of the angles, but of the [lengths of the inclined] lines measured to a horizontal intersecting line by which they [the inclined lines] partake equally of the vertical [i.e., to a line which would cut off equal projections of these lines on the vertical]. Hence let weight e be on DC and h on DA. Let e be to h as DC is to DA. I say that they (the weights) are of the same force *(virtutis)* in this position. For let line DK be of the same obliquity as DC and [similarly] let the weight g on it be equal to weight e. Then, if it is possible, let e descend to L drawing h to M, and let GN be equal to HM which is also equal to EL. Pass a perpendicular to line DB through G and H and let it be GHY, and [another perpendicular] from L, [it being] LT. Erect also on GHY [perpendiculars] NZ and MX and on LT [perpendicular] ER. Then since the proportion of NZ to NG is as DY is to DG and hence as DB is to DK, and since similarly MX will be to MH as DB is to DA, MX will be to NZ as DK is to DA. This is as g is to h. But since e is not sufficient to raise g to N [the effective weights of e and g being the same from proposition 9], neither will it be sufficient to raise h to M. Therefore, the weights will remain [as they are, in equilibrium].

COMMENTARY

1. The form of the proof of proposition 6 is indirect. Weights inversely proportional to the arm lengths are either in equilibrium about c or they are not. If they are not, they are displaced, with, say, b descending. Then b descending through the vertical distance HE lifts a through the vertical distance DG. Suppose further then that weight l is equal to b and is placed at L on LC equal to CB or CE. It is then shown by simple geometry that a is to l inversely as DG is to LM. Hence the principle of work applies, or as Jordanus says, "that which suffices to lift a to D would be sufficient to lift l through LM." But since l and b are equal in weight and at equal distances from the fulcrum, they are in equilibrium and thus b does not have sufficient effective weight to lift l through LM. Hence it is not of sufficient weight to lift a through DG as was proposed. The proof implies but does not state the following additional steps: DG is any potential vertical displacement of a at all. The same reasoning could be used for any potential displacement of b. Hence, if no displacement of either a or b could take place, equilibrium must obtain when the arm lengths are inversely proportional to the suspended weights.

2. So far as I know, this is the first correct analysis with proof of the case of the bent lever, although Hero had correctly applied the law of the lever to irregular levers but without proof. We have already pointed out its importance in its stressing of the necessity of using the *horizontal* distance to the vertical running through the fulcrum, rather than the distance to the fulcrum itself, for determining the effective force of a weight suspended on one of the bent arms of the lever.

As in the case of the straight lever, the principle of virtual work lies at the bottom of the proof of this proposition. If we assume the disequilibrium of the bent lever with equal weights suspended at equal distances from the vertical through the fulcrum, then we will have a greater work output than work input. Or as this proof more specifically holds, weight b would be lifted by weight a equal to it a greater vertical distance than that through which a descends, which is impossible.

3. In suppositions 4 and 5 and proposition 9, none of which has been included in these selections, Jordanus already laid the basis for the solution of the problem of the inclined plane. In supposition 4 he declared that something is heavier according to position (we might say effectively

heavier by reason of its position) when in that position its descent is less oblique. He then proceeded to give the measure of the obliqueness, declaring in supposition 5 that the "descent is more oblique by the amount it takes less of the vertical." By this he meant that obliqueness is greater, by the amount that the length of the trajectory's projection on the vertical is less. When the trajectory is a straight line, this is equivalent to saying that the inclination is measured by the sine of the angle of inclination. Hence by suppositions 4 and 5 together, positional or effective weight is measured by what is equivalent to the sine of the angle of inclination. Proposition 9 assures us that as long as the inclination of the plane or planes is constant, the effective or positional weight will be the same. As a proof, he shows that the vertical intercepts of any two equal segments of the plane or planes will be the same.

Now in proposition 10 Jordanus has proved that $W_1/L_1 = W_2/L_2$, where W_1 and W_2 are the natural weights of weights resting on inclined planes of lengths L_1 and L_2 respectively. This can be easily shown to be equivalent to the correct modern formulation $W_1 \sin a = W_2 \sin b$, where a and b are the angles of inclination of the planes. The substance of Jordanus' proof is as follows: If we assume that e moves down to L, i.e. through a vertical distance ER, and e and h were connected, then h would move to M. But that which is sufficient to move h to M can move g to N, since it can be shown geometrically that MX is to NZ as g is to h. And since g is equal to e and at the same inclination, by proposition 9 (not given in our selections, but noted above) its resistance to ascent is equal to e's power for descent. Hence e is not sufficient to move g to N and thus not sufficient to move h to M. Hence, there is no movement of h. Although our author omitted it (perhaps because it would be similar to the refutation of the descent of e), the proof ought to have the refutation of the ascent of e. Thus with both the descent and ascent of e refuted, the proposition would follow.

Document 2.2

An Anonymous Commentary on the Elements of Jordanus on Weights

PROPOSITION 8. If the arms of a balance are proportional to the weights, suspended in such a manner that the heavier weight is suspended on the shorter arm, the suspended weights will be of equal positional gravity.

Let a and b (see Fig. 2.9) be two unequal weights; and let b be the

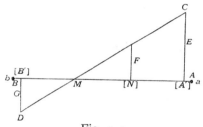

Fig. 2.9

greater and hung on the shorter arm, while a [is hung] on the longer arm, so that the arms of the balance are proportional to the weights; or, the proportion of weight b to [weight] a is as that of arm AM to arm BM. I say that a and b are equally heavy in position. If it can be done, let b descend to D and a ascend to C; and from C let a perpendicular be dropped to line AB, and similarly [let there be a perpendicular] from D. These [two perpendiculars] we let be E and G. Thus we have two similar triangles EM and GM, since they have right and thus equal angles, and opposite and thus equal angles; hence the third [angle] is equal to the third [angle]. Therefore, the proportion of CM to DM is as AM to BM and E to G. But CM is greater than DM by hypothesis. Therefore $A'M$ is greater than $B'M$. Hence let AM be cut at a distance equal to MB' by a line erected perpendicularly from AM and meeting CM, which line we let be F. Thus we have two similar triangles for the aforesaid reason. Hence the proportion of side $M[N]$ to side $B'[M]$ is as F to G, just as if C were closer to D. But $M[N]$ is equal to $B'[M]$. Therefore F is equal to G, and C [in this situation] is closer to D. Thus immediately the proportion of CM to DM is as E to F. But CM is to DM as weight b is to weight a. Therefore, the proportion of weight to weight is as ascent to ascent, because if b had ascended in the same time as the ascension of a, it would have ascended through F. Therefore, what will be sufficient for lifting a through E would be sufficient for lifting b through F. But b is sufficient for lifting a through E. Therefore, it is sufficient for lifting itself substituted in place of a through F. But placed there, it would be on a longer arm than is b. Therefore by the fifth [proposition] of this [work] it would produce descent, which is against the hypothesis.

COMMENTARY

This commentary to the *Elementa de ponderibus* is the one which I have called the "Corpus Christi version" after its earliest manuscript, (Oxford, Corpus Christi 251, ff. 10r–12r). I assume that this work, like its earliest extant manuscript, dates from the late thirteenth century. On the whole it is not so penetrating as the commentary which I have called in the body of the chapter *Aliud commentum* and from which I have extracted a long passage.

The lever proof whose text I have here translated is quite different in its concluding section from any other commentary or version of Jordanus' proof. The proof quickly reaches the point of Jordanus' proof, where the work principle is to be applied; but in so doing, the commentator does things rather peculiar for mathematicians of the period, as for example labeling a triangle by using a single letter for one side joined with a single letter for the opposite angle, e.g. EM and GM. It should be added that in the early part of this proof, the letters are somewhat corrupt in both manuscripts; and I have had to reconstruct the text. Now the proof gives the work principle rather neatly: "Therefore, the proportion of weight to weight is as ascent to ascent.... what is sufficient for lifting a through E would be sufficient for lifting b through F. But b is sufficient for lifting a through E. Therefore, it is sufficient for lifting itself substituted in place of a through F." But after this clear-cut statement of the heart of the proof, the proof then falters by placing b not at the position where it would in fact be lifted through a vertical distance F, but rather at the position where a is. This has not entered into the proof. Hence his conclusion that b placed at C would produce descent—while true—is beside the point.

I have had to reconstruct the diagram somewhat. I have added B' and

A' and [N], and I have distinguished weights a and b from positions A and B, although no such distinction is made in the manuscripts.

Commentum in Elementa Jordani de ponderibus*

Propositio 8. Si fuerint brachia libre proportionalia ponderibus appensorum, ita ut in breviori gravius appendatur, eque gravia erunt secundum situm appensa.

Sint a et b duo pondera inequalia, et b sit maius et appendatur breviori brachio, a longiori, ita quod brachia libre sint proportionalia ponderibus, vel que sit proportio b ponderis ad a ea sit AM brachii ad BM brachium. Dico quod a et b sunt eque gravia situ. Si fieri potest quod b descendat usque ad D et a ascendat usque ad C et a C protrahatur perpendicularis ad AB lineam, et a D similiter, et sint ille E et G. Habebuntur itaque duo trianguli similes, EM et GM, quia habent angulos rectos et ita equales, et contrapositos et ita equales; quare tertius tertio. Ergo que est proportio CM ad DM ea que AM ad BM et E ad G. Sed CM est maior DM ex ypothesi. Ergo $A'M$ est maior $B'M$. Resecatur ergo AM ad equalitatem MB' linea erecta perpendiculariter ab AM et ducta ad CM, que sit F. Habemus itaque duos triangulos similes predicta ratione. Ergo que est proportio $M[N]$ lateris ad $B'[M]$ latus ea est F ad G, etsi C propinquioris ad D. Sed $M[N]$ est equale $B'[M]$. Ergo F est equale G et C propinquioris D. Ergo a primo que est proportio CM ad DM ea est E ad F. Sed que est CM ad DM ea est b ponderis ad pondus a. Ergo que est proportio ponderis ad pondus ea ascensus ad ascensum, quia si ascendisset

---

* MSS Oxford, Corpus Chirsti College 251, f. 11v (A); Cambridge Univ. Library Mm. 3.11, f. 152v (B); cf. Erfurt, Stadtbibliothek, Amplon. Q. 348, ff. 1r–4v (this prop. 4r), a fourteenth-century copy. Weights have been left in lower-case letters, but I have capitalized points and magnitudes, thus following the procedure adopted by E. A. Moody and me in the *Medieval Science of Weights*.

---

7 vel A ut B
9 potest A probatur B / a *om.* B
10 perpendicularis A perpendiculariter B
11 EM et GM A cum M B
12 et² *om.* B
12–13 et ita equales *om.* B
13 Ergo A igitur B
13–14 ea que A est B
14 ex ypothesi A ypothenusa B
15 Resecatur B resecare A / ergo A igitur B
15 MB' A BM B
16 CM B EM A
17 Ergo A igitur B
18 B' [M]: B *corr.* B *ex* V; D A / etsi B et A / propinquioris B partialiter (pti^{lit}) A
19 Ergo A igitur B / propinquioris *correxi ex* ptile
20 Ergo A igitur B / E A G B
21 pondus *om.* B / Ergo A igitur B
22 ea ascensus *iter.* A

b per tempus ascensionis a, ascendisset per F. Quod ergo sufficiet ad
attollendum a per E sufficeret ad attollendum b per F. Sed b sufficit
ad attollendum a per E. Ergo sufficit ad attollendum se positum in
loco a per F. Sed ibi positum esset in longiori brachio quam sit b.
Ergo per quintam istius faceret nutum, quod est contra preconcessum.

23 ascendiscet B / sufficiet *corr. ex* sufficienter

25, 27 Ergo A igitur B
27 istius A *et* B *supra;* ibi B

Document 2.3

Johannes de Muris
The Four-Parted Work on Numbers

Book IV, Second Tract

These are Propositions or Conclusions

1. THE WEIGHT of any body in air exceeds its weight in water by the weight of water equal to it in volume.

Let the given body be gold. It weighs less in water than in air by the second postulate [i.e., every body is of greater weight in air than it is in water]. But a part of the gold as it enters the water expels the same amount of water, since the "containing" and the "contained" are equal in volume, that is, in surface. And so it is the same for all parts of the gold. Therefore, since water weighs nothing in itself [i.e., in water] by the first postulate [i.e., no body is heavy in relationship to itself—so that water is not heavy in water], but only has weight outside of itself, hence if the gold weighs less in water than in air, this is because of the expelled water. But the expelled water is equal to the immersed gold [in volume]. Therefore the gold weighs less in water by the weight of water equal to it in volume, which is that which was proposed. Understand that the *calculi* are in air, since if they were weighed in water like the gold to be weighed, their weights or the weight of gold would not (! delete?) be varied. From this it can be known how to distinguish which of two "waters" is lighter or rarer. For if weights equal in air [and of the same specific weight] are placed in two waters [at the same time] and they are still equal, the two waters are of equal [specific] weight and rarity. If, however, inclination of the balance takes place, that water in which the inclination takes place is rarer and lighter [according to species]. For resistance to descent takes place only because of the density of the medium. Hence if there were no medium, a heavy body would descend in an instant.

An example of the aforementioned: Let the gold be of one cubic foot and its weight in air 8 pounds. Weigh it in water and it would weigh 7 pounds. The difference, therefore, in air and in water is one pound. I say that this is the weight of water equal in volume to that of gold, i.e., the weight of one [cubic] foot of water. For, by induction through the parts of gold, the proposition is evident. Since, if 1/4 foot of gold is submerged, i.e., 2 pounds, the amount of water displaced is similarly one quarter of a foot, i.e., 1/4 of a pound, hence the proportion of 2 pounds to 1/4 is as 8 is to 1. Therefore, etc. Similarly for [any] other parts. Hence the conclusion remains for the whole.

2. Of any two bodies, whether the same or different in kind, the ratio in volume of one to another is as the ratio of the difference between the weight of one [of them] in air and its weight in water, to the difference between the weight of the other in air and its weight in water.

Let a quantity of gold be weighed in air and in water and its difference be 1 pound. Similarly another quantity of gold is weighed in air and in water and let its difference be 2 pounds. I say that just as [the one] difference is double [the other] difference, so the second quantity of gold is double the first quantity of gold in volume, which is clearly evident.

Also let a quantity of gold be weighed in air and in water—with the difference of 1 pound, and a quantity of something else of different kind, e.g., silver, is weighed in air and water—with a difference of 4 pounds. Then the proportion will be a quadruple one, as stated. And so that this might be understood briefly, I adduce an example using volumes and numbers.

Let there be a quantity of gold figured in the manner of a cubic foot, for this is reputed to be more apt for [making] divisions. Let this foot of gold weigh 12 pounds. And let a foot of silver of the same quantity weigh 9 pounds in air, but a foot of lead 8 pounds, and further a foot of copper 6 pounds. But a foot of oil of equal volume to the foot of gold would weigh 2 pounds in air; while a similar foot of water would weigh 1 pound. In addition a foot of gold in water would weigh 11 pounds. Therefore the difference in weight of that [gold] in air and in water will be 1 pound. In regard to all of these, sensible experience by means of a balance will restore you [who are in] doubt. The foot of silver is also less [heavy] in water than in air by the weight of water equal in volume to it, i.e., by 1 pound. Therefore, it will weigh 8 pounds in water. The same holds for the others.

Hence let there be two bodies, gold and silver, in accordance with the proposed theorem. Let the gold be 1 [cubic] foot, the difference of whose

[Johannes de Muris, *On Numbers* 115]

weight in air and water is 1 pound. Let there also be some silver and let the [number of] feet be unknown. Then the difference of its weights in air and water will be 4 pounds. The experience of the balance proves this, and it is evident from the suppositions. For 4 feet of silver weigh 36 pounds in air and 32 pounds in water. The difference is 4 pounds. Now moreover as 4, the second difference, is to 1, the first difference, so the quantity of silver in volume is to the quantity of gold, i.e., [it is] a quadruple [proportion]. For just as there was then [in that case] 1 foot of gold and 4 of silver, so you will see [the same thing] happen in all cases. Therefore, it is clear that the whole proposed theorem is proved. You can prove it by letters of the alphabet, but to me numbers are clearer.

And it is clearly evident in another way. For the difference of the weight of gold in air and in water is the weight of water equal in volume to it, and this is 1 pound. Similarly the difference of silver in air and water is the weight of water equal to it in volume, and this is 4 pounds. Now the ratio of volumes is the same as [the ratio] of weights in bodies of the same specific gravity, and [this is true of] water. But [one quantity of] water is of the same species as [another quantity] of water. And [so] there is between the volumes a quadruple proportion, because between their weights there is a quadruple proportion. Therefore, between equivalents of these same [quantities of water], i.e., between gold which is equal [in volume] to water of 1 pound weight and silver which is equal [in volume] to water of 4 pounds in weight, there will be a quadruple proportion. Therefore the silver is quadruple the gold [in volume], which is that which was proposed.

3. The ratio of the specific weights of any two bodies is equal to the ratio of the weight of the [specifically] heavier body to its volume, multiplied by the ratio of the volume of the [specifically] lighter body to its weight.

For example: Let the two bodies be gold and silver and let the volume of the gold be half the volume of the silver, and the weight of the gold be in subsesquialterate proportion to the weight of the silver, which is written 2/3. Therefore, multiply 2 by 2/3 and the result will be $1\frac{1}{3}$, which is a sesquitertiate proportion. Say therefore that the gold for an equal volume is related in specific weight to the silver as 4 is to 3—and this is what was proposed—or as 12 is to 9, as was stated before. And thus it is evident that the ratio of volumes multiplied by the [inverse] ratio of weights from the balance yields the ratio of specific weights which you seek. Proceed by [particular numerical] proportion and you will arrive at the same conclusion.

Let the determined weight of gold be 12 and the weight of silver 18.

Since, therefore, the silver was double the gold in volume, the ratio was as 2 is to 1. Hence the ratio of the volume of the gold to its weight is as 1 is to 12, and consequently of the weight to its volume as 12 is to 1. Similarly the ratio of the volume of the silver to its weight was as 1 is to 9. Hence multiply 12 by 1/9 and the result is 1⅓, which is the ratio of the specific weight of gold to that of silver for an equal volume—i.e., a sesquitertiate proportion, as before.

4. If the specific and numerical weights of two bodies are known, the ratio of their volumes will be known.

Divide the numerical weight of the specifically heavier body, i.e. gold, by the ratio of the specific weights of the said bodies, [the latter relationship] being had from the preceding. The result is divided into the numerical weight of the specifically lighter body, i.e. silver, and that which results is the ratio of the volume of the body specifically lighter to [that] of the body [specifically] heavier, which is what was proposed.

Example: Let the numerical weight of gold be 1—this is found by means of a balance; this is the [weight of the] body specifically heavier. Let the weight of the silver be 2, and let the ratio of specific gravities be 4 to 3. Hence divide 1 by 4/3 and the result is 3/4, which is then divided into 2, and the result [of that division] is 2⅔. Say, therefore, that the volume of silver is 2⅔ the volume of gold, i.e., as 8 is to 3.

5. Given three bodies equal in volume, two of which are simple bodies of diverse species, while the third is a mixture of both [the simple bodies], and given that one of the simple bodies is specifically heavier than the other, then the ratio of that part [by volume] of the mixture which is of the same species as the heavier body is to the other part [by volume], which is of the same species as the lighter body, as the ratio of the excess of weight of the mixture over the weight of the lighter body to the excess of the weight of the heavier body over that of the mixture.

Let there be three equal bodies: gold, silver, and the third a mixture (i.e. alloy) of these two. Let the gold be heavier specifically than the silver, as is in fact so. And let the weight of gold be 36 pounds, that of silver 27, and that of the mixture 30. Furthermore, the difference in weight of the mixture and the heavier body is 6; this is the amount by which the mixture is exceeded [in weight]. Further, the difference in weight of the mixture and the lighter body is 3. Therefore, just as 3 is to 6, so is that part [by volume] of the mixture which is of the same species as the heavier body to the other part [by volume] of the mixture which is of the same species as the lighter—i.e., the part of gold is half the part of silver in that mixture, which is that which was proposed. There was in the mixture, moreover, one foot of gold, following the measures of weights mentioned earlier. And these three [feet of the mixture] weigh 30 pounds, and three feet of gold 36, and three feet of silver 27. From these statements the aforementioned proposition is evident. And continue to suppose with me that a foot of gold weighs 12, [one] of silver 9, and [one] of lead 8 [pounds], as was designated earlier.

Proceed in the same way in regard to a mixture of three bodies. Let there be four bodies: gold, silver, lead, and the fourth a mixture of these simple bodies. Let them be equal in volume. Also let the weight of the mixture be 54 pounds, the weight of gold 72, the weight of silver 54, and that of lead 48. Then the difference in weight between the mixture and the heavier weight is 18 pounds. But the difference between the mixture and the lightest weight is 6. Say, therefore, that as 6 is to 18, so the part [by volume] of gold existing in the mixture is to the part [by volume] of lead existing there, and [thus] the lead is double [in weight] the gold in the mixture. Since, therefore, there are 24 pounds of lead, there will be 12 of gold, which joined together give 36. With these taken away from the total [weight], which is 54, 18 pounds remain for the part of silver. You will have, therefore, [by weight] one part of gold, one and one-half parts of silver, and two parts of lead. The numbers are these: of gold 12, of silver 18, of lead 24, of the whole mixture 54; which was useful to explain.

Proceed in the same way with a liquid mixed of several simple liquids. With their specific and numerical weights known, it is shown in the same way as with solid bodies. For the technique is similar in both cases. I shall speak briefly [of liquids].

6. The ratio of the specific weight of any liquid to that of another [liquid] of a different kind is as the ratio of the difference of weight of some body in air and in the one liquid to the difference [in weight] of the same body in air and in the other liquid.

Example: Let water and oil be the given liquids and let the heavy body be gold, whose difference of weight in air and in water is a and whose difference in air and in oil is b. Therefore a and b will be the weights of water and oil, each of which is equal [in volume] to the body of gold. Therefore the ratio of the weight of water and oil is as the ratio of a to b, and you will find it a double proportion, following the numbers assigned previously. And we call this a proportion in species when two bodies are of the same volume, but one of them is found to be naturally heavier than the other. Concerning this the balance gives assurance.

7. To assign conveniently a volume of one body which is equal to that of any given body and whose specific gravity is different from that of the given body.

Let the given body be iron and the body which we wish to equate to it, lead. And so we weigh the iron and [some] lead in air and in water and note the difference of weight of each in air and in water. For the ratio [of the volumes] of the given iron and lead will be as the ratio of one difference to the other, by one of the preceding [conclusions]. Then if they [i.e. the differences] are equal, we have what was intended. If, however, the lead is less, add—or if more, substract—according as the differences demand, until the differences are equal.

Example: Let the difference of weight of the gold in air and water be 1 pound and the difference in weight in air and in water of the silver to be equated to it, 2 pounds. Say that the volume of silver is double the volume of gold. Therefore take one-half of it, and then the volume of silver will be made equal to the volume of gold, which was that which we proposed to explain.

8. In a body mixed of two simples, of which one is specifically heavier than the other, to find out how much there is of each of them.

Let there be a body mixed of gold and silver. Make a body of gold equal to it in volume; and also [make] a body of silver equal in volume to that mixture—by the preceding [conclusion]. Then you will have three bodies, of which two are simple bodies and the third is a mixture of them—[and all] of the same volume. Proceed, therefore, as was said before, and it will be obvious to you how much of gold and how much of silver there is in the mixture. And if that mixture is of different liquids, you will be able to know how much there is of each. For let that mixture be of water and milk. Take a quantity of water equal [in volume] to that of the mixture and similarly the same quantity of milk—and this you can easily do by measuring the said liquids with an equal measure. And thus you have three bodies, two of which are simples and the third a mixture of them—[and all] of the same volume. Therefore proceed as was said before.

In another way [it can be done] more artificially. Consider the ratio of specific weights of water and milk, as was said in regard to diverse liquids, and make the body of milk equal in volume to the mixture and similarly [make] the body of water [equal in volume to the mixture] by the aforesaid method. Then proceed as was said in regard to three bodies.

9. The ratio of the specific gravity of any solid body to [that of] a given

liquid is as the ratio of the weight of the same body in air to [its loss of weight] in that liquid.

Example: Let it be silver and water whose ratio of specific gravities you wish to know. Weigh the silver in air and water, and the difference is that which shows the ratio. Whence silver would weigh in air 9 pounds and in water 8 pounds. The difference is one pound, which is the weight of water equal in volume to that of the silver. Therefore, the ratio of the specific weight of silver to the specific weight of water is as 9 to 1, i.e., nine times greater. So it is with the others, each in its own way.

10. If a body were an alloy of three metals, it is possible to find out how much there is of each one [in it].

This was explained before in the fifth proposition. But if these three metals are unknown [as to quantity], how will they be found, particularly if the mass of the alloy is covered by a patina (? *serico*)? It ought to be said however that this is a mass mixed of three metals. Respond as follows: *(Here all manuscripts omit any explanation of the procedure for the determination of the components by volume or weight of an alloy of three metals.)*

11. Every body floating in water displaces in the water a volume of water of its own weight.

For example: If half of the floating body were in the water and half outside of it, the water equal in volume to half of that body will be equal in weight to the whole body. And it is proportionally the same in regard to other fractions. Thus if 2/3 of the floating body is in the water and 1/3 of it outside of the water, the volume of water equal to 2/3 of that body will be equal in weight to the whole body. And if that body is equal in weight to water equal to it in volume, the whole body is submerged in the water; but it will not descend to the bottom, since it is not superior in weight to the water. Rather the [upper] surface [or the body] will be at the surface [of the water]. If, however, the body is superior in weight to the water which is equal to it in volume, it will descend immediately to the bottom.

12. If there is a vessel made of submergible material (i.e. specifically heavier than water), to find out whether or not it will float in water.

Let there be an iron vessel. I wish to know whether it will float in water. I take a quantity of iron which weighs the same as the said vessel. I also take as much water as fills up the said vessel. I further take water equal in volume to the said iron, and I join together these two quantities of water. These two waters denote the weight of the said vessel if the whole vessel is [just] submerged in the water. Leaving it in air, I shall consider whether the said iron body weighs more than the said waters joined together, or less. For if more, the vessel will be submerged; if not, it will float; which is that which was proposed.

An example in numbers: Let the vessel be iron—and cubical if you wish. Its capacity is 64 feet and the weight in air is 8 pounds. The ratio of iron to water in specific weight would be 8 to 1. So the iron weighs more than the water in an eightfold proportion. Thus let there be taken a quantity of iron equal in weight to the said vessel, namely, of 8 pounds. And let a quantity of water be taken which fills up the said vessel, and it will contain 64 feet, and its weight will be known by a balance; let that weight be 15 pounds. Take another quantity of water equal in volume to the said quantity of iron; and so it will weigh—by hypothesis—1 pound. Join these quantities of water together; they make 16 [pounds]. And since the quantity of iron weighs 8 and the waters joined together 16, it is clear that the vessel will float; since the weight of a quantity of iron equal in weight to the said vessel is not superior to that of the waters but is inferior to it, hence the whole vessel will not enter the water but will float. And say that half of it is in the water with the other half floating when the weight of the waters joined together is double the weight of the quantity of iron. If they were equal in weight, the whole vessel would enter the water up to the surface, but it would not be immersed [further]. If, however, the waters joined together weighed 16 and the quantity of iron 12, 3/4 of the vessel would be in the water and 1/4 would float, and so in other cases proportionally; this it was proposed to explain.

COMMENTARY

The *Quadripartitum numerorum* of the eminent Parisian mathematician and astronomer Johannes de Muris was completed in 1343, we are told in the Paris manuscript (BN lat. 7190, f. 100v). The fourth book concerns itself with physical problems. The first tract of the fourth book, which we have not translated, contains simple problems of motion. The second tract, entitled in the margins of the same manuscript "De ponderibus et metallis," concerns problems of hydrostatics. The full Latin text of this tract is published here for the first time. In the translation I have omitted only the definitions and postulates, since they were taken over from the *De insidentibus in humidum* and have already been translated in Moody and Clagett, *Medieval Science of Weights* (pp. 41, 43), and furthermore the translation of the most significant of them has been included above in the body of this chapter. There their significance for the distinction between specific and numerical weight has been pointed out.

1–2. The propositions and conclusions were also largely drawn from the *De insidentibus*—but with an important change. For most of the propositions Johannes gives numerical explanations rather than attempting mathematical proofs. As Johannes himself pointed out in the second proposition, "You can prove it by letters of the alphabet, but to me numbers are clearer." The first two enunciations are drawn verbatim from the *De insidentibus*. Needless to say, the first is important, since it states Archimedes' principle and thus became a common source for that principle in the fourteenth century. There is an effort to give a rhetorical proof of the proposition by reference to the first and second postulates. There seems to be an error here in the text, which I have corrected in my translation. The text appears to say that the counterweights are weighed in air, for they would *not* vary in weight in water. It is clear, on the contrary, that they would vary in weight in water. Following the "proof," there is a brief diversion, which says that by using this principle we can distinguish which of two waters is the lighter or rarer. This passage is somewhat confused. But what I believe he means to say is this: If we have equal weights of the same specific weight that balance each other when weighed in air, and if the pan holding one weight is placed in one "water" and the other pan with the other weight is placed in the second water, then, if equilibrium remains, the waters are of the same specific weight; but if one sinks, the water in which it sinks is lighter. The "proof" of the first proposition is concluded by a numerical example which is empirically ridiculous since it leads to a specific gravity of 8 for gold.

3. Proposition 2 relates the ratio of the volumes of two bodies to their differences in weight in air and water. This is obvious, since the differences in weight are equal to the weights of water displaced, the weights of water are of course proportional to the volumes of water, and the bodies displace their own volumes of water. This proposition is of considerable importance later when Johannes comes to determine the parts by volume and weight of the components of a mixture or alloy, as we shall see.

4. Proposition 3 is a rhetorical expression of the following formula:
$$\frac{S_{W_1}}{S_{W_2}} = \frac{W_1}{W_2} \cdot \frac{V_2}{V_1}.$$
In this formula S_{W_1} and S_{W_2} are specific weights (densities), W_1 and W_2 weights, and V_1 and V_2 the volumes of the two bodies in question. The next proposition stems directly from this one and merely says that if

S_{W_1} and S_{W_2}, as well as W_1 and W_2, are known, then the ratio V_1/V_2 can be found.

5. Propositions 5, 8, and 10 directly concern the "crown" problem, that is the problem of finding the component parts (by volume and then by weight) of a mixture* or an alloy of two and then three components. This was to be done apparently without chemical action on the unknown mixture. Proposition 5 concludes that $V_1/V_2 = (W_c - W_s)/(W_g - W_c)$ where V_1 and V_2 are the partial volumes of the components g and s, and W_c, W_g, and W_s are the weights respectively of a volume V of mixture c, of a volume V of component g, and of a volume V of component s. There is no proof, only a numerical example. But a proof in terms of the material of this treatise is easily given.

(1) Let W_c, W_g, and W_s be given.

(2) Note that $S_{Wg} = (W_1/V_1) = (W_g/V)$, where S_{Wg} is the specific weight of substance g, and thus $W_1 = W_g(V_1/V)$ and similarly $W_2 = W_s(V_2/V)$. W_1 and W_2 are the weights of the component parts.

(3) Then $W_c = W_1 + W_2$.

(4) Multiplying by V and substituting $V_1 + V_2$ for V, the conclusion of proposition 5 follows.

(5) Further, with V_1/V and V_2/V known, then W_1 and W_2 can be found from (2).

Johannes also attempts to apply this conclusion to a body which is a mixture of three simple bodies. It happens to work for his example because he takes a specific case where the weight of the mixture (an alloy of gold, silver, and lead) happens to equal the weight of an equal volume of one of the three components (silver), so that one of the components is eliminated, thus: $VW_c = V_1W_g + V_2W_s + V_3W_p$, and $V = V_1 + V_2 + V_3$. Hence if $W_c = W_s$, then V_2W_s is eliminated, and so

$$\frac{V_1}{V_3} = \frac{W_c - W_p}{W_g - W_c}.$$

The enunciation concerns itself only with volumes, but the composition by weight still has to be found. He only partially indicates this solution for the mixture of two simples. However in his case of a mixture of three simples, he indicates the complete solution by weight.

It ought to be noted in connection with this method of the solution of

---

* The Latin word *mixtum* is used by Johannes for mixtures of all kinds, including alloys of metals. Generally I have followed Johannes by simply translating it as "mixture," although in proposition 10 I have translated it as "alloy," because the proposition is specifically limited to metals.

the crown problem that it depends upon the volume of the mixture being equal to the sum of the volumes of the components. As I indicated in the text of the chapter, the volumetric changes in mixtures (including alloys) are manifold, and hence the assumption is often a not too accurate approximation.

The crudity of the experimental data used in his numerical examples should be further remarked. His ratio of the specific weights of gold and silver is 4/3 (instead of 19.3/10.5). He considers lead to be less dense than silver, which is contrary to fact. His ratio for the specific weights of gold and lead works out to 3/2 (instead of 19.3/11.3). We noted earlier his hypothetical and absurd value of 8 for the specific gravity of gold (proposition 1). Later in proposition 9, he relates the specific weight of silver to that of water as 9/1. Somewhat better is his figure of 8/1 for the ratio of the specific weights of iron and water. Johannes strikes us in his numerical examples as being concerned solely with illustrating how to perform the arithmetical operations involved rather than in the collection of the empirical data, although to be sure he often says that something can be tested or found out by the use of the balance.

6. Proposition 6 (similar to proposition 3 of the *De insidentibus*) relates the specific weights of liquids by comparing the differences in weight of some one body in air and in each of the two liquids. This method is an alternative to the use of a hydrometer, which, of course, compares the relative displacement by a standard floating body of each of the liquids.

7. We have said that Johannes starts his determination of the component parts of any alloy or mixture by taking equal volumes of the mixture and the simple substances composing the mixture. In proposition 7 he suggests how we might find a volume of each of the components equivalent to the volume of the mixture, although actually all he tells here is how to find the volume of a body that will be equal to the volume of a given body which is of different specific gravity. He uses iron as the given body and lead as the other body. He takes both specimens and weighs them in air and in water and notes the difference of weight of each in air and water. This gives the ratio of the volumes of the specimens. One then sees what fraction of the present volume of the other body (lead) will make the differences in weight in air and water of the two bodies equal. We are not told how one can easily take a fraction of a solid body like lead. He simply says, for example, "take one-half..." etc. Perhaps he reduces the lead to a molten state. But if he does do this, why worry about the differences in weight in air and water? By a spill-over method the volume of the given

body could be determined, and then an attempt could be made to pour a volume of the molten material that would be equivalent to it (although this might be difficult because of the volumetric changes involved in the different states of the metal). He might have had in mind some method of casting using wax patterns as that mentioned in the *Carmen de ponderibus* and other technological treatises. Be that as it may, he certainly does not allude to any such method, and one is forced to conclude that Johannes has not faced, and indeed is not interested in, the practical problems involved in this volumetric method.

8. Proposition 8 is, as we have said, related directly to proposition 5. We are to find volumes of the component substances equal to that of the mixture and then proceed as in proposition 5, first finding the volumes of the components and then presumably weights.

9. Proposition 9 is not found in the *De insidentibus*. It tells us that the ratio of the specific weight of a submersible body (i.e., one specifically heavier than a liquid) and that of the liquid is as the ratio of the weight of that body in air and [the loss of weight] in the given liquid. Later Albert of Saxony was also to compare solids specifically heavier than the liquids in which they are submersed (see Doc. 2.4, passage 7); but Albert was not interested in the quantitative relationship but only in knowing that the solid body was indeed heavier than the liquid.

10. Proposition 10 relates to the second half of the explanation of proposition 5. As before, we are to make volumes of each substance represented in an alloy of three metals equal to the quantity of alloy being tested. Once we have done so, we are to proceed as in proposition 5. The only difficulty about this is that in proposition 5 we are given only a special case where the weight of the alloy is equivalent to the weight of an equal volume of one of the constituents. The general procedure is not delineated in either proposition 5 or proposition 10 of the text as we have it. Possibly some attempt was made by Johannes (or perhaps he hoped to add this procedure later), for our earliest Parisian manuscript leaves a large lacuna following the brief statements given.

11. Johannes' proposition 11 is missing in all but one copy of the *De insidentibus*, namely in Paris, BN lat. 7377B (see Moody and Clagett, *Medieval Science of Weights*, pp. 355, 318), where it appears as the second proposition (but without proof). It was perhaps this manuscript then that Johannes used in incorporating it into his work. The proposition tells us that a floating body displaces its weight of water. It is thus equivalent to Book I, proposition 5, of the genuine *Floating Bodies* of Archimedes. In the ex-

planation we are told (as in the genuine *Floating Bodies*, props. 3, 4, and 7) of the three cases of bodies plunged into fluids. The first is that of a body specifically lighter than the fluid, i.e., a body which floats with a fraction emergent; the second that of a body equal in specific weight to that of the fluid, i.e., a body which floats but is completely submerged with the upper surface of the body at the surface of the water; and the third that of a body specifically heavier than the fluid, i.e., a body which sinks to the bottom. The juxtaposition of these cases without proof is like the brief treatment in a fragment attached to the aforementioned copy of the *De insidentibus* (MS Paris, BN lat. 7377B; Moody and Clagett, *op. cit.* pp. 320–21), except that this latter fragment merely says for the case where the body is equally heavy in species with the liquid that it is totally submerged and does not descend; and thus, unlike Johannes, it says nothing of the surface of the body being at surface of the water. The inclusion of these three cases without proof was also found in the Arabic fragment of the genuine *Floating Bodies* of Archimedes (see Doc. 1.4). The three cases appear together at least twice more in the fourteenth century, once in Albert of Saxony's treatment of the *De caelo* (see Doc. 2.4, passage 7) and again in Book III, proposition 3, of Blasius of Parma's *Tractatus de ponderibus* (Moody and Clagett, *op. cit.* p. 276).

12. Johannes' last proposition (prop. 12) gives a method of determining whether a vessel made of a material specifically heavier than the fluid in which it is immersed will float. Johannes' procedure is to take a quantity of the material equal in weight to the vessel and see how much water is displaced by it. This water is added to the quantity of water which fills up the vessel. If the weight of the two quantities of water joined together is equal to that of the material, the vessel will just float with no part emergent. If the waters are heavier, the vessel will float, and the proportion of the part emergent to that submerged depends on how much heavier the waters are than the vessel. Finally, if the waters weigh less than the weight of the vessel, it sinks.

Quadripartitum numerorum Johannes de Muris*

Liber quartus. Incipit secundus tractatus. Diffinitiones

[1] Statera est instrumentum examinis ponderum quo gravia corpora ponderamus, quod fit per virgulam rectam in cuius medio est foramen recipiens perpendiculum cum quo sustinetur virgula cum ponderibus in extremitatibus ipsius appensis, cum debet pondus alicuius corporis deprehendi.

[2] Calculus est minima pars ponderum ad quem omnes mensure ponderum referuntur.

[3] Calculi equari dicuntur quando corpore in una extremitate virgule appenso et calculis in alia, virgula in neutram partem nutum facit.

[4] Illius ponderis esse calculi, quorum pariter acceptorum pondus illi ponderi adequatur.

[5] Scitum pondus est cuius calculorum numerus est scitus.

[6] Corpus naturaliter descendens grave dicitur.

[7] Duorum gravium unius ad aliud relatio duplici modo potest considerari: uno modo secundum speciem, alio modo secundum numerositatem.

[8] Secundum speciem ut si volumus gravitatem auri et gravitatem argenti comparare. Hoc debet fieri supposita duorum corporum auri et argenti equalitate magnitudinis corporalis.

[9] Secundum numerositatem quando volumus discernere per pondus an massa auri sit gravior quam massa argenti cuius magnitudinis masse sint date. Et huius rei iudicium est in statera.

---

* I have used three manuscripts: Paris, BN lat. 7190, ff. 81r–84v (P); Vienna, Nat.-bibl. 4770, ff. 314r–324r (VV); Vienna, Nat.-bibl. 10954 (V). The first is of the fourteenth, the second of the fifteenth, and the third of the sixteenth century. All manuscripts are close to each other. V is copied from VV. Although later, sometimes V and VV have a better reading than P. All manuscripts on occasion use lb. as an abbreviation for all forms of *libra*. I have here represented it merely by lb. Note this tract is entitled on the margins of P, *De ponderibus et metallis*.

3–4 corpora P om. V et VV
8 calcus P
9 ponderum P om. V et VV
11 calculis P calculi V et VV / nutum P natum V et VV
14 adequatur V et VV et vide Moody et Clagett, *Weights*, p. 40, linea 21; adequetur P
16 naturaliter P similiter V et VV / grave PVV genere V
18 consyderari V
19, 23 numerositatem V et VV, et P supra; diversitatem P in textu

[10] Corpus dicitur alio gravius secundum numerositatem, cuius virgula statere nutum facit eisdem corporibus in extremitatibus statere appensis, aut cuius pondus plurium calculorum ponderi coequatur.

[11] Corpora eiusdem generis sunt inter que est nulla differentia substantialis: ut auri ad aurum, argenti ad argentum.

[12] Differentia duorum corporum in magnitudine est magnitudo qua maius excedit minus.

[13] In pondere vero pondus in quo gravius excedit levius.

Iste sunt petitiones

[1] Nullum corpus in seipso grave est.

2^a. Omne corpus in aere quam in aqua maioris esse ponderis.

3^a. Duorum equalium corporum alterum altero gravius esse specie, cuius pondus maiori numero calculorum equatur.

4^a. Corporum eiusdem generis magnitudinum et ponderum eandem esse proportionem.

5^a. Omnia pondera suis calculis esse proportionalia.

6^a. Eque gravia in specie dicuntur, quorum equalium pondus est equale.

Iste sunt propositiones sive conclusiones

Prima. Omnis corporis pondus in aere quam in aqua maius est per pondus aque sibi equalis in magnitudine.

Sit corpus datum aurum. Ponderat autem ipsum minus in aqua quam in aere per secundam petitionem. Sed pars auri subintrans aquam tantum expellit de aqua, quoniam continens et contentum sunt equalis magnitudinis, scilicet, superficiei, et ita de omnibus partibus auri. Cum igitur aqua in seipsa non ponderat per primam petitionem, sed extra se ponderat, igitur si aurum minus ponderat in aqua quam in aere, hoc est propter aquam egressam. Sed aqua egressa equalis est auro immerso. Ergo minus ponderat aurum in aqua per pondus aque equalis sibi in magnitudine, quod est propositum. Intellige quod calculi sint in aere, quoniam si in aqua ponderentur sicut et aurum ponderandum, non (!) essent pondera sive pondus

---

26 alio gravitus *tr. V et VV*
27 facit *P om. V et VV*
30 est nulla *tr. V et VV*
35 Iste sunt *PVV om. V*
37 2^a *P; om. V et VV hic et omnes* numeros petitionum et propositionum
45 Iste sunt *om. V*
49 in aere *om. V*
57 quoniam *PVV* etiam *(?) V*

auri variatum. Ex hoc notatur discretio duarum aquarum, que sit
60 levior sive magis rara. Nam si equalia pondera in aere apponantur
in duabus aquis et restent equalia, ille due aque sunt equalis ponderis
et raritatis. Si autem fiat nutus statere, illa aqua in qua fit nutus est
rarior et magis levis. Non enim fit resistentia descensus gravis nisi
propter dempsitatem medii. Igitur si nullum esset medium, grave
65 descenderet in instanti.

Exemplum de predictis: Sit aurum unius pedis cubici, cuius pondus
in aere 8 lb. Pondera ipsum in aqua et ponderet 7 lb. Est ergo differentia ipsius in aere et in aqua una libra. Dico quod hoc est pondus
aque equalis in magnitudine ipso auro, scilicet, pondus aque unius
70 pedis. Nam inducendo per partes auri patet propositum; quoniam
si de auro immergatur quarta pedis, scilicet, 2 lb, pars egressa aque
est similiter quarta pedis, hoc est, 1/4 lb, erit igitur proportio 2 lb
ad 1/4 sicut 8 ad 1; quare etc. Et ita in aliis partibus; quare et in toto
conclusio manet.

75 2^a. Omnium duorum corporum eiusdem sive diversi generis est
unius ad alterum proportio in magnitudine tanquam proportio differentie ponderis unius in aere et in aqua ad differentiam ponderis
alterius in aere et in aqua.

Sit auri massa ponderata in aere et in aqua et sit differentia 1 lb.
80 Similiter alia massa auri ponderata in aere et in aqua et sit differentia
2 lb. Dico quod sicut differentia est dupla ad differentiam, ita ultima
massa auri est dupla ad primam massam auri in magnitudine, quod
clare patet. Item sint (! sit?) iterum [differentia] masse auri ponderate
in aere et in aqua 1 lb, et alterius generis masse ponderate in aere et
85 in aqua, sicut argentum, sit differentia 4 lb. Et tunc erit proportio
quadrupla, sicut dicta. Et ut brevis capiatur, adduco magnitudines
et numeros in exemplum.

Sit auri massa ad modum pedis cubici figurata, eo quod ad divisiones
aptior reputatur. Et ponderet in aere iste pes aureus 12 lb. Sitque pes

60 Nam P om. V et VV / de pondera
  mg. scr. P sic: Intellige etiam quod
  sunt equalia in magnitudine —
  sunt diverse speciei ita esset. Iudicium
  ex, — constat, — esse equalia, —.
62 ^{1,2} nutus PVV intus V
64 densitatem V et VV / Igitur V et VV
  item P
67 Pondera P ponderemus V et VV
72 est P erit V et VV

73 quare P om. V et VV
79 sit differentia tr. V et VV
83 patet P patet intuenti V et VV /
  [differentia] scripsi. supra scr. P scilicet
  differentia
86 brevis capiatur P breviter capiam
  V et VV
88 massa P om. V et VV / cubici
  V et VV cubi P
89 ponderetur V

90 eiusdem quantitatis argenteus ponderans in aere 9 lb. Sed pes plumbeus 8 lb; pes quoque cupreus 6 lb. Sed pes olei equalis magnitudinis pedi aureo ponderet in aere 2 lb, pes vero aque consimilis ponderet 1 libram, preterea pes aureus in aqua ponderet 11 libras. Erit igitur differentia ponderis illius in aere et in aqua 1 lb. Et de hiis omnibus
95 sensibilis experientia per stateram te reddit in dubium. Pes etiam argenteus in aqua minus quam in aere per pondus aque sibi equalis in magnitudine, hoc est, per unam libram. Ponderabit ergo in aqua 8 lb; et sic de aliis consequenter.

Sint igitur iuxta propositum theoreuma duo corpora aurum et
100 argentum. Sit aurum pes unus, cuius ponderis in aere et in aqua differentia est una libra. Sit quoque argentum et sint pedes ignoti. Tunc erit differentia ponderum eius in aere et in aqua 4 lb. Ex hoc experientia statere probat et patet ex suppositis. Nam quatuor pedes argenti ponderant in aere 36 libras et in aqua 32. Differentia est 4 lb.
105 Nunc autem sicut 4, que est differentia ultima, ad 1, que est differentia prima, ita massa argenti ad massam auri in magnitudine, hoc est, quadrupla. Et ita iam enim fuit pes unus de auro et quatuor de argento, ita in cunctis evenire videbis. Quare constat theoreuma propositum declaratum. Posses declarare per litteras alphabeti; sed
110 mihi sunt numeri clariores.

Et patet aliter manifeste. Nam differentia ponderis auri in aere et in aqua est per pondus aque sibi equalis in magnitudine, et illud est 1 lb. Similiter differentia argenti in aere et in aqua est per pondus aque sibi equalis in magnitudine, et illud est 4 lb. Nunc autem que
115 est proportio magnitudinum, eadem est ponderum in corporibus eiusdem speciei, et aqua. Sed aqua aque eiusdem speciei est. Et est inter magnitudines proportio quadrupla, quia inter earum pondera est quadrupla. Ideo et inter equalia eorundem, scilicet, inter aurum, quod est equale aque unius libre, et argentum, quod est equale aque
120 4 lb, erit proportio quadrupla. Est ergo argentum quadruplum auri, quod est propositum.

3ª. Omnium duorum corporum est ponderum eorum in specie proportio secundum proportionem ponderis maioris corporis ad magnitudinem eiusdem corporis ducta in proportionem magnitudinis

---

90 argenteus... Sed *P om. V et VV*
93–94 Erit... 1 lb *P om. V et VV*
102 Tunc *P* tunc ibi *V et VV*
113–14 1 lb... est *P om. V et VV*
119 est equale *tr. V*
120 erit *om. P*

corporis minoris ad pondus eiusdem corporis.

Verbi gratia. Sint duo corpora aurum et argentum. Sitque magnitudo auri subdupla ad magnitudinem argenti et pondus auri sit ad pondus argenti in proportione subsesquisecunda, que scribitur per 2/3. Duc igitur 2 in 2/3; exibit $1\frac{1}{3}$, que est proportio sesquitertia. Dic igitur aurum ponderare in specie contra argentum in equali magnitudine sicut 4 contra 3, et hoc est propositum, aut sicut 12 ad 9, ut prius dictum est. Et sic patet quod proportio magnitudinum ducta in proportionem ponderum per stateram reddit proportionem ponderum in specie quam querebas. Fac secundum proportionem et in idem redibit.

Sit inventum pondus auri 12 et pondus argenti 18. Cum igitur argentum fuit duplum in magnitudine auri, fuit proportio sicut 2 ad 1. Erit ergo proportio magnitudinis auri ad suum pondus 1/12, et per consequens pondus ad magnitudinem in 12^a proportione. Similiter proportio magnitudinis argenti fuit ad suum pondus 1/9. Duc ergo 12 in 1/9 et provenit $1\frac{1}{3}$, que est proportio in specie ponderis auri ad pondus argenti in equali magnitudine, hoc est, proportio sesquitertia, sicut prius.

4^a. Si fuerint pondera duorum corporum in numero et specie nota, erit proportio magnitudinum eorum nota.

Divide pondus in numero gravioris corporis in specie, hoc est, auri, per proportionem ponderum dictorum corporum in specie; habitum per precedentem. Per id autem quod inde exibit divide pondus in numero levioris corporis in specie, scilicet, argenti; exibit proportio in magnitudine levioris in specie ad gravius, quod est propositum.

Exemplum: Sit pondus auri in numero 1, hoc est per stateram; quod est corpus gravius in specie. Sitque pondus argenti 2; et sit proportio in specie eorundem sesquitertia. Divide ergo 1 per proportionem sesquitertiam; exit 3/4, per quas divide 2; exit $2\frac{2}{3}$. Dic

---

125 minoris *V et VV* brevioris *P / de eiusdem mg. scr. P. sic:* scilicet minoris
128 scribitur *PVV* est sive scribitur *V*
134 Fac *PV* reffac (?) *VV / de proportionem mg. scr. P sic:* id est, vice versa sicut si diceretur: Omnium duorum corporum est ponderum eorum in specie proportio secumdum proportionem

138 ergo *P* igitur *V et VV / supra* proportio... 1/12 *scr. P* ponderis ad suam magnitudinem sicut 12 ad unum, i.e. sicut unum ad 12
141 et *om.* P / $1\frac{1}{3}$ *correxi ex* 1/13 *in MSS*
149 exibit *om. V*
152 1 *P om. V et VV*
155 exit *PVV* exeunt *V*

ergo quod magnitudo argenti est dupla superbipartiens ad magnitudinem auri, hoc est, 8 ad 3.

5ª. Si fuerint tria corpora equalia in magnitudine, quorum duo sunt simplicia generum diversorum, aliud vero mixtum ex utrisque, fueritque simplicium unum altero gravius in specie, erit partis mixti que in ipso est de genere gravioris proportio ad aliam sui partem que in ipso est de genere levioris, tanquam proportio differentie ponderis mixti ad pondus levioris, ad differentiam ponderis gravioris ad pondus mixti.

Sint tria corpora equalia: argentum, aurum, tertium ex hiis mixtum. Et sit aurum argento gravius in specie, sicut est in rei veritate. Et sit pondus auri 36 lb, argenti 27, compositi vero 30. Erit autem differentia ponderis mixti ad pondus gravioris 6, hoc est, in tantum exceditur mixtum. Item differentia mixti ad levius in pondere est 3. Sicut ergo 3 ad 6, ita pars que est in mixto de genere gravioris ad aliam partem mixti que est de genere levioris, hoc est, pars auri est subdupla partis argenti. Argentum ergo est duplum ad aurum, quod est propositum.

Erat autem in mixto pes auri et duo pedes argenti secundum mensuras ponderum prius dictorum. Et isti tres ponderant 30 libras, et tres pedes auri 36, et tres pedes argenti 27. Ex hiis patet propositum antedictum. Et suppone semper mecum, quod auri pes ponderet 12, et argenti 9, plumbi 8, sicut fuit anterius designatum.

Eodem modo fac de corpore ex tribus mixto. Sint quatuor corpora: aurum, argentum, plumbum, quartum vero ex hiis simplicibus compositum. Sint autem hec in magnitudine equalia. Sit quoque pondus mixti 54 libre, pondus auri 72, argenti vero 54, plumbi quoque 48. Est autem differentia ponderis mixti ad gravioris pondus 18 libre. Sed differentia mixti ad levius est 6. Dic ergo quod sicut 6 ad 18, ita se habet pars auri existens in mixto ad partem plumbi ibidem existentis, et hoc est duplum esse plumbum ad aurum in mixto. Quoniam igitur sunt 24 lb de plumbo, erunt de auro 12, quibus iunctis exit

---

159 sunt *PV* sint *VV*
162–63 levioris... mixti *Moody et Clagett, Weights, p. 50 lineae 93–94;* levioris ad differentiam ponderis mixti ad pondus gravioris *P* gravioris *V et VV*
169 est *om. P*
172–74 quod... mixto *V et VV* id est *P*
174–75 mensuras: numeros *P*
179 mixto *tr. P ante* ex
184 levius *P* levioris *V et VV*
186 mixto *P* magnitudine *V et VV*

36. Demptis quoque ex summa, que est 54, remanent pro parte argenti 18 libre. Habes ergo de auro partem unam et partem cum dimidia de argento duasque partes de plumbo. Numeri sunt isti: auri 12, argenti 18, plumbi 24, summe corporis mixte 54, quod fuit utile declarare. Eodem modo fac de uno liquore mixto ex pluribus liquoribus simplicibus; cognitis eorum ponderibus in specie et numero, sicut de fixis corporibus est ostensum. Ars enim consimilis est utrobique; breviter dicam.

6ª. Omnis liquoris ad alium alterius generis est proportio ponderis in specie secundum proportionem differentie ponderis alicuius corporis in aere et in altero liquore ad differentiam eiusdem corporis in aere et in alio liquore.

Exemplum: Sint aqua et oleum liquores dati; et sit corpus grave aurum, cuius differentia ponderis in aere et in aqua sit a. Et differentia eiusdem in aere et in oleo sit b. Erunt igitur a et b pondera aque et olei quorum utrumque est equale corpori auri. Ergo erit proportio ponderis aque et olei secundum proportionem a ad b, et invenies eam duplam secundum numeros anterius assignatos. Et hanc proportionem in specie appellamus, quando duo corpora eiusdem magnitudinis sunt, quorum alterum altero naturaliter ponderosius experitur; de quo equilibris facit fidem.

7ª. Quocunque corpore dato sibi equale in magnitudine alterius generis conveniter (?) assignare.

Sit corpus datum ferrum et corpus plumbi quod volumus adequare. Ponderetur itaque ferrum et plumbum in aere et in aqua, et nota differentiam ponderis utriusque in aere et in aqua. Nam secundum proportionem unius differentie ad aliam erit proportio ferri et plumbi datorum per unam precedentium. Et tunc si fuerint equalia, habetur intentum. Si autem plumbum minus fuerit, adde, vel si maius fuerit, deme secundum quod exigunt differentie usquequo differentie sint equales.

Exemplum: Sit differentia ponderis auri in aere et in aqua 1 lb et argenti adequandi sit differentia ponderis in aere et in aqua 2 lb. Dic quod magnitudo argenti dupla est ad magnitudinem auri. Summe

188 quoque V et VV que P / ex P de V et VV / summa P summa totali V et VV
191 summe: summa P
192 liquore mixto *tr.* V et VV
194 consimilis est *tr.* V et VV
195 *ante* breviter *habet* P *unam verbam*

quam non legere possum
204 ad P et V et VV
207 naturaliter P similiter V et VV
207–8 de quo equilibris P et conclusio V et VV
219 2 P 20 V et VV

igitur eius medietatem et erit magnitudo argenti ad magnitudinem auri coequata, quod fuit nobis propositum declarare.

8ª. In corpore ex duobus mixto, quorum unum est altero gravius in specie, quantum sit de utroque declarare.

Sit enim corpus mixtum ex auro et argento. Fac corpus equale sibi in magnitudine de auro, et item corpus argenti equale illi mixto in magnitudine, per precedentem. Et tunc habes tria corpora, quorum duo simplicia et tertium ex hiis mixtum eiusdem magnitudinis. Operare ergo ut dictum est prius, et patebit tibi quantum est de auro in illo mixto et quantum de argento. Et si illud corpus mixtum fuerit ex diversis liquoribus, poteris scire quantum erit de utroque. Sit enim illud mixtum ex aqua et lacte. Sume de aqua secundum quantitatem liquoris mixti et similiter de lacte secundum eandem quantitatem, quod facile scies eosdem liquores equali mensura mensurando. Et sic habes tria corpora, quorum duo sunt simplicia, tertium ex hiis mixtum, eiusdem magnitudinis. Ergo fac ut dictum est.

Aliter artificialius. Considera proportionem ponderum aque et lactis in specie, ut dictum fuit de diversis liquoribus. Et fac corpus lactis equale in magnitudine illi mixto, et similiter corpus aque, per modum antedictum. Et tunc perfice sicut de tribus corporibus dictum est.

9ª. Cuiuslibet corporis generis fixi est proportio ad liquorem datum in specie secundum proportionem ponderis eiusdem corporis in aere et in illo liquore.

Exemplum: Sit argentum, et aqua, quorum vis scire proportionem in pondere secundum speciem. Pondera argentum in aere et in aqua, et differentia est que proportionem ostendit. Unde ponderet argentum in aere 9 lb, in aqua 8 lb. Differentia est una libra, que est pondus aque equalis in magnitudine ipsi argento. Est ergo proportio ponderis argenti ad pondus aque secundum speciem sicut 9 ad 1, hoc est, proportio noncupla. Ita de aliis suo modo.

10ª. Si fuerit corpus ex tribus metallis datis mixtum, quantum sit de utroque reperire.

Ista exposita est prius in propositione quinta. Sed si illa tria metalla sint ignota, quo modo erunt reperta ac si illa massa mixti esset aliquo

224 in *P om. V et VV* / *post* sit *scr. V et VV* ibi
228 et *om. P*
234 eosdem *P* dictos *V et VV*
235 duo *tr. V et VV post* simplicia
239 et *correxi ex* ut (?) *in MSS*
243 *post* et *add. P et V* in aqua seu
246 differentie *V* / est que *tr. V* / ostendit *V et VV* erit (?) *P*
251 sit *P* est *V et VV*
253 Ista *P* ita *V et VV*

serico cooperta. Diceretur tamen hoc est massa ex tribus metallis composita, que sunt illa. Responde sic.

11ª. Omne corpus supernatans aque occupat in ea locum aque sui ponderis.

Verbi gratia: Si fuerit medietas corporis supernatantis in aqua et medietas supra, erit aqua equalis in magnitudine medietati huius corporis equalis in pondere ipsi toti corpori, et sic de aliis partibus proportionaliter; ut si 2/3 corporis supernatantis sit infra aquam et 1/3 eius extra aquam, erit aqua equalis 2/3 eiusdem corporis equalis in pondere ipsi toti corpori. Et si fuerit pondus ipsius corporis equale in pondere aque equali et in magnitudine, totum corpus mergitur in aqua, sed non descendet ad fundum, cum non vincat aquam in pondere, sed superficies superficiei applicabitur. Si tamen corpus illud aquam vincat in pondere equalem ei in magnitudine statim descendet ad fundum.

12ª. Si fuerit vas datum ex materia mergibili, utrum supernatabit aque vel non invenire.

Sit vas ferreum. Volo scire utrum supernatabit aque. Sumam ferrum in massa eiusdem ponderis cum dicto vase. Sumam etiam aquam replentem dictum vas. Sumam quoque aquam equalem in magnitudine dicto ferro, et iungam istas aquas simul, que aque denotant gravitatem dicti vasis in aqua si totum vas demergatur. Remanente eo in aere considerabo utrum dictum corpus ferreum ponderet plus dictis aquis simul iunctis vel minus. Quod si plus, mergetur vas in aqua; si non, supernatabit, quod est propositum.

Exemplum in numeris: Sit vas ferreum et cubicum si vis, cuius capacitas 64 pedes, et pondus in aere 8 lb. Ponderetque ferrum ad aquam in specie sicut 8 ad 1. Itaque ferrum ponderet plus quam aqua in octupla proportione. Sumatur itaque massa ferri equalis in pondere dicto vasi, scilicet, 8 librarum. Sumaturque aqua replens dictum vas et continebit 64 pedes; et erit eius pondus notum per stateram, sitque 15 lb. Recipiatur etiam alia aqua equalis in magnitudine dicte masse

256 *post* sic *habet* P *magnam lacunam*
263 2/3 eiusdem corporis equalis P *om.* V et VV
266 descendet P descendit V et VV
267 sed *om.* P / superficiei *om.* P *et habet lacunam* | Si V et VV sicud P
278 vas *tr.* V et VV *post* aqua
281 ad P *om.* V et VV
282 ad 1 P *om.* V et VV / *post* quam *scr.* P et V in
284 scilicet P *om.* V et VV / aqua V et VV que aque P
285 sitque 15 P *om.* V et VV
286 Recipiatur P recipit V et VV / etiam V et VV etc. P / in magnitudine P magnitudinis V et VV

ferri, ponderabit utique per ypothesim 1 libram. Iunge has aquas simul; exeunt 16. Et quoniam massa ferri ponderat 8 et aque simul iuncte 16, palam est vas supernatare; quoniam pondus masse ferri equalis in pondere dicto vasi non vincit aquas sed vincitur, igitur vas non intrabit aquam totum sed natabit. Et dic quod medietas eius erit infra aquam, alia natabit, sicut [quando] pondus aquarum simul iunctarum est duplum ad pondus masse ferri. Et si erunt equales in pondere, intraret totum vas usque ad superficiem in aquam sed non mergeretur. Si autem aque simul iunte ponderent 16 et massa ferri 12, 3/4 vasis erunt infra aquam et 1/4 nataret, et ita in aliis proportionaliter, quod fuit propositum declarare.

287 hypothesim *V*
291 totum *PVV* totam *V*

292 natabit *V et VV* natat *P*
296 erunt *P* essent *V et VV*

Document 2.4

Albert of Saxony, *Questions on the [Four] Books on the Heavens and the World* of Aristotle*

1. BOOK III, QUESTION 2.... Whether any body at all which is heavier than another body in air is heavier than the same body in water.... In connection with this question I wish, in the first place, to posit three conclusions on the topic under investigation. Secondly, I wish to posit in regard to it certain ways by which we can without weighing find the proportion of the [specific] weights of heavy and light bodies of different specific weights.

As regards the first category, let the first conclusion be this: Not every body whatsoever which is heavier than another in air is heavier than the same body in water. This is obvious from the case of a large piece of wood and a small piece of lead.... The second conclusion: Not every body whatsoever is equally heavy in air and water. This is obvious concerning wood, which descends in air to the bottom of the air, but in water does not. This could only be so if it were heavier in air and less heavy in water. Third conclusion: Not every body whatsoever is equally heavy in water and oil. This is obvious from the case of a piece of wood which descends not so far in water as it does in oil. From which follows finally a fourth conclusion: Not every body whatsoever which is heavier than another in water is heavier than the same body in oil and conversely. Not only is this true concerning water and oil but also concerning all bodies of different specific weights.

2. As regards the second category, it is to be supposed first that there are some things which are called "equally heavy according to species" and some things which are "unequally heavy." Accordingly, those things are called "equally heavy according to species" which are so selected that equal portions of them in volume weigh the same. On the other hand, those things are called "unequally heavy according to species" which are

* Rendering *mundus* by "world" rather than by "universe."

so related that equal portions in volume do not weigh the same. Whence, if two portions are taken—one of wood and the same amount of lead—the portion of lead weighs more.

3. Then let this be the first conclusion: With two solid bodies given, it is possible without weighing [them] in a balance to find out [1] whether they are of the same or of different specific weights and [2] which of them is heavier. For let these two bodies be *a* and *b* and let equal volumes of *a* and *b* be taken—if *a* and *b* are unequal in volume; and let these portions be released so that they fall in the same water. Then if they descend equally fast toward the bottom of the water, or if just as large a part of one as of the other [is submerged in the water], say that the said bodies are of equal specific weight. If, however, they descend to the bottom unequally fast, or more of one of these portions is submerged in the water and less of the other, say that the one is heavier according to species whose portion descends more quickly or whose portion is submerged further in the water.

4. Second conclusion: With two solid bodies of different specific weights given—both of them, however, floating above the water—not only can one find out without weighing them in a balance which of them is heavier, but also the proportion in which the one is heavier than the other. This can be found out as follows: For let the two bodies of this kind be different kinds of wood of different specific weights. Take two equal portions of them and divide each of them [by means of markings] on the side into some equal number of parts, e.g., into 12, which parts we let be called "points." When this has been done, let them be released to fall in the water, and observe which of them has more of its points submerged and which fewer, and finally observe how many points of each of them are submerged. By this would be judged the proportion of their [specific] weights. For example let 2 points of one of them be submerged and 4 points of the other. Then say that the one which has 4 points submerged is twice as heavy according to species as the one which has only two points submerged.

5. Third conclusion: With two liquid bodies of different specific weights given, it is possible to find out without weighing them which of them is the heavier. For let water and oil be two bodies of this kind and let a piece of wood be released first into the water and then into the oil. If less of the wood is submerged in the water and more in the oil, say that the water is heavier and the oil is lighter [according to species]. This is obvious, for always the same weight descends more deeply in the lighter than in the heavier [medium].

6. *Fourth conclusion*: With two liquid bodies of different specific weights given, it is possible to find out without weighing them not only which of them is heavier than the other but also in what proportion the one is heavier than the other. For let the liquid bodies of this kind be water and oil and take a piece of wood, dividing it in the said manner into 12 equal parts. Let it be released first into one of them, e.g., into water, and observe how many of its parts are submerged in the water—let this be, for example, 4 of its parts. Then let it be released into oil and see how many of its parts are submerged in oil. Let this, for example, be 3 points. With the results thus considered, say that the water is heavier according to species than the oil in the same proportion that the number of points of that body submerged in water has to the number of points of the same body submerged in oil. And because the proportion of those numbers is sesquitertiate, i.e., 4 to 3, say that the water is in sesquitertiate proportion heavier according to species than the oil.

7. *Fifth conclusion*: With two bodies given—one solid and the other liquid—it is possible to find out without weighing them whether they are equally heavy according to species or one of them is heavier than the other according to species. For let the liquid body be water and the solid body be lead or wood or something of this sort. Let the latter be released into the water. Then if it descends in the water to the bottom, say that it is heavier than the water according to species. If, however, it descends until its upper surface is equally high as the upper surface of the water, say that the solid body and the water are equally heavy according to species. If however it descends in water partially and floats partially, say that it is lighter than the water according to species—as in the case of wood.

8. *Sixth conclusion*: With two bodies of different specific weights given —one a solid and the other a liquid—and with the solid body moreover a body floating in the liquid and not descending to the bottom, it is possible to find out in what proportion one of them is heavier than the other according to species. For let the liquid body be water and the solid body be wood, and let the wood be divided in the said manner into 12 equal parts. When this has been done, let it be released into the water and observe how many of its points are submerged in the water—and, for example, let them be 4. With this done, observe what proportion the total number of points—12—has to the number of points submerged in water, namely, 4. Say that that liquid is heavier according to species than that solid in such a proportion, i.e., in a triple proportion. This is evident, for if that wood descended in the water completely through all of its points until its upper

surface became equally high as the upper surface of the water (i.e. until the surfaces coincided), it would be just precisely as heavy as the water. Therefore, the less of the wood that descends in the water, the heavier is the water than the wood. Hence, if half of that wood, i.e., 6 points, should descend in the water, the water would be precisely twice as heavy. But if less than half, say, a third—4 points—the water would be three times as heavy—which is just as the proportion of 12 to 4.

Here there have been proposed six conclusions. By means of the first two, solids are compared to solids; then by the next two, liquids [are compared] to liquids; and by the last two, solids [are compared] to liquids....

9. Question 3.... Whether some element is heavy in its very own [natural] place.... For the response to this question, I first posit a distinction; and it is this, that gravity can be taken in two ways: In the first way it is a certain habitual and potential disposition which follows the form of the heavy body—whether it actually tends to motion or not—and this we call "habitual" or "potential gravity"; in another way [the term] gravity is used for a disposition that actually tends to motion, and this we call "actual gravity." A similar distinction can be posited for levity.

This, then, we let be the first conclusion: Wherever a heavy element is situated, it has habitual gravity. This is proved, for gravity of this kind is a certain quality which follows the substantial form of the heavy body.... The second conclusion: A heavy element *does* have habitual gravity in its very own place.... The third conclusion: No element has actual gravity in its very own place.

COMMENTARY

1. The document here presented is representative of the fourteenth-century treatment of the subject of "hydrostatics" in the schools. As in other areas of physical thought, it represents a juncture of Aristotelian philosophical ideas with the Hellenistic-Arabic mechanical tradition, although the latter tradition plays a very minor role in Albert of Saxony's hydrostatics. In general, Aristotle, whose views on weight and movement we shall describe briefly in Chapter 7 below, treated the problem of specific gravity without precision (cf. C. Thurot's summary in the *Revue archéologique*, New Series, Vol. *18* [1868], 397–99). But it ought to be recognized immediately that Aristotle in the *De caelo* (III. 1, 299b 8–9) notes that the "dense" (πυκνόν) differs from the "rare" (μανόν) "in containing more (πλεῖον) in an equal bulk (ὄγκος)." It is from this discussion of "dense"

and "rare" that scholastic discussions of density take their point of departure. But let us develop Aristotle's general discussion of weight further. It is well known that Aristotle held that gravity is the tendency for heavy bodies to move to their natural places and lightness the tendency for light bodies to move to their places. The natural places of the elements then become ordered about the earth; the arrangement becomes one of concentric shells the central sphere of earth, these "shells" being of water, air, and fire (*De caelo*, IV. 3 and 4, 310a–312a). Bodies when out of their natural places tend toward them and, granting no obstacles, will move to them. Neither fire nor earth tends to move naturally from its place; earth never ascends and fire never descends naturally. However, it is conceivable that air and water could move in both directions, if some parts of their adjoining elements were withdrawn or added. Another way of putting this idea is that earth is "heavy" everywhere; water is heavy everywhere but in the place of earth. Air is heavy in its own place and in the place of fire but light in the place of earth and water (thus tending to ascend). The fire is heavy nowhere. In the same discussion we are told (*ibid.*, IV. 4, 311b) that air is heavy in its own place, and the experimental proof adduced is that an inflated skin weighs more than a deflated one. This leads to a much discussed question among the scholastics as to whether air has weight in air, or, more generally, whether any element has weight in its own place. According to Aristotle, it is because air has weight in its place that a piece of wood of one talent weight, which is heavier in air than a piece of lead of one mina in weight, will be lighter in water. It is heavier in air because it contains more air than the lead and it is lighter in water for the reason that the air tends to return to its natural place.

Aristotle vaguely discusses phenomena connected with specific gravity in the *Meteorology*. Thus he distinguished (II. 3, 359a) the weight of a liquid from its "thickness," recognizing that salt water is "heavier" than sweet water. The difference in consistency is given as the reason that ships with the same cargo "very nearly sink in a river when they are quite fit to navigate in the sea." More interesting but less well known were the views of Strato (who succeeded to the headship of the Lyceum at the death of Theophrastus in 287 B.C.). He held that all bodies have weight and that the heavier (presumably, the "specifically" heavier) displaced the lighter (cf. Clagett, *Greek Science in Antiquity*, p. 69). Among the medieval authors who held such a view of weight was Marcus Trivisano of the fourteenth century (see Doc. 2.5).

Now the sections of Albert's work included here are still very much in the tradition of the Aristotelian discussions. But at the same time Albert had become aware of the tradition of the specific gravity problem originating ultimately with Archimedes' genuine *Floating Bodies* but introduced into the West with Pseudo-Archimedes' *De insidentibus in humidum*.

2–8. In submitting himself to the influence of the tradition of Pseudo-Archimedes' *De insidentibus* (perhaps through the restatement of the conclusions of that treatise by Johannes de Muris; see Doc. 2.3), Albert makes two important changes: (1) Instead of employing the principle of Archimedes to compare the specific weights of bodies heavier than the fluid in which they are immersed, he takes over from Aristotle and the *Liber de ponderoso et levi*—attributed to Euclid—the procedure of comparing the speeds of fall of the heavy bodies through their fluids. (2) He adds (possibly from the tradition of *Carmen de ponderibus*, with certain influence from proposition 11 of Johannes de Muris' *Quadripartitum numerorum*) the idea of a hydrometer to compare the specific weights of liquids and of solids and liquids. He plunges a graduated floating body into the liquids to be measured, in each case seeing how many points are submerged.

It is clear that Albert's experiments were "thought" experiments and were not actually performed by Albert. For example, it is all very well to say, as he does, that one takes equal portions [by volume] of two solids to be compared in specific weight but quite unpractical in actuality to do so. Similarly his plunger for measuring the density of liquids is quite impractical. Had he actually used such a hydrometer, he could hardly have failed to mention the important consideration of the stability of the plunger, unless he employed a slim-necked vessel for the liquid being tested. Note also that in the fourth conclusion Albert has oil specifically heavier than water, at least so far as the figures he gives.

It is clear on comparing these passages of Albert with Book IV of Blasius of Parma's *De ponderibus* that Blasius drew heavily from Albert or some similar scholastic discussion. The reader wishing to make this comparison may consult the edition and translation of Blasius' text which I have published (Moody and Clagett, *Medieval Science of Weights*, pp. 272–79).

9. I have introduced brief sections from question 3, which takes up the problem of whether an element has weight in its own place. Aristotle is "saved" by introducing a distinction between habitual (or potential) and actual gravity. By its very nature, i.e., its substantial form, a heavy body has habitual weight wherever it is. But actual weight, i.e., the actual inclination that produces motion, is present only when a body is outside of

its natural place. Blasius of Parma briefly alludes to the question, saying that Aristotle affirms that an element has weight in its place while Archimedes (actually, Pseudo-Archimedes) denies it (*ibid.*, p. 272).

*Questiones [subtilissime] Alberti de Saxonia in libros de celo et mundo Aristotelis**

[1] Liber tertius. Questio secunda.... Utrum quodlibet corpus quod est alio gravius in aere sit eodem gravius in aqua.... Iuxta istam questionem primo volo ponere conclusiones circa quesitum; secundo iuxta hoc volo ponere quosdam modos quibus possumus invenire sine ponderatione proportionem corporum gravium et levium adinvicem, in gravitate diversarum gravitatum secundum speciem.

Quantum ad primum sit prima conclusio: Non quodlibet corpus quod est gravius alio in aere est eodem gravius in aqua. Patet de magno ligno et parvo plumbo.... Secunda conclusio: Non quodlibet corpus est eque grave in aere et in aqua. Patet de ligno, quod in aere descendit usque ad profundum aeris et in aqua non, quod non esset nisi in aere esset gravius et in aqua minus grave. Tertia conclusio: Non quodlibet corpus est eque grave in aqua et in oleo. Patet de ligno, quod non ita profunde descendit in aqua sicut in oleo. Ex quo ulterius sequitur quarta conclusio: Non quodlibet corpus quod est gravius alio in aqua est gravius eodem in oleo et econverso. Et non solum istud verum est de aqua et oleo sed de omnibus corporibus diversarum gravitatum secundum speciem.

[2] Quantum ad secundum supponendum est primo, que dicantur equaliter gravia secundum speciem et que inequaliter. Unde illa dicuntur equaliter gravia secundum speciem que sic se habent quod equales portiones eorum in magnitudine equaliter ponderant. Illa vero dicuntur inequaliter gravia secundum speciem que sic se habent quod portiones eorum equales in magnitudine inequaliter ponderant. Unde si capiantur due portiones, una ligni et tanta plumbi, portio plumbi plus ponderat.

[3] Tunc sit prima conclusio: Datis duobus corporibus solidis, an

---

* Edition of Venice, 1492, sign. h. 1v–h. 2r (ff. 42v–43r). Punctuation and capitalization have been altered; paragraph numbers have been added. The word "subtilissime" in the title is not used in the manuscripts of this work that I have seen. Notice the early publisher vacillates between "quatuor" and "quattuor."

sint eiusdem gravitatis in specie vel diversarum, et quod eorum sit gravius, possibile est sine ponderatione in equilibra invenire. Nam sint ille duo corpora *a* et *b* et capiantur de *a* et *b* portiones equales in magnitudine, si *a* et *b* sint inequalia in magnitudine, et dimittuntur ille portiones cadere in eandem aquam. Tunc si equevelociter descendunt in profundum aque vel tanta pars de una quanta de alia [submergitur in aqua], dic quod dicta corpora sunt equalis gravitatis in specie. Si autem ineque velociter descendunt usque ad profundum vel plus de una illarum portionum submergitur in aqua et minus de alia, dic quod illud est gravius secundum speciem cuius portio velocius descendit vel cuius portio* submergitur plus in aqua . . . .

[4] Secunda conclusio: Datis duobus corporibus solidis diversarum gravitatum in specie, ambobus tamen supernatantibus in aqua secundum aliquas sui partes, non solum sine ponderatione in equilibra contingit invenire quod illorum sit gravius, sed etiam proportionem in qua alterum est altero gravius. Et hoc potest inveniri sic. Nam sint duo talia corpora duo ligna diversarum specierum et diversarum gravitatum in specie. Capiantur de eis due portiones equales et quelibet illarum dividatur in latus in aliquot partes equales, verbi gratia, in 12, que partes vocentur puncta. Quo facto dimittantur cadere in aqua et videatur de quo plura puncta submerguntur et de quo pauciora et ulterius videatur quot puncta submerguntur de uno et quot de alio. Et secundum hoc iudicetur proportio eorum adinvicem in gravitate. Verbi gratia, sit quod unius submergantur duo puncta et alterius quatuor, tunc dic quod illud cuius quatuor puncta submerguntur est in duplo gravius secundum speciem quam illud cuius solum duo puncta submerguntur.

[5] Tertia conclusio: Datis duobus corporibus liquidis diversarum gravitatum adinvicem secundum speciem, est possibile sine ponderatione† invenire quod illorum sit gravius. Nam sint duo talia aqua et oleum et dimittatur unum lignum in aquam primo et deinde in oleum. Si de illo ligno minus submergitur in aqua et magis in oleo, dic quod aqua est gravior et oleum est levius. Patet, nam semper idem grave in leviori profundius descendit quam in graviori.

[6] Quarta conclusio. Datis duobus corporibus liquidis diversarum gravitatum in specie, possibile est sine ponderatione† invenire non solum quod eorum sit altero gravius sed etiam in qua proportione alterum est altero gravius. Nam sint duo talia corpora liquida aqua et

---

* *Edit. habet* portionis.   † *Edit. habet* pondere.

oleum et capiatur unum lignum, quod dicto modo sit divisum in 12 partes equales, et dimittatur illud primo in unum eorum, verbi gratia, in aquam, et videatur quot de illis partibus submergantur in aqua et sit, verbi gratia, quod quattuor eius puncta, et deinde dimittatur in oleum, et videatur quot puncta eius submergantur in oleo, et sint, verbi gratia, tria. Hoc considerato dic quod in illa proportione aqua est gravior oleo secundum speciem in qua numerus punctorum illius corporis submersorum in aqua se habet ad numerum punctorum eiusdem corporis submersorum in oleo. Et quia proportio illorum numerorum adinvicem est sexquitertia, quia 4 ad tria, dic quod aqua in sexquitertia proportione est gravior secundum speciem quam sit oleum.

[7] Quinta conclusio: Datis duobus corporibus, uno solido et altero liquido, possibile est sine ponderatione* invenire an sint eque gravia secundum speciem vel unum eorum sit altero gravius secundum speciem. Nam sit corpus liquidum aqua et corpus solidum sit plumbum vel lignum vel aliquid huiusmodi. Et dimittatur illud in aquam. Tunc si descendit in aqua usque ad profundum, dic quod est gravius aqua secundum speciem. Si autem descendit usquequo superficies eius superior fiat eque alta cum superficie superiori aque, dic quod illud corpus solidum et aqua sunt eque gravia secundum speciem. Si autem secundum aliquas eius partes descendit in aquam et secundum aliquas supernatat, dic quod est levius aqua secundum speciem, sicut est lignum.

[8] Sexta conclusio: Datis duobus corporibus, uno solido et altero liquido, diversarum specierum in gravitate, corpore tamen solido natante in liquido et non descendente ad profundum, possibile est invenire in qua proportione unum illorum sit alio gravius secundum speciem. Nam sit corpus liquidum aqua et corpus solidum sit unum lignum, et dividatur illud lignum dicto modo in 12 portiones equales. Quo facto dimittatur in aquam et videatur quot eius puncta submergantur in aqua, et sint, verbi gratia, quattuor. Quo facto videatur in qua proportione totalis numerus punctorum, duodecim, habeat se ad numerum punctorum submersorum in aqua, scilicet, ad quatuor. Dic quod in ea, scilicet, tripla proportione illud liquidum est gravius secundum speciem illo solido. Patet, nam si illud lignum secundum omnia eius puncta descenderet in aquam usquequo eius superficies superior fieret eque alta cum superficie aque superiori, esset precise

* *Edit. habet* pondere.

eque grave cum aqua. Ideo in quanto minus de ligno descendit in aquam in tanto aqua est gravior ligno. Ideo si medietas illius ligni, scilicet, sex puncta, descenderet in aquam, aqua esset precise in duplo gravior. Sed si minus quam medietas, scilicet, tertia, ut puta quatuor puncta, aqua esset in triplo gravior; qualis est proportio 12 ad quattuor.

Sic ergo posite sunt sex conclusiones: et iuxta primas duas conferuntur solida ad solida, secundum sequentes duas conferuntur liquida ad liquida, secundum ultimas duas solida ad liquida....

[9] Questio tertia.... Utrum aliquod elementum in suo proprio loco sit grave.... Pro respondendo ad istam questionem pono primo unam distinctionem, et est ista, quod dupliciter potest capi gravitas: uno modo pro dispositione quadam habituali et potentiali consequente formam gravis, sive actualiter inclinet ad motum sive non, et illa vocetur gravitas habitualis seu potentialis; alio modo accipitur gravitas pro tali dispositione actualiter inclinante ad motum et illa vocatur gravitas actualis. Consimilis distinctio potest poni de levitate.

Tunc sit prima conclusio: Ubicunque sit elementum grave ipsum habet gravitatem habitualem. Probatur, nam talis gravitas est qualitas quedam consequens formam substantialem ipsius gravis.... Secunda conclusio: Elementum grave in suo proprio loco habet gravitatem habitualem.... Tertia conclusio: Nullum elementum in suo loco proprio habet gravitatem actualem.

Document 2.5

Marcus Trivisano of Venice
On the Macrocosm or Greater World

BOOK I. (8v).... For the major confirmation of this position which holds that no body is light, some rational arguments proving it are put forth. For just as elements have a certain proportion in volume *(magnitudo)*, so it is fitting that they have in a contrary way a proportion in gravity *(gravitas)*.... In the third place [if we posited both heavy and light as absolutes], we would have a case of that which could be done by fewer causes being done in vain by a multitude of causes. But God and Nature do nothing in vain. However, in positing only gravity without levity*—gravity which is increased or decreased—both upward and downward motions, and [in fact] other things that take place according to those who posit levity, can be saved *(salvari)*, as is obvious to him who understands our position. Therefore, it is not necessary to posit such levity.... But gravity [rather than lightness] is posited, since we perceive it by our senses. Hence levity is to be rejected, and consequently nothing is light.

If, moreover, it is many times argued by famous and illustrious authors that "fire is lighter than air" and similar propositions, we answer that they [should] understand by "lighter," "less heavy," and by "absolutely light" *(leve simpliciter)* that which is "least heavy" among elementary bodies.

From these things it is apparent that greater or lesser weight follows a unit or a group of units of matter. So that wherever more units subsist in less or equal volume, there it is continually necessary that the one thing is heavier than the other. If however they are in a greater [volume], it is not always necessarily so, but sometimes so and sometimes not, since the more powerful and more effective unit force *(virtus)* is [sometimes] itself dispersed. By the amount there is greater perfection (9r) and form in some body—even indeed where it has less matter—by that same amount that body

* I have altered the text from *sive levitatem* to *sine levitate* in my translation.

[Marcus Trivisano, *On the Macrocosm*   147]

seems heavier.* Speaking in this way, earth is said to be heavier than water, water than air, air than fire. This method of understanding "heavier" is the method by which something is said to be "intensively heavier" *(gravius intensive)* than something else. There is another way, by which something is said to be heavier than another "extensively" *(extensive)* where volume is not considered but only units and numbers are considered. Thus that body where more units are grouped together would be called heavier. And wherever we have the same number of units we have the same weight.

COMMENTARY

Johns Hopkins University possesses a manuscript of this interesting treatise, and Professor George Boas of the Philosophy Department has transcribed the complete manuscript. He kindly provided me with a copy of this transcription, and my extract has been drawn and translated from that transcription. The author of this work (see note 46) was Marcus Trivisano who died in 1378 as a member of the "contrata Sancti Martialis Venetiarum" before completing Part VI of the treatise. Incidentally, it ought to be noticed that Part VI, written at a later time, is interesting in that it is in the form of a dialogue—a somewhat rare form for philosophical works of the fourteenth century.

Before commenting on the brief passage translated above, it should be noted that somewhat later in the work (f. 10v), the author in a discussion of "number" briefly treats of center of gravity: "Centrum gravitatis in figuris regularibus sequitur centrum numerorum. Et sic centrum gravitatis erit idem cum centro magnitudinis in lineis sive planis sive solidis regularibus." The substance of these references is that center of gravity in regular geometrical magnitudes is identical with the center of magnitude. We have discussed scholastic considerations of center of gravity in Chapter 10 below.

As for the passage translated as Document 2.5, several points are worth noting. (1) Trivisano assumes an anti-Aristotelian position regarding "heavy" and "light." He argues that all bodies are heavy, and we only call bodies or elements "light" which are "less heavy." (2) One of the arguments to support this position is the principle of economy. The phenomena of terrestrial motion can be *more simply* explained by positing that bodies are only heavy. Positing both heaviness and lightness as entities is unnecessarily multiplying entities. (3) The distinction between specific weight and gross weight is made as one between "intensive" and "extensive" weight.

* I have altered *minus grave* in the transcription to *gravius*.

As in the case of the Merton schoolmen, gravity is discussed in terms of the number of units of matter in a certain volume. As in the discussion of other qualities among the schoolmen, we find the fundamental division is between "intension" and "extension."

Liber Aurelii Auctoris Mar[ci] Trivisano Veneti
De macrocosmo, i.e., de maiori mundo*

[Liber primus.] Sextum corollarium est unitas per se in quantum est subiecta formae et qualitatibus alicuius elementi est maximum grave. Expondendo lib[ere] maximum negative et capiendo grave intensive, probatur consequentia quia sub nullo corpore tanto datur gravitas maior. Ergo est maximum grave intensive.

Pro maiori confirmatione huius argumenti quod supponit quod nullum corpus sit leve, ponuntur aliquae rationes idem probantes. Nam sicut elementa servant proportionem certam in magnitudine, ita conveniens est modo contrario servare in gravitate. Verbi gratia, si detur gratia exempli proportio elementorum secundum magnitudinem in decuplo, tunc magnitudo elementi terrae se habebit in proportione ad magnitudinem ignis, sicut unum ad mille. Modo versa vice erit proportio gravitatum. Verbi gratia, capiatur de quolibet elemento aliqua portio sub aequali magnitudine. Dicimus quod proportio gravitatum ignis ad terram econtra se habebit sicut unum ad mille. Sed omnes naturales ponunt quod magnitudinem (! magnitudinum?) inter elementa certa sit proportio, igitur et gravitatum. Et si sic, ergo si sub uno pugillo aquae est gravitas ut 100, in uno pugillo aeris erit ut 10. Et primo ignis gravitas ut 1, ergo ignis erit gravis de quo minus videtur inesse....

Tertio, frustra fit per plura quod potest fieri per pauciora. Deus etenim et natura nihil frustra faciunt. Sed ponendo solummodo gravitatem sive (! sine?) levitatem (! levitate?) secundum intensum et remissum possunt salvari motus sursum et motus deorsum et cetera quae contingunt secundum eos qui ponunt levitatem, sicut patet intelligenti nostram positionem. Ergo non est necesse ponere talem levitatem.

Quarto, probatum est quod unitas subiecta et cetera est corpus

---

* Transcription of the Johns Hopkins manuscript by George Boas, ff. 8v–9r (Boas transcription, pp. 28–30). I have made an occasional punctuation change from the Boas transcription.

gravissimum. Et per eandemmet rationem est levissimum. Sed summa gravitas et summa levitas in eodem subiecto non sunt compossibiles. Ergo non sunt ambae ponendae; sed cum una ponitur, debet altera removeri. Sed gravitas ponitur cum eam per sensum nostrum percipimus. Ergo levitas est abiicienda et per consequens nihil est leve.

Si autem ab auctoribus illustribus et famosis pluries proferatur quod ignis est levior aere et propositiones consimiles, dicimus quod ipsi intelligunt per levius minus grave, et per leve simpliciter quod inter corpora elementaria est minus grave.

Ex quibus apparet quod gravitas maior et minor sequitur unitatem vel unitates congregatas materiae sic quod ubicumque suberint plures unitates sub minori vel aequali magnitudine, semper illud altero gravius esse necesse est. Si autem sub maiori, non semper necesse, immo aliquando sic, aliquando non, quoniam virtus unita potentior ac efficacior est seipsa dispersa. Et quanto in aliquo corpore plus existit de perfectione (9r) et forma, minus vero de materia, tanto illus (! illud?) minus grave videtur esse. Et isto modo terra dicitur gravior aqua, aqua aere, aer igne. Et hic modus capiendi gravius dicitur modus quo aliquid aliquo est gravius intensive. Est alius modus quo quod alio gravius dicitur extensive, ubi non magnitudo sed solum unitates considerantur et numeri sic quod ubicumque plures existent unitates aggregatae, illud gravius diceretur et ubicumque tot aequegrave.

Document 2.6

Galileo Galilei, *Mechanics**

1. (PP. 163–65) HAVING demonstrated how the moments *(momenti)* of unequal weights come to be equal when the weights are suspended at distances inversely proportional [to the weights], it does not appear to me that I should pass over in silence another congruence and probability by which the same truth can be rationally confirmed.

Accordingly, consider the balance AB (see Fig. 2.10) divided into

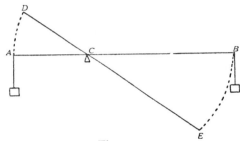

Fig. 2.10

unequal parts in point C and weights suspended at A and B which are inversely proportional to distances CA and BC. It is already evident how the one [weight] balances the other, and consequently, how, if to one of these [weights] there were added a minimal moment of gravity, it would descend and lift the other. Thus if we add an insensible weight to heavy body B, the balance will move, with point B descending to E and the other extremity A ascending to D. And, since to make the weight B descend any minimal increase in gravity to it is sufficient, hence we shall not take into account this insensible amount, and so will not make any

---

* Translated from the Italian text of *Le Opere*, Ed. Naz., Vol. 2 (Florence, 1891), 181–96. An early English translation of this work was made by Thomas Salusbury in his *Mathematical Collections and Translations*, Vol. 2 (London, 1665), 271–303. The sections equivalent to those I have here translated occupy pages 294–96, 297–98. Mr. Stillman Drake has recently prepared a new English translation, which he hopes to publish soon.

distinction between the power to sustain a weight and the power to move it. Now consider the motion which the heavy body B has in descending to E and which the other body A has in ascending to D. And we shall find without any doubt that by the amount that space BE exceeds space AD, distance BC exceeds CA, because there are formed at center C two opposite and thus equal angles DCA and ECB, and consequently the two arcs BE and AD are similar and have the same proportion between themselves as the radii BC and CA by means of which they are described. It then follows that the velocity of the motion of heavy body B in descending surpasses the velocity of the other moving body A in ascending by the amount that the gravity of the latter exceeds the gravity of the former. Since the weight A can only be lifted (although slowly) to D if the other heavy body B is moved rapidly to E, it is not surprising nor unnatural that the velocity of the motion of the heavy body B compensates for the greater resistance of the weight A, as long as it (A) is moved slowly to D and the other (B) descends rapidly to E. And so, contrariwise if we place the heavy body A in point D and the other body in point E, it will not be unreasonable that the former (D) by falling slowly to A can quickly lift the latter (E) to B, restoring with its gravity that which it comes to lose as the result of the slowness of [its] movement. And from this discourse we can recognize how the velocity of motion is able to increase the moment in the mobile by the same proportion as the velocity of motion itself is increased.

2. Before proceeding further, one other thing needs to be considered, and this is in regard to the distances at which the heavy bodies are suspended. For it is very important to know how to consider equal and unequal distances and, briefly, in what manner they ought to be measured. Now, if from the extremities of the straight line AB (see Fig. 2.11) there are

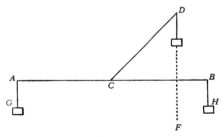

Fig. 2.11

suspended two equal weights and point C [the support] is in the middle of this line, equilibrium will prevail because the distance AC equals the distance *CB*. But if by elevating line *CB* and rotating it about point *C*, it is transposed into line *CD* so that the balance *(libra)* lies in accordance with the two lines *AC* and *CD*, two equal weights suspended from the termini *A* and *D* no longer equilibrate about point *C*, because the distance of the weight placed in *D* has been made less than it was when it was at *B*. Thus if we consider the lines by means of which the said weights produce impetus *(impeto)* and through which they would descend if they moved freely, there is no doubt that they would be the lines *AG*, *DF*, and *BH*. Hence the weight hanging from point *D* produces moment *(momento)* and impetus *(impeto)* according to line *DF*. But when it was hanging from point *B*, it was producing impetus in line *BH*. And because this line *DF* lies closer to the support *C* than does the line *BH*, therefore we must understand that the weights suspended from points *A* and *D* are not equal in distance from point *C*, as they are when they are constituted according to the straight line *ABC*. And finally one must be mindful to measure the distances with lines which fall at right angles on those in which the weights are hanging and in which they would move were they to descend freely.

3. (pp. 181–83) The present speculation has been undertaken anew by Pappus Alexandrinus in Book VIII of his *Mathematical Collections;* but in my opinion he has not achieved his goal and is deceived in the assumption that he makes when he supposes that the weight ought to be moved in the horizontal plane by a given force. This is false, no sensible force being required to move the given weight in the horizontal plane (with all the accidental impediments removed, impediments which in the theoretical discussion are not being considered). So that afterwards he seeks in vain with what force [the weight] is to be moved on an elevated plane. It will be better then, given the force which moves the weight up perpendicularly (which equals the gravity of the weight), to seek what would be the force which moves it on the inclined plane. This we shall attempt to do with a different kind of attack than that of Pappus.

Suppose the circle *AIC* (see Fig. 2.12), and in it the diameter *ABC*, and the center *B*, and two weights of equal moments *(momenti)* on the extremities *A*, *C*, so that, the line *AC* being a beam or a balance movable around the center *B*, the weight *C* will be supported by the weight *A*. But if we imagine the arm of the balance *BC* to be depressed downward to line *BF*, in such a way, however, that two lines *AB*, *BF* remain fixed together in point *B*, then the moment *(momento)* of weight *C* will no longer be equal to the moment of weight *A*, because the distance of point

F from the vertical *(linea delle direzione)* which goes through the support *B* along *BI* to the center of the earth is decreased. But if we draw from point *F* a perpendicular to *BC*, which is *FK*, the moment of the weight in *F* will be as if it were hung from line *KB*; and by the amount the distance *KB* is decreased with respect to the distance *BA*, so the moment of weight

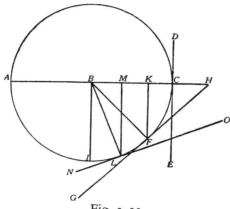

Fig. 2.12

F is decreased with respect to the moment of weight *A*. And so likewise when the weight is depressed more, as in the line *BL*, its moment will be decreased; and it will be as if it were hung from the distance *BM* along the line *ML*. In this point *L* it can be balanced by a weight placed at *A* which is less than it (at *L*) by the amount by which the distance *BA* is greater than the distance *BM*. Regard, then, how in inclining downward through the circumference *CFLI*, the weight placed on the extremity of the line *BC* successively decreases its moment and impetus *(impeto)* of descent, because it is supported more and more by the lines *BF*, *BL*. But to consider this heavy body descending and supported by the radii *BF*, *BL*, now less and now more, and constrained to traverse the circumference *CFL*, is no different than imagining the circumference *CFLI* to be a surface curved in the same way and placed under the same body so that if it [the body] were resting on it, it [the body] would be forced to descend along it [the curve].... Therefore, we can undoubtedly affirm that, while the heavy body is descending from point *C* through the circumference *CFLI*, in the first point *C* its moment of descent *(momento di descendere)* is total and complete because in that first point *C* it is not supported at all by the circumference and is not disposed to move any differently than if it should act freely along the tangent and perpendicular *DCE*. But if the moving body is in point *F*, then its gravity is supported in part by the circular path, which is placed under it, and its moment in descending downward is diminished in the proportion that line *BK* is exceeded by line *BC*. But when the moving body is in *F*, in the first point of such [movement] its motion is as if it were in the inclined plane along the tangent line *GFH*, because the inclination of the circumference in point *F* differs from the inclination of the tangent only by the insensible angle of contact. And in this same way we shall find in point *L* that the moment of the same moving body is decreased by the same ratio that the line *BM* is decreased with respect to *BC*; so that in the plane tangent to the circle in point *L*, i.e., along the line *NLO*, the moment of descent downward in the moving body decreases in the same proportion. If then the moment of the moving body on plane *HG* is less than its total impetus *(impeto)* which it has in the perpendicular *DCE* according to the proportion of line *KB* to line *BC* or *BF*—and by the similarity of the triangles *KBF* and *KFH*, $KF/FH = KB/BF$—then we will conclude that the whole and absolute moment, which the moving body has in the perpendicular to the horizon, has the same proportion to the moment it has on the inclined plane *HF* as the line *HF* has to the line *FK*, i.e., as the length of the inclined plane has to the perpendicular dropped from it to the horizon.

So that, passing to the more distinct figure here presented (see Fig. 2.13),

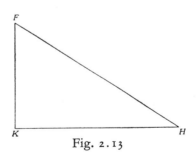

Fig. 2.13

the moment of descent *(momento di venire al basso)* which the moving body has on the inclined plane *FH* is to its total moment, with which it gravitates in the perpendicular to the horizon *FK*, as this line *KF* is to *FH*. . . .

4. (pp. 185–86) Finally one cannot pass over in silence this consideration which in the beginning has been said to be necessary in all the mechanical instruments: namely, that by the amount *(quanto)* there is a gain in force *(si guadagna di forza)* by means of them (the machines) by the same amount *(altrettanto)* there is a loss in time and velocity *(si scapita nel tempo e nella velocità)*. Perchance this would not appear in the present inquiry as true and manifest to someone. Rather it appears as if the force is multiplied

[Galileo, *Mechanics*

without the motor being moved through a longer path than the moving body. If this is what we understand (i.e., increase in force without the motor being moved through a longer path) then in triangle *ABC* (see Fig. 2.14) let the line *AB* be the plane of the horizon, *AC* the inclined

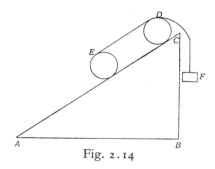

Fig. 2.14

plane whose height is measured by the perpendicular *CB*, [and let there be] a moving body placed upon the plane *AC* to which is tied the cord *EDF*, and [let there be] placed at *F* a force or a weight which has the same ratio to the gravity *(gravità)* of weight *E* that line *BC* has to *CA*. By what has been demonstrated, the weight *F* will fall downward drawing the moving body *E* over the inclined plane, but the said heavy body *F* will measure no more space in falling downward than that which the moving body *E* measures over the line *AC*.

But here, however, one ought to be warned that even if the moving body *E* will have traversed the whole line *AC* in the same time that the other heavy body *F* will have been lowered through the same space, nevertheless the heavy body *E* will not have been removed from the common center of gravity of heavy things more than by the length of the perpendicular *CB*, while the weight *F* descending perpendicularly will be lowered through a space equal to the whole line *AC*. And because heavy bodies do not offer any resistance to transversal movements, if by such movements they will not come to be any amount [more] distant from the center of the earth, therefore, since the moving body *E* in the whole movement *AC* will be lifted no more than by the distance *CB* while the other body *F* is lowered perpendicularly by the magnitude of the length *AC*, we can justly say: the distance of the force *F* *(viaggio della forza F)* to the distance of force *E* is the same as the ratio which line *AC* has to *CB*, i.e., as weight *E* to weight *F*.

COMMENTARY

1. In his *Mechanics* Galileo shows himself as the heir of both antique traditions of statics, the statical and the dynamic. In the section just prior to where this translation begins Galileo gives an Archimedean type of statical demonstration of the law of the lever. Now in this first section in the document at hand he applies the dynamic approach to the same law. His method is that which goes back to the *Mechanica* attributed to Aristotle, a method which indicates that the law of the lever follows from the fact that the velocities (i.e., the arcs swept out by the extremities of the lever arms in the same time) are in inverse proportion to the weights or forces on those extremities. As in the case of the *Liber karastonis*, Galileo's proof depends on the simple geometrical relationship that the arcs swept out are in the same proportion as the lever arms. It should be underlined that Galileo does not use the preferable method of Jordanus of employing vertical rather than arcal distances, although the vertical distances are used later in his analysis of the work principle as applied to weights on inclined planes (see section 3 below). Galileo of course knew of the *Mechanica* attributed to Aristotle, which would have been sufficient as a starting point of his analysis. Whether he was acquainted with the *Liber karastonis* of Thābit ibn Qurra is not known; the probability is that he was not, since he does not mention such a work when speaking of the Roman balance.

One interesting and important part of this first section is Galileo's bold declaration that since it would take only a minimal, insensible surplus of weight to move a weight on a balance over the weight necessary to maintain it in equilibrium, he is not going "to take in account this insensible amount" and so he will not "make any distinction between the power to sustain a weight and the power to move it." This idea had of course been implicit in both the *Mechanica* of Pseudo-Aristotle and Hero's *Mechanics*. Both apply the same proportionality of weights and distances (or times) regardless of whether they are dealing with balances in equilibrium or machines in motion. It is to Galileo's credit that he makes this idea explicit.

The principle of virtual work or displacements (or velocities) occupied Galileo later in 1612 when he briefly gives an account of its application to balances and levers in his *Discorso che stanno in su l'acqua* (in *Le Opere*, Ed. Naz., Vol. 4, 68–69) and once more in his famous *Dialogo sopra i due massimi sistemi del mondo*. I give this latter passage in the translation of Salusbury (as revised by G. de Santillana, [Chicago, 1953], pp. 228–31):*

---

* I have added certain corrections and some Italian words in parentheses from the text of the *Dialogo* given in the *Opere*, Ed. Naz., Vol. 7, 240–42.

SALV.: Therefore, tell me again, do you not think that the inclination *(inclinazione)*, for example, of grave (i.e., heavy) bodies to move downwards is equal to the resistance of the same to the motion of projection upwards?

SAGR.: I believe that it is exactly the same. And for this reason I see that, if two equal weights are put into a balance, they stand still in equilibrium, the gravity of the one resisting its being raised by the gravity wherewith the other pressing downwards would raise it.

SALV.: Very well; so that if you would have one raise up the other, you must increase the weight *(peso)* of that which depresses, or lessen the weight of the other. But if the resistance to ascending motion consists only in gravity, how does it happen that in a balance of unequal arms, to wit, in the steelyard *(stadera)*, a weight sometimes of a hundred pounds, with its urge downwards *(gravare in giù)*, does not suffice to raise up one of four pounds that shall counterpoise (i.e. balance) with it, nay, this of four, descending, shall raise up that of a hundred? For such is the effect of the sliding poise upon the weight which we would weigh. If the resistance to motion resides only in the gravity, how can the arm with its weight of only four pounds resist the weight of a sack of wool, or a bale of silk, which shall be eight hundred or a thousand weight; yea, more, how can it overcome the sack with its moment *(momento)* and raise it up? It must therefore be confessed, Sagredus, that here it makes use of some other resistance, or other force *(forza)*, besides that of simple gravity.

SAGR.: It must be so; therefore, tell me what this second virtue *(virtù)* should be.

SALV.: It is that which is not in the balance of equal arms; see then what variety there is in the steelyard, and upon this doubtless depends the cause of the new effect.

SAGR.: I think that your putting me to it a second time has made me remember something that may be to the purpose. In both these beams the business is done by the weight *(peso)* and by the motion *(moto)*. In the balance [of equal arm lengths] the motions *(movimenti)* are equal, and therefore the one weight must exceed the other in gravity, before it can move it; in the steelyard, the lesser weight will not move the greater, except when this latter, being hung at the lesser distance, moves but little, while the former moves a long way, as hanging at a greater distance [from the lacquet or cock]. It is necessary to conclude therefore that the lesser weight overcomes the resistance of the greater by moving more, while it (the greater) moves less.

SALV.: Which is to say that the velocity *(velocità)* of the less heavy body compensates the gravity of the heavier and less swift body.

SAGR.: But do you think that the velocity fully makes good the gravity? That is, that the moment *(momento)* and force *(forza)* of a body of, for example, four pounds weight is as great as that of one of a hundredweight, when the first has a hundred degrees *(gradi)* of velocity, and the latter but four only?

SALV.: Yes, doubtless, as I am able by many experiments to demonstrate....

Now fix it in your belief *(fantasia)* as a true and manifest axiom *(principio vero e notario)* that the resistance which proceeds from the velocity of motion compensates that which depends on the gravity of another body, so that a body of one pound that moves with a velocity equal to a hundred resists all obstructions as much as another movable of a hundredweight whose velocity is only equal to one. And two equal bodies will equally resist their being moved, if they shall be moved with equal velocity; but, if one is moved more swiftly than the other, it shall make greater resistance according to the greater velocity that shall be conferred on it.

The principle of virtual velocities is also found in Galileo's famous *Discorsi e dimostrazioni matematiche intorno a due nuove scienze* (Ed. Naz., Vol. *8*, 310–11), and in his last work *Della forza della percossa* that remained unpublished until 1718 (Ed. Naz., Vol. *8*, 329–30).

2. The second passage illustrates that Galileo was well aware that in the case of the bent lever, static moment is measured by the weight and the *horizontal* distance to the vertical through the center of motion. It is possible that this came to him from the Jordanus tradition that existed in the sixteenth century (see section 5 of this commentary below). It is most doubtful that it could have come to him directly from Hero, since the *Mechanics* of Hero does not appear to have been known in the late sixteenth century.

3. The third section I have translated is most interesting to the historian of medieval statics, for Galileo is doing more correctly what the various Jordanus treatises tried to do, namely to apply the principle of the resolution of forces to both lever and inclined planes, and his is certainly a most ingenious way to reduce the inclined plane problem to the lever. It will be recalled that, although the authors of the medieval treatises correctly used the principle of gravity according to position (identical, we have suggested, with $F = W \sin a$) to inclined planes, they misapplied it to arcs, while at the same time making a fertile suggestion that would lead to its correct application to arcs. Galileo then has made the correction by asserting that the moment of descent on every point F of the circumference is the same as if it were on an inclined plane tangent to the circumference at that point.

It must also be pointed out that the first paragraph of this passage contains an early reference by Galileo to his inertial concept when he says, "no sensible force being required to move the given weight in the horizontal plane." He many times in later works enunciates a principle of inertia, which in certain physical problems he applies to rectilinear paths and in other problems to circular paths (see Chapter 11, note 132).

4. I believe there has been some misunderstanding of what Galileo is trying to do in this fourth section (pp. 185–86). He is not proving the inclined plane principle as Stevin has done. This Galileo has already done in the third section by reducing it to the lever. Here he is trying to show that the principle of work (force × distance = force × distance) is valid even in a case where some might feel it does not apply, namely, in the case when a smaller weight moving through a perpendicular distance moves a larger weight in an incline through what seems to be the same distance on that incline. His solution, of course, is that we must not consider the length of the movement on the inclined plane, but only its vertical component. If we do so, then the principle "force × distance = force × distance" will hold. This passage, then, justifies the principle as applying only to the inclined plane. Jordanus, on the other hand, had used the principle, which he assumed, as justification for the proof of the inclined plane problem. Compare an analysis similar to this in a section of the *Discorsi* elaborated by Viviani (*Le Opere*, Ed. Naz., Vol. *8*, 216–17).

5. It can be reasoned that Galileo knew of the *De ratione ponderis* in the Tartaglia edition. We are told by his disciple Bardi that Galileo "with Jordanus" uses the expression specific weight: "Gravitas, de qua hic agitur, ea est, quam nonnulli a pondere distinguunt, Galileus vero cum Iordano, gravitatem in specie appellat" (Giovanni Bardi, *Eorum quae vehuntur in aquis experimenta*... [Rome, 1614], p. 6). If "with Jordanus" means that Galileo took it from Jordanus, we can then ask from which work of Jordanus he could have taken it? So far as I know, no work attributed in the Middle Ages to Jordanus has a definition of specific weight. But just following the *De ratione ponderis* in the Tartaglia edition of 1565, there is, without indication of authorship and occupying folios 16v–19v, the *De insidentibus in humidum* attributed to Archimedes elsewhere, which of course does have a definition of specific weight. A reader unfamiliar with the medieval manuscripts might well conclude from the edition that the *De insidentibus* was also by Jordanus. If such was the case with Galileo, we would then have to conclude that he was familiar with the Tartaglia edition as a whole, and thus with the best of the medieval statical treatises.

Part II

MEDIEVAL KINEMATICS

Chapter 3

Gerard of Brussels and the Origins of Kinematics in the West[1]

THE mechanics of large bodies comprises two main divisions. The first is *kinematics*, which studies movements taken in themselves, i.e., the spatial and temporal aspects or dimensions of movement without any regard for the forces which engender changes of movement. The second part of mechanics considers movement from a different point of view. Called *dynamics*, it studies movements in relationship to their causes, or as is more commonly said, in relationship to the *forces* associated with movement. Although they did not have the Newtonian concept of force, the mechanicians of the high and late Middle Ages nevertheless evolved this fundamental division in the manner of treating movement. Perhaps the first to deal exclusively with kinematic problems for their own sake and not as a part of some wider objective was a little-known geometer of the first half of the thirteenth century, Gerard of Brussels. Although Gerard's *Liber de motu* was the first entirely kinematic treatise in the Latin West, he, of course, fell heir to the kinematic ideas of Greek antiquity.

Like statics and dynamics, kinematics had solid beginnings in Greek antiquity. This does not mean that it was a separate and independent branch of mechanics at that time, but only that some fundamental kinematic definitions, theorems, and points of view can be traced back to the fourth, and perhaps even the fifth century B.C. Hence it will be instructive for us to see in what way kinematics was nourished in antiquity and particularly how much of this earlier work was available in Latin translation to Gerard of Brussels in the thirteenth century.

[1] The first section of this chapter is largely drawn from my text and study of Gerard of Brussels in *Osiris*, Vol. *12* (1956), 73-175.

Kinematics was fostered in antiquity by three distinguishable currents of scientific activity: (1) the geometrization of astronomy, (2) the emergence of a geometry of movement, or generative geometry, and (3) the development of physical and mechanical treatises whose theoretical parts had a geometrical character. We do not suggest this order as a strictly chronological one, for almost certainly the preliminary steps in the development of a geometry of movement took place before either of the other two events; and the kinematic statements of Aristotle, which belong to the last group, happen to be our earliest *extant* kinematic definitions. However, we find this order a more convenient way to present the kinematic ideas available to Gerard.

While there is clearly some usage of kinematics in the astronomy of Mesopotamia, I shall confine myself to Greece. Greek tradition would have us believe that the Pythagoreans were the first to apply mathematics to astronomy in a systematic way. It is said, for example, that the Pythagorean astronomers were concerned with the speeds of planets and that they declared that "those bodies move most quickly which move at the greatest distance, that those bodies move most slowly which are at the least distance, and that the bodies at intermediate distances move at speeds corresponding to the sizes of their orbits."[2] In the course of this concern with the speed of the heavenly bodies, no doubt the Pythagoreans began to apply geometry to the problems of spherical motions. As a result of this application of geometry to astronomy, certain basic kinematic ideas connected with the movement of astronomical spheres were fashioned—and no doubt were fully formed by the time that Eudoxus in the fourth century B.C. constructed his intricate system of homocentric spheres to account for the irregular movements of the heavens. But the basic kinematic ideas developed were in their way so simple that they remained pretty much unexpressed in the Greek astronomical works available to medieval authors.

Such, however, was not the case of the widely popular *De spera* mota* of Autolycus of Pitane, who flourished about 310 B.C.[3] I have purposely given the title in the form found in the medieval translation of Gerard of Cremona, and my citations will be to that translation. Autolycus sets

---

* This is, of course, the common medieval form for *sphaera*.

[2] M. R. Cohen and I. E. Drabkin, *A Source Book in Greek Science* (New York, 1948), p. 96.

[3] For general literature on Autolycus, see Paul-Henri Michel, *De Pythagore à Euclide* (Paris, 1950), pp. 84–85, 225.

out initially to define uniform velocity, or equal movement as he calls it.[4] "A point is said to be moved with equal movement when it traverses equal and similar quantities in equal times. When any point on an arc of a circle or on a straight line traverses two lines with equal motion, the proportion of the time in which it traverses one of the two lines to the time in which it traverses the other is as the proportion of the one of the two lines to the other."

There are several interesting points to observe about this early definition of uniform velocity. In the first place, notice how the approach to movement is entirely geometrical. We are not dealing with a gross body in movement but with a geometric point. It was, of course, the geometrization of movement that produced kinematics and lies at the base of all of its progress. Hence it was an important step when Greek geometers disregarded all of the unessential properties of matter in movement and substituted therefor the geometric point in movement, an event that no doubt had already taken place by the time of Zeno's anlysis of movement in the fifth century B.C. And naturally, since Autolycus is using a point in movement, the paths described are geometric lines, either arcs of circles or straight lines.[5]

A second and important point concerns itself with the terminology of the Latin translation, for that terminology is a key to the words used by Gerard. The word *motus* is employed here and in the treatise of Gerard as the word *velocitas* comes to be employed later, and furthermore the word *equalis* is used as the word *uniformis* is used more frequently later; and so the expression "an equal movement of a point" is equivalent to "a uniform velocity of a point."

Keeping the meaning of this terminology in mind, Autolycus' statement can be restated in modern fashion as follows: (1) the velocity of a point is uniform when that point traverses equal linear distances in equal periods

---

[4] MS. Paris, BN lat. 9335, f. 19r; cf. the excellent edition of J. Mogenet in *Archives internationales d'histoire des sciences*, Numéro 5 (1948), 146, lines 4-9: "Punctum equali motu dicitur moveri cum quantitates equales et similes in equalibus pertransit temporibus. Cum aliquod punctum super arcum circuli aut super rectam existens lineam duas pertransit lineas equali motu, proportio temporis in quo super unam duarum linearum pertransit ad tempus in quo transit super alteram est sicut proportio unius duarum linearum ad alteram."

[5] A later astronomer, Geminus (1st century. B.C.?), stresses the importance of geometry for the study of astronomy, including its kinematic aspects, when he says: "...the astronomer will prove them (i.e., the problems of astronomy) by the properties of figures or magnitudes or by the amount of movement and the time appropriate to it." (Cohen and Drabkin, *Source Book*, p. 90).

of time, (2) and furthermore if the velocity of a point is uniform and it traverses two distances with that uniform velocity, then the distances traversed are directly proportional to the periods of time in which they were traversed: Or, for a uniform velocity:[6]

$$\frac{S_1}{S_2} = \frac{T_1}{T_2}.$$

Following upon this definition of uniform velocity, Autolycus makes some further kinematic observations regarding the movement of points on the surface of a rotating sphere. These constitute the first three theorems of the *De spera mota*, of which we quote the first two without their proofs. The third is omitted as being scarcely more than a repetition of the second:[7]

1. When a sphere turns on its axis in uniform revolution, the points on its surface except the polar points describe parallel circles, whose poles are the poles of the sphere on its axis....
2. When a sphere is moved in uniform movement on its axis, all the points on its surface in equal times describe similar arcs of the parallel circles they traverse....

The author now has introduced the basic concepts connected with the sphere in motion. All of these theorems describe the effect of a uniform rotation of the sphere on the movements of the points on the surface of a sphere. Thus we are told in the first theorem that when a sphere rotates on its axis uniformly, all except the polar points describe parallel circles. Theorem 2 emphasizes the fact that when uniform rotation takes place, the surface points describe similar arcs in equal periods of times. Hence, the author is describing the conditions of movement of the surface points of the sphere which attend the uniform rotation of the sphere. These theorems constitute a definition of the uniform rotation of a sphere in terms of the movement of the points on the surface of the sphere.

I quote these theorems of Autolycus regarding the sphere in motion mainly because much of the kinematic development of the Middle Ages

---

[6] I have generally followed the practice of using capital letters when a formula is illustrative of a simple Euclidean proportion, even in the case of a variable which is customarily represented by a small letter in metrical formulae.

[7] *MS. cit. loc. cit; edit. cit.* in note 4, p. 146, lines 13–19; and pp. 147–48, 1. 25–2. 4: "1. Cum spera super suum meguar revolutione voluitur equali, puncta que sunt supra ipsius superficiem, preter ea que sunt supra meguar, circulos designant equidistantes quorum poli sunt poli spere et sunt erecti super meguar.... 2. Cum spera equali motu supra suum meguar movetur, omnia puncta que sunt supra spere superficiem in equalibus temporibus transeunt supra similes arcus circulorum equidistantium supra quos transeunt."

is tied in with the attempt to find the best way to measure the velocities of rotating geometrical figures and in particular that of the sphere. In fact this is, as we shall see, the basic objective of Gerard's treatise: to find some uniform linear velocity of a point which can be said to represent the multitude of different punctual velocities arising from the rotation of a geometric figure about an axis. But Autolycus, as an astronomer, is only interested in the movements of the points on the surface of a rotating sphere, while Gerard is interested in all the points in motion when different geometric figures are rotating. Gerard does not cite Autolycus' treatise in any fashion, but he may very well have been familiar with it, since it was quite popular and had been translated as early as the twelfth century by Gerard of Cremona, as we have already observed. Gerard of Brussels has something of the same geometric spirit as Autolycus, but this he undoubtedly got from studying the works of Euclid and Archimedes.

One final point concerning Autolycus' kinematics ought to be observed, namely that Autolycus and most Greek mathematicians give comparative rather than metric definitions. That is to say, they compared the *distances* traversed in two uniform movements when the times are assumed to be the same, or the *times* when the distances are the same, or the ratios of the distances on the one hand and the times on the other. Hence we find no metric definition like $V = k\,(S/t)$. The comparisons of Autolycus and other Greek authors were thus true proportions in the Euclidean sense as being between like quantities. It is not surprising, then, that none of the Greek authors arrived at the idea of velocity itself as a number or a magnitude representing a ratio of "unlike" quantities, namely distance and time. On the other hand, Gerard in his treatise of the thirteenth century, while not yet defining velocity as a ratio of unlike quantities, seems to assume that speed or motion can be assigned some magnitude not simply identical with either the quantity of time or the distance alone, although it is measured by either one with the other considered constant. At least he tells us in supposition 8 of the first book that "the proportion of the movements of points is that of the lines described in the same time" (see Doc. 3.1, supposition 8). This must mean that some magnitude or number V_1 representing one motion is to another magnitude or number V_2 representing another motion, as the distance S_1 is to the distance S_2; otherwise it would make no sense to use the language of proportions. We shall see in the next chapters how considerations of instantaneous or qualitative velocity independent of the total velocity measured by the total distance in a given time brought kinematicists in the fourteenth century one step closer to the definition of velocity as a ratio and of instantaneous velocity as the limit of a ratio.

Now let us look at the second current of Greek scientific activity in which kinematics played an important role. This second field fertile for kinematics lay within the development of pure geometry itself; it was, as I have said, the evolution of a geometry of movement or a kinematic geometry opposed to the static geometry so firmly established by Euclid and his predecessors. To illustrate the difference of viewpoint between a static and kinematic approach to geometry, let us describe two possible ways to define the circumference of a circle. Statically, the circumference of a circle can be defined as the locus of all points in a plane a given distance from some point in the plane which we designate as the center. On the other hand, we could define the circumference kinematically as the curve generated by a point moving in a plane in such a way that it is always the same distance from some point in the plane designated as the center and it returns to the position of departure.

There is evidence that a geometry of movement began to emerge as early as the middle of the fifth century B.C. But I hesitate, in contrast to some historians, to link this kinematic geometry too surely with Heraclitean concepts of reality epitomized in the phrase πάντα ῥεῖ, "everything flows," since we know so little about the details of pre-Socratic writings. But it does make an attractive cogent theory to establish on the one hand a harmony of viewpoint between the Parmenidean idea of reality (which relegates movement to the plane of illusion) and a static geometry of pure form with locus definitions, and on the other hand a coincidence of approach between the Heraclitean view of a dynamic reality and a geometry where the figures are generated by the movements of points, lines, and figures. That the connection between views of reality and geometry is not completely illusory is illustrated in the paradoxes of Zeno, which serve to reduce to absurdity the existence of movement and thus to defend the earlier ideas of Parmenides on reality, and which at the same time reveal clearly a knowledge on Zeno's part of the kinematic factors of space and time in the definition of velocity.

But we are on more solid ground when we seek the immediate origin of kinematic geometry in the attempts made to solve the three classical problems of Greek geometry, the trisection of an angle, the squaring of a circle, and the duplication of a cube. These problems led to the invention of higher curves, which were often curves defined in terms of movement.

[the Origins of Kinematics in the West   169]

This chapter is hardly the place to discuss these curves in detail, but for purposes of illustration we can point to the *quadratrix*, whose invention tradition assigns to Hippias of Elis (born about 460 B.C)[8] This curve was employed in the solution of two of the three classical problems, the trisection of an angle (or more generally, the dividing of an angle into any desired ratio) and the squaring of a circle (or more generally, in finding the length of any arc of a circle). To show the kinematic aspects of this curve, I shall repeat a brief description of it from the Greek mathematician Pappus:[9]

Suppose that $ABCD$ (see Fig. 3.1) is a square and BED a quadrant of a circle with Center A. Suppose (1) that a radius of the circle moves uniformly about A from the position AB to the position AD, and (2) that *in the same time* the line

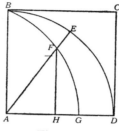

Fig. 3.1

BC moves uniformly, always parallel to itself and with its extremity moving along BA, from the position BC to the position AD. Then, in their ultimate positions, the moving straight lines and the moving radius will coincide with AD; and at any previous instant during the motion the moving line and the moving radius will by their intersection determine a point, as F or L. The locus of these points is the *quadratrix*. The property of that curve is that $<BAD : <EAD =$ (arc BED) : (arc ED) $= AB : FH$. [This property then allows us not only to trisect an angle, but to divide it into any number of parts. It can also be shown that, granted we can find point G, arc $BED/AB = AB/AG$. But arc BED is a quadrant of a circle. Hence with this curve we can square the circle.]

I give this section on the *quadratrix* only for illustration of the growth of kinematics. It had no direct influence on Gerard, since Pappus was unknown in the medieval Latin West. Of course, Gerard did employ the

---

[8] See Michel, *De Pythagore*, pp. 245–47. Cf. T. L. Heath, *A History of Greek Mathematics*, Vol. 1 (Oxford, 1921), 225–26.
[9] Heath, *History*, Vol. 1, 226–27. See

Pappus of Alexandria, *Mathematical Collection*, Bk. IV, prop. 30 (Edition of F. Hultsch, Vol. 1 [Berlin, 1876], 252).

concepts of rotary motion and of the movement of translation of a line, but he did not join them together to generate curves like the *quadratrix*. Needless to say, both concepts had become commonplace by the time of Gerard.

The continuing study by Greek geometers of problems of the quadrature of curves like the circle and the parabola led to the development of the celebrated method of exhaustion, possibly invented by Eudoxus in the fourth century, B.C., utilized on occasion by Euclid, and then employed with complete understanding and great skill by Archimedes. But although this method of exhaustion was in a real sense the product of a geometry of movement if only as a reaction thereto, its exposition belongs more to the history of mathematics than it does to our immediate study of early kinematic definitions and theorems. Let us merely say that in the hands of Archimedes it produced techniques and results that were employed by Gerard. Thus, for example, Gerard depends heavily on the *Dimensio circuli* of Archimedes, at least three times translated in the Middle Ages, possibly first by Plato of Tivoli and then almost certainly by Gerard of Cremona in the twelfth century, and then again by William Moerbeke in 1269 as a part of his general translation of the works of Archimedes.[10] I believe it most probable that Gerard of Brussels used Gerard of Cremona's translation. It was the first proposition of the *Dimensio* that Gerard of Brussels cited (see Doc. 3.1, prop. 1), the proposition which equates a circle to a right triangle whose sides including the right angle are the radius and circumference of the circle respectively. We ought to notice that in addition to using the proposition itself, Gerard made considerable use of the Archimedean type of proof by reduction to absurdity that is employed in this proposition, a type of proof always called by Gerard the proof *per impossibile*. The reader can examine the proofs given by Gerard in the propositions included in Document 3.1 below to see how he has employed this form of demonstration.

At the same time, Gerard used the propositions contained in a work based on Archimedes' *Sphere and Cylinder*, the *Commentum in demonstrationes Archimenidis* of Johannes de Tinemue, a work known to many medieval authors as *De curvis superficiebus Archimenidis* and to Gerard as *De pyramidi-*

---

[10] M. Clagett, "Archimedes in the Middle Ages: The 'De Mensura circuli,'" *Osiris*, Vol. *10* (1942), 591–618. See also Clagett, "The Use of the Moerbeke Translations of Archimedes in the Works of Johannes de Muris," *Isis*, Vol. *43* (1952), 236–42.

bus.[11] For example, Gerard used the first proposition of that work, a proposition which equates the lateral surface of a cone to a right triangle, one of whose sides including the right angle is equal to the circumference of the base of the cone and the other of which is equal to the slant height of the cone.

While there is no doubt, then, that Archimedes' general geometric procedures influenced Gerard, less evident is any specific use of Archimedes' strictly kinematic ideas as found in his *Spiral Lines*. In fact, it is clear, that no translation of the *Spiral Lines* had been made at Gerard's time. But since an abbreviated version of this latter treatise seems to have circulated in the fourteenth century and thus to have been available to at least the French kinematicists of that century,[12] we ought to examine the kinematics of Archimedes in some detail. The version of the *Spiral Lines* available in the fourteenth century was produced from William Moerbeke's translation of 1269, and my citations are to that translation. The initial kinematic considerations of the *Spiral Lines* are evident in the introduction, where Archimedes defines in general terms one of the spirals that is going to be the special object of his work:[13] "If a straight line one of whose extremities is fixed turns with uniform speed in a plane, reassuming the position from which it started, and at the same time a point of that rotating line is moved uniformly fast on that line, starting from its fixed extremity, the point will describe a spiral in the plane." Now Gerard, of course, makes no use of this definition, since it was not directly available to him. He is concerned only with a rotating radius in his first proposition, and he does not consider curves that arise from two different but simultaneous movements. Later authors of the fourteenth century, following William Heytesbury, became interested in the spiral type of line, although only one of them, Nicole Oresme, in his highly original *De configurationibus qualitatum*, writes as if

---

[11] Clagett, "The *De curvis superficiebus Archimenidis*: A Medieval Commentary of Johannes de Tinemue on Book I of the *De sphaera et cylindro* of Archimedes," *Osiris*, Vol. *11* (1954), 295–358. For the citation of this treatise as *De pyramidibus* by Gerard, see p. 290, note 4 in the above article.

[12] A hybrid work entitled *De quadratura circuli*, consisting of thirteen of the first eighteen propositions of the *Spiral Lines* together with the first proposition of the *Dimensio circuli*, was put together, probably at Paris, in 1340. I have discussed this hybrid treatise in my article, "The Use of the Moerbeke Translations," already cited in note 10.

[13] MSS Vat. Ottob. lat. 1850, 13c f. 11r, c. 2; Vat. Reg. lat. 1253, 14c, f. 15r: "Si recta linea in plano manente altero termino equevelociter circumdelata restituatur iterum unde incipit, simul autem linee circumdelate feratur aliquod signum equevelociter ipsum sibi ipsi per recta[m] incipiens a manente termino, signum helicem describet in plano."

he might have been familiar with the treatise of Archimedes, at least in its abbreviated version.[14]

Of more importance to us in our study of early kinematics are the first two theorems of this work on spirals of Archimedes:[15]

---

[14] Nicole Oresme, *De configurationibus qualitatum*, Bk. I, chap. 21 (MS London, Brit. Mus. Sloane 2156, f. 166v). Cf. *Isis*, Vol. *43* (1952), 239, note 7.

[15] MSS Vat. Ottob. lat. 1850, 13c, f. 12r, c. 1; Vat. Reg. lat. 1253, 14c, ff. 15v–16r. "1. Si per aliquam lineam feratur aliquod signum equevelociter ipsum sibi ipsi motum, et accipiantur in ipsa due linee accepte eandem habebunt proportionem adinvicem quam quidem tempora in quibus signum lineas perambulavit. Feratur enim aliquod signum penes lineam AB equevelociter et accipiantur in ipsa due linee CD, DE. Sit autem tempus in quo lineam CD signum perambulavit ZH, in quo autem lineam DE, HT. Ostendendum quod eandem habent proportionem CD linea ad lineam DE quam tempus ZH ad tempus HT. Componantur enim ex lineis CD, DE linee AD, DB, et secundum quamcunque compositionem ut excedat linea AD lineam DB, et quotiens quidem componitur linea CD in linea AD totiens componatur tempus ZH in tempore LH. Quotiens autem componitur DE linea in DB totiens componatur HT tempus in tempore KH. Quoniam igitur supponitur signum equevelociter delatum esse per lineam AB, palam quod in quanto tempore per lineam CD delatum est in tanto [per] unamquamque [delatum est] equalem ei que est CD. Manifestum igitur quod et per compositam lineam AD in tanto tempore delatum est quantum est LH tempus, quoniam totiens componitur CD linea in AD linea et ZH tempus in tempore LH; propter hoc itaque et per lineam BD in tanto tempore signum motum est quantum est KH tempus. Quoniam igitur maior est AD linea quam BD, palam quod in pluri tempore signum per lineam AD motum est quam per BD. Quare tempus LH maius est tempore KH. Similiter autem ostendetur et si ex temporibus ZH, HT componantur tempora secundum quamcunque compositionem ut excedat alterum alterum quia et compositam ex lineis CD, DE secundum eandem compositionem excedet proportionalis excedenti tempori; palam igitur quod eandem habebit rationem CD ad lineam DE quam tempus ZH ad tempus HT.

"2. Si duobus signis utroque per aliquam lineam moto non eandem equevelociter ipsum sibi delato accipiantur in utraque linearum due linee, primeque in equalibus temporibus a signis permeentur et secunde, eandem habebunt proportionem adinvicem accepte linee.

"Sit per lineam AB motum aliquod signum equevelociter ipsum sibi ipsi et aliud per lineam KL. Accipiantur autem in linea AB due linee CD, DE, et in linea KL, ZH, HT. In equali autem tempore id quid per AB lineam delatum signum per lineam CD moveatur in quanto alterum per KL delatum per eam que ZH. Similiter autem et per lineam DE in equali moveatur signum in quanto alterum per HT. Ostendendum quod eandem habet proportionem CD ad DE quam ZH ad HT. Sit itaque tempus in quo per lineam CD movetur signum MN. In hoc itaque tempore et alterum signum movetur per ZH. Rursum itaque et in quo per lineam DE movetur signum, sit tempus NX. In hoc itaque et alterum signum movetur per HT; eandem itaque proportionem habebunt CD ad DE lineam quam tempus MN ad NX, et ZH et HT quam tempus MN ad NX. Palam igitur quod eandem habeat proportionem CD ad DE quam ZH ad HT." I have substituted capital letters for the lower-case letters of the figures. For bracketed words, see margin of MS Vat. Ottob. lat. 1850, f. 12r, c. 1. Cf. the Greek text in J. L. Heiberg, edition, *Archimedis Opera omnia*, Vol. *2* (Leipsig, 1913), 12–16 and the brief Eng-

1. If some point is displaced with a uniform velocity along a certain line, and if upon this latter line we take two lines, they will have the same ratio between them as the times during which the point has traversed these lines.

For let us transport some point with uniform velocity along line AB (see Fig. 3.2), upon which we take two lines CD and DE. Moreover let ZH be the

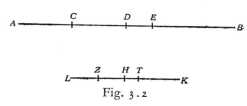

Fig. 3.2

time during which the point has traversed the line CD and let HT be the time in which the point has traversed line DE. It is necessary to demonstrate that the proportion of line CD to line DE is the same as that of time ZH to time HT. Let us make up the lines AD and DB out of lines CD and DE in any way such that AD exceeds the line DB and such that time ZH is contained in time LH as many times as line CD is contained in line AD, and furthermore such that time HT is contained in time KH as many times as DE is contained in line DB. Then since it is supposed that the point is carried in a uniform velocity along line AB, it is evident that it traverses each of the lines equal to CD in a time equal to that in which it traverses CD. It is then clear that the point traverses the composite line AD in a time equal to LH, because ZH is contained in time LH as many times as line CD is contained in line AD. For the same reason, therefore, the point traverses the line BD in the time KH. Hence, since the line AD is greater than line BD, it is clear that the point will traverse the line AD in a greater time than that in which it traverses the line BD. Consequently, the time LH is greater than the time KH. It is similarly demonstrated that even if one composes times out of ZH and HT in any such fashion that one exceeds the other, then among the lines composed in the same way out of CD and DE, one will similarly exceed the other, and it is that one which will be homologous to the greater time. It is clear, therefore, that CD will have the same ratio to line DE as the time ZH has to the time HT.

2. If two points are displaced with uniform velocity, each along a different line, and if on each of these lines one takes two lines such that the first line segment on the one line is traversed by its point in the same time as the first segment on the second line is traversed by the other point, and similarly the second segments on the lines are traversed in equal times, then the line segments on each line respectively are in the same proportion.

Let there be a point moved uniformly along line AB (see Fig. 3.3) and another point along line KL. Further, let us take two lines CD and DE on the line AB

---

lish paraphrase of T. L. Heath, *The Works of Archimedes* (Cambridge, 1897), p. 155.

and the lines ZH, HT on the line KL. Let the point carried along line AB traverse the line CD in the same time as the other point carried along line KL traverses the line ZH. Similarly let the one point traverse line DE in the same time as the other point traverses the line HT. It is necessary to demonstrate that the ratio of CD to DE is the same as that of ZH to HT. Therefore, let MN be the time in which

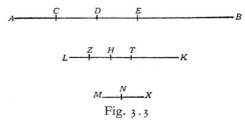

Fig. 3.3

the one point has traversed the line CD. And so in this time the other point traverses the line ZH. Further, let NX be the time in which the first point traverses the line DE; the other point traverses the line HT in the same time. And so CD will have the same proportion to DE as time MN to time NX [from the first proposition], and the ratio of ZH to HT will be as time MN to time NX. Therefore it is clear that the ratio of CD to DE is the same as that of ZH to HT.

Now Archimedes' first theorem here is precisely what Autolycus had premised in the second sentence of his introductory paragraph in the *De spera mota*, namely, that if a point is moving with uniform velocity, any two distances described by it are in the same ratio as the times. Since he gives it as a theorem, Archimedes feels it necessary to add a proof, and this proof rests on the assumption which Autolycus had made in the first sentence of his opening paragraph to the effect that uniform velocity means that equal spaces are traversed in equal times. The substance of this theorem, then, is that for uniform velocity $S_1/S_2 = T_1/T_2$. The proof is based on the sixth proposition of the fifth book of Euclid's *Elements*.

The second theorem poses two points, each moving with a uniform velocity on different lines. The point on the first line describes distances S_1, s_1, which according to the first proposition are related to the times of description T_1, t_1 as follows: $S_1/s_1 = T_1/t_1$. The other point moving on the second line similarly describes distances S_2, s_2 in times T_2, t_2 so that $S_2/s_2 = T_2/t_2$. Then if $T_1 = T_2$ and $t_1 = t_2$, $S_1/s_1 = S_2/s_2$, as the theorem states.

With this second proposition, we have a case of the comparison of the kinematic factors of two uniform velocities, i.e., a comparison of the ratio of two distances covered by one point moving with a uniform velocity to the ratio of two distances covered by the second point moving with uniform velocity and traversed in the same times as the distances traversed by the first point. But even though this is a comparison of the effects of uniform velocity, it is not the same thing that Gerard is doing in his eighth supposition and elsewhere throughout the treatise, because all that Archimedes is doing is throwing more light on what constitutes uniform velocity, while Gerard is giving us a general formula whereby we know that, if the times are the same, the *velocities* can be compared by comparing the distances traversed. It is true that it is quite simple to go from the basic understanding of uniform velocity as given by Autolycus or Archimedes to the comparison of velocities given by Gerard. But it seems to me as if Gerard gets his basic suppositions rather from the third general current of Greek scientific activity, namely, that of the physical-mechanical treatises, and in particular from the *Physics* of Aristotle, and perhaps from the *Mechanica* attributed to Aristotle. Let us look then at this third current of scientific activity which helped nourish kinematics.

While not strictly speaking a mechanical treatise in the same sense as the *Mechanica* or Hero's *Mechanics*, the *Physics* nevertheless belongs at the head of this tradition because of its obvious influence on the *Mechanica*. Needless to say, the *Physics* of Aristotle was the most influential single physical work during the later Middle Ages. There appear to have been at least five efforts to translate or revise old translations of the *Physics*:[16] (1) the so-called *Physica Vaticana*, an early version from the Greek dating from the middle of the twelfth century, (2) a second version from the Greek also made in the twelfth century and perhaps done by James of Venice; this version is known as the *physica veteris translationis*, (3) a version from the Arabic by Gerard of Cremona, before 1187, (4) another version from the Arabic accompanying the commentary of Averroës, dating from the early thirteenth century, and perhaps made by Michael Scot, and (5) a revision (without much change) of the *vetus translatio*, possibly done by William Moerbeke between 1260 and 1270, and known as the *nova translatio*. Presumably any one or more of the first four versions could have been known to Gerard of Brussels. But whether or not Gerard was acquaint-

[16] L. Minio-Paluello, "Iacobus Veneticus Grecus: Canonist and Translator of Aristotle," *Traditio*, Vol. 7 (1952), 265, 284–87. G. Lacombe, *Aristoteles latinus* (Rome, 1939), pp. 51–52. M. Grabmann, *Guglielmo di Moerbeke O.P., Il traduttore delle opere di Aristotele* (Rome, 1946), pp. 90–91. E. Franceschini, "Ricerche e studi su Aristotele nel medioevo latino," *Aristotele nella critica e negli studio contemporanei* (*Revista di Filosofia Neoscolastica*, supplemento speciale al volume XLVIII, 1956), p. 150.

ed with the *Physics* of Aristotle, we must examine the principal kinematic passages from that work because of their obvious influence on fourteenth-century authors whose works we shall have occasion to examine in the succeeding chapters. There are two principal passages in the *Physics* bearing on the formation of kinematic rules. The first occurs in VI.2, where these rules are used in connection with Aristotle's arguments regarding the nature of *continua*. The second is in VII.4, where Aristotle discusses the comparison of movements. Let us examine the first of these passages using a medieval Latin text as our point of departure (the Latin text is that of the translation accompanying the commentary of Averroës, i.e., the fourth one listed above):[17]

[17] Edition with Averroës, Venice, 1495, ff. 91v–92r. Cf. MSS Paris, BN lat. 6505, 14c, ff. 164v–165v; BN lat. 16141, ff. 130r–v. "11. Et cum omnis magnitudo est divisibilis in magnitudines, nam declaratum est quod impossibile est ut magnitudo componatur ex indivisibilibus, et omnis magnitudo est continua, necesse est ut velocius in equali tempore cum motu tardioris moveatur per maius et in minori tempore [per maius, et in minori tempore] per equale, ut quidam determinant velocius.

"Sit igitur A velocius quam B; quoniam velocius est illud quod citius mutatur et tempus in quo mutatur A de C in D, et sit ZH, non potest in eo B pervenire ad D. Ergo in equali tempore velocius pertransit per maius spacium.

"12. Et etiam in minori tempore pertransit maius.

"Ponatur igitur quod in tempore in quo A pervenit ad D, pervenit B, quod est tardius, ad K. Et quia A pervenit ad D in toto tempore ZH, perventus ergo eius ad T erit in minori hoc tempore. Sit igitur in tempore ZL. B igitur pertransit CK in tempore ZH et A pertransit CT in ZL. Ergo illud quod pertransit A est maius quam CK et tempus ZL est minus quam totum ZLH. Ergo in minori tempore pertransit maius.

"13. Et apparet etiam ex hoc quod velocius in minori tempore pertransit spatium equale.

"Quoniam cum pertransit spatium longius in minori tempore quam sit tempus in quo pertransit illud tardius, cum accipitur per se, tunc pertransibit spatium longius in maiori tempore quam tempus in quo pertransivit brevius spatium. Verbi gratia, quod spatium LM est longius spatio LN et tempus ZH in quo pertransit [tardius] spatium LN est maius tempore ZQ in quo [velocius] pertransit spatium LM. Ex quo sequitur quod si tempus ZQ fuerit minus tempore ZH in quo tardius pertransit spatium LN ut ZF sit minus tempore ZH, quoniam est minus tempore ZQ, et quod est minus minore est etiam minus. Ergo necesse est ut moveatur in tempore minori equaliter." I have altered the punctuation somewhat. The passage numbers are those employed by Averroës and after him by the medieval Latin schoolmen.

Among the manuscripts of the *Physics* which I have used, the most interesting is Paris, BN lat. 16141, which contains in three parallel columns: the *physica veteris translationis*, the translation from the Arabic probably of Michael Scot, and the translation of Gerard of Cremona. So far as the passages under question are concerned, I find almost no difference between the *physica veteris translationis* represented by MSS BN lat. 16141, 16142, and 12592 and the so-called new translation appearing in MSS BN lat. 6320 and 16672. It will be noticed that I have rendered the Latin text accompanying the Averroës commentary. I am, of course, not sure which of the translations Gerard might have used;

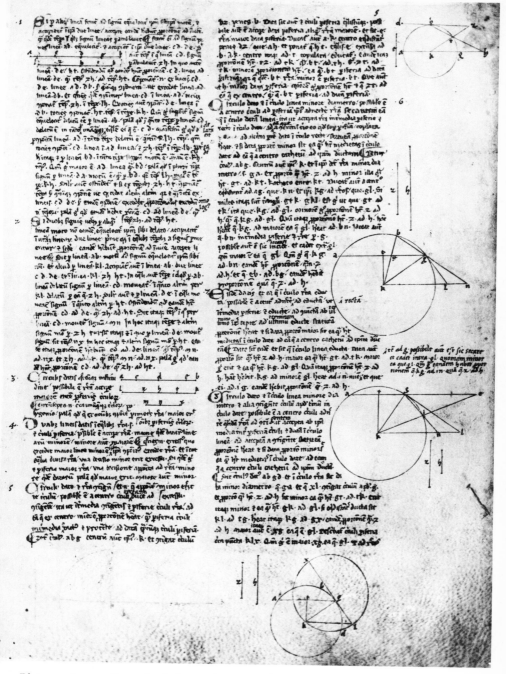

Plate 2: A page from the holograph copy of William Moerbeke's translation of the works of Archimedes. MS Vat. Ottob. lat. 1850, f. 12r.

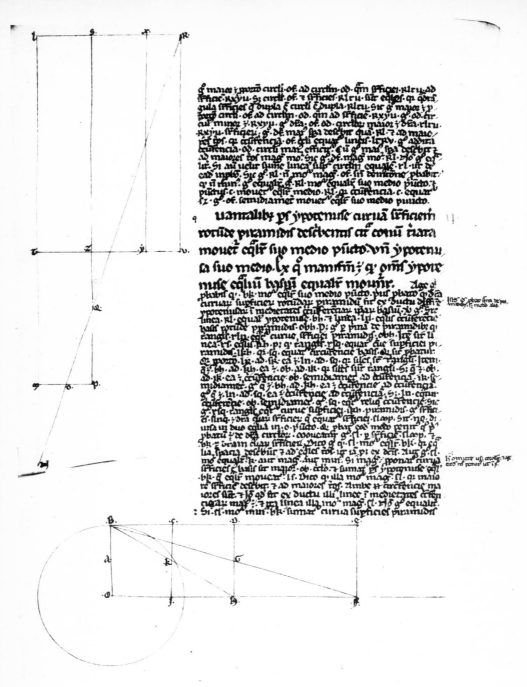

Plate 3: A page from a thirteenth-century manuscript of the *Liber de motu* of Gerard of Brussels. MS Bodleian Library, Auct. F. 5.28, f. 117v.

11. And since every magnitude is divisible into magnitudes, for it was demonstrated that it is impossible for a magnitude to be composed of indivisibles, and every magnitude is continuous, it is necessary that the "quicker" [of two bodies] is moved through more in the same time, and it is moved through more in less time, and through an equal amount in less time, as some define the "quicker."

For let A be quicker than B. Since the quicker is that which is changed more swiftly and the time in which A is changed from C to D is ZH, in that time B cannot arrive at D. Therefore, the "quicker" traverses more space in the same time.

12. And it [the quicker] also traverses more space in less time.

For let it be posited that in the time in which A (see Fig. 3.4) arrives at D,

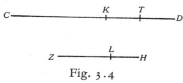

Fig. 3.4

B, which is slower, arrives at K. And because A arrives at D in the whole time ZH, hence its arrival at T will be in less than this time. Therefore, let it be in time ZL. Hence B traverses CK in time ZH and A traverses CT in ZL. Therefore, that which A traverses is more than CK and time ZL is less than the whole time ZLH. Therefore, it traverses more in less time.

13. And it is also apparent from this that the "quicker" traverses an equal space in less time. For since it traverses a longer space in less time than that of the "slower," and since it [the "quicker"] is accepted *per se*, then it will traverse a longer space in a greater time than that in which it itself traversed a shorter space. For example, the space LM is longer than the space LN and the time ZH in which the "slower" traverses the space LN (see Fig. 3.5) is more than the time

---

however, the text I have given here was quite popular among the fourteenth-century schoolmen whom we treat later. I have rendered this medieval text rather than the Greek one, because of the interesting divergencies from the Greek text, W. D. Ross, *Aristotle's Physics* (Oxford, 1936), 232a.23—232b.15 (cf. pages 404, 641). The actual Latin text reproduced here in the note is basically that of the Venice, 1495 edition, but the manuscripts are quite close in readings to the edition. However, the bracketed phrase in 11. "[per maius et in minori tempore]" is in the manuscripts but is omitted in the edition. It is interesting to note that this omitted material is a statement of the third case of the "quicker," as given in the Greek text. The early medieval translation from the Greek correctly renders the Greek text (see BN lat. 16141, f. 130r, c. 1): "necesse est velocius in equali tempore maius et in minore (!) equale, et [in] minori plus moveri." The translation of Gerard of Cremona is similar to that from the Greek rather than to the translation accompanying Averroës' commentary (see BN lat. 16141, f. 130r, c. 3). For an English translation, see P. H. Wicksteed, and F. M. Cornford, *Aristotle, The Physics*, Vol 2 (Cambridge, Mass., London, 1934), 102–5.

ZQ in which the "quicker" traverses space LM. From this it follows that if time ZQ is less than the time ZH in which the "slower" traverses space LN, ZF the time in which the "quicker" traverses LN is less than the time ZH, since it is less than time ZQ and that which is less than something less is also less. Hence it is necessary that it [the "quicker"] is moved equally in less time.

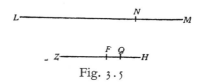

Fig. 3.5

The substance of this chapter of Aristotle's *Physics* can be summarized in three statements regarding the "quicker." (1) The "quicker" traverses more space than the "slower" in the same time. (2) It can traverse more space in less time. (3) It can traverse the same space in less time. The modern reader cannot but feel the lack of mathematical economy in these expressions, which can all be contained in the single formula, $V = S/T$, where V is velocity, S is distance, and T is the time. Thus if we start from this formula, these three statement of Aristotle can be easily illustrated. (1) The first statement says that if $V_1 > V_2$ and $T_1 = T_2$, then $S_1 > S_2$. For values V_1, V_2, S_1, S_2, T_1, T_2, the general formula $V = S/T$ is rewritten $V_1/V_2 = (S_1 T_2)/(S_2 T_1)$ and for $T_1 = T_2$, $V_1/V_2 = S_1/S_2$. Thus it is clear that if $V_1 > V_2$, then $S_1 > S_2$. (2) The second statement indicates that if $V_1 > V_2$, it is possible for both $T_1 < T_2$ and $S_1 > S_2$. Assuming $V_1/V_2 = (S_1 T_2)/(S_2 T_1)$ as before, then we let $V_1/V_2 = a > 1$ and $T_2/T_1 = b > 1$. Thus we get $S_1/S_2 = a/b$ and so $S_1 > S_2$ when $T_1 < T_2$ so long as $a > b$. (3) Finally Aristotle says that if $V_1 > V_2$, it is possible for $S_1 = S_2$ and $T_1 < T_2$. From (2) we take the expression $S_1/S_2 = a/b$. Hence $S_1 = S_2$ when $T_1 < T_2$ so long as $a = b$. It ought to be observed, however, that this use of V_1 and V_2 which I have made here is most anachronistic because it assumes that Aristotle has some magnitude exclusively representing velocity, whereas we have already said the Greek authors do not give a numerical value to motion different from space on the one hand or time on the other.

Now we ought to observe that these statements of Aristotle are the oldest *extant* kinematic statements from the Greek antiquity, for the earlier, fifth-century kinematics we know of only at second or third hand. It should also be pointed out that while there is evidence of the geometrical approach in these passages, they are much less precise than those of later mathematical writers like Archimedes. Loose and uneconomical as the statements are, they are historically important for a number of reasons: (1) they constitute the first *extant* statement of the comparison of velocities; (2) they influenced later Greek treatises like the *Mechanica* attributed to Aristotle; and (3) they were the point of departure for kinematic discussions in the Middle Ages.

I feel that Gerard probably knew of them or of the *Mechanica*, which they influenced. For it is evident that the first statement of Aristotle—to the effect that the "quicker" traverses more space than the "slower" in the same time—is contained within the general statement of Gerard's eighth supposition, which we can write in modern form $V_1/V_2 = S_1/S_2$ when $T_1 = T_2$. But Gerard's statement not only contains the definition of $V_1 > V_2$, but also of $V_1 < V_2$ and of $V_1 = V_2$. From Aristotle's definition of $V_1 > V_2$ once we have decided V_1 and V_2 are magnitudes of some kind, we easily arrive at the case of $V_1 < V_2$, as his passages imply and as Averroës in his commentary on chapter 2 (comment 11 to the passage cited in note 17 indicates when he says: (1) "velocius enim dicitur in respectu tardius," i.e., "something is said to be quicker in respect to something slower"; and (2) "diffinitio tardioris... est quod movetur in tempore in quo movetur velocius per spatium minus, et... diffinitio velocius... est quod movetur in tempore equali per maiorem magnitudinem," i.e., "the slower is defined as that which is moved through less space in the same time as the quicker moves, and the quicker is defined as that which is moved through a greater magnitude in the same time [as the slower]."

Now as for the third case of $V_1 = V_2$, this is evident in all that Aristotle has to say about the comparison of movements, particularly in VII.4, parts of which are of interest to our historical introduction:[18]

[18] Edition of Venice, 1495, ff. 117v, 119r–v. MSS Paris, BN lat. 6505, ff. 208r, 208v, 211r; Paris, BN lat. 16141, ff. 167r–v. "21. Et queritur utrum motus coniungetur cum motu vel non. Quoniam si omnes motus coniunguntur sibi et convenientia in velocitate sunt illa que in tempore equali moventur per equale, tunc aliquis motus circularis erit equalis recto. Ergo aliquis motus circularis erit maior aut minor illo et alteratio et translatio erunt equales, cum in tempore equali hoc alteratur et hoc mutat locum. Ergo passio erit equalis longitudini. Sed hoc est impossibile. Quoniam tunc velocitas est equalis quando duo pertranseunt in equali tempore aliquod equale. Sed passio et longitudo non sunt equales; ergo neque alteratio et translatio sunt equales. Neque una minor est alia. Unde necesse est ut non omnis motus coniungatur cum omni motu....

"30. Si igitur aliquis dixerit quod quando in ista longitudine demonstrata hoc alteratur, hoc vero transfertur, ergo ipsa alteratio convenit in velocitate cum translatione, dicemus quod est improbabile. Et causa in hoc est quia motus habet species. Ergo necesse est si illa que mutant in tempore equali longitudines equales sint equalis velocitatis, ut motus rectus sit

21. And it is sought whether or not [every] movement is compared with [every] movement. For if all movements are mutually comparable and those equal in velocity are the ones which in equal time are moved through an equal amount, then some circular movement will be equal to a rectilinear movement. Hence some circular movement would be more or less than that motion, and also an alteration and a motion of transport would be equal, since in the same this is altered and that changes place. Hence a passion [qualitative reception] would be equal to a length. But this is impossible. For velocity is equal when two things traverse something equal in the same time. But a passion and a length are not equals. Hence alteration and transport are not equals; nor is one more than another. Hence it is necessary that not every movement is comparable with every movement....

30. If someone says this is altered in the same magnitude when that is transported, and so this alteration is equal in velocity with that transportation, we answer that this is impossible because motion has species. So if things which traverse equal length in equal times are of equal velocity, then a rectilinear motion must necessarily equal a circular one. Is this because the motion of transport is a genus or because the line is a genus? For time is indivisible in species. Hence we answer in this matter that the two take place at the same time. But there is a diversity of species. For the movement of transport has species because that through which it moves has species. But perhaps it will be said that the diversity comes from that which produces the movement, e.g., a thing with feet has a motion of ambulation, a thing with wings one of flying. But this is not so because movement of transport is not diverse in species from any instrumental diversity. But the motion of transport has diversity here because of diversity according to figure. And so it is necessary that those which are moved through the *same space* in equal time are of equal velocity. And I define the "same" as that in which there is no diversity in species, and [so] not in motion.

---

equalis motui circulari. Utrum igitur est causa quod translatio est genus aut quod linea est genus. Temporis enim semper due (est?) species indivisibilis. Dicamus igitur in hoc quod hec duo sunt insimul, sed diversitas in specie. Translatio enim habet species, quia illud per quod fit iste motus habet species. Et forte erit hoc cum fuerint eius per quod fit motus, verbi gratia, quoniam quod habet pedes, motus eius est ambulatio; et quod habet alas, motus eius est volatio. Sed non est ita, scilicet quoniam motus translationis non diversatur in specie propter diversitatem instrumenti. Sed translatio erit in hoc diversa secundum figuras. Ergo necesse est ut illa que in tempore equali moventur per idem spatium sint equalis velocitatis et dico idem illud in quo non est diversitas in specie neque in motu." Cf. the Greek text in the Ross edition cited in note 17, 248a.10–18, 249a.8–20; and see pages 425–27, 680–81. The English translation of Wicksteed and Cornford, is in *op. cit.*, Vol. 2, 240–43, 248–51. The argument and notes with this translation really do not represent what Aristotle is getting at. For they suggest that Aristotle is comparing *angular* and *rectilinear* velocities, while in actuality he is comparing *curvilinear* and *rectilinear* velocities.

Aristotle then is concerned in this chapter with the problem of the equality of movements and how movements can be compared. It is not so much a mathematical discussion as a philosophical one. He points out initially that if one accepts the definition of equal velocity as being the moving of an equal amount in a equal time, then it would appear that an alteration, i.e., a qualitative mutation, could be equal to a local movement, i.e., a movement involving change of place. But this turns out not to be so, for the qualitative amount through which the alteration takes place cannot be equated to a space length. Hence alteration and transport cannot be equal. But then he takes up a more difficult case, that of the possible equality of velocity between something traversing a circular path and something describing a rectilinear path, i.e., the problem of comparing curvilinear and rectilinear velocities. One could certainly seem to compare the lengths traversed along these paths in a given time and thus establish, according to Aristotle's kinematic rules, whether one is traversing a greater length in the same time, or the same length in less time, or an equal length in the same time as the other. But, answers Aristotle, the curvilinear and rectilinear movements are of different species because their paths are different in species. And their paths are specifically different because they are different in figure (and perhaps because of the difficulty or impossibility of finding any straight line equal to a given curvilinear line). Hence the conclusion of these passages is that things which are moved through the *same space* in the same time are said be equal in velocity; but the expression "same space" means that the spaces are not different in species, i.e., figure.

As Averroës points out in his commentary to this passage (comment 30 to the passage cited in note 18, f. 119v, c. 1), the circle and the straight line are the same in genus and are different in species; hence the curvilinear movement is of the same genus as the rectilinear movement but of different species. It is undoubtedly the similarity in genus that makes it seem possible to compare the velocity of the one to that of the other. But, as we have said, Aristotle would prefer to limit comparisons of velocities to cases where there is a specific identity of figure in the spaces traversed. This distinction, one might say, between the genus of movement and species of movement as applied to the case of comparing curvilinear and rectilinear velocities reminds us a little of the modern distinction between the scalar quantity of speed and the vector quantity of velocity. We could say, of course, that the curvilinear and rectilinear movements would have the same speed if the scalar values of the lengths traversed, irrespective of direction, were equal in the same time. But these movements could never be equal in velocity vectorwise because of the fact they are so differently directed.

These last comments of Aristotle I believe to be of importance for Gerard. Not only do they give authority to the establishment of his all-important supposition 8, as we have already pointed out, but they discuss the question that lies at the heart of most of Gerard's propositions, namely, the comparison of rectilinear and curvilinear velocities. Gerard does not concern himself with whether these movements are different in species, but rather he accepts readily the comparison of these movements which Aristotle outlined only to reject. For example, in the first proposition Gerard specifically declares in the course of his proof that a point moving through a given circumference in a given time is equally moved as another point moving in the same time through a straight line equal to the circumference; and this is, of course, exactly the example given by Aristotle in paragraph 30.

The influence of Aristotle's *Physics* and his Lyceum generally on the history of mechanics is undeniable. Among the important results of the work he initiated was the mechanical investigation of Strato, who became director of the Lyceum in 287 B.C. We know that Strato composed a work *On Motion*, now lost, that may very well have had important kinematical passages, for Simplicius quotes directly from this work an extremely interesting passage discussing the acceleration of falling bodies which seems to suggest a kinematic analysis of accelerated movement (see Chapter 5 below for a discussion of the importance of this passage).

It may be that Strato was also the author of the *Mechanica* attributed to Aristotle, the statical and dynamical ideas of which we have already discussed in Chapter 1. It is also of some importance for our survey of Greek kinematic ideas. Chapter 1 of the *Mechanica*, we recall, is a strange eulogy to the circle and circular movement.

After his praise of the circle, the Peripatetic mechanician takes up the case of the rotating radius, of which he notes that none of the points is moved equally fast as any other, "but of two points that which is further from the fixed center is the quicker."[19] Here, then, in the second part of this statement we have the first supposition of Gerard stated very much as Gerard has given it (see Doc. 3.1, supposition 1). The only difference is that Gerard does not specify that it is a point which is farther removed from the center and thus moves more quickly.

[19] See Chapter 1, p. 5 for the full quotation.

In the second chapter the author of the *Mechanica* returns to this question of circular movement. The basic problem he is pursuing, we remember, is to find out why larger balances are more accurate than smaller ones. In order to solve this, he suggests, we have to solve the more basic question of why a line in a circle more distant from the center is moved more swiftly than a line closer to the center, although it is moved by the same force. This question leads him to define the "quicker." "Something is said to be the 'quicker' in two ways, that which traverses an equal space in less time or that which in an equal time traverses more space. But the longer line describes a larger circle, for that which is exterior is greater than that which is interior."[20] The terminology in Greek is as ambiguous as my English. It seems therefore a moot question as to whether the "lines in the circles" of which he is speaking are radial lines or lines in the circumferences. The whole context of the first two chapters where he is continually talking about circles in movement suggests the latter interpretation. Similarly Gerard, in his first supposition, speaks quite generally: "those which (*que*) are farther from the center or immobile axis are moved more quickly."

The only other strictly mechanical treatise extant from antiquity is the *Mechanics* of Hero. The pertinent sections on kinematics have been included in Document 1.3, appended to the first chapter. Passages I.2–6 (I.5 is omitted in our translation) are extensions of the directional analysis of circular movement found in the *Mechanica* of Pseudo-Aristotle. They are elementary but largely kinematic. Perhaps the most important observation in these chapters is the recognition in chapter 3 of the completely relative nature of the directions of movements. These chapters throw some light in retrospect on the terminology employed by the *Mechanica*, particularly on the expression "circles in movement." Here in chapters 2–5, circles in movement are analyzed on the basis of curvilinear velocity (as they are with Gerard also). And hence it seems likely that when the author of *Mechanica* (on which Hero bases these chapters) spoke of the movements of lines in circles in movement, he was referring to the movement of the circumferences themselves.

On the other hand, it is true that in chapter 6 Hero is clearly talking about two different kinds of velocity. When he speaks of the movement of larger circles being greater than the movement of smaller ones on the same axis, he is talking about the greater curvilinear velocity on the circumferences in movement farther from the center. This is, of course, the twice-

---

[20] Aristotle (Pseudo-), *Mechanica*, 848b, 1–9 (Apelt edition [Leipzig, 1888], p. 98).

repeated axiom of the *Mechanica* and the first supposition of Gerard. However, Hero's example of the wagon with different-sized wheels, of which he says the smaller wheels revolve more rapidly than the larger ones, can only be explained in terms of angular rather than curvilinear velocity. *Gerard does not speak anywhere of angular velocity.*

Hero's passage I.7 is the so-called paradox of the wheel of Aristotle, contained also in the *Mechanica*, chapter 24. I included it in my translation because of its statement of the basic theorem defining equality of velocity, which Aristotle had already noted, and which, of course, lies at the heart of Gerard's treatise. Hero succinctly tells us that "things which traverse equal distances in equal times have movements of equal swiftness." Passage I.8 is the parallelogram of velocities. It is of interest to us because it gives us an example of the straight line moving in uniform translation. Gerard, it will be noted, in his second supposition tells us: "When a line is moved equally, uniformly, and equidistantly (i.e. uniformly parallel to itself—*equidistanter*), it is moved equally in all of its parts and points." Line *AB* in Hero's passage I.8 was just such a line, as was the line in movement in Hippias' description of the generation of a *quadratrix*. The reader will observe that in the proof of proposition 1 Gerard employs the straight line moving uniformly to sweep out a rectangle equal to the circle swept out by a rotating radius equal in length to this line.

This brings us to the end of our discussion of the principal Greek kinematic passages in the three currents of scientific activity which we proposed at the beginning of the chapter. It would appear, in summary, that so far as his knowledge of the geometry of surfaces and volumes is concerned, as well as his basic geometrical method of procedure, Gerard drew most heavily from the second current of activity, namely, the development of a geometry of movement, and within that current from the *Dimensio circuli* and *De sphaero et cylindro* of Archimedes (or rather from the *Commentum* of Johannes de Tinemue based on the *De sphaero et cylindro*). On the other hand, Gerard's interest in comparing velocities and his preoccupation with curvilinear velocities and their equation with rectilinear velocities was probably more directly the result of the third current, that of Aristotle and the mechanical tradition.

Not long after Hellenic learning began passing into Latin in the twelfth and early thirteenth century, Gerard of Brussels composed his tract consisting of thirteen kinematic propositions and ordinarily circulating with the title of *Liber de motu*. I have published elsewhere the text of the *Liber de motu,* discussed its dating, and analyzed its contents in detail.[21] My conclusions there regarding the general dating and milieu of Gerard's treatise I repeat here now:[22] (1) The *De motu* of Gerard of Brussels was composed between the late twelfth and mid-thirteenth century, i.e., between 1187 and 1260. (2) It is a work that continually appears with the mechanical and mathematical works of Jordanus, and thus may be related in some close fashion to the latter's activity. (3) It is a work which definitely cites and uses Euclid and Archimedes and thus is obviously in the geometrical tradition revived in the last half of the twelfth and the first half of the thirteenth centuries. (4) Finally, it is at least possible that our "Gerardus" is identical with the unknown "Gernardus" who was the author of the mathematical work entitled *Algorithmus demonstratus*.

The *Liber de motu* is divided into three short "books." The first attempts to find uniform punctual rectilinear velocities which will make uniform the varying curvilinear velocities possessed by different lines (and perimeters of regular polygons) as they rotate or revolve. The second discovers similar uniform movements for surfaces in rotation, and the last seeks uniform movements for solids in rotation. Gerard's work, then, is probably the first Latin treatise to take the fundamental approach to kinematics that was to characterize modern kinematics, namely, to see in kinematics the basic objective of reducing variations in velocity to uniform velocity.

I have included as Document 3.1 parts of two propositions of the *De motu,* namely, proposition I.1 and proposition II.1, as well as the suppositions that precede these propositions. In the commentary to Document 3.1 I have analyzed the proofs of these propositions in detail. Hence at this point I would like merely to characterize the propositions in a general fashion to illustrate Gerard's basic objectives in the treatise.

In proposition I.1 Gerard says that a segment of a rotating radius or the radius is moved "equally" as its middle point. By this he means that if we take the segment or the radius itself and allow it to move, not in a movement of rotation, but rather in one of translation always parallel to itself so that all of its points are moving rectilinearly with the velocity which the middle point had when it was rotating, then the line segment or radius so moved traverses the same area in the same time as it did in revolution. And when Gerard says that the rotating lines move "equally" as its

[21] Clagett, "The *Liber de motu* of Gerard of Brussels and the Origins of Kinematics in the West," *Osiris,* Vol. 12 (1956), 73–175.

[22] *Ibid.,* p. 107.

middle point, he means thereby that it is the velocity of the middle point which can convert or transform the varying motion of rotation into uniform motion of translation. We shall see in Chapters 4 and 5 how this theorem may have stimulated by analogy the discovery at Merton College of the fundamental theorem of uniform acceleration—a theorem which held that with respect to the traversal of space in a given time a uniform acceleration is equal to a uniform motion at a velocity equal to the velocity at the middle instant of the time of acceleration. At any rate, Thomas Bradwardine, the founder of kinematic studies at Merton College, in his *Tractatus de proportionibus velocitatum* of 1328, mentions Gerard's proposition I.1 and uses the treatise as a point of departure for his chapter on kinematics, even while he disagrees with Gerard's conclusions (see Doc. 4.1 and its commentary).

In proposition II.1 Gerard attempts much the same sort of thing that he accomplished in proposition I.1. But now he is allowing a circle to rotate about one of its diameters, and he seeks the rectilinear velocity such that if the circle moved in translation, all of its points moving with that rectilinear velocity, it would sweep out the same volume as that of the sphere it produced in rotation. His conclusion is that $V_1/V_2 = 4/3$ where V_1 is the equalizing motion of the circle, i.e., the uniform punctual velocity which converts the rotating circle to uniformity and V_2 is the equalizing velocity of the radius of diameter, i.e., the uniform velocity which converts the rotating diameter or radius to uniformity. This conclusion, it turns out, is erroneous, and should in fact be $V_1/V_2 = 4/3\pi$. But in the course of "proving" his incorrect theorem (see commentary to Doc. 3.1, prop. II.1), Gerard uses an ingenious technique of considering surfaces as composed of line elements in a manner not unlike that of Archimedes in his *Method*, a work which Gerard certainly did not have access to. This was a technique which assumed that if equality of movement or velocity was true for any pair of corresponding line elements of two rotating surfaces, then the equality also held for the whole surfaces.

Finally, we should notice once more that, unlike the Greek authors, Gerard throughout his treatise seems to be assigning some given magnitude to velocity, a magnitude distinct from, although proportional to, the distance traversed in a given time. He thereby opens the way for a similar but more varied treatment of the proportionality of velocities in the fourteenth century, and it is this fourteenth-century study of velocity that next demands our attention.

Document 3.1

Gerard of Brussels, *On Motion**

Book I

[SUPPOSITIONS:]

[1] Those which are farther from the center or immobile axis are moved more [quickly]. Those which are less far are moved less [quickly].

[2] When a line is moved "equally," "uniformly," and "equidistantly" (i.e., uniformly parallel to itself), it is moved equally in all of its parts and points.

[3] When the halves [of a line] are moved equally and uniformly to each other, the whole is moved equally [fast] as its half.

[4] Of equal straight lines moved in equal periods of time, that which traverses greater space and to more [distant] termini is moved more [quickly].

[5] [That which traverses] less [space] and to less [distant] termini is moved less [quickly].

[6] That which does not traverse more space nor to more distant termini is not moved more [quickly].

[7] That which does not traverse less space nor to less distant termini is not moved less [quickly].

[8] The proportion of the movements (i.e. speeds) of points is that of the lines described in the same time.

[Proposition] I.1: Any part as large as you wish of a radius describing a circle, which part is not terminated at the center, is moved equally as its middle point. Hence the radius is moved equally as its middle point. From this it is clear that the radii and the speeds are in the same proportion.

Proceed therefore: I say that CF (see Fig. 3.6) is moved equally as its middle point, it having been proved that the [annular] difference

---

* For the Latin text, see my study cited in note 21, p. 185. For the first part of proposition I.1 here translated I have followed Manuscript Tradition I as given in the text used in the *Osiris* article.

[188   Origins of Kinematics in the West: 3.1]

between [concentric] circles is equal to the product of the difference of the radii into half the sum of the circumferences.

For let lines OF and RL be equal, and let line LN be equal to the circumference of the circle with OF radius. It is evident by the first proposition of the *De quadratura circuli* [of Archimedes] that the circle

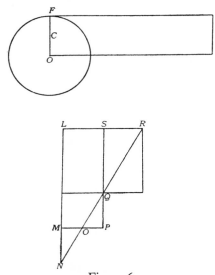

Fig. 3.6

OF and the triangle RLN are equal. Also, let lines SL and CF be equal, as well as lines ON and OQ. And let lines SL and MP be parallel. It is necessary, therefore, that triangle RSQ be equal to circle OC and line SQ be equal to the circumference of that circle. For, since triangles RLN and RSQ are similar, the proportion of LR to SR is as the proportion of LN to SQ. But the proportion of LR to SR is as that of OF to OC, and the proportion of OF to OC is as that of the circumference of OF circle to the circumference of OC circle, because the diameters and circumferences [of circles] are in the same proportion. Since, therefore, the circumference of OF circle is equal to the line LN, SQ will be equal to the circumference of circle OC. And the surface $SLNQ$, which is the difference of the triangles, will be equal to the difference of the circles OF, OC. Furthermore, the surface $SLNQ$ is equal to the quadrangular surface $SLMP$. This is proved as follows: Triangles OMN and OPQ are equal and similar because M and P are right angles, MP being parallel to line SL, and the opposite angles at Q are equal. Hence angle N is equal to angle Q. Therefore, the sides are proportional. But ON is equal to OQ, so that NQ has been divided into two equal parts at point O. Hence OM and OP are equal, as are MN and PQ. Therefore, the triangles are equal. Consequently the surfaces $SLNQ$ and $SLMP$ are equal. But line LN plus SQ is equal to LM plus SP because MN and PQ are equal. But lines LN and SQ are equal to the circumferences of circles OF and OC. Hence the lines LM and SP together are equal to these circumferences together. Therefore, the surface $SLMP$ is equal to the product of the difference of the radii by half of the sum of the circumferences, and this is equal to the difference of the circles. This same thing can be proved in another way. But the above proof suffices for the present.

Let SL, then, be moved through the surface $SLMP$ and the line CF through the [annular] difference of the circles OF and OC. I say, therefore, that lines SL and CF are equally moved, for they traverse equal spaces and to equal termini, as is now clear from the [following] statements: SL is either equally moved as CF or it is moved more or less [rapidly]. It is not moved more [rapidly] because it does not describe more space to more [distant] termini [by supposition 5]. Nor is it moved less rapidly, for it does not describe less space to less [distant] termini [by supposition 6].... Since, therefore, SL is not moved more [rapidly], and is not moved less [rapidly], it must be moved equally as CF.

But SL is moved equally as any of its points by the second supposition because it is moved equally and uniformly in all of its parts and points. Hence it is moved equally as its middle point. But the middle point of SL is moved equally as the middle point of CF because these points describe equal lines in equal times [by supposition 7]. That, moreover, these lines are equal you will prove in the same way that it was proved that line SQ is equal to the circumference of circle OC. And so, therefore, SL is moved equally as the middle point of CF. But SL is moved equally as CF, as was proved. Hence CF is moved equally as its middle point. The demonstration is the same for any part you wish of the radius OF not terminated at O..... [and also for the radius itself].

Book II

[Suppositions:]

[1] Of equal squares, the one whose sides are moved more [quickly] is said to be moved more [quickly].

[2] The one [whose sides are moved] less [quickly is said to be moved] less [quickly].

[3] The one whose sides are not moved more [quickly] is not moved more [quickly].

[4] The one [whose sides are not moved] less [quickly is not moved] less [quickly].

[5] If surfaces are equal and all lines taken in the same proportion are equal, the one none of whose lines so taken is moved more [quickly] is not moved more [quickly].

[6] The one none [of whose lines is moved] less [quickly] is [not moved] less [quickly].

[Proposition] II.1 An equinoctial circle is moved in four-thirds proportion to its diameter. Whence it is evident that the proportion of circles is as the proportion of the movements squared.

Let squares $BDFH$ and $OLMN$ (see Fig. 3.7) be equal and let surface

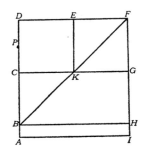
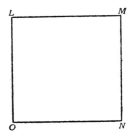

Fig. 3.7

$ADFI$ be moved to describe a cylinder around axis AI. And let $OLMN$ be moved uniformly and equally in all of its parts and points by moving in a direction perpendicular to itself, so that a side of it would be moved equally as point C of line CG, which divides square $BDFH$ into two equal parts. It is obvious, therefore, that square $LMNO$ is moved equally as line CG, because line CG is moved equally in all of its parts and points, and $LMNO$ is also. I say, therefore, that squares $BDFH$ and $LMNO$ are equally moved. For they are either moved equally or one or the other is moved more [quickly].

If [surface] $BDFH$ is moved more [quickly], it is argued in the negative, for its sides are not moved more than the sides of $OLMN$ because BD and FH are moved equally as OL and MN, and DF and BH are moved equally as LM and NO, and so on. Hence this square is not moved more [quickly] than that one [by supposition 3]. By the same reasoning it is not moved less [using supposition 4]. Hence it is moved equally.

[Gerard of Brussels, *On Motion*

Let there be proposed another equal square which is designated as Z. Let it be moved so much more [quickly] than $OLMN$ as $BDFH$ is moved more [quickly] than $OLMN$, and let it proceed perpendicularly like $OLMN$. Since Z is moved equally in all of its parts and points, it is obvious that its side is moved more [quickly] than point C and thus more [quickly] than side BD, because BD is moved equally as point C. And so it is also moved more [quickly] than side FH and, in the same way, than all the intermediate lines between BD and HF. And so by the given description [and supposition 1], the square Z is moved more [quickly] than $BDFH$ and not equally as it.

If moreover the sophist objects that side DF is moved more [quickly] than the side of Z, we answer that sides DF and BH taken together are moved equally as sides BD and FH taken together. For by the amount in its movement that DF exceeds the movement of DB, by that same amount the movement of DB exceeds the movement of BH. And so the sides DF and BH taken together are moved less than [a pair of] the sides of Z. It is obvious therefore that $BDFH$ is not moved more [quickly] than $LMNO$. If less, it will be refuted in the same way, a square having been proposed which moves less [quickly] than $LMNO$ [using supposition 4]. The response indicates it is moved equally as $BDFH$. It is obvious, therefore, that $BDFH$ is moved equally as $LMNO$. But $LMNO$ is moved equally as any of its sides. Therefore, it is moved equally as $B[D]$, and hence equally as point C. So $BDFH$ is moved equally as point C. By the same proof it will be completely proved that square $CDEK$ is moved equally as point P and $KEFG$ also as point P, and so rectangle $CDFG$ is moved equally as point P, and rectangle $BCGH$ is moved equally as the middle point of CB. It is similarly demonstrated for other rectangles.

Then *per impossibile* you will prove in the same way as was proved concerning the radius, that square $BDFH$ is moved equally as point C, if it is moved around the immobile axis BH to describe a cylinder. Therefore let it be so moved. Since then it it moved equally as point C and the rectangle $BCGH$ is moved equally as the middle point of BC, the square $BDFH$ is moved twice as fast as the rectangle $BCGH$. But the rectangle $CDFG$ is moved equally as point P, which is in three-halves proportion to point C. Hence the rectangle $CDFG$ is moved with respect to the rectangle $CBGH$ in the proportion composed of 3/2 and 2, i.e., 3. The space described by rectangle $CDFG$ is in this same proportion to the space described by rectangle $BCGH$. For the cylinder described by rectangle $BDFH$ is to the cylinder described by rectangle $BCGH$ as the base is to the base, since the rectangles are between parallel lines. But the proportion of the bases is quadruple because the radius BD is double the radius BC. Hence one circle is four times the other. Thus the circles are in quadruple proportion. So the cylinders are in quadruple proportion. Hence that which remains of the larger cylinder after the lesser cylinder has been substracted is three times the lesser cylinder. And this is what is described by the rectangle $CDFG$. Hence the proportion of the figure described by $CDFG$ to that described by $BCGH$ is as the proportion of movement to movement.

From this it is obvious that the proportion of the figure described by triangle BDF to the figure described by the triangle BFH is as the proportion of the movement to the movement. For the excess of the movement of $CDFG$ to the movement of BDF is the same as the excess of the movement of KFG to the movement of BCK, since the movement of $CDFK$ is common to each. And the excess of the movement of BFH to the movement of $BCGH$ is the excess of the movement of FGK to the movement of BCK, for the movement of $BKGH$ is common to each. Therefore the excess of the movement of the superior rectangle to the movement of the superior triangle is the same as the excess of the movement of the inferior triangle to the inferior rectangle. Therefore, the movements of the triangles and the rectangles are equal. Furthermore, the space described by the movement of the superior rectangle exceeds that described by the superior triangle, as that described by the movement of the inferior triangle exceeds that described by the movement of the inferior rectangle. This will be clear afterwards. Therefore, the excess of the movements and the excess of the amounts described are the same, and the total movements are equal and the total amounts described are equal. But the proportion of space described by the superior rectangle to that described by the inferior rectangle is as their movements. Hence that space which is described by the superior triangle is to the space described by the inferior triangle as the movement [of the one] is to the movement [of the other]. But the space which is described by the superior triangle is to the space described by the inferior triangle as two is to one, because a cone is a third part of a cylinder. Hence the remainder [of the cylinder] is double [that cone] because the former is described by the superior triangle. Therefore, the movement of the superior triangle is twice that of the inferior triangle. Since, therefore, the movements of the triangles and the movements of the rectangles are equal and the movement of the superior rectangle is triple the movement of the inferior rectangle and the movement of the superior triangle is double the movement of the inferior triangle, these movements are related as 3/9 and 8/4. Hence the movement of the superior rectangle is to the movement of the superior triangle as 9/8, i.e., in sesquioctaval proportion. And the movement of the inferior triangle is to that of the inferior rectangle as 4/3, i.e., in sesquitertial proportion. Since, therefore, the superior rectangle is moved in sesquialteral (3/2) proportion to BD and is moved in sesquioctaval (9/8) proportion to the superior triangle, the superior triangle is moved in sesquitertial (4/3) proportion to BD, for if a sesquialteral proportion is divided by a sesquioctaval proportion, a sesquitertial proportion remains, as is evident with the numbers 9, 8, 3, 2. Also since the space described by the superior rectangle is triple that described by the inferior and the space described by the superior triangle is double the space described by the inferior, and the total spaces described are equal, then the excess of the particular spaces described will be the same, as is evident with the numbers 9, 3, 8, 4.

Now look at the next figure (see Fig. 3.8). I shall prove that the triangles

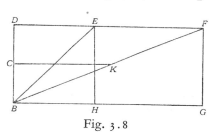

Fig. 3.8

BDE and BDF are equally moved [around axis BHG], for the rectangles $BDEH$ and $BDFG$ are equally moved by the present proof. And in whatever proportion $BDEH$ is moved with respect to its superior triangle, $BDFG$ is moved in the same proportion with respect to its superior triangle. And this is obvious from the adjacent proof. The proportion is evidently 3/4. And since these rectangles are moved [equally], rectangle $BDEH$ is moved in a proportion of 3/4 with respect to either triangle. Hence these triangles are moved equally; and in the same way the inferior triangles are moved equally.

Therefore let line DF equal the circumference of a circle whose radius is BD. It is obvious therefore by the first [proposition] of the *De quadratura circuli* [of Archimedes] that triangle BDF is equal to the circle of radius BD. Let this circle be "applied." Therefore the lines of the triangle and circle C taken in similar proportion are equal, for the line DF is equal to the circumference of the circle C. But the proportion of the radius

BD to the radius BC is as the proportion of circumference to circumference. And the proportion of BD to BC is that of DF to CK, because these are similar triangles, CK being parallel to DF. But DF is equal to the larger circumference. Hence CK is equal to the smaller [circumference]. And so the lines are taken in the same proportion, because just as is the proportion of BD to BC squared so is the proportion of circle to circle. And thus the square of the proportion of BD to BC is as the proportion of the larger triangle to the smaller one. Therefore, just as the larger circle and the larger triangle are equal, so the lesser circle and the lesser triangle are equal. Therefore, all the lines taken in the same proportion are equal and are equally moved, because just as DF is moved equally as the larger circumference, so CK is moved equally as the lesser circumference. It is the same with respect to all the lines. I say, therefore, that the circle and the triangle are moved equally. This will be proved *per impossibile* .... But the triangle is moved in the four-thirds proportion to BD. Therefore, the circle is moved in four-thirds proportion to BD. And hence [it is so moved in relation to the movement of its] diameter [which is moved with the same velocity as its radius, BD].

COMMENTARY

I.1 The basic intent of this proposition we have already indicated. But we must say something more about the lack of precision in the terminology used. I have purposely left some of the ambiguities, since it would be a mistranslation to give precision where precision is lacking. In the first place, the word "motion" *(motus)* is continually used as we would use the more precise term "speed." Thus the verb "to move" is used with the meaning of "to have a speed of motion." And "to move more" means "to have a greater speed of motion."

Similarly there is a lack of precision in the use of the word "equal" as applied to movement. Not only is it used to equate one speed to another, i.e., to identify the space traversed in the same time by one line or point with that traversed by another line or point. But it is also used as we use the term "uniform," namely, to describe the character of a single movement. Here it is employed to identify the space traversed by a line or point in any one part of the duration of a movement with the space traversed by a line or point in any other equal part of the duration of the movement. There is an element of both usages in the statement of the first proposition. Thus to say a segment of a radius is moved "equally" as its middle point means that were the segment to move "uniformly" with a speed "equal" to that of its middle point when rotating so that all of its points were moving uniformly with the speed of the middle point, then it would traverse the same total area in its uniform movement as when rotating.

The proof is simple enough. He finds a rectangular area $SLMP$ equal to the area which the rotating segment CF describes. Then he lets SL equal to CF move uniformly over $SLMP$ in the same time that CF sweeps out its annular area. The areas being equal, the "movements" or speeds of the two lines, SL and CF, are by definition equal. But what is the punctual speed of all the points of SL? Since SL is moving uniformly, the speeds of all of its points are the same. Hence the speed of all of its points is the speed of any one of them, and thus of the speed of its middle point. But the speed of the middle point of SL is the same as the speed of the middle of CF, since these points describe equal lines in the same time. Therefore, SL and CF being called equal in movement or speed because they describe equal spaces in equal times, CF is said to move equally as its middle point.

II.1 In the second book Gerard turns to finding the rectilinear uniform punctual velocity that will convert rotating surfaces to uniformity. In this first proposition he claims (erroneously) that the desired [punctual] speed which makes a rotating great circle uniform is equal to 4/3 of the [punctual] speed which makes the rotating diameter uniform. The first thing he does is to show that a square rotating about an external axis or one of its sides is made uniform by the speed of the middle point of one of its sides perpendicular to the axis, or, in the example he gives, that the desired velocity for the square $BDFH$ is that of point C, the midpoint of BD. As in the first book, Gerard makes use of the indirect method, *per impossibile*, to show that $BDFH$ in rotation traverses the same volume as its equal $LMNO$ moving in translation uniformly with the speed of point C. The suppositions preceding the proposition are stated in the negative as well as the positive form for just the reason that Gerard is going to use the proof *per impossibile*.

Then by a series of comparisons of the figures traversed by the rectangles and triangles into which $BDFH$ is divided by line CG and diagonal BF, he ultimately comes to the conclusion that the [punctual] speed which makes uniform the rotating triangle BDF is 4/3 the [punctual] speed that makes BD uniform. We can show quickly by modern formulation that[23] Gerard is quite correct in this conclusion:

---

[23] I have used periods in my formulae to indicate multiplication only when a geometrical magnitude is given as two letters, e.g. BD.

(1) Volume of Tri. BDF rotating,
$$\text{Vol.} = \frac{2\pi \cdot BD^2 \cdot DF}{3}$$

(2) Volume of Tri. BDF in translation is the same volume by supposition, and to traverse this volume, it must move a distance S in time t, and
$$S = \text{Vol. Tri. } BDF \text{ rotating}/\text{Area Tri. } BDF$$
$$= \frac{\frac{2}{3}\pi \cdot BD^2 \cdot DF}{\frac{1}{2} BD \cdot DF}$$
$$= \tfrac{4}{3}\pi \cdot BD$$

(3) Thus the desired speed
$$V_1 = S/t$$
$$= \frac{4\pi \cdot BD}{3t}$$

(4) And the desired speed of BD rotating is the speed of point C, and it is computed from the circumference of C, $V_2 = \dfrac{\pi \cdot BD}{t}$

(5) And thus $V_1/V_2 = 4/3$. Q.E.D.

So far Gerard is on the right track. He further shows that, regardless of the length of line DF, the speed of triangle BDF is 4/3 the speed of BD. Hence he hypothesizes a triangle BDF in which DF is equal to the circumference of a circle of radius BD; and thus triangle BDF is equal in area to the circle of radius BD. Hence any line parallel to DF, say line CK, is the circumference of a smaller circle of radius BC, and thus there is a corresponding and equal line CK for any circumference of any circle of radius BC within the larger circle of radius BD. Then Gerard asserts that line DF is not only equal to the circumference of BD, but that, as well, it is *equally moved* as that circumference, and similarly for any other line CK and its equivalent circumference at C. It is here that Gerard makes his crucial error, an error he repeats in other propositions later. For DF is *not* moved equally as the circumference of BD, since DF traverses the surface of a cylinder, namely, $4\pi^2 r^2$, while the circumference traverses the surface of a sphere, namely, $4\pi r^2$, and they are obviously not equal. Consequently the whole triangle BDF does not traverse the same volume in rotation around the axis BG as does a circle of radius BD around one of its diameters. Rather it traverses Volume $= 4\pi^2 \int_0^r r^2 \, dr = \frac{4}{3}\pi^2 r^3$ instead of Volume $= \frac{4}{3}\pi r^3$, as does the circle. Thus the punctual speed that makes the triangle uniform will not make the circle uniform, but believing that the triangle is moved equally as the circle, and knowing from his earlier proof that the triangle is moved uniformly with a [punctual] speed 4/3 that of the radius, he concludes that the equinoctial or great circle is also moved uniformly with a speed 4/3 that of the radius. Furthermore, since the radius is made uniform by the same [punctual] velocity as the diameter, Gerard asserts that he has proved his proposition, namely, $V_1/V_2 = 4/3$. But it can easily be shown that $V_1/V_2 = 4/3\pi$, since in actuality the circle of area πr^2 must move in time t through a distance $4r/3$ in order to traverse a cylinder equal to the sphere $4\pi r^3/3$ and we recall the radius in becoming uniform moves in time t through a distance πr.

But in spite of Gerard's error, it must be admitted that his technique of equating the movements of any pair of corresponding lines of two figures is most interesting. Actually the technique worked well when he was comparing, line for line, the rotation of a square and its movement in translation.

Chapter 4

The Emergence of Kinematics at Merton College

WE indicated in the preceding chapter that Gerard of Brussels' excursion into kinematics proved to be the starting point for an important development of kinematic ideas almost a century later at Oxford (see page 186 above). This development in mechanics centered at Merton College and was fostered by four able logicians and natural philosophers: Thomas Bradwardine, William Heytesbury, Richard Swineshead, and John Dumbleton. The activity of these men at Merton spans the period from about 1328 to about 1350.

Only the meagerest biographical details of these Merton scholars remain. But there is no doubt that the mechanical activity of Bradwardine precedes that of the other scholars. Bradwardine appears in the Balliol College records for 1321 and in the Merton College records for the first time in 1323, is a proctor of the university in 1325–26 and 1326–27, and composed his influential *Tractatus de Proportionibus* in 1328.[1] It was this last work which did much to stimulate the succeeding activity at Merton. We cannot be absolutely sure that another work in this logico-physical tradition, *De continuo*, is in fact Bradwardine's composition.[2] But it is

---

[1] So dated for example in MS Paris, BN lat. 14576, ff. 255r–261v. A recent biographical summary is in H. Lamar Crosby Jr., *Thomas of Bradwardine, His Tractatus de Proportionibus* (Madison, 1955), pp. 3–7. Consult also George Sarton, *Introduction to the History of Science*, Vol. 3 (Baltimore, 1947), 668–71; G. C. Brodrick, *Memorials of Merton College* (Oxford, 1885), p. 188; "Bradwardine, Thomas" in the *Dictionary of National Biography* (hereafter cited as *DNB*). I am indebted to Mr. A. B. Emden of Oxford for the early reference of 1321; see his *A Biographical Register of the University of Oxford to A.D. 1500*, Vol. *1* (Oxford, 1957), 244.

[2] There are two main manuscripts of the *De continuo*: Erfurt, Stadtbibliothek, Amplon. Q.385, 14c, 17r–48r; and Thorn, R 4° 2, 14c, pp. 153–92. It is in the latter

probable, and if so, it shows that Bradwardine was further preoccupied with kinematic problems. That the latter was composed after the former is obvious since the one cites the other (see Doc. 4.3 below). Neither his purely mathematical works, nor his ecclesiastical career is of concern to us here. Suffice it to say, he left Merton in the middle of the 1330's,[3] was chancellor at St. Paul's cathedral in 1337, was appointed a royal chaplain, and is said to have been in Flanders with the royal court in 1338. Consecrated in Avignon in 1349 by the Pope as Archbishop of Canterbury, he returned to England, only to fall victim to the plague and die in that same year.

The second of the foursome at Merton, William Heytesbury,[4] is mentioned in the records of Merton for 1330 and 1338–39. He is probably the Hoghtelbury or Heightilbury who was one of the original fellows named by the founder of Queen's College in 1340. He was, however, still a fellow of Merton in 1348. There is some evidence that Heytesbury became chancellor of Oxford in his old age, in 1371. His work which is of most importance for the study of fourteenth-century mechanics is a *Regule solvendi sophismata*, which is dated 1335 in an Erfurt manuscript (Stadtbibliothek, Amplon. F. 135, 171),[5] and which includes the important section entitled *Tria predicamenta de motu* that often circulated as a separate work. There is also a *Probationes conclusionum*, which consists of proofs of conclusions

manuscript that the work is unequivocally identified as being by Bradwardine. On page 192 the tract terminates: "Explicit tractatus bratwardini de continuo." A third manuscript (Paris, BN Nouv. Acquis. lat. 625, f. 71v) contains a fragment of the *De continuo*. On the margin we read: "Bradwardin." This treatise has been examined by E. Stamm, "Tractatus de continuo von Thomas Bradwardina," *Isis*, Vol. *26* (1936), 13–32, and M. Curtze, "Über die Handschrift R. 4° 2: Problematum Euclidis Explicatio, des Königl. Gymnasial Bibliothek zu Thorn," *Zeitschrift für Mathematik und Physik*, Vol. *13* (1868), Suppl., pp. 85–91. A text of this work based on all of the manuscripts was started by John Murdoch in my seminar in medieval mechanics. Its publication is awaited soon. (See Document 4.2).

[3] See the biographical works listed in note 1.

[4] On Heytesbury, see C. Wilson, *William Heytesbury: Medieval Logic and the Rise of Mathematical Physics* (Madison, 1956), pp. 7–8; Sarton, *Introduction*, Vol. *3*, 565–66; Brodick, *Memorials*, p. 207; *DNB*, Vol. *26*, 327–28. Mr. Emden writes that Heytesbury was a doctor of theology by 1348 (cf. *Munimenta Academica, or Documents Illustrative of Academical Life and Studies at Oxford*, edited by H. Anstey, Roll Series Vol. *1* [Oxford, 1868], 167).

[5] See A. Maier, *An der Grenze von Scholastik und Naturwissenschaft*, 2d ed. (Rome, 1952), p. 266, note 26. For the first time Miss Maier identifies the so-called *Loyca Hesbri* with the *Regule*. Wilson mentions a fairly extensive list of the manuscripts of the *Regule* in his *William Heytesbury*, Appendix B, to which we can add MSS Bruges, Stadsbibliotheek 500, 14c, ff. 33r–71v, and 497, 14c, ff. 46r–59v; and Venice, Bibl. Naz. San Marco Lat. VIII, 38, ff. 66v–72r.

arrived at in the *Regule*. There has been some tendency to doubt that the *Probationes* is by Heytesbury,[6] but we have here tentatively accepted it as being his. The question of its genuineness is of some importance, for if it is by Heytesbury then its proof of the famous acceleration theorem, which we shall discuss in the next chapter, is probably the first proof of that theorem which we have, just as Heytesbury's statement of the theorem without proof in the *Regule* in 1335 seems to be the first extant statement of the theorem.

Contemporary with Heytesbury was Richard Swineshead, in many ways the subtlest and most able of the four logicians. Richard is confused (apparently in his own times as well as now) with John Swineshead, also a member of Merton College, whose legal career can be clearly traced.[7] Hence the reference for 1340 in the Merton Bursar rolls to Swineshead (without praenomen) may be either to John or to Richard, or even to an entirely different third person, Roger Swineshead.[8] However Richard can be clearly dated at Merton in 1344, and he was still a fellow of Merton in 1355.[9] The name of "Roger" further clouds the picture. Although no specific reference to a Roger at Merton College can be found, an Erfurt codex includes a work entitled *De motibus naturalibus* which is attributed in the colophon to "Roger," although in the title it is attributed to "William."[10] That work is specifically noted as having been done at

---

[6] Wilson, in his *William Heytesbury*, p. 210, so doubts Heytesbury's authorship of the *Probationes*. As Wilson notes, the *Probationes* appears in Vat. lat. 2189, ff. 13v–38r under the title *Anonymi conclusiones*.

[7] Mr. A. B. Emden of Oxford has gathered considerable evidence of John's career in the 1340's, 1350's and 1360's, down to the time of his death in 1372. It is doubtful whether John had any of the logical interests of Richard, although it is of some interest that a manuscript dated 1378 (Paris, BN lat. 14715, ff. 86v–90v) attributes the short logical tract known as *Obligationes* to John—f. 90v: "Explicit. Et in hoc terminantur obligationes Reverendi Magistri Jo. Swiinsed (*or* Swinised) de anglia doctoris in sacra theologia."

[8] I am indebted to personal correspondence from Mr. Emden, once more, for the 1340 reference, this constituting a correction of the reference of 1339 in Brodrick's *Memorials*, p. 213.

[9] Brodrick, *Memorials*, pp. 212–13. For other information on Swineshead, see my *Giovanni Marliani and Late Medieval Physics* (New York, 1941), Appendix I; and L. Thorndike, *A History of Magic and Experimental Science*, Vol. *3* (New York, 1934), chap. 23. Mr. Emden supplied me the information for 1355.

[10] The *De motibus naturalibus* is contained in MS Erfurt, Stadtbibl., Amplon. F. 135, ff. 25v–47r. On f. 25v we read: "Incipit tractatus Magistri Wilhelmi Swineshep datus Oxonie ad utilitatem studencium." However, on f. 47r the colophon reads: "Explicit tractus de motibus naturalibus datus a magistro Rogero Swyneshede." Parts of the same work are also included in MS Paris, BN lat. 16621, where it is entitled *De primo motore* from its incipit. This manuscript is a very poor one and it is difficult to locate all of the parts of the Swineshead treatise given there. A collation with the Erfurt manuscript is

Oxford "for the utility of students," and presumably before 1337 when the codex was begun.[11] Two short logical tracts, *Obligationes* and *Insolubilia*, while generally without praenomen in the manuscripts, have been attributed to Roger, to John, and to Richard.[12] Hence, the identity of Richard and Roger is still an open question.

The most important of Richard's works, at least so far as mechanics is concerned, is his collection of logico-physical tracts which came to be known as the *Liber calculationum*. None of the many manuscripts carries the date of composition,[13] although a partial copy not previously identified appears in a codex that has been dated about 1346, although probably erroneously.[14] We know that it was composed after the *Tractatus propor-*

necessary. The sixth and seventh *differentie* of the *De motibus naturalibus* are of particular interest to mechanics. The sixth on local motion occupies folios 41r–43v of the Erfurt manuscript, while the seventh *de proportionibus velocitatum in motu* runs from 43v–44v.

[11] Erfurt, Stadtbibl., Amplon. F. 135, f. 1r; the scribe notes that "inceptum est hoc opus in festa Magdalene a.D. MºCCCºXXXVIIº...." For the reference to Oxford, see note 10 above.

[12] For example, in MS Bruges, Stadsbibl. 500, 14c, f. 150r, we read: "Expliciunt obligationes magistri Rogeri Suincet anglici." Cf. f. 157v for the attribution of the *Insolubilia* to "Domino Rogero Seveneset." Cf. MS Vienna, Dominik 160/130, f. 122v: "Explicit Magistri Rudegerii Swenishaupt (Sweinshaupt?) anglici de obligatoriis." For the attribution of the *Obligationes* to John, see note 7 above. The above mentioned Erfurt codex, Amplon. F. 135, contains on folios 17–20 a treatise entitled *Insolubilia Swynesheyft data Oxonie*, but which at the end is said to be given by "mag. Thomas Bradewardyn" of Oxford. I am indebted to Father J. A. Weisheipl, O.P., for the following attribution to Richard: Vat. lat. 2154, 15c., f. 1r: "Richardi Suiset Anglici obligaciones," f. 6rb: "Expliciunt obligaciones Ricchardi suiset anglici doctoris," f. 12 va: "Expliciunt insolubilia subtilissimi viri et in tota philosophia profundissimi Ricchardi Suiset doctoris anglici."

[13] I know of the following manuscripts of the *Liber calculationum*: Worcester Cathedral F 35, 15c., ff. 3–124 (I have not seen this manuscript); Cambridge, Gonv. and Caius 499/268, 14c, ff. 165r–203v; Paris, BN lat. 6558 in toto, dated 1375 (MCCCLXXV, f. 1r); Rome, Bibl. Angelica 1963 in toto, 15c; Rome, Bibl. Naz., Vittorio Emanuele 250, 15c, 1r–81v; Pavia, Bibl. Univ., Aldini 314, 15c., 1r–83r; Vatican, Vat. lat. 3064 in toto, 15c; Vatican, Vat. lat. 3095 in toto, 15c; Vatican, Chigi E IV 20 (I have not seen this manuscript; see Maier, *An der Grenze*, p. 269, note 35); Venice, Bibl. Naz. San Marco, Lat. VI, 226, 15c., ff. 1r–98v; Erfurt, Stadtbibl., Amplon. Oct. 78, ca. 1346 (?), ff. 1r–36v. Editions of Padua, ca. 1477; Pavia, 1498; Venice, 1520. The there fourteenth century copies, i.e., the Cambridge, Paris, and Erfurt manuscripts do not contain the *Regule de motu locali*.

[14] W. Schum, *Beschreibendes Verzeichniss der Amplonianischen Handschriften-Sammlung zu Erfurt* (Berlin, 1887), p. 373, dates this codex "um 1346." However he takes this date from the colophon of the last item in the codex, which reads: "Expl. reportata libri de reg. san. missa Allexandro regi ab Aristotile (!) principe philosophorum a.D. MºCCCºXLVIº feria quinta post festum Dyonesii in Bremis." It would appear to me that this is the date of the "reportata" rather than that of the copying date of the codex, for the codex includes (ff. 41–132) a copy of the *Abbreviata* on

tionum, which it cites and uses,[15] and we have already seen that this work of Bradwardine was written in 1328. Thus the *Liber calculationum* was written after 1328 and probably not later than 1350.

For mechanics we are particularly interested in a tract entitled *Regule du motu locali* that appears in the editions and fifteenth-century copies of the *Liber calculationum*. Although I have not found this tract in any fourteenth-century copy of the *Liber calculationum*,[16] there is a fourteenth-century Cambridge manuscript, Gonville and Caius 499/268, which contains two tracts (or one tract of two parts?), the first of which (ff. 212r–213r) is a short tract on motion which is not specifically attributed to Swineshead, although it appears between two Swineshead items.[17] It is similar to, but shorter than, the second tract (ff. 213r–215r), which appears to make a direct reference to the first tract; it bears the colophon: "Explicit

the *Physics* by Marsilius of Inghen, which is certainly not early enough to be included in a codex of 1346. It could be, of course, that Schum has misidentified this item. In such a case, 1346 would not then be an improbable date for the codex. It should be noted that Schum has erroneously attributed the Swineshead work (ff. 1r–36v) to "Clymiton." While I have not seen the codex as a whole, I do have photographs of the Swineshead item and so have established without any doubt that it is a partial copy of the *Liber calculationum*.

[15] Cf. M. Clagett, "Richard Swineshead and Late Medieval Physics," *Osiris*, Vol. 9 (1950), 158, note 51. Thorndike, *Magic and Experimental Science*, Vol. 3, 376, argues concerning the dating of the work from its citation of Bradwardine thus: "The use of the present tense in this citation of Calculator and the calling Bradwardine master instead of archbishop is no sure indication that it was written during his lifetime and before he became archbishop, but it would be an appropriate enough form to employ under such circumstances." However, it should be noted that early manuscripts do not use the present tense *(declarat)* but rather the perfect *(declaravit)*. Thus the citation reads: "...ut venerabilis Magister Thomas Braduardini in suo tractatu de proportionibus liquide declaravit."

[16] I do not mean to imply that the *Regule de motu locali* is not genuinely by Swineshead. There are too many citations to it in the course of the other tracts of the *Liber calculationum* to support such a statement. Still its relationship to the other two tracts on motion in the Gonv. and Caius manuscript would be interesting to explore. It is obviously a much more fully developed tract, and thus perhaps it post-dates these shorter tracts.

[17] Ff. 204r–211v contain a series of subtle problems that are at least in part suggested by the *De caelo* of Aristotle. This piece begins (f. 204r): "In primo de de celo philosophus in commento 39 arguit corpus infinitum circumvolui non posse...." Its explicit and colophon read (f. 211v): "...cum igitur aer aut b tantam latitudinem deperdet quantam ipsa aqua, scilicet, 3 illius latitudinis. Explicit tractatus de Swynyshed." The short tract itself begins (f. 212r): "Cum omnis effectus naturalis per motum ad esse producitur...." It ends (f. 213r): "...Sed si per tantum tempus vel per tantum continaret talem motum pro quolibet instanti talis motus attenditur. Ecce finis." MS Oxford, Bodl. Digby 154, 13c, ff. 42r–46v has the same incipit. It is the short tract that we used as Document 4.5.

tractatus de Swynyshed de motu locali."[18] Another work is falsely attributed to Swineshead, a *lectura* on the *Sententie* of Peter Lombard.[19] We can note finally that there appears to be no contemporary evidence for an oft-repeated assertion that Swineshead joined the Cistercian Order about 1350.[20] The source of the Cistercian association probably arises from the fact that there was a monastery of Swineshead.

The last of our Merton quadrumvirate, John Dumbleton, is certainly not identical with the Thomas de Dumbleton mentioned in the Merton College accounts of 1324.[21] He is, I am sure, the John Dumbleton appear-

---

[18] This longer tract is in many respects like the shorter tract preceding it. Both have two main divisions, namely, the treatments of motion or velocity *quo ad causam* and *quo ad effectum*. The first section on dynamics in each tract explores the Bradwardine proportionality law (see Chapter 7 below), while the second section on kinematics in each case defines and explores the kinematic factors in uniform and difform (i.e., nonuniform) movements. The longer tract begins (f. 213r): "Multe possunt elici conclusiones sive regule super variationem proportionum et motus ex variatione potentie motive ad suam resistentiam et econtra. Et quia omnis variatio potentie motive seu resistentive vel est uniformis vel difformis...." It ends (f. 215r): "...et quamadmodum variata fuerit ipsa proportio ita et velocitas ideo et cetera. Et ecce finis. Explicit tractatus de Swynyshed de motu locali." The title given in this colophon is misleading since the longer tract includes a brief treatment of the motion of alteration. There is some evidence that the two tracts are merely parts of a loosely constructed longer treatise. For the longer tract in giving the rule for the traversal of distance in a uniformly accelerated motion notes that the rule was stated earlier (f. 213v): "Et iuxta illud satis faciliter poteris elicere illam prius dictam, quod universaliter in omni intensione motus uniformi a non gradu ad aliquem gradum et similiter in omni uniformi remissione motus ab aliquo gradu ad non gradum triplum precise pertransietur per medietatem illius inten-siorem quam per medietatem remissiorem eiusdem latitudinis, et utrumque satis diffuse est alibi demonstratum." It should be observed that indeed the rule is stated "earlier" in the shorter tract (f. 213r): "Ex istis patet universaliter quando a non gradu ad certum gradum uniformiter intenditur motus vel ab aliquo gradu ad non gradum uniformiter remittitur motus, triplum pertransietur precise per medietatem talis motus intensiorem quam remissiorem...."

[19] In my *Giovanni Marliani*, pp. 172–73, I discussed and refuted the possibility of Swineshead's having composed this *Lectura*. I suggested rather that this work is by the English Franciscan Roger Royseth in spite of its attribution to Swineshead on almost every page of Oriel College MS 15, ff. 235r–279v. The catalogue of Vat. lat. MSS *(Codices Vaticani Latini,* Vol. 2, Part 1 [Vatican, 1931], 719) mentions a manuscript of this work dated at Norwich in 1337.

[20] Swineshead is called a Cistercian by the Cistercian historian Ch. de Visch, *Bibliotheca scriptorum sacri ordinis cisterciensis* (Duaci, 1649), under "Rogerius Suisetus." See also C. G. Jöcher, *Allgemeines Gelehrten-Lexicon*, Vol. *4* (Leipzig, 1751), c. 933, who claims Swineshead went into the Cistercian Order about 1350.

[21] On Dumbleton, see Brodrick, *Memorials*, pp. 190, 343, 346; *DNB*, Vol. *16*, 146–47; P. Duhem, *Études sur Léonard de Vinci*, Vol. *3* (Paris, 1913), 410–12; Sarton, *Introduction*, Vol. *3*, 564–65; Emden, *Biographical Register*, Vol. *1*, 603. Mr. Emden

ing in the Merton College records for 1331(?), 1338, 1344, 1347–48. He seems to have been one of the fellows of Merton who had a part in the foundation of Queen's College in 1340, but who probably did not actually take up fellowships there. His principal work of interest to mechanics is his *Summa de logicis et naturalibus*. We can say little that is definite about the date of the *Summa's* composition, but presumably it dates from the time he was active at Merton. Of minor interest is an *Expositio cap. 4 Bradwardini de proportionibus*, since it deals only with the geometrical conclusions of the fourth chapter rather than the kinematic conclusions.[22]

From the discussions of these four men at Merton emerged some very important contributions to the growth of mechanics: (1) A clear-cut distinction between *dynamics* and *kinematics*, expressed as a distinction between the *causes* of movement and the spatial-temporal *effects* of movement. (2) A new approach to speed or velocity, where the idea of an instantaneous velocity came under consideration, perhaps for the first time, and with it a more precise idea of "functionality." (3) The definition of a uniformly accelerated movement as one in which equal increments of velocity are acquired in any equal periods of time. (4) The statement and proof of the fundamental kinematic theorem which equates with respect to space traversed in a given time a uniformly accelerated movement and a uniform movement where the velocity is equal to the velocity at the middle instant of the time of acceleration. It was this last theorem in a somewhat different form that Galileo states and which lies at the heart of his description of the free fall of bodies. We shall treat it at greater length in succeeding chapters.

It is worth noting that, while the *mathematical* work of Gerard of Brussels proved to be the principal point of departure for the kinematic section of the earliest of the Merton treatises, namely, the *Proportions of Velocities in Movements* of Thomas Bradwardine, of perhaps equal importance and influence was a *philosophical* current. And indeed it was philosophers trained in mathematics who were responsible for the investigations into kinematics at Oxford and Paris. The philosophical problem which gave

mentions the date of 1347–48 as the last citation of Dumbleton's name; this presumably is a correction of the 1349 date in the *Memorials* of Brodrick. Mr. Emden further suggests in personal correspondence that the subsequent silence regarding Dumbleton might be accounted for by supposing him a victim of the plague of 1349.

[22] MS Paris, BN Nouv. Acquis. 625, ff. 70v–71v. I am grateful to Father J. A. Weisheipl, O.P., for calling my attention to this work and sending me a transcription of it.

stimulus to kinematics was the problem of how qualities (or other forms) increase in intensity, e.g., how something becomes hotter, or whiter. In the technical vocabulary of the schoolmen, this was called the problem of "the intension and remission of forms," intension and remission simply meaning the increasing and decreasing of the intensity of qualities or other forms.[23] The solution of this problem worked out by the philosopher Duns Scotus during the early years of the fourteenth century assumed a *quantitative* treatment of variations in the intensities of qualities suffered by bodies. It was accepted by the successors of Scotus, and it perhaps influenced the view held by some of the men of Merton that the increase or decrease of qualitative intensity takes place by the addition or subtraction of degrees *(gradus)* of intensity. With this approach to qualitative changes accepted, the Merton schoolmen applied various numerical rules and methods to qualitative variations and then by analogy to kindred problems of motion in space *(motus localis)*. Needless to say, the discussion of variations of quality and velocity by the Oxford schoolmen (and their medieval successors) was almost entirely hypothetical and *not rooted in empirical investigations;* nor was it framed in such a way as to admit of such investigations. In fact, in medieval kinematics as well as dynamics all of the quantitative statements relative to pretended physical variables are in terms of general proportionality expressions; *and the proportionality constants, which can only be determined by experiment, are never found.* Nevertheless, in stimulating the abstract kinematic definitions and descriptions here described in Chapters 4, 5, and 6, this essay into hypothetical problems of change was not unproductive of important results, as we shall show.

In the preceding chapter on Gerard of Brussels, we have already noted that Gerard presented exclusively kinematic problems. From this we might infer that he distinguished such problems from those of dynamics. But it was at Merton College that the implicit distinction between *dynamics* and *kinematics* became explicit. It is true that on a philosophical level the distinction had already emerged clearly in attempts by William Ockham, the celebrated Oxford nominalist who was a contemporary of Bradwardine, to resolve an old argument on the cause of the temporal aspect of movement. This argument was between those who stressed the necessity of a *forceful resistance* of the medium in which motion takes place and those who said that the temporal nature of movement demands only *extension* or *space*

---

[23] The best summary of the problem of the intension and remission of forms is given by Maier, *Zwei Grundprobleme der scholastischen Naturphilosophie*, 2d ed. (Rome, 1951). Cf. Clagett, "Swineshead," *Osiris*, Vol. *9* (1950), 132–38.

through which the movement takes place. Put in a somewhat more modern terminology, this was an argument between those who would define movement primarily in terms of forces and those who would define it kinematically. The argument had come to the fore in the commentary on the *Physics* of Aristotle by the Spanish Moslem philosopher Averroës. He argued for the dynamic definition against the kinematic approach of his fellow Spanish Moslem Avempace. The argument was renewed in the Latin West in the thirteenth century when Aquinas stressed the kinematic aspect, while Aegidius Romanus emphasized the dynamic. Ockham in his treatment of the problem recognized that which is valid in each of the two approaches. Ernest Moody has summarized Ockham's contribution as follows:[24]

> ...although he (Ockham) defends St. Thomas' thesis that local motion is temporal because the medium is extended, he denies that this kinematic definition of what motion is, constitutes a definition and measure of force.... Ockham's discussion and resolution of the controversy between Averroes and Avempace... is of great theoretical interest in relation of (! to?) the development of modern mechanics. For Ockham introduces the distinction fundamental to our own mechanics, between the condition of "being in motion," and the condition of "being moved" in the sense of being acted upon by a force. The kinematic problem is clearly distinguished from the dynamic one. Avempace had seen that motion is kinematically defined as traversal of extended magnitude in time, irrespective of material resistance; but he had constructed this kinematic analysis of motion as equivalent to a definition and measure of motive power or force. Averroes had clung to what is essentially a sound conception of force or motive power, as that which does work on materially resistant body; but he had assumed that motion under the action of no force is impossible. Ockham was the first to separate what was true in the positions of these two Arabian philosophers, from what was erroneous, and to combine the sound kinematic insight of Avempace and St. Thomas, with the sound dynamic principle of Aristotle and Averroes.

But important as the theoretical considerations of Ockham undoubtedly were, the oft-repeated distinction between dynamics and kinematics in the mechanical treatises of the fourteenth century had its origins in the clear-cut separation of dynamics and kinematics in Thomas Bradwardine's *Treatise on the Proportions of Velocities in Movements* of 1328. In the third chapter of the work, Bradwardine treats of the "proportion of velocities in movements in relationship to the *forces (potentias)* of the movers and the things

---

[24] E. A. Moody, "Ockham and Aegidius of Rome," *Franciscan Studies*, Vol. 9 (1949), 436–38.

moved," in short, to the dynamic considerations of velocity.[25] On the other hand, the fourth chapter treats of velocities "in respect to the magnitudes of the thing moved and of the *space traversed (spatii pertransiti)*," i.e., to the kinematic measure of movement. Furthermore, in discussing his dynamic law of movement (which law we shall consider in Chapter 7), the English Master kept the two aspects of movement distinct even while he related them as a law. Bradwardine's junior contemporary, Richard Swineshead, if indeed he is the author of the short tract here quoted, in like manner explicitly made the distinction between the measure of velocity dynamically *(quo ad causam)* and kinematically *(quo ad effectum)*. We read in a fragment on motion following his *Liber calculationum*:[26]

With slowness set aside, it ought to be investigated diligently as to how velocity in any motion is measured[27] *(penes quid attendatur)*, both with respect to cause and with respect to effect.... In the first place the measure of velocity in motion with respect to cause should be noted. For this it should be remarked generally that the succession of velocity in every movement is measured causally by the proportion of the motive force to the resisting force.... (212v). Having spoken generally regarding the measure of velocity in any motion with respect to cause, now we should see its measure with respect to effect. For this it should be known that of local motions some are uniform, some nonuniform. For uniform local motion it should be known that its velocity is measured simply by the line described by the fastest moving point in such and such time....

The author of this statement has explicitly added time as a kinematic factor. Time was, of course, implicit in the statement of Bradwardine, as

---

[25] Thomas of Bradwardine, *Tractatus de proportionibus*, edition of Crosby, p. 64: "Tertium capitulum veram sententiam de proportione velocitatum in motibus, in comparatione ad moventium et motorum potentias, manifestat.... Capitulum autem quartum de proportione velocitatum in motibus, in comparatione ad moti et spatii pertransiti quantitates, pertractat...."

[26] MS Cambridge, Gonv. and Caius 499/268, f. 212r: "Dimissa tarditate penes quid quo ad causam et quo ad effectum in quocunque motu attendatur velocitas diligentius est inquirendum.... Primo notandum est penes quid attenditur velocitas in motu quo ad causam. Pro quo notandum est quod generaliter in omni motu successio velocitatis quo ad causam attenditur penes proportionem potentie motive supra potentiam resistentivam, sic quod si aliquod movens ad suum motum habuerit maiorem proportionem quam aliud, motus proveniens ab illa proportione velocior est et velocitas maior." The Latin for the rest of the quote is found at the beginning of the Latin text of Document 4.5.

[27] Here and troughout the translations I have rendered the verb *attendo* with "is measured." I do not, of course, use "measured" in the sense of an act involving an instrument. I use it only in the sense of "is mathematically determined" or "is a function of." In some cases in the Latin texts *mensuratur* is used instead of *attenditur*, and thus I have followed the lead of the fourteenth-century authors themselves (see Doc. 4.2, lines 42–43).

[at Merton College

we can tell from his succeeding remarks. Now whether or not Swineshead was the author of this short treatise, we have a similar statement in the longer *Tractatus de motu locali* which is definitely assigned to Swineshead:[28]

> Since, as it was stated before when we defined the velocity of motion causally as following the proportion of motive power to resisting power, a corresponding effect uniformly follows its sufficient cause, we will now see therefore with what ought to be measured the velocity of motion with respect to effect.... Speaking about local motion, in regard to the measure of its velocity, [we say that it is measured] by the maximum line which the most rapidly moving point in such a velocity would describe in [some] time.

Numerous other authors made a similar distinction between the measure of velocity in terms of force and its measure in terms of the distance traversed in some given time. One of these authors, who may be identical with Marsilius of Inghen (a master at Paris in the 1360's), but who at any rate is speaking as an Ockhamist, states that, of the two ways to measure velocity, the kinematic method is preferable, because the space traversed is a *quasi-effect*, and in natural things effects are more readily knowable than causes.[29]

In their analysis of kinematic problems, the men of Merton established a vocabulary of kinematics whose influence on Galileo is clearly discerni-

---

[28] MS Gonv. and Caius 499/268, f. 214r: "Quia, prius dictum est, diffinito de velocitate motus quo ad eius causam, videlicet, quod sequitur proportionem potentie motive ad potentiam resistentivam, et causam sufficientem uniformiter sequitur suus effectus correspondens, videtur igitur modo penes quid attendi debeat velocitas motus quo ad eius effectum.... Loquendo igitur de motu locali, penes quid attenditur eius velocitas, penes maximam lineam quam describeret punctus velocissime motus uniformiter contenta tali velocitate per tempus...."

[29] Marsilius of Inghen (?), *Quaestiones in octo libros physicorum Aristotelis* [*secundum nominalium viam*], in *Opera omnia* of Duns Scotus, Vol. *2* (Lyon, 1639), 366, c. 2, Bk. VI, quaest. V: "Item, notandum quod duo concurrent ad velocitatem motus, per quorum quodlibet potest cognosci quanta est velocitas, scilicet proportio potentiae motoris ad resistentiam moti; & qualiter penes hoc attendatur velocitas, dictum fuit superius. Secundo, concurrit spatium pertransitum a mobili, & magis proprie attenditur velocitas penes spatium descriptum a mobili, quam penes proportionem potentiae ad resistentiam. Primo, quia illud spatium est nobis notius; sed proportio potentiae ad resistentiam non cognoscitur nisi arguitive, & et ex consequenti, eo quod talis proportio non sentitur. Secundo, quia spatium ponitur in definitione *velocioris;* & ideo spatium videtur magis de intensione velocitatis, quam proportio potentiae ad resistentiam. Tertio, quia huiusmodi proportio potentiae ad resistentiam est causa velocitatis; sed spatium pertransitum est quasi effectus: modo in naturalibus effectus sunt nobis notiores causis; igitur potius devenimus in notitiam velocitatis per spatium, quam per proportionem potentiae ad resistentiam...."

ble, but which has archaic elements that make its clarification by Galileo and his successors important. We can summarize in tabular form some of the more important terms and their modern approximations:

Scholastic Term	Modern Approximation
1. Motion *(motus)*. (See Doc. 4.4, Prologue; Doc. 6.1, passage II.1; Doc. 7.1, passages 1–3.)	Motion. On occasion, speed or velocity, as used loosely without vectorial implications.[30]
2. Velocity *(velocitas)*. (See Doc. 4.5, passage 1.)	Speed or velocity (without vectorial implications). It is not defined as a ratio of two unlike quantities, although it is considered as capable of quantification.
3. Quality of motion or intensity of velocity *(qualitas motus, intensio motus, intensio velocitatis.)* (See notes 33 and 34; cf. Doc. 6.1, passages II.1, II.3.)	Velocity without consideration of its continuation or duration in time. In the cases of nonuniform motion, instantaneous velocity (although, of course, not defined in the modern manner as the limit of a ratio).
4. Quantity of motion or total velocity *(quantitas motus, quantitas totalis velocitatis)*. (See the same passages as in 3.)	The velocity over a definite period of time, measured by the distance traversed in that time.
5. Degree of motion or velocity *(gradus motus, gradus velocitatis)*. (See commentary to Doc. 4.3; Doc. 4.5, passage 2.)	Generally, the numerical designation of the magnitude of the quality or intensity of motion. In nonuniform movements, the magnitude of an instantaneous velocity.
6. Instantaneous velocity *(velocitas instantanea)*. (See Doc. 4.4, passage 2.)	Instantaneous velocity (with the limitations mentioned in 3). Note that the magnitude of instantaneous velocity is said by Heytesbury to be determined by the path which would be described in a given period of time if a moving point were moved uniformly at the degree of velocity with which it is moved in the assigned instant.
7. Latitude of motion, or latitude of velocity *(latitudo motus, latitudo velocitatis)*. (See Doc. 4.4, passage 3, *et passim* in the other documents of Chapters 4–6).	Properly, positive or negative increment of velocity. Improperly, the integral of all degrees of velocity between two terminal velocities. For the "proper" and "improper" usages, see the commentary to Doc. 6.3, second paragraph. For latitude as the "distance" between two degrees of quality

[30] In the documents of Chapters 4 and 5 we often find *motus* being used in the sense of the "speed or velocity of motion." See, for example, Document 5.3, passage 3.

8. Uniform motion (or velocity) *(motus uniformis)*. Measured by the traversal of an equal amount of space in every equal period of time. (See Doc. 4.5, passage 3; Doc. 4.6, passage 1.)

or velocity, see Doc. 6.2 and the commentary to Doc. 7.1, passage 4.
Uniform speed or velocity.

9. Intension and remission of motion *(intensio et remissio motus)*. (See Doc. 4.5, passages 4–6.)

When used in connection with an increment of velocity, acceleration. When used statically to distinguish intension and extension, the degree of intensity.

10. Uniformly difform motion *(motus uniformiter difformis)*. Defined as the acquisition of equal increments of velocity in any equal periods of time. (See Doc. 4.4, passage 3; Doc. 4.5, passage 6; Doc. 4.6, passages 3 and 4. Cf. also Oresme's alternate expression for acceleration, *velocitatio*, Doc. 6.1, passage II.5.)

Uniform acceleration.

11. Uniformly difformly difform motion *(motus uniformiter difformiter difformis)*.

By analogy with uniform motion and uniform acceleration, this should mean uniformly changing acceleration, i.e., a movement in which there are equal increments of acceleration in any equal periods of time. Perhaps Heytesbury, Swineshead, and Oresme would understand it in this way. Presumably such motion would fall under the categories given by Swineshead in Doc. 5.3, passage 3, and would constitute one kind of difform *velocitatio* with Oresme (see Doc. 6.1, passage II.5). However, other authors like the author of the *Tractatus de latitudinibus* (commentary to Doc. 6.3, third paragraph) seem to hold in the case of uniformly difformly difform motion, that, in any two successive equal periods of time, the ratio of the excesses of velocity in those times is a constant (other than one, which would be motion uniformly difform). Galileo has a similar view of a quality uniformly difformly difform, as can be noted in the passage given in the commentary to Doc. 4.7.

For the confirmation, clarification, and extension of these definitions, I have presented in the documents appended to this and the succeeding chapter, several important passages of the Merton College works. But a few preliminary considerations are necessary, particularly as to how the fourteenth-century authors arrived at a concept of instantaneous velocity from their view of "intension and extension of motion." This consideration is of the greatest importance, for their concept of instantaneous velocity lay at the basis of their kinematic achievements.

We have already suggested that the Merton kinematics developed out of the discussions of the problem of the intension and remission of forms. These discussions went deep into the thirteenth century, and in fact ultimately to antiquity.[31] Out of them emerged a clear-cut distinction between the measure of the intensity of a quality and its quantitative extension in some subject. This distinction was first expressed in terms of one between intensive or "virtual" quantity *(quantitas virtualis)* and "dimensive" or "corporeal" quantity *(quantitas dimensiva, corporalis)*. One of the first examples developed of the distinction between "intensive" and "extensive" factors in the treatment of qualities was the rather common distinction made between intensity of heat (temperature) and quantity of heat.[32] A somewhat similar distinction between total weight (extensive weight) and specific weight (intensive weight) was recognized (see Chapter 2 p. 97). And further, "extensive" force or power was distinguished from "intensive" force, a distinction something like the more modern one of force and pressure (although their treatment of the interaction of forces was not by *vectorial additions* but rather by the use of an exponential function involving the *ratio* of the motive force to the resistance, as we shall see in Chapter 7). But observe now how Bradwardine distinguishes between the qualitative and quantitative aspects of force:[33]

---

[31] Clagett, "Swineshead," p. 134.

[32] *Ibid.*, p. 137. Cf. Clagett, *Marliani*, pp. 34–39.

[33] *Proportionibus*, edition of Crosby, p. 118: "Pro primo istorum dicendum non esse inconveniens idem habere eandem proportionem qualitative (scilicet in virtute agendi) ad totum et ad partem sed quantitative non; quia licet totum et pars sint inaequalia in quantitate, possunt tamen esse aequalia in qualitate resistendi. Et ideo sicut non differunt in qualitate resistendi sed in quantitate, sic nec motus per media differunt in qualitate motus (quae

[at Merton College   213]

In answer to the first of these [objections], one should say that it is not inconsistent for the same [agent] to have an identical proportion *qualitatively*, i.e. in its power of acting, with both the whole and the part [of its resistance], but not so *quantitatively*. For although the whole and the part are unequal in quantity, they can however be equal in the quality of resisting. And therefore just as they do not differ in their quality of resisting, so their movements through media do not differ in quality of motion (which is swiftness and slowness), but rather in quantity of motion, which is [in] length or brevity of time.

Notice that at the end of this passage Bradwardine also refers to a distinction between the "quality of motion" and the "quantity of motion." The same distinction is repeated in a passage immediately following:[34]

In answer to the second [objection], one should say that agents can be proportional to their resistances *qualitatively*, that is, in their power of acting. From such proportionality arises equality of motions *qualitatively*, that is, in swiftness and slowness. Or [the agents can be proportional to their resistances] *quantitatively*, that is, in regard to their action throughout the whole quantity of their resistance. And from such a proportionality follows in similar fashion the equality of speeds *quantitatively*. By this is meant [equality] in [consideration of] the length and brevity of the time of movement.

The distinction of quantitative from qualitative velocity[35] so obscurely expressed by Bradwardine only becomes completely clear with the system of applying geometric figures to qualities and movements some twenty years after Bradwardine's treatise (see Chapter 6). For in Fig. 4.1 we represent two uniform movements by rectangles whose altitudes are the same and whose bases are of different length. The altitudes are representa-

est velocitas et tarditas) sed in quantitate motus (quae est longitudo vel brevitas temporis)." The italics in the English translation of this and the succeeding passage are my own.

[34] *Ibid.*, "Pro secundo dicendum quod aliqua agentia possunt esse proportionalia suis passis qualitative (scilicet in virtute agendi) et ex ista proportionalite sequitur aequalitas motuum qualitative (scilicet in velocitate et tarditate); vel quantitative, quod agendum in suum passum per totam suam quantitatem, et ex tali proportionalitate sequitur aequalitas motuum correspondenter (videlicet quantitative), et hoc est in longitudine et brevitate temporali motus."

[35] Let me remind the reader once more that the medieval authors do not distinguish between "velocity" as a vectorial quantity and "speed" as a scalar quantity. I have not attempted to decide in each case whether direction is implied; I have in fact used "velocity" as a translation of *velocitas* regardless of its particular context. A start toward vectorial distinction was made by Swineshead, when he declared that "the way *(via)* from *a* to *b* and the converse way from *b* to *a* are the same according to thing *(rem,* i.e., magnitude), but different according to nature *(rationem,* i.e., directionally)." See Clagett, "Swineshead," *Osiris*, Vol. *9* (1950), 158, note 51.

tive of the "qualities" of the two velocities; and since they are equal, these velocities can be said to be equal in "quality." But since the total areas (i.e. the total spaces traversed, or the sum of all the velocity perpendiculars through definite periods of time) are different, the velocities are different in quantity. In this case, the ratio of the "quantities" of velocity is as the

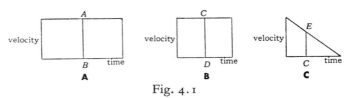

Fig. 4.1

ratio of the bases or the periods of times. It can also be seen that qualitative velocity is at the same time representative of instantaneous velocity. AB is not only the measure of the quality of velocity of the total movement (fig. A) but also of the instantaneous velocity at instant B. In a uniform acceleration (fig. C) there is obviously no single "quality" of velocity; rather it varies as the instantaneous velocity represented by EC varies from instant to instant. The distinction between qualitative and quantitative velocities in non-uniform motion becomes then an intuitive appreciation of the distinction between instantaneous and average velocities, or rather between instantaneous velocity and the total space traversed in some given time. I hardly need say that in view of their lack of a clear-cut definition of velocity as S/t and of infinitesimal calculus, the fourteenth-century authors *never defined* instantaneous velocity as the limit of a ratio, i.e. by dS/dt.

In the case of all "intensive" measures as opposed to "extensive" measures, the Merton kinematicists were of course faced with the problem of how the magnitude of intensity was to be determined. In the particular case of intensive or instantaneous velocity (as in the case of specific gravity), they made the measure of intensity depend ultimately on units of extension. Thus Heytesbury in Doc. 4.4 (passage 3) tells us that "the velocity at any given instant will be determined by the path which *would** be described by the most rapidly moving point if, in a period of time, it were moved uniformly at the same degree of velocity with which it is moved in that given instant, whatever given instant be assigned."[36]

* The italics are mine here and in the next two quotations.

[36] The reader might find it interesting to compare this statement of Heytesbury with the following remarks from a modern treatment of calculus (*The New International Encyclopaedia*, 2d, ed., Vol. *4* [1914], p. 311): "Studying the motion of a ball thrown up in the air, we consider infinitely small intervals of time dt merely in order to think of the motion as uniform; for within any finite interval the motion is

[at Merton College   215]

Similarly, Richard Swineshead says (Doc. 4.5, passage 2), "To every degree of velocity (i.e., qualitative or instantaneous velocity), there corresponds a lineal distance which *would* be described, assuming a movement throughout the time at this degree." It will be noticed that Galileo speaks in the same way regarding the measure of instantaneous velocity (Doc. 4.7), when he says "if a body were to continue its motion with the same degree or moment of velocity it acquired in the first time interval, and continue to move uniformly with that degree of velocity, then its motion would be twice as slow as that which it *would* have if its velocity had been acquired in two time-intervals." Thus he compares two instantaneous velocities in a uniform acceleration by positing two uniform movements over some time period and then comparing these uniform velocities. Some authors, however, without the intuition of the Merton logicians would say that "instantaneous motion is neither swift nor slow, because swift or slow are defined in time."[37]

We are now prepared to turn to our documents which contain more extended kinematic passages from the works of the Merton authors. The first document (Doc. 4.1) consists of select passages from chapter 4 of Bradwardine's *De proportionibus*. This section is the one devoted to kinematics in general, but more specifically to the problem of assigning a particular punctual velocity by which we can represent rotating bodies.

variable. But if at any given instant the motion should actually become uniform and continue so, we might think of our differential dt as representing any finite length of time, be it 5 minutes, or 10 minutes, or 500 minutes. For when a body moves with perfectly uniform speed, that speed may be readily determined by ascertaining the distance traversed during any interval of time whatever; the result is the same whether we divide the distance traversed in 5 minutes by 5, or that distance traversed in 10 minutes by 10. We may, accordingly, define the differential or distance dl as the distance that *would* be traversed by the ball in an arbitrary, finite interval of time, dt, beginning at a given instant, if at that instant the motion became uniform. In this manner we may avoid thinking of infinitely small quantities."

[37] Marsilius of Inghen (?), *In libros physicorum*, edit. cit. in note 29, p. 366, c. 1: "Secundo, notandum quod sicut in qualitatibus est extensio qualitatis, & intensio, ita est etiam in motu: nam extensio subiecti in qualitatibus vocatur extensio qualitatis; & multitudo graduum in eadem parte subiecti vocatur intensio qualitatis, sed duratio illius vocatur eius extensio; & ideo sicut differunt qualitas maior, & qualitas intensior, ita etiam differunt motus maior, & motus velocior: quia ille motus est maior, qui maiori tempore durat, & ille motus velocior, quo in minori tempore maius pertransitur. Ex quo patet, quod velox, & tardum definiuntur in comparatione ad tempus, & ad spatium, quod acquiritur tanto, vel tanto tempore. Et ex isto sequitur, quod motus instantaneus non est velox, neque tardus: quia velox & tardum definiuntur tempore."

In contrast to the theory of Gerard of Brussels, which used the velocity of the middle point of a rotating radius, Bradwardine advances the opinion that we should use the speed of the fastest moving point.

The main reason for this change on Bradwardine's part is that he has basically a different objective. He is not interested in finding a motion of translation where the area or volume swept out is equivalent to that swept out in rotation. Rather, he is interested in assigning arbitrarily a significant punctual curvilinear velocity representing a line, surface, or body in rotation. He has in the back of his mind the astronomical problem of rotating spheres and the comparison of their movements. Hence, he is interested primarily in the velocity of the points on the circumferences of great circles, which points are, of course, the fastest moving points.

The opinions of both Bradwardine and Gerard were mentioned by the author of the treatise composed about 1350 by someone intimately familiar with the Oxford schools and entitled *Tractatus de sex inconvenientibus* in a rather undistinguished treatment of the problem of rotary motion.[38] Over at Paris at about the same time John Buridan, while accepting Bradwardine's dynamic description of motion (see below Chapter 7, footnote 40), rejects his idea that we should use the speed of the fastest moving point as the measure of the speed of a revolving sphere and appears rather to accept Gerard's view, saying that it is by using mean values that we reduce nonuniformity to uniformity.[39] A very neat summary of the

---

[38] See Chapter 5, footnote 8, for the Latin citation of the position on rotary motion of the *Auctor de sex inconvenientibus*.

[39] John Buridan, *Questiones super octo physicorum libros Aristotelis* (Paris, 1509), f. 15c, c. 1: "...suppono etiam quod si columna fuerit eque longa ex uno latere sicut ex alio ita quod ex utroque latere sit decem pedum et alia columna fuerit difformiter (!) scilicet ex uno latere decem pedum et ex alio latere novem pedum, prima columna erit longior quam secunda in semipede, quia longitudo corporis non est solum in eius dextro vel solum in sinistro vel in medio sed est in dextro, medio, et sinistro. Ideo non debet simpliciter dici longum vel tante longitudinis ex dextro latere solum vel ex sinistro sed coniunctim ex dextro, sinistro, et medio. Et si non sit uniformitas longitudinis, oportet inferre longius ad minus longum auferendo de longiori latere et apponendo minus longo ut inveniatur medium. Et hoc etiam probatur per simile, quia si a lucido illuminaretur aer ad certam distantiam, puta usque ad leucam, continue difformiter ita quod propre lucidum sit lumen intensum et ad finem sit in infinitum remissum lumen ut omnino nichil ultra sit de lumine et vocetur ille aer totalis *b* et millesima pars eius propinqua lucido vocetur *a* que est valde intense illuminata, ergo tunc non est dicendum quod aer *b* debeat dici eque intense illuminatus sicut aer *a* nisi tu velis denominare totum ab infinita parva parte eius dimittendo denominationem secundum omnes alias partes, quod videtur absurdum. Est etiam aliud simile, si baculus sit uniformiter difformiter coloratus secundum album et nigrum secundum processum de cono *a* ad conum *b*, ab albissimo ad nigerrimum, tunc non est

problem raised by the English kinematicists was given further treatment in the second half of the century by Buridan's disciple, Albert of Saxony, whose contributions to dynamics we shall have an opportunity to treat later. An examination of Document 4.2 will show that, while following Bradwardine, Albert at least distinguished curvilinear and angular velocities, much as his contemporary at the University of Paris, Nicole Oresme, did (see Doc. 6.1, passage II.4).

The third document appended to this chapter (Doc. 4.3) consists of extracts from what is believed to be another work by Bradwardine, the *Tractatus de continuo*. We cannot date this work accurately, although we may presume that it postdates Bradwardine's *Tractatus de proportionibus* (1328), since it makes clear reference to that work (see Doc. 4.3, conclusion 23). It is probable that, if the work is truly by Bradwardine as one of the three extant manuscripts asserts and another implies, then it probably predates 1335 when Bradwardine left Merton College for less academic pursuits. The brief passages we have extracted from the *De continuo* show that already the new vocabulary arising from the quantitative approach to forms was taking hold, particularly as applied to problems of local motion. This included the use of *gradus motus*, with a definition that clearly links it with the measure of the "quality" of motion. It will also be noted that in his suppositions eight and nine, we have concise statements to the effect that when[40] $T_1 = T_2$, then $V_1/V_2 = S_1/S_2$, and that when $S_1 = S_2$, then $V_1/V_2 = T_2/T_1$. We have already seen in the preceding chapter that such statements had antecedents in Greek antiquity. However, there is some difference in the medieval statements as compared to the antique ones. In both the definitions of Autolycus and the theorems of Archimedes from the *Spiral Lines*, we are assuming uniform speeds and then are relating spaces and times; the proportionality statements do not involve "velocity" itself as some kind of magnitude, but rather relate distances to each other or times to each other. On the other hand, Gerard of Brussels in his *De*

dicendum quod a cona *a* debeat denominari albissimum, pari enim ratione diceretur a cono *b* nigerrimum et sic simpliciter loquendo dicetur albissimum et nigerrimum, quod est impossibile. Ergo ad simpliciter denominandum oportet recompensare inter partes, ut a medio fiat simpliciter denominatio. Et ideo manifestum est quod mensurantes superficiem quanta sit vel corpus quantum sit reducunt difformitates ad uniformitatem. Propter quod videtur michi concludendum corelarie (!) quod non a velocitate puncti velocissime moti in circumferentia debet denotari simpliciter velocitas totalis spere, sicut multi communiter dicunt dimittendo denominationem ab omni alia parte...." I have altered punctuation and capitalization somewhat.

[40] As noted in the preceding chapter, I use capital letters when expressing Euclidean proportionality statements.

motu (Doc. 3.1, supposition 8) tells us that "the proportion of the movements (i.e., velocities) of points is that of the lines described in the same time." Here we are relating "motions" or "velocities" as if they themselves are magnitudes mutually comparable. Gerard's supposition is essentially Bradwardine's seventh supposition. Bradwardine's eighth supposition extends the treatment of velocities as magnitudes in a sense distinct from spaces by relating the velocities inversely as the times. Now it appears to me that such a shift from the ancient identification of velocity with space to the medieval conception of velocity as a magnitude in its own right was necessary before we could have the modern idea of a velocity as a ratio of the two unlike magnitudes of space and time. As I have said before, this shift of viewpoint was absolutely necessary before we could arrive at the correct view of instantaneous velocity as ds/dt, i.e., as a limit of a ratio.

Our next three documents (Docs. 4.4, 4.5, and 4.6) serve to confirm and elaborate the definitions of kinematic terms which we presented in tabular form earlier in this chapter. The first of these documents is a selection from Heytesbury's *Regule solvendi sophismata*. It is particularly noteworthy for its concern with, and definition of, instantaneous velocity, with its definitions of uniform velocity and uniform acceleration. The selection from Heytesbury is followed by one from a fragment probably by Richard Swineshead (Doc. 4.5). It will be noticed that this document includes much the same distinctions and definitions; its interesting distinction of dynamics from kinematics we have already quoted earlier in this chapter. Document 4.6, a selection from the *De motu* of John of Holland, is an illustration of how the English kinematics had spread even to Eastern Europe (Prague) by the 1360's or 1370's. How even earlier the definitions influenced the Italians, by 1350 or so, is illustrated in the kinematic passages in Document 6.2, taken from Giovanni di Casali's *De velocitate motus alterationis*. The way in which these same definitions reached France and were utilized by Nicole Oresme about the same time is treated in Chapter 6 and illustrated by Document 6.1. Further definitions are included in the anonymous summary of mechanical ideas appearing with the 1505 edition of Bradwardine's *Proportions* (see Doc. 7.1).

Finally, I have attempted in the documentary section of this present chapter to extract certain definitions and statements from Galileo's *Two New Sciences* (Doc. 4.7) which serve to show the persistence of the Merton vocabulary and definitions. In the commentary to that document, I have included a brief paragraph from one of Galileo's youthful works wherein he makes the usual Merton distinctions between uniform and difform qualities, and between uniformly difform and difformly difform qualities. This paragraph demonstrates that he was conversant at that time with some account of the scholastic doctrine of the latitude of forms.

Document 4.1

Thomas Bradwardine, *Treatise on the Proportions of Velocities in Movements**

CHAPTER IV, PART 2. A number of different views concerning the proportion of velocities in motions in relationship to space [traversed] are held by different people. For it seems to some that the proportion of local motions in respect to velocity is just as that of the corporeal (i.e., three-dimensional) volumetric spaces described in the same time. This is refuted easily. For if it were so, then the whole of a body would be moved twice as rapidly as its half. And also it would follow that, if any body would traverse a corporeal, volumetric space of a foot in length in an hour, and half of the body would traverse a corporeal, volumetric space of double length in an hour, the two would be moved equally fast *(aequevelociter)*. Also the movement of a point or a line could not be compared in speed to the movement of a body, because there would be no way to compare quantitatively the spaces described by them.

Others, indeed, posit that the proportion of local movements in respect to velocity is as the proportion of the surfaces described in the same time. This opinion, moreover, could be refuted in the same way as the first one.

The author of the treatise *On the Proportionality of Movements and Magnitudes* (i.e., the *De motu* of Gerard of Brussels), much more subtle than the others, posits that in the case of equal straight lines moved in equal times, that which traverses more superficial (i.e., two-dimensional) space and to greater termini is moved more swiftly; and that [which traverses] less [area] and to lesser termini [is moved] more slowly; and that [which traverses] an equal area and to equal termini he supposes to be moved equally fast. He understands by "greater" termini final termini *(termini ad quos)* which are a greater distance from the initial termini *(termini a quibus)*.

This position, however, seems to be contrary to reason in some respect.

* Translated from the Latin edition of H. Lamar Crosby, Jr. (Madison, 1955), pp. 128–30.

[Bradwardine, *Proportions of Velocities* 221]

For according to it, any part at all of a rotated radius not terminated at the center, as well as the whole radius itself, would be moved equally as its middle point, as the first conclusion of the first [part] of this treatise posits. Consequently [the segment] would be slower than its end point (i.e., the point farthest from the center). Then the equinoctial circle would be moved in a four-thirds proportion more swiftly than its diameter, as the first conclusion of the second part wishes. And accordingly the equinoctial radius would neither be moved more swiftly, nor more slowly, nor equally fast as some moving body, no point of which is at rest. For it (the radius) does not traverse some space to certain *(aliquos)* termini, but rather to a unique terminus, since one end of the radius is not moved.

Therefore, it can be said more reasonably that the speed of local motion is measured *(attenditur)* by the speed of the fastest moving point of a body moved locally.... Now that space by whose magnitude the speed of local motion is to be measured is neither corporeal nor superficial, as was proved before; it is, therefore, linear space....[Hence this is a first supposition:] [1] In any local motion at all the speed is assumed with respect to the maximum linear space described by some point of the body. [2] In the case of any two local motions, their speeds are directly proportional to the maximal lines described in the same time by two points, belonging respectively to the two bodies in motion....

COMMENTARY

It should be pointed out that Bradwardine does seize upon an ambiguity in Gerard's definition of the relative velocities of lines. Gerard has said one line moves more quickly than another when it traverses a greater area *et ad maiores terminos*. Bradwardine thinks this means "to more distant termini." He then asks how we can apply such terminology to a rotating line where one end of the line (at the axis) does not move to any terminus at all. We can extend this criticism one step further by saying that when we compare a rotating line to its equivalent movement in translation, the phrase "to equal termini" does not seem to apply to the movements of both rotation and translation even though they sweep out equal areas in the same time. For only the midpoints of the two lines move to equally distant termini. Gerard no doubt realized this but assumed that "on the average" the two lines go to equal termini, for all the less distant termini of the points on the bottom half of the radius are just compensated for by all the "farther" distant termini of the points of the top half of the radius. It appears to me that Bradwardine's selection of the most rapidly moving point as the measuring point is even more arbitrary. Although useful for measuring the movement on the surfaces of spheres, i.e., useful for astronomy, it makes less kinematic sense than Gerard's attempt to reduce the complete nonuniformity of the curvilinear velocities of the points of the rotating figures to uniformity.

Document 4.2

Albert of Saxony, *Questions on the Eight Books of the Physics of Aristotle*

1. BOOK VI, QUESTION 5. It is sought, in the fifth place, how *(penes quid)* the velocity of circular movement is measured *(attendatur)*.... In the preceding question it was seen how the velocity of rectilinear local movement is measured in regard to effect. We should like now to investigate the same thing with respect to circular movement. In this matter there are two things to be considered: namely, "movement" and "making a circuit" *(circuitus)*. Although they may be the same thing, yet they are different according to connotation and nature *(rationem)*. For "making a circuit" adds to the connotation of movement [the idea of] describing an angle around the center or the axis according to which the movement takes place. Hence, it is quite possible for two things to make a circuit *(circuire)* equally fast and yet to be moved unequally fast, and conversely. Whence, if two points are designated, namely, *a* on the equinoctial circle and *b* near the pole, *a* and *b* make a circuit equally fast around the axis of the world, and yet they are not moved equally fast, for the points near the equinoctial move much more swiftly than the points near the pole. Hence, it could be conceded that some movement is swifter than another [with respect to curvilinear velocity], but that the same movement is not a swifter circuit *(circuitionem)* than the other.

In the first place, it should be observed how the velocity of movement simply in circular motion is measured; in the second place, how the velocity of "making a circuit" is measured.

2. [Conclusions relative to the linear velocity of circular movement:]

With respect to the first [division of my treatment] there is this first conclusion: Velocity of movement in circular motion is measured by the space described in such or such a time, as is obvious by the description of velocity posited by Aristotle in the sixth book of this work (i.e., the *Physics*).

The second conclusion: [Velocity of movement in circular motion] is not measured by corporeal space.

Third conclusion: It is not measured by superficial (i.e., two-dimensional) space. These [last] two conclusions are proved in the same way that they were proved with regard to rectilinear local movement.

Fourth conclusion. [The velocity] is measured by linear space. This is proved as follows: By the first conclusion [it is proved that] the velocity of circular movement is measured by space, and by the second conclusion [it is shown that it is] not [measured] by corporeal space, nor with superficial (two-dimensional) space by the third conclusion. Therefore, etc. The implication holds from [the argument of] "sufficient division."

But [granting that velocity is measured by linear space], there is a doubt as to whether it is measured by the linear space described by a point in the convex [surface], i.e., in the circumference of a great circle of the spherical or orbicular body circularly moved, or by a point in the middle, or as to just where the point is.

Concerning this matter there are a number of opinions. One opinion imagines that just as a movement uniformly difform corresponds to its mean degree, as when the velocity of a movement of alteration uniformly difform is measured by the uniform quality corresponding to the mean degree of the quality uniformly difform acquired through a uniformly difform alteration, so also in a local movement uniformly difform the velocity is measured *(mensuratur)* by the line described by the middle point among the points moved uniformly difformly. This occurs in such a way that this opinion accordingly wishes that if there were two spherical bodies circularly moved, and if equal linear spaces were described by the middle points of their radii in the same time, these bodies would be equally swiftly moved. This opinion is reasonable enough. Nevertheless, the velocity of an orbicular (i.e., shell-like) body could not be very well assumed as in this opinion because the middle point of its radius would lie outside of the orbicular body. Now it does not seem that the velocity of any body ought to be measured *(mensurari)* by something which is not within it. For this reason, some say that the velocity of an orbicular body ought to be measured by the linear space described by a point midway between the concave and convex (i.e., innermost and outermost) surfaces. But we can argue against this opinion, for it would thus follow that if a condensation of the orbicular body took place toward the convex surface without the convex surface receding from the center, the orbicular body would by this [theory be said to] be moved faster. This seems to be inconsistent. The implication *(consequentia)* is proved, for by such condensation the middle point between the convex and concave [surfaces] would become more distant from the center, and consequently a greater circumference would be described. Hence, if the velocity is measured by the linear space described by such a point, what was said [above] follows.

Another more common and more famous opinion posits the conclusion that the velocity of a movement of a circular body or of an orbicular body moved circularly is measured by the linear space described by the most swiftly moving point of the body. Such a point is in the circumference of one of the great circles of that circular or orbicular body; e.g., in the heavens it is in the equinoctial [circle]. It (the conclusion) is proved from this, that the whole is moved as swiftly as any of its parts; hence, the spherical body is moved with just as great a velocity as any of its parts. Therefore, the velocity of the whole and that of the most swiftly moving part ought to be measured by the same thing. In the second place, since the movement of the most swiftly moving point is more knowable *(notior)*, hence it seems quite reasonable that we should measure the movement and velocity of such a body by the space described by such a point. [This opinion] is confirmed by the fact that if the velocity is not measured by the space described by the middle point of the radius, nor by the point midway between the concave and convex [surfaces], then it seems that [it is measured] by the space described by a point located in the convex [surface]; therefore, etc.

3. With respect to the second [division of my treatment], I say that the velocity of "making a circuit" is measured by the angle described around the center or axis of the moved body [i.e., the axis] about which the circuit takes place. So that if two moving bodies would circuit the same center or axis and would describe about the center or axis equal angles in equal periods of time, then these moving bodies would be circuiting equally fast. If, however, they would describe unequal angles about the aforementioned [center or axis] in equal periods of time, then they would be circuiting unequally fast. The conclusion is obvious. For otherwise point b designated as near the pole would not revolve equally fast around the axis of the universe as point a designated as on the equinoctial [circle].

From these arguments it follows that "making a circuit" is in no way comparable with a rectilinear local movement, since an angle is in no way comparable to a linear space. Now the velocity of making a circuit is measured by an angle, while the velocity of rectilinear local movement is measured by a linear space. Nor is the velocity of making a circuit comparable to circular movement, since an angle is not comparable in any way to a circular line....

COMMENTARY

1. The purpose of this question is to determine the kinematic measure of circular motion, i.e., its measure in terms of "effect." First, according to Albert, we must distinguish "movement" and "making a circuit." This is a distinction between curvilinear and angular velocities.

2. The curvilinear velocity of a rotating sphere or shell is measured by the traversal of space in time, but not by either the volumetric or areal surface described, but rather by the linear space described by some point of the rotating body. But which point is the proper point? It would appear reasonable to measure velocity by the linear space traversed by the middle point of the radius (as Gerard wished), but this presents some difficulty in the movement of a shell, since the middle point of the radius lies outside of the shell, and it does not appear proper to measure the velocity of a body by the motion of a point lying outside of the body. If one uses the point midway between the innermost and outermost circumferences of the shell, Albert makes this objection: If condensation toward the outer surface takes place without the outer surface being any farther from the center, the velocity of the shell would be measured continually by a faster and faster moving point. This seems to him absurd because the outer dimension of the shell has not changed. This is an objection which certainly seems without substance. We must conclude, according to Albert, that the velocity of the rotating body is measured by the linear (i.e. curvilinear) space traversed by its most rapidly moving point (namely, a point on the outer surface). This, of course, is the opinion of Bradwardine, presented in Document 4.1. It is extraordinary that these authors did not realize (or at least did not point out) the completely arbitrary nature of their selection of one or another points as the proper point to measure the speed of the rotating body.

3. Albert holds that angular velocity is measured by the angle swept out in a given time. He correctly observes that angular velocity is not directly comparable to linear velocity, for there is no way to compare an angle and a line. As we pointed out in the text, this distinction of velocities was well known at Paris. The popularity of Albert's treatise makes his the most significant of the accounts of angular and curvilinear velocities. It is for this reason it is included here, even though chronologically it dates from almost a generation later than the Merton documents.

Questiones Alberti de Saxonia in octo libros physicorum Aristotelis*

Liber Sextus. Questio quinta. Queritur quinto penes quid attendatur velocitas motus circularis.... In ista questione precedenti videbatur de motu locali recto penes quod (!) attenditur eius velocitas tanquam penes effectum. Nunc videndum est de hoc quantum ad motum circularem, in quo duo sunt consyderanda, scilicet motus et circuitus, que, licet sint eadem res, tamen differunt secundum connotationem et rationem. Circuitio enim addit super connotationem motus describere angulum circa centrum vel axem secundum quem fit motus. Et ideo bene possibile est aliqua duo eque velociter circuire, ineque velociter tamen moveri, et econverso. Unde signatis duobus punctis, scilicet a in equinoctiali et b prope polum, a et b eque velociter circuunt axem mundi, non tamen eque velociter moventur. Puncta enim circa equinoctialem multo velocius moventur quam puncta circa polum. Et ideo posset concedi aliquem motum esse velociorem alio et eundem motum non esse velociorem circuitionem alio.

Primo ergo videndum est penes quid attenditur velocitas motus simpliciter in motu circulari. Secundo, penes quid attenditur velocitas circuitionis.

Quantum ad primum sit ista prima conclusio, quod velocitas motus in motu circulari attenditur penes spatium pertransitum in tanto vel in tanto tempore, sicut patet per descriptionem velocitatis positam ab Aristotele sexto huius.

Secunda conclusio, quod non attenditur penes spatium corporale.

Tertia conclusio, quod non attenditur penes spatium superficiale. Sic due conclusiones probantur sicut probabantur de motu locali recto.

Quarta conclusio, quod attenditur penes spatium lineale. Probatur, quia attenditur penes spatium per primam conclusionem et non penes corporale per secundam nec penes superficiale per tertiam; ergo et cetera. Consequentia tenet a sufficienti divisione.

Sed tunc est dubium ex quo velocitas attenditur penes spatium lineale, utrum penes spatium lineale descriptum a puncto existente in convexo, scilicet in circumferentia maximi circuli corporis spherici vel orbicularis circulariter moti, vel in medio, vel ubi.

* Edition of Paris, 1518, ff. 67v–78r (i.e. 68r). I have altered the punctuation.

Circa quod sunt opiniones. Una opinio imaginat* quod sicut motus uniformiter difformis correspondet suo gradui medio, taliter quod sicut velocitas motus alterationis uniformiter difformis attenditur penes qualitatem uniformem correspondentem medio gradui qualitatis uniformiter difformis acquisite per alterationem uniformiter difformem, sic etiam in motu locali uniformiter difformi velocitas mensuratur penes lineam descriptam a puncto medio inter puncta uniformiter difformiter mota, ita quod secundum hoc hec opinio vult quod si essent duo corpora spherica circulariter mota a quorum punctis mediis semidiametrorum in equali tempore equalia spatia linealia describerentur, quod illa corpora essent eque velociter mota. Et ista opinio est satis rationalis. Verumtamen secundum eam non posset bene capi velocitas corporis orbicularis propter hoc quod punctus medius sui semidiametri est extra illud corpus orbiculare. Modo non videtur quod velocitas motus alicuius corporis debeat mensurari penes aliquid quod non est in seipso. Et ideo alii dicunt quod velocitas corporis orbicularis debeat attendi penes spatium lineale descriptum a puncto medio inter superficiem concavam et superficiem convexam. Sed contra arguitur, nam sic sequeretur quod si fieret condensatio corporis orbicularis versus convexum, convexo non recedente ab eis centro, quod ex hoc illud corpus orbiculare moveatur velocius. Hoc videtur esse inconveniens. Probatur consequentia, nam per talem condensationem punctus medius inter convexum et concavum fieret magis distans a centro et per consequens maior circumferentia describeretur. Ergo si penes spatium lineale descriptum a tali puncto medio attenditur velocitas, sequitur quod dictum est.

Alia opinio communior et famosior ponit illam conclusionem, quod velocitas motus corporis circularis vel orbicularis circulariter moti attenditur penes spatium lineale descriptum a puncto velocissime moto† talis corporis, et talis punctus est in circumferentia unius maximi circuli illius corporis circularis vel orbicularis, verbi gratia, in celo est in equinoctiali. Probatur ex eo quod aliquod totum movetur ita velociter sicut aliqua eius pars; ideo tanta velocitate movetur corpus sphericum quanta velocitate movetur aliqua eius pars; et ideo penes idem debet attendi velocitas totius et partis velocissime mote.

Secundo, quia motus puncti velocissime moti est notior, et ideo bene rationale est quod penes spatium descriptum a tali puncto motum et velocitatem talis corporis mensuremus. Et confirmatur ex quod non

---

* Imaginatur *in edit*.   † modo *in edit*.

intenditur (attenditur?) penes spatium descriptum a puncto medio semidiametri, nec a puncto medio inter concavum et convexum, videtur quod penes spatium a puncto existente in convexo; quare et cetera.

Quantum ad secundum dico quod velocitas circuitionis attenditur penes angulum descriptum circa centrum vel axem corporis moti circa quod vel quem fit circuitio, ita quod si duo mobilia circuirent aliquod centrum vel axem et in equalibus partibus temporis equales angulos describerent circa illud centrum vel axem, tunc illa mobilia eque velociter circuirent. Si autem describerent inequales angulos circa predicta in partibus equalibus temporis, tunc ineque velociter circuirent. Patet conclusio, quia aliter b punctus signatus prope polum non eque velociter circuiret axem mundi sicut punctus a signatus in equinoctiali.

Ex his sequitur quod circuitio nullo modo est comparabilis motui locali recto, ex eo quod angulus nullo modo est comparabilis spatio lineali. Modo velocitas circuitionis attenditur penes angulum, velocitas autem motus localis recti attenditur penes spatium lineale. Nec velocitas circuitionis est comparabilis motui circulari, ex eo quod nullo modo angulus est comparabilis linee circulari; ergo [et cetera].

Document 4.3

Thomas Bradwardine, *On the Continuum*

[Definitions]

1. A CONTINUUM is a quantity whose parts are mutually joined.

2. A permanent continuum is a continuum whose individual parts persist at the same time.

3. A successive continuum is a continuum whose parts are successive with respect to "before" and "after"....

7. An indivisible is that which can never be divided.

8. A point is an indivisible in position.

9. Time is a successive continuum measuring succession.

10. An instant is a certain atom of time.

11. Motion is a successive continuum measured by time.

12. A "moved being" is the indivisible end of a motion.

13. The matter *(materia)* of motion is that which is acquired through motion.

14. A grade *(gradus)* of motion is that part of the matter of motion susceptible to "more" and "less"—which matter is acquired through some "moved being"....

23. An infinite taken absolutely and categorematically is a quantity without end.

24. An infinite taken syncategorematically or relationally *(secundum quid)* is a finite quantity, and a finite quantity greater than that finite quantity, and a finite quantity greater than that greater finite quantity, and so on without any final term; that is, it is a quantity, yet one not so great but that it can be greater.

[Suppositions]

6. Every body, surface, line, and point can be moved uniformly and continually.

7. In the case of two local motions which are continued in the same

or equal times, the velocities and distances traversed by these [movements] are proportional, i.e., as one velocity is to the other, so the space traversed by the one is to the space traversed by the other.

8. In the case of two local motions traversing the same or equal spaces, the velocities are inversely proportional to the times, i.e., as the first velocity is to the second, so the time of the second velocity is to the time of the first.

9. A [given] moving body can be moved with any whatsoever quickness or slowness or a [given] space can be traversed by any [body] at all.

[Conclusions]

22. In the case of any finite straight line one of whose termini is at rest, the other terminus can be made to revolve uniformly and continually, so that the whole line and any part of it as large as you wish but terminated at the immobile terminus describes a circle and any of its points in motion forms a circumference of a circle....

23. If a finite straight line is moved circularly with one of its termini at rest, then any two line segments, terminating on the one end at the immobile point and at the other end at movable points, you most certainly know to be proportional to the velocities of the movable points. Although this is clear enough by the first conclusion of the second part of the fourth chapter of the *De proportione velocitatum in motibus*, it can however, as there, be briefly demonstrated....

24. There can be found a uniform and continuous local motion swifter or slower in any proportion of finite straight line to finite straight line than any designated local motion at all. Hence it is manifest that any finite space at all can be traversed uniformly and continuously in any finite time at all....

26. If anything is moved locally and continuously, it does not acquire multiple positions *(situs)* in the same instant, nor can it be in the same position in diverse instants....

COMMENTARY

As I pointed out in the text, we can conclude that the work is later than 1328 since in conclusion 23 (and in fact in several other places) it cites the *Tractatus de proportionibus* of that date. While this treatise is of more importance for the study of the history of ideas concerning continuity than for mechanics, at least a few citations from it are of interest to us. The definition of *gradus motus* (definition 14) is of considerable interest.

The substance of this definition is that the *gradus* is the measure of the intensity of motion, for it is to the intensive aspects of a quality that the words "more" and "less" belong, as distinct from "great" or "little" (see Clagett, "Swineshead," *Osiris*, Vol. 9 [1950], 132). It, then, is the measure of "qualitative" velocity rather than "quantitative" velocity. The significance of suppositions 8 and 9 for the development of the idea of velocity as a magnitude representing a ratio of unlike quantities has already been discussed in the body of the chapter. Conclusions 22 and 23 hearken back to the discussions of Gerard of Brussels and Bradwardine on velocities of rotating lines and need no further discussion. Conclusion 22 merely confirms that velocities can be compared geometrically, i.e., as straight lines. In short, uniform velocities are comparable in every way that lines are comparable.

*Liber Th. Bradwardini de contimo**

[Definitiones]

1. Continuum est quantum cuius partes adinvicem copulantur.
2. Continuum permanens est continuum cuius partes singule manent simul.
3. Continuum successivum est continuum cuius partes succedunt secundum prius et posterius.
4. Corpus est continuum permanens, longum, latum, at profundum.
5. Superficies est continuum permanens, longum, latum, sed non profundum.
6. Linea est continuum permanens, longum, non latum nec profundum.
7. Indivisibile est quod numquam dividi potest.
8. Punctus est indivisibile situatum.
9. Tempus est continuum successivum successionem mensurans.
10. Instans est certus athomus temporis.
11. Motus est continuum successivum tempore mensuratum.
12. "Motum esse" est indivisibile finis motus.
13. Materia motus est quod per motum acquiritur.

* MS Thorn R 4° 2, pp. 153–92 (*T*); Q. 385, ff. 17r–48r (*E*). cf. MS Erfurt, Stadtbibliothek, Amplon.

3 continum *T*
5 manent simul *tr. E*
6 succedunt *T* successive *E*

9–10 Superficies ... profundum *om. E*
14 situatum *T* situ *E*
15 successionem *T* successive *E*

14. Gradus motus est illud materie motus suscipientis magis et minus, quod acquiritur per aliquod "motum esse"....

23. Infinitum cathegorematice et simpliciter est quantum sine fine.

24. Infinitum synkathegorematice et secundum quid est quantum finitum, et finitum maius isto, et finitum maius isto maiori, et sic sine fine ultimo terminante; et hoc est quantum et non tantum quin maius....

[Suppositiones]

6. Omne corpus, superficiem, lineam atque punctum uniformiter et continue posse moveri.

7. Omnium duorum motuum localium eodem tempore vel equalibus temporibus continuatorum velocitates et spatia illis pertransita proportionales existere, i.e., sicut una velocitatum ad aliam ita spatium per unam velocitatem pertransitum ad spatium per aliam pertransitum.

8. Omnium duorum motuum localium super idem spatium vel equalia deductorum velocitates et tempora proportionales econtrario semper esse, i.e., sicud velocitas prima ad secundum ita tempus secunde velocitatis ad tempus prime.

9. Quacunque velocitate vel tarditate potest unum mobile moveri vel unum spatium pertransiri potest quocunque....

[Conclusiones]

22. Cuiuslibet recte linee finite uno termino quiescente potest reliquus eius terminus circulariter uniformiter et continue circumferri, tota recta et qualibet parte eius magna ad terminum eius immobilem terminata circulum describente et quolibet eius puncto moto

---

20 magis: maius *E*
22 kathegorematice *E* / *ante* quantum *add. E* tantum
23 synkategorematice *E*
24 maiori *om. E*
25 ultimo non *E* / non tantum quin *T* termino terminatum quam *E*
28 lineam *om. T*
29 et continue *om. T*
31 *post* pertransita *add. E* eodem tempore
32 i.e. sicut *E* semper *T* / ita *E* vel *T*
33 unam velocitatem *E* istam *T* / per... pertransitum *T* quod per aliquod

pertransitur *E*
34 vel *E* simul *T*
35 deductorum *E* deditorum *T* / velocitates et tempora *om. T*
36–37 i.e.... prime *om. T*
38 quacunque *T* quantum *E*
39 unum *om. E* / pertransiri potest *om. T* / quocunque *E* quodcumque *T*
42 reliquus *T* alius *E*
42–43 circumferri *E* transferri *T*
43 qualibet *T* quelibet *E*
44 circulum *om. E* / describentem *T* describente *E* / quolibet *T* quodlibet *E*

circumferentiam circuli faciente....

23. Si recta finita super unum eius terminum quiescentem circulariter moveatur, omnes duas rectas terminatas ad punctum immotum et alia puncta mota et velocitates istorum punctorum proportionales certissime scias esse. Licet ista per primam conclusionem secunde partis quarti capituli de proportione velocitatum in motibus satis apparet, potest tamen ut ibi breviter demonstrari....

24. Quocunque motu locali signato potest motus localis uniformis et continuus in omni proportione recte finite ad rectam finitam velocior et tardior inveniri. Unde manifestum est, quodcunque spacium finitum quocunque tempore finito posse uniformiter et continue pertransiri....

26. Si quid continue localiter moveatur, in eodem instanti non acquirere multos situs, nec in eodem situ in diversis instantibus esse posse....

45 circumferentiam E circumstantiam T / faciente T facientem E
46 quiescentem E quiescente T
46–47 circulariter moveatur T circummovetur E
47 rectas *om.* T
48 mota T immota E / proportionales *om.* T
49 scias T habeas E / ista E prima T /
primam T proximam E / conclusionem E conclusionis T
50 quarti T aut E / satis *om.* E
55 quocunque E quodcunque T / finito T finitum E
55–56 pertransire E
57 continue *om.* E
58 aquirit T / nec *om.* E

Document 4.4

William Heytesbury, *Rules for Solving Sophisms**

[Part VI. Local Motion]

[Prologue]

THERE are three categories or generic ways in which motion, in the strict sense, can occur. For whatever is moved, is changed either in its place, or in its quantity, or in its quality. And since, in general, any successive motion whatever is fast or slow, and since no single method of determining velocity is applicable in the same sense to all three kinds of motion, it will be suitable to show how any change of this sort may be distinguished from another change of its own kind, with respect to speed or slowness. And because local motion is prior in nature to the other kinds, as the primary kind, we will carry out our intention in this section, with respect to local motion, before treating of the other kinds.

[1. Measure of Uniform Velocity]

Although change of place is of diverse kinds, and is varied according to several essential as well as accidental differences, yet it will suffice for our purposes to distinguish uniform motion from nonuniform motion. Of local motions, then, that motion is called uniform in which an equal distance is continuously traversed with equal velocity in an equal part of time. Nonuniform motion can, on the other hand, be varied in an infinite number of ways, both with respect to the magnitude, and with respect to the time.

In uniform motion, then, the velocity of a magnitude as a whole is in all cases measured *(metietur)* by the linear path traversed by the point which is in most rapid motion, if there is such a point. And according as the position of this point is changed uniformly or nonuniformly, the complete motion of the whole body is said to be uniform or difform (nonuniform). Thus, given a magnitude whose most rapidly moving point is

* The translation kindly supplied to me by Ernest Moody has been slightly altered.

moved uniformly, then, however much the remaining points may be moving nonuniformly, that magnitude as a whole is said to be in uniform movement....

[2. Measure of Nonuniform Velocity]

In nonuniform motion, however, the velocity at any given instant will be measured *(attendetur)* by the path which *would* be described by the most rapidly moving point if, in a period of time, it were moved uniformly at the same degree of velocity *(uniformiter illo gradu velocitatis)* with which it is moved in that given instant, whatever [instant] be assigned. For suppose that the point A will be continuously accelerated throughout an hour. It is not then necessary that, in any instant of that hour as a whole, its velocity be measured by the line which that point describes in that hour. For it is not required, in order that any two points or any other two moving things be moved at equal velocity, that they should traverse equal spaces in an equal time; but it is possible that they traverse unequal spaces, in whatever proportion you may please. For suppose that point A is moved continuously and uniformly at C degrees of velocity, for an hour, and that it traverses a distance of a foot. And suppose that point B commences to move, from rest, and in the first half of that hour accelerates its velocity to C degrees, while in the second half hour it decelerates from this velocity to rest. It is then found that at the middle instant of the whole hour point B will be moving at C degrees of velocity, and will fully equal the velocity of the point A. And yet, at the middle instant of that hour, B will not have traversed as long a line as A, other things being equal. In similar manner, the point B, traversing a finite line as small as you please, can be accelerated in its motion beyond any limit; for, in the first proportional part of that time, it may have a certain velocity, and in the second proportional part, twice that velocity, and in the third proportional part, four times that velocity, and so on without limit.

From this it clearly follows, that such a nonuniform or instantaneous velocity *(velocitas instantanea)* is not measured by the distance traversed, but by the distance which *would* be traversed by such a point, *if* it were moved uniformly over such or such a period of time at that degree of velocity with which it is moved in that assigned instant.

[3. Measure of Uniform Acceleration]

With regard to the acceleration *(intensio)* and deceleration *(remissio)* of local motion, however, it is to be noted that there are two ways in which a motion may be accelerated or decelerated: namely, uniformly, or nonuniformly. For any motion whatever is *uniformly accelerated (uniformiter intenditur)* if, in each of any equal parts of the time whatsoever, it acquires an equal increment *(latitudo)* of velocity. And such a motion is uniformly decelerated if, in each of any equal parts of the time, it loses an equal increment of velocity. But a motion is *nonuniformly accelerated or decelerated*, when it acquires or loses a greater increment of velocity in one part of the time than in another equal part.

In view of this, it is sufficiently apparent that when the latitude of motion or velocity is infinite, it is impossible for any body to acquire that latitude uniformly, in any finite time. And since any degree of velocity whatsoever differs by a finite amount from zero velocity, or from the privative limit of the intensive scale, which is rest—therefore any mobile body may be uniformly accelerated from rest to any assigned degree of velocity; and likewise, it may be decelerated uniformly from any assigned velocity, to rest. And, in general, both kinds of change may take place uniformly, from any degree of velocity to any other degree.

COMMENTARY

Notice in this passage the definitions of uniform velocity, uniform acceleration, and instantaneous velocity. In the definition of uniform velocity, Heytesbury speaks of the traversal of an equal space in an equal part of the time. Thus he failed to say in *any* equal parts of the time. That this would appear to be understood by him is clear, however, from his definition of uniform acceleration in terms of the acquisition of an equal increment of speed in *any* equal parts of the time. Heytesbury's contemporary, Richard Swineshead, was careful to specify that uniform velocity is to be defined by the traversal of an equal distance in *every (omni)* equal period of time (see the succeeding document of Swineshead). Hence Swineshead, at least, was anticipating Galileo's admonition to include the word "any" in a proper definition of uniform motion (see the Galileo selection, Doc. 4.7, below).

One final important point should be noticed about this selection from Heytesbury's *De motu*. For him instantaneous velocity is to be measured or determined by the path which *would* be described by a point if that point were to move during some time interval with a uniform motion of the velocity possessed at the instant. It will be noticed in the passage from Galileo's *Two New Sciences* quoted below as Document 4.7 that he did much the same thing as Heytesbury.

Regule solvendi sophismata Guillelmi Heytesberi*

Tria sunt predicamenta vel genera in quorum quolibet contingit proprie motum esse; mutatur enim localiter, quantitative, aut qualitative, quodlibet quod movetur. Et cum universaliter motus quilibet successivus velox sit vel tardus, nec aliquid est idem univocum penes quid attendi poterit velocitas in hiis tribus, conveniens erit ostendere qualiter quecunque mutatio huiusmodi, quo ad eius velocitatem seu tarditatem, ab alia sui generis distinguatur. Et quia motus localis naturaliter precedit alios tanquam primus, circa ipsum in hac parte transcurrens saltem intentio ceteris premittatur.

Loci autem mutatio, quamvis diversas habeat species, et tam essentialibus quam etiam accidentalibus differentiis pluribus varietur, ad propositum tamen sufficiet motum uniformem distinguere a difformi. Motuum igitur localium dicitur uniformis quo equali velocitate continue in equali parte temporis spacium pertransitur equale. Difformis quidem in infinitum variari potest, et respectu magnitudinis et etiam quo ad tempus.

In uniformi itaque, penes lineam a puncto velocissime moto descriptam, si quis huiusmodi fuerit, quanta sit totius magnitudinis mote velocitas universaliter metietur. Et penes hoc quod punctus talis uniformiter seu difformiter mutat situm, totius totus motus uniformis

---

* MS Bruges, Stadsbibliotheek 497, f. 56r (*A*); MS Bruges, Stadsbibliotheek 500, ff. 56v–57r (*B*); MS Vat. lat. 2136, ff. 24v–25r (*V*); Edition of Venice, 1494. ff. 37r–39v (*Ed*). Some differences in orthography between these various copies have been omitted in the variant readings. I have on occasion changed lower-case letters to capitals, e.g., motus *a, b, c* to motus *A, B, C*. I have included somewhat more in the Latin text than in the translation.

---

2 quorum quolibet: quibus *Ed*
3 proprie... esse: motum fieri proprium *Ed* / mutatur: movetur *B* / qualitative aut quantitative *A* / aut: vel *B*
4 universaliter *BVEd* uniformiter *A*
5 idem *om. AV*
6 erit: est *Ed*
7 huiusmodi: prius in *A* / seu: seu eius *A*
8 ab: ab ad *B* / alia: altera *A*
12 etiam *om. Ed*
13 sufficiet motum *tr. B*
14 igitur: ergo *Ed* / motuum... localium *AEd* motus igitur localis *B* / dicitur
om. Ed. ille dicitur *V* / quo *BV* que *A* est quo *Ed*
15 parte temporis *ABEd* tempore seu temporis parte *V* / pertransiretur *Ed*
16 in *om. B* / variari potest *tr. V* / potest: poterit *Ed*
18 In... itaque: in motu itaque uniformi *V*
19–20 mote velocitas *tr. B*
20 metietur *V* mestietur *B* mentietur *A* mensuratur *Ed* / Et *om. B*
20–21 talis uniformiter *tr. A*
21 seu: vel *A* / situm: scitum *B* / totus *om. Ed*

[*Regule solvendi sophismata Heytesberi*

dicitur vel difformis. Unde data magnitudine cuius punctus velocissimus uniformiter moveatur, quantumcunque difformiter residua omnia differantur, uniformiter moveri conceditur tota proposita magnitudo.

Posito nempe casu quo mote magnitudinis nullus sit punctus velocissime motus, penes lineam quam describeret punctus quidam qui indivisibiliter velocius moveretur, quam aliquis in magnitudine illa data tota, totius velocitas attendetur: Sicut posito quod continue incipiant corrumpi puncta extrema, aut quod nulla sint ultima puncta illius sicut accidit in linea girativa que ponitur infinita. Movetur enim omnis magnitudo, mota localiter, ita velociter sicut aliqua pars ipsius, aut sicut aliquis eius punctus; unde in casu isto indivisibiliter velocius movetur talis magnitudo quam aliquis eius punctus. Et ideo, iuxta illud, conceditur tanquam possibile quod continue tardius et tardius movebitur mobile *A* per horam, et tamen continue per eandem horam erit ita quod quilibet punctus illius qui movetur, velocius movetur quam prius, et intendit motum suum. Et similiter quod *B* magnitudo per horam continue eque velociter movebitur, et uniformiter saltem quo ad tempus, et tamen quilibet punctus illius continue per eandem horam tardabit motum suum. Primum, per continuam corruptionem punctorum extremorum ipsius mobilis, satis poterit verificari; et secundum absque inconvenienti concedi poterit de linea girativa. Potest etiam concedi tanquam imaginabile similiter, quod *A* magnitudo per horam continue velocius et velocius movebitur, et tamen continue erit ita per eandem horam, quod quilibet punctus motus ipsius *A*

---

22 *post* difformis *tr. A* dicitur
23 *ante* uniformiter *add. A* motus / residua omnia *tr. Ed* reliqua omnia *V* residua *A*
24 conceditur: contra *B*
24–25 proposita magnitudo *tr. VEd*
26 mote: note *Ed* / magnitudinis motus *A*
27 penes: tunc penes *V* / describeret: difformis et *B*
28 movetur *A* / *ante* quam *add. Ed* ac si esset
29 illa *om. B* / totius velocitas: tota velocitas totius *A* / attendatur *Ed*
30 incipiat *B* / puncta *om. Ed* / ultima puncta *tr. V*
31 illius: istius *A* / accidet *Ed* / enim: autem *A*
32 mota *om. B* / pars ipsius: eius pars *Ed*
33 isto *AB* illo *VEd*
34 eius punctus *tr. BV*
35 illud: idem *A*
36 mobile A *tr. V Ed*
37 illius *BV* istius *A* eius *Ed* / qui *om. B* / movetur *tr. B post* prius
38 *ante* quod *add. Ed* conceditur
40 illius: eius *Ed*
42 ipsius *VB* illius *Ed* istius *A*
44 etiam concedi *tr. Ed*
46 puncta *B* / motus ipsius A *BV* istius A illius *Ed*

remittit motum suum; et hoc per adventum continuum novarum partium secundum extremum velocius motum.

Et quamvis universaliter omnis magnitudo mota localiter, ita velociter moveatur sicut aliqua eius pars, ex hoc tamen non sequitur etiam ita tarde moveatur sicut aliquid eiusdem. Tarditas enim est quasi privatio velocitatis, non obstante quod de virtute sermonis omnis tarditas sit velocitas et econverso—sicut quelibet parvitas est magnitudo—et universaliter, ab habitu et perfectione sua denominanda est res quecunque habens illam. Unde omne continuum, quamvis quamcunque parvitatem habeat quam habet aliqua sui pars, non est tamen aliquid quod infinite parvitatis dicitur. Consimiliter, quamvis omni tarditate qua movetur pars alicuius, moveatur et totum; non tamen requiritur quod ita tarde moveatur totum sicut aliqua eius pars.

In motu autem difformi, in quocunque instanti attendetur velocitas penes lineam quam describeret punctus velocissime motus, si per tempus moveretur uniformiter illo gradu velocitatis quo movetur in eodem instanti, quocunque dato. Posito enim quod A punctus per horam continue intendet motum suum, non oportet quod in aliquo instanti totius hore attendatur velocitas illius penes lineam quam describet punctus ille in illa hora. Ad hoc enim quod aliqua duo puncta seu aliqua alia duo mobilia eque velociter moveantur, non requiritur quod in equali tempore equales pertranseant magnitudines, sed stat quod inequales pertranseant in quacunque proportione volueris imaginari. Posito enim quod A punctus continue uniformiter

---

47 remittit: continue remittit V / continuum: continue B
49 universaliter *om.* Ed *tr.* A ante localiter / mota localiter *tr.* A / localiter *om.* B
50 moveatur *tr.* A *post* pars / ex: sed ex V / ex hoc *om.* A / tamen non *tr.* A
51 etiam AV *om.* Ed totum B / moveatur *om.* A / aliquid: aliqua eis pars B
52 quasi *om.* V / virtute: utate (?) V
53 sit: est Ed / econverso AEd econtra BV
54 ab *om.* B / sua *om.* B
54–55 denominanda est: nominanda A est nominanda B
55 est *om.* A / quecunque: quod met B / illam: istam A
56 habet: habeat A / sui: eius Ed

57 dicitur: dicuntur Ed / Similiter V
58 et: ipsum V
59 tamen *om.* B / eius pars: pars ipsius B
60 *post* autem *add* Ed locali / attendetur AB attenditur V Ed
63 *ante* dato *add.* Ed instanti
64 intendet AB intendat V intenderet Ed / *ante* motum *add.* A moveri / aliquo: quocunque V
65 *ante* hore *add.* Ed illius / attendetur AB / illius BV istius A eius Ed
66 describit Ed / enim *om.* B / aliqua: sic B
67 alia *om.* AV / duo *om.* AB / moveantur *tr.* A *post* duo
68 pertranseant magnitudines *tr.* Ed
70 continue uniformiter *tr.* A
70–71 uniformiter moveatur *tr.* Ed

[*Regule solvendi sophismata Heytesberi* 241]

moveatur *C* gradu velocitatis per horam, pertranseundo pedalem quantitatem, et *B* punctus a quiete incipiat uniformiter intendere motum suum, in prima medietate illius hore acquirendo illum *C* gradum, et in secunda medietate uniformiter remittat motum suum ab eodem gradu usque ad quietem; tunc notum est quod in medio instanti totius hore movebitur *B* punctus *C* gradu velocitatis, et eque velociter omnino cum ipso *A* puncto; et tamen in medio instanti illius hore non erit tanta linea pertransita per *B* punctum, sicut per *A*, ceteris paribus.

Similiter, pertranseundo lineam solummodo finitam quantumcunque modicam volueris, potest *B* punctus in infinitum velocitare motum suum; quia super primam partem proportionalem aliqua velocitate, super secundam dupla, et super tertiam quadrupla, et sic in infinitum. Ex quo manifeste sequitur quod huiusmodi velocitas difformis seu instantanea, non attenditur penes lineam pertransitam, sed penes lineam quam describeret punctus talis, si per tantum tempus vel per tantum uniformiter moveretur illo gradu velocitatis quo movetur in illo instanti dato.

Est autem circa intensionem et remissionem motus localis advertendum, quod motum aliquem intendi vel remitti dupliciter contingit: uniformiter scilicet aut difformiter. Uniformiter enim intenditur motus quicunque, cum in quacunque equali parte temporis, equalem acquirit latitudinem velocitatis. Et uniformiter etiam remittitur motus talis, cum in quacunque equali parte temporis, equalem deperdit latitudinem

---

71 C: ad *B* / gradu: gradum *B* / per horam *tr. A ante* pedalem / transeundo *Ed* / pedalem: precise *(?) B*
72 B: E *B* / a quiete *tr. A post* incipiat / uniformiter *om. A*
73 prima *VEd om. AB* / illius: istius *A* / illum *om. Ed*
74 secunda: 4ª *(?) A* instanti *B* / motum *om. A* / suum *om. AB* / *ante* ab *tr. A* usque
75 usque *om. V*
76 totius: ipsius *B* / B: E *B*
77 medio instanti *tr. AB*
78 erit: est *B* / pertransita *V* pertransiri *AEd* in pertransiri *B* / B: C *B*
79 ceteris: cum *A*
80 solum *B*
81 B: A *A* / punctum *B* / in *om. B*

83 *ante* super *add. V* et / et¹ *om. A* / quadrupla *tr. BV post* et
84 *post* quo *add. BV* satis
85 instantanea: distantia *B* / attenderetur *AB*
86 describet *Ed* / talis *om. B*
88 illo: isto *A*
90 vel: aut *Ed* / *post* remitti *add. E* est / contingit *om. Ed*
91 uniformiter scilicet *tr. V* / scilicet *om. A* / aut: vel *V* / enim *om. A* autem *V*
92 quacunque: aliqua *(?) A*
93 *post* velocitatis *add. Ed* motus / Et uniformiter *tr. A*
94 cum: esse *B* / *post* equalem *add. A* partem *(?)* / latitudinis *A*

velocitatis. Difformiter vero intenditur aliquis motus, vel remittitur, cum maiorem latitudinem velocitatis acquirit vel deperdit in una parte temporis quam in alia sibi equali.

Iuxta illud sufficienter apparet, quod cum latitudo motus seu velocitatis sit infinita, non est possibile aliquod mobile ipsam uniformiter acquirere in aliquo tempore finito. Et quia quilibet gradus velocitatis per latitudinem tantummodo finitam distat a non gradu, seu termino privativo totius latitudinis, qui est quies; ideo a quiete ad gradum quemcunque datum, contingit aliquod mobile uniformiter intendere motum suum; et consimiliter, a gradu dato contingit motum uniformiter remittere ad quietem; et universaliter, a quocunque gradu ad quemcunque alium contingit utramque mutationem fieri uniformen.

95 vero *AV* enim *B* autem *Ed* / aliquis motus *tr. V*
96 *post* maiorem *add. B* velocitatem et *om.* velocitatis
97 sibi equali *tr. Ed*
98 sufficienter: etiam *Ed* / seu: vel *A*
98–99 velocitas *V*
99 non *Ed* quod non *ABV* / ipsam: illam *Ed*
100 *ante* acquirere *add. A* autem / Et: Sed *B*
101 *ante* per *add. V* finitus / *post* seu *add. V* a
102 privativum *B*
103 quemcunque *om. A* quecum (?) *B* / quecun[que] datum *tr. V* / datum *om. B*
104 motum *om. V* / dato *om. A* / contingit *tr. A ante* a / motum *tr. A ante* ad
105 universaliter: uniformiter *B*
106 *post* alium *add. Ed* gradum

Document 4.5

On Motion (A Fragment)
Attributed to Richard Swineshead

1. HAVING spoken generally regarding the measure of velocity with respect to cause, now we should see its measure with respect to effect. For this it should be known that some local motion is uniform and some is difform (nonuniform = *difformis*). For uniform local motion, it should be known that its velocity *(velocitas)* is measured *(attenditur)* simply by the line described by the fastest moving point in some given time *(in tanto tempore et in tanto)*, if there is any such fastest moving point in the magnitude in motion....

2. The cause of why velocity of this motion is measured by a described line is this: To every degree of velocity in local motion (i.e., of qualitative or instantaneous velocity) there corresponds a linear distance which *would* be described in some given time, assuming a movement throughout the time at this degree. Just as in the manner described above, there is a certain degree of movement corresponding to some degree of a proportion between agent and patient (i.e., resistance), so there is a certain distance traversed in some given time which corresponds to a certain degree of speed.

3. Furthermore it should be known that uniform local motion is one in which in every *(omni)* equal part of the time an equal distance is described.

Difform (nonuniform) movement is that in which more space is acquired in one part of the time and less in another equal part of the time.

4. Similarly for acceleration *(intensio localis motus)* and for deceleration *(remissio localis motus)*, it should be known that local motion can be accelerated *(intendi)* in two ways, namely, uniformly and difformly (nonuniformly); and similarly with regard to its deceleration. Difform (nonuniform) acceleration is of two kinds, either where the rate of acceleration is increasing *(velocius et velocius)* or where it is decreasing *(tardius et tardius)*.

Furthermore, wherever there is increase *(intensio)* or decrease *(remissio)* of speed—either as uniform or difform increase, or uniform or difform decrease—we have difformity of speed.

5. It ought to be known also that just as acceleration *(intensio motus)* is related *(se habet)* to speed, so speed is related to space; for just as space is acquired in speed, so speed is acquired in acceleration. Hence just as in the case of uniform local motion where the velocity is measured by the maximum line described by some point [in some given time], so in acceleration the rate of acceleration is measured by the maximum increment *(latitudo)* of velocity acquired in some given time.

6. Wherever there is uniform increase *(intensio)* of local motion, the local motion is uniformly difform motion. Since local motion uniformly difform corresponds to its mean degree [of velocity] in regard to effect, so it is evident that in the same time so much is traversed by means of [a uniform movement] at the mean degree as by means of the uniformly difform movement.

7. Furthermore, any difform motion—or any other quality—corresponds to some degree....

COMMENTARY

It is clear that Swineshead's definitions are very similar to those of Heytesbury, and indeed on the basis of extant evidence it is impossible to say which of the two wrote down these definitions first. Like Heytesbury, Swineshead made it clear that to determine the magnitude of instantaneous or qualitative velocity *(gradus velocitatis)* there must be sought a corresponding distance which *would* be traversed in some given time, assuming a uniform movement through that time at that instantaneous velocity (passage 2).

Also of interest is passage 5, in which he sets up a proportion between acceleration and speed on the one hand, and speed and distance on the other, with the statement that these proportions are equal, i.e., $A/V = V/S$. It is quite clear that the concept of acceleration *(intensio motus)* has been constructed from the analogy to speed itself; but in place of the acquisition of space we have the acquisition of speed. This type of analogy also appealed to Galileo, as the selection below reveals (Doc. 4.7).

De motu Ricardi Swineshead*(?)

Dicto generaliter penes quid habet attendi velocitas in quocunque motu quo ad causam, videndum est penes quid attenditur in motu quo ad effectum. Pro quo sciendum quod motuum localium quidam est uniformis et quidam difformis. Unde pro motu locali uniformi est illud sciendum, quod velocitas in motu locali uniformi attenditur simpliciter penes lineam descriptam a puncto velocissime moto in tanto tempore et in tanto, si aliquis fuerit talis punctus velocissime motus in tali magnitudine....

Causa autem quare penes lineam descriptam velocitas illius motus attenditur, est hoc: cuicunque gradui in motu locali correspondet certa distantia linealis que in tanto tempore et in tanto cum partibus tali gradu describeretur. Consimiliter in toto [modo] superius aliquo gradui proportionis agentis ad passum correspondet certus gradus motus, sicut certo gradui motus correspondet certa distantia que in tanto tempore et in tanto tali gradu pertransitur.

Unde sciendum quod motus localis uniformis est quo in omni parte temporis equali equalis distantia describitur.

Difformis motus quo in una parte temporis plus adquiritur et in alia parte temporis equali minus.

Similiter pro intensione motus localis et remissione est sciendum quod motus localis potest intendi dupliciter, scilicet uniformiter et difformiter, et etiam remitti.

Difformiter dupliciter, quia aut velocius et velocius, aut tardius et tardius. Unde ubicunque est intensio et remissio—sive fuerit intensio uniformis sive intensio difformis sive remissio uniformis sive difformis—est motus difformis.

Sciendum est etiam quod consimilter se habet intensio motus ad motum sicut se habet motus ad spatium, quia sicut per motum pertransitur spatium sic per intensionem motus adquiritur motus. Unde sicut in motu locali uniformi attenditur velocitas penes maximam lineam descriptam ab aliquo (f. 213r) puncto, sic in intensione motus attenditur velocitas penes maximam latitudinem motus adquisitam in tanto temporis vel in tanto. Ubicunque est intensio motus localis uniformis est motus localis uniformiter difformis. Cum motus localis uniformiter difformis correspondet quo ad effectum suo medio gradui, sic patet quod tantum per idem tempus ponitur pertransiti per medium

* MS Cambridge, Gonv. and Caius 499/268, ff. 212v–213r.

[gradum] sicut per illum motum localem uniformiter difformem. Unde quiscunque motus difformis, et etiam quecunque alia qualitas, alicui
40 gradui correspondet....

Document 4.6

John of Holland,* *On Motion*

.. OF local motions some are uniform, some are difform (nonuniform = *difformis*). Uniform local motion is of two kinds. Some is uniform as to magnitude and some is uniform as to time. Local motion uniform as to magnitude is the motion of some magnitude whereby the whole magnitude is moved equally fast *(equevelociter)* as any part at all and so the movement of a stone downward is moved uniformly as to magnitude. Motion uniform as to time is the motion of some moving body *(mobilis)* whereby the body traverses an equal space in every equal part of the time *(in omni parte equali temporis)* in which it is moved. From this it follows that the ninth sphere is moved uniformly as to time, but nonuniformly as to magnitude, because the parts of the ninth sphere nearer to the equinoctial circle are moved more quickly than the parts nearer the poles.

2. Difform motion is of two kinds. Some is difform as to magnitude and some is difform as to time. Motion difform as to magnitude is the motion of a magnitude where some part of the magnitude is moved more slowly than the whole magnitude (*actually*, than another part), as in the case of the movement of the ninth sphere. Similarly the motion of any sphere at all when rotated is difform as to magnitude.

3. Movement difform as to time is the motion of a mobile body whereby the body traverses more space in one part of the time than in some other equal part of the time. Motion difform as to time is twofold: some uniformly difform, some difformly difform. Local motion uniformly difform is described by the [English] Calculators as follows: "Local motion uniformly difform is a difform movement in *any* designated part of which the middle speed of that part exceeds the minimum terminal speed by the same increment *(latitudo)* as the middle is exceeded by the maximum terminal speed." Motion difformly difform occurs when uniformly difform movement is nonexistent.

* Another of John's works, his *Liber de instanti*, is dated 1369 at Prague in MS Oxford, Bodl., Canon. Misc. 177, f. 61v.

4. *To be moved as to time* is twofold, either uniformly or difformly. *To be moved uniformly as to time* is to traverse an equal space in every equal period of time.... *To be moved difformly as to time* is twofold, to increase speed *(intendere motum)* or to decrease speed *(remittere motum)*. *To increase speed* is twofold, uniformly or difformly. *To increase speed uniformly* is to acquire an equal increment of speed *(latitudo motus)* in every equal part of the time....

COMMENTARY

My principal purpose in including this passage from John of Holland's *De motu* is to show how the English ideas spread quickly to other parts of Europe. This document reveals their presence at Prague in the 1360's. Succeeding chapters will show how they spread to France and Italy by about 1350.

Worth special notice is John's quotation of a common English definition of uniform acceleration, namely that in any designated part of the time of that accelerated movement, $V_f - V_m = V_m - V_i$, where V_f, V_m, V_i are respectively the final, middle, and initial instantaneous velocities of the designated time period.

Finally, it is of some interest to observe that John of Holland called his Merton College predecessors the "Calculators." This was a title that was to stick, particularly to Richard Swineshead, who was known in Italy and elsewhere in the last part of the fourteenth and throughout the fifteenth century as the Calculator. The whole technique of treating qualities and motion in the Merton manner became known as "the calculations."

*De motu Johannis de Hollandia**

Motuum localium quidam est uniformis, quidam est difformis. Motus localis uniformis est duplex; nam quidam est uniformis quo ad magnitudinem et quidam est uniformis quo ad tempus. Motus
5 localis uniformis quo ad magnitudinem est motus alicuius magnitudinis quo ipsa magnitudo movetur equevelociter cum qualibet parte sua, et sic lapis motus deorsum movetur uniformiter quo ad magnitudinem. Motus uniformis quo ad tempus est motus alicuius mobilis

---

* MS Oxford, Bodl. Canon. Misc. 177, f. 100v, (*B*); MS Venice, Bibl. Naz. San Marco Lat. VIII, 19, f. 1r. (*V*); MS Vat. lat. 1108, f. 144r (*C*).

2 quidam²: et quidam *C* / est² *om. VC*
5 uniformis *tr. B post* magnitudinem
6–7 parte sua *tr. BC*
7 movetur *om. B*

quo ipsum mobile in omni parte equali temporis pro quo illud movetur pertransit spatium equale.

Ex isto sequitur quod nona spera uniformiter movetur quo ad tempus, difformiter quo ad magnitudinem, quia partes none spere circulo equinoctiali propinquiores velocius moventur partibus existentibus circa polos.

Motus difformis est duplex; nam quidam est difformis quo ad magnitudinem et quidam quo ad tempus. Motus difformis quo ad magnitudinem est motus magnitudinis quo movetur eiusdem magnitudinis aliqua pars tardius ipsa tota magnitudine (i.e., alia parte magnitudinis) mota, ut motus none spere; et similiter motus cuiuslibet spere mote circulariter est difformis quo ad magnitudinem.

Motus difformis quo ad tempus est motus mobilis quo ipsum mobile in una parte temporis plus pertransit quam in alia parte temporis sibi equali. Motus difformis quo ad tempus est duplex; nam quidam est uniformiter difformis, quidam difformiter difformis.

Motus localis uniformiter difformis describitur a calculatoribus sic: Motus localis uniformiter difformis est motus difformis, cuius quacunque parte signata medius gradus illius partis per equalem latitudinem excedit extremum remissius eiusdem sicud ipse ab extremo intensiori illius partis exceditur.

Motus difformiter difformis est motus difformis non existens uniformiter difformis. Moveri quo ad tempus est duplex, vel uniformiter vel difformiter. Uniformiter moveri quo ad tempus est in omni parte temporis equali equale spatium pertransire, vel secundum rem vel secundum equevalentiam. Difformiter moveri quo ad tempus est duplex, vel est intendere motum vel est remittere motum. Intendere motum est duplex, vel uniformiter intendere vel difformiter intendere. Uniformiter intendere motum est in omni parte temporis equali equalem latitudinem motus acquirere. Verbi gratia, quiescat Sortes

9 equali *om. VC* / pro quo *VB* quod *C*
10 pertransit spatium *tr. C*
11 quod *om. C* / uniformiter movetur *tr. B*
12 et difformiter *C*
13 propinquius *V*
16 *post* quidam *add. C* est difformis / difformis *om. C*
17 magnitudinis¹ *VC* difformis magnitudinis *B*
18 ipsa *V* ipsa scilicet *B* movetur ipsa *C*
19 ut *V* et *B* / motus *tr. C post* circulariter *in linea 20*
22 plus *VC* non *B* / quam *VC* tantum quantum *B*
23 difformis *om. V*
24 quidam² *V* et quidam *BC*
25 sic *tr. C ante* calculatoribus
28 remissius *om. V*
29 exceditur *tr. C ante* ab *in linea 28*
33 pertransire *BC* pertransiri *V*
38 equalem *VC* equalis *B*

(i.e. Socrates) iam et incipiat ipse moveri acquirendo latitudinem
40 motus a non gradu usque ad gradum uniformiter, sic quod totam
latitudinem acquirat in tota ista hora sequenti, sic quod ipsius medietatem acquirat in medietate eiusdem hore, quartam in quarta parte
hore, et sic correspondenter de aliis, et tunc Sortes uniformiter
intendet motum suum....

40 gradum *VB* & *C* / sic quod *V* sic
    quod sicut B quod sicut *C*
41 ipsius *VC* ipsum *B*
41–42 medietatem *VC* medietate *B*

42 acquirat *om. C* / quartam *VB* et
    quartam *C*
44 intenderet *C*

Document 4.7

Galileo Galilei, *The Two New Sciences**

THE THIRD DAY.—On equable (i.e., uniform) motion *(De motu aequabili)*. In regard to equable or uniform *(uniformis)* motion, we have need of a single definition, which I give as follows. I understand by equal *(equalis)* or uniform movement one whose parts *(partes)* gone through *(peracte)* during any *(quibuscunque)* equal times are themselves equal. We must add to the old definition—which defined equable motion simply as one in which equal distances *(spatia)* are traversed in equal times—the word "any" *(quibuscumque)*, i.e., in "all" *(omnibus)* equal periods of time.... (p. 197). On movement naturally accelerated *(accelerato)*.... If we examine the matter, we find no addition *(additamentum)* or increment *(incrementum)* more simple than that which always increases in the same way. This we readily understand when we consider the intimate relationship between time and motion; for just as equality, and uniformity, of motion is defined by, and conceived through, equalities of the time intervals and of the spaces.... so through the same equal periods of time we can conceive of increments of speed *(incrementa celeritatis)* simply added. Thus we may conceive that a motion is uniformly and continually accelerated when in any equal time periods equal increments of swiftness are added.... To put the matter more clearly, if a moving body were to continue its motion with the same degree or moment of velocity *(gradus seu momentum velocitatis)* it acquired in the first time-interval, and continue to move uniformly with that degree of velocity, then its motion would be twice as slow as that which it would have if its velocity *(gradus celeritatis)* had been acquired in two time-intervals. And thus, it seems, we shall not be far wrong if we assume that increase in velocity *(intentio velocitatis)* is proportional to *(fieri iuxta)* the increase of time *(temporis extensio)*.

* *Le Opere*, Ed. Naz., Vol. *8* (Florence, 1898), 191, 197–98. Cf. Translation of Henry Crew and A. de Salvio, (Chicago 1946), p. 148; although I have used the Crew translation in part, in some places I have made the translation very literal to reveal more clearly Galileo's degree of dependence on the Merton vocabulary.

COMMENTARY

Anyone paying due attention to the Latin terminology of this passage will see how dependent Galileo still was on the Merton vocabulary. Thus *equalis motus, uniformis motus, gradus velocitatis*, and *intentio velocitatis* were all part of both vocabularies. Note once more that Galileo in this passage compared the instantaneous velocities at the end of the first time-period and at the end of the second time-period (in a uniformly accelerated movement) by imagining that the bodies were moving uniformly over some time-period with these respective instantaneous velocities. This, as we have seen, was precisely what Heytesbury and Swineshead recommended in their treatment of instantaneous velocity. Needless to add also, the definitions of uniform velocity and uniform acceleration given by Galileo have their almost exact Merton counterparts.

The irrefutable proof that Galileo was familiar with the Merton vocabulary is given by a juvenile work, which bears the title *De partibus sive gradibus qualitatis* (Ed. Naz. Vol. *1*, 119–22). There we read (p. 120):

It should be noted in the third place that, since a quality is always in a quantitative subject, in addition to having proper (i.e., intensive) degrees it also participates in a latitude of quantity and can be divided into quantitative parts. If the degrees or parts of the quality are compared throughout the parts of the quantity, either [1] the degrees of quality will be equal in any part at all and then the quality will be called "uniform," or [2] the degrees will be unequal and then it will be called "difform." If the excesses of the parts will be equal so that in the first part there are two degrees, in the second part 4, in the third 6, and so on, so that the excess is always two, the quality will be called "uniformly difform." If however the excesses will be unequal, it will be called "difformly difform." Again if they will be unequal in such a way that in the first part, for example, there will be 4 degrees, in the second 6, in the third 9, and so on, then the quality is called "uniformly difformly difform." If in truth the excesses will not be proportional, it will be called "difformly difformly difform."[41]

---

[41] Advertendum est, 3°, cum qualitas semper sit in subiecto quanto, praeter proprios gradus participare etiam latitudinem quantitatis, et dividi posse in partes quantas. Quod si comparentur invicem gradus sive partes qualitatis cum partibus quantitatis, vel in qualibet parte quantitatis erunt aequales gradus qualitatis, et tunc dicetur qualitas uniformis; vel erunt inaequales gradus, et tunc dicetur difformis. Quod si excessus illarum partium erunt aequales, ita ut, si in prima parte sint duo gradus, in secunda sint 4, in tertia 6, et sic deinceps, ut excessus sit semper per duo, qualitas dicetur uniformiter difformis; si vero excessus erunt inaequales, dicetur difformiter difformis. Rursus, si excessus inaequales erunt ita, ut in prima parte, verbigratia, sint 4 gradus, in secunda 6, in tertia 9, et sic deinceps, tunc dicetur

[Galileo, *The Two New Sciences*

It is of interest to note that Galileo's conception of uniformly difformly difform motion, if based on his use of the terms for qualities, would not be that of uniformly changing acceleration, but rather of an acceleration such that in succeeding equal periods of time the *ratio* of the velocity increments is some constant (other than 1). This is the definition found in the *Tractatus de latitudinibus formarum* (see the commentary to Doc. 6.3). This leads us to suspect that Galileo was familiar with this treatise, no doubt in one of its published editions.

Finally, we can remark that Galileo was much preoccupied with the whole problem of the intension and remission of forms and more than once he had occasion to write about it (cf. Ed. Naz., Vol. *1*, 111–19, 133–57).

qualitas uniformiter difformiter difformis; si vero excessus non erunt proportionales, dicetur difformiter difformiter difformis."

Chapter 5

The Merton Theorem of Uniform Acceleration

WE have already singled out as one of the most important results of the Merton studies in kinematics the discovery of the theorem giving the measure of uniform acceleration in terms of its medial velocity, i.e., its velocity at the middle instant of the period of acceleration.[1] The significance of this theorem in the history of mechanics lay in the fact that once it was recognized that the free fall of bodies is an example in nature of uniform acceleration, the Merton theorem automatically expressed one form of the law of free fall. A Spanish-Parisian scholastic in the sixteenth century, Domingo de Soto, when discussing uniform acceleration and the Merton theorem, casually suggested as an example of uniform acceleration the free fall of bodies.[2] It was Galileo, however, who believed that he had confirmed experimentally the fall on an incline to be a movement of uniform acceleration and who specifically applied the acceleration theorem to free fall to derive his famous law. As we shall see in the next chapter, Galileo's demonstration of the theorem was quite similar to the fourteenth-century geometric proof associated with the names of Nicole Oresme and Giovanni di Casali.

The Merton theorem can be expressed in modern symbols as (1) $S = 1/2\, V_f t$, for acceleration from rest, where S is the distance traversed, V_f is the final velocity, and t is the time of acceleration; or as (2)

---

[1] It is this theorem that is understood when, for the sake of economy, I speak of "the mean speed theorem" or "the acceleration theorem."

[2] See Chapter 9 below, particularly footnote 21. See also footnote 4 of this present chapter.

$S = \left[(V_o + \frac{(V_f - V_o)}{2}\right]t$, for acceleration from some velocity V_o. Since, in the first case, $V_f = at$ (a being the acceleration), we can reduce (1) to the familiar formula $S = 1/2\, at^2$. In the second case $V_f - V_o = at$. And thus (2) reduces to the general formulation $S = V_o t + 1/2\, at^2$.

Now in the fourteenth and fifteenth centuries there were about twenty attempts (many of them mere repetitions of earlier efforts) to give a formal proof of the theorem. These proofs were basically of two kinds: arithmetical and geometrical. The former were the earlier and arose out of the Merton College activity. The latter may have had roots in Merton College, but they do not reach their perfection until between 1350 and 1360 at Paris. In either case the medieval proofs reveal an intuitive if incomplete grasp of the infinitesimal analysis necessary for a proper treatment of instantaneous velocity and acceleration.

We cannot date exactly the first discovery of the acceleration theorem at Merton. But since it appears alike in the works of Heytesbury, Swineshead, and John Dumbleton, all of whom were active at Merton in the 1330's and 1340's,[3] we would probably be not too far wrong to date its discovery somewhere close to the early years of the 1330's.

It is tempting to speculate on the origin of the theorem. It certainly did not come directly from any Greek source now known to us. But in a sense the whole acceleration problem, and particularly the problem of the measure of free fall, which we shall return to in Chapter 9, started with two passages from Aristotle's *Physics*.

The first of these, chapter 2 of Book VI, is that which we have examined at some length above in Chapter 3 and in which Aristotle gives three definitions of "the quicker." The second significant passage is that of chapter 7 of the same book.[4] Its main objective is to show that neither (1)

---

[3] See Chapter 4 above for the dating of these Merton scholars.

[4] In this chapter I have chosen to cite the Latin translation of the *Physics* known as the *nova translatio* from the Greek, since it was quite widely known by the schoolmen of the fourteenth century. Edition of Venice, 1495, ff. 103v–104v: "60. Quoniam autem omne quod movetur in tempore movetur, et in pluri maior magnitudo, in infinito tempore impossibile est moveri magnitudinem finitam.... Quod quidem igitur si aliquid moveatur eque velociter, necesse est finitum in infinito (! *but read* finito) moveri, manifestum est.... 61. Sed si non sit eque velociter, differt nihil. Sit enim in quo *AB* spatium finitum quod motum sit in infinito tempore et tempus infinitum in quo *CD*. Si igitur necesse est prius alterum altero motum esse, hoc autem manifestum, quod temporis in priori et posteriori alterum est motum. Semper enim in pluri alterum est motum esse, sive eque velociter mutet sive non eque velociter mutet, sive intendatur motus sive remittatur sive maneat, nihil minus. Accipiatur enim aliquid *AB* spatii quod sit *AH* (! *AE*), quod mensurat *AB*. Hoc itaque

can a finite movement occupy an infinite time, nor (2) can an infinite movement take place in a finite time. In pursuing these conclusions, Aristotle takes up and categorizes movement as uniform and nonuniform; and nonuniform movement is further subdivided into that in which the velocity is increasing or decreasing: "...it makes no difference... whether the movement (ἡ κίνησις, *motus*) is increased in intensity (ἐπιτείνῃ, *intendatur*), decreased in intensity (ἀνίῃ, *remittatur*) or is constant (μένῃ, *maneat*)." It is apparent that when he attempts to analyze the case of nonuniform movement to show that a finite movement cannot occupy an infinite time, his measure of the nonuniform movement is the varying time period for the traversal of equal spaces. On the other hand, it is equally obvious that when he proves the second conclusion, that infinite motion cannot take place in a finite time, his measure of nonuniform velocity is the varying space traversed in equal periods of time.

---

infiniti in quodam factum est tempore, in infinito enim non potest esse, omne enim in infinito est. Et iterum alterum iam si accipiamus quantum est *AH* (! *AE*) necesse est in finito tempore esse; omne enim est in infinito. Et sic accipiens, quoniam infiniti quidem nulla pars est que mensuret, impossibile enim infinitum esse ex finitis et equalibus et inequalibus, propter id quod mensurantur finita multitudine et magnitudine a quodam uno, sive equalia sive sint inequalia, finita autem magnitudine nihil minus, spatium autem finitum tantis que sunt *AB* mensuratur, ergo in finito tempore *AB* movetur.... 62. Eadem autem ratio et quod neque in finito tempore infinitum possibile est moveri, neque quiescere, neque quod regulariter movetur, neque quod irregulariter. Accepta enim quadam parte, que metietur totum tempus in hac aliquid tantum transibit magnitudinis, et non totam; in omni enim totam. Et iterum in equali aliam, et in unoquoque similiter, sive equalis est, sive inequalis ei, que est a principio. Differt autem nihil, si solum sit finita unaqueque. Manifestum enim, quod divisio (! diviso) tempore infinitum non aufertur, finita divisione facta et quanto et eo quot tot modis. Quare non transibit in finito tempore infinitum. Nihil autem differt magnitudinem in altera aut in utraque esse infinitam." I have changed the punctuation and capitalization somewhat. The better readings in parentheses I have taken from the copy of this translation given in Walter of Burley's *Super Aristotelis libros de physica auscultatione...commentaria* (Venice, 1589), cc. 797–802 and from the Greek text of W. D. Ross, *Aristotle's Physics* (Oxford, 1936), 237b.23–238a.32. It is of interest to note that the new medieval translation from the Greek contains as translations for ἐπιτείνω and ἀνίημι *intendere* and *remittere*, the terms that were to become standard in the fourteenth century to represent increasing and decreasing speed. The old translation from the Greek (or at least that copy of it in MS Paris, BN lat. 16141, f. 146r, c.1) has "extendatur" and "minuatur" for "intendatur" and "remittatur." Domingo de Soto, commenting on this same passage, suggests much the same thing he takes up later, namely that natural motion is an example of movement always increasing in swiftness: *Super octo libros physicorum Aristotelis commentaria* (Salamanca, 1582), p. 96, c.2: "Et in quidem sive mobile moveatur velocius, semper atque velocius, ut in motu naturaliter: sive tardius, ac tardius, ut in motu violento: sive maneat in eadem velocitate."

Now historically what seems to have happened is that Aristotle's definition of the quicker as the traversal of equal spaces in less time was applied to the parts of a uniformly accelerated movement, i.e., the Aristotelian *first* measure of nonuniform movement was made specific for the case of uniform acceleration by the application of one of his definitions of the quicker. This application appears to have produced the conclusion that in uniform acceleration, and more specifically in the acceleration of falling bodies, the velocity is *directly* proportional to the distance traversed. As we shall see here and in Chapter 9, this conclusion was supported in one way or another by Strato, Alexander of Aphrodisias, Simplicius, Albert of Saxony, and by Galileo before he hit upon the correct solution. On the other hand, another of Aristotle's definitions of the quicker—the quicker traverses greater space in the same time—when applied to uniform acceleration (i.e., when his *second* measure of nonuniform movement was made specific), appears to have led to the more fruitful conclusion that velocity in a movement uniformly accelerated is directly proportional to the time elapsed. As we shall see shortly, this conclusion appeared in a primitive way in the thirteenth-century *Liber de ratione ponderis* attributed to Jordanus. Its more mature expression is found in the acceleration theorem that was produced at Merton College and which is the object of our study in this chapter. But now let us backtrack and see specifically how acceleration was treated by Strato in antiquity and Jordanus in the thirteenth century.

The first explicit kinematic treatment of acceleration appears to have been that of Strato, who in 287 B.C. became the successor to Theophrastus as the head of the Lyceum. His account of acceleration appeared in a treatise *On Motion*, which is now lost; but fortunately the commentator Simplicius (*fl.* 519) quotes from this lost work in his *Commentary on the Physics of Aristotle*.[5] We are told that Strato had asserted in the treatise *On Motion* that a falling body as it accelerates "completes the last part of its trajectory (τὴν ἐσχάτην τοῦ τόπου) in the shortest time." Strato had then added, according to Simplicius, "In the case of bodies moving through the air under the influence of their weight, this is clearly what happens. For if one observes water pouring down from a roof and falling from a considerable height, the flow at the top is seen to be continuous, but the

---

[5] Simplicius, *In Aristotelis physicorum libros...commentaria*, edition of H. Diels, *Commentaria in Aristotelem graeca*, Vol. *10* (Berlin, 1882–95), 916, lines 4–30. Cf. M. R. Cohen and I. E. Drabkin, *A Source Book in Greek Science* (New York, 1948), p. 211.

water at the bottom falls to the ground in discontinuous parts. This would never happen unless the water traversed each successive space (τὸν ὕστερον τόπον) more swiftly (θᾶττον)." Possibly Strato gave a more extended kinematic analysis in his work, just as it is equally possible that Hipparchus, the famous astronomer of the second century B.C., gave a kinematic description of acceleration in a work (also lost) entitled *On Bodies Carried Down by Weight*. But these works were unknown to both the Arabic and Latin traditions, and we must be satisfied with these few remarks given by Simplicius.

We can represent these two closely akin descriptions of acceleration attributed to Strato as follows. The statement describing acceleration by the completion of the last part of its trajectory in the shortest time assumes that (1) for $S_1 = S_2 = S_3 = \ldots = S_n$, $t_1 > t_2 > t_3 > \ldots > t_n$. On the other hand, the last phrase characterizing acceleration in terms of the traversal of each successive space more quickly assumes (2) that for $S_1 = S_2 = S_3 = \ldots = S_n$, $V_1 < V_2 < V_3 < \ldots < V_n$. One can get from the description of (1) to that of (2) by applying one of Aristotle's definitions of the "quicker," namely that which defines the "quicker" as that which traverses the same space in less time. As we said before, Aristotle in defining the quicker was thinking of the comparison of two movements, while Strato was applying this kind of comparison to the parts of a single continuous movement which was not uniform. In summary, then, the definition of acceleration understood by Strato would be "a movement such that equal spaces are traversed in succeeding periods of less time, i.e., to say, at continually greater speed." Alexander of Aphrodisias must have had a similar definition of acceleration, as we shall show in Chapter 9, where the problem of free fall is discussed.

The natural question to ask at this point was whether the account of Simplicius was known to the medieval authors. It should be pointed out, in the first place, that no extant medieval translation of the complete commentary of Simplicius on the *Physics* is known. But recently I discovered a Latin fragment of Simplicius' commentary.[6] Furthermore, it is possible that knowledge of Simplicius' *Physics*, and perhaps even of this specific passage on acceleration, is reflected in the thirteenth-century *Liber de*

---

[6] I have published the text of this in "The *Quadratura per lunulas*: A Thirteenth Century Fragment of Simplicius' Commentary on the *Physics* of Aristotle," *Essays in Medieval Life and Thought Presented in Honor of Austin Patterson Evans* (New York, 1955), pp. 99-108.

ratione ponderis attributed to Jordanus.[7] Jordanus has given the example of the falling water, submitting it to elementary kinematic analysis:

> A liquid which is continuously poured forms a narrower stream at its lower end, to the degree that it falls further.
>
> Let the aperture through which the liquid escapes be AB (see Fig. 5.1), and

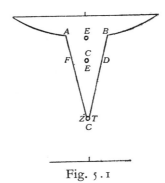

Fig. 5.1

> let the first portion of the liquid be C. When C shall have descended as far as DF, let the portion E be at the aperture. And then, when E has reached DF, let C be at ZT. Since, therefore, a body becomes heavier the more it descends, C will be heavier at DF than at AB. Hence it will be heavier at DF than E is at AB. But because, when E arrives at DF, C reaches ZT, the distance FZ will be longer than the distance AF and hence [the stream will be] more slender. Thus [the stream] will be continuously more slender, because the first parts are faster *(velociores)*, and thus, finally, it will break into drops.

One can readily see that the passage implies that when $t_1 = t_2 = \ldots = t_n$, then $S_1 < S_2 < \ldots < S_n$, for it specifically says "when E arrives at DF, C reaches ZT, [and] the distance FZ will be longer than the distance AF." Thus the quicker *(velociores)* parts of the accelerating movement are defined in terms of Aristotle's first definition of the "quicker," namely, the traversal of more space in an equal time. So we can deduce the following definition of acceleration for Jordanus: "Acceleration occurs when, in equal periods of time, greater and greater space is traversed, i.e., when in equal periods of time the speed increases."

Thus for Strato-Simplicius and Jordanus, acceleration reduces to a continuous movement one of whose parts is "quicker" than another, the "quicker" being defined as either the traversal of an equal space in less time, or more space in an equal time. The approach of Jordanus in terms

---

[7] Proposition R4.16. See the text in E. A. Moody and M. Clagett, *The Medieval Science of Weights* (Madison, 1952), pp. 224–27.

of more space in equal times was the more fruitful one in the fourteenth century. It will be noticed from the various definitions given of acceleration in the documents of the preceding chapter that the Merton College kinematicists describe the conditions of acceleration in terms of equal time-periods.

The main reason for the unsatisfactory nature of the discussions of Strato-Simplicius and Jordanus is that no clear concept of instantaneous velocity has yet developed. It is this concept, which we have already discussed in the preceding chapter, which differentiates decisively the Merton analysis of acceleration from all preceding treatments of the problem. And, indeed, it was the distinction between "qualitative" and "quantitative" velocities which probably led directly to the theorem of the mean speed. For from the concept of instantaneous velocity came, no doubt, the proper definition of uniform acceleration, and from the proper definition of uniform acceleration the mean speed theorem is immediately derivable.

There is, however, the interesting possibility that the mean speed theorem of uniform acceleration came by analogy from the theorem of Gerard of Brussels which equated the movement of a rotating line with the speed of its middle point. Our passage from John of Holland in the preceding chapter (Doc. 4.6) will help us understand how this might have happened. Following the English authors, John gave nonuniform motion a twofold division: as to magnitude and as to time. Now it is clear that Gerard in his theorem is talking about a movement uniformly nonuniform as to magnitude, i.e., the curvilinear velocities of the points on that rotating radius vary uniformly from zero velocity at the axis to a maximum velocity at the other end of the radius. The Merton kinematicists in their acceleration theorem were talking about a motion uniformly nonuniform as to time—where the instantaneous velocities vary uniformly, for example, from zero instantaneous velocity at the first instant of time to a maximum instantaneous velocity at the last instant of time. Now, since Gerard worked out the theorem that allowed him to reduce the nonuniformity of the motion uniformly nonuniform as to magnitude to uniformity [in terms of space described in a given time] by the use of the middle punctual velocity of the rotating magnitude, it seems at least possible that a later author drew the analogy in discussing the reduction of motion uniformly nonuniform as to time to uniformity (also in terms of space described in a given time) by trying the instantaneous velocity of the middle instant of the acceleration. At any rate, the analogy between the two theorems was remarked on by later authors of the fourteenth century.[8]

While we are in doubt as to the exact origin of the mean speed theorem, the earliest statement of the theorem that we can pin down to a specific date occurs in the *Regule solvendi sophismata* of William Heytesbury, dated 1335, as we have seen in Chapter 4. In that statement of the theorem (see Doc. 5.1, passages 1, 7), we see it stated as an obvious truth without proof.

For whether it commences from zero degree or from some [finite] degree, every

---

[8] The best example occurs in the treatment by Albert of Saxony of the question of the proper measure for circular motion: *Questiones in octo libros physicorum* (Paris, 1518), Bk. VI, quest. 5, f. 67v. For the Latin text of the pertinent passage, see Document 4.2 above, lines 37–48. The mysterious author of the *Tractatus de sex inconvenientibus*, written under the influence of Oxford about 1350, was also interested in comparing uniformity and nonuniformity of movements with respect to the subject and with respect to time. He explains (but not clearly) why it is possible to maintain with Bradwardine that a rotating body is measured by the speed of its fastest moving point, while a uniform acceleration in time is measured by the speed of the middle point of time, i.e. the mean speed (*Sex inconvenientes*, edition of Venice, 1505, no. pag. but see sig. h, col. 1): "Dico etiam ut ille magister Ricardus [i.e. Gerard of Brussels] demonstrat quod velocitas motus spere attendatur penes punctum medium: nec hoc tenendum est: sed forte videatur dicere quod tota latitudo motus localis correspondeat suo medio gradui: sicut consequenter conceditur. nec hoc repugnat huic, quod motus localis attenditur penes punctum velocissime motum. unde stant simul quod in omni motu sperali vel locali quocunque motu istius attendatur penes punctum velocissime motum et tamen in intensione motus ubi partes motus non remanent quod tota latitudo motus correspondeat suo gradui medio: sed hoc non oportet nisi motus extendatur: et per partes motus remaneret motum: sicut patet in motu spere: et illud videtur dicere magister guielmus eceberi (i.e. Heytesbury) in tractatu suo de motu: ...unde motus talis spere non attenditur penes latitudinem motus a centro spere usque ad circumferentiam que est uniformiter difformis: sed penes gradum quo movetur punctus velocissime motus qui in casu supposito manebit continue uniformis. nec hoc repugnat huic quod dictum est supra, quod tota latitudo movetur uniformiter difformis suo medio gradui: et responderet quod in motu spere extenso correspondet suo gradui ultimo et supremo ubi vero motus continue intenditur etiam extensus ibi habet opinio illa locum."

In the next question treated, namely, the mean speed theorem, the author conversely notes that one cannot argue that because one accepts that a circular movement is measured by the fastest moving point, that an acceleration in time is to be measured by the fastest or terminal speed of the acceleration. He would hold with the rest of the Oxford school that we must use the speed of the middle instant as the measuring speed (*ibid.*, sign. h 2 v, c.2): "et concedo quod in motu locali uniformiter difformiter incipiente a non gradu tota latitudo motus est equalis suo medio gradui. et hoc loquendo de motu qui continue est intendi, cuius nullus gradus acquisitus manet similis (? simul) cum alio, et per tempus; quod dico pro tanto: quia in motu extenso spere uniformiter et revolute: cuius quilibet gradus motus manet cum alio in tali motu extenso tota latitudo motus correspondet gradui intensissimo et extremo: sed in motu continue intenso et non extenso non oportet, nec est verum."

latitude (i.e., increment of velocity or velocity difference), as long as it is terminated at some finite degree (i.e., of velocity), and as long as it is acquired or lost uniformly, will correspond to its mean degree. Thus the moving body, acquiring or losing this latitude (increment) uniformly during some assigned period of time, will traverse a distance exactly equal to what it would traverse in an equal period of time if it were moved uniformly at its mean degree of velocity.... For every motion as a whole, completed in a whole period of time, corresponds to its mean degree—namely, to the degree which it would have at the middle instant of the time.

Other early statements of the mean speed theorem are found in the two short treatises on motion by Swineshead which are attached to the *Liber calculationum* in a Cambridge manuscript and which we have quoted from in the previous chapter. In the first tract we read (see Doc. 4.5, passage 6): "Since local motion uniformly difform corresponds to its mean degree [of velocity] in regard to effect, so it is evident that in the same time so much is traversed by means of [a uniform movement at] the mean degree as by means of the uniformly difform movement." In the longer tract much the same sort of thing is said:[9] "Motion uniformly difform always corresponds to its mean degree, and we are saying this everywhere in regard to motion nonuniform as to time.... Whence I affirm universally that uniform acquisition of motion corresponds to the mean degree of that latitude of motion uniformly acquired, and the same thing is correspondingly true concerning the loss of a latitude of motion."

However the mean speed theorem might have arisen at Merton, a number of interesting proofs of it were given by the Merton College kinematicists. It is difficult to date the relative order of the various Merton proofs. For a number of reasons, I have adopted (but with little surety) the following order for the proofs: (1) the proofs of Heytesbury (i.e., of the author of the treatise entitled *Probationes conclusionum*, who may not actually be Heytesbury), (2) those of Swineshead, (3) those of Dumbleton, (4) those of an anonymous author of a treatise entitled the *Tractatus de sex inconvenientibus*. While this relative order is by no means certain, we should not be too far wrong, as I have already said, in supposing that the theorem was discovered sometime during the early 1330's and that all of these

---

[9] MS Cambridge, Gonv. and Caius 499/268, ff. 214r–v: "Motus autem uniformiter difformis semper correspondet suo medio gradui et hoc loquendo utrobique de motu difformi quo ad tempus.... Unde dico universaliter quod uniformis motus correspondet medio gradui illius latitudinis motus sic uniformiter adquisite et idem correspondenter est verum de deperditione latitudinis motus."

proofs were completed before 1350. Of the various proofs, I have omitted here any extended discussion of those proofs appearing in the *De Sex inconvenientibus*, since they are the least worthy of analysis.[10]

I have attempted to analyze in some detail the various proofs in the commentaries following the translations below. But it would certainly

---

[10] See the *Sex inconvenientes*, Venice, 1505 (no pag., but sig. h–h2): "Utrum velocitas omnis motus localis uniformiter difformis incipiens a non gradu sit equalis suo medio gradui.... (h2r) Propter illa et similia argumenta que possent fieri, dicitur a quibusdam quod in latitudine motus localis terminata ad non gradum tota latitudo motus non est equalis suo gradui medio, nec sibi correspondet, sed solum gradui intensissimo, sic quod denominatio latitudinis totius sit a denominatione gradus intensissimi in illa latitudine, et proportio motuum secundum proportionem intensissimorum illorum motuum. Sed hoc totum est falsum, sicut arguitur in argumento ad oppositum articuli, quia improbare articulum est improbare illam positionem et ideo contra utrumque arguitur simul. Ad oppositum istius articuli arguitur. Et probatur (! ponatur?) quod in omni motu locali uniformiter difformi incipiente a non gradu est tota latitudo *a*. Tunc arguo sic: [in] *a* latitudine est aliquis gradus medius, puta *b*, qua est alia latitudo intensior et qua est alia latitudo remissior. Igitur in *a* latitudine est aliqua latitudo sibi equalis, et aliqua non; totum (! solum?) igitur tota est sibi equalis. Consequentia ultima est necessaria, sicut patet, et prima similiter est formalis, ut patet per commenta 2 physicorum commentatore, ideo (! id est,) ubi dicit quod in instanti, continuo, et dissimili, ubicunque est invenire maius et minus et equale. Et probatur argumentum, quia latitudo a *b* ad extremum sui intensius est intensior *b* gradu et latitudo a *b* usque ad non gradum est remissior *b* gradu, ut patet satis etc. Sequitur ergo quod in *a* sit aliqua sibi equalis et nulla nisi tota, quia accepta quacunque latitudine que est pars illius latitudinis totius illa vel est intensior *b* vel remissior *b*, et per consequens tota latitudo totalis est sibi equalis; igitur etc. Ex quo sequitur ultra, quod tota latitudo nec est equalis nec contraria (! correspondet?) suo gradui intensissimo. Secundo arguitur, sit Sortes qui nunc incipit a non gradu uniformiter difformiter intendere motum suum usque ad *b* gradum et Plato ab eodem gradu vel consimili incipiat remittere motum suum uniformiter difformiter ad non gradum. Tunc ille latitudines motuum Sortis vel (! et?) Platonis sunt equales et aliquibus gradibus sunt equales, et non extremis gradibus, ut patet; igitur mediis quia non videtur quibus aliis conveniunt ut forent equales; igitur etc. Item si aliqua latitudo motus localis sit uniformiter difformis incipiens a non gradu et signata per *a* et eius medius gradus per *b* et incipiat Sortes moveri *a* latitudine, et incipiat Plato moveri *b* gradu medio eiusdem, tunc sic Plato pertransiens aliquod spacium *b* gradu in aliquo tempore tantum spacium precise pertransiet quantum pertransiet Sortes tota *a* latitudine in eodem tempore vel equali. Igitur *a* latitudo est equalis *b* gradui. Consequentia est manifesta, et probo antecedens. Ad cuius probationem sumo quod gradus terminans in extremo suo intensiori sit *c* signatus per *a* (! 8?), gradus medius inter *b* et *c* signatus per *gb* (! 6?), vero gradus medius latitudinis totalis per 4 gradus, etc., et [gradus medius inter *b* et] non gradum sit *e* signatus per 2; et sit *f* tempus per quod pertransiret aliquod spacium, quod sit *g*. Et arguo tunc sic: per totam primam medietatem *f* temporis Sortes et Plato movebuntur, Sortes continue tardius Platone; igitur Plato continue velocius Sorte sit; igitur quod in duplo velocius, et quod Sortes per latitudinem que est a non gradu usque ad gradum medium extensive, qua solum movebitur [in] prima medietate temporis, pertranseat unam quartam tantum; et

be worthwhile to characterize them briefly at this point. The proof in the *Probationes* (Doc. 5.2) is a direct arithmetical proof. It is based on the equal, symmetrical gain and loss of velocity beyond the middle or mean degree of velocity. It says, then, that every *excess* in distance traversed by a body moving uniformly with the mean velocity during the first half of the time *over* the distance traversed by the body uniformly accelerating

sequitur tunc in prima medietate *f* temporis Sortes pertransiet solum unam quartam et Plato solum duas. Et tunc ultra sic, in prima medietate *f* temporis Sortes pertransiet unam quartam et in secunda medietate temporis pertransiet Sortes 3 quartas; igitur intensio (! in toto?) *f* temporis (! tempore?) pertransiet 4 quartas, et tantum precise in eodem tempore pertransitum a Platone. Et [cum] in prima medietate *f* temporis pertransiet Plato 2 quartas et per totum *f* tempus movebitur uniformiter, igitur in secunda medietate *f* temporis tantum pertransiet quantum in prima. Sed duas pertransivit in prima medietate; igitur 2 alias pertransibit in ea; igitur etc. quartas in toto; igitur tantum quantum Sortes. Et [probo] quod Sortes in secunda medietate *f* temporis movebitur in triplo velocius quam in prima. Primo signetur latitudo motus qua movebitur Sortes in secunda medietate *f* temporis, que erit latitudo terminata ad *b* et *c* gradus, cuius etiam medius gradus est *d*, sicut patet. Tunc sic Sortes movebitur velocius Platone per totam secundam medietatem *f* temporis, vel igitur secundum proportionem graduum mediorum illarum latitudinum, que est a *b* ad *c* et a *b* ad non gradum, vel secundum proportionem graduum extremorum latitudinum earundem, quia non videtur penes quas alias velocitates sex (! sed illam?) in secunda medietate *f* temporis supra velocitatem eius in prima medietate *f* temporis. Si penes proportionem graduum mediorum: cum gradus medius acquisitus in secunda medietate sit triplus ad acquisitum in prima, quia *d* ad *c* est proportio tripla etc. ut patet, igitur Sortes movebitur in triplo velocius in secunda medietate quam in prima, et habetur propositum. Si detur alia pars, quod in secunda medietate *f* temporis movebitur Sortes non secundum proportionem graduum mediorum sed secundum proportionem graduum extremorum: cum unus precise sit duplus ad alium, ut est *c* ad *d*, igitur solum pertransiet Sortes in secunda medietate 2 quartas; et sic in toto *f* tempore plus pertransietur a Platone, quia in toto tempore pertranseuntur a Platone 4 quarte et a Platone nunc nisi 3, quod est falsum...." The arguments of this author do not represent sound proofs. The first is much too syncopated. The author does not actually show that, because the second half of the acceleration is equivalent to a velocity greater than the mean and the first half to a velocity less than the mean, only the whole accelerated movement is equivalent, with respect to space traversed, to the mean. In the second argument he does not truly exclude all the possibilities. He merely states that since the latitudes are not equivalent to the terminal velocities, they must be equivalent only to the mean velocities. Furthermore, although he rests his third argument on the sound theorem, that a body uniformly accelerated traverses three times as much space in the second half of the time as in the first, his proof of that distance relationship is completely inadequate. On the whole, it is obvious that the author of the *Sex incovenientes* is attempting to give the Merton arguments but that he does not have a very perceptive understanding of them. On those who held the theory that a latitude uniformly acquired corresponds to its most intense degree, see M. Clagett, *Giovanni Marliani and Late Medieval Physics* (New York, 1941), pp. 103–5.

during that same first half of the time is just compensated for in the second half of the time by a corresponding *deficiency* in distance traversed by the body moving uniformly with respect to the distance traversed by the body uniformly accelerating during that second half. Thus the gain in velocity in the first half is precisely compensated for by the loss of velocity in the second half. The two bodies, therefore, *in the whole time* will move the *same distance*. Thus the uniform acceleration is said to be equivalent to the mean velocity for the traversal of the same space in a given time.

From this "mean speed theorem," the author of the *Probationes* then goes on to prove the "distance theorem," namely, that a body uniformly accelerating from rest traverses three times as much space in the second half of the time as in the first. This is the *first step* toward the establishment of the *distance* relationship emphasized by Galileo, who generalized it further by saying that in succeeding periods of time the distances increase as the odd numbers 1, 3, 5, 7 . . . . As we shall show in Chapter 6, Nicole Oresme generalized the "distance theorem" as applied to qualities in the next generation after Heytesbury and thus two and one-half centuries earlier than Galileo.

Passing on to Swineshead, we should notice that in his tract in the *Liber calculationum* entitled *Regule de motu locali* the Calculator gives four proofs. Three of them are direct proofs based on symmetry and are thus very much like that in the *Probationes*, except that the implications of instantaneous velocity are more clearly present in Swineshead's proofs. But the fourth proof (actually his third in order of presentation) is the most original, as the detailed analysis given below reveals (Doc. 5.3). We can say here only that first Swineshead proves the 3:1 distance theorem by the simultaneous comparison and summation of two infinite series. Once he has established the distance theorem, he proceeds to show by indirect proof that that theorem would be violated unless the mean speed theorem were also true. That is to say, he states that the velocity to which the acceleration is equivalent either is equal to the mean velocity or is greater than or less than that mean velocity. In either of the cases of "greater than" or "less than" the distance theorem is violated and a contradiction ensues. Hence we conclude that the equivalent speed is, indeed, the mean speed.

John Dumbleton likewise gives an interesting indirect proof of the mean speed theorem. Following Document 5.4 I have given a detailed analysis of the proof. But I should like to say a word here about the structure of the proof. As an indirect proof, it first supposes that if the desired equivalent speed is not the mean speed, it must be either greater than or less than the mean. Let us suppose with Dumbleton the second alternative. If the equivalent velocity is said to correspond to a speed less than the mean, Dumbleton goes on to show that, because of the nature of uniform acceleration, the velocity to which the whole accelerated movement corresponds would have to be less than the mean by the same amount (k) that the velocity corresponding to the first half of the accelerated movement is less than the mean of that first half. If this is so, then let us divide the first half of the time into proportional parts: $1/4$, $1/8$, $1/16$ ... By the same line of argument as before, the equivalent degree of every smaller and smaller proportional part of the acceleration would be less than its mean by the constant amount k. But soon we must arrive at a part so small that its equivalent speed, which is less than its mean by the amount k, will either be zero or less than zero; which is, of course, an absurdity. This whole can be shown symbolically as follows:

(1) We let $P/2^n$ be the value of the mean speed of the nth successive proportional part of the original acceleration from initial speed o to to final speed P.

(2) We further suppose that the speed to which the nth successive proportional part corresponds does not equal $P/2^n$, but rather equals $V_m < P/2^n$.

(3) But $V_m = (P/2^n) - k$, where k is the constant amount by which the mean degree of any proportional part exceeds the speed to which that proportional part corresponds.

(4) If we increase n towards infinity, $P/2^n$ approaches o.

(5) Before it becomes o it must attain some magnitude $P/2^q$ which either equals k or is less than k (if we increase n further).

(6) At the point where $P/2^q = k$, obviously $V_m = $ o; or where $P/2^q < k$, then $V_m < $ o. Either of which conclusions is absurd.

In a similar fashion it was shown before this proof by Dumbleton that if we assume that the uniform acceleration is measured by some velocity greater than the mean, we will arrive at the absurdity that an acceleration could be measured by its final velocity or even by a velocity greater than its final velocity. Since, then, the acceleration is not measured by a speed greater than the mean nor less than the mean, it must be measured by the mean.

Although we have stressed the mean speed and distance theorems as the most fruitful results of the Merton kinematic activity, other kinematic theorems were developed. For example, there were theorems involving problems of a changing rate of acceleration or deceleration. Both Heytesbury and Swineshead gave several of these theorems. We have limited our selection (Doc. 5.3, passage 3) to a single example from Swineshead's *Liber calculationum*. This theorem asserts that in a movement where the rate of deceleration is increasing, the movement is equivalent—so far as space traversed in a given time is concerned—to a speed greater than the mean speed between the initial and final speeds of the whole velocity decrement lost in the increasingly decelerated movement.

These various English theorems regarding acceleration passed with the Merton logic and kinematics to France and Italy, and then to other parts of Europe. In the next chapter we shall see how the passage to France and Italy was followed almost immediately by the invention of a graphic system to represent intension and remission problems, and in Chapter 11 we shall trace how the Merton kinematics continued to flourish at the end of the fourteenth century and into the fifteenth and even sixteenth centuries. It will be noticed that there was very little change in the form of the Merton kinematics when it was adopted by later scholastics. Thus when John of Holland takes up the English kinematics at Prague in the late 1360's, he parrots the English "Calculators" (see Doc. 4.6). And in fact when he presents a proof of the mean speed theorem, it obviously comes directly from Swineshead's *Liber calculationum*.[11] Similarly, the

[11] John of Holland, *De motu*, MS Venice, Bibl. Naz. San Marco, Lat. VIII, 19, ff. 5r–5v (cf. MS Oxford, Bodl. Canon. Misc. 177, f. 114v, *MS B*): "*Conclusio quinta. (Marg.)* Alia conclusio: omnis latitudo motus uniformiter acquisita vel deperdita suo gradui medio correspondet, id est, si aliquis motus uniformiter acquiret vel deperdet latitudinem motus tantum spacium precise pertransibit in illo tempore quantum pertransiret si continue per idem tempus moveatur gradu medio illius latitudinis. Probatur et suppono quod omne compositum ex duobus inequalibus est duplum ad medium inter ista, quod sic probatur: Sit A compositum ex B et C inequalibus, et sit B maius, C vero minus, et D medium inter ista, et decrescat B uniformiter in ista hora ad equalitatem D et sit E instans terminans istam horam, et maneat medium D invariatum. Tunc arguitur sic: In E instanti A totum erit duplum ad D medium, et A iam est tantum quantum erit in E instanti, et similiter D est tantum quantum erit in E instanti. Ergo A iam est duplum ad D. Patet consequentia, et antecedens pro prima parte arguitur sic: A in E instanti erit adequate compositum ex B et C, et B et C in E instanti erunt equalia inter se, et similiter quodlibet istorum in E instanti erit equale ipsi D; ergo in E instanti A erit duplum ad D. Patet consequentia et similiter antecedens, et quod A non augmentabitur neque diminuetur probatur. Quantum deperdet A per decrementum B tantum acquiret per *(MS B)* crementum C, et A nichil acquiret nec deperdet nisi per

numerous Italian scholastics take their proofs of this and other kinematic theorems from Heytesbury and Swineshead, as, for example, an anonymous *questio* of the late fourteenth century entitled *Utrum omnis motus uniformiter difformis correspondeat suo gradui medio*.[12] And at Pavia in the fifteenth century Giovanni Marliani takes over the proof of Dumbleton with little change.[13]

crementum vel decrementum alicuius illorum; ergo et cetera. Ista suppositione admissa, arguitur sic: Et suppono quod Sortes acquirat latitudinem motus a non gradu usque ad gradum ut 8 uniformiter in ista hora, et deperdat Plato consimiliter eandem per omnia in eadem hora usque ad non gradum, et sit D gradus medius, scilicet ut 4, latitudinis acquiribilis a Sorti. Tunc sic arguitur, per totam istam horam compositum ex motu Sortis et motu Platonis equivalebit gradui duplo ad D, qui est ut 4. Patet, quia illud compositum erit duplum ad D, et motus Sortis eque intenso gradui correspondebit in ista hora sicut motus Platonis et econverso. Ergo quilibet illorum motuum per se correspondebit gradui D. Patet consequentia et antecedens pro secunda parte patet, et pro prima parte probatur sic: Pro quolibet instanti istius hore future preter quam in medio instanti D erit medium inter motum Sortis et motum Platonis, et in medio instanti erit ita quod D, motus Sortis, et motus Platonis sunt equales. Ergo per totam istam horam compositum ex motu Sortis et motu Platonis equivalebit gradui duplo ad D, et per consequens quilibet illorum motuum per se equivalet gradui D qui est medius utriusque latitudinis, quod fuit probandum."

[12] See the Latin text in Document 5.5 (Venice, Bibl. Naz. San Marco, Lat. VIII, 19, 230v–231v). It will be noticed that this document includes the usual Merton definitions of uniform and uniformly difform motion, which are then followed by suppositions and conclusions leading to the proof of the mean speed theorem. I have not included any English translation because of its similarity with documents already translated.

[13] For an analysis of Giovanni Marliani's treatment of the problem of the mean speed theorem, see my *Giovanni Marliani*, chapter 5.

Document 5.1

William Heytesbury, *Rules for Solving Sophisms**

1. PART VI. Local Motion (continued[14]). In this connection, it should be noted that just as there is no degree of velocity by which, with continuously uniform motion, a greater distance is traversed in one part of the time than in another equal part of the time, so there is no latitude (i.e., increment, *latitudo*) of velocity between zero degree [of velocity] and some finite degree, through which a greater distance is traversed by uniformly accelerated motion in some given time, than would be traversed in an equal time by a uniformly decelerated motion of that latitude. For whether it commences from zero degree or from some [finite] degree, every latitude, as long as it is terminated at some finite degree, and as long as it is acquired or lost uniformly, will correspond to its mean degree [of velocity]. Thus the moving body, acquiring or losing this latitude uniformly during some assigned period of time, will traverse a distance exactly equal to what it would traverse in an equal period of time if it were moved uniformly at its mean degree [of velocity].

2. For of every such latitude commencing from rest and terminating at some [finite] degree [of velocity], the mean degree is one-half the terminal degree [of velocity] of that same latitude.

3. From this it follows that the mean degree of any latitude bounded

---

* English translation by Ernest Moody—slightly modified.

[14] This follows directly after the selection from the *Regule* given as Document 4.4 and thus continues the section on the measure of uniform acceleration. The changes I have made in Moody's translation are those which bring it more in line with my own translations of the succeeding documents and which follow the modifications I have made in the text by using manuscripts in addition to the edition. However, I have left Moody's translation of "latitude" for *latitudo*, although generally in the other documents I have rendered it as "increment" or "velocity increment." I have done this because in some parts of this document Heytesbury is speaking of latitudes in a more general sense than velocity increments. Of course, when I use the word "increment," I understand that it may apply to either positive or negative quantities, that is, to velocity gains or to velocity losses.

by two degrees (taken either inclusively or exclusively) is more than half the more intense degree bounding that latitude.

4. From the foregoing it follows that when any mobile body is uniformly accelerated from rest to some given degree [of velocity], it will in that time traverse one-half the distance that it would traverse if, in that same time, it were moved uniformly at the degree [of velocity] terminating that latitude. For that motion, as a whole, will correspond to the mean degree of that latitude, which is precisely one-half that degree which is its terminal velocity.

5. It also follows in the same way that when any moving body is uniformly accelerated from some degree [of velocity] (taken exclusively) to another degree inclusively or exclusively, it will traverse more than one-half the distance which it would traverse with a uniform motion, in an equal time, at the degree [of velocity] at which it arrives in the accelerated motion. For that whole motion will correspond to its mean degree [of velocity], which is greater than one-half of the degree [of velocity] terminating the latitude to be acquired; for although a nonuniform motion of this kind will likewise correspond to its mean degree [of velocity], nevertheless the motion as a whole will be as fast, categorematically, as some uniform motion according to some degree [of velocity] contained in this latitude being acquired, and, likewise, it will be as slow.

6. To prove, however, that in the case of acceleration from rest to a finite degree [of velocity], the mean degree [of velocity] is exactly one-half the terminal degree [of velocity], it should be known that if any three terms are in continuous proportion, the ratio of the first to the second, or of the second to the third, will be the same as the ratio of the difference between the first and the middle, to the difference between the middle and the third; as when the terms are 4, 2, 1; 9, 3, 1; 9, 6, 4. For as 4 is to 2, or as 2 is to 1, so is the proportion of the difference between 4 and 2 to the difference between 2 and 1, because the difference between 4 and 2 is 2, while that between 2 and 1 is 1; and so with the other cases.

Let there be assigned, then, some term under which there is an infinite series of other terms which are in continuous proportion according to the ratio 2 to 1. Let each term be considered in relation to the one immediately following it. Then, whatever is the difference between the first term assigned and the second, such precisely will be the sum of all the differences between the succeeding terms. For whatever is the amount of the first proportional part of any continuum or of any finite quantity, such precisely is the amount of the sum of all the remaining proportional parts of it.

Since, therefore, every latitude is a certain quantity, and since, in general, in every quantity the mean is equidistant from the extremes, so the mean degree of any finite latitude whatsoever is equidistant from the two extremes, whether these two extremes be both of them positive degrees, or one of them be a certain degree and the other a privation of it or zero degree.

But, as has already been shown, given some degree under which there is an infinite series of other degrees in continuous proportion, and letting each term be considered in relation to the one next to it, then the difference or latitude between the first and the second degree—the one, namely, that is half the first—will be equal to the latitude composed of all the differences or latitudes between all the remaining degrees—namely those which come after the first two. Hence, exactly equally and by an equal latitude that second degree, which is related to the first as a half to its double, will differ from that double as that same degree differs from zero degree or from the opposite extreme of the given magnitude.

And so it is proved universally for every latitude commencing from zero degree and terminating at some finite degree, and containing some degree and half that degree and one-quarter of that degree, and so on to infinity, that its mean degree is exactly one-half its terminal degree. Hence this is not only true of the latitude of velocity of motion commencing from zero degree [of velocity], but it could be proved and argued in just the same way in the case of latitudes of heat, cold, light, and other such qualities.

7. With respect, however, to the distance traversed in a uniformly accelerated motion commencing from zero degree [of velocity] and terminating at some finite degree [of velocity], it has already been said that the motion as a whole, or its whole acquisition, will correspond to its mean degree [of velocity]. The same thing holds true if the latitude of motion is uniformly acquired from some degree [of velocity] in an exclusive sense, and is terminated at some finite degree [of velocity].

From the foregoing it can be sufficiently determined for this kind of uniform acceleration or deceleration how great a distance will be traversed, other things being equal, in the first half of the time and how much in the second half. For when the acceleration of a motion takes place uniformly from zero degree [of velocity] to some degree [of velocity], the distance it will traverse in the first half of the time will be exactly one-third of that which it will traverse in the second half of the time.

And if, contrariwise, from that same degree [of velocity] or from any

Plate 4: A page from a fourteenth-century manuscript of Swineshead's *Calculationes*. Note that it bears date of 1375 and that on the margin of column 1 the Oresme configuration system is employed to illustrate the text. MS Paris, BN lat. 9558, f. 1r. See also Plate 6.

due regule τ ꝙ īter istos tres gradus sūt due distantie intermedie quar vna est dupla ad aliā:pꝫ:naꝫ octo distat a quatuoꝛ per qua ternariū īntermediuꝫ:τ quatuoꝛ distat a duo per binariū intermedi um:pꝫ ꝙ quaternarius est duplus ad binariū τ sic dicas de triplis τ sexquialteris Deīde dīc.

¶ Signato igitur aliquo termino sub quo infiniti alii tꝰ mini sunt continue pꝛopoꝛtionales ꝓpoꝛtione dupla. quilibet ad sibi pꝛoximum coniugatur. et quanta fue rit differentia pꝛimi termini dati ad secūdū tantum ꝑ cise erit aggregatū ex omnibus differentiis terminoꝝ sequentiū. quanta enī est pꝛima pars ꝓpoꝛtionalis ali cuius continui vel alicuius q̄ti finiti tm̄ erit ꝑcise ag gregatum ex omnibus partibus pꝛopoꝛtionalibus re siduis eiusdem.

¶ Ex dictis ponit hanc cōclusionē calculatoriā ꝙ posito vno ter mino seu quāto vel cōtinuo vel discreto et diuidatur in oēs suas ptes ꝓpoꝛtionales duplas que sūt infini te īfra seu versus nō gradū īteīsiōis:puta. 8. 4. 2. 1. ½. ¼. ⅛. 1/16. τ sic in infinitum:dicit mō ꝙ quanta fuerit differētia seu per q̄tū distabit vnus signatus a suo īmediate subsequente vt quatuoꝛ duo. vnū: ꝙ per tantū precise erit distantiā oīum subsequentium simul sūptā. pꝫ. Nā octo distat a quatuoꝛ per quaternariū:modo dico ꝙ quaternariꝰ ꝑcise est mēsura distantie alioꝝ graduuꝫ īfra quatuoꝛ:pꝫ:nam a quatuoꝛ ad duo est binariꝰ in medio:a duo ad vnum est vnitas:que cum binario facit ternarium τ ab vno ad ½ est semivnitaꝫ que facit tres. ½. τ similiter distantia vſ qꝫ ad non gradum est precise faciens ½. τ sic erunt omnes si mul sumpte precise vt quatuoꝛ sicut fuit pꝛima distantia. τ sic facias de aliis terminis inmediatis vt a primo ad ½. Nam di stantia est semiunitas:mō dico ꝙ omnis distantia simul sumpta citra illam vſqꝫ ad non gradum faciet precise semi vnitatem:τ sic postea dicit.

¶ Cum ergo quelꝫ latitudo finita sit q̄ dā q̄titas. τ vniuer

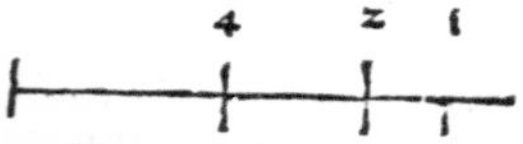

saliter sicut in oī q̄to medium equaliꝰ distat ab extre mis. ita cuiuslibet latitudīs finite medius gradus eq̄li ter distat ab vtroqꝫ extremoꝛum. siue illa duo extrēa sint duo gradus aut vnum illoꝛum fuerit aliquis gra dus alterū oīo pꝛiuatio siue nō gradus illius.

Sed sicut iam ostē sum est dato gradu aliq̄ sub quo eēnt i finiti alii cōtinue ꝓ poꝛtiōales τ glꝫ ad sibi pꝛoximū signe tur equalis erit dif

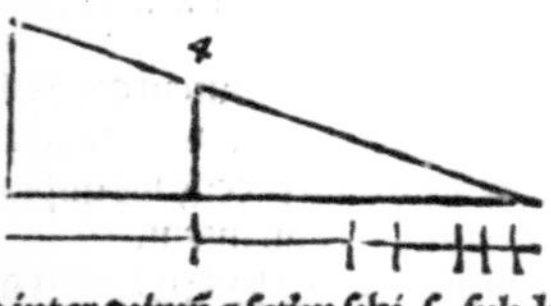

ferentia seu latitudo inter pꝛimū τ scōm sibi. s. subdu plum sicut latitudo compoſita ex oībꝰ differētiis seu latitudinibus īter omnes gradus residuos seqnentis videlicet duos pꝛimos. ergo eq̄liter precise τ per eq̄lē latitudinem distabit ille gradus secundus subduplus ad ꝑmū duplū ab illo duplo sic distabit idē scōs. a. nō gradu seu ab extrēo opposito illiꝰ magnitudis date.

¶ His verbis dīc ꝙ cū latitudo īteīsionis sit quātitas quedā τ ī omni quanto tā cōtinuo q̄ discreto finito tamē:sit assignare vnuꝫ mediū. puta ī linea pedali ē dare suuꝫ mediuꝫ ꝙ equalr distat ab extremis ita ī latitudīe:cū sit quātitas ē dare suū mediū ꝙ equalr distabit ab extremis:siue extrema:sint positiua siue vnū positiuuꝫ aliud ꝑuatiuū. vt nō gradus īteīsionis:τ ꝙ latitudo sit q̄titas. su pra declaratum ꝑ. Hic ꝯsiderat latitudinē abstractā a subō. τ vult ꝓbare ꝙ cuiuſꝫ latitudinis dꝫ assignari mediū:τ addit rōnē:qꝛ ē quantitas:τ ꝑmo recapitulat qđ est dictū. s. qꝛ vt ꝙ illō sit me dium ad ꝓbādū τ ad īueniēdū mediū latitudinis. Nā. s. ꝓbaui mus ꝙ dr̄ia seu latitudo inter media inter octo τ quatuoꝛ est q̄ ternarius:τ ꝓbauimꝰ ꝙ dr̄ie seu latitudines sūpte sil̄ a quatuoꝛ vſqꝫ ad nō gradū erāt ꝑcise quaternarii:pꝫ mō ꝙ ab octo ad q̄t tuoꝛ est dr̄ia ītermedia quaternariꝰ:τ a quatuoꝛ ad nō gradū est dr̄ia ꝑcise quaternarius. Patet ergo ꝙ q̄tuoꝛ distat ꝑcise tm̄ ab octo sicut a nō gradu. τ iō quartus gradus est mediꝰ seu mediuꝫ inter octo. τ non gradum:qđ erat ꝓbandum. Et iō ē tm̄ intelli gendum de huiusmodi latitudine cuius extrema est non gradus vnde ipē subdit.

¶ Et sicut vniuersaliter ꝓbaꝝ de omni latitudine in cipiente a nō gradu τ terminata ad aliquē gradū fini tū cōtinente ēt gradum aliquem τ subduplum τ subq̄ druplū. τ sic in infinitū ꝙ eius gradus medius est pꝛe cise subduplus ad gradum ipsam terminantē. Nō nō solū est hoc verū de latitudine velocitatis motus inci pientis. a. nō gradu sed etiā de latitudie caliditatis fri giditatis. luminis τ aliarū similiū qualitatum.

¶ Hic dicit ꝙ pꝛobatio ētsūp facta est ad demonstrandū ꝙ ī lati tudine cū sit q̄titas est assignare suū mediū. Sed ipē dicit ꝙ ta le mediuꝫ in latitudine incipiente a nō gradu τ termīante ad ali quem gradum finitū ē ꝑcise subduplū ad gradū termīantem ī lā latitudinē:qđ ē clarū. vt supꝛa deduxi d. 8. 4. τ ꝓpoꝛtiōibꝰ du plis ad ñ gradū:qꝛ quatuoꝛ duplus ad 8. τ est medium īter octo τ nō gradum τ dicit in lā terminata ad gradū habentem subdu plum subquadruplū τ sic in infinitū p gradū infinitū q nō habet gradum subduplū nec subquadruplum. qꝛ alr̄ eēt finitū.

¶ Consimiliter argui poterit τ pꝛobari:q̄tū aūt ad ma gnitudinē ꝑtranseundā vnifoꝛmiꝰ acqrendo talem la titudinē motus incipientē a nō gradu τ terminatā in aliquē gradū finitū dictū est pꝛiꝰ qđ totꝰ ille motꝰ seu illa acq̄sitio coꝛrespondebit gradui suo medio.

¶ Hic dicit ꝙ ex habitis pōt ꝓbari τ patere ꝙ mobile motū aī gradu intēdēdo motū suū:vꝫ. ad certū gradū vnifoꝛmiter difoꝛ miter ꝙ pertrāsibit tantam magnitudinē:ac si fuisset motū p idē tp̄s tm̄ gradui medio illiꝰ latitudinis acq̄site vnifoꝛmiter difoꝛ miter a nō gradu vſqꝫ ad gradū terminantem illū. Et ille magi ster non ꝓbat sed dicit ꝙ pōt pꝛobari:τ est verum:ꝙ ꝓbo sic. naꝫ gradus mediꝰ inter non gradū τ quatuoꝛ est duo vt ꝓbatum ē supꝛa. adde modo oēs gradꝰ supꝛa duo partibus non hēntibus duo τ sic erunt duo. Deinde dicit.

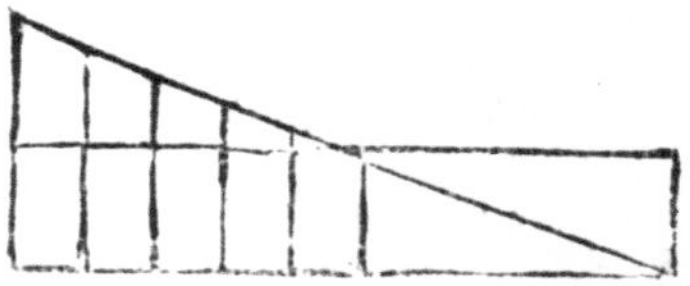

¶ Et consimiliter etiam si ab aliquo gradu exclusiue vnifoꝛmiter acquiratur latitudo motus ad aliquē gra dum terminata finitum.

¶ Hic dicit si ille gradus fuerit exclusiue terminās illā latitudinē ꝙ gradꝰ mediꝰ erit exclusiue talꝭ. Deīde dīc τ īfert ꝯclōnē

¶ Ex quo satis cognosci poterit in huiusmodi vnifoꝛ mi intensione vel remissione quāta erit magnitudo ꝑ trāsita ceteris pibꝰ in pꝛima medietate τ quāta a scōa

¶ Hic dicit ꝙ ex dictis patebit ī tali motu vnifoꝛmiter difoꝛmi quanta fuerit acquisitio vel deperditio ī pꝛma medietate tp̄is:τ quanta in scōa eiusdē tp̄is ī q̄ fiet talis alteratio τ dicit.

¶ Cum enim a nō gradu ad aliquem gradum vnifoꝛ mis fiat alicuius motus intensio subtriplū pertrāsibit ꝑcise in pꝛima medietate tēpoꝛis ad illud quod ꝑtrā sibit in secunda. τ si alias ab codē gradu aut ab aliquo alio quocunqꝫ ad non gradū vnifoꝛmis fiat remissio triplum pꝛecise ꝑtransiet in pꝛima medietate tempo ris ad illud qđ pertransiretur in secunda.

Plate 5: A page with the statement of the Merton uniform acceleration theorem. From William Heytesbury's *Regule solvendi sophismata*, edition of Venice, 1494, f. 40v. The smaller print is the commentary of Gaetano de Thienis.

[Heytesbury, *Rules for Solving Sophisms*

other degree whatsoever, there is uniform deceleration to zero degree [of velocity], exactly three times the distance will be traversed in the first half of the time, as will be traversed in the second half. For every motion as a whole, completed in a whole period of time, corresponds to its mean degree [of velocity]—namely, to the degree it will have at the middle instant of the time. And the second half of the motion in question will correspond to the mean degree of the second half of that same motion, which is one-fourth of the degree [of velocity] terminating that latitude. Consequently, since this second half will last only through half the time, exactly one-fourth of the distance will be traversed in that second half as will be traversed in the whole motion. Therefore, of the whole distance being traversed by the whole motion, three-quarters will be traversed in the first half of the whole motion, and the last quarter will be traversed in its second half. It follows, consequently, that in this type of uniform intension and remission of a motion from some degree [of velocity] to zero degree, or from zero degree to some degree, exactly three times as much distance is traversed in the more intense half of the latitude as in the less intense half.

8. But any motion can be uniformly accelerated or decelerated from some degree [of velocity] to another degree in an endless number of ways, because it may be from some degree to a degree half of that, or to a degree one-fourth of it, or one-fifth, or to a degree two-thirds of that degree, or three-quarters of it, and so on. Consequently there can be no universal numerical value by which one will be able to determine, for all cases, how much more distance would be traversed in the first half of this sort of acceleration or deceleration than in the second half, because, according to the diversity of the extreme degrees [of velocity], there will be diverse proportions of distance traversed in the first half of the time to distance traversed in the second half.

But if the extreme degrees [of velocity] are determined, so that it is known, for instance, that so much distance would be traversed in such or such a time by a uniform motion at the more intense limiting degree [of velocity], and if this is likewise known with respect to the less intense limiting degree [of velocity], then it will be known by calculation how much would be traversed in the first half and also how much in the second. For, if the extreme degrees [of velocity] are known in this way, the mean degree [of velocity] of these can be obtained, and also the mean degree between that mean degree and the more intense degree terminating the latitude. But a calculation of this kind offers more difficulty than advantage.

[274 }   Merton Theorem of Uniform Acceleration: 5.1]

And it is sufficient, therefore, for every case of this kind, to state as a general law, that more distance will be traversed by the more intense half of such a latitude than by the less intense half—as much more, namely, as would be [the excess of distance] traversed by the mean degree [of velocity] of this more intense half, if it moved in a time equal to that in which this half is acquired or lost uniformly, over that [distance which] would be traversed by the mean degree [of velocity] of the less intense half, in the same time.

9. But as concerns nonuniform acceleration or deceleration, whether from some degree [of velocity] to zero degree or *vice versa*, or from one degree to some other degree, there can be no rule determining the distance traversed in such or such time, or determining the intrinsic degree to which such a latitude of motion, acquired or lost nonuniformly, will correspond. For just as such a nonuniform acceleration or deceleration could vary in an infinite number of ways, so also that motion as a whole could correspond to an infinite number of intrinsic degrees [of velocity] of its latitude—indeed, to any intrinsic degree whatsoever, of the latitude thus acquired or lost.

In general, therefore, the degree [of velocity] terminating such a latitude at its more intense limit is the most remiss degree [of velocity], beyond the other limit (i.e., the most remiss extreme) of the latitude, to which such a nonuniformly nonuniform motion as a whole *cannot* correspond; and the degree [of velocity] terminating that latitude at its more remiss limit is the most intense degree [of velocity] beneath the upper limit of the same latitude, to which such a nonuniformly nonuniform motion *cannot* correspond. Consequently, it is not possible for such a motion as a whole to correspond to such a remiss degree (as that of the lower limit); nor to such an intense degree (as the upper limit).

COMMENTARY

1. An able commentary on this whole section, comparing this passage to similar passages in the *Probationes conclusionum* attributed to Heytesbury, to the section of Swineshead's *Liber calculationum* entitled *Regule de motu locali*, and to the *De motu* of John of Holland, is given by Curtis Wilson in his *William Heytesbury* (Madison, 1956), Chapter 4.

The substance of this first passage is that it makes no difference whether a body is uniformly accelerating from V_o to V_f or uniformly decelerating from V_f to V_o in the same time, the distance traversed, ΔS, will be the same. The reason given (without proof) is that regardless of whether ΔV

is positive or negative, the equivalent uniform speed with which a body would traverse the same distance in the same time is the mean velocity, i.e., the velocity at the middle instant of the period of acceleration. This would appear to be the earliest definitely established statement of the mean speed theorem, as we have already observed above in the body of the chapter. No formal proof of this theorem is given. The succeeding documents in this chapter present several different proofs of it.

2. Furthermore, it can be shown that for uniform acceleration, if $V_o = 0$, then $V_m = V_f/2$, where V_f is the final velocity of the acceleration, V_m is its mean velocity and V_o is the initial velocity.

3. Hence if $V_o > 0$, then $V_m > V_f/2$, whether we include or exclude the terminal velocities in ΔV.

4. $S_a = S_m = S_f/2$, S_a being the distance traversed in the course of the uniform acceleration from rest, S_m the distance traversed by a body moving uniformly during the same time with the mean velocity, and S_f being the distance traversed by a body moving uniformly in the same time with the final velocity.

5. When $V_o > 0$, $S_a > S_f/2$.

6. For the next conclusion, Heytesbury feels it necessary to give a fuller treatment. It is clear that Heytesbury here characterizes a uniform acceleration from rest not only by equal increments of velocity in equal time periods (see Doc. 4.4), but also by the fact that if any degree of speed within such an acceleration is taken, we shall always find, on dividing the time period (to that degree of velocity) into continually proportional parts, that the degrees of velocities at the ends of those periods are continually proportional in the same way. We shall see that such definition of uniform acceleration was also assumed by Swineshead in one of his proofs of the

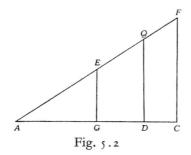

Fig. 5.2

mean speed theorem (see commentary to Doc. 5.3). The meaning of this alternate definition of uniform acceleration can be made clearer if we pretend for a moment that the geometric system of representing movements developed later at Paris by Nicole Oresme was in existence. In the accompanying figure (see Fig. 5.2), AC represents the time of acceleration and the perpendicular lines on AC making up *in toto* the triangle represent velocities at successive instants along AC. Heytesbury's alternate definition is illustrated by taking any velocity QD. Then at time $AD/2$ there is a velocity $QD/2$, at $AD/4$ a velocity $QD/4$, etc. The same would hold true if we took the final velocity FC instead of QD, namely, there would be a velocity $FC/2 (= EG)$ at time $AC/2 (= AG)$.

In Heytesbury's elaboration he indicates that EG is the mean or middle velocity, and thus there is equal distance in velocity increment (latitude) from the two extremes, namely FC and zero. Hence $FC = 2EG$. While this would appear to be obvious, Heytesbury would like to show this is so without directly reckoning with zero velocity. Assume

$$\Delta V_1 = \Delta V_2, \text{ where } \Delta V_1 = FC - EG, \text{ and}$$
$$\Delta V_2 = \left(EG - \frac{EG}{2}\right) + \left(\frac{EG}{2} - \frac{EG}{4}\right) + \cdots + \left(\frac{EG}{2^{n-1}} - \frac{EG}{2^n}\right) + \cdots \text{ and}$$
$$\text{rewriting } \Delta V_2 = \frac{EG}{2} + \frac{EG}{4} + \cdots + \frac{EG}{2^n} + \cdots.$$

It is clear then that, as n goes to infinity, ΔV_2 sums to EG, and thus $FC = 2EG$.

7. From the special case of uniform acceleration from rest, the mean speed theorem can be proved for uniform acceleration commencing from a finite velocity. Furthermore, from the special case of the theorem for acceleration from rest, it can be proved that the distance traversed in the first half of the uniform acceleration from rest is 1/3 of that traversed during the second half of the period of acceleration. Conversely, the distance traversed during the first half of the time of uniform deceleration to rest is three times that traversed in the second half. The proof is indicated for the case of deceleration. The deceleration of each half corresponds to its mean velocity (for the equivalent traversal of space). Thus in the second half, the mean velocity is 1/4 the initial velocity and in the first half it is 3/4 the initial velocity. Using modern formulas, $S_1 = (3\ V_f/4)(t/2)$ and $S_2 = (V_f/4)(t/2)$ and thus $S_1/S_2 = 3/1$.

8. No single universal relationship between the distances traversed in the first and second half of the time of uniform acceleration can be established if the acceleration commences with a finite velocity rather than rest. But, if we know the terminal velocities, we can, by using the mean speed theorem, determine what the relationships are, for the pro-

portion will be as that of the mean velocities of the respective halves of the acceleration.

9. In the case of nonuniform accelerations, no general rule is applicable for determining the equivalent uniform speed allowing for an equal traversal of space in the same time. The only evident conclusion is that the equivalent speed cannot be either of the terminal velocities. (See examples for a uniformly increasing or decreasing acceleration drawn from Swineshead's treatment in Doc. 5.3, passage 3.)

*Regule solvendi sophismata Guillelmi Heytesberi**

Circa quod est advertendum, quod sicut nullus est gradus velocitatis quo uniformiter continue movendo plus pertransitur in aliquo tempore equali, quam in alio; sic nec est aliqua latitudo velocitatis incipiens a non gradu aut ab aliquo gradu certo, per quam complete ipsam uniformiter acquirendo in aliquo tempore assignato, plus pertranseatur quam per eandem pertransiretur in aliquo tempore equali, ipsam uniformiter deperdendo. Omnis enim latitudo sive a non gradu incipiat, sive a gradu aliquo, dum tamen ad gradum aliquem terminetur finitum, et uniformiter acquiratur seu deperdatur, correspondebit equaliter gradui medio sui ipsius, sic scilicet quod mobile illud, ipsam uniformiter acquirens seu deperdens in aliquo tempore dato, equalem omnino magnitudinem pertransibit sicut si ipsum per equale tempus continue moveretur medio gradu illius; cuiuslibet etiam talis latitudinis incipientis a quiete et terminante ad aliquem gradum, est gradus suus medius subduplus ad gradum eandem latitudinem terminantem.

Ex quo sequitur quod cuiuslibet latitudinis terminate ad duos

---

* MSS cited in Doc. 4.4 *A*, ff. 56r–v; *B*, ff. 57r–58r; *V*, ff. 25r–26r; *Ed.*, ff. 39v–41v. As before, in the variant readings I have not indicated all word transpositions and spelling differences.

---

2 quod *om. A*
3 movendo: moto *V* / pertransiretur *Ed*
4 *post* alio *mg. add. A* sibi equali / sic: sicut *V*
5 gradu certo *tr. Ed*
6 ipsam *AB* illam *Ed* propriam *V*; sed *A scr.* ipsam *post* uniformiter
7 pertransitur *Ed*
8 ipsam: illam *Ed* / sive: seu *A*
9–10 terminatur *B*

10 correspondet *B*
11 ipsius *om. A* / scilicet *om. Ed*
12 seu *AB* vel *VEd*
13 pertransetat (?) *A* / si: se *A*
14 illius: istius *A*
14–15 talis latitudinis: latitudinis illius *Ed*
15 a: a scilicet *B* / *post* aliquem *add. Ed* certum
17 Ex... terminate *om. B in textu*; sed *B scr. in marg.* et cuiuslibet latitudinis

gradus, inclusive vel exclusive, est gradus medius maior quam subduplus ad gradum intensiorem eandem latitudinem terminantem. Ex precedenti sequitur quod cum mobile aliquod a quiete uniformiter intendat motum suum ad aliquem gradum datum, quod ipsum in duplo minus pertransibit in tempore illo quam si ipsum uniformiter per idem tempus moveretur gradu illo ipsam latitudinem terminante; quia totus ille motus correspondebit gradui medio istius latitudinis, qui est precise subduplus ad gradum illum qui terminus est eiusdem.

Sequitur etiam consimiliter quod cum aliquod mobile ab aliquo gradu exclusive ad alium gradum inclusive vel exclusive intendet uniformiter motum suum, quod ipsum plus pertransibit quam subduplum ad illud quod ipsum uniformiter pertransiret in equali tempore secundum istum gradum ad quem stabit intensio sui motus; quia totus ille motus correspondebit gradui suo medio, qui maior est quam subduplus ad gradum terminantem illam latitudinem acquirendam; quamvis autem motus huiusmodi difformis correspondebit equaliter gradui suo medio, ita velox tamen erit totus ille motus cathegorematice, sicut est aliquis motus uniformis sub aliquo gradu intrinseco istius latitudinis acquirendo, et consimiliter ita tardus.

Ad probandum autem quod cuiuslibet latitudinis incipientis a quiete et terminantis ad aliquem gradum finitum, est medius gradus precise subduplus ad gradum eundem latitudinem terminantem: Est sciendum quod si aliqui sint tres termini continue proportionales, qualis est proportio primi ad secundum, aut secundi ad tertium, talis

20 precedenti: predictis *B*
21 *post* aliquem *add. V Ed* certum / datum *om. Ed*
22 illo: isto *A* / uniformiter *tr. Ed. post* tempus
23 movetur *B* / terminante *Ed* terminantem *ABV*
24 istius latitudinis *A* illius latitudinis *B* totius latitudinis *V* totius latitudinis illius *Ed*
25 est precise *tr. B* / terminus est *tr. A*
26 consimiliter *om. Ed* / aliquod mobile *tr. VB* / ab: sub *Ed*
27 alium *AV* aliquem *BEd* / intendat *B*
28 ipsum *om. Ed*
29 pertransibit *B*
30 istum: illum *B*
31 *post* quia *add. A* ut *et add. Ed* ut

dictum est / ille: ipse *B* / suo *BV* sui *AEd*
32 terminantem *tr. B post* latitudinem *in linea 32 et tr. A post* acquirendam / illam: istam *A* illum *B*
33 huiusmodi *AEd* huius *BV*
34 medio suo gradui *AV* / tamen *om. A* / ille: iste *A*
36 istius *A* totius *Ed* illius *BV* / *post* latitudinis *add. Ed* istius
37 cuiuslibet: cuiuscunque *B*
38 terminante in *B* / medius gradus *tr. Ed*
39 gradum eundem *A tr. Ed.* gradum eandem *V* gradum illam *B*
39–40 Est sciendum *tr. Ed*
40 sint: sunt *V*
41 aut *AEd* et *BV*

erit proportio differentie primi et medii, ad differentiam medii et tertii; ut 4, 2, 1; 9, 3, 1; 9, 6, 4. Qualis enim est proportio 4 ad 2, aut 2 ad 1, talis est proportio differentie 4 ad 2, ad differentiam 2 ad 1. Est enim differentia 4 ad 2 binarius, et differentia vero duorum ad unum est unitas; et consimiliter est in aliis.

Signato igitur aliquo termino sub quo infiniti alii termini sunt continue proportionales, proportione dupla, quilibet ad sibi proximum coniugatur, et quanta fuerit differentia primi termini dati ad secundum, tantum precise erit aggregatum ex omnibus differentiis terminorum sequentium. Quanta enim est prima pars proportionalis alicuius quanti finiti, tantum erit aggregatum precise ex omnibus partibus proportionalibus residuis eiusdem. Cum ergo quelibet latitudo sit quedam quantitas, et universaliter sicut in omni quanto medium equaliter distat ab extremis, ita cuiuslibet latitudinis finite medius gradus equaliter distat ab utroque extremorum, sive illa duo extrema sint duo gradus, aut unum illorum fuerit aliquis gradus, et alterum omnio privatio sive non gradus illius.

Sed sicut iam ostensum est, dato gradu aliquo sub quo etiam infiniti alii continue proportionales, quilibet ad sibi proximum signetur, equalis erit differentia seu latitudo inter primum et secundum, suum scilicet subduplum, sicut latitudo composita ex omnibus differentiis seu latitudinibus inter omnes gradus residuos, sequentes videlicet duos primos; ergo equaliter precise et per equalem latitudinem distabit ille gradus secundus subduplus sibi ad primum duplum, ab

---

42 erit AEd est BV / et medii: ad secundum V / medii²: secundi V
43 4³ *om. B*
43–44 aut... differentiam *om. B*
44 post 1² *add. V* 4 enim 2 est dupla proportio et 2 est dupla proportio ad 1
45 enim: etiam V / et *om. BEd* / vero *om. AB*
46 est *om. AB* / *post* unitas *add. V* inter quos est consimiliter proportio dupla
47 igitur *om. V* ergo A / aliquo *om. A* / termini *om. A*
48 continue proportionales *tr. B* / quilibet: et quilibet V quibus B / ad sibi AEd ad sui V sibi ad B
50 precise erit *tr. V*
51 subsequentium A / *post* proportionalis *add. Ed* alicuius continui vel

52 aggregatum precise *tr. Ed*
53–54 *post* latitudo *add. Ed* finita
54 universaliter: uniformiter B
56 *post* utroque *add. A* gradu / illa *om. V*
57 sint: fuerint V / illorum: istorum A / aliquis *om. V* / et *om. Ed*
58 sive: seu A
59 Sed *om. B* / etiam: essent *Ed.*
60 *ante* quilibet *add. Ed* et / sibi A Ed sui V suum B / signaretur B
61–62 suum scilicet A scilicet suum B sibi V sibi scilicet Ed
62 subduplum: duplum B / *post* composita *add. A* scilicet
63 sequentes AV sequentur B sequentis Ed
65 ille: iste A / secundus subduplus: duplus B / sibi ad A scilicet ad V ad BEd

illo duplo, sicut distabit idem secundus a non gradu seu ab extremo opposito illius magnitudinis date.

Et sicut universaliter probatur de omni latitudine incipiente a non gradu et terminata in aliquem gradum finitum, continente etiam gradum aliquem et subduplum et subquadruplum et sic in infinitum, quod eius gradus medius est precise subduplus ad gradum ipsam terminantem. Unde non solum est hoc verum de latitudine velocitatis motus incipientis a non gradu, sed etiam de latitudine caliditatis, frigiditatis, luminis, et aliarum similium qualitatum, et consimiliter argui poterit et probari.

Quantum autem ad magnitudinem pertranseundam, uniformiter acquirendo talem latitudinem motus incipientem a non gradu et terminatam in aliquem gradum finitum, dictum est prius quod totus ille motus seu tota illa acquisitio correspondebit gradui suo medio. Et consimiliter etiam, si ab aliquo gradu exclusive uniformiter acquiratur latitudo motus ad aliquem gradum terminata finitum.

Ex quo satis cognosci poterit in huiusmodi uniformi intensione vel remissione, quanta erit magnitudo pertransita, ceteris paribus, in prima medietate temporis, et quanta in secunda. Cum enim a non gradu ad aliquem gradum uniformis fiat alicuius motus intensio, subtriplum precise pertransibit in prima medietate temporis, ad illud quod pertransibit in secunda. Et si alias ab eodem gradu, aut ab aliquo alio quocunque ad non gradum, uniformis fiat remissio, triplum precise pertransietur in prima medietate temporis, ad illud quod pertransiretur in secunda.

Totus enim motus factus in toto tempore correspondebit medio

---

67 magnitudinis *AEd* latitudinis *BV*
68 sicut: sic *B*
69 terminate *AV* / in: ad *Ed* / ante finitum *add. V* certum / continente *Ed* continentem *ABV*
70 et¹ *om. B* / in *om. B*
71 eius: eiusdem *V* ille *B* / ipsam: illam *B*
71–72 ipsam terminantem: terminantem illam latitudinem *V*
73 incipiente *B* / etiam *om. V*
74 vel luminis *V* / consimilium *B* / et *om. VEd*
78 terminante *B* / totus: tertius *(?) B*
79 ille: iste *A* / tota *om. Ed*

81 terminata *tr. V ante* ad
82 huiusmodi: huius *V*
84 temporis *om. Ed* / enim: autem *B*
85 gradum *om. B* / uniformis *BV* uniformiter *AEd; cf. lineam 88* / intensio *tr. V ante* alicuius
86 pertransibit: pertransietur *B*
87 pertransibit *AEd* pertransietur *B* pertransiret *V* / alias: aliquis *V*
88 *post* quocunque *add. V* gradu / fiat: fiet *A*
89 pertransiet *Ed*
90 pertransiretur *VEd* pertransiet *B* transietur *A*
91 enim *om. B*

[*Regule solvendi sophismata Heytesberi* 281]

gradui suo, illi scilicet quem habebit in instanti medio istius temporis. Et secunda medietas istius motus correspondebit gradui medio secunde medietatis eiusdem motus, qui est subquadruplus ad gradum istam latitudinem terminantem. Ideo cum ipsa secunda medietas durabit solum per subduplum tempus, in quadruplo minus precise pertransietur per ipsam secundam medietatem quam per totum motum. Igitur totius magnitudinis pertranseunde toto motu, pertransientur tres quarte per primam medietatem totius motus, et ultima quarta pertransietur per secundam medietatem eiusdem. Sequitur igitur quod in huiusmodi remissione uniformi vel intensione alicuius motus ab aliquo gradu usque ad non gradum, vel a non gradu in aliquem gradum, in triplo plus precise pertransietur per medietatem istius latitudinis intensiorem quam remissiorem.

Motum autem aliquem uniformiter intendi vel remitti ab aliquo gradu ad alium gradum, infinitis modis contingit, quia ab aliquo gradu ad suum subduplum vel ad suum subquadruplum aut subquintuplum, aut suum subsexquialterum aut subsexquitertium, et sic deinceps. Ideo nullus potest esse terminus universalis penes quem universaliter cognosci poterit quanto plus pertransiretur per primam medietatem huius intensionis vel remissionis, quam per secundam; quia respectu diversorum graduum extremorum diversa erit proportio magnitudinis pertransite in prima medietate temporis ad magnitudinem pertransitam in secunda. Cognitis tamen gradibus extremis, ita scilicet quod cognoscatur quantum pertransiretur uniformiter in tanto tempore vel in tanto, per gradum extremum intensiorem terminantem, et consimiliter respectu gradus extremi remissioris, cognosci poterit per calculatio-

92 illi scilicet: sive illi *V* / instanti medio *VEd* / istius: illius *Ed*
93 Et: Et ita *V* / istius: illius *V*
95 istam: illam *Ed* / terminantem *tr. BV post* gradum *in linea 94* / ipsa: ista *Ed*
96 solum *AV* solummodo *Ed B* / precise *tr. BEd ante* per
97 pertransietur *om. B* pertransibit *V* / ipsam: istam *V* illam *Ed*
99 *ante* totius *add. Ed* istius *et add. V* et / et *om. V* / ultima: una *B*
100 pertransiretur *V*
101 huiusmodi *AEd* huius *BV* / intensione vel remissione uniformi *Ed*
102 in: usque in *V* non *B*
103 pertransiretur *BV*
104 istius: illius *V* / quam: quam per *V*
105 aliquem *om. B*
106 ad alium: in aliquem *V* / modis: motis *B*
107 *post* suum *add. Ed* gradum / subduplum: duplum *B* / vel *VB om. A* et *Ed* / aut: ad *A*
108 subsexquialterum *BV* sesquialterum *AEd* / deinceps: subdeinceps *B* / Et ideo *B*
109 nullus potest: non poterit *V* / universalis *om. B*
110 poterit: potest *(?) A* / pertransietur *Ed*
111 intensio vel remissio *V*

nem quantum pertransiretur in prima medietate et etiam quantum in secunda; quia cognitis isto modo extremis gradibus, haberi potest etiam gradus medius inter istos, et etiam gradus medius inter illum gradum medium et gradum intensiorem illam latitudinem terminantem. Sed huiusmodi calculatio maiorem sollicitudinem ageret quam profectum.

Et ideo sufficit quod in omni huiusmodi casu respondeatur generaliter, quod plus pertransietur per medietatem talis latitudinis intensiorem quam remissiorem; tantum, scilicet, quantum per gradum medium istius medietatis intensioris plus uniformiter pertransiretur in equali tempore, sicut si ipsa medietas uniformiter acquiritur vel deperditur, quam per medium gradum medietatis remissioris in eodem pertransiretur.

De difformi autem intensione vel remissione, sive ab aliquo gradu usque ad non gradum vel econtra, vel ab aliquo gradu ad aliquem alium, nulla potest esse regula quantum in tanto tempore vel in tanto pertransiretur, aut cui gradui intrinseco istius latitudinis correspondebit talis latitudo motus difformiter deperdita seu acquisita; quia sicut infinitis modis contingit talem remissionem seu intensionem difformem variari, ita etiam infinitis gradibus intrinsecis eiusdem latitudinis, ymmo cuilibet gradui intrinseco istius latitudinis sic acquisite vel deperdite, poterit totus ille motus correspondere.

Unde universaliter gradus terminans talem latitudinem secundum extremum intensius est remississimus supra aliud extremum eiusdem

118 pertransietur *A* / etiam *BEd*; *om.* *AV* / quantum *om.* *B*
120 etiam¹ *om.* *A* / inter... medius *om.* *B* / istos: illos *V* / illum: istum *B*
121 et gradum: et econtrario gradum medium inter *A*
122 huiusmodi *AEd* huius *V* illa *B* / maiorem *ABV* minorem *Ed*
123 profectum *A* proficuum *BV* prefectum *Ed*
124 omni *om.* *A* / huiusmodi: huius *B* / omni... casu: omnibus casibus *V* / de respondeatur *scr.* *V in marg.* vel ostendatur
125 pertransietur *AB* pertransitur *V* pertransitetur *Ed* / talis latitudinis *tr.* *A ante* per
126 quam per *V*
127 istius *AEd* illius *BV* / plus *Ed om.* *ABV*
128 si *om.* *ABV*
129–30 quam... pertransiretur *Ed om.* *ABV*
131 sive: seu *A*
132 econtrario *Ed* / vel *A* dum *B* seu *VEd*
133 alium: alium gradum *V* / in² *om.* *V*
134 pertransiretur: pertransietur *B* / istius: illius *V*
134–35 correspondebit *AB* correspondeat *V* corresponderet *Ed*
135 disperdita *B* / sicut: si *A*
136 modis: motus *B*
137 etiam *om.* *V*
140 gradus *om.* *B* / talem: illam *V*
141 remississimus *Ed* remissius *AV* remitissimus *B* / supra *Ed* citra *ABV*

latitudinis cui non potest totus huiusmodi motus difformiter difformis correspondere; et gradus terminans illam latitudinem secundum extremum remissius, est remississimus citra aliud extremum eiusdem
145 latitudinis, cui non potest huiusmodi motus difformiter difformis correspondere. Unde nec est possibile quod totus huiusmodi motus ita remisso gradui correspondeat, sicut idem motus poterit correspondere, nec ita intenso.

142 huiusmodi *AEd* ille *V* huius *B*
143-46 et... correspondere *Ed om. ABV*
146 nec: non *V* / huiusmodi *AEd*

huius *BV*
147 gradui: gradu *B* / motus *om. B* / poterit: potest *(?) A*

Document 5.2

Proofs of Propositions Posited in the Rules for Solving Sophisms Attributed to William Heytesbury

EVERY increment of velocity *(latitudo motus)* uniformly acquired or lost will correspond to its mean degree [of velocity]. This means that a moving body uniformly acquiring or losing that increment will traverse in some given time a magnitude completely equal to that which it would traverse if it were moving continuously through the same time with the mean degree [of velocity].

Proof. I assume a total [velocity] increment from zero degree to a degree of eight. And I take three bodies, namely, a, b, c, all now moving with the mean degree of that velocity increment (i.e., with a velocity of 4).... Let a move through an hour uniformly at that degree. Let b accelerate uniformly from that degree to a degree of eight in half of the hour and let c decelerate uniformly from that mean degree to zero degree in the same half hour. Furthermore, I take a body d which uniformly acquires the whole designated increment (i.e., from 0 to 8) in the whole hour. Then it is argued as follows.

Body d will traverse just as much space in the whole hour as do b and c in one-half of the hour. But b and c will traverse in the half hour just as much as a will traverse in the whole hour. Therefore d will traverse just as much in the whole hour as a. The [syllogistic] consequence is evident.[15] From the consequent it is [further argued] as follows: d will traverse in the whole hour as much as a in the same hour. But a will traverse neither more nor less in the whole hour than that which moves with the mean

---

[15] For fourteenth century logicians, syllogisms were only one species of the generic, logical form called a consequence, and thus this author's arguments are syllogistic consequences. See P. Boehner, *Medieval Logic* (Chicago, 1952), pp. 52–53; E. A. Moody, *Truth and Consequence in Medieval Logic* (Amsterdam, 1953), p. 65; and I. M. Bocheński, *Formale Logik* (Munich, 1956), pp. 219–43.

degree of the increment. Therefore d will traverse, in acquiring the velocity increment, just as much space as if it were continually moved with the mean degree of the increment. The inference is obvious, and the major is argued.

Body d will traverse as much in the first half of the hour as c in the same half. But d will traverse in the second half of the hour as much as b in the first. Therefore d will traverse so much in the whole hour as b and c joined together in half of that hour. The consequence is obvious and also the major premise. But that a will traverse in the whole hour as much as b and c in half of the hour, which is the minor premise of the argument [of the first syllogism], is argued thus. If b were not increased and c were not decreased, they would, joined together, in one-half hour traverse as much as a in the whole hour. But b and c are as much now as they were then. Therefore, b and c will traverse in half an hour as much as a in the whole hour. The consequence is obvious; and the minor premise is argued.

In the half hour a will traverse more space than c by the same amount that b traverses more space than a in the half hour. Therefore, in comparing the excess of a over c, or b over a, in that half hour, both of them, namely, b and c, taken together would traverse the same amount as if neither b nor c were increased or diminished. And therefore b and c will traverse the same amount now as then. Therefore, etc. And just as it is argued concerning moving body d, so it can be argued for any moving body acquiring that increment in an hour; therefore, etc.

If any moving body uniformly increases its motion from zero degree up to any degree in an hour, it will traverse in the second half hour precisely three times the distance that it will traverse in the first half of the same [hour].

Proof: The degree to which corresponds the whole motion made in the first part of the hour is one-third the degree to which the whole motion completed or made in the second half hour corresponds. Therefore the distance traversed in the first half is one-third the distance traversed in the second half hour. The consequence is proved. Just as the degree to which corresponds the motion in the first half hour is related to the degree to which corresponds the motion in the second half, so is related the distance traversed in the first part of the hour to the distance traversed in the second. But from this—that the degree to which corresponds the motion in the first half hour is one-third the degree to which corresponds the motion in the second half hour—it follows that the distance traversed [in the first half will be one-third that traversed in the second half].

And the minor is argued thus. The degree to which corresponds the motion completed in the first half hour is one-third, etc., since I posit that the moving body *a* uniformly increases its motion in an hour from zero degree up to eight. And then it is argued thus: a degree of two is one-third of a degree of six. But to a degree of two corresponds the motion in the first half, and to a degree of six corresponds the motion in the second half. Therefore, etc. The minor is proved: the whole motion made in the first half hour corresponds to its mean degree, which is two. And the whole motion made in the second half hour corresponds to its mean degree, which is six, as is obvious to one who attends closely. Therefore, etc.

COMMENTARY

This selection consists of two theorems—the mean velocity theorem and the theorem comparing the distances traversed in the first and second half of the time by a body uniformly accelerating. It will be noticed that the form of the arguments in the proof of the first theorem is completely syllogistic (see note 15). Before analyzing the syllogisms given, we should notice the data of the proof:

1. A body *a* moves uniformly through an hour with a velocity of 4.
2. A body *b* accelerates uniformly from 4 to 8 in one-half hour.
3. A body *c* decelerates uniformly from 4 to 0 in one-half hour.
4. A body *d* accelerates uniformly from 0 to 8 in the whole hour.
5. Bodies *a*, *b*, *c*, and *d* move respectively through distances S_a, S_b, S_c, and S_d.

Then Heytesbury states the first syllogism.

Syllogism I

If (1) $S_d = S_b + S_c$
and (2) $S_b + S_c = S_a$
then (3) $S_d = S_a$

He follows this by the statement of an obvious syllogism not requiring demonstration (S_{dm} in (4) being the distance traversed at mean speed of the increment acquired by *d*):

Syllogism II

If (3) $S_d = S_a$
and (4) $S_{dm} = S_a$
then (5) $S_d = S_{dm}$

Having outlined the principal syllogisms, he then goes on to demonstrate the veracity of the major and the minor premises, (1) and (2), of Syllogism I. The proof of the major is rather obvious, since b and c are precisely equivalent to the two halves of d. The minor premise [I (2)] is the crux of the whole demonstration. This form of the proof is as follows (S_{bi}, S_{ci} being the distances traversed if b and c were held at the mean degree rather than accelerating and decelerating):

$$\text{Since (6) } S_{bi} + S_{ci} = S_a$$
$$\text{and (7) } S_{bi} + S_{ci} = S_b + S_c$$
$$\text{therefore (8) } S_b + S_c = S_a$$

The crucial part of this supplementary syllogism is the minor premise (7). This is never really proved. It is simply intuitively grasped that the $S_a/2$ covered in the first half (or its equivalent S_{bi}) exceeds S_c by the same amount that $S_a/2$ traversed in the second half (S_{ci}) is exceeded by S_b. Now with both (1) and (2) assumed to be proved, (3) follows; and from (3) and the defined fact of (4) follows the mean speed theorem, at least as applied to bodies accelerating from rest. A succeeding proof, not included in our translation here, generalizes the law for all bodies uniformly accelerating from any given velocity. The reader may examine proofs of this more generalized theorem by Swineshead and Dumbleton in the two succeeding documents.

The three-to-one distance theorem is proved immediately from the mean velocity theorem. Its proof needs no additional commentary.

Probationes conclusionum tractatus regularum solvendi sophismata Guilelmi Heytesberi[*]

Omnis latitudo motus uniformiter acquisita vel deperdita correspondebit gradui medio ipsius, i.e., quod mobile idem ipsam latitudinem uniformiter acquirens seu deperdens in aliquo tempore dato equalem omnino magnitudinem pertransibit ac si ipsum continue per equale tempus moveretur medio gradu. Istud probatur sic: Accipio totam latitudinem a non gradu ad gradum ut octo. Et capio tria mobilia, scilicet, a, b, c, sub medio gradu illius latitudinis. Quorum a ponatur pertransire [4?] pedale in hora. Et moveatur per horam illo

---

[*] Edition of Venice, 1494, ff. 198v–199r *(Ed)*; cf. MS Venice, Bibl. Naz. San Marco Lat. VI, 71, f. 129v *(V)*. The manuscript is quite defective and it proved of no profit to collate the two. Cf. Oxford, Bodl. Canon. Misc. 376, ff. 23–32 (contains only chapters 3 to 5, and hence not the relevant part).

gradu. Et intendat *b* uniformiter motum suum ab isto gradu usquam ad gradum ut octo in medietate illius hore. Et remittat *c* uniformiter motum suum ab isto medio gradu ad non gradum in eadem medietate hore. Et capio *d* mobile quod uniformiter acquirat totam latitudinem iam assignatam in tota ista hora. Tunc arguitur sic:

Tantum pertransibit *d* in tota hora quantum pertransibunt *b*, *c* in medietate hore. Sed tantum pertransibunt *b*, *c* in medietate hore quantum pertransibit *a* in tota hora. Igitur tantum pertransibit *d* in tota hora quantum pertransibit *a* in tota hora. Consequentia patet et ex consequenti arguitur sic: tantum pertransibit *d* in tota hora quantum pertransibit *a* in eadem hora. Sed nec maius nec minus pertransibit *a* in tota hora quam a movendo medio gradu illius latitudinis. Igitur tantum pertransibit sic acquirendo istam latitudinem ac si pertransiret continue movendo medio gradu illius latitudinis *d*. Consequentia patet et maior arguitur.

Tantum pertransibit *d* in prima medietate hore quantum *c* in eadem medietate hore. Sed tantum pertransibit *d* in secunda medietate quantum *b* in prima. Igitur *d* pertransibit tantum in tota hora quantum *b*, *c* coniuncti in medietate illius hore. Consequentia patet et maior similiter. Sed quod tantum pertransibit *a* in tota hora quantum pertransibunt *b*, *c* in medietate hore que est minor illius argumenti arguitur sic. Si nec *b* intenderetur nec *c* remitteretur pertransirent coniuncti in medietate hore tantum quantum *a* in tota hora. Sed tantum pertransibunt *b*, *c* nunc sicut tunc. Igitur tantum pertransibunt *b* et *c* in medietate hore quantum *a* in tota hora. Consequentia patet et minor arguitur.

Quia quanto minus pertransiretur a *c* quam ab *a* in medietate hore tanto plus pertransiretur a *b* quam ab *a* in eadem medietate. Igitur comparando excessum quo pertransiretur plus ab *a* quam a *c* vel a *b* quam ab *a* in illa medietate hore ab utroque, scilicet *b*, *c* equaliter pertransiretur coniunctim a *b*, *c* si nullum illorum intenderetur nec aliquod remitteretur, et ultra igitur tantum pertransibit *b*, *c* nunc quantum tunc. Igitur, etc. Et sic arguitur de *d* mobili sic potest argui de quocumque mobili uniformiter acquirendo istam latitudinem in hora. Igitur, etc. . . .

Si aliquod mobile uniformiter intendat motum suum a non gradu usque ad aliquem gradum in hora triplum precise pertransibit in secunda medietate hore ad hoc quod pertransibit in prima medietate eiusdem. Probatur: nam gradus cui correspondet totus factus motus in

50 prima parta hore est subtriplus ad gradum cui correspondet totus motus finitus vel factus in secunda medietate hore. Igitur pertransitum in prima medietate est subtriplum ad pertransitum in secunda medietate hore. Consequentia probatur: Quia sicut se habet gradus cui correspondet motus in prima medietate hore ad gradum cui correspon-
55 det motus in secunda medietate sic se habet pertransitum in prima parte hore ad pertransitum in secunda. Sed ex quo gradus cui correspondet motus in prima medietate hore est subtriplus ad gradum cui correspondet motus in secunda medietate hore sequitur quod erit pertransitum, etc.
60 Et minor arguitur sic: quod gradus cui correspondet motus factus in prima medietate hore est subtriplus etc. quia posito quod *a* mobile uniformiter intendat motum in hora a non gradu usque ad octo: Et tunc arguitur sic: gradus ut duo est subtriplus ad gradum ut vi (6) sed gradui ut duo correspondet motus in prima medietate et
65 gradui ut vi correspondet motus in secunda medietate. Igitur, etc. Minor probatur: totus motus factus a prima medietate hore correspondet medio suo gradui qui est ut duo. Et totus motus factus in secunda medietate hore correspondet medio suo gradui qui est ut vi ut patet intuenti. Igitur, etc.

Document 5.3

Richard Swineshead, *The Book of Calculations:*
Rules on Local Motion

1. [SECOND SUPPOSITION FOR CONCLUSION 38.] The second is [that] every increment of velocity uniformly acquired or lost corresponds to its mean degree [of velocity]. In the first place this is demonstrated as follows. By an "increment acquired corresponds to its mean degree [of velocity]" I mean the following: Just as much space would be traversed by means of that increment so acquired as by means of the mean degree [of velocity] of that increment, assuming something were to be moved with that mean degree [of velocity] throughout the whole time.

Initially it is argued that every composite of two unequal numbers is twice their mean. For example, that which is a composite of 8 and 4 is twice 6. If it is argued to the contrary, let it be posited that a is more than b and c is their mean. Let a be diminished toward c equally fast as b is increased toward c. Then in the end the composite of a and b will be twice c because then a and b will be equal to each other [and to c]. But the composite of a and b will continually be as in the end because one will acquire just as much as the other will lose. Therefore, the composite of a and b is *now* twice c, which was to be proved.

Then this second supposition is argued in the first place as follows: Let a acquire some velocity increment uniformly, which increment is terminated in the upper end at a degree [of velocity] of c. And b will lose the velocity increment in every way just as a now acquires it. And e is the mean degree [of velocity] of that increment. It makes no difference whether this increment is terminated at zero degree or at some other degree. Then the velocities [*motus*] of a and b together continually will be equal to a degree twice that of e because [by the argument of the previous paragraph] the composite of two unequal degrees is always equal to twice their mean, and e will always be the mean between a and b, when they will be unequal. And these two together will continually correspond to the same degree be-

[Swineshead, *Book of Calculations* 291]

cause the lesser is continually increased equally fast as the greater is decreased. Therefore, they will continually correspond to a degree of twice e, and one of them will correspond to the same degree [as the other]. Hence the two of them together will correspond to a degree twice that to which one by itself corresponds. Therefore, the movement of either of them by itself during the whole time will correspond to the mean degree e, which was to be proved. The consequence follows, for let d be moved with a degree of twice e. It follows that a and b will traverse as much as d and one will traverse just as much as the other. Hence it will traverse half of that which d traverses. Consequently, it traverses just as much as if it were moved with e degree, which is one half of the degree with which d is moved, which degree is the mean of the increment to be acquired. Therefore the supposition follows....

2. [Third argument for the second supposition.] The same position (i.e., the mean speed theorem) is argued in a third way. In the first place it is argued that whenever a power accelerates *(intendit motum suum)* uniformly from zero degree, it will traverse three times as much space in the second half of the time as in the first. For let it be posited that a decelerates uniformly to zero degree. Then in every instant of the first proportional part [of the time] it will be moved twice as swiftly as in the corresponding instant *(instanti correspondente)* of the second proportional part [of the time], and similarly for the succeeding [proportional parts of the time], as is evident. Since, therefore, the first proportional part of the time is twice the second, it is evident that a will traverse four times as much space in the first proportional part as in the second, and four times as much space in the second as in the third, and thus to infinity *(in infinitum)*. Therefore, in all the proportional parts, it will traverse four times as much as in those parts remaining [after the first part]. The consequence follows from the fact that if two sets of quantities be compared term for term so that from each comparison arises the same proportion, then, if the sum of the first set is compared with the sum of the terms of the second set, the sums will have the same proportion as that of the individual term for term comparisons. It follows, therefore, in the whole time [a] will traverse four times as much space as in the second half of the time. Hence, in the first half of the time it will traverse three times as much space as in the second half of the time—evidently where there is uniform deceleration to zero degree. Conversely, therefore, in the case of acceleration, it will traverse three times as much in the second half as in the first. From this conclusion we shall argue to our principal objective.

For let *a* be uniformly accelerated from zero degree to 8. Then if the motion in the first half of the time will correspond to a degree that is less than 2, let that degree be *d*. And let the degree to which the movement in the second half corresponds be *c*. Then from the fact that the movement is uniformly accelerated in the second half in the same way as in the first, and the fact that in the middle instant [of the whole time] the degree is 4, it follows that *c* equally exceeds 4 as *d* exceeds zero, and 6 equally exceeds *c* as 2 exceeds *d*. Hence 6 equally exceeds 2 as *c* exceeds *d*. Hence, since the ratio of 6 to 2 is 3, and *c* degree is less than 6, it is evident that the ratio of *c* to *d* is greater than 3. Consequently, it (*a*) will traverse more than three times as much in the second half as in the first. But this conclusion is refuted by the initial conclusion.... If, therefore, the movement in the first half will correspond to a degree less than the mean, it is evident that the motion in the second half will correspond to a degree more than triple that [to which the motion] in the first [half corresponds]. But this consequent is false and refuted. Now if the motion corresponds to a degree greater than the mean, it follows that, since the motion in the second half equally exceeds the motion in the first half as 6 exceeds 2, the motion in the second half would not be triple to the motion in the first half. The consequent is false by the previous arguments. It is evident, therefore, that the motion in the first half will correspond to a degree of 2 and in the second half to one of 6. It is argued this way for every increment terminated at zero degree.

The same thing follows if the acceleration starts from some degree other than zero. For a body so moving will acquire a velocity increment equivalent to an increment immediately acquired starting from rest, which latter increment [uniformly acquired] corresponds to its mean degree. And so [the final degree of speed] exceeds the initial degree by that increment immediately acquired, and the degree midway between the initial and final degrees is greater than the initial degree by one-half of the whole increment immediately acquired. Therefore, just as the increment immediately acquired from rest corresponds to its mean degree, so this uniform acceleration starting from some speed corresponds to its mean degree, which was to be proved....

3. ... This conclusion is proved, namely, that every motion whose deceleration is increasing *(motus velocius et velocius deperditus)* corresponds, with respect to the traversal of space [in the same time], to a degree greater than the mean [between the initial and final velocities]. If this is denied, [argue as follows]. Let *a* be decelerated at an increasing rate and let

[Swineshead, *Book of Calculations* 293]

b uniformly decelerate through the same velocity increment *(latitudo)*, starting from the same initial degree and ending at the same final degree. This is possible in the same way that one body moving uniformly traverses a certain space in an hour and another body by accelerating traverses the same space in the hour, it having, however, begun its movement more slowly.

With this case admitted, then argue as follows. [In the beginning] the velocities *(motus)* of *a* and *b* are equal, and [then] the movement of *a* will be continually faster and faster than the movement of *b*; therefore, the movement of *a* will correspond, so far as the traversal of space is concerned, to a greater *(intensiori)* degree than the movement of *b*. The over-all argument is clear *(consequentia patet)*. Its antecedent is argued as follows. If the velocity of *a* is not continually greater than the velocity of *b*, let an instant in which it is not swifter be *c*. Then it is argued, either the motion of *a* at instant *c (tunc)* will decrease more swiftly than that of *b*, or equally swiftly, or more slowly. If at that time the motion of *a* is decreased more swiftly, and afterwards continually more swiftly, and [if] in the end *a* and *b* will be at the same degree, therefore in instant *c* the velocity of *a* will be greater than that of *b*. The consequence is evident. For, from the fact that the speed of *a* is decreased more rapidly after instant *c* than before, it follows that *a* loses a greater increment [of velocity], and this increment lost will only be that which is between the degree it has at *c* and the degree it has in the end. Hence *a*'s increment from *c* to the end will be greater than *b*'s, from *c* to the end. Consequently, the velocity of *a* is greater than that of *b*.

If the speed at instant *c (tunc)* is assumed to be decreased exactly equally as fast as that of *b*, and continually afterwards more rapidly than that of *b*, and [if] in the end they will be equal, therefore at *c (tunc)* the speed of *a* will be greater—by the previous argument.

[Finally,] if the velocity of *a* at instant *c (tunc)* is decreased more slowly than that of *b*, and continually before the same instant *a* is decreased more slowly than *b*, and [if] in the beginning they are equal—by assumption, therefore, at *c* instant *(tunc)* the velocity of *a* is greater than that of *b*. The consequence is evident in itself *(de se)*.

Since it was posited, therefore, that in instant *c* the speed of *a* will be decreased, either more swiftly or more slowly than, or equally fast as, that of *b*, it follows that at *c (tunc)* the speed of *a* will be greater than that of *b*. And it ought to be argued in the same way for every other instant *(de omni alio instanti)*....Therefore, the speed of *a* will be continually greater than that of *b*. Therefore, the [whole] motion of *a* will correspond to a degree greater than that to which the motion of *b* corresponds. The mean degree of the increments of velocity lost by *a* and *b*, (i.e., the increments between the initial and final degrees) is the same. But the movement of *b* will correspond to the mean degree of its increment, as was proved in the [second] supposition to the thirty-eight conclusion. For it proved there that every movement uniformly acquired or lost corresponds to its mean degree. Therefore it follows that the movement of *a* corresponds to a greater degree than the mean, which was to be proved.... And similarly every movement whose rate of deceleration is decreasing *(motus tardius et tardius deperditus)* corresponds to a degree less than the mean.... And wherever the rate of acceleration is increasing *(velocius et velocius motus acquiritur)* it corresponds to a degree less than the mean.... Wherever the rate of acceleration is decreasing *(motus tardius et tardius acquiritur)*, it corresponds to a degree greater than the mean....

COMMENTARY

1. The first Swineshead proof should be compared with the Heytesbury proof (Doc. 5.2), as should Swineshead's second proof, not here translated but given in the Latin text. Logically, it seems somewhat tighter than the proof of Heytesbury, but basically it is of the same symmetrical type. Swineshead posits a body uniformly accelerating through a velocity increment and another body uniformly decelerating through the same increment in the same time. At every instant of time, the sum of their velocities will be twice the mean. Hence, throughout the whole time their measurement together will be equivalent, for the traversal of space, to a uniform movement at twice their mean speed. But the movement of acceleration traverses precisely as much as the movement of deceleration. Hence for the whole time each will traverse half of the total space traversed by them both together. To do this each must be equivalent to a degree half of that to which they both correspond; or, in short, each must correspond to the mean degree. It should be noticed that the proof is based on the mathematical assumption that the sum of two unequal numbers is twice their mean. Needless to say, the passage from the *continual* comparison of the two velocities with twice their mean to the comparison for the *whole* time is an intuitive grasp of the summation of infinitesimals that will be used later in integral calculus.

2. The next proof (Swineshead's third proof) is the most interesting of the various Merton proofs. I have attempted below to reconstruct the

proof, using modern symbolization and making explicit those statements which are only implicit in Swineshead's vocabulary. Notice, first, that Swineshead sets out to prove the distance formula, namely, that a body uniformly accelerating traverses three times as much distance in the second half of the time as in the first. Actually, Swineshead gives his detailed proof for a body decelerating instead of accelerating. But having proved the theorem for deceleration, he assumes that it holds also for acceleration.

In the beginning of the proof the author speaks of dividing the time into proportional parts, which we can represent as follows:

(1) $T = t + t/2 + t/4 + \ldots + t/2^{n-1} + \ldots$, T being the total time, and t being $T/2$. He then uses the expression "corresponding instants" of succeeding proportional parts of the time. By this he clearly means that we have in each proportional part of the time an "infinite" number of infinitesimal time intervals or "instants." Like most of his contemporaries, Swineshead in all probability considered "infinite" and "infinitesimal" from a potential or syncategorematic standpoint.[16] Thus for him an infinite number would be a number greater than any assignable quantity and an instant would be a time interval shorter than any assignable interval. Keeping this in mind we can now deduce from the phrase "corresponding instants" two more formulations:

(2) $t = t_{11} + t_{12} + t_{13} + \ldots + t_{1m}$, where t is the total time of the first proportional part, and m goes toward infinity, and $t_{11} = t_{12} = t_{13} = \ldots = t_{1m}$;

(3) $t/2 = t_{21} + t_{22} + t_{23} + \ldots + t_{2m}$, where $t/2$ is the total time of the second proportional part, and $t_{21} = t_{22} = t_{23} = \ldots = t_{2m}$.

And using the term "corresponding," Swineshead further implies that there are just as many terms in series (2) as in series (3). But since $t : t/2 = 2/1$, then

(4) $t_{11}/t_{21} = t_{12}/t_{22} = t_{13}/t_{23} = \ldots = t_{1m}/t_{2m} = 2/1$.

Let us assume further that there are velocities $V_{11}, V_{12}, \ldots, V_{1m}$ for times $t_{11}, t_{12}, \ldots, t_{1m}$ and velocities $V_{21}, V_{22}, \ldots, V_{2m}$ for times $t_{21}, t_{22}, \ldots, t_{2m}$. Then, according to Swineshead, the decelerating body moves such that "in every instant of the first proportional part of the time, it will be moved twice as swiftly as in the corresponding instant of the second proportional part." In short,

(5) $V_{11}/V_{21} = V_{12}/V_{22} = \ldots = V_{1m}/V_{2m} = 2/1$.

He apparently assumes that this follows from the definition of uniform

[16] Clagett, "Richard Swineshead and Late Medieval Physics," *Osiris*, Vol. 9 (1950), 144-45.

acceleration. He then appears to assume the well known kinematic proportions:

(6) $V_{11}/V_{21} = (S_{11}/S_{21})(t_{21}/t_{11}) = \ldots = V_{1m}/V_{2m} = (S_{1m}/S_{2m})(t_{2m}/t_{1m})$.

Now applying the values of (4) and (5) to (6), we arrive at values for the distances traversed in the corresponding instants of the first and second proportional parts:

(7) $S_{11}/S_{21} = \ldots = S_{1m}/S_{2m} = 4/1$.

Now suppose that we designate the whole distances traversed in the first and second proportional parts of the time as S_1 and S_2, so that

(8) $S_1 = S_{11} + S_{12} + \ldots + S_{1m}$ and $S_2 = S_{21} + S_{22} + \ldots + S_{2m}$.

Then by the basic theorem of ratios relating the sums as the individual terms,

(9) $S_1/S_2 = 4/1$.

By a similar line of reasoning, we can show that the same relationship holds for all the succeeding proportional parts,

(10) $S_1/S_2 = S_2/S_3 = S_3/S_4 = \ldots = S_{n-1}/S_n = \ldots = 4/1$.

Now if S_T is the total distance traversed in the whole time, Swineshead goes on to assume

(11) $S_T = S_1 + S_2 + S_3 + \ldots + S_{n-1} + \cdots$;

and if S_R is the distance traversed in the remaining parts of the time, after the first part, i.e., in the second half of the time, then

(12) $S_R = S_2 + S_3 + S_4 + \ldots + S_n + \cdots$.

Therefore, by (10), (11), (12) and the basic theorem equating the ratio of the sums to the ratio of the parts,

(13) $S_T/S_R = 4/1$.

From this it is evident that in the case of uniform deceleration, the distance traversed in the whole time to that traversed in the second half is 4 to 1, and thus the distance traversed in the first half of the time to that traversed in the second half is as 3 to 1. Conversely, for acceleration the ratio is 1 to 3.

With the distance formula proved, Swineshead shows by indirect proof that with respect to distance traversed in the same time, the uniformly accelerated movement is equivalent to a uniform movement at the mean velocity. He does this by showing that if we assume that the equivalent uniform velocity is not equal to the mean velocity, but rather is greater or less than that mean velocity, in either case the distance relationship of 3 : 1 established in the first part of the proof is violated; hence the equivalent velocity must be the mean velocity.

This selection is finally concluded by a proof of the more general

[Swineshead, *Book of Calculations* 297]

expression of the mean speed theorem, namely, when the initial velocity is not zero. It is simple and offers no difficulties.

3. To explain this theorem, let us graph the problem (see Fig. 5.3) in the manner developed at Paris by Oresme, which we shall describe in some detail in the next chapter. Now we assume the following data:

(1) Body a, beginning from speed V_i, decelerates *at an increasing rate* until it reaches speed V_f. Its instanteneous speed at any instant c is V_a. Body a traverses in the whole time (t) the distance S_a.

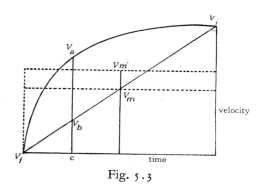

Fig. 5.3

(2) Body b beginning from V_i decelerates *uniformly* to V_f in time t. It traverses S_b in so decelerating.

(3) S_m is the distance traversed in time t by a body moving uniformly with $V_m = (V_i + V_f)/2$.

To prove: $S_a > S_m$.

(1) At any instant c, $V_a > V_b$, for if we assume otherwise, then the assumption violates the datum that a is decelerating at an increasing rate and yet starts and ends at the same speeds as a.

(2) If $V_a > V_b$ for any instant c, it is true for the whole time (other than the initial and final instants), and thus V_a is always greater than V_b (except at first and last instants). Hence, the times being the same $S_a > S_b$. (This is essentially an operation of infinitesimal calculus that has been here performed. If any $dS_a/dt >$ the corresponding infinitesimal dS_b/dt, then $(S_a/t) > (S_b/t)$, and with t the same, $S_a > S_b$.)

(3) But $S_b = S_m$ by the mean speed theorem.

(4) Therefore, $S_a > S_m$.

Swineshead also gives similar theorems for decreasing deceleration, and for both increasing and decreasing acceleration.

Liber calculationum Ricardi Swineshead*

[1] Regule de motu locali.... [Secunda suppositio pro conclusione 38ᵃ.] Secundum est: omnis latitudo motus uniformiter acquisita vel deperdita suo gradui medio correspondet. Hoc demonstratur primo sic—tunc voco latitudinem ut acquiretur correspondere suo gradui medio sic: tantum pertransiretur precise mediante illa latitudine sic acquisita sicut mediante gradu medio eiusdem, si per totum tempus illo gradu medio moveretur.

Sed primo arguitur quod omne compositum ex duobus inequalibus est duplum ad medium inter illa, sicut compositum ex 8 et 4 est duplum ad 6. Nam si non, ponatur quod a sit maius b et sit c medium inter illa. Et diminuatur a ad c et maioretur b equevelociter versus c. Tunc in fine compositum ex a, b erit duplum ad c, eo quod a, b inter se tunc erunt equalia. Sed compositum ex a, b continue erit tantum sicut in fine, quia quantum unum acquiret tantum aliud deperdet. Ergo compositum ex a, b nunc est duplum ad c, quod fuit probandum.

[Primum argumentum pro suppositione secunda.] Tunc arguitur illa

---

* It is of interest that, of the many copies of the *Liber calculationum* that I have examined, only the Italian manuscripts contain the *Regule de motu locali* (although, to be sure, Worcester Cathedral F. 35, which I have not seen, appears to contain the *Regule*). I have no doubt of the genuineness of the *Regule*, however, since there are several cross references to it in the other parts of the *Liber calculationum*. The manuscripts I have employed for this section of the *Regule* follow: Vat. lat. 3064, ff. 102r–102v (*Va*) (passage 3 in the selection below is missing in this manuscript); Vat. lat. 3095, ff. 113–114r, 118r–v (*Vb*); Pavia, Bibl. Univ., Aldini 314, pp. 158, 153 (pag. out of order), 164–65 (*A*); Rome, Bibl. Angelica 1963 (*L*); the second edition, Pavia, 1498 (*Es*). *Va*, *Vb*, *L*, and *Es* are fairly close to each other, but *L* is quite careless and I have used it only on occasion to confirm one or another reading from the other manuscripts. *A* seems to be of a different tradition from the other manuscripts, but at least in some cases it gives better readings. No effort has been made to reproduce differences in spellings. In a few ambiguous cases letters italicized in text are given in Roman type in variant readings (e.g., d in line 77, b in line 120, a^2 in line 132).

---

2 Secunda suppositio *mg.* A secunda *mg.* Vb / pro conclusione 38ᵃ *supplevi*
3 Secundum est: secundo A
4 gradui *om.* A / primo: et primo A
5 tunc LEs tamen A cum (?) VaVb / acquiretur AVaVb acquiritur EsL / suo A *om.* VaVbLEs
6 tantum: per tantum A / pertransiretur A pertransiret L pertransietur Es transiretur VaVb
7 mediante A *om.* VaVbLEs
8 medio *om.* LVb
11 ad *om.* A
12 equevelociter: equaliter A
13 erit: et A
13–14 inter se *tr.* A post equalia
15 aliud: reliquum A
17 Primum... secunda *supplevi* Probatur secunda suppositio *mg.* Va / illa: quod illa Vb

[*Liber calculationum Ricardi Swineshead* 299]

suppositio secunda sic primo: acquirat *a* aliquam latitudinem uniformiter terminatam in extremo intensiori ad *c* gradum; et deperdat *b* eandem latitudinem consimiliter omnino sicut *a* acquirit illam. Et *e* gradus medius illius latitudinis, sive latitudo terminetur ad gradum sive ad non gradum non est cura. Tunc *a*, *b* motus continue equivalebunt gradui duplo ad *e*, eo quod gradus compositus ex duobus gradibus inequalibus semper est duplus ad medium inter illos, et *e* semper erit medius inter *a*, *b*, quando erunt inequales. Et illi duo coniunctim continue eidem gradui correspondebunt, eo quod brevior continue equevelociter intendetur sicut maior continue remittetur. Ergo continue correspondebunt gradui duplo ad *e* et unus eidem gradui correspondebit sicut alius. Ergo illi duo gradus in duplo intensiori gradui correspondebunt quam unus illorum per se. Ergo motus utriusque per se *e* gradui medio in totali tempore correspondebit, quod fuit probandum.

Consequentia tenet, nam moveatur *d* gradu duplo ad *e*, et sequitur quod *a*, *b* pertransibunt tantum sicut *d*, et unum tantum pertransibit sicut aliud. Ergo pertransibit in duplo minus quam *d*, et per consequens tantum sicut si moveretur *e* gradu subduplo ad gradum quo movetur *d*, qui gradus est medius latitudinis acquirende. Ergo sequitur suppositio.

[Secundum argumentum pro secunda suppositione.] Secundo sic, moveatur aliquid gradu uniformi ut 4. Cum ergo ibi sint due latitudines uniformiter difformes a 4 usque ad non gradum, ergo erit uniformis acquisitio unius latitudinis et uniformis deperditio alterius.

18 suppositio secunda *tr. Va* / sic primo *tr. A*
19 deperdet *A*
20 acquireret *Vb* / illam: iam *A* / e: c *A hic et ubique*
21 illius: huius *A*
22 non[1] *tr. A post* ad[1] *in linea 21*
24 illos: illa *A* / e: a *A* / e semper *tr. Vb*
25 medius: medium *A* / illi *A* illa *VaVbLEs* / coniunctim *AVb* coniuncta *VaLEs*
26 brevior continue *A* unus *VaVbLEs*
27 intendetur *ALEs* intentitur *VaVb* / maior continue *A* alius *VaVbLEs* / remittetur *ALEs* remittitur *VaVb*
27, 29 Ergo: igitur *A*
29 illi duo gradus *A* illa duo gradui *VaVbLEs*

30 gradui *om. VaVbLEs hic* / illorum *A* eorum *VaVbLEs* / Ergo: igitur *A*
31 utriusque: unius movetur de *Vb* unus utrius *Va* / gradui: gradui medio *A*
33 d *om. A*
34 pertransibunt: pertransibit *Es*
35 Ergo *VaVb* ergo unus *Es* ergo unum *L* igitur *A*
37 movetur: movebitur *LEs* manebitur *Va*
37–38 Ergo... suppositio *om. A*
39 Secundum... suppositione *mg. A* secundum argumentum *mg. Va*
40 Cum ergo *om. A* / sint: sunt *A*
41 ergo: ubi ergo *Es* igitur *A*

Et mediantibus istis duabus pertransibunt tantum quantum mediante gradu ut 4 in illo tempore poterit pertransiri, quia ex illis duabus resultabunt quatuor. Ergo uterque per se gradui ut duo correspondebit, quia illi duo motus gradui in duplo intensiori quam unus illorum per se.

[2. Tertium argumentum quod sepe pro secunda suppositione sumptum est.] Tertio arguitur idem. Et primo arguitur quod quecunque potentia a non gradu uniformiter intendit motum suum, in triplo plus pertransibit in secunda medietate temporis quam in prima, ponatur enim quod *a* remittat motum suum uniformiter ad non gradum. Tunc in omni instanti prime partis proportionalis in duplo velocius movebitur quam in instanti correspondente secunde partis proportionalis, et sic deinceps, ut patet. Ergo cum prima pars proportionalis sit dupla ad secundam, patet quod in quadruplo plus pertransibit in prima parte proportionali quam in secunda, et in quadruplo plus in secunda quam in tertia, et sic in infinitum. Ergo in omnibus partibus in quadruplo plus quam in residuis partibus *a*.

Prima consequentia tenet per hoc, si alique quantitates ad alias comparentur in eadem proportione sicut unum illorum se habet ad sui compar ita aggregatum ex omnibus comparatis se habet ad omnia ad que sit comparatio. Sequitur ergo quod in toto tempore pertransibit in quadruplo plus quam in secunda medietate, et per consequens in prima medietate in triplo plus pertransibit quam in secunda, ubi, scilicet, est uniformis remissio motus ad non gradum. Ergo econverso intendendo in triplo plus pertransiretur in secunda medietate temporis quam in prima.

43 duobus *Vb* / pertransibitur *LEs*
44 poterit pertransiri *om. LEs* / illis: istis *A* / duabus *LEs* duobus *VbA*
45 resultant *A* / Ergo: igitur *A* / uterque *VaLEs* utrumque *A* utrumque de *Vb* / correspondebunt *A*
47 *post* se *add. Vb* correspondent
48–49 Tertium... est *mg. A* tertium argumentum per quod probat quod in quadruplo plus pertransibitur in toto quam in medietate, si movetur uniformiter difformiter *mg. Va*
49–50 quecunque: ubicunque aliqua *A*
50 uniformiter intendit *tr. Vb* / triplo: duplo (?) *Vb*
53 *post* instanti *add. A* correspondente
54 secunde *om. Vb*
55 Ergo cum: cum igitur *A*
56 sit: temporis sit *EsVb*
57 parte proportionali: prima medietate temporis *Vb*
58 Ergo: igitur *A*
59 partibus *om. Vb*
60 Prima *om. A* / si: quod si *A*
61 comparentur *tr. Vb ante* ad *in linea.* 60 / illorum: eorum *A*
62 compar *LEs* comparationem *A* partem *VaVb*
63 ergo: igitur *A* / pertransibit *tr. Es post* plus *in linea* 64
65 triplo: duplo *A* / pertransibit *om. A*
66 motus *om. Es* / Ergo: igitur *A*
67 pertransibit *Es*
67–68 temporis *VaA om. VbLEs*

[*Liber calculationum Ricardi Swineshead* 301]

Ex illo arguitur intentum, intendat enim *a* uniformiter a non gradu usque ad 8. Tunc si motus in prima medietate correspondebit gradui remissiori quam duo, sit ille gradus *d*, et sit gradus cui correspondebit motus eius in secunda medietate *c*. Tunc ex quo motus equevelociter intendetur in secunda medietate sicut in prima et in instanti medio habet quattuor, sequitur quod *c* equaliter distat a quatuor sicut *d* a non gradu, et 6 equaliter distant a *c* sicut duo a *d*. Ergo 6 equaliter distant a duobus sicut *c* a *d*. Ergo cum sit tripla proportio, 6 ad duo, et gradus *c* est remissior 6, patet quod inter *c*, *d* est maior quam tripla proportio, et per consequens plus quam triplum pertransibit in secunda medietate quam in prima. Consequens est improbatum. Et quod sequatur quod *c* equaliter excedit *d* sicut 6 duo patet per hoc: Si sint quattuor termini, quorum primus et maximus equaliter excedit secundum sicut tertius quartum, primus equaliter excedit tertium sicut secundus quartum ut sint *a*, *b*, *c*, *d* quatuor termini ita quod *a* sit maximus equaliter excedens *b* sicut *c*, *d* minimum, tunc distantia inter *a*, *c* est maior distantia inter *a*, *b* solum per distantiam inter *b*, *c*, et excessus *b* supra *d* est maior quam excessus *c* supra *d* solum per latitudinem inter *b*, *c*. Cum ergo inter *a*, *b*, et *c*, *d* est distantia equalis, patet quod inter *a*, *c* et *b*, *d* est distantia equalis, quod est probandum. Si ergo motus in prima medietate gradui remissiori medio correspondebit, patet quod motus in secunda medietate plus quam in triplo intensiori gradui correspondebit quam in prima. Consequens est falsum et improbatum. Si motus gradui intensiori medio correspondebit, sequitur quod cum motus in secunda medietate equaliter excedit

69 Ex: et ex *A* / intendit *A*
70 8: aliquem gradum *A*
71 duo: ut duo *A*
72 c *tr. A post* sit *in linea 71* / equevelociter intendetur *tr. LVb*
74 habebit *LVa* / sequitur *bis Vb*
75 distant *AEs* distat *VaVb* / Ergo: igitur *A*
76 distant *VbAEs* distat *Va* / sicut: et sicut *Es* / Ergo: igitur *A* / cum: cum illa *A*
77 6: d *Vb*
79 medietate: medietate temporis *Vb* / est *om. A*
80 sequatur: sequitur *A* / excedit: excedere *LEs* / 6: a *A* / hoc: hoc quod *LEs*

81 Si *om. L* / excedit *AVa* excedunt *Vb* excedat *LEs*
82–83 primus... quartum *om. A*
82 equaliter *om. Vb*
84 sit: ut *VaVb* / excedens *AEs* excedat *VaVb* / sicut: sic *Es*
85 est *om. Va*
86 quam excessus: excessu *A*
87 Cum ergo *VbEs* ergo *Va* cum igitur *A* / d *om. Vb*
88 patet: patet ergo *Vb* / est²: erat *A* fuit *L*
89 ergo: igitur *A* / prima medietate *tr. Vb* / remissiori: intensiori *Vb*
91 intensiori: remissiori *Vb* / est *om. Va*
92 motus: vero *A*
93 excedit: excedat *Vb*

motum in prima medietate sicut 6 excedunt 2, patet quod motus in secunda medietate non foret triplus ad motum in prima. Consequens falsum per prearguta. Patet ergo quod motus in prima medietate correspondebit gradui ut duo et in secunda gradui ut 6, et sic arguitur de omni latitudine ad non gradum terminata.

Idem sequitur si intendat a gradu quia acquiret latitudinem de novo a non gradu que correspondet suo gradui medio et excedit gradum a quo intendet per latitudinem istam de novo acquisitam et gradus medius inter gradum in principio habitum et gradum in fine habitum est intensior illo gradu nunc habito per medium totius latitudinis de novo acquirende. Ergo sicut illa de novo acquisita correspondet medio gradui, ita ista cum toto motu prehabito suo gradui medio correspondet, quod fuit probandum....

[3. Conclusio 51^a]. Pro qua consequentia probanda probatur illa conclusio, quod omnis motus velocius et velocius deperditus quantum ad pertransitionem spatii gradui intensiori medio correspondet, quia, si non, remittatur a motus velocius et velocius et b uniformiter per idem tempus et per eandem latitudinem ab eodem gradu ad eundem gradum; sicut est possibile quod aliquod moveatur uniformiter aliquod spatium in hora pertranseundo et quod aliquod aliud mobile idem spatium transiret in hora velocius et velocius movendo illud tamen tardius inciperet moveri. Isto casu admisso arguitur sic: motus a et b sunt equales et motus a continue erit velocior et velocior motu b; ergo gradui intensiori correspondebit motus a quam motus b quantum ad pertransitionem spatii.

Consequentia patet et antecedens arguitur sic: si a motus non est

---

94 patet *om. Es* et patet *Vb*
95 Consequens: consequens est *Es*
96 ergo: igitur *A*
99 Idem: item *AVb* / si: quod si *A* / quia: quod *A*
100 a... que: quomodo gradus *A* / medio *om. Vb*
101 intenditur *A*
102 habitum²: habendum *A*
103 medium: mo^m *A*
104 Ergo: igitur *A*
104–5 medio gradui *tr. VbL*
105 ista *AEs om. VaVb*
107–8 consequentia... conclusio: probanda ponatur ista regula et probetur scilicet *Vb*
109 quia: probatur quia *Vb*
110 *b*: *b* motus *Vb* / uniformiter: uniformiter remittatur *Vb*
112 aliquod: aliquid *Vb* / moveatur *EsL* moveretur *Vb* motum *A*
113 et: remittit motum suum et *A*
114 transiret *AEs* pertransiret *VbL* / hora: eadem hora *Vb*
114–15 illud tamen *A* tamen idem *Es* idem tamen *L*
115 incipet *Vb*
116 velocior et *om. Vb*
117 ergo: igitur *Es*
119 patet: tenet *Vb* / est: erit *Es*

[*Liber calculationum Ricardi Swineshead* 303]

velocior motu *b* continue, sit *c* illud instans in quo non. Et arguitur sic: vel *a* motus tunc velocius remittetur quam *b* vel equevelociter vel tardius. Si velocius tunc remittetur *a* motus quam *b* et continue plus velocius et in fine erunt *a*, *b* sub eodem gradu, ergo in *c* instanti erit *a* motus intensior *b*. Consequentia patet, ex quo post *c* instans velocius remittetur quam ante sequitur quod maiorem latitudinem deperdet *a* et non deperdet aliquam nisi illam que est inter gradum habendum in *c* et gradum habendum in fine. Ergo tunc per plus distabit a gradu habendo in fine quam *b* ab eodem et per consequens erit intensior *a* motus *b* motu. Si tunc equevelociter precise remittetur *a* motus sicut *b* et continue post velocius quam *b* et in fine erunt equales, igitur tunc erit *a* motus intensior, ut prius argutum est. Si tardius tunc remittetur *a* motus quam *b* et continue ante illud instans remittetur illud *a* tardius quam *b* et in principio sunt equales per casum, ergo tunc *a* motus est intensior *b*. Consequentia patet de se.

Sive ergo ponatur quod in *c* instanti velocius remittetur *a* quam *b* sive tardius sive equevelociter, sequitur quod *a* tunc erit intensior *b*. Et sic de omni alio instanti est arguendum, sed continue velocius remittetur *a* motus quam *b* vel tardius vel equevelociter. Ergo continue erit *a* motus intensior *b*. Ergo gradui intensiori correspondebit motus *a* quam motus *b*. Et idem est gradus medius latitudinis deperdite ab *a* in illo tempore sicut a *b*, et *b* motus correspondebit gradui medio, ut probatum est in suppositione conclusionis 38^{ve}. Ibi enim probatur

120 motu b *tr. Vb* / continue *tr. Vb* post non *in linea 119* / illud instans: id *Es*
122–23 *b*... velocius: *b* motus et *a* motus continue velocius remittetur quam tunc et *b* uniformiter ergo continue post illud instans velocius remittetur *a* motus quam *b Vb*
123 *a*, *b* : *a*, *b* motus *Vb* / ergo: igitur *Es* / instanti *Vb om. AEs*
124 ex quo: quia *Es*
125 remittet *A* / ante: in *c* instanti *Vb* / quod: etiam quod *Vb* / deperdet *a*: motus deperdet *Vb*
127 Ergo: igitur *Es*
128 erit: tunc erit *Vb*
128–29 intensior *a* motus: motus intensior *Vb*
129 motus: motu *Es* / tunc *om. L* autem tunc *Vb*
130 et... equales *om. Vb* / igitur *Vb Es* ergo *A*
132 illud[1]: idem *Es* / remittetur illud *a*: remittetur *A Es* / post *a*[2] *add. Vb* motus
133 *b* : *b* motus *Vb*
135 ergo: igitur *Es* / instanti: instans *Es* / velocius remittetur *tr. Vb* / *a om. VbEs*
136 *a* : *a* motus *Vb*
138 Ergo: igitur *Es*
138–39 continue... motus: *a* continue erit *A*
139 Ergo: igitur *Es*
140 motus *om. Vb*
141 sicut: sic *A* / *b* motus *tr. Vb* / medio: medio istius latitudinis *Es* intensiori medio illius latitudinis *Vb*
142 38^{ve}: 33^{e} *Vb*

quod omnis motus acquisitus uniformiter vel deperditus suo gradui medio correspondet. Ergo sequitur quod motus *a* gradui intensiori medio correspondet, quod fuit probandum.... Et consimiliter omnis motus tardius et tardius deperditus gradui remissiori medio correspondet.... Ubicunque velocius et velocius motus acquiritur gradui remissiori medio correspondet.... Ubicunque motus tardius et tardius acquiritur correspondebit gradui intensiori medio....

143 quod *om. A* / acquisitus uniformiter *tr. Vb*
144 Ergo: igitur *Es*
145 correspondebit *LVb et tr. Vb post* motus *a* / Et consimiliter: consimiliter arguitur quod *Vb*
146–47 correspondebit *LVb*
147–48 Ubicunque... correspondet: Alia regula ubicunque motus velocius et velocius intenditur correspondet gradui remissiori medio *Vb*
148 Ubicunque: Alia regula est ubicunque *Vb*
149 acquiritur: inducitur *Vb*

Document 5.4

John Dumbleton, *The Summa of Logical and Natural Things*

1. PART III, CHAPTER 10. Consequently it is proved that a [velocity] latitude *(latitudo)** corresponds to its mean degree [of velocity] *(gradus medius)*. It is demonstrated in the first place, however, that if some latitude of velocity *(latitudo motus)* terminated at rest [and uniformly acquired] is equivalent to a degree [of velocity] *(gradus)* greater *(intensior)* than its mean, then it is refuted that the lesser *(remissior)* half of the latitude terminated at rest corresponds to [a degree of velocity] less than the mean of the same half.

This is evident as follows (see Fig. 5.4). Let us posit AF as a latitude of velocity, B the mean degree [of velocity] of the same latitude, D the mean degree [of velocity] of the half of the latitude terminated at rest, and C the degree *(gradus)* to which the whole latitude AF uniformly acquired corresponds. And let us posit R, a degree corresponding to the half BF. And let F be rest. I say that if C is greater *(intensior)* than B, then R is greater than D. For I posit that Sortes (Socrates) acquires latitude AF uniformly in an hour. It is obvious that Sortes will traverse just as much in the first half as if he had moved uniformly at degree R. But since Sortes in the second half of the hour will have [acquired] increment BA terminated short of rest, hence Sortes will traverse in the second half just as much as two moving bodies, one moving [uniformly] at degree B only in the same half hour and the other moving at degree R, because the latitude FB is the same as latitude BA.... Consequently, just as much will be traversed precisely in an hour by [a mobile experiencing] latitude AF as by two [mobiles], one of which is moved through the [whole] hour

---

* In this document *latitudo* appears sometimes in the improper sense of an integral of velocities and sometimes in the proper sense as a difference of velocities. Because of this ambiguity, I have retained the single word "latitude," which is equally as ambigious, in my translation.

at degree R only and the other at degree B only through the half hour, as is evident to one who is attentive. But since B, lasting through one half hour, corresponds with respect to the traversal of space *(quantum ad transitum spacii)* to its half D, lasting through the whole hour, it follows that latitude AF, acquired uniformly through the hour, is equivalent to R and D degrees [together], through the same hour. But since R and D degrees [together], lasting through the hour, are precisely as one velocity composed of these two, it follows that G (D plus R), which is just as far

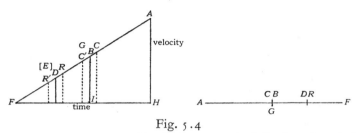

Fig. 5.4

Oresme's system is represented in the figure on the left; that given in MS G is shown on the right. It is obvious that in the one-dimensional system of Dumbleton (represented by the single line AF on the right), all the degrees of velocity are laid out on the same line so that both degrees and latitudes (i.e., differences of degree) can be directly represented on that line. Thus if point F is zero degree and hence the referent point, degree B, is represented by line FB and latitude AB (the difference between degree A and degree B) is represented by the line AB. In the Oresme system on the left, each degree is represented by a perpendicular to the base line (i.e., velocity B is represented by perpendicular BI), and a latitude "properly speaking" is in this system the difference between two perpendiculars (i.e., latitude AB is not the segment AB of the hypotenuse but is rather the vertical component of AB, or, in other words, a linear segment equal to AH–BI). Latitude AB "improperly speaking" is the sum of all the velocity perpendiculars between BI and AH, i.e., the area $BAHI$. For a discussion of Oresme's system, see Chapter 6.

removed [in velocity increment] from degree D as R is from F, will be equivalent to latitude AF, uniformly acquired in the hour. The consequence holds from the fact that G in one movement through the hour is equivalent to D and R in separate movements through the hour.

But since the increment *(distantia)* between R and F is less than the increment between D and B, and G exceeds D only by the increment RF, hence it follows that G is less *(remissior)* than B. Therefore the degree

[Dumbleton, *Logical and Natural Things*

equivalent to the whole latitude AF acquired uniformly through an hour is less than B. Hence C is less than B, which was to be proved.

In the same way it is proved that if the degree corresponding to the half BF, acquired uniformly, is greater *(intensior)* than D, the degree corresponding to the whole latitude AF is greater than B....

2. Chapter 11. There follows the proof of this, that every latitude etc. (i.e., corresponds to its mean degree of velocity—see Fig. 5.5). For if

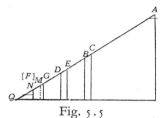

Fig. 5.5

As graphed using the Oresme system (for Dumbleton everything is measured along a single line AQ; see Fig. 5.4).

some latitude beginning from rest and uniformly acquired does not correspond to its mean degree with respect to the traversal of space, let there be given such a latitude AQ, and let Q be rest and A the degree terminating the latitude, and let B be the mean of the same total latitude and D the mean [degree] of the BQ half of the same latitude. If, therefore, latitude AQ uniformly acquired does not correspond to its mean with respect to the traversal of space, let it be given in the first place that degree C, greater than B, is equivalent to latitude AQ uniformly acquired. And let E be the degree corresponding to the half BQ. Because, therefore, C is greater than B, it is obvious by the preceding conclusion that E is greater than D. I say, therefore, that it follows from this supposition that the latitude acquired uniformly in an hour is equivalent to the degree terminating the latitude, which is impossible.

This conclusion is argued as follows. Let us posit that Sortes acquires latitude AQ uniformly in an hour. It is obvious by the things proved before that Sortes will traverse just as much space in the first half hour as if he had continually moved through that half hour at a degree of E, and in the second half hour just as much space as two bodies, one moving uniformly through that same half hour at degree B only and the other moving at degree E through the same half hour. Therefore, Sortes will traverse just as much space acquiring the latitude AQ as two moving bodies, one of which would move through the whole hour at E degree only and the other through half [of the hour] at B degree only. Consequently, Sortes will traverse just as much space acquiring latitude AQ uniformly as would be traversed in an hour at degree E and at degree D, which is one-half B. The consequences follow, from the preceding conclusion, to one who is attentive. Since, therefore, D and E degrees through an hour are equivalent to latitude AQ produced in the same hour, it follows that the degree which is equivalent to the latitude AQ in an hour exceeds in [velocity] increment the [velocity] D by the same amount that E exceeds Q, which is rest. But since C by hypothesis is equivalent to latitude AQ in the hour, it follows that C exceeds E by the same amount that D exceeds Q. These consequences hold by the supposition. So the increment (*distantia*) between D and Q is equal to that between E and C. But since B exceeds D as D exceeds Q, hence C exceeds E as B exceeds D. The consequence holds by the identical middle: whatever are equal to a third etc. (i.e., are equal to each other). If, therefore, increment BD is equal to EC, with the common term, i.e., BE, subtracted, the terms which remain, namely, increments ED and CB, will be equal. Consequently, the velocity corresponding to AQ exceeds the mean B as E exceeds the mean of BQ, which is itself half of AQ. In the same way it is proved that the velocity corresponding to $1/4$ AQ exceeds the mean of that $1/4$ as degree E exceeds degree D, and in the same way [dividing] to infinity through $1/8$, $1/16$.

Then let us take GQ an aliquot part of the latitude AQ, the mean degree of which part is N, and N is as far [in velocity increment] from G as E is from D, or let it be a smaller increment, [i.e.,] let the increment between its mean degree [N] and the degree to which it corresponds be as the increment between C and B. That such can be given is obvious, because latitude AQ is infinitely divisible into aliquot parts terminating at degree Q, and of any aliquot part at all such is the increment between the degree corresponding to that part and its mean degree as between C and B, as is obvious by the argument we have made. Then let us take the degree corresponding to latitude GQ acquired uniformly in time and let it be M. By the argument made, it follows that the increment between M and N degrees is the same as between D and E. Therefore, M is as far removed [in velocity increment] from N as G from degree N. Consequently, degree M is identical with G, [and G thus] is the degree to which the latitude GQ acquired uniformly in an hour corresponds, which was to be proved, [G being the terminal degree of the latitude].

But if it is given that there is no aliquot part of AQ, of which the mean degree is as distant [in velocity increment] from its terminal [velocity] as C is distant from B, there will be a part of which the mean velocity is less distant from the terminal [velocity] than C is distant from B. Let this part be GQ. From the denomination made, it is evident that GQ corresponds to a degree greater than the terminal degree, which is impossible.

If then we assume the converse possibility, namely, that the latitude uniformly acquired is equivalent to a degree less than the mean of the latitude, it follows by the preceding conclusion that any lesser latitude at all which is terminated at rest corresponds to a degree less than its mean.

We assume for the aforementioned conclusion that C is the mean of AQ, and B is the lesser degree to which the latitude AQ uniformly acquired in an hour corresponds, and D corresponds to latitude CQ. From this position it follows that *nothing* (i.e., no space) is acquired in the course of some acceleration *(latitudo)*. This is proved as before. In that conclusion it is evident that latitude AQ acquired uniformly in an hour will be equivalent to D, which corresponds to half of it through the hour, and C through half the hour; and consequently latitude AQ acquired uniformly in an hour is equivalent to degree D through an hour together with degree E through the same hour, E being one-half C. But since it is given that B is equivalent to latitude AQ acquired uniformly in the hour, therefore degree B exceeds D by the same amount E exceeds Q. The consequence holds by the supposition. But since the distance [in velocity increment] between E and Q is the same as that between C and E, it follows that B is the same distance from D as E is from C. With the common part BE subtracted, the remaining increments, namely, DE and CB, will be equal. Therefore, let there be some part of the half CQ in which the mean degree is just as distant [in velocity increment] from rest as D is distant from E, the mean of the half CQ. Let that latitude be GQ, the mean degree of which latitude let be N, and let F be the degree corresponding to that latitude, a degree less than the mean, as is clear by the preceding conclusion. Since, therefore, the increment between D and E is equivalent to that between B and C, notwithstanding the fact that CQ is double EQ, and GQ is either a fourth part or an eighth part of latitude EQ (so that we will not have to go further toward infinity in our division), it follows that degree F corresponding to latitude GQ is the same distance [in velocity increment] from N, the mean of the increment GQ, as degree D corresponding to latitude CQ is distant from E, the mean degree of latitude CQ. But since rest is just as distant from N, the mean degree of latitude GQ, as D is distant from E, therefore degree F, which corresponds to latitude GQ, is rest, and thus it follows that nothing (i.e., no space) is traversed in the course of the uniform acquisition of latitude GQ, which was to be proved.

3. It is proved in another way that a latitude of velocity terminated at rest and uniformly acquired etc. (i.e., is equivalent to its mean degree of velocity with respect to the traversal of space in the same time). This is evident, once the following supposition is posited along with the previous proofs: The proportion between the terminal degrees of some latitudes starting at rest is the same as that of the degrees corresponding to these same latitude in respect to the traversal of space. Thus if AB and AC (see Fig. 5.6) are two latitudes terminated at rest, AB being the

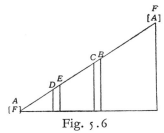

Fig. 5.6

As graphed using the Oresme system. Letters in brackets are those used in proof. As before, Dumbleton simply puts all of his letters on a single line, such as AF.

greater, and E and D the the uniform velocities by which the same spaces are traversed in an hour as in the course of the uniform acquisition of velocity latitudes AB and AC, then B, the degree terminating latitude AB, is to degree E as C is to D, just as is so argued from that supposition. And we posit latitude AF uniformly acquired by Sortes in an hour. I say that so much will be traversed in the course of the uniform acquisition of latitude AF in an hour as by C, the mean degree of that latitude. For if not, let it be given that so much will be traversed in the course of the acquisition of AF latitude as by [a uniform movement of] a degree greater than the mean degree of the same latitude. Let a foot be the maximum traversed by Sortes in an hour. And we assume that F is zero velocity *(quies)* of the latitude AF. I then take a degree of one-half B and let it be E, which is greater than D, the mean degree of the lesser half of the latitude AF, as is obvious. Then $A/C = B/E$, because the ones are double the others. Therefore by permutation $A/B = C/E$. The consequence is evident from a supposition of Book V of Euclid; thus $A/B = C/E$. Hence, by the given supposition, just as B corresponds to latitude AF uniformly acquired, so E corresponds to the lesser half of the

[Dumbleton, *Logical and Natural Things* 311]

same latitude [uniformly acquired]. The consequence holds because there is a unique degree in latitude AF which is related to C as B is to A. But because E is one-half B and Sortes will move in the first half hour in acquiring latitude CF, corresponding to degree E, hence it will traverse one-fourth of the total space to be traversed by the degree corresponding to the total latitude AF. The consequence is obvious to a person of intelligence. Consequently, Sortes will traverse in the first half hour one-fourth of a foot. But since Sortes will acquire latitude AC uniformly in the second half hour, hence Sortes will traverse as much in the second half hour as two mobiles in the same time, one of which uniformly accelerates *(intendet)* from rest to degree C, and the other of which is moved continually with degree C. The consequence is obvious by the preceding conclusion of chapter seven.* But because one-half foot will be traversed by [a body moving] at degree B in the second half hour, since the whole latitude AF corresponds to it (B), and C is a degree less than B by assumption, therefore Sortes will traverse in the second half hour, if moved at degree C and the latitude CF, less than three-fourths of a foot. Therefore, Sortes will not traverse a foot in the whole time, which is contrary to that which we have posited. The consequence holds, because if Sortes is moved at degree B plus that latitude, then he would traverse three-fourths of a foot. But now [at degree C velocity plus the latitude, he is moved] less than then [when moved at degree B plus the latitude]. Therefore, et cetera, as before.

If the degree corresponding to the latitude is less than the mean of the latitude, let us posit such a latitude AF. We intend B to be the mean degree of the latitude, C the degree corresponding to the whole latitude AF, C being less than the mean degree of AF, as is given. And let E be the mean degree of latitude BF, and let D be one-half C. By the argument previously made, it is obvious that $A/C = B/D$, a proposition of [Book] V of Euclid being assumed. We posit that a foot is the maximum which will be traversed by Sortes in an hour when he is acquiring the latitude AF uniformly. Since, therefore, C is $2D$ and precisely one foot is traversed by velocity C in an hour, it being equivalent to the latitude AF [in space traversed], it follows according to the prior kind of argument that Sortes will traverse in the first half hour one-fourth of a foot and that in the second half hour he will traverse precisely as much space as two mobiles in that same half hour, one moving with [a uniform movement at velocity

---

* This concluson occurs in chapter 6   lat. 16146, 27v c. 1.
(but marked chapter 5) in MS Paris, BN

C], the mean of the whole latitude *AF*, and the other moving so as to acquire latitude *BF* uniformly, as is obvious by the preceding conclusion. Therefore, Sortes will traverse in the second half hour one-fourth of a foot plus more than three-fourths of a foot. The consequence holds, because by the last supposition [the mean velocity] is more than 2*D* velocity. And so it is obvious that more than a foot will be traversed by Sortes in the hour when acquiring latitude *AF*. But this is contradictory. By the argument previously made, the consequences for this part are obvious to anyone who is attentive.

Here it is noted, however, that this demonstration is founded upon this, that, if a latitude begins a rest and is uniformly acquired in some time, it is necessary that one-fourth of the total space is traversed in the first half of the time, i.e., during the acquisition of the lesser half of the latitude.

Therefore, [a uniformly accelerated motion] in which that latitude is acquired corresponds, with respect to the traversal of space, neither to a velocity greater than the mean nor to one less than the mean.

4. From these two conclusions [of chapters 10 and 11] follows a third to the effect that every finite latitude uniformly acquired [of which the initial terminal velocity is] short of rest (i.e., has some value greater than zero) corresponds to its mean degree [of velocity].

Let there be given latitude *AB* terminated short of rest, and let its lesser *(remissius)* terminus be *A*. I say, therefore, that, when it is uniformly acquired, it corresponds to its mean degree. For just as much space is traversed by [a mobile] acquiring latitude *AB* uniformly in an hour as by one moving uniformly and continuously through the hour with velocity *A* together with one acquiring a latitude equal to *AB* uniformly, [assuming, however, that the increment is] acquired from rest, as is evident by one conclusion previously proved. But since every latitude beginning from rest and uniformly acquired corresponds to its mean degree [of velocity], as is evident from the preceding conclusion, hence just as much will be traversed in the course of the acquisition of latitude *AB* in an hour as by two mobiles, one moving uniformly through the hour with velocity *A* only and the other moving with the mean of the latitude, [if the latitude is assumed as] starting at rest. The consequence holds by things previously proved. But since the mean velocity of *AB* latitude is composed of the uniform velocity *A* and a velocity equal to the mean of such a latitude, [thought of] as beginning at rest, hence just as much is traversed in the course of the uniform acquisition of latitude *AB* as by [a mobile moving uniformly with] the mean degree of that latitude, which was to be proved.

COMMENTARY

1. Having illustrated the nature of Dumbleton's interesting proof with algebraic symbols in the body of the chapter, we now wish to summarize the proof in greater detail.

Before actually proving that a uniform acceleration is equivalent to its mean velocity in regard to the traversal of an equal space in an equal time, Dumbleton first proves that if we assume that the equivalent velocity of such an acceleration from rest is greater than the mean velocity, i.e., the velocity at the middle instant of time, then the velocity equivalent to the uniform acceleration through the first half of the time cannot be less than its mean. This is proved inversely by showing that if the equivalent velocity R of the first half of the total acceleration is less than the mean D of that first half of the acceleration, then the whole acceleration through increment A–F corresponds to a velocity G less than its mean B rather than to a velocity greater than that mean (see Fig. 5.4), which we assume.

For the exposition of the proof of this conclusion let us assume the following definitions:

(1) S_{af} = the distance traversed in the course of the uniform acceleration from velocity F (zero) to velocity A in the whole hour.

(2) S_{bf} = the distance traversed in the course of the uniform acceleration from velocity F to velocity B in the first half hour.

(3) $S_{r/2}$ = the distance traversed by a body moving uniformly for one-half hour with velocity R.

(4) S_{ab} = the distance traversed in the course of the uniform acceleration from velocity B to velocity A in the second half hour.

(5) S_r = the distance traversed by a body moving uniformly with velocity R through the whole hour.

(6) S_b = the distance traversed by a body moving uniformly with velocity B for an hour.

(7) $S_{b/2}$ = the distance traversed by a body moving uniformly with velocity B for one-half hour.

(8) S_d = the distance traversed by a body moving uniformly with velocity D for one hour.

(9) S_g = the distance traversed by a body moving uniformly with velocity G for one hour.

(10) S_c = the distance traversed by a body moving uniformly with velocity C for one hour.

The proof that if $R < D$, then $G < B$ and hence $C \not> B$, but rather $C < B$, follows:

(1) $S_{bf} = S_{r/2}$, by the definition of R.
(2) $S_{ab} = S_{b/2} + S_{r/2}$, since $A - B = B - F$.
(3) Hence, $S_{af} = S_r + S_{b/2}$.
(4) But $S_{b/f_2} = S_d$, since $B/2 = D$.
(5) Hence, $S_{af} = S_r + S_d$.
(6) Let us suppose that $S_r + S_d = S_g$ (and thus $G = R + D$).
(7) But $R - F < B - D$, since $R < D$ and $D - F = B - D$.
(8) And so $G < B$, since $R + D = G$ and $F = 0$.
(9) And hence $C < B$, since $C = G$.

Dumbleton has proved, therefore, that as long as the equivalent velocity of the first half of the acceleration is less than the mean velocity of that first half, then the equivalent velocity of the whole acceleration must be less than its mean. It can also be shown in the same way that, if the equivalent velocity of the first half is greater than the mean velocity of that half, the equivalent velocity of the whole acceleration must be greater than its mean.

2. Dumbleton now passes on to the special case of the mean speed theorem for acceleration commencing from rest (see Fig. 5.5). Supposing we have a uniform acceleration from velocity Q (which is zero) to velocity A, and velocity B is one-half A, D is one-half B. Then if we assume either that the equivalent velocity C is greater than or less than B, an absurdity will follow. For the exposition of this proof assume the following definitions:

(1) S_{aq} = the distance traversed by a body accelerating uniformly from velocity Q (zero velocity) to velocity A in an hour.
(2) S_{bq} = the distance traversed by a body accelerating from velocity Q to velocity B in the first half hour.
(3) S_{ab} = the distance traversed by a body accelerating from velocity B $(= A/2)$ to velocity A in the second half hour.
(4) S_c = the distance traversed by a mobile moving uniformly with velocity C through the hour, the velocity C being such that $S_c = S_{aq}$. In the first part of the proof C is assumed to be greater than B.
(5) S_e = the distance traversed by a body moving uniformly for an hour with velocity E, E being such that $S_{e/2} = S_{bq}$.
(6) $S_{e/2}$ = the distance traversed by a body moving uniformly for a half hour with velocity E.
(7) $S_{b/2}$ = the distance traversed by a body moving uniformly with velocity B for one half hour.

[Dumbleton, *Logical and Natural Things* 315]

(8) S_d = the distance traversed by a body moving uniformly for an hour with velocity D, D being $B/2$.

Now let us detail the proof:

(1) Since $C > B$, then $E > D$, by the preceding proof.

(2) $S_{aq} = S_d + S_e$, since $S_{bq} = S_{e/2}$, $S_{ab} = S_{b/2} + S_{e/2}$, $S_{aq} = S_{bq} + S_{ab} = S_e + S_{b/2}$, and $S_d = S_{b/2}$.

(3) Hence $C - D = E - Q$, since $S_{aq} = Ct$, $S_d = Dt$, and $S_e = Et$, inserting in (2) and dividing throughout by t. Q of course is zero.

(4) Therefore, $D - Q = C - E$, by transposing terms in (3).

(5) But $B - D = D - Q$, by definition.

(6) Hence $B - D = C - E$.

(7) But $B - D = (B - E) + (E - D)$ and $C - E = (B - E) + (C - B)$.

(8) Therefore, $E - D = C - B$ (and this equals k in the discussion given of this proof in the chapter above). This conclusion means that the velocity corresponding to the whole latitude exceeds its mean by the same amount that the velocity corresponding to the first half of the latitude exceeds its mean. Dumbleton then goes on to say that precisely the same reasoning can be used to prove that the velocity which corresponds to 1/4 of the original latitude exceeds its mean by this same amount (k), and similarly for the remaining proportional parts of the latitude, 1/8, 1/16, etc.

(9) In the process of dividing the original latitude $A - Q$ into proportional parts, Dumbleton now supposes we reach an aliquot part of the original latitude, a part which we call $G - Q$; this part is such that its mean is N, and $G - N = C - B = E - D$.

(10) But since $G - Q$ is a part arrived at by progressive division of the original latitude, the velocity corresponding to it (velocity M) exceeds its mean (velocity N) by the constant amount k (equal to $C - B$ or $E - D$).

(11) Therefore $M - N = G - N$, or $M = G$, i.e., the velocity corresponding to an acceleration is equal to the final velocity of the acceleration, which is absurd.

(12) Since the conclusion of (11) is absurd, the assumption from which it follows is erroneous, namely, that the equivalent velocity is greater than the mean velocity. Of course it may be that in step (9) there is no aliquot part where $G - N = C - B$; but if not, one takes a part where $G - N < C - B$. If this is assumed, one also arrives at an absurdity, namely, that the equivalent velocity for $G - Q$ is greater than the final velocity G.

(13) By similar reasoning, Dumbleton shows that if we assume that the equivalent velocity for the whole acceleration is less than the mean, we can conclude that the equivalent velocity of a certain acceleration is the initial velocity of the acceleration or zero velocity, which is equally absurd. I have outlined this part of the proof in algebraic symbols in the text of the chapter.

(14) Our conclusion, therefore, is that the desired velocity by which the same space is traversed as in the course of the acceleration is the mean velocity.

3. Dumbleton's first ingenious proof of the mean speed theorem is followed by another one. The second proof is based on an assumption he feels to be self-evident. This assumption is that $V_{f_1}/V_{f_2} = V_{m_1}/V_{m_2}$, where V_{f_1} and V_{f_2} are the final velocities of two accelerations commencing from rest and V_{m_1} and V_{m_2} are the uniform velocities equivalent to the accelerations in the traversal of space in the same time.

For the exposition of the proof, accept the following data (see Fig. 5.6): $C = A/2$, $B =$ the velocity equivalent to the whole acceleration from F (zero velocity) to A, $D = C/2$, and $E = B/2$. The details of the proof follow:

(1) $A/C = B/E$, from the data.

(2) $A/B = C/E$, by permutation of the proportions.

(3) But since B "corresponds" to the acceleration from F to A as regards the traversal of space in the same time, then E corresponds to the acceleration from F to C. This follows from the fact assumed as our basic supposition, namely, that for two uniform accelerations from rest the final velocities are in the same proportion as the equivalent velocities, and B we have assumed to be the equivalent velocity of the acceleration from F to A.

(4) Hence a body, uniformly accelerating from F to A, in its first half hour traverses 1/4 of the total distance traversed in the whole hour. This is true because a body moving uniformly with velocity B for the whole hour traverses the same distance as a body uniformly accelerating from F to A velocity. But E is $B/2$, and thus a body moving with E through the whole hour would traverse 1/2 the distance of a body moving with B for the whole hour. And so a body moving with E in the first half hour only would traverse 1/4 the distance traversed by a body moving with B throughout the hour. But in (3) we showed that E was equivalent to the acceleration from F to C, namely, the acceleration in the first half hour.

(5) Now in the second half of the hour Sortes (accelerating from C to F) traverses the same distance as two bodies, one moving uniformly through one-half hour with C velocity and the other accelerating uniformly from rest to C velocity in one-half hour. This conclusion Dumbleton had already proved in an earlier chapter.

(6) But since $C < B$ by assumption and B as the equivalent velocity would traverse 1/2 the distance in the second half of the hour, then from (5) Sortes, traversing the same distance as two bodies, one moving with C (less than B) uniformly and the other accelerating from rest to C and thereby traversing 1/4 distance, will traverse in the second half hour less

[Dumbleton, *Logical and Natural Things* 317]

than 3/4 of the total distance. And since from (4) Sortes in the first half hour traverses 1/4 the whole distance, we arrive at the absurd conclusion that Sortes does not traverse the whole distance in the whole time.

(7) By precisely the same kind of reasoning, Dumbleton shows that a contradiction will follow if we assume that the velocity equivalent to the acceleration from rest is less than the mean.

(8) If the equivalent velocity is not greater than the mean and not less than the mean, it must be equal to the mean. Q.E.D.

As Dumbleton points out, this proof rests on the assumption that in a uniform acceleration from rest, 1/4 of the total space is traversed in the first half of the time. A complete proof would demand a proof of this assumption, perhaps in the manner of Swineshead's proof.

4. Our last section from Dumbleton's work gives a proof of the general mean speed theorem, namely, that any uniform acceleration from one finite velocity to another is equivalent with respect to space traversed in a given time to its mean velocity. This follows immediately from the simultaneous assumption of (1) the conclusion previously referred to in step (5) of the proof above, namely, that a body accelerating from one finite velocity to another will traverse just as much space as two bodies moving in the same time, one uniformly with the initial velocity and the other accelerating from rest to a velocity equal to the difference of the final and initial velocities, and (2) the special case of the mean speed theorem for an acceleration from rest.

*Summa Johannis Dumbletonis de logicis et naturalibus**

Pars III. Cap. X. Consequenter probatur latitudinem suo medio gradui correspondere. Primo tamen monstratur quod si aliqua latitudo motus terminata ad quietem intensiorem gradum suo medio

---

* MS Paris, BN lat. 16146, 14c, ff. 29v–30r (*B*). Cf. MS Cambridge, Peterhouse No. 272, 14c, (no pag.) (*C*); and Cambridge, Gonv. and Caius 499/268, 14c, (no pag.) (*G*). MS Merton College 306 was not available to me at the time of the preparation of this text. I have constructed the text from these three, MSS *B*, *C*, *G*. Occasional reference has been made to Vat. lat. 954, 15c, ff. 23v–24v (*V*). I have maintained spellings of *B* such as "sicud" and "set," although "sicut" and "sed" appear in other manuscripts. I have not noted all the transpositions in words, nor when *C* has "ista" for "illa" in *BG*. Nor have I noted the variant forms of "ergo" and "igitur." I have capitalized letters standing for degrees and latitudes. I have not indicated all deletions made by the various scribes. The capital letters italicized in the text are given in Roman type in the variant readings.

---

2 Cap. X: Cap. IX *B*   gradui *B*
4 intensiorem gradum *CG* intensiori

valet, medietatem remissiorem eius ad quietem terminatam remissiori medio eiusdem medietatis correspondere improbatur.

Hoc sic patet. Ponamus AF latitudinem motus, B medium gradum eiusdem, D medium gradum medietatis ad quietem terminate, et C gradum correspondere toti AF latitudini uniformiter acquisite. Et ponamus R gradum correspondere BF medietati. Et sit F quies. Dico quod si C est intensior B, ergo R est intensior D. Nam pono Sortem adquirere uniformiter AF latitudinem in hora. Patet quod Sortes tantum pertransibit in prima medietate acsi movisset [uniformiter] R gradu. Set cum in secunda medietate hore habebit Sortes BA latitudinem citra quietem terminatam, ergo Sortes tantum pertransibit in secunda medietate sicud duo mota, quorum unum movetur B gradu solum [in] eadem medietate hore et aliud movetur R gradu, eo quod latitudo FB est tanta latitudo sicud est BA. Consequentia tenet per secundam conclusionem capituli septimi illius partis. Et per consequens vero tantum pertransietur precise in hora AF latitudine sicud a duobus, quorum unum movetur per horam R gradu solum et aliud solum [in] medietate hore B gradu, ut patet intuenti. Sed cum B tantum valet per medietatem hore quantum ad transitum spacii sicud D subduplus valet per horam, sequitur quod AF latitudo adquisita uniformiter per horam valet [sicud] R, D gradus per eandem horam. Set cum D, R gradus tantum valent per horam precise sicud unus gradus compositus ex istis precise, sequitur quod G gradus, qui tantum precise distat a D gradu qua R ab F, equevalebit AF latitudini uniformiter adquisite per horam. Consequentia tenet per hoc, quod G in uno motu per horam valet D, R in diversis motibus per horam. Set cum minor est distantia inter R, F quam inter D, B et G non addit supra D nisi [per] latitudinem RF, sequitur quod G est remissior B. Ergo gradus valens totam latitudinem AF uni-

5 valet: videlicet B / remissiorem *om.* G
6 improbatur: non posse C
9 latitudini uniformiter CG *tr.* B
10 BF CG toti BF B
14 hore *om.* BG
18 latitudo... latitudo: tanta est latitudo BF C / est BA *tr.* C
19 conclusionem CG suppositionem B / secundam... partis: conclusionem partis illius capituli septimi C / septimi: primi B ?G
20 vero *om.* C / precise *om.* C
21 movebitur C
22 solum^2... hore *om.* C
23 transitionem C
24 suus subduplus C
25 adquisita uniformiter *tr.* C
26 horam^1 *om.* C / tantum valent CG valent B *tr.* C *post* horam
27 gradus^2 *om.* C
28 precise distat *tr.* C / gradu qua: quam C / equevalent G
30 R, D CG
31 R et F C
31–32 D et B C

formiter adquisitam per horam est remissior B. Et per consequens C est remissior B, quod est probandum.

Per idem probatur quod si gradus correspondens BF medietati uniformiter adquisite sit intensior D, qui gradus correspondens totali AF latitudini est intensior B.

Pro conclusione precedenti et aliis providendum in illa materia ponenda est suppositio, quod si sint duo gradus uniformes motus, ut R, D inequales sive equales, quod tantum valent precise D et R per horam quantum unus qui tantum addit de latitudine super D quantum R distat a quiete. Hec suppositio patet per conclusionem primam.

Cap. XI. Sequitur probatio illius, quod omnis latitudo et cetera, nam si aliqua latitudo incipiens a quiete uniformiter adquisita, in transitu spacii suo medio gradui non correspondet, detur AQ latitudo talis, et sit Q quies et A gradus eandem terminans, et sit B medius totalis latitudinis eiusdem, et D medius BQ medietatis eiusdem. Si ergo AQ latitudo uniformiter adquisita suo medio in transitu spacii non correspondet, detur ergo primo quod C gradus intensior B equevalet AQ latitudini uniformiter adquisite. Et sit E gradus correspondens medietati BQ. Quia ergo C est intensior B, patet per precedentem conclusionem quod E est intensior D. Dico ergo ex illa impositione sequi quod latitudo adquisita uniformiter in hora gradum ipsam latitudinem terminantem equevalet, quod est impossibile.

Istud sic arguitur. Ponamus Sortem adquirere AQ latitudinem in hora uniformiter. Patet per prius probata, quod in prima medietate hore Sortes tantum pertransibit acsi per illam medietatem hore continue movisset E gradu et in secunda medietate tantum sicut duo mota, quorum unum movetur solum B gradu uniformiter per eandem medietatem et aliud E gradu per eandem medietatem. Ergo Sortes

---

35 B: DB *G*
37 qui: quia *(?) G*
39 illa: ista *C*
40 est: est hec *C*
42 *post* tantum *tr.* B addit
43 hec: et *(?) B*
45 Cap. XI: Cap X *B*
47 non *om. G* / AQ *CV* A *BG*
47–48 latitudo talis *CV* latitudo *BG*
48 Q: A *G*
51 correspondet B equeposset *CGV* /

ergo *om. C*
53 QB *C*
54 E *om. C* / BD: D *BG*
55 ista positione *C*
56 equevalet BC esse valet *G*
58 AQ *CG* A *B*
60 hore *CV om. BG* / illam *B* istam *C* primam *G*
60–61 continue *om. C*
61 sicut *om. G*

tantum pertransibit AQ latitudine sicut duo mota, quorum unum per totam horam moveret E gradu solum et aliud per medietatem solum B gradu et per consequens Sortes tantum pertransibit illa AQ latitudine uniformiter adquisita quantum pertransietur in hora E gradu et D qui est subduplus ad B. Consequentie tenent intuenti per precendentem conclusionem. Cum ergo D et E gradus per horam facti valent AQ latitudinem per eandem [horam], sequitur quod gradus qui valet AQ latitudinem in hora tantum addit de latitudine supra D quantum distat E a Q qui est quies. Set cum C per positionem equevalet AQ latitudinem in hora, sequitur quod tantum C distat a E quantum distat D a Q. Iste consequentie tenent per illam suppositionem. Tunc sic equalis est distantia inter D et Q sicud inter E et C. Set cum equaliter distet B a D sicud D distat a Q, ergo equaliter distat E a C sicud B a D. Consequentia tenet per idem medium: quecunque sunt equalia tertio et cetera. Si ergo equalis est latitudo DB sicud EC, dempto ergo communi, scilicet BE, illa que relinquuntur, scilicet ED et CB latitudines, erunt equales. Et per consequens tantum distat gradus correspondens AQ a B medio sicud distat E a medio BQ qui est medietas AQ. Et per idem probatur quod per tantum distat gradus correspondens quarte AQ a medio gradu eiusdem quarte sicud E gradus a D gradu et sic in infinitum per 8as et 16as. Capiamus inde GQ partem alicotam AQ latitudinis, cuius medius gradus sit N et tantum distat N a G quantum distat E a D vel quod minor. Sit distantia inter gradum eius medium et gradum sibi correspondentem, qua sit distantia inter C et B; quod talis sit danda patet, quia AQ latitudo est divisibilis in infinitum in partes alicotas terminatas ad Q gradum et cuiuscunque partis alicote dande, tanta est distantia inter gradum sibi correspondentem et eius gradum medium sicud inter C et B; sicud patet per argumentum factum. Capiatur gradus correspondens GQ latitudini uniformiter adquisite per tempus et sit M. Per argumentum factum sequitur quod

64 *post* pertransibit *add.* B acsi
64–67 sicut... latitudine *CV om. BG*
70 valent *CG* valeant *B*
72 *E C D BG*
73 *C om. C*
74 *E C B BG* / D a *tr. B* / Q: Q latitudine *C* / illam *om. CV*
76 distet: distat *CG* / B a D: D a B *C*
77 *D B C C*
78 tertio *BG* alicui tertio *C*

81 B medio: B medio euisdem *C*
85 per 8as et 16as: per BG et per *G* / inde *om. C* / aliquotam *C hic et alicubi*
86 N^1 *C F B* / et *om. G* / N^2 *C A BG*
87 E: *C (?) BG*
87–88 Sit... distantia *in marg. B; correxi ex. GVC*
89 AQ *B* GQ *C* / in^1 *om. B*
91 gradum *om. C*
93 factum et *C*

[*Summa Dumbletonis de logicis et naturalibus* 321]

95 tanta est distantia inter M et N gradum sicud inter D et E. Ergo tantum distat M ab N sicut G ab N gradu. Et per consequens idem est M gradus et G, cui GQ latitudo uniformiter adquisita in hora correspondet, quod est probandum ex dato.

Si ergo detur quod nulla est pars alicota AQ cuius medius gradus 100 precise tantum distat ab extremo quantum C distat a B, capiatur una cuius medius minus distat ab extremo quam C distat a B, et sit illa GQ. Ex denominatione facta patet quod GQ correspondet intensiori gradui extremali, quod est impossibile.

Si ergo detur pars conversa, quod latitudo uniformiter adquisita 105 valet remissiorem gradum medio eiusdem, sequitur per conclusionem precedentem, quod quelibet latitudo minor illa terminata ad quietem remissiori medio suo correspondet.

Intendimus pro predicta conclusione, C esse medium AQ et B remissiorem gradum correspondentem AQ latitudini uniformiter ad-110 quisite in hora et D correspondere CQ latitudini. Ex ista positione sequitur per aliquam latitudinem uniformiter adquisitam nichil adquiri. Iuxta prius probatur; in illa conclusione patet quod AQ latitudo uniformiter adquisita in hora valebit D correspondentem sue medietati per horam et C per medietatem hore, et per consequens valebit 115 AQ latitudo adquisita uniformiter in hora D gradum per horam et E gradum per eandem horam, qui E est subduplus ad C. Set cum B per datum valeat AQ latitudinem in hora uniformiter adquisitam, ergo B gradus tantum continet ultra D quantum distat a E, Q. Consequentia tenet per suppositionem. Set cum equalis est distantia inter E et Q 120 sicud inter C et E, sequitur quod equaliter distat a B, D sicud E a C. Dempto ergo communi BE, relicte latitudines, scilicet DE et CB, erunt equales. Capiatur ergo aliqua pars CQ medietatis, cuius medius gradus tantum precise distat a quiete sicud distat D ab E medio medietatis CQ. Et sit latitudo illa GQ, cuius latitudinis gradus 125 medius sit N et gradus correspondens eidem sit F, qui est remissior

97–98 uniformiter... dato *om.* C
99 AQ C AQ que BG
100 distat² *om.* CG
105 eiusdem BC eius G
106 *post* illa *scr.* B *et del.* latitudine
108 predicta conclusione: ista probatione CV / AQ: AC G
109 gradum B C gradum CG
110 CQ C GQ GB

111 aliquam: aliam *(?) in MSS*
112 probatum C / latitudo *om.* G
117 valet CG
118 a E *tr.* G
120 a B *tr.* C a A BG / D: D gradu C A BG
122 aliqua CG prima B
123 precise *om.* G

medio, ut patet per conclusionem precedentem. Cum ergo tanta est latitudo inter D et E sicud inter B et C, non obstante quod CQ sit dupla latitudo ad EQ et latitudo GQ vel est quarta pars, vel octava latitudinis EQ, ita quod non in infinitum exceditur ab eadem, sequitur quod tantum distat F gradus correspondens GQ latitudini ab N medio gradu latitudinis GQ quantum distat D gradus correspondens latitudini CQ ab E medio gradu CQ latitudinis. Set cum quies tantum distat ab N medio GQ latitudinis quantum D distat ab E, ergo F gradus qui correspondet GQ latitudini est quies, et sic sequitur per GQ latitudinem uniformiter adquisitam nihil pertransiri, quod est probandum.

Aliter probatur quod omnis latitudo motus terminata ad quietem et uniformiter adquisita et cetera. Hoc patet, posita hac suppositione, cum predictis probatis, quod omnium graduum terminantium aliquas latitudines incipientes a quiete ad gradum correspondentem eisdem in pertransitione spacii eadem est proportio. Ut si AB, AC sint due latitudines ad quietem terminate, quarum AB sit maior, et D, E sint gradus uniformes quibus tantum pertransietur in hora sicud AB, AC latitudinibus uniformiter adquisitis, quod eadem est proportio B gradus terminantis AB latitudinem ad E gradum sicud C ad D, ex illa suppositione sic arguitur. Et ponamus AF latitudinem uniformiter adquisitam in hora a Sorte. Dico quod tantum pertransietur AF latitudine in hora sicud a C eiusdem medio gradu, quia si non, detur quod tantum pertransietur AF latitudine sicut ab gradu intensiori gradu medio eiusdem latitudinis, et sit pedale maximum pertransitum a Sorte in hora. Et intendimus F esse quietem AF latitudinis. Capio tunc gradum subduplum ad B et sit E qui intensior est D gradu medio medietatis remissioris AF latitudinis, ut patet. Tunc eadem est proportio A ad C sicut B ad E, quia dupla. Ergo permutatim qualis est

127 B: D et B *G* B et D *B*
129 in *om. C*
131 gradu *CV om. BG*
133 GQ: GA *B*
134 latitudinis *B*
135 pertransiri *CGV* pertransiri potest *B*
136 probandum: probandum ex dato *C*
137 *ante* aliter *scr. C* capitulum 12^m
138 Hoc patet *CGV* hec patent *B*
139 cum predictis probatis *C* ad predictum probatur *BG* cum priori seu cum predictis probatis *V* / aliquas

CG reliquas *B*
142 terminate *CGV* terminantes *B* / sint: sunt *B*
143 pertransietur *CV* pertransiatur *BG* / sicud *om. C*
145 terminantes *B* / C: a C G / *post* D *add.* BC sicut
148–49 a C . . . sicut *om. B*
150 pedale *CGV* precise *B*
151 intendimus *BG* intelligamus *CV*
153 eadem est: sit eadem *C*

[*Summa Dumbletonis de logicis et naturalibus*

proportio A ad B gradum medium, eadem est C ad E gradum. Consequentia patet per suppositionem quinti Euclidis. Tunc sic eadem est proportio A ad B gradum sicud C ad E gradum. Ergo per suppositionem datam sicud B correspondet AF latitudini uniformiter adquisite, sic E gradus correspondet medietati remissiori eiusdem latitudinis. Consequentia tenet, quia unicus est gradus in AF latitudine qui se habet ad C sicut B ad A. Set quia E est subduplus ad B gradum et Sortes movebitur in prima medietate hore CF latitudine correspondenti E, ergo pertransibit quartam totalis spacii pertranseundi ab gradu correspondenti toti AF latitudini. Consequentia patet intelligenti. Et per consequens Sortes pertransibit in prima medietate hore quartam pedalis. Set cum Sortes in secunda medietate hore habebit uniformiter AC latitudinem, ergo Sortes tantum pertransibit in secunda medietate hore sicud duo mota in eodem tempore, quorum unum uniformiter intendet a quiete ad C gradum et aliud movetur continue C gradu. Consequentia patet per precedentem conclusionem capituli septimi. Set quia B gradu pertransietur in secunda medietate hore semipedale, quia sibi tota AF latitudo correspondet et C est gradus remissior B per datum, ergo Sortes minus pertransibit in secunda medietate hore per C gradum et CF latitudinem quam tres quartas pedales. Ergo Sortes non pertransibit in toto tempore pedale, quod est contra impositioni. Et consequentia tenet per hoc, quod si Sortes movetur B gradu et illa latitudine, tunc tres quartas pertransiet; set iam minus quam tunc; ergo et cetera ut prius.

Si gradus correspondens latitudini est remissior medio eiusdem, ponamus ergo AF latitudinem talem. Et intendimus B esse medium gradum eiusdem, C correspondentem toti AF latitudini, qui C gradus sit remissior medio gradu AF, ut datur. Et sit E medius gradus BF latitudinis et sit D subduplus ad C. Per prius argumentum factum, patet quod eadem est proportio A ad C qualis est B ad D, capta pro-

155 gradum medium *om.* C / C: proportio C C
156 patet: tenet C
161 E: A C
164–65 intelligenti BG intuenti C
169 moveatur C
171 pertransietur CG pertransiet B
175 pedales B pedes C ponendo G / Ergo C Ergo quod BG
176 Et: sed C / si *om.* G
177 movebitur G / tunc *om.* B / pertransiret CV
178 ut prius BG *om.* C
180 talem *om.* G / Et: ut primus et C / intendimus: intelligamus C
181 C¹: et C esse gradum C quarte G / gradus *om.* C
182 Et *om.* C
183 subduplus GV subduplum BC
184–85 propositione CG suppositione B
184–86 capta… pertransietur *bis* B

positione quinti Euclidis. Ponamus quod pedale sit maximum quod pertransietur a Sorte in hora adquirendo uniformiter AF latitudinem. Quia ergo C est duplus ad D, et per C gradum pedale precise pertransietur in hora quia latitudinem AF equivalet, sequitur iuxta formam priorem, quod Sortes precise pertransibit in prima medietate hore quartam pedalis et quod in secunda medietate hore Sortes tantum pertransibit precise quantum duo mota in eadem medietate, quorum unum movetur sub gradu medio totius AF latitudinis et aliud movetur adquirendo uniformiter latitudinem BF, ut patet per conclusionem precedentem. Ergo Sortes pertransibit in secunda medietate hore quartam pedalis et plus quam duplum ad quartam pedalis. Consequentia tenet per hoc, quod iuxta ultimum suppositum hoc est maius quam in duplo intensior D gradu et sic patet quod plus quam pedalis quantitas est pertransita a Sorte in hora AF latitudine, quod est contra dictum. Consequentie patent pro illa parte intuenti per argumenta prius facta.

Hic tamen notatur quod hec demonstracio fundatur super hoc, quod si latitudo motus incipiens a quiete et uniformiter sit adquisita in aliquo tempore, necessario in prima medietate eiusdem temporis quarta totalis spacii pertransietur et hoc medietate remissiori eiusdem. Et ideo neque intensiori medio neque remissiori eiusdem illa latitudo in pertransitione spacii correspondet.

Ex istis duobus sequitur tertia, quod omnis latitudo finita citra quietem terminata suo medio gradui uniformiter adquisita correspondet.

Detur latitudo AB citra quietem terminata, cuius remissius extremum sit A. Dico ergo eam uniformiter adquisitam suo medio gradui correspondere. Nam tantum pertransietur AB latitudine uniformiter adquisita in hora sicud A gradu continuato per horam uniformiter et latitudine equali AB incipiente a quiete uniformiter adquisita, ut patet per unam conclusionem prius probatam. Sed cum omnis latitudo incipiens a quiete et uniformiter adquisita suo medio gradui cor-

---

185 sit CG est B
187 et *om.* G
187–88 et... AF CGV *om.* B
188 pertransitur C
192 sub *om.* C b G
193 moveatur C / ut *om.* B
195 secunda CGV prima B / pedalis *om.* G

197 maius: magis C
199 est CGV *om.* B
211 eam CGV *om.* B
212 latitudine *om.* G
214 incipiens B
215 unam: secundam P
215–17 prius... conclusionem *om.* B in textu; forte in mg.

respondet, ut patet per precedentem conclusionem, ergo AB latitudine tantum pertransietur in hora sicud a duobus motis per horam, quorum unum movetur A gradu solum per eandem et aliud gradu medio tante latitudinis incipientis a quiete. Consequentia tenet per prius probata. Sed cum gradus medius AB latitudinis sit compositus ex A gradu uniformi et gradu medio tante latitudinis incipientis a quiete, ergo tantum pertransietur AB latitudine uniformiter adquisita sicud gradu medio eiusdem, quod fuit probandum.

219 movebitur CG
220 *ante* incipientis *scr.* B et / tenet: patet C
221 probatum C
222 *post* latitudinis *scr.* B incipientis / *ante* incipientis *scr.* B et

Document 5.5

*De motu incerti auctoris**

(230v) UTRUM omnis motus uniformiter difformis correspondeat suo gradui medio. Quod sic probo quia omnis motus uniformiter difformis correspondet illi gradui quo gradu tantum precise pertransietur de spatio sicut illo motu. Sed gradus medius est huiusmodi. Igitur et cetera.

Ad oppositum: Motus semidyametri in circulo non correspondet suo gradui medio; sed motus illius semidyametri est motus uniformiter difformis; igitur et cetera.

Ad evidentiam conclusionis ponende sunt primo alique divisiones; secundo alique suppositiones; tertio ex illis concludende sunt alique conclusiones.

Prima divisio est quod quidam est motus uniformis, quidam difformis. Motus uniformes sunt cuius omnes gradus sunt equales vel eque intensi vel eque remissi; motus difformes sunt cuius non omnes gradus sunt equales.

Secunda divisio est quod quidam est motus uniformiter difformis, quidam difformiter difformis. Motus uniformiter difformis est cuius quilibet gradus magis accedens ad terminum ad quem est intensior aliquo gradu qui minus illo accedit ad eandem terminum ad quem, et similiter cum hoc quod quibuscunque gradibus acceptis inter illos gradus non est accipere aliquem gradum qui deberet esse inter illos qui ille gradus sit inter illos dispositus sicut deberet. Motus difformiter difformis est cuius gradus magis accedens ad terminum ad quem est remissior tali gradu qui minus accedit ad eundem gradum ita quod gradus non ordinatur sic deberet ordinari.

Alia divisio est quod tam motus uniformiter difformis, quam difformiter difformis, vel incipit a gradu vel a non gradu, et terminatum

---

* Venice, Bibl. Naz. San Marco Lat. VIII, 19, ff. 230v–231v. I have capitalized letters designating bodies and movements, although no such capitalization exists in the manuscript.

ad gradum vel ad non gradum. Et si terminatum ad gradum, vel ad gradum finitum vel [ad gradum] infinitum, nam latitudo uniformiter difformis potest acquiri uniformiter vel difformiter. Et accipiatur difformiter, vel hoc est uniformiter difformiter vel difformiter difformiter. Tunc latitudo acquiritur uniformiter quando in omni parte temporis (231r) equali equalis pars acquiritur de latitudine. Tunc acquiritur difformiter quando non in omni parte temporis equali equalis pars latitudinis acquiritur. Tunc acquiritur uniformiter difformiter quando gradus illius latitudinis non acquiruntur sicut deberent. Tunc difformiter difformiter quando non acquiruntur sicut deberent et hoc de divisionibus sufficiant.

Prima suppositio est ista, quod omnis motus uniformiter difformis sive omnis latitudo uniformiter difformis potest acquiri in omni tempore sive in omni parte temporis, i.e., quandocunque latitudo est terminata ad aliquem gradum, si illa latitudo potest acquiri in aliquo tempore, ipsa potest acquiri in medietate temporis et sic in omni parte temporis.

Secunda suppositio, quod cuiuscunque latitudinis uniformiter difformis incipientis a non gradu et terminate ad aliquem gradum, gradus eius medius est precise in subduplo intensior ad gradum summum intensissimum.

Tertia suppositio, quod cuiuscunque latitudinis incipientis a gradu vel a non gradu gradus equaliter distans a duobus extremis est gradus medius eiusdem latitudinis.

Quarta est, quod omnis latitudo motus uniformiter difformis potest deperdi in omni tempore et in omni parte temporis, sicut dicebatur de acquisitione eiusdem in eadem prima suppositione.

Quinta est, quod omnis talis acquisitio et omnis talis deperditio sunt equales quibus per equale tempus equalis latitudo precise acquiritur.

Sexta est, quod omnis motus uniformiter difformis motui uniformi correspondet, quia illi motus corresponde[n]t quibus equalia spatia per equale tempus precise acquiruntur; sed quocunque motu uniformiter difformi dato tanto (! tantum?) precise pertransibitur vel saltem poterit pertransiri aliquo motu uniformi per equale tempus. Ideo et cetera.

Prima conclusio est hec, cuiuscunque latitudinis uniformiter difformis incipientis a gradu et terminate ad gradum, gradus medius est plus quam in [sub]duplo intensior ad gradum suum intensissimum.

Probatur sic: Sit A aliqua latitudo motus uniformiter difformis incipiens a non gradu et terminata ad gradum. Sit B tota latitudo excepta prima quarta illius latitudinis. Sit C gradus medius totius latitudinis A. Sit D gradus medius B latitudinis. Sit E gradus intensissimus totius latitudinis.

Tunc probatur sic., D gradus est plus quam in subduplo intensior ad E gradum. Probatur quia C gradus est precise in subduplo intensior [E] et D (231v) intensior C. Ergo patet conclusio.

Secunda conclusio, quod ex omni latitudine motus uniformiter difformis tantum precise per equale tempus pertransibitur sicut suo gradu medio.

Ista conclusio probatur per 6^{am} suppositionem que ponit quod omnis motus uniformiter difformis correspondet alicui motui uniformi, sed nullo nisi suo gradui medio; igitur suo gradui medio correspondet. Et si sic tantum precise pertransibitur illo toto motu per equale tempus sicut suo gradu medio et econverso, et sic patet conclusio. Minorem probo sic, quia si aliquod mobile moveatur uniformiter gradu medio alicuius motus uniformiter difformis tantum precise pertransibit per equale tempus sicut mobile quod movetur illa tota latitudine et nec magis nec minus pertransibitur quam tota latitudine. Igitur sequitur alter gradus correspondeat toti latitudini. Sed accipiatur alium [gradum] illius latitudinis, quecunque volueris vel maius vel [minus] pertransibitur illo nisi sit gradus medius, sicut potest probari inductive, et ad hoc erit ymaginari duo mobilia quorum unum movetur uniformiter super medio et aliud movetur tota illa latitudine et statim apparebit conclusio.

Tertia conclusio est, quod omni latitudine data motus uniformiter difformis plus pertransitur in secunda medietate quam in prima. Probatur sic: Sit A aliquod corpus quod moveatur uniformiter difformiter per unum diem continue intendendo motum suum et sit tota latitudo que acquiritur per illum diem B et sit C aliud mobile quod moveatur super medio uniformi illius latitudinis. Tunc sic A et C moventur per equale tempus precise et equalia spatia pertranseunt precise in toto tempore. Sed A continue per primam medietatem temporis movetur tardius C; igitur A in prima medietate temporis minus pertransit de spacio quam C. Sed C in prima medietate temporis precise pertransit medietatem spatii. Igitur [A] in prima medietate temporis non pertransit medietatem spatii cum [in] toto tempore pertransit totum spatium. Igitur in secunda medietate plus pertransit quam in prima. Igitur sequitur consequentia. Et quod C movetur velocius continue per primam medietatem quam A probatur, quia C continue movetur super medio et A nunquam habebit illum gradum ante medietatem istius temporis; igitur et cetera....

Chapter 6

The Application of Two-Dimensional Geometry to Kinematics

NOT long after the Merton kinematics had reached its maturity, its distinctive vocabulary and principal theorems began to spread throughout Europe. Their passage to Italy and France took place about A.D. 1350. In the course of this passage a significant event took place, the application of graphing or coördinate techniques (or more exactly two-dimensional figures) to the English concepts dealing with qualities and velocities. In the sense that this new system represented the functions implicit in the concepts of uniform velocity and acceleration, it resembled the analytic geometry of the seventeenth century; but it did not translate algebraic expressions as such into geometric curves, and vice versa. Hence we cannot yet call it analytic geometry.

The basic idea of the system is simple. Geometric figures, particularly areas, can be used to represent the quantity of a quality. Extension of the quality in a subject is to be represented by a horizontal line, while the qualitative intensities at different points in the subject are to be represented by perpendiculars erected on the extension or subject line (see Fig. 6.1). In the case of motion, the line of extension represents time, and the line of intensity, velocity.

It has been customary to assume that this application of coördinates or at least two-dimensional figures to the English kinematics was invented by the Parisian scholastic, Nicole Oresme. The date of Oresme's principal expositions of this technique probably lies somewhere between the years 1348 and 1362, while he was at the College of Navarre in the University of Paris; and their titles are *Questions on Euclid's Elements* and *On the Configurations of Qualities* (*De configurationibus qualitatum*). The composition of these treatises is presumably closer to the later date than the earlier; but even if one or the other were composed around 1350, it would have to yield precedence, so far as the application of a two-dimensional geometric system is concerned, to a treatise of an Italian Franciscan named Giovanni di Casali. For we can date the latter's work, *On the Velocity of the Motion*

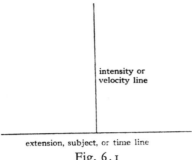

Fig. 6.1

of Alteration (De velocitate motus alterationis), in the year 1346, as one manuscript has clearly stated in its colophon.[1] A recent historian has questioned the date of this manuscript but apparently without examining it, for she gives an argument against this date in terms of Arabic numerals, when in fact the date is in Roman numerals.[2] Its early date is further suggested by the existence of a copy made in 1355.[3] We are told in the colophon to this copy that Casali composed his work in the Franciscan monastary of Bologna. Giovanni's work is thoroughly dependent on the Merton College texts, and, in fact, it centers on a question disputed at Oxford. Giovanni's interest in the English kinematics was not unique in Italy at this early date. For at the University of Padua in 1352, one Francischus de Ferraria composed a reworking of Bradwardine's *Proportions of Motions* which shows close acquaintance with the Merton kinematics as well as dynamics (see Doc. 7.4). And so if Casali was already using the method of graphing in Bologna in 1346, we can hardly say that Oresme invented the technique.

[1] MS Florence, Bibl. Riccard. 117 (L.I.33), ff. 135r–144v (colophon with date "anno domini millesimo CCCXLVI°" on f. 144v). The table of contents of the codex on the verso of the flyleaf gives: "Tractatus philosophicus, Auct. Fratre Joanne de Casali ordinis minorum, anno 1346 elaboratus."

[2] A. Maier, *An der Grenze von Scholastik und Naturwissenschaft*, 2d ed. (Rome, 1952), p. 360, n. 8. Although the manuscript is here mentioned erroneously as a Laurentian manuscript, it is the Riccardian manuscript under discussion. As I indicated in footnote 1 the date is in Roman numerals, which makes Miss Maier's discussion of the confusion of "5" and "4" beside the point.

[3] *Ibid.*, p. 360, where the explicit of Vat. lat. 2185 (ff. 61v–71r) with the date of 1355 is given.

Nevertheless, Oresme's exposition so far surpasses in clarity and further development the efforts of Casali that we should like to concentrate our attention on Oresme's discussion, leaving the treatment of Casali's few statements for later exposition.

But the system, whether invented by Oresme or Casali, had antecedents. It would not be out of place to examine those antecedents briefly before examining Oresme's system closely. Now, of course, outside of the study of qualities, the graphing concept had been applied from antiquity to cartography and astronomy; the very terms used by Oresme and his contemporaries to graph qualities and movements—"latitude" and "longitude"—were geometrical terms that long since had become the property of geography and astronomy, and thus their use by the scholastics might reflect not simply a basic understanding of geometry but rather an intentional transfer of the terms from one type of graphing to another.

Oresme himself claims as justification for his representation of qualities and movements by geometric figures (1) the fact that Aristotle employs lines for representing periods of time, (2) their use by Witelo and Lincoln (Grosseteste) to imagine the "intension of light," and (3) a statement by the Euclid Commentator (Campanus) that anything that is of the nature of a continuum can be imagined by a line, surface, or body. (See note 14 for Latin passage.)

But the immediate antecedents for Oresme's system lie more surely in the discussion of qualities and their measurements. I think in fact we may single out several basic steps in this qualitative study that lead to Oresme's system.

1. The first step was the general distinction made between "extension" and "intension" in measuring qualities (at least hypothetically) and comparing them. This distinction was evident in the flourishing discussion that the thirteenth century inherited from antiquity, of the problem of the intension and remission of qualities, a problem we have already briefly noted in the preceding chapters. This distinction had even been made numerical in medical treatises since at least the time of Galen. In the medical context it concerned itself largely with the distinction between heat intensity (temperature) and heat quantity.[4]

2. The next step came with the comparison of intensities or velocities by the use of line lengths, a one-dimensional system. As A. Maier[5] has

---

[4] For a discussion of the temperature concept in medicine, see my *Giovanni Marliani and Late Medieval Physics* (New York, 1941), pp. 34–36.

[5] Maier, "La Doctrine de Nicolas d'Oresme sur les configurationes inten-

already pointed out, this usage is apparent in the treatise entitled *De graduatione medicinarum compositarum* and attributed (doubtfully in my opinion) to Roger Bacon.[6] We read in the beginning of the treatise:[7]

> Every inherent form receives intension and remission, on account of which it becomes understandable when set forth as a line that is called the line of intension and remission. And since every inherent form has a contrary and a mean, that same line will be imagined containing contrary forms. Suppose a hotness is placed in any place whatever on the aforesaid line. Through the intellect it is understood that it may be increased above that point and similarly be remitted until it comes to the first point of the mean between hot and cold; that very same mean, since it is understood to have longitude, can be remitted or, by another consideration, be increased until it comes to a point equidistant from the contraries. And similarly it is understood that the mean, through intension, recedes from the middle point until it comes to the first point of the contrary, and the contrary is understood to be able to be increased until it comes to whatever point you wish of intension.

Quite obviously this unilinear system of representing intensities gives a method of illustrating the basic Aristotelian idea of qualitative alteration as a movement from one contrary to the other (see Chapter 7 below). Nowhere is there any idea of two geometrical dimensions or of a coördinate system. In short, this system is graphically immature, as is illustrated later when the author works out some problems of mixtures, as for example,

---

sionum," *Revue des sciences philosophiques et théologiques*, Vol. *32* (Jan.-Apr., 1948), 52–67, particularly p. 57. Cf. Maier, *Zwei Grundprobleme der scholastischen Naturphilosophie*, 2d ed. (Rome, 1951), p. 97 (pp. 89–109 discuss the system of Oresme; cf. *An der Grenze*, pp. 289–384 for a long discussion of the system and its propagation, and see particularly p. 290).

[6] A. G. Little and E. Withington, editors, *Opera hactenus inedita Rogeri Baconi*, fasc. IX (Oxford, 1928) p. xxiii, accept the work as genuinely by Bacon, although admittedly on no external evidence. My idea is that it has the ring of the early fourteenth-century English vocabulary after the intension and remission problem had had fairly extensive consideration.

[7] *Ibid.*, p. 144. The English translation is the unpublished one of Thomas Smith. The Latin text follows: "Omnis forma inherens recipit intensionem et remissionem: propter quod intelligitur tanquam exposita in linea que dicitur linea intensionis et remissionis. Et quia omnis forma inherens habet contrarium et medium, erit eadem linea intellectualis continens formas contrarias. Puta calor in quocunque loco dicte linee ponatur. Per intellectum intelligitur posse intendi supra illum punctum et similiter remitti donec venerit ad primum punctum medii inter calorem et algorem; ipsum quoque medium quia longitudinem habere intelligitur remitti potest sive per aliam considerationem intendi, donec venerit ad punctum eque distantem a contrariis. Et similiter intelligitur quod medium per intensionem recedit a puncto medio donec pervenerit ad primum punctum contrarii, et contrarium intelligitur posse intendi donec perveniat ad quemcunque punctum intensionis."

the mixture of a water of two weights hot in the sixth degree with a water of one weight hot in the twelfth degree:[8]

For example, let there be given water of two weights hot in the sixth degree, in respect to some point contained in the same line; let there be given again another water of one weight hot in the twelfth degree with respect to the same point; a mixture of the two waters having been made, the hotness of the mixture will be raised in a line intension through eight degrees, with respect to the aforesaid point, since the distance that is between six and eight is one-half the distance that is between eight and twelve, just as the water of one weight is half the water of two weights.

The author is of course arithmetically taking into account extension in his computation when he considers the relative weights going into the mixture. But he has not as yet represented both extension and intension geometrically. In fact, he has computed the intensions arithmetically and then applied the results to single lines.

The same sort of unilinear procedure is used by the Merton College authors (and this is the principal reason I believe the treatise belongs to the fourteenth century). For example, the system is applied to motion in the selection from John Dumbleton's *Summa* given in the preceding chapter (see Fig. 5.4). There the degrees of velocity are laid out along a single line.

3. There is, however, one English author, Richard Swineshead, who gives a two-dimensional geometrical *analogy* to explain intension and extension and their relationships. In short, he compares geometrical length and breadth in an area with extension and intension:[9]

[8] *Ibid.*, p. 148. The English translation is again that of Thomas Smith. The Latin runs: "Cum igitur duorum commixtorum quantitates sunt diverse et intensiones formarum confusarum in diversis gradibus elevate in linea, sive fuerint ille forme eiusdem speciei sive contrarium specierum, sive medie et contrarie, que est proportio quantitatis unius partis commixti ad aliam, eadem erit proportio differentie que est inter gradum forme confuse et gradum maioris partis commixti corporis ad distantiam que est inter eundem gradum commixti et gradum minoris partis. Exempli gratia, detur aqua duo librarum calida in 6° gradu linee sumpto in respectu alicuius puncti in eadem linea contenti; detur etiam alia aqua unius libre calida in 12° gradu respectu eiusdem puncti; facta commixtione istarum aquarum, caliditas commixti erit elevata in linea intensionis per 8 gradus respectu predicti puncti, quia distantia que est inter 6 et 8 est subdupla ad distantiam que est inter 8 et 12, sicut aqua unius libre est subdupla ad aquam duarum librarum. Et e converso vero, si aqua duarum librarum fuerit elevata in linea per 12 gradus et aqua unius libre per 6 gradus, erit ipsum commixtum calidum in gradu 10 per predictam regulam."

[9] The translation is a joint translation of Mr. Smith and myself. The passage was variable from copy to copy, and so Mr. Smith and I prepared a text from the following manuscripts of the *Liber calculationum* of Richard Swineshead: *BN*, Paris,

If a length of a foot is taken, and if there were added to the side of it a length as great or smaller, the whole length would not be increased, because [the second length] is added in such a way that it cannot increase that dimension (i.e., length) as such. And so if a body with a length of a foot be given, to which is added breadthwise a length of half a foot, still the whole is said to be of a length of a foot, as it was before. And if that length of half a foot were enlarged until it contained three-fourths of a foot, and the length of a foot were reduced to the same degree of length, the whole length is decreased as to length, even though one acquires as much length as the other loses.... Thus in the question at hand, that acquired quality is acquired in such a way that it does nothing intensively to the whole quality, but only [affects it] quantitatively (i.e., extensively). Nay rather, in order that it would affect intensively the whole quality, it would be necessary that it be coextended with it, just as if the acquired length were to increase another length [extensively] or if the other were to become greater on account of it, [then] it is necessary that it be acquired at the extreme of the other length; and just as the depth of a body is not increased intensively by the advent of another depth unless their coextension is effected, or at least their addition according to depth. Although there would be acquisition at the extreme, it would not make the whole deeper, but would only affect it quantitatively, by occupying a greater space. So neither would this acquired quality affect something qualitatively, but only quantitatively with respect to the whole quality.

BN. lat. 6558, ff. 5r–5v; *Ca*, Cambridge, Gonv. and Caius MS. 449/268, f. 167v, c.2; *Ald*, Pavia, Bibl. Univ. Aldini 314, f. 5r. The last manuscript has only been used for an occasional variant. The text and variants follow (one might see a similar but vaguer analogy in the *Questions on the Eight Books of the Physics* of Buridan [see above, Chapter 4, footnote 39]):

Et ideo tota qualitas remaneret eo quod pars deperdita in medietate remissiori nichil facit qualitative ad totam qualitatem. Sicud si capiatur una longitudo pedalis et si ad latus eius adderetur tanta longitudo vel brevior, tota longitudo non maioraretur eo quod illo modo additur quod non potest maiorare illam dimensionem ut sic. Et ideo si detur corpus pedalis longitudinis cui sit additum semipedale secundum latitudinem, totum adhuc dicitur pedalis longitudinis, ut ante fuit. Et si illa semipedalis longitudo maioraretur quousque contineret tres quartes pedalis et longitudo pedalis minoraretur ad eundem gradum longitudinis, totum longum minoratur quo ad esse longum; et tamen tantam longitudinem adquirit unum sicut aliud deperdit. Ymo longitudo acquisita illo modo acquireretur quod intensive nichil faceret ad intensionem totius longitudinis vel maiorationem, ymo extensive. Et illa deperditio nichil faceret ad deperditionem longitudinis intensive, ut est cuius dimensio. Sic in proposito, illa qualitas acquisita illo modo acquiritur quod ad totam qualitatem intensive nichil facit, sed solum quantitative. Ymo ad hoc quod ad totam qualitatem faceret intensive oporteret quod coextenderetur cum illa, sicud longitudo acquisita si maioraret aliam vel fuerit alia maior propter illam, oportet quod ad extremum alterius longitudinis acquiratur, et sicud profunditas corporis non maioratur intensive propter aliam profunditatem adve-

[Geometry to Kinematics   337]

It is, of course, still a crucial step to go from a geometrical analogy to a full graphing system. That step was taken by Giovanni di Casali and Nicole Oresme, and it is their system we wish to analyze.

A word should be said about the life and activity of Nicole Oresme. Without sure evidence, his birth is conjectured to have occurred about 1323, possibly in the neighborhood of Caen.[10] The first date in his academic career is 1348, when his name is mentioned as one of a number of theological students in the College of Navarre at the University of Paris. Presumably he was a theological student for no longer than eight years, since he was clearly a Master in 1356. It is presumed that he heard the lectures or had some fairly close contact with the greatest of the Parisian Masters, John Buridan, whose important ideas in dynamics we shall explain below in the eighth chapter. In 1356 Oresme became Grand Master of the Parisian College of Navarre. It is said that at about this time he was *precepteur* to the future Charles V and wrote, at Charles' father's request,

nientem nisi fiat coextensio illarum vel saltim additio secundum profunditatem. Licet ad extremum acquireretur non faceret totum profundius sed 45 solum quantitative maiorem locum occupando. Sic nec et illa qualitas acquisita aliquid faceret qualitative sed quantitative quo ad totam qualitatem.

4 Sicud *BN* ut *Ca*   5 si *BN* om *Ca*   6 eius om *Ca*   6–7 vel...longitudo *BN* illa pedalis *Ca*   8 *ante* eo *add Ca* in longitudine illo *om Ca*   8 quod *BN* aliquid et *Ca*   9 non *BN* nec *Ca*   10 detur *BN* capiatur *Ca*   10–11 additum semipedale *BN* addita semipedalis *Ca*   13 totum *Ald* totam *BN Ca*   14 ut...fuit *Ca* ut dictum est *BN* illa *om Ca*   15 maioraretur *BN* maioretur *Ca*   17 minoraretur *BN* minoretur *Ca*   18 longitudinis *Ca Ald* latitudinis *BN*   19 minoratur...esse *Ald* minorabitur esse *Ca* diceretur esse brevius *BN*   20 tantam longitudinem *Ca Ald* tanta longitudo *BN*   20–21 adquirit...deperdit *Ca* acquireret sicud deperderet *BN*   23 quod *BN* et *Ca*   faceret *om BN*   24 vel *BN* vel ad *Ca*   24–25 maiorationem: minorationem *Ca*   25 *ante* ymo *add Ca* faciet extensive *BN* ad extensionem *Ca*   illa *BN* ita illa *Ca*   25–26 deperditio *BN* deperdita *Ca Ald*   26 nichil *om Ald* faceret: faciet *Ca*   27 longitudinis intensive: intensionis vel longitudinis *Ca* cuius: talis *Ald*   33 faceret intensive: intensio faceret *Ca et tr ante* ad   34 coextenderetur *BN Ald* ex quo extenderetur *Ca*   illa *om Ca*   34–35 longitudo *Ca* latitudo *BN*   35 si *om Ca*   38 *post* sicud *add Ca Ald* etiam   41 illarum *Ca Ald* illa *BN*   42 saltim: saltem *Ca*   42–43 profunditatem: profundum *Ca*   43 Licet: set si *Ca*   44 non...profundius *om Ald*   46 nec et: etiam nec *Ca*

[10] The biography of Oresme has been treated often, but still a new and complete biography needs to be done. We have generally followed A. D. Menut, *Le Livre de Ethiques d'Aristote* (New York, 1940), and A. D. Menut and A. J. Denomy, "Maistre Nicole Oresme Le Livre du ciel et du monde, etc.," *Mediaeval Studies*, Vol. 4 (1942), 239–43. Consult also E. Bidrey, *Nicole Oresme; Etude d'histoire des doctrines et des faits economiques, etc.* (Paris, 1906); F. Meunier, *Essai sur la vie et les ouvrages de Nicole Oresme* (Paris, 1906); and L. Thorndike, *A History of Magic and Experimental Science*, Vol. *3* (New York, 1934), 398–99. These various works have citations to the extensive bibliography on Oresme.

his famous treatise on money, *De mutationibus monetarum*. Oresme remained at the College of Navarre until 1362 in spite of an appointment as archdeacon of Bayeux in 1361. It has been speculated that all of his Latin works date from the period at the College of Navarre.[11] If so, we should have to

---

[11] I suggest in regard to the dating of Oresme's scientific works the following tenuous argument. In the course of his translation of Plato of Tivoli's Latin version of the *Quadripartitum* with an attendant commentary of ʿAlī ibn Riḍwān, Oresme mentions Charles as "hoir de France, a present gouverneur du royalme." This according to Menut and Denomy (*Mediaeval Studies*, Vol. *4* [1942], 241) "must signify that he was writing during the period of King John's absence in England, between 1356–1360, when the dauphin Charles was acting as regent." Now in Oresme's French work on astrology, *Livre de divinacions*, he tells us in the proemium (edition of G. W. Coopland [Cambridge, Mass., 1952], p. 50) "et supplie que on me ait pour excuse de la rude maniere de parler, car je n'ay pas aprins de (estre) acoustume de riens baillier ou escripre en francois." From which statement that "I have never learned or been used to set forth or write anything in French," one might deduce that this is his first French effort. If so, then the work precedes the translation of the *Quadripartitum* and thus was written before the end of the period 1356–60. But the *De divinacions* (*edit. cit.*, pp. 60, 92), twice cites the *De configurationibus qualitatum*, and thus it too would appear to have been composed before 1360. In turn the *De configurationibus* (III.6, MS London, Brit. Mus., Sloane 2156, f. 192r) cites the *Algorismus proportionum*, which thus would also put that work back into the fifties. This is further confirmed by the fact that the *Algorismus* is dedicated to Philip of Meaux (MS Florence, Bibl. Laurent., Ashbur. 210, f. 172r), who was originally of Vitry and did not assume the attribution "Meldensis" until he became bishop of Meaux in 1351, in which position he continued until his death in 1361. But the dedication is being made to him as a living man. The *De divinacions* (*edit. cit.*, p. 54) also cites by title his *De commensurabilitate motuum celestium* which puts this work also in the fifties. Furthermore, the *De commensurabilitate* (BN lat. 7281, f. 267r) cites by title the brilliant *De proportionibus proportionum* (a critical text of which is in preparation by an ex-student of mine, Edward Grant). Here Oresme is also interested in part in irrational proportions. At any rate in the *De proportionibus* he appears in one place (MS Cambridge, Peterhouse College 277—Pepys 2329—f. 93v) to be correcting specifically an earlier statement on incommensurability. One might judge on internal evidence that the more complex and difficult *De proportionibus* followed the simpler *Algorismus proportionum*, but such reasoning is always hazardous. However, references to *agorismus* (i.e. *algorismus*) in the *De proportionibus* (*MS cit.*, ff. 94r, 95r), if to his own *Algorismus proportionum*, would confirm an earlier dating of the latter. But in all probability these references are simply to general treatises on algorism rather than to his own treatise. At any rate, we do know that Oresme's Latin *Questiones de spera* (MS Florence, Bibl. Riccard. 117, f. 134r) quotes an opinion on the probability of any two quantities or times being incommensurable from the *Liber de proportionibus*. Turning to that work we find the quoted opinion in chapter 4 (*MS cit.*, f. 110r). But here again we run into difficulty, for we do not know when the *Questiones* were prepared, although their scholastic form suggests that Oresme was still at the College of Navarre, i.e., before 1362. The scholastic form of the *Questiones in libro Euclidis* makes a similar dating plausible, although he does not cite any specific previous work therein. A work entitled *De proportionibus velocitatum in motibus* (MS

[Geometry to Kinematics

place there his *Questions on Euclid's Elements* as well as his treatises on proportions written under the stimulus of Bradwardine's work, his Latin *Questions* on the *De caelo* of Aristotle, and finally his fine work which we are studying in this chapter, the *De configurationibus qualitatum*.[12]

Paris, Bibl. de l'Arsenal 522, ff. 126r–168v) is also attributed (f. 168v) to Oresme, and was compiled in part of tracts of the author's masters at Paris, as he tells us in the introduction (f. 126r). Unfortunately, this work cites no other work of Oresme. Nor in fact does he use some characteristic doctrines of Oresme (like his configuration technique) when dealing with problems where we might expect him to. From this we might deduce either that this work is very early or that it is not by Oresme. The upshot of this discussion is that, if sound, it confirms the older idea that all of Oresme's Latin scientific works were composed while at the College of Navarre.

[12] The question as to what was the original title of the work we have called *De configurationibus qualitatum* is a difficult one to answer. Two Parisian manuscripts (BN lat. 14580 ff. 37r–60v, and Bibl. de l'Arsenal 522, ff. 1–29r) give in their colophons (f. 60v and f. 29r respectively) the title as *Tractatus de configurationibus qualitatum*. Erfurt, Stadtbibl. Amplon. Q. 350, ff. 1r–14r (incomplete) has (f. 1r) "*Tractatus de configuracionibus.*" The very good British Museum manuscript, Sloane 2156, ff. 159r–193v, which we have used in this volume for our text, has *Tractatus de configurationibus qualitatum et motum* (f. 159r). The handsome Laurentian manuscript (Florence, Bibl. Laurent., Ashbur. MS 210, ff. 101v–129v) has *Tractatus de configuratione qualitatis* in the beginning (f. 101v) and *Tractatus de uniformitate et difformitate intensionum seu configuratione qualitatum* at the end (f. 129v). Similarly the subject matter as mentioned in the incipits and explicits of almost all the manuscripts is referred to as *De uniformitate et difformitate intensionum*. (It is for this reason we have made this a subtitle after choosing *De configurationibus qualitatum* as the main title.) And *De uniformitate et difformitate intensionum* is in fact the title given in the colophons of two Parisian manuscripts (BN lat. 7371, ff. 214r–266r and BN lat. 14579, ff. 18r–40v) and the Basel manuscript (Univ. Bibl. F. III. 31, ff. 2r–29r). No title is given in one of the Vatican manuscripts (Chigi lat. E. IV. 109, ff. 87r–149r), while in the other (Vat. lat. 3097, ff. 1r–22v) we read in the colophon: "*Tractatus de latitudinibus etc.*" The Bruges manuscript (Stadsbibliotheek 486, ff. 159r–173r) and the manuscript of the Bibl. Naz. at Florence (J. IX. 26, ff. 13r–35r) are without specific title, although they have in *incipit* and *explicit* the subject matter indicated above. One further possible title appears in the first section just prior to the list of titles for parts and chapters in certain of the manuscripts. It runs in MSS Paris, BN lat. 7371, f. 241r; Paris, Bibl. de l'Arsenal 522, f. 1r; Erfurt, Stadtbibl. Amplon. Q. 350, f. 1r, and London, Brit. Mus., Sloane 2156, f. 159r, as follows: *Tractatus de figuratione potentiarum et mensura difformitatum;* and in Vat. Chigi lat. E.IV. 109, f. 87r: *Tractatus de figuratione, potencia, et de qualitate per predicta causa* (!) *quorundam et mensura difformitatum*. But the title is missing in the other manuscripts (except perhaps Basel F.III.31, which was temporarily unavailable to me). The other manuscripts have some phrase like the following (Florence, Bibl. Laurent. Ashbur. 210, f. 101v): "Huius autem tractatus tres sunt partes...." This title might well have been a scribal addition in an early manuscript. However, I cannot be sure and hence I have given it here as a further possibility. It is of interest to note that Oresme twice cites this treatise in his translation of the *Politics* of Aristotle with the title *De difformitate qualitatum* (see Thorndike, *History of Magic*, Vol. *3*, 425, n. 2). He also twice cites it in his *Livre de divinacions* (edition of G. W. Coopland, 1952), once (p. 60) as

Oresme perhaps left Navarre in 1362 to become a canon at Rouen, shortly after which he became canon at La Sainte Chapelle, and then when the later Charles V became regent again in 1364, he helped Oresme secure the appointment as dean of the cathedral at Rouen. That same year, the Dauphin assumed the crown as Charles V, and Oresme while dean of Rouen continued as a counsellor to him and later undertook at the King's request the translation into French of four Aristotelian treatises, the most important for our purposes in this volume being the French text of and commentary on Aristotle's *De caelo*, the approximate date of which was 1377, the time of Oresme's appointment to the bishopric of Lisieux. Thus, toward the end of his career, his interest in scientific matters was as pronounced as it was in the early years. Only five years later, in 1382, Oresme died.

The object of Oresme's *On the Configurations of Qualities* is, as the title suggests, to represent by figures, that is geometrically, variations in qualities. It must be understood immediately, however, that Oresme's representations are entirely "ymaginationes," as he calls them. They are concerned with a figurative presentation of hypothetical quality variations and thus are totally unrelated to any empirical investigations of actual quality variations. But his importance for us is, that in treating of qualities (however "imaginatively") he follows his English predecessors by treating also of velocity changes; and so he gives us a geometrical method of representing velocity changes.

As examples of Oresme's technique let us consider the accompanying rectangle and right triangle (see Fig. 6.2). Each measures the quantity of some quality. Line AB in either case represents the extension of the quality in the subject. But in addition to extension, the intensity of the quality from point to point in the subject has to be represented. This Oresme did by erecting lines perpendicular to the base line, the length of the lines varying as the intensity varies. Thus at every point along AB there is some intensity of the quality, and the sum of all these lines is the figure representing the quality. Now the rectangle $ABCD$ is said to represent a uniform quality, for the lines AC, EF, BD representing the intensities of the quality at points A, E, and B (E being any point at all

La configuracion des qualites et des mouvemens, and the second time (p. 92) as *Livre de la Figuracion des Qualitez*. After Mr. Smith and I have established the text and thus know better the relationships of the various manuscripts listed with the Latin text of Document 6.1, we perhaps can better choose the original title. For now we can be satisfied with the title and subtitle here employed as indicating very nicely the objectives of the work.

[Geometry to Kinematics

on AB) are equal, and thus the intensity of the quality is uniform throughout. In the case of the right triangle ABC it will be equally apparent that the lengths of the perpendicular lines representing intensities uniformly decrease in length from BC to zero at point A. Hence, the right triangle is said to represent a uniformly difform (nonuniform) quality. Of course the

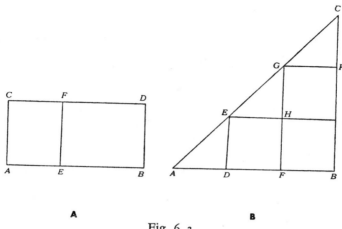

Fig. 6.2

intensities could vary in an infinite number of ways and we would have a limitless variety of figures to represent other kinds of nonuniform qualities. It is worth pointing out that Oresme designated the limiting line CD (or AC in the case of the triangle) as the "line of summit" or the "line of intensity." This is comparable to a "curve" in modern analytic geometry. And thus the figures themselves in Oresme's system are comparable to "the areas under curves." The "curve" or summit line of Oresme is representing a "function" expressed verbally instead of by algebraic formula, the verbal expressions of the functions being "a uniform quality," "a uniformly nonuniform quality," etc. The variables in these functions of Oresme are the quantity of a quality and the extent and intensity of the quality.

In one interesting passage, Oresme foreshadows the definition of a function in terms of the kind of curve drawn, i.e., in the form and disposition of his so-called summit line (see Doc. 6.1, passage I.13).

...the superior line of the figures by which we let the quality be imagined we let be called the line of intension or the summit line.... For example, let there be a [base] line AB in quadrangle $ABCD$ [with a summit line CD]. Hence, if the summit line of a figure of this kind by which a quality is imagined is parallel to the base, as for example, to the base AB, then the quality imaginable by such a figure

[342   The Application of Two-Dimensional]

is simply uniform. If, however, it is not parallel to the base and yet is a straight line, then the quality is uniformly difform. So that if the aforementioned line joins the base at one extreme, that uniform difformity is terminated at zero degree.... But if the line of intension is a curve or composed of more than one line, then the quality imaginable by that figure will be difformly difform. It can be terminated at some degree in both extremes or at zero degree in both extremes or at some degree at one extreme and at zero degree in the other.

It can be pointed out that Oresme earlier (*ibid.*, passage I.11) had already given one definition of a quality uniformly difform that is not unlike the definition of a straight line in analytic geometry:

...a quality uniformly difform is one in which, when any three points [of extension] are taken, that proportion of the distance between the first and second to the distance between the second and the third is the same as the proportion of the excess in intension of the first over the second to the excess of the second over the third.... This statement is clarified as regards a quality uniformly difform which is terminated at zero degree and which we let be signified or imagined by triangle *ABC* (see Fig. 6.2B). After three perpendicular lines *BC*, *FG*, and *DE* have been erected, *HE* parallel to line *DF* and *GK* parallel to *FB* are protracted. Hence two triangles are formed, namely *GHE* and *CGK*, which triangles are of equal angles. Hence by the fourth [proposition] of the sixth [book of the *Elements of Euclid*], the proportion of *GK* to *EH* is as the proportion of the excess *CK* to the excess *GH*. And since *GK* is equal to *FB* and similarly *EH* equals *DF*, the proportion of *FB* to *DF*, which lines indeed are the distances between the three points of the base, will be as the proportion of *CK* to *GH*, which lines are the excesses of altitude proportional to the intension at these same points.... The same thing holds for any other three points.

Now it may be asked by the reader, what has all of this about qualities to do with kinematics? The answer, as I have said, lies in the fact that like his English predecessors Oresme passed from variation in the intensity of permanent qualities to variations in velocities. Thus, if we look again at our Figure 6.2, Oresme held that the line *AB*, the base line in each figure, could also represent instead of the extension of a permanent quality, the duration of some local movement. The perpendiculars *AC*, *EF*, *DB* in the rectangular figure would than represent instantaneous velocities at instants *A*, *E*, and *B*. Since *EF* would have the same value regardless of what instant *E* was taken, the rectangle represents a uniform movement. *CD* then instead of being the line of intension would now be the line of velocity. In Figure 6.2B, the right triangle would now represent in local motion a uniformly difform movement, i.e., a uniform accelera-

[Geometry to Kinematics

tion, the velocity line BC being the maximum velocity and the velocity at instant A being zero. It is clear from at least one of his theorems that Oresme understood that the areas of the figures in the case of local motion represent the distances traversed in the times represented in each case by AB.

Now since the basic kinematic acceleration theorem discovered by the English equates a uniform acceleration with a uniform speed equal to its mean *in so far as the same space is traversed in the same time*, the geometric proof of this theorem using Oresme's system must show that a rectangle whose altitude is equal to the mean velocity is equal in area to a right triangle whose altitude represents the whole velocity increment, i.e., a line equal to twice that of the altitude of the rectangle. Oresme in effect demonstrates this in the selection below (see Doc. 6.1, passage III.7). True, the proof is given in terms of uniform and uniformly difform qualities. But after completing the proof for qualities, he states that it also follows for velocities.

In addition to giving Oresme's proof of the mean speed theorem, we have included several other interesting chapters from Oresme's *De configurationibus qualitatum* with the object of further characterizing his system and his knowledge of kinematics. Part I defines the system and applies it to qualities. In Part II, chapter 1, Oresme applies his graphing system to velocity, and there emerges a system of tridimensional coördinates—one coördinate (the ordinate) being velocity, and the other two orthogonal abscissas being respectively time and the physical extent of the moving body. Such a system would allow us to represent the variations in speed from point to point in a moving body not only at any given instant but throughout some specified period of time. In II.3 Oresme suggests the fundamental idea of quantity of velocity which arises from considering both speed and the time through which the movement continues—for Oresme quantity of velocity is measured by the *area* of the figure. This was also known as "total velocity"; it is, as we have already said, dimensionally equivalent to distance. In II.4 the author distinguishes various kinds of velocity, and particularly noteworthy is his distinction of angular velocity from curvilinear velocity, a distinction repeated carefully by Oresme's contemporary at the University of Paris, Albert of Saxony (see Doc. 4.2). In the succeeding chapter (II.5) a description of acceleration ensues; for acceleration he uses the word *velocitatio* as distinguished from the word *velocitas* used for speed. In II.8 Oresme applies his coördinate system to velocity changes occurring in time; and a justification for his system is

[344] The Application of Two-Dimensional]

given. Part III of the *De configurationibus* concerns itself with the measure and comparison of qualities and velocities. In III.7 occurs the mean speed theorem, treated geometrically, which, of course, is a comparison between the quantity of a uniform quality or velocity with a uniformly difform quality or velocity; the comparison, as we have pointed out, is between a rectangle and right triangle raised on the same base. In III.8 he gives an example of how a convergent infinite series is summed geometrically by the use of his system. This and other series had also been employed by the English to problems of variations in quality and velocity, as our selections from Heytesbury, Swineshead, and Dumbleton in the preceding chapter illustrate.

It should be remarked at this point that Oresme in his *Questions on the Elements of Euclid* not only describes the coördinate system in much the same manner as he does in his *De configurationibus*, but he has one interesting addition, namely, the generalization of the Merton distance rule for uniform acceleration from rest.[13] It will be recalled from the preceding chapter that many of the Englishmen had noted that in the second half of the time of acceleration a body traversed three times the distance traversed in the first half of the time. Now Oresme, while not specifically applying the generalization to uniform acceleration but rather to a uniformly difform quality, says that if we divide the subject line into equal parts (equivalent to dividing the time into equal periods in problems of velocity), the successive

[13] MS. Vat. Chis. F.IV.66, f. 29rb: "Hic est notandum quod quedam sunt superficies equales, alie inequales. Ita quedam sunt superficies similes, alie dissimiles.... Secundo notandum quod eodem modo quedam sunt qualitates equales et alie inequales. Et similiter quedam sunt qualitates similes et alie dissimiles.... Qualitates similes in uniformitate aut difformitate sunt que ymaginande sunt per similes superficies...omnes qualitates uniformes sint sibi in (!) invicem similes in intensione...quia ymaginande sunt per quadrangula similia.... non omnes uniformiter difformes sunt similes. Patet quia una ymaginanda est per unum triangulum et alia per triangulum dissimilem... (29va).... Proportio qualitatum vel extensionum est sicud proportio subiecti ad subiectum proportione duplicata....Patet exemplo si *a* movetur uniformiter per horam et *b* per eandem horam uniformiter incipiendo a gradu duplo et terminando ad non gradum, tunc pertranseunt equalia spacia (*cor. ex* spacium) sicud possem faciliter probare. Igitur per diffinitionem velocitatis equevelociter movebantur per totam horam... (30ra)... omnis qualitas uniformiter difformis ad non gradum, cuius subiectum per aliquas partes equales ymaginatur dividi, signanda est per numerum quadratum, cuius numerus partium [subiecti] esset radix... proportio qualitatis totius ad qualitatem illius partis est sicud proportio subiecti ad illam partem duplicatam...subiecto taliter diviso—et vocetur semper pars remissior prima—proportio parcialium qualitatum et habitudo earum ad invicem est sicut series imparium numerorum, ut prima est 1, secunda 3, tripla 5, etc."

Plate 6: A page from a fourteenth-century manuscript illustrating the use of Oresme's configuration techniques to interpret text of Swineshead's *Calculationes*. MS Paris, BN lat. 9558, f. 6r.

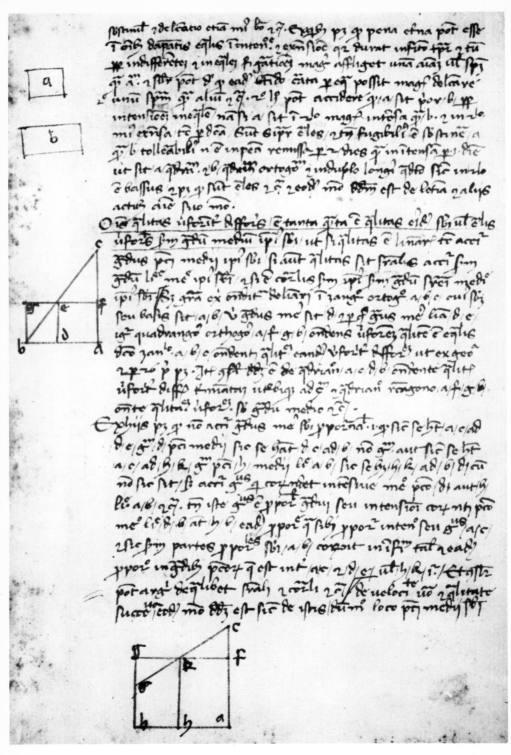

Plate 7: A page from a fifteenth-century manuscript of Oresme's *De configurationibus qualitatum*, illustrating his geometric proof of the Merton uniform acceleration theorem. MS Florence, Bibl. Naz., Conv. Soppr. J. IX. 26, f. 32v.

partial qualities (equivalent to the successive distances) in those parts are as a series of odd numbers. Galileo was later to repeat this statement for uniform acceleration (see Doc. 6.5).

After studying the detailed exposition of Oresme's geometry of qualities and velocities, we are particularly struck by its superiority to the few casual statements of Giovanni di Casali in his *De velocitate motus alterationis* (Doc. 6.2, passage 2). There we find Casali briefly utilizing the system (1) to represent a uniformly intense quality by a rectangle, (2) to represent a uniformly difform quality beginning at zero degree by a right triangle, and (3) to represent a uniformly difform quality beginning and ending at some degree by a quadrangle having two right angles on the base and two other unequal angles, one acute and the other obtuse. We should note further (4) that he equates a rectangle with a right triangle of twice the altitude in order to prove the theorem that the quantity of a uniformly difform quality is equivalent to the quantity of quality uniform in the degree mean between the initial and final degrees of the latitude uniformly difform. This is, of course, the mean speed theorem as applied to qualities. One final interesting feature of his system is (5) that he appears to be using latitude as a horizontal line and longitude as the vertical line. In modern parlance, we would say that his coördinates are rotated 90° from the positions adopted by Oresme.

Following the initial accounts of Oresme and Casali, the geometry of qualities and velocities seems to have become fairly popular in the second half of the fourteenth century. We have attempted to mention the incidences of its usage in Chapter 11, where we trace the course of the English and French mechanics in the late medieval period. Here we can be satisfied with comparing two accounts composed toward the end of the fourteenth century with the prior expositions of Casali and Oresme. The first is found in a tract entitled *De latitudinibus formarum* (see Doc. 6.3). This tract, while attributed to Oresme, was probably composed in the last quarter of the fourteenth century by one Jacobus de Sancto Martino (perhaps of Florence or Naples) following "the doctrine of Oresme" (see commentary to Doc. 6.3). The treatise concerns itself largely with a description of the various kinds of qualities and the figures used to represent them: uniform, uniformly difform, difformly difform, uniformly difformly difform, etc. The treatise does not compare the quantities of qualities or total velocities as does Oresme in the third part of his *De configurationibus*. Hence it is not surprising to find missing the mean speed theorem with its proof by the equation of rectangle and right triangle. As we point out in the commentary to Document 6.3, the author was thoroughly confused as to how to represent a quality uniformly difformly difform. He represented it by a quadrant of a circle instead of a segment of a parabola. However, he does have one interesting statement (Doc. 6.3, passage 10) to the effect that in the case of a quadrant or segment of circle, the rate of change of intension with respect to extension (i.e., the slope) is a maximum at zero extension (and thus minimum intension) and a minimum as we proceed to the maximum intension.

The second tract of the late fourteenth century, one inspired initially by the *De latitudinibus formarum*, is a short *Questiones super tractatu de latitudinibus* by the famous scholastic Blasius of Parma, who taught at various Italian universities from 1377 to 1411, as we shall see in Chapter 11. It will be noticed (Doc. 6.4) that Blasius presents the equation of rectangle and right triangle to prove the mean speed theorem.

The final fruition of the doctrine of latitude of forms with its geometric representation was with Galileo and Isaac Beeckman. Galileo was in all probability not acquainted with Oresme's *De configurationibus*, but he could have easily read the published accounts of the mean speed theorem with geometric proof found in the treatises of Casali and Blasius of Parma. For the sake of comparison I have given in Document 6.5 Galileo's geometric proof of the mean speed theorem. I think it can hardly be doubted that Galileo obtained both the theorem and the essentials of its proof from the medieval Oxford-Paris tradition, although the exact sources he used are not known. I have given the geometric proof of Isaac Beeckman in Document 6.6. It can be remarked that Beeckman's proof contains the first real improvement over Oresme's treatment, since his treatment of infinitesimals is more explicit.

It was, then, in the area of kinematics, and particularly in the geometrical analysis of uniform acceleration, that the medieval tradition was to be most significant for the development of modern mechanics. Of course, side by side with this growing interest in kinematics was a corresponding interest in dynamics, and although the results of medieval dynamical investigations were less significant than those of medieval statics and kinematics, we would have a poor idea indeed of medieval mechanics without some treatment of medieval dynamics. It is to dynamics, then, that I have devoted the third section of this work.

Document 6.1

Nicole Oresme, *On the Configurations of Qualities**

I.O. WHEN I began to set in order the views of earlier people on the uniformity and difformity of intensions, certain additional things occurred to me which I introduced into the proposed matter, so that this treatise might be useful not only for training but also for organized study *(discipline)*. In this treatise I have striven to treat clearly and skillfully those matters which others seem to perceive in a confused manner, to express in an obscure way, and to apply illogically; and [I have also tried] to apply them in a useful way to other matters. The treatise on figuration of powers and the measure of difformities has three parts: [1] on the figuration and the power of the uniformity and difformity of permanent things, [2] on the figuration and power of successive things, and [3] on the acquisition and measure of quality and velocity....

I.1. (Part I). On the continuity of intension. Every measurable thing except numbers is conceived in the manner of continuous quantity. Hence it is necessary for the measure of such a thing to imagine points, lines, and surfaces—or their properties—in which, as Aristotle wishes, measure or proportion is originally found.... Hence every intension which can be acquired successively is to be imagined by means of a straight line erected perpendicularly on some point or points of the [extensible] space or subject of the intensible thing.

For example, whatever kind of a proportion is found between one intension and another—among intensions of the same kind—a similar proportion is found between some one line and another, and conversely. For just as one line is commensurable to another line and incommensurable to [still] another, so it is the same with respect to intensions that, because they are continua, some are mutually commensurable and some incommensurable in any way you wish. Therefore, the measure of intensions

---

* Or *On the Uniformity and Difformity of Intensions*. Still another alternate title is: *The Treatise on the Figuration of Powers and the Measure of Qualities*, as given in the first paragraph. For a discussion of the alternate titles, see note 12 of this chapter.

can be imagined in a manner congruent to the measure of lines, since intension can also be diminished and increased *ex se* toward infinity in the same way as a line.

And again, intension, according to which something is said to be more or less in some way—such as more white or more rapid—is indeed that according to which the intension or extension of a point is divisible by a certain amount in one way (i.e., in one dimension) and it is divisible to infinity in the manner of a continuum. Therefore, nothing can be more fitting *(convenientius)* than having intension represented by that species of a continuum which is first of all divisible—and [furthermore] is divisible in one way by a certain amount—namely, a line. And since the quantity or proportion of lines is better known and is conceived more easily by us—indeed a line is in the first species of continua—hence such an intension is to be imagined by lines; and, most conveniently and truly, by lines applied to the subject, which are erected perpendicularly on it. The consideration of these lines is of assistance and leads naturally to the knowledge of any intension whatsoever, as will be more fully apparent in the fourth chapter below. Hence equal intensions are designated by equal lines, a double intension by a double line, and so continually in a proportional manner as we proceed. And this is to be understood universally with respect to every intension which is divisible in the imagination, whether it is an intension of an active or of a nonactive quality, or of a sensible or an insensible subject or object, or for a medium....

A "line of intension" of this sort concerning which we have just spoken is not extended outside of the point or the subject in actuality *(secundum rem)*, but only in imagination *(secundum ymaginationem)*. It could be imagined as going in any direction you wish, but it is more convenient to imagine it as standing perpendicularly on the subject which is informed with this quality.

I.2. On the latitude of quality. Every intension designated by the aforesaid line ought, therefore, to be called properly the "longitude" of that quality. [Why?] First, because in continuous alteration there is not essentially demanded a succession according to extension or according to the parts of the subject, since the whole subject can begin to be altered intensively at one time. But succession according to intension is required there. Hence just as in local motion this dimension would be called the "longitude" (i.e. length) of space or way according to which the succession is required, so in the same way the intension of this kind (that is, of a quality) according to which succession is required ought to be called the "longitude" of that quality. Also, just as the velocity in local motion is measured *(mensuratur)* according to the length of space, so in alteration velocity is measured *(attenditur)* by intension. Therefore such intension ought to be called "longitude." Also, no quality acquirable by alteration can be imagined without intension, or diversity according to intension; but it can be readily imagined without extension. Indeed the quality of an indivisible subject, such as a soul or an angel does not have extension.

Since then length (longitude) can be imagined without breadth (latitude) mathematically, but not conversely, and since intension should be referred to some dimension, as is obvious in the preceding chapter, then intension should be referred to longitude and not to latitude, and it ought more properly to be called by the name of "longitude." Whence it is obvious that the quality of an indivisible subject does not properly have latitude. But many theologians speak improperly of the "latitude" of charity, because if by "latitude" they understand "intension," then we would find latitude without longitude, and so the transposition of these [terms] appears to be incongruous. Nevertheless, I shall call intension of this kind "latitude" of quality, as I shall more fully explain in the chapter immediately following.

I.3. On the longitude of a quality. The extension of an extended quality of any kind ought to be called its "latitude," and the aforesaid extension can be designated by a line described in the subject upon which the line of intension of the same quality is erected perpendicularly. For since every such quality has intension and extension, and its measure is a function of both of them, hence, if its intension were called "longitude," then extension, which would be the second dimension, would be called "latitude." Conversely, if the intension is called "latitude," extension will be called "longitude." Just as the lines of longitude and latitude of a surface or of a body cut each other perpendicularly, so also the extension of a quality, which ought to be called its "latitude," is to be imagined by a line perpendicularly adjacent to the line of longitude of the same quality. And just as extension in "permanent" things ought to be called "latitude" of quality and intension "longitude," so in the same fashion the extension in time of "successive" things, such as motion and sound and the like, ought to be called "latitude" and the intension "longitude." It is true, however, that intension is more clear and palpable as it is commonly used and so is prior in our cognition to extension, and perhaps [it is prior even] with respect to nature. Hence, notwithstanding the aforementioned remarks, extension, according to the common way of speaking, is assigned to the first dimension, namely longitude, and intension to latitude. And since the difference of an assumption [of terms] of this kind or an impropriety of speech makes no difference in the actual thing *(ad rem)*, which in fact can be expressed in either way, I wish to follow the common method, so that those things which I say will not be less easily understood because of the unaccustomed locution.

Therefore, let extension of a quality be called nominally its "longitude" and intension be called its "latitude" or "altitude." However this may be, it is obvious from some assertions that certain moderns call (but not correctly) the whole quality itself "latitude"—just as there is an incorrect use of words when one understands by "latitude" (breadth) of surface the whole surface or figure. It is an abuse of terminology, because just as some latitudes of unequal surfaces or figures are equal, so similarly, as will be obvious from the following, many latitudes of unequal qualities are equal or conversely so.

I.4. On the quantity of qualities. The quantity of any linear quality at all is to be imagined by a surface whose longitude or base is a line protracted in some way in the subject, as the preceding chapter says, and whose latitude or altitude is designated by a line erected perpendicularly upon the protracted base line, as the first chapter posits. And I understand by "linear quality" the quality of some line in the subject "informed" with that quality. For that the quantity of such a quality can be imagined by a surface of this kind is obvious, because there can be found a surface relatable to that quantity, a surface equivalent in length or extension and whose altitude is similar to the intension, as will be obvious afterwards....

I.8. Concerning a right triangular quality. Every quality imaginable by a triangle having a right angle upon the base can be imagined by every triangle having a right angle upon the same base and by no other figure. For that such a quality is imaginable by such a triangle is obvious from the preceding chapter, because some quality can be proportional in intension to such a right triangle in altitude [i.e., some quality can be related to a triangle whose varying altitude represents the varying intension]. This quality is commonly called a "quality uniformly difform terminated at zero degree" *(non gradum)* [of intension]. More properly, however, it could be called a "quality uniformly unequal in intension," just as the triangle to which it is proportionally related is uniformly unequal in altitude. Similarly, it would be better to say that it is terminated at the privation [of the quality] rather than at zero degree. But since the other

[Oresme, *Configurations of Qualities*   351]

terminology is more widely used among the moderns and is adaptable enough, hence I accept it and follow it in this tractate....

I.10. On quadrangular quality. Some quality is imaginable by a rectangular quadrangle, in fact by any such rectangular quadrangle constructed on the same base, and it can be designated by no other figure. This last part is obvious by the statements in chapter 6. And thus let there be a rectangular quadrangle *ABCD* (see Fig. 6.3A). It is possible,

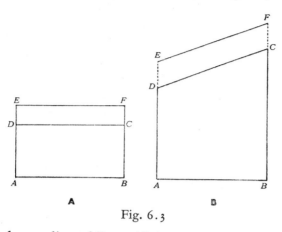

Fig. 6.3

therefore, that the quality of line *AB* is proportional in intension to this quadrangle in altitude. Therefore, it will be proportional to any rectangular quadrangle constructed on *AB*, because evidently all such [rectangles] are of proportional—even if of unequal—altitude. Therefore by the ninth chapter, this quality is imaginable by the [rectangular] quadrangle *ABCD* and similarly by the larger quadrangle *ABEF* or also [any] smaller [quadrangle]. Moreover, any such quality is said to be uniform or of equal intension in all of its parts. Again it ought to be known that some quality is imaginable by a quadrangle having two right angles on the base and its other angles unequal, as for example by quadrangle *ABCD* (see Fig. 6.3B). And [it is imaginable also] by every quandrangle of proportional altitude constructed upon base *AB*, whether it is larger or smaller, as is obvious in chapter 9. Moreover such a quality is called "uniformly difform terminated on both sides at some degree," so that the more intense extreme is designated at acute angle *C* and the more remiss extreme at obtuse angle *D*. The superior line, *DC*, is called the "line of the summit" (*linea summitatis*) or in relation to quality it can be called the "line of intension" because the intension [of the quality] is varied according to its variation.

I.11. On uniform and difform quality. Thus every uniform quality is imagined by a [rectangular] quadrangle, and every quality uniformly difform terminated at zero degree is imaginable by a right triangle. Every quality uniformly difform terminated at both ends at some degree is to be imagined by a quadrangle having right angles on the base and the other angles unequal. Moreover, every other linear quality is called "difformly difform" and is to be imagined by figures disposed in other and considerably varying ways. These other ways will be examined later.

While the aforementioned differences of intensions cannot be known better, nor more clearly, nor more easily, than by such representations and relations to figures, still there can be given certain other descriptions or notations, which are also known through figurative representations of this kind. For example, if it were said that a uniform quality is one which is equally intense in all of its subject parts, and a quality uniformly difform is one in which, when any three points [of extension] are taken, that proportion of the distance between the first and the second to the distance between the second and third is the same as the proportion of the excess in intension of the first over the second to the excess of the second over the third, with the proviso that I call the first of these three points the most intense point. This statement is clarified as regards a quality uniformly difform which is terminated at zero degree and which we let be signified or imagined by triangle ABC (see Fig. 6.2B). After three perpendicular lines BC, FG, and DE have been erected, HE parallel to line DF and GK parallel to FB are protracted. Hence two triangles are formed, namely GHE and CGK, which triangles are of equal angles. Hence by the fourth [proposition] of the sixth [book of the *Elements* of Euclid], the proportion of GK to EH is as the proportion of the excess CK to the excess GH. And since GK is equal to FB and similarly EH equals DF, the proportion of FB to DF, which lines indeed are the distances between the three points of the base, will be as the proportion of CK to GH, which lines are the excesses of altitude proportional to the intension at these same points. Since, therefore, the quality of line AB is such that the proportion in intension of the points on the line is as the proportion of the lines perpendicularly erected in altitude on these same points, the proposed statement is clearly obvious, namely that the proportion of the excess in intension of the first point beyond the second to the excess of the second point beyond the third is the same as the proportion of the distance between the first point and the second to the distance between the second and the third. The same thing holds for any other three points.... By the same method the aforementioned description or property can be demonstrated with respect to a quality uniformly difform terminated on both sides at [some] degree. And let there be one [such quality] which is imagined by quadrangle *ABCD* (see Fig. 6.4). Let line *DE*, parallel to the base *AB*, be constructed to form

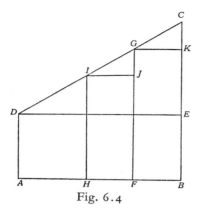

Fig. 6.4

triangle *DEC*. Then lines of altitude [*HI* and *FG*] are protracted in the quadrangle and [also] transversal distances [*IJ* and *GK*] parallel to the transversal distances in the base [*HF* and *FB*]. Thus small triangles [*IJG* and *GKC*] are formed in the large triangle [*DEC*]. Then it can easily be argued concerning those excesses and distances in this triangle as it was argued in the seventh (? i.e., in the demonstration above). This will be readily apparent to the attentive person. Moreover, every quality which is different from the above mentioned qualities we let be called "difformly difform," and can be described in a negative manner, namely as a quality which is not equally intense in all of the parts of its subject nor a quality where the proportion of the excess [of the intension] of the first [point] over that of the second to the excess of [the intension of the] second point over the third is as of the proportion of their distances....

I.13. On the same [qualities, considered] in another way. Furthermore, the aforesaid qualities can be distinguished in another way by calling the superior line of the figure by which we let the quality be imagined the "line of intension" or the "summit line," just as was said in the tenth chapter. For example, let there be [base] line *AB* in quadrangle *ABCD* (see Fig. 6.2). Hence, if the summit line of a figure of this kind by which a quality is imagined is parallel to the base, as for example to the base *AB*, then the quality imaginable by such a figure is simply uniform. If, however, it is not parallel to the base and yet is a straight line, then the quality is uniformly difform, so that if the aforementioned line joins the base in one extreme, that uniform difformity is terminated at zero degree. And if it is joined to the base at neither extreme, the quality or difformity is terminated at both ends at [some] degree. Since such a line cannot be joined to the base in either extreme—for it is a straight line, and the base line is a straight line, and thus the [two lines] would be a single line—hence it is obvious that it cannot be a quality uniformly difform terminated at zero degree. But if the line of intension or summit line is a curve or composed of more than one line, then the quality imaginable by that figure will be difformly difform. It can be terminated at some degree in both extremes or at zero degree in both extremes or at some degree in one extreme and zero degree in the other.

II.1. Part II. On the difformity of successive [entities]. Chapter 1. On the difformity of motion. Every successive motion of a divisible subject has parts. It is divisible in one way according to the extension and division or continuity of the moving body. It is divisible in another way according to the divisibility of time and its duration or continuity. In a third way it is divisible, at least in imagination, according to the degree *(gradus)* and intension *(intensio)* of velocity. Speaking in the first way, motion is called "great" *(magnus)* or "small" *(parvus)*. In the second way, "short" *(brevis)* or "long" *(longus)*. In the third way, [it is called] "swift" *(velox)* or "slow" *(tardus)*. Thus movement has two kinds of extension—one an extension in the subject *(subjectiva)* and the other an extension in time *(temporalis)*. It has [only] one intension.

The two extensions, moreover, can be imagined in a certain way as orthogonally and mutually intersecting like a cross, so that the extension of duration would be called "longitude" *(longitudo)* and the subjective extension would be called "latitude" *(latitudo)*. The intension then can be called the altitude *(altitudo)* of the motion or velocity. But if we follow the premises of the third chapter of Part I and [for the sake of convention] call the intension of velocity "latitude," then either of the extensions in comparison to the intension can be called the "longitude." Thus velocity will have longitude of two kinds just as it has extension of two kinds. The intension of velocity can be multiply varied throughout either of these extensions. And since difformity arises from the fact that intension is variously extended, hence it follows that movement or velocity can have two kinds of difformity, or even two kinds of uniformity. The one is according to the part, or [subjective] extension of the moving body—which is properly called "uniformity" or "difformity." The other is according to the parts or duration of time. This is properly called "regularity" or "irregularity." Hence movement has difformity or uniformity by reason of the subject and has regularity or irregularity by reason of, or according to, time. According to this way of speaking, the movement of the heavens is called "difform" and "regular." Conversely, the movement downward of a heavy body can be "uniform," and "irregular" or "regular." ... But it is not possible for circular (i.e., rotary) movement to be "uniform" [in the sense that all parts of the body have the same velocity].

Nevertheless, in following the customary way of speaking I shall call any regularity by the name of uniformity and any irregularity by the name of difformity....

II.3. On the quantity of the intension of velocity. Since each uniformity of motion, according to the first chapter, consists in the equality of intension, and each difformity arises from inequality [of intension], one ought to premise the measure of the quantity of the intension of velocity in degrees. However, with regard to velocity, three closely and mutually related things are to be considered. One is the quantity of the total velocity itself, dependent on both the intension and the extension. This will be spoken of in the third part of this treatise, which will concern the measure of qualities and velocities. A second thing can be considered in this matter, namely, the denomination by which a subject is said to come into such a condition *(tale)* more quickly or more slowly. This will be treated in the next chapter. The third thing to be considered is the gradual intension itself, which leads to the subject of this chapter and of which we now speak. Hence I say universally that a degree of velocity is simply more intense or greater [than another] when in an equal period of time more of the perfection with respect to which the motion takes place is acquired or lost. For example, in local motion, that degree of velocity is more intense or greater by which more space or distance is traversed....

II.4. On different kinds of velocity. It should not be overlooked that the same movement or flux is called by many terms connoting different things. Hence velocity denominated in different ways is attended or measured accordingly as its quantity of gradual intension is assigned in different ways.... For example, in the first way, in circular motion a body is said "to be moved" *(moveri)*, and it is [also] said "to revolve" *(circuire)*. Now the intension of a velocity of *motion* (i.e., rectilinear or curvilinear motion) is measured by the linear space which will be traversed at that degree [of speed]. But the intension of a degree of a rotary velocity *(velocitas circuitonis)* is measured by the angles described about the center. Hence it happens that one body moved circularly in comparison to another is *moved* more quickly but *revolves* less rapidly. Thus perhaps Mars is *moved* more rapidly in its proper motion than the sun because of the magnitude of the circle described, and yet the sun makes a quicker circuit [in terms of angular velocity] and *revolves* more swiftly around the center.... Astronomers *(astrologi)* actually pay more attention to the velocity of revolution than to the velocity of motion.

Also in rectilinear motion—for example in motion of descent—the *velocity of motion* is measured by the space traversed. However, the *velocity of descent* is measured by proximity to the center [of the earth]. Hence it is possible that A and B are *moved* equally fast, and both downward, and yet they do not *descend* equally fast. So that A will be moved over a straight line toward the center and B over a transverse line. Therefore A descends more quickly than B, and yet B is moved equally as fast....

However, universally in all these cases, that degree of velocity is more intense or greater by which in equal time the subject becomes somehow or so much greater according to that denomination by which the velocity is said to be acquired, whatever that may be. For example, a degree of velocity of descent is greater where the subject moving body descends more or would descend more if the descent were continued....

II.5. .... Furthermore, another succession can be imagined. For every velocity is intensible or remissible. Continuous increase [of velocity] is called acceleration *(velocitatio)*. And indeed this acceleration or augmentation of velocity can take place either more quickly or more slowly. Thus sometimes it happens that velocity is increasing while acceleration is decreasing; sometimes both are increasing at the same time. And similarly acceleration of this sort can sometimes take place uniformly or difformly in various ways....

II.8. On the difformity of velocities of motion with respect to time. Every velocity endures in time. And so the time or duration of that velocity will be the longitude. The intension of the velocity will be its latitude. And although time and a linear magnitude *(linea)* are not [strictly] comparable in quantity, yet there is no ratio found between one time [interval] and another that is not also found between lines, and vice versa.... The same thing holds for intension of velocity, namely, that every ratio which is found between one velocity intension and another is also found between one line and another.... Therefore, we can arrive at a knowledge of the difformities of velocities by the imagination of lines and also figures. Let us designate, for example, a line, which is AB; and let body D be moved through time EF in whatever way you wish.

[Oresme, *Configurations of Qualities*   357]

Either it will be moved in a circular movement or a rectilinear movement, or [it will be moved] in alteration; and it will be moved over line AB or some other line. And let perpendicular lines be erected upon the whole of line AB to form a surface. Let this surface or figure have an altitude proportional to the intension in velocity of D. I say, therefore, that the velocity of the moving body D can be assimilated to that surface or figure, and [hence] can be suitably imagined *(ymaginari)* by means of it, so that line AB, which is the longitude of a figure of this sort, will designate the duration of this velocity and the altitude of this figure will designate the intension of this velocity. For example, if in all instants of time EF the velocity is equally intense, then on any point of line AB the altitude will be the same and the figure will be uniformly high. [It will] evidently be a rectangle designating this velocity, which is simply uniform. But if in the first instant of the time, the body has some velocity, and in the middle instant of the whole time it has a velocity one-half [of the first velocity], and in the middle instant of the last half [of the time] it has a velocity one-fourth [of the first velocity], and so on proportionally in other instants, so that in the last instant it has no velocity, then there will be [erected] on AB proportional lines of altitude in the aforementioned way, and the figure will be a right triangle designating that velocity. Indeed that velocity was one uniformly difform terminated at zero velocity in the last instant. In short, I say, every uniformity and every difformity of velocities can be described and made known by those same techniques, techniques which were described in Part I of this [treatise] . . . .

III.1. Here begins the third part, on the acquisition and measure of quality and velocity. Chapter 1. How the acquisition of a quality is to be imagined. Succession in the acquisition of a quality can take place in two ways, according to extension and according to intension, as was stated above in the fourth chapter of the second part. And so extensive acquisition of linear quality is to be imagined by the motion of a point flowing over that subject line, so that the part [of the line] traversed has the quality *(sit qualificata)* and the part not yet traversed has not the quality. Example: if point C were moved over line AB, whatever part was traversed by that point would be white and whatever was not yet traversed would not yet be white. Moreover, the extensive acquisition of a surface quality would have to be imagined by the motion of a line dividing the part of the surface altered from the other part which has not yet been altered. In the same manner, the extensive acquisition of corporeal [quality] is to be imagined by the motion of a surface dividing the part altered from the part not yet altered.

Furthermore, the intensive acquisition of a punctual quality is to be imagined by the motion of a point continually ascending above the subject point, while the intensive acquisition of a linear quality is to be imagined by the motion of a line perpendicularly ascending above the subject line; and by its flux or ascent it describes the surface by which the acquired quality is designated. For example: Let AB be the subject line (see Fig. 6.5 A). Hence I say that the intension of point A is imagined by the motion

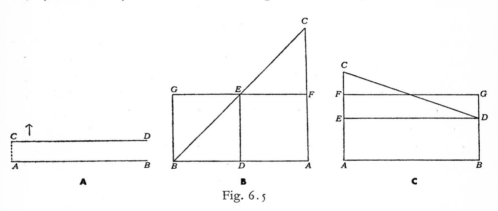

Fig. 6.5

or perpendicular ascent of point C. And the intension of line AB or the acquisition of intension is imagined by the ascent of line CD. The intensive acquisition of a surface quality in the same way is to be imagined by the ascent of a surface, and by its imagined motion it describes the body by which the quality is designated. And similarly the intensive acquisition of a corporeal quality is imagined by the motion of a surface quality, because by its imagined flux a surface describes a body. And it is not necessary to give a fourth dimension, as was said in the fourth chapter of the first part.

And what we have just now said about the acquisition of quality ought to be said *and imagined* in the same way about the loss [of quality]... for such loss is imagined by movements opposite to the aforementioned motions [of acquisition]. What has been just now said about the acquisition or loss of quality is to be imagined as applying in the same way to the acquisition or loss of velocity both in intension and in extension....

III.7. On the measure of difform qualities and velocities. Every uniformly difform quality [in a subject] is just as great as would be a quality in the same or equal subject uniform at the degree [of intensity] of the middle point of the same subject; and I understand this [to be so] if the quality is linear. If it is a surface quality, [it would be equal to a quality uniform] at the degree of the mean line; if corporeal, [to one uniform] at the degree of the mean surface, all of them being understood in the same way.

In the first place this is demonstrated for a linear [quality]. Let there be quality imaginable by a triangle ABC, which is uniformly difform, and is terminated at zero degree in point B (see Fig. 6.5B); and let D be the middle point of the subject line. The degree of this midpoint, or its intension, is imagined by the line DE. Hence the quality which is uniform at degree DE throughout the whole subject is imaginable by a quadrangle $AFGB$, as is clear from the tenth chapter of Part I. And it is evident by the twenty-sixth [proposition] of the first [book] of Euclid, that the two small triangles EFC and EGB are equal. Therefore, the larger triangle BAC, which designates the quality uniformly difform, and the quadrangle $AFGD$, which would designate the quality uniform at the degree of the middle point, are equal. Hence the qualities imaginable by a triangle of this kind and a quadrangle are equal; and this was proposed.

In the same way it can be argued with respect to a uniformly difform quality terminated in both extremes at some degree. This quality would be imaginable by a quadrangle $ABCD$ (see Fig. 6.5C); for let there be drawn a line DE parallel to the subject base [line] and let triangle CED be formed. Then there is protracted through the degree of the middle point line FG, equal and parallel to the subject base [line]. Also another line GD is drawn. Then, as before, it will be proved that triangle CED is equal to quandrangle $EFGD$. Therefore, with quadrangle $AEDB$ common to both, the two total [areas] are equal, namely, the quadrangle $ACDB$, which designates the uniformly difform quality, and the quadrangle $AFGB$, which designates the quality uniform at the degree of the middle point of the subject AB. Therefore, by the tenth chapter of the first part, the qualities representable by these quadrangles are equal.

It can be argued in the same way with respect to surface and even corporeal [quality]. We ought to speak of velocity completely in the same way as linear quality, except that in the place of the middle point [of the subject] the middle instant of time is taken as the measure of this [uniformly difform] velocity....

III.8. On the measure and intension to infinity of certain difformities. A finite surface can be made as long and as high as you wish without an [over-all] increase in the area by varying the extension. For such a surface has both longitude and latitude, and so it is possible for one dimension of the surface to be increased at will without increasing the over-all area, so long as the second dimension is decreased proportionally. And it is thus for bodies as well.

Let there be taken, for example (see Fig. 6.6), a rectangular surface

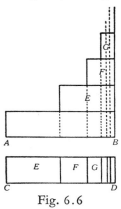

Fig. 6.6

whose [altitude] is one foot and whose base line is AB, and let there be another surface similar and equal to it whose base is CD. This [latter figure] is pictured as being divided into parts continually proportional to infinity, according to a double proportion, on its base CD, divided in the same way. Let E be the first part, F the second, G the third, and so on for the others. Take the first of these parts, namely E, which is one-half of the whole [subject], and place it on top of the first surface toward the end B. Then on top of both of them put the second part F. Then again on top of all of them put the third part G, and similarly with the other parts to infinity. When this has been done, the base line AB is imagined as being divided into proportional parts continually according to a double proportion and working toward B. Then it is immediately clear that above the first proportional part of line AB stands a surface with altitude of one foot, upon the second part a surface altitude of two feet, upon the third a surface with altitude of three feet, [upon the fourth part a surface with altitude of four feet], and so forth to infinity. And yet the whole surface as originally conceived with only an altitude of two feet is in no way augmented overall [by this proportional division]. Consequently, the total surface which stands over line AB is precisely four times the area of the surface of the part of it which stands over the first proportional part of the same line AB. Hence, that quality or velocity which will be proportional in intension as this figure is in altitude would be precisely quadruple to the part of it which would be in the first part of the time or subject according to a dimension of this kind....

Similarly, if any moving body were moved with some velocity in the first proportional part of a period of time so divided [into proportional parts] and were moved with a double velocity in the second [proportional part], with a triple velocity in the third [proportional part], and with a quadruple velocity in the fourth [proportional part], and continually increasing velocity in this way to infinity, the "total velocity" would be precisely quadruple to the velocity of the first part, so that a body in the whole hour would traverse four times as much distance as it traversed in the first half of that hour. And if in the first half or proportional part it traversed one foot, in the whole remaining part it would traverse three feet, and in the whole time it would traverse four feet.

COMMENTARY

I.O. In the body of Chapter 6 (see notes 11 and 12) I have discussed the date and title of this treatise. One certainly infers from the introduction that Oresme utilized the ideas and works of his predecessors—in all probability the Merton scholars but also possibly Giovanni di Casali. Of course, we cannot infer that he specifically drew his idea of a geometry of qualities from these predecessors. The question of the invention of the system remains unanswered, as I have already suggested in the body of the chapter.

I.1. Here Oresme sets out the fundamental idea of his system. Since natural things are continuous, we can appropriately apply the continuous quantities of geometry to represent the relationships of extension and intension that exist between such natural things. He well knows that other natural philosophers have already applied geometrical quantities to make not dissimilar representations of natural entities.

As we have noted in the chapter above, in his *Questions on Euclid's Elements* he gives as justification for representing intensions by geometry such previous representations of natural phenomena as that of time by Aristotle, optical phenomena by the opticians, etc.[14] The clinching argu-

[14] MS Vat. Chis. F.IV.66, f. 27v (MS *C*); cf. Vat. lat. 2225, f. 94r. (MS *V*). The following passage is generally from *C*: "Oppositum patet in precedenti questione. Respondeo quod questio (*V*; quo *C*) est vera et posset confirmari per perspectivos, qui ita ymaginantur de intensione luminis sicud Vitelo (Utilo *V*; Ultio *C*) et Lyncolniensis, et per Aristotelem quarto physicorum qui ymaginatur tempus ad modum linee, et per commentatorem quinto huius (*i.e.*, elementorum) ubi dicit quod omne habens naturam continui potest ymaginari sicud linea aut superficies aut corpus sicud declarat de proportionibus." Aristotle clearly treats times by the use of lines in numerous places in the fourth, the sixth, and the seventh books of the *Physics*. The

ment here in chapter I.1 in support of the suitability of using geometrical quantities to represent intensity relationships is that "whatever kind of a proportion is found between one intension and another—among intensions of the same kind—a similar proportion is found between some one line and another, and conversely. For as one line is commensurable to another line and incommensurable to [still] another, so it is the same with respect to intensions...." He cautions at the end of these preliminary remarks that the line of intension which we erect perpendicularly to represent the magnitude of intension at a particular point in a body "is not extended outside of the point in actuality, but only in imagination." In short, there is nothing "real" about his system. It is purely figurative.

I.2 and I.3. In these chapters Oresme makes the point at some length that it would be preferable to call intension by the name of "longitude" and extension by that of "latitude" in discussing the quantity of qualities. But conventionally the terms are used in the converse manner—i.e., "latitude" for "intension," and "longitude" for "extension." For this reason, and since it makes no difference what terms we use so long as we employ the basic idea of the system itself, Oresme will himself adopt the conventional terms in this treatise.

It should also be noted, Oresme tells us, "that certain moderns call (but not correctly) the whole quality itself 'latitude'—just as there is an incorrect use of words when one understands by 'latitude' of surface the whole surface or figure." In short, some call the whole quantity of quality "latitude," while it is more proper to apply it to one dimension alone (conventionally, intension). A similar difference in terminology had been noted by John Dumbleton in his *Summa de logicis et naturalibus*.[15] As John points out, it is only in the improper sense that we can speak of latitude being "equivalent to some properly intrinsic degree." Thus when the authors we quoted both in Chapters 5 and 6 speak of a "latitude uniformly difform corresponding to the mean degree," it is obvious that it is

remarks of Campanus appear in his comment on the third definition of Book V (see Euclid's *Elements*, edition of Basel, 1546, pp. 103–4).

[15] John Dumbleton, *Summa de logicis et naturalibus*, Part II, chap. 35, MS Cambridge, Peterhouse 272, f. 20r, col. 1: "Post hec tertiam opinionem, que est vera, exprimamus. Pro qua monstranda dupliciter latitudinem esse in qualitate difformi intel[l]igimus iuxta usum, scilicet, proprie et improprie. [1] Proprie secundum quod intel[l]igimus eam tantum intensive continere non referendo eam ad aliquam quantitatem sive extensionem in subiecto, sed ut est tanta distantia qualitativa inter gradus, penes quam attenditur motus alterationis, sicut linea est pedalis quia extrema tantum distant, et sic tota illa latitudo considerata est gradus summus sue speciei. [2] Improprie dicitur latitudo qualitatis cuius partes qualitative in sub-

latitude in the sense of the whole uniformly difform quality or movement (i.e., the whole area of the figure used to represent the quality or movement).

I.4. Here Oresme tells us what is meant by quantity of quality as far as his system is concerned, namely, the whole figure, i.e., the summation of all the intension lines erected in all the points of the extension line.

I.8. In this chapter Oresme tells us that a uniformly difform quality terminated at one end at zero degree is represented by a right triangle. In fact, it is represented by any right triangle erected on a given base (a point he proves later in a long section we have not translated). He believes it would be better to call such a quality, a "quality uniformly unequal in intension," and one would better say instead of "terminated at zero degree," "terminated at the privation [of the quality]." But again he will follow the conventional terminology.

I.10. Furthermore, a uniformly difform quality terminated at each end at some degree is to be represented in Oresme's system by any quadrangular figure erected on a given base, such that on the base there are two right angles and of the other two angles one is obtuse and the other acute (see Fig. 6.3B). Notice, as we have indicated in the body of the chapter, that line DC is called "the line of summit." We have said that it is similar in modern coördinate geometry to the "curve."

I.11. Any quality that is not uniform (represented by a rectangle), not uniformly difform terminated at zero degree (represented by a right triangle), and not uniformly difform terminated at both ends at some degree (represented by the quadrangle described in I.10), is said to be a quality "difformly difform." The figures of such difformly difform qualities vary in manifold ways. Oresme goes on to say that we can define in other ways the three basic qualities (uniform, uniformly difform terminated at zero degree at one end, uniformly difform terminated at both ends at some degree). The most interesting description here is the one in which he defines uniformly difform quality by saying in substance that for equal increments of extension we find (in a uniformly difform quality) equal increments of intension. This is equivalent in modern parlance to saying that, in the representation of a uniformly difform quality, the slope of the summit line is a constant. He proves that such a description fits both kinds

---

iecto sunt diverse intensionis et inequalis, penes quem motum solum qualitatem difformem ponentes intensam, dicunt eam ex diversa coextensione intensionem sibi adquireri vel alicui gradui proprie intrinseco equevaleri." Cf. MS Paris, BN lat. 16146, f. 23r, c. 2 (here given as chap. 33). I have altered spellings of *intencionis*, *extencionem*, *coextencione*, and *intencionem* which appear in MS.

of uniformly difform qualities. One can say further that such cannot be the case for any quality difformly difform.

I.13. Still another way of delineating the various kinds of qualities is to describe their summit lines. A uniform quality is characterized by a summit line parallel to the base, a uniformly difform quality terminated at one end at zero degree by a summit line that is a straight line intersecting the base in an acute angle, the uniformly difform quality terminated at both ends at some degree by a straight line not parallel to the base and not intersecting the base line, a difformly difform quality by any other kind of curved line or line composed of straight and/or curved lines. It is in this section that the author makes a first step towards describing functions in terms of "curves."

II.1. The most interesting part of this chapter is that in which Oresme's coördinate system is extended to three dimensions. It will be noticed that he bows to convention and calls an increment of velocity by the term "latitude." As the result, his other two dimensions become "longitudes." Although he posits this three-dimensional system, he does not discuss it in any detail. Most of his work is concerned with the application of a two-dimensional system. His remark that velocity can vary through two extensions, namely, through the parts of the body and through some duration of time, is a repetition of the distinction customarily made by the English (see Doc. 4.6). Just as Oresme yielded to convention in the case of the terms "latitude" and "longitude," so he follows customary usage in talking of "uniformity" and "difformity," applying them indifferently to constancy or variation in velocity from one part of a moving body to another, or from one moment to another. But he adopts the conventional usage only after saying that the terms "uniformity" and "difformity" properly should be restricted to describing the uniformity or variation in velocity from one part to another; while for uniformity or variation of velocity from one moment to another, he would prefer the terms "regularity" and "irregularity."

II.3. This chapter indicates that the measure of a specific grade or degree of velocity must be the amount of space traversed in some period of time. We would compare one degree with another by comparing the spaces traversed in equal periods of time. This, of course, is completely in the English tradition. Notice further that "total" velocity is distinguished from "gradual" velocity. Total velocity is velocity through the whole time and thus is dimensionally equivalent to distance. It is actually what the whole geometric figure represents.

[Oresme, *Configurations of Qualities* 365]

II.4. This chapter brings out clearly the distinction between angular velocity and curvilinear velocity. Albert of Saxony, Oresme's contemporary at Paris, in Document 4.2 makes the same distinction between *movere* (to move with curvilinear or rectilinear speed) and *circuire* (to rotate or revolve with angular speed).

In the last paragraph of this chapter, Oresme appears at first glance to be maintaining that the velocity of descent increases as the time. Actually, all Oresme is saying here is that a velocity of descent would be greater if more space were traversed. Elsewhere in his Latin *De caelo* Oresme does hold that velocity of descent is directly proportional to the time of descent (see Chapter 9 below).

II.5. In the brief section which I have drawn from this chapter, Oresme notes that "intending" or increasing velocity is called *velocitatio*, i.e., acceleration. It is possible, he says, for velocity to be increasing while acceleration is decreasing, or for both velocity and acceleration to be increasing. Furthermore, we are told that acceleration can be uniform or difform. However, Oresme does not further categorize difform acceleration as uniformly difform and difformly difform, as the author of the *De latitudinibus formarum* does in distinguishing motion uniformly difformly difform and motion difformly difformly difform. One would gather, however, by the parallel that Oresme draws between velocity and acceleration, that he would consider uniformly difform acceleration as one where equal increments of acceleration are acquired in equal periods of time, rather than in the peculiar way that the author of the *De latitudinibus* considers motion uniformly difformly difform (see Doc. 6.3, passage 9).

II.8. Here Oresme distinguishes between uniform and uniformly difform movements in the manner we have already described in the body of the chapter. The figures representing these movements are clearly being considered as the summation of all the latitude or velocity lines erected on the base-longitude line, and each perpendicular represents an instantaneous velocity. The reader should compare the remarks of Galileo to the same effect in Document 6.5 below.

III.1. In this chapter Oresme has summarized briefly his theory of how the acquisition of qualities can be imagined by the flow of geometric magnitudes. He here makes an interesting distinction developed elsewhere between punctual, linear, surface, and corporeal qualities. Punctual quality is the intensity of a quality at a given point; a linear quality is the quality extended over a given line, with its intensity being measured at every point along the line. Surface quality is quality spread over a surface, with its intensity being measured at every point over the surface. Finally, corporeal quality is quality spread throughout a whole body. Now Oresme is saying here that we can imagine the *acquisition* of the extension of a linear quality by the flow of a point over the line being infused with the quality. Similarly the acquisition of intension of quality can be imagined by a point ascending perpendicularly from some point on the line being infused with the quality. We can in the same way imagine the acquisition of the extension and intension of a surface quality—the extension by a line flowing over the surface area, the intension by a line ascending perpendicularly from the surface. Finally, the extension and intension of a corporeal quality are representable by the flow of a surface through the body. He makes a brief allusion to the fact that it is not necessary to employ a fourth dimension to represent corporeal qualities. This arises from the consideration that a punctual quality is representable by a line, a linear quality by a surface, a surface quality by a body. Hence should not a corporeal quality be represented by a four-dimensional figure? But elsewhere he says that such is not necessary, for if we use the three dimensions of the body to represent the extension of the quality, we can imagine a whole series of perpendiculars raised at every point in the body. The final figure would be an interlacing of the original three dimensional figure with a new three-dimensional figure whose shape is dependent on the variations in intensity throughout the original body.

III.7. We have already discussed this geometric proof of the mean speed theorem in the introductory material. Its resemblance to the proofs presented in Documents 6.2, 6.4–6 should be emphasized—all involve the equation of a rectangle and a right triangle.

III.8. This chapter is presented by Oresme as a kind of paradox, the paradox being that it is possible to imagine the intensity of a quality (or a velocity) increasing toward infinity while the whole quantity of the quality (or total distance traversed) remains finite. Essentially he has presented us with the summation of the following convergent series:

$$1 + \tfrac{1}{2} \cdot 2 + \tfrac{1}{4} \cdot 3 + \ldots + \frac{n}{2^{n-1}} + \ldots = 4.$$

This same convergent series, as well as others, had already been discussed by the Merton College group. Particularly noteworthy are the discussions by Richard Swineshead in his *Liber calculationum* (e.g. see his chapter *De difformibus*, in the Pavia edition of 1498, with no pagination, but pp. 14–15 in the text). Oresme's discussion is somewhat different from those of the English kinematicists. He works backwards, starting with the sum as a geometric area and then breaking up that sum into proportional parts. Thus he incidentally gives a proof of the summation of the series we have already noted. It is of interest to note finally that for motion, the surface (or area under the curve) represents distance or "total velocity," its dimensional equivalent.

*De configurationibus qualitatum Nicolai Oresme**
(sive *De uniformitate et difformitate intensionum*)

I.0. Cum ymaginationem veterum de uniformitate et difformitate intentionum ordinare cepissem, occurrerunt mihi quedam alia que huic
5   proposito interieci, ut iste tractatus non solum exercitationi prodesset sed etiam discipline. In quo ea que aliqui alii videntur circa hoc confuse sentire et obscure eloqui ac inconvenienter aptare, studui dearticulatim et clare tradere et quibusdam aliis materiis utiliter applicare. Tractatus de figuratione potentiarum et mensura difformitatum
10 (159v) habet tres partes: Prima pars de figuratione et potentia uniformitatis et difformitatis qualitatum permanencium. Secunda pars

---

* These selections were prepared by consulting the following manuscripts: London, Brit. Mus., Sloane 2156, ff. 159r–193v (*S*); Paris, BN lat. 7371, ff. 241r–266r (*E*; a poor, partial edition was made from this copy by H. Wieleitner, *Bibliotheca Mathematica*, 3rd Series, Vol. *14* [1913–14], 193–243); Paris, BN lat. 14580, ff. 37r–60v (*P*); Florence, Bibl. Laurent. Ashbur. 210 (cat. no. 136), ff. 101v–129v (*F*). In general *S* was followed, but corrected fairly extensively by the use of *EPF* in the first chapter and on occasion by the use of *EP* in all the succeeding chapters. I hope to publish in the future a fully critical text, which will take into account, in addition to the manuscripts already noted, the following manuscripts: Paris, BN lat. 14579, ff. 18r–40v; Basel, Univ. Bibl. F.III.31, ff. 1r–28r; Bruges, Stadsbibliotheek 486, ff. 159r–173r; Erfurt, Stadtbibliothek, Amplon. Q. 150, ff. 1r–14r (incomplete); Florence, Bibl. Naz. Conv. sopp. J. IX. 26, ff. 13r–35r; Paris, Bibl. de l'Arsenal 522, ff. 1r–29r; Vat. lat. 3097, ff. 1r–22v; Vat. Chigi lat. E.IV.109, ff. 87r–149r. I have films of all of these manuscripts. As usual, in these selections here presented I have freely capitalized the letters used for qualities and movements where they were lower-case letters in the manuscripts.

Part I

3 veterum *ES* veterum vel meam *P* meam *F*
4 cepissem *ESF* incepissem *P*
5 interieci *PF lac. S* intento sunt consona *E* / exercitationi *EF* ignorancie exercitanti *P* cognitationi *S*
6 ea *ESF* illa *P*
7 confuse *om. P* / sentire *om. S*
8 tradere: ostendere *F*
9–10 Tractatus... partes *S* Huius (huius autem *F*) tractatus tres sunt partes principales *PF* Tractatus de figuratione potentiarum et mensura difformitatum *E*
11 qualitatum permanencium *F* (*P*?) *om. ES* / pars *om. F*

de figuratione et potentia successivorum. Tertia pars de acquisitione et mensura qualitatis et velocitatis.... [Then follows a complete list of the chapters in most manuscripts.]

I.1. Pars Prima (160r) [Capitulum primum]. Omnis res mensurabilis exceptis numeris ymaginatur ad modum quantitatis continue. Ideo oportet pro eius mensuratione ymaginari puncta, lineas, et superficies aut istorum proprietates, in quibus, ut vult Aristoteles, mensura seu proportio per prius reperitur. In aliis autem cognoscitur in similitudine qua per intellectum referuntur ad ista; et si nichil sunt puncta indivisibilia, aut linee, tamen oportet ea mathematice fingere pro rerum mensuris et earum proportionibus cognoscendis. Omnis igitur intensio successive acquisibilis ymaginanda est per lineam rectam perpendiculariter erectam super aliquod punctum aut aliquot puncta spacii vel subiecti illius rei intensibilis. Verbi gratia, qualiscunque proportio reperitur inter intensionem et intensionem, de intensionibus que sunt eiusdem rationis, similis proportio invenitur inter lineam et lineam et econtra. Quemadmodum enim una linea alteri linee est commensurabilis et alteri incommensurabilis, ita est conformiter de intensionibus quod quedam sunt commensurabiles adinvicem et quedam incommensurabiles quomodolibet, propter continuitatem earumdem. Ergo mensura intensionum potest ymaginari congrue sicut linearum mensura, cum etiam intensio possit eodem modo sicut linea in infinitum diminui et quantum est ex se in infinitum augeri. Et rursum intensio secundum quam aliquid dicitur magis tale vel minus, ut magis album aut magis velox, ipsa quidem secundum quod in-

---

12 successivorum ES successivarum PF / pars *om.* F
13 qualitatum et velocitatum F
15 omnes res mensurabiles P
17 lineas et superficies: superficies et lineas E
18 Aristoteles PE philosophus SF
20 qua: dum F / per... ista: ad ista referuntur per intellectum P
21 mathematice *om.* E / fingere: sumere P
24 aliquod... puncta E aliquot punctum P aliquota puncta S aliquod punctum F
25 spacii... intensibilis PSF intensibilis spacii vel subiecti E / rei *om.* F / Verbi gratia P verbi gratia nam E ut

verbi gratia S ymaginatur verbi gratia ut qualitas nam F / qualiscunque: quecunque F
26 et intensionem *om.* F
27 que: quod quedam F / inter lineam *om.* F
28 econtra: econverso F / enim *om.* E / linee *om.* S
30 intensionibus: intensis P / et *om.* F
34 diminui: dividi et diminui F / quantum est: quantitas F / ex PSF de F
35 aliquod E / vel minus *om.* PF
35–36 vel... magis[1]: aut minus vel (ut?) E
36 album: album aut niger F

tensio vel extensio puncti est tantum uno modo divisibilis et in infinitum ad modum continui. Igitur non potest convenientius ymaginari quam per illam speciem continui que primo est divisibilis et uno modo tantum, scilicet, per lineam. Et quoniam linearum quantitas sive proportio notior est et facilius a nobis concipitur, ymmo linea est in prima specie continuorum, ideo per lineas ymaginanda est intensio talis—maxime vero et convenientissime per illas que subiecto applicate super ipsum perpendiculariter eriguntur, quarum consideratio ad cuiuslibet intensionis noticiam naturaliter iuvat et ducit, prout in IIII° capitulo sequenti plenius apparebit. Ideoque intensiones equales per equales lineas designantur et dupla intensio per duplam lineam et sic semper proportionaliter procedendo, et istud est universaliter intelligendum de omni intensione ad ymaginationem divisibili, sive sit intensio qualitatis active sive non active, sive sensibilis sive insensibilis subiecti aut obiecti aut medii, ut de luce corporis solis et de lumine medii, vel de specie in medio, vel influentia aut virtute diffusa, et sic de aliis, excepta forsitan intensione curvitatis, de qua dicetur ad partem in capitulis 20mo et 21mo huius partis. Huiusmodi vero linea intensionis de qua nunc dictum est non extenditur extra punctum vel extra subiectum secundum rem sed solum secundum ymaginationem et ad quamvis partem nisi quod convenientius ymaginatur in sursum perpendiculariter stare super subiectum qualitate formatum.

I.2. Capitulum secundum de latitudine qualitatis. Omnis igitur intensio per predictam lineam designata proprie vocari deberet longitudo illius qualitatis. Primo igitur quidem quia in alteratione continua essentialiter non exigitur successio secundum extensionem sive secundum partes subiecti, quia totum potest simul incipere alterari, sed ibi requiritur successio secundum intensionem. Igitur sicut in motu locali ita dimensio diceretur longitudo spacii seu vie secundum quam exigitur successio, ita conformiter huiusmodi enim intensio secundum quam requiritur successio deberet dici longitudo ipsius qualitatis. Item sicut velocitas in motu locali secundum longitudinem spatii mensuratur, ita in alteratione velocitas attenditur penes intensionem.

37 vel extensio puncti *om.* F
39 primo est *tr.* F / divisibilis: modo divisibilis F
42 ymaginanda est *tr.* P
44 ipsum: ipsam S / eriguntur quarum *tr.* S

47 per... intensio *om.* E
50 intensio *om.* F
53 diffusa: infusa F
54 ad partem... 20mo: postea in 2° P
56 solum *om.* E
58 in *om.* E

Ergo talis intensio deberet dici longitudo. Item (160v) nulla qualitas alteratione acquisibilis potest ymaginari sine intensione seu diversitate secundum intensionem. Sed bene potest ymaginari sine extensione, ymmo qualitas subiecti indivisibilis, ut anime vel angeli, non habet extensionem. Cum igitur ymaginatur mathematice longitudo sine latitudine et non econverso et cum intensio sit referenda ad aliquam dimensionem, ut patet in precedenti capitulo, ipsa referenda est ad longitudinem et non ad latitudinem et nomine longitudinis magis proprie appellanda. Unde patet quod qualitas subiecti indivisibilis proprie non habet latitudinem. Sed multi theologi loquuntur improprie de latitudine caritatis, quia si per latitudinem intelligunt intensionem, tunc continget invenire latitudinem sine longitudine, et sic eorum transsumptio videtur incongrua. Verumtamen huiusmodi intensionem vocabo latitudinem qualitatis sicut plenius dicam in capitulo immediate sequenti.

I.3. Capitulum tertium de longitudine qualitatis. Cuiuslibet qualitatis extense sua extensio deberet vocari sua latitudo et predicta extensio designari potest per lineam in subiecto descriptam super quam linea intensionis qualitatis eiusdem perpendiculariter erigitur. Nam cum omnis talis qualitas habeat intensionem et extensionem que in eius mensura sit attendenda, ideo si eius intensio diceretur longitudo, tunc extensio que esset secunda dimensio vocaretur latitudo. Et etiam econverso, si intensio dicatur latitudo, extensio vocabitur longitudo. Sicut igitur corporis vel superficiei linea longitudinis et linea latitudinis perpendiculariter sibi invicem dividunt, ita etiam extensio qualitatis que deberet dici ipsius latitudo ymaginanda per lineam perpendiculariter adiacentem linee longitudinis qualitatis eiusdem. Et quemadmodum in permanentibus extensio dici debet latitudo qualitatis et intensio longitudo, ita conformiter in successivis, cuiusmodi sunt motus et sonus et similia, extensio eorum in tempore vocaretur latitudo et intensio longitudo. Verum est tamen quod intensio est manifestior et palpabilior ut ita loquitur et prior cognitione quo ad nos quam sit extensio et forsan quo ad naturam. Ideo non obstantibus predictis ipsa extensio secundum communem usum loquendi attribuitur primo dimensioni, scilicet, longitudini, et intensio latitudini et quoniam differentia huiusmodi impositionis seu improprietas voca-

84 vocabo *ES* aliquociens vocabo *P*
94 et *EP om. S*
103 extensio *P* intensio *ES*

106 quoniam differentia *P tr. S* quia differentia *E*

[*De configurationibus qualitatum Nicolai Oresme* 371]

tionis nichil facit ad rem, sed utroque modo potest idem exprimi, volo sequi modum communem, nec propter locutionem inconsuetam illa que dicam maius leviter intelligantur. Et extensio igitur qualitatis in nomine dicti vocetur eius longitudo et intensio eius vocatur latitudo sive altitudo. Sed qualitercunque sit, patet ex dictis quod quidam moderni non bene vocant latitudinem qualitatis ipsam totam, sicut abusio est per latitudinem superficiei intelligere totam superficiem vel figuram, nam quodammodo alique latitudines superficierum sive figurarum inequalium sunt equales, ita similiter, sicut postea videbitur, multe latitudines qualitatum inequalium sunt equales aut etiam e converso.

I.4. Capitulum quartum de quantitate qualitatum. Cuiuslibet qualitatis quantitas linearis ymaginanda est per superficiem cuius longitudo seu basis est linea in subiecto quali protracta, ut dicit precedens capitulum, et cuius latitudo seu altitudo designatur per lineam super basim productam perpendiculariter erectam secundum quod ponit capitulum primum, et intelligo per qualitatem linearem qualitatem alicuius linee in subiecto informato qualitate. Quod enim quantitas qualitatis talis per huiusmodi superficiem possit ymaginari patet, quoniam continget dare superficiem illi quantitati equalem in longitudine seu extensione et similem in altitudine eidem quantitati intensione, ut patebit post [ea?] . . . .

I.8. (161v) Capitulum 8^m de qualitate triangulari rectangula. Omnis qualitas ymaginabilis per triangulum habentem rectum angulum super basim potest ymaginari per omnem triangulum habentem rectum angulum super eandem basim et per nullam aliam figuram (162r) potest ymaginari. Quod enim talis qualitas sit ymaginabilis per talem triangulum patet ex capitulo precedenti, eo quod aliqua potest esse proportionalis in intensione tali triangulo rectangulo in altitudine et illa est que vocatur communiter qualitas uniformiter difformis terminata ad non gradum, que tamen magis proprie posset dici qualitas uniformiter inequalis intensive, sicut triangulus cui ipsa proportionatur est altitudinis uniformiter inequalis. Similiter ipsa magis deberet terminari ad privationem quam ad non gradum. Sed quoniam alia locutio magis est apud modernos assueta et est satis transibilis, ideo in hoc tractatu eam recipio et admitto . . . .

I.10. (162v) Capitulum 10^m de qualitate quadrangulari. Quedam qualitas ymaginabilis est per quadrangulum et per quamlibet talem super eandem basim constitutum et per nullam aliam figuram designari potest. Hec ultima pars patet per dicta in capitulo VI⁰. Sit itaque quadrangulus rectangulus $ABCD$. Possibile est igitur quod qualitas linee AB sit proportionalis in intensione huic quadrangulo in altitudine. Igitur erit proportionalis cuilibet quadrangulo rectangulo super AB constituto, eo quod omnes scilicet tales sunt proportionalis altitudinis quamvis tamen inequalis. Igitur per capitulum 9^m ipsa qualitas est ymaginabilis per quadrangulum $ABCD$ et similiter per quadrangulum $ABEF$ maiorem sive etiam per minorem. Quelibet autem talis qualitas dicitur uniformis seu equalis intensionis in omnibus partibus eius. Rursum sciendum est quod aliqua qualitas est ymaginabilis per quadrangulum habentem duos rectos angulos super basim et alios inequales sicut per quadrangulum $ABCD$ et per omnem quadrangulum proportionalis altitudinis super basim AB constitutum, sive fuerit maior sive minor, ut patet in 9⁰ capitulo. Quelibet autem talis qualitas dicitur uniformiter difformis utrinque terminata per gradum, ita quod extremum intensius designatur in angulo C acuto et extremum remissius in angulo D obtuso. Superior vero linea sicut est CD dicitur linea summitatis vel in relatione ad qualitatem potest vocari linea intensionis, quia secundum varietatem ipsius variatur intensio.

I.11. Capitulum 11^m de qualitate uniformi et difformi. Omnis itaque qualitas uniformis ymaginatur per quadrangulum et omnis qualitas uniformiter difformis terminata ad non gradum ymaginabilis est per triangulum rectangulum. Omnis vero qualitas terminata utrinque ad gradum ymaginanda est per quadrangulum habentem rectos angulos super basim et alios inequales. Omnis autem alia qualitas linearis dicitur difformiter difformis et est ymaginanda per figuras aliter dispositas secundum multifariam variationem, cuius aliqui modi postea videbuntur. Predicte vero differentie intensionum non melius nec clarius neque facilius notificari possunt quam per tales ymaginationes et relationes ad figuras, quamvis quedam alie descriptiones seu notificationes possunt dari, que etiam per huiusmodi figurarum ymaginationes fuerint note, ut si diceretur qualitas uniformis est que in omnibus partibus subiecti est equaliter intensa. Qualitas vero uniformiter difformis est cuius omnium trium punctorum proportio distantie inter primum et secundum ad distantiam inter secundum et tertium est sicut proportio excessus primi supra

170 utrinque ad gradum E ad utrumque gradum PS 181 proportio EP om. S

[*De configurationibus qualitatum Nicolai Oresme* 373]

secundum ad excessum secundi supra tertium in intensione, ita quod punctum intensiorem illorum trium voco primum. Istud ergo declaratur de ea qualitate uniformiter difformi que terminatur ad non gradum que signetur seu ymaginetur per triangulum *ABC* erectum. Itaque tribus perpendicularibus lineis *BC*, *FG*, et *DE*, protrahatur *HE* equedistans (163r) linee *DF* et similiter *GK* equedistanter *FB*. Fiunt ergo duo trianguli *GHE* et *CKG* qui sunt equeanguli. Igitur per quartam 6[1] proportio *GK* ad *EH* est sicut proportio *CK* excessus ad *GH* excessum et quoniam *GK* est equalis *FB* et similiter *EH* est equalis *DF*, erit proportio *FB* ad *DF*, que quidem linee sunt distantie trium punctorum ipsius basis, sicut proportio *CK* ad *GH*, que sunt excessus altitudinis proportionalis intensionum eorundem punctorum. Cum igitur qualitas linee *AB* sit talis quod proportio punctorum linee in intensione est sicut proportio linearum in altitudine super eadem puncta perpendiculariter erectarum, patet evidenter propositum, scilicet quod que est proportio excessus primi puncti supra secundum ad excessum secundi supra tertium in intensione, eadem est proportio distantie inter primum punctum et secundum ad distantiam inter secundum et tertium et ita de quibuscunque aliis tribus punctis. Igitur qualitati sic difformi recte convenit, quod premittebatur et ita per talem triangulum bene designabatur. Per eundem modum predicta descriptio sive proprietas potest ostendi de qualitate uniformiter difformi terminata utrobique ad gradum et sit una que ymaginetur per quadrangulum *ABCD* in quo protrahatur linea *DE* equedistans basi *AB* et fiet triangulus *DEC*. Deinde protrahantur linee altitudinis in quadrangulo et alie transversales equedistantes basi in isto triangulo faciendo parvos triangulos et tunc facile poterit argui de illis excessibus et distantibus in isto triangulo sicut superius arguebatur in 7°, prout intuenti potest leviter apparere. Omnis autem qualitas se habens alio modo a predictis dicatur difformiter difformis et potest describi negative, scilicet qualitas que non est in omnibus partibus subiecti equaliter intensa nec omnium trium punctorum ipsius proportio excessus primi supra secundum ad excessum secundi super tertium est sicut proportio distantiarum eorum....

I.13. Capitulum 13^m. De eisdem alio modo. Adhuc autem aliter possunt predicta distingui ita quod linea superior figure per quam

194 proportionalis intensionum *PE om. S*
199 in intensione *EP om. S*
210 et... triangulo *EP om. S*
211 Omnis *EP* quamvis *S*
217 aliter *EP om. S*

ymaginetur qualitas vocetur linea intensionis seu linea summitatis sicut dicebatur in capitulo 10°. (163v) Verbi gratia, linea AB in quadrangulo $ABCD$. Si igitur huiusmodi linea summitatis figure per quam ymaginatur qualitas fuerit equedistans basi, sicut basi AB, qualitas per talem figuram ymaginabilis est simpliciter uniformis. Si autem non fuerit equedistans basi et fuerit recta, tunc qualitas est uniformiter difformis, ita quod si predicta linea coniungatur basi in uno extremo illa difformitas uniformis terminatur ad non gradum; et si in neutro extremo coniungatur basi, qualitas sive difformitas utrobique ad gradum terminatur. Et quoniam talis linea non potest coniungi basi in utroque extremo, quoniam ipsa est recta et basis recta, et sic esset linea una, inde patet quod non potest esse aliqua qualitas uniformiter difformis terminata usque ad non gradum.

Si vero linea intensionis sive summitatis fuerit curva aut ex multis lineis composita et non una, tunc qualitas per illam figuram ymaginabilis erit difformiter difformis et potest etiam terminari utrinque ad gradum vel utrinque ad non gradum vel ad gradum in uno extremo et ad non gradum in alio....

II.1. Secunda pars huius tractatus de difformitate successivorum. Capitulum primum de difformitate motus. (172v) Omnis motus successivus subiecti divisibilis habet partes. Et est divisibilis uno modo secundum extensionem et divisionem seu continuitatem mobilis; alio modo secundum divisibilitatem temporis et durationem seu continuitatem eiusdem; tertio modo saltem ymaginative secundum gradum et intensionem velocitatis. A prima autem continuitate dicitur motus magnus vel parvus; a secunda brevis vel longus; a tertia velox vel tardus. Habet itaque motus duplicem extensionem, unam subjectivam et aliam temporalem; et habet unam intensionem.

Due autem extensiones possunt ymaginari quodam modo orthogonaliter se invicem ad modum crucis intersecare, ita quod extensio durationis diceretur longitudo et extensio subiectiva vocaretur latitudo. Intensio vero potest vocari altitudo ipsius motus seu velocitatis. Sed si iuxta premissa in tertio capitulo prime partis intensio velocitatis

228 non E om. PS
230 linea EP linea recta S
234–35 utrinque... utrinque E om. S

PART II
3 subiecti EP sive S
7 intensionem S intensionem E intensionis P
12 se EP seu S / intersecare EP intersecate S

appelletur eius latitudo, tunc utraque extensionum ad intensionem comparata poterit dici longitudo, et sic velocitas habebit duplicem longitudinem sicut habet duplicem extensionem. Et in utraque istarum extensionum potest intensio velocitatis multipliciter variari. Et quoniam difformitas oritur ex eo quod intensio varie extenditur, inde sequitur quod motus sive velocitas potest habere duplicem difformitatem, vel etiam uniformitatem duplicem, unam secundum partem vel extensionem mobilis que proprie dicitur uniformitas aut difformitas, aliam vero secundum partes vel durationem temporis, que proprie dicitur regularitas vel irregularitas. Habet igitur motus ratione subiecti uniformitatem vel difformitatem, et ratione temporis sive secundum tempus regularitatem et irregularitatem. Et secundum hoc dicitur quod motus celi est difformis et est regularis. Motus vero gravis deorsum potest esse econverso uniformis et irregularis et etiam potest esse uniformis et regularis, vel difformis et irregularis. Sed non est possibile quod motus circularis sit uniformis. Verumptamen in sequendo modum loquendi consuetum vocabo quamcunque regularitatem nomine uniformitatis et irregularitatem nomine difformitatis....

II.3. (173r) Capitulum tercium de quantitate velocitatis intensionis. Cum utraque uniformitas motus ex primo capitulo consistat in intensionis equalitate et utraque difformitas ex inequalitate proveniat, premittendum est penes quod attendatur quantitas gradualis intensionis ipsius velocitatis. Verumptamen circa velocitatem tria sibi invicem propinqua possunt considerari: unum est quantitas ipsius velocitatis totalis pensatis intensione et extensione, et de hoc dicetur in tertia parte huius tractatus que erit de mensuris qualitatum et velocitatum. Aliud etiam potest ibi considerari, scilicet denominatio qua subiectum dicitur tale fieri velocius vel tardius, de quo etiam dicetur in capitulo sequenti. Tertium est ipsa gradualis intensio que facit ad illud propositum de qua nunc dicendum est. Dico igitur universaliter quod ille gradus velocitatis est simpliciter intensior sive maior quo in tempore equali plus acquiritur vel deperditur de illa perfectione secundum quam fit motus. Verbi gratia, in motu locali ille gradus velocitatis est intensior et maior quo plus pertransitur de spacio vel de distantia, et in alteratione similiter ille gradus velocitatis est maior quo plus

29 irregularis *E* irregularis seu regularis *PS*
30 uniformis et regularis *E* regularis et uniformis *PS*
44 tale fieri *EP* calefieri *S*
46 qua *EP* quo *S*

acquireretur vel deperderetur de intensione qualitatis....

II.4. Capitulum quartum de diversis modis velocitatis. Non est pretermittendum quod idem motus vel fluxus multis nominibus diversimode connotationibus appellatur. Et secundum hoc velocitas denominata diversimode attenditur sive mensuratur, ita quod quantitas intensionis gradualis multis modis assignatur, quibus tamen convenit descriptio prius dicta in capitulo precedenti. Verbi gratia, primo modo in motu circulari mobile dicitur moveri et dicitur circuire. Intensio autem velocitatis motionis attenditur penes spacium lineare quod illo gradu pertransiretur. Sed intensio gradus velocitatis circuitionis attenditur penes angulos circa centrum descriptos. Inde contingit quod aliquod mobile ad aliud comparatum circulariter motum velocius movetur, et tamen minus velociter circuit, sicut forsan mars velocius movetur quam sol et hoc motu proprio propter magnitudinem circuli descripti, et tamen sol velocius circuit et velocius (173v) revolvitur circa centrum.... Astrologi vero magis attendunt ad velocitatem circuitionis quam ad velocitatem motionis.

Item in motu recto, verbi gratia, in motu descensus, velocitas motus attenditur penes spacium pertransitum. Velocitas autem descensus attenditur penes appropinquationem ad centrum. Ideo possibile est quod A et B equevelociter moveantur, et utrumque deorsum, et tamen non equevelociter descendunt, sic quod A movebitur per lineam rectam ad centrum et B per lineam transversalem, et ideo A descendit velocius quam B et tamen B equevelociter movetur....

Verumptamen universaliter in omnibus ille gradus velocitatis est intensior sive maior quo in tempore equali subiectum fit magis tale aut tantum secundum illam denominationem qua dicitur velocitas acquiri, quecunque sit illa. Verbi gratia, gradus velocitatis descensus est maior quo subiectum mobile magis descendit vel magis descenderet si continuaretur descensus....

II.5.... Adhuc autem potest ymaginari una alia successio. Omnis enim velocitas est intensibilis et remissibilis. Eius vero continua intensio vocatur velocitatio, et hec quidem velocitatio seu augmentatio velocitatis potest fieri velocius aut tardius. Unde quandoque contingit quod velocitas intenditur et velocitatio remittitur, quandoque vero utraque simul intenditur. Et similiter huiusmodi velocitatio

---

54 idem EP idem est S
62 circuitionis P circuitonis S
71 appropinquationem EP propinqui-
tatem S
74 transversalem ES circularem vel transversalem P

[*De configurationibus qualitatum Nicolai Oresme* 377]

quandoque fit uniformiter et aliquando difformiter et diversimode....

II.8. (1751) Capitulum octavum de difformitate velocitatum motus quo ad tempus. Omnis velocitas tempore durat, tempus itaque sive duratio erit ipsius velocitatis longitudo, et eius velocitatis intensio erit sua latitudo. Et quamvis tempus et linea sint incomparabiles in quantitate, nulla tamen proportio reperitur inter tempus et tempus, que non invenitur in lineis et econtra, et per prius reperitur in lineis secundum Aristotelem 6 physicorum; et similiter est de intensione velocitatis, videlicet, quod omnis proportio que reperitur inter intensionem et intensionem velocitatis reperitur etiam inter lineam et lineam, sicut de aliis intensionibus dicebatur in secundo capitulo prime partis. Ideoque in noticiam difformitatum velocitatum possumus devenire per ymaginationem linearum ac etiam figurarum. Signetur gratia exempli linea ita que sit AB et moveatur D mobile per tempus EF quomodolibet. Sive sit motu circulari sive recto seu alteratione, et sive sit super lineam AB sive super aliam. Eriganturque super lineam AB per totum linee perpendiculares in superficie, que quidem superficies vel figura sit proportionalis in altitudine velocitati ipsius D in intensione. Dico igitur quod illi superficiei vel figure poterit velocitas D mobilis assimilari, et per eam congrue ymaginari, ita quod linea AB que est huiusmodi figure longitudo designabit longitudinem durationis huius velocitatis et altitudo huius figure designabit intensionem huius velocitatis. Verbi gratia, si in omnibus instantibus temporis EF velocitas sit equaliter intensa, tunc super quemlibet punctum linee AB erit altitudo equalis ubique et figura uniformiter alta, scilicet, quadrangulus rectangulus designans hanc velocitatem simpliciter uniformem. Si vero in primo instanti illius temporis fuerit aliquanta velocitas et in instanti medio totius temporis fuerit velocitas subdupla et in instanti medio ultime medietatis fuerit subquadrupla, et sic proportionabiliter de aliis instantibus, et per consequens erit in ultimo instanti nulla, tunc super lineam AB erunt linee altitudinis proportionales secundum modum predictum et erit figura trianguli rectanguli illam velocitatem designantis, que quidem velocitas erat uniformiter difformis terminata ad non gradum in ultimo

95 6 P 8vo ES
109–10 longitudinem... designabit *om. S*
110 huius E alibi extensionem ipsius S
  in alio extensam ipsius P
112 ubique EP *om. S*

115 totius EP *om. S*
119 predictum EP predictum et quadranguli S / et EP *om. S*
120 *post* figura *add. S* sic / rectanguli EP *om. S*

instanti suo. Et ut breviter dicam, omnis uniformitas et omnis difformitas velocitatum potest eisdem modis notificari et describi, qui modi fuerunt positi in prima parte huius [tractatus]....

III.1. (190r) Incipit tertia pars de acquisitione et mensura qualitatis et velocitatis. Capitulum primum per quod ymaginanda est acquisitio qualitatis. Duplici modo potest contingere successio in acquisitione qualitatis, secundum extensionem et secundum intensionem, sicud superius fuit dictum quarto capitulo secunde partis. Acquisitio itaque extensiva qualitatis linearis ymaginanda est per motum puncti fluentis super ipsam lineam subiectivam, ita quod pars pertransita sit qualificata, et pars nondum pertransita non qualificata. Sicud si punctus C moveretur super lineam AB et quidquid esset ab eo pertransitum, esset album; et quidquid nondum esset pertransitum, nondum esset album. Acquisitio autem extensiva qualitatis superficialis ymaginanda esset per motum linee dividentis partem superficiei alteratam ab altera parte nondum alterata. Et acquisitio extensiva corporee [qualitatis] conformiter ymaginanda est per motum superficiei dividentis partem alteratam a parte nondum alterata.

Acquisitio autem intensiva qualitatis punctualis ymaginanda est per motum puncti continue ascendentis super punctum subiectivum. Acquisitio vero intensiva qualitatis linearis ymaginanda est per motum linee perpendiculariter ascendentis super lineam subiectivam, et suo fluxu vel ascensu derelinquentis superficiem, per quam designatur qualitas acquisita. Verbi gratia: Sit AB linea subiectiva; dico igitur quod intensio puncti A ymaginatur per motum vel per ascensum perpendicularem puncti C et intensio linee AB vel acquisitio intensionis ymaginatur per ascensum linee CD. Acquisitio intensiva qualitatis superficialis conformiter ymaginanda est per ascensum superficiei, motu suo ymaginato derelinquentis corpus per quod illa qualitas designatur, et similiter acquisitio intensiva corporee qualitatis ymaginatur per motum superficiei qualitatis, quia superficies fluxu suo ymaginato derelinquit corpus, et non contingit dare quartam dimen-

122 suo *EP om. S*

    Part III

6 extensiva *E* extensive *PS*
7 motum *EP* modum *S* / subiectivam *EP om. S*
9 esset *tr. S post* pertransitum

10 esset[2] *EP* est *S*
11–12 superficialis *EP* linearis *S*
13 alteratam *om. S*
17 subiectivum *EP* subiectum *S*
19 subiectivam *EP* summam *S*
21 subiectiva *EP* subiecta *S*
27 acquisitio *E om. PS*

[*De configurationibus qualitatum Nicolai Oresme* 379]

30 sionem, sicud dictum est quarto capitulo prime partis. Et sicud nunc dictum est de acquisitione qualitatis, ita conformiter dicendum est et ymaginandum de deperditione.... Ymaginatur enim talis deperditio per motus oppositos motibus predictis. Sicud enim nunc dictum est de acquisitione aut deperditione qualitatis, ita conformiter ymaginandum
35 est de acquisitione aut deperditione velocitatis, tam in intensione quam in extensione....

III. 7. (192r) Capitulum septimum de mensura qualitatum et velocitatum difformum. Omnis qualitas, si fuerit uniformiter difformis, ipsa est tanta quanta foret qualitas eiusdem subiecti vel equalis uniformis
40 secundum gradum puncti medii eiusdem subiecti. Et hoc intelligo, si qualitas fuerit linearis. Et si fuerit superficialis, secundum gradum linee medie; si vero fuerit corporalis, secundum gradum superficiei medie, secundum hoc conformiter intelligendo.

Illud primo ostenditur de lineari. Sit igitur una qualitas ymaginabilis
45 per triangulum ABC, que est uniformiter difformis terminata ad non gradum in puncto B; et sit D punctus medius linee subiective. Cuius quidem puncti gradus vel intensio ymaginatur per lineam DE. Igitur qualitas que est uniformis per totum subiectum secundum gradum DE ymaginabilis est per quadrangulum $AFGB$, ut patet per 10^m
50 capitulum prime partis. Constatque per 26^{am} primi Euclidis, quod duo parvi trianguli EFC et EGB sunt equales. Igitur maior triangulus BAC qui designat qualitatem uniformiter difformem et quadrangulus $AFGB$ qui designaret qualitatem uniformem secundum gradum puncti medii sunt equales. Igitur qualitates per huiusmodi triangulum
55 et quadrangulum ymaginabiles sunt equales, et hoc est propositum. Eodem modo potest argui de qualitate uniformiter difformi terminata utrobique ad certum gradum, sicud esset qualitas ymaginabilis per quadrangulum $ABCD$; protrahatur enim linea DE equedistans (193v) basi subiecte et fiat triangulus CED. Deinde protrahatur per
60 gradum puncti medii linea FG equalis et equedistans basi subiecte. Protrahatur etiam linea GD. Tunc sicut prius probabitur quod triangulus CED et quadrangulus $EFGD$ erunt equales. Ergo utrobique communi quadrangulo $AEDB$ fient duo tota equalia, scilicet, quadrangulus $ACDB$ qui designat qualitatem uniformiter difformem et

57 utrobique P utriusque S
58 protrahatur EP pertrahatur S / DE EP *om.* S
59–60 et... subiecte EP *om.* S

59 fiat E fiet P
62 erunt E fient S sunt P / equales *om.* E
62–63 Ergo... fient P quare S
63–64 quadrangulus... et EP *om.* S

quadrangulus $AFGB$ qui designat qualitatem uniformen secundum gradum puncti medii ipsius subiecti AB. Igitur per capitulum 10^m prime partis qualitates per huiusmodi quadrangulos designabiles sunt equales. Conformiter potest argui de qualitate superficiali ac etiam de [qualitate] corporali. De velocitate vero omnino dicendum est sicut de qualitate lineari, dum tamen loco puncti medii [subiecti] instans medium temporis capiatur velocitatem huius mensurantis....

III.8. Capitulum octavum de mensura et intensione in infinitum quarundam difformitatum.

Superficies finita potest fieri quantumlibet longa vel alta per variationem extensionis absque eius augmento. Nam talis superficies habet longitudinem et latitudinem, et possibile est ipsam secundum unam dimensionem quantumlibet augeri, ipsa non augmentata simpliciter dummodo secundum aliam dimensionem proportionaliter minuatur, et ita etiam de corpore.

Verbi gratia de superficie. Accipiatur superficies quadrata [altitudinis] pedalis, cuius basis sit linea AB, et sit alia superficies similis et equalis, cuius basis sit linea CD que ymaginetur in infinitum dividi per partes continue proportionales, continue secundum proportionem duplam, super basim CD eodem modo divisam. Et sit E prima pars, et F secunda, et G tertia, et sic de aliis. Sumatur ergo prima istarum partium, scilicet E, que est medietas sui totius. Et ponatur super primam superficiem versus extremum B. Deinde super totum hoc ponatur secunda pars, scilicet F, et iterum super totum hoc ponatur tertia pars, scilicet, G, et ita de aliis in infinitum. Quo facto ymaginetur basis AB dividi per partes continue proportionales secundam proportionem duplam eundo versus B. Et statim patebit quod super primam partem proportionalem linee AB stat superficies alta per unum pedem, et super secundam partem per duos pedes et super tertiam partem per tres pedes, [et super quartam per 4] et ulterius in infinitum. Et tamen totalis superficies non est nisi duo pedalia prius dicta in nullo augmentata, et per consequens totalis superficies que stat super lineam est precise quadrupla ad illam sui partem que stat super primam partem proportionalem eiusdem linee AB. Illa ergo qualitas sive velocitas que proportionaretur in intensione huic figure

69 De E om. PS
70 instans P instantis S (E?)
75 Nam talis S om. E qualis P
93 duos pedes EP duas partes S
94 [et... 4] S om. EP

95–96 Et... argumentata EP om. S
95 non est P om. E
96 per consequens EP cum S
98–100 Illa... quadrupla bis S

100 in altitudine esset precise quadrupla ad partem sui que foret in prima parte temporis vel subiecti secundum huiusmodi dimensionem....
(194r) Eodem modo si aliquod mobile moveretur in prima parte proportionali alicuius temporis taliter divisi aliquali velocitate, et in secunda moveretur in duplo velocius, et in tertio triplo, et in quarta
105 quadruplo, et sic consequenter in infinitum semper intendendo, velocitas totalis esset precise quadrupla ad velocitatem prime partis, ita quod illud mobile in tota hora pertransiret precise quadruplum ad illud quod pertransivit in prima medietate illius hore. Et si in prima medietate vel parte proportionali transivit unum pedem, in toto
110 residuo pertransiret tres pedes, et in toto tempore pertransiret quattuor pedes.

108 medietate *ES* parte *P*

Document 6.2

Giovanni di Casali
On the Velocity of the Motion of Alteration

1. ...LATITUDE of calidity is [1] calidity uniformly difform, or [2] the qualitative distance between degrees of calidity, by which distance the motion of alteration of calidity is measured, or [3] the distance from a summit degree to zero degree. A summit degree *(gradus summus)* of calidity is calidity which throughout is equally and maximally distant from zero degree. An "intended" degree of calidity is calidity which throughout, or at its more intense extreme, is distant from zero degree by a latitude. A "remitted" degree of calidity is [1] a calidity which in its most intense part is truly distant from its summit degree by [some] indivisible latitude, or [2] a remitted degree is spoken of where in its latitude there is, or can be, another degree more intense than it, or [3] [it] is that which contains less intension or qualitative distance, or [4] [it is] that which is less distant from zero degree. A degree in latitude is nothing else than a part of the latitude which is more intense or more remiss than another part of it. A degree of calidity is calidity which is uniform, [difform], or uniformly difform. A degree of uniform calidity is a calidity no part of which is more intense or more remiss than another part of it. A difform calidity is [for example] one whose first part is hot, whose second part is hotter than the first, whose third part is hotter than the second, and so on. A degree of calidity uniformly difform is a difform calidity, wherein, when any two immediate parts are taken, the most intense uniform degree which is not in the one part is the most remiss degree which is not in the other part. According to some, calidity uniformly difform is one whose second part exceeds its first by [some] degree and whose third exceeds its second [by the same degree], and so on. The most intense uniform degree which is not in A is the uniform degree not in A such that any degree more remiss than any such uniform more intense degree is in A. And in such a degree [there is] no degree more remiss or as remitted [as any one] in A. The most

[Giovanni di Casali, *On Velocity of Motion* 383]

remiss degree which is not in A is the uniform degree not in A such that any degree more intense than any such more remiss degree is in A. And in such a degree [there is] no degree as intense [as any one] in A....

2. The third conclusion is that any uniformly difform latitude is precisely as much as a latitude uniformly intense at the mean degree. Or, anything at all uniformly nonuniform in hotness is precisely as hot [quantitatively] as something uniformly hot at the mean degree of the uniformly nonuniform hotness.... [Then he defines uniformly difform hotness as well as uniform hotness.]

One can exemplify these qualitative things [by Fig. 6.7]. For something

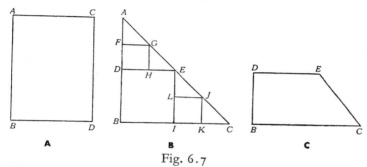

Fig. 6.7

These diagrams have been reconstructed in accordance with the text. They bear letters I have added for purposes of explanation.

uniformly hot is throughout like a rectangular parallelogram constructed between two parallel lines [AB and CD].* Then any part of such a rectangle is equally wide *(lata)* as any other part, because the latitude of any such part is measured by the base.

Similarly a uniformly difform hotness is in every way like a right triangle [ABC]. This would be a uniformly difform hotness terminated in one extreme [point A] at zero. [This correspondence of uniformly difform hotness and a right triangle] holds because one quarter [ADE] of such a triangle has a line [FG] which is just as distant from one extreme of the quarter as from the other (i.e., the latitude at A is 0, and the latitude at D is DE, and there is a mean latitude FG), and it is just as distant from either extreme as the middle line [LJ] of another small quarter [EIC] is distant from its extremes. But when we speak of distance [we speak in terms] of arithmetic proportion. So the smaller any latitude of some part toward the base is, the more distant is the part it is in, [i.e.,] in any more

* All of the specific letters included in brackets I have added to make the concepts clear; they do not appear in Casali's text.

distant part the latitude is smaller, and no part is uniformly wide. Similarly, any latitude at all is imaginable between the latitude corresponding to the base and a point corresponding to some (! the highest?) point of that triangle, because it is possible to find in such a triangle any line that is less than the base of the triangle.

Any latitude uniformly difform terminated at both extremes at some degree is similar to a quadrangle [*BDEC*] produced by a line [*DE*] cutting off the apex of the triangle mentioned above, so that it would be terminated at a degree in the more remiss extreme which would be continually less as the line approaches the apex. [Thus *DE* forms the less intense extreme of the uniformly difform quality represented by *BDEC*, and obviously as we let *DE* be closer to *A* it becomes smaller.]

[Casali then gives a proof of the mean speed theorem—applied to qualities—which is rhetorical and in the Heytesbury tradition without specific reference to his geometric figures (see Latin text, pp. 389–90, lines 66–99). But at the conclusion of this proof, he suggests what is equivalent to the Oresme proof, namely, taking a right triangle representing the uniformly nonuniform quality and equating it to a rectangle representing the quality uniform at the mean degree.] Therefore, because any triangle having two equal sides is equal to some rectangle, it is evident that the latitude of the rectangle will be uniform and measurable by a line passing through the midpoint of the same triangle. This is evident, for let there be given such a triangle and let it be *A* (see Fig. 6.8). Its mean line, i.e., the line passing

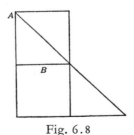

Fig. 6.8

through the middle point, we let be *B*. Let this triangle be placed between two parallel lines, and [so] become a parallelogram [constructed] between these parallel lines—whose base is equal to line *B*, and let it be a rectangle. [Q.E.D.]

COMMENTARY

1. We have already noted in the body of the text that this treatise apparently dates from 1346. There can be little doubt that the author

draws his inspiration from the English treatises that discuss latitudes. However, in this first section in using the term *gradus* he is not thinking only of the magnitude of an intensity at a particular point (or the magnitude of instantaneous velocity, if he were thinking about motion), but rather he uses the term to mean simply the magnitude of intensity of some extended part. Thus he can talk about a uniform degree and a uniformly difform degree—meaning that the degree is uniform throughout an extended part of the subject or uniformly difform throughout that part. His definition of uniformly difform should be compared to the definition intimated by Francischus de Ferraria (see Doc. 7.4) and that given in the anonymous summary on motion prefixed to Bradwardine's *Proportions* (see Doc. 7.1). For the definition of "immediate parts" the reader is referred to Chapter 11, note 4. Note also that his second definition of latitude is verbally like one mentioned by John of Dumbleton in his *Summa* (see note 15).

2. The most interesting feature of Casali's use of coördinates, which distinguishes it from Oresme's technique, is that he has apparently rotated the coördinates. Instead of intensity (or velocity) being measured by a vertical line perpendicular to the base line, as Oresme wishes, Casali measures intensity by a horizontal line parallel to the base. This allows him to maintain the pristine meaning of latitude as width. It should be observed, however, that while the text is quite clear in its representation of degrees of intensity by lines parallel to the base line, two of the manuscripts which I have used actually draw the figures according to the conventional system of Oresme, and thus represent intensity by vertical lines. The third manuscript (*F*—see the Latin text) does not have any figures. The edition sometimes has figures representing intensity by horizontal lines (as the text suggests) and sometimes figures representing intensity by vertical lines. It would seem that the scribes of the first two manuscript copies merely added figures that followed the accepted system of their day.

It is of some interest to note that Galileo in using the coördinate geometry for his proof of the mean law (see Doc. 6.5) employs the coördinates of Casali rather than those of Oresme.

*De velocitate motus alterationis Johannis de Casali**

.... Latitudo caliditatis est caliditas uniformiter difformis sive distantia qualitativa caliditatis inter gradus, penes quam distantiam attenditur velocitas motus alterationis [caliditatis, sive distantia a gradu summo usque ad non gradum].

Gradus summus caliditatis est caliditas secundum se totam equaliter maxime distans a non gradu.

Gradus intensus caliditatis est caliditas secundum se totam vel secundum extremum intensius distans a non gradu per latitudinem.

Gradus remissus caliditatis est caliditas secundum partem intenssissimam indivisibiliter per latitudinem vere distans a summo gradu, vel gradus remissus dicitur quo in eius latitudine est alius intensior vel esse potest, vel qui minus continet de intensione sive de distantia qualitativa, vel qui minus distat a non gradu.

Gradus in latitudine nihil aliud est quam pars latitudinis alia parte eius intensior vel remissior existens.

---

* My text is based on the following copies: MSS Florence, Bibl. Riccard. 117, ff. 135r–144v (*F*; but this selection, ff. 137r–v); Oxford, Bodl. Canon. Misc. 376, ff. 32r–47r (*B*; this selection, ff. 34r–35r); Vienna, Nat.-bibl. 4217, ff. 154r–172r (*V*; this selection, ff. 154v–155r); and the edition of Venice, 1505 (*Ed;* without pagination, but this selection, sign. J2v–J 3v). I have in addition consulted Vat. lat. 2185, ff. 61v–71r, but it was not available at the time of preparing this text. As usual I have capitalized letters representing figures and have punctuated freely. In the variant readings I have not indicated all orthographic variations; thus only *F* spells *parallelogrammum* in the conventional way, while the other manuscripts have variations not always indicated in the variant readings. The additions to the text in brackets are all from *F* except in the case of line 17, where the addition is my own. The addition in parentheses in line 28 is also my own. I have bracketed the material from *F* because, although *F* may be the earliest of the manuscripts I used, it nevertheless has a number of obviously erroneous readings, and so it may well be that the bracketed material is a scribal addition. By examining the variant readings, the reader will notice that *F* has rearranged the initial definitions. I have not followed *F* in this rearrangement, although it may be that *F* is correct in so doing.

---

3 caliditatis *om. B*
4 velocitas *F om. BEd*
4–5 [caliditatis... gradum] *F om. BVEd*
6–14 Gradus... gradu *tr. F post* difformis *in linea 18*
8 caliditatis *om. B*
10 caliditatis...caliditas *F* est caliditas *B* caliditatis *Ed*
10–11 intensissimam *Ed* remissimam *F*

intensivam *B*
11 divisibile *F*
12 remissus: remissus caliditatis *F*
13 intensione *FEd* latitudine intensive (!) *B* / de² *om. B*
15 aliud est *tr. F*
16 eius: eiusdem *F* / vel... existens: existens vel remissior *B*

[*De velocitate motus Johannis de Casali* 387]

  Gradus caliditatis est caliditas uniformis vel [difformis vel] uniformiter difformis.
  Gradus uniformis caliditatis est caliditas, cuius nulla pars alia parte eiusdem intensior vel remissior existit.
  [Caliditas difformis est cuius prima pars est calida, secunda magis calida quam prima, et tertia magis quam secunda, et sic deinceps.]
  Gradus caliditatis uniformiter difformis est caliditas difformis, cuius quarumlibet duarum partium sibi invicem immediatarum gradus uniformis intensissimus qui non est in una parte est remississimus qui non est in alia. [Vel secundum aliquos caliditas uniformiter difformis est cuius secunda pars excedit primam per gradum, et tertia secundam (per idem gradum), et sic deinceps.]
  Intensissimus gradus uniformis qui non est in A est gradus uniformis qui non est in A, quo quolibet gradu uniformi intensiori gradus remissior est in A et in eodem gradu nullus gradus remissior vel ita remissus est in A.
  Remississimus gradus uniformis qui non est in A est gradus uniformis qui non est in A, quo quolibet gradu uniformi remissiori gradus intensior est in A et in eodem gradu nullus gradus ita intensus est in A....
  Tertia conclusio est quod quelibet latitudo uniformiter difformis est tanta precise quanta est latitudo uniformis sub gradu suo medio, vel sic quodlibet uniformiter difforme calidum est precise ita calidum sicut aliquod uniforme calidum sub gradu medio illius uniformiter difformis calidi....
  Exemplum ad ista in qualitatibus potest haberi. Sic enim est per omnia de calido uniformi sicut est de parallelogrammo rectangulo

17–20 Gradus... existit *om. B*
17 [difformis vel] *supplevi*
19 pars: pars eius *F*
21–22 [Caliditas... deinceps] *F om. BEd*
23 Gradus caliditatis *B* Gradus *Ed* Caliditas *F*
25 intensivus *B*
26–28 [Vel... deinceps] *F om. BEd*
28 (per idem gradum) *supplevi*
30 quo: est quorum quodlibet *B*
31 in[1] *om. B*
31–32 vel... remissus *om. B*
33 uniformis[1] *om. B*
34 A quo: aliquo *B*
35 in[2]... gradu: quo *B* / gradus *om. B* / ita intensus: intensior *B*
37 est *FEd* est hec *B* est ista *V*
40 sicut: sicud est *V* / aliquod: aliquod precise *B* aliquid *Ed* / calidum *om. F* / illius: illius precise *B*
41 calidi *om. Ed*
42 ista: omnia ista *Ed* / in qualitatibus *om. Ed et tr. F post* haberi / in quantitatibus *B* / Sic enim est *FV* sicut enim est *B* sit enim *Ed*
43 calido uniformi *tr. F* / est *om. B* / paralelogramo *mg. B* / rectangulo *om. F*

inter duas lineas equedistantes constituto, cuiuslibet talis quelibet pars est eque lata cum alia, quia cuiuslibet partis unius talis latitudo mensuratur post basem illius parallelogrammi.

Et sic est per omnem modum de uniformiter difformiter calido sicut est de triangulo rectangulo, et hoc si illud uniformiter difformiter calidum terminetur ad non gradum in uno extremo, quia cuiuslibet talis trianguli una quarta habet lineam que tantum distat ab uno extremo eius quarte sicut ab alio et a quolibet suo extremo precise tantum distat sicut una alia linea media alterius quarte minoris distat a suis extremis et in eadem proportione. Sed hoc loquendo de distantia proportionis arismetice. Similiter quecunque sit latitudo alicuius partis versus basem trianguli talis minor est in quacunque parte distanti, et in parte magis distanti minor latitudo, et nulla pars est uniformiter lata. Similiter quelibet latitudo ymaginabilis inter latitudinem correspondentem basi et punctum correspondentem alicui (! altissimo?) puncto illius trianguli, quia quelibet linea mundi que est minor quam basis trianguli continget in tali triangulo reperire minorem. Sed latitudo quelibet uniformiter difformis terminata in utroque extremo ad certum gradum similis est quadrangulo, qui causaretur

44 lineas equedistantes *F tr. BV* lineas distantes *Ed* / talis: termio (?) *B*
45 latitudo *om. Ed*
46 post *BVEd* per *F* / illius: scilicet *V*
47 sic est *F* sic est etc. *V* sic *B* sit *Ed* / per omnem modum *F* pro toto modo *B* eodem modo *V* per omnia *Ed*
48 rectangulo *mg. B om. FVEd* / hoc si illud: sic id *Ed* / difformiter: difforme *B*
49 terminatur *Ed*
51 eius: eiusdem *V* / quolibet: quocunque *Ed*
51–52 tantum distat *tr. B*
52 media *om. F* / minoris *Ed* minus *F* uniformiter *BV*
53 extremis: extremitatibus *V* / Sed *VEd* et *B*
53–54 Sed... arismetice *om. F* / distantia... arismetice *VEd* proportione distantia arismetica *B*
54 Similiter *BVF* sic *Ed* / sit *om. Ed* / alicuius *VEd* illius *F* sit alicuius *B*
55 minor *V* brevior *F* basis *Ed* modi *B*
55, 56 distanti[1,2] *FEd* distante *BV*

56 minor: brevior *F* / nulla *FV Ed* nūma (?) *B*
57 ymaginabilis *FB* qualitatis (?) *Ed* triangulis vel ymaginabilis *V* / inter *BFV* et inter *Ed*
57–59 *mg. V* Nota quod basis latitudinis est linea correspondens vel designans gradum intensiorem latitudinis sicud est linea AB
58 alicui *FEd* alii vel alicui *V* alt'i (?) *B*
59 illius: istius *V*
59–60 quia... minorem: quelibet linea data illa *mg. B et inferius in mg. add. B* alia littera habet: quelibet linea data est minor quam basis 3[11] trianguli in 3[10] et quacunque minori data continget minorem reperire, et ista est melior littera; *in textu om. B* quia... que *in linea 59*
60 continget: continget *B* / in tali triangulo *F* trianguli tali *B* triangulo tali *VEd* / minorem: breviorem *F*
62 quadrangulo: triangulo *B* / qui *VF* quod *B* quid *Ed*

[*De velocitate motus Johannis de Casali* 389]

per lineam abscindentem conum trianguli supradicti, ita quod consequenter terminaretur ad gradum minorem in extremo remissiori quantum linea talis plus appropinquatur cono.

His visis probatur conclusio proposita sic. Sit A aliquod calidum uniforme per totum, cuius una medietas sit B et alia C. Et applicentur puncto medio A duo agentia, vel extremis, non curo, ita quod unum illorum uniformiter difformiter remittat ipsum C, ita quod in fine illius hore terminetur C ad non gradum, et ad gradum sub quo fuit uniforme prius. Et intendat illud agens B uniformiter difformiter in eadem proportione in qua C est remissum, et hoc semper loquendo de proportione arismetica, ita quod si gradus sub quo A fuit uniforme calidum signetur per 4, tunc quicunque punctus C medietatis remittitur ad 2, quod punctus sibi correspondens in B medietate intenditur ad 2 ultra illud quod habuit prius, ita tamen quod ille punctus sit proportionaliter signabilis per 6, et quando punctus aliquis C remittitur ad unum, vel gradum signatum per unitatem, quod tunc punctus sibi correspondens in B intendatur sic quod ille punctus signetur per 7, et sic de aliis. Sed remittatur C per omnem gradum inter gradum sub quo fuit et non gradum in diversis punctis.

Et arguitur tunc sic: Si illud precise quod auferretur de quolibet puncto ipsius C adderetur consimili puncto in B, et illud sic additum in eadem proportione intenderet punctum cui additur sicut ablatum remittit punctum cui aufertur, et hoc loquendo de proportione arismetica.

63 abscendentem *Ed*
64 minorem: breviorem *F* / remissiori: remotiori *Ed*
65 quantum *FEd* cum *BV* / appropinquatur *F* appropinquat *BVEd*
66 proposita *FB om. VEd* / conclusio... sic: sic conclusio *V* / sic: ipso sic *F*
67 una medietas *tr. V* / et *BVEd om. F* / alia *FEd* alia sit *BV* / applicetur *B*
69 illorum *VB* istorum *Ed* illarum *F*
70 illius *BEd om. F* huius *V* / fuit: fit *F*
71 intendatur *V* / illud: aliud *B*
73 si *om. F* / A: illud A *F* / fuit: fuerit *B* / uniforme: uniformiter scilicet *V*
74 signaretur *B*
74–75 remittetur *F*
75, 76 ad 2: ad'o *F*
76 prius: primo *V* / tamen *om. F* tunc *BV* / ille punctus: unus talis *V* ille *B*
77 proportionalis *VB* / 6: 8 *EdB* (?) /

quando punctus aliquis *F* quod punctus aliquis *B* alia *et lac. Ed* alius sit proportionaliter signabilis per 2 et quod quecunque punctus *V*
77–78 remittur *VEd* remittetur (?) *F* remittatur *B*
78 quod: et *B* / tunc *om. F*
79 sic: ita *F* / ille: iste *V* / signatur *BV*
80 7 *mg. V* 8 *BFEd* / omnem: communem *Ed*
82 arguo *F* / Si: sic *B* / illud: istud *BV* / aufertur *BV* (?) / quolibet *FEd* quocunque *BV*
83 ipsius C: in C *V*
84 eadem proportione: consimili proportione eadem *Ed* / intenderet *BF* intendat *V* intendet *Ed*
84–85 remitteretur *B*
85 punctum *om. Ed* / cui *FBEd* a quo *V* / auferretur (?) *B* / hoc *om. B*

Tunc esset precise in fine talis alterationis tanta latitudo quanta fuit in principio, quia eadem et nihil deperditur de ea. Sed modo tali intensione et remissione facta in partibus A sic est per omnia sicut esset, si illud precise quod aufertur de quocunque puncto ipsius C, et remittit punctum istum, adderetur puncto simili in B et tantum proportionaliter intendet sicud illud remisit punctum a quo auferebatur, quia iam est medietas A ita remissa sicut foret per talem ablationem et alia medietas ita intensa sicut foret per talem additionem. Igitur modo in fine alterationis non est latitudo illa intensior nec remissior quam fuit in principio, sed erit tanta quanta fuit in principio. Sed in principio fuit illa latitudo uniformis sub gradu medio illius latitudinis iam uniformiter difformis. Igitur modo illa latitudo est tanta quanta est una latitudo uniformis sub gradu medio suo, quod erat probandum.

Consimiliter per omnia arguitur de latitudine uniformiter difformi terminata ad certum gradum in extremo remissiori, sicut potest patere cuilibet intuenti.

Ergo quod quilibet triangulus mundi habens duo latera equalia equalis est alicui parallelogrammo rectangulo, cuius parallelogrammi, patet quod, latitudo erit uniformis et mensurabilis a linea transuente per medium punctum eiusdem trianguli; patet, quia detur unus triangulus talis et sit A, cuius linea media vel transiens per punctum

---

86 precise in fine: in fine precise BV
87 fuit om. F / et: vel V / perditur V / modo FEd non B solum modo V
88 remissione et intensione Ed
89 esset si om. B / illud: istud B / precise add. F post aufertur / quocunque: quolibet F
90 remittit $FVEd$ remittat B / istum: illum B / simili V consimili B scilicet FEd / in FB om. VEd
91 intendet BV intendetur F intenderetur Ed / sicud illud: istud sicud istud V / a quo: cui F
93 et... additionem om. F
94 est del. B et scr. erat vel erit / latitudo illa tr. V / illa BV om. F ita Ed / intensior: ita intensior Ed / nec VFB vel Ed
95 fuit: fuerit Ed / sed... principio FEd om. BV
96 illa: ista V / uniformis BEd om. F

difformis V / medio: medio uniformi F / illius: istius V
97 illa: ista V / est FBV erit Ed
98 medio suo: illo medio F
99 erat: esset vel erat F
100 per omnia arguitur FBV arguitur per omnia Ed
102 intuenti: intelligenti hoc Ed (?F)
103 Ergo quod FEd tenet consequentia quia B Habeo igitur quod V / quilibet: quecunque V / mundi om. Ed
104 rectangulo VEd lac. F om. B / parallelogrammi BV parallelogrammo FEd
104–9 cuius... parallelogrammum om. B
105 patet om. F / quod V om. $FBEd$ / a linea FB alia VEd
106 eiusdem: illius F / quia FV quod Ed
107 linea media tr. V / vel om. F
107–8 punctum medium tr. VEd

medium sit B, et ponatur iste triangulus inter duas lineas equedistantes, et fiat inter easdem lineas parallelogrammum, cuius basis sit equalis linee B et sit rectangulum.

108 iste VF ille BEd 110 sit rectangulum BEd fiat triangulus etc. F sit rectiangulum V

Document 6.3

Jacobus de Sancto Martino
*On the Latitudes of Forms**
(According to the Doctrine of Nicole Oresme)

1. BECAUSE the latitudes of forms are varied in many ways, this multiplicity is discerned with difficulty unless reference is made to geometric figures. . . .

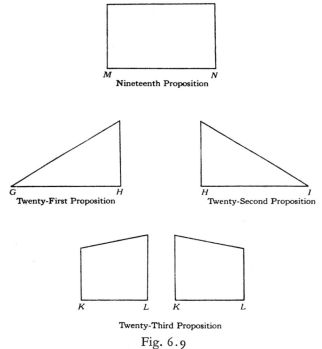

Fig. 6.9

2. Nineteenth proposition. Every latitude uniform throughout is to be imagined by a rectangle. This is proved because since every uniform

* Translated by Thomas Smith from his own unpublished text—see Latin text.

[Jacobus de Sancto Martino, *Latitudes of Forms* 393]

latitude is of the same degree throughout, it is to be imagined by a figure which is of the same latitude throughout....

3. Twenty-first proposition. Every uniformly difform latitude beginning from zero degree is to be imagined by a rectilinear [right] triangle beginning from an acute angle and terminated at a right angle....

4. Twenty-second proposition. Every uniformly difform latitude beginning from a certain degree and terminated at zero degree is to be imagined by a triangle beginning from a right angle and terminated at an acute angle....

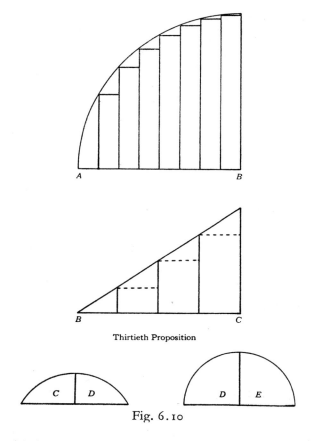

Thirtieth Proposition

Fig. 6.10

5. Twenty-third proposition. Every uniformly difform latitude beginning from a certain degree and terminated at a certain degree is to be imagined by a quadrangle of which the two angles on the base are right angles and of the remaining (angles), one is acute, the other obtuse....

6. Twenty-fourth proposition. No latitude beginning from zero degree

[394]   Two-Dimensional Geometry: 6.3]

and terminated at zero degree is uniform or uniformly difform, although it could have parts that are uniform or uniformly difform....

7. Twenty-sixth proposition. No latitude difformly difform throughout is to be imagined by a rectilinear figure....

8. Twenty-seventh proposition. Every latitude difformly difform throughout is to be imagined by a figure of which the altitude (i.e., the continuous height) is bounded by a curved line or by curved lines....

9. Thirtieth proposition. Every latitude uniformly difformly difform is to be imagined by a [recticurvilinear] triangle having upon the base one angle right and rectilinear, the remaining two indeed acute and curvilinear.... For the purpose of visualizing, moreover, by what way in such figures the same proportion may be maintained among [intensive] increments of degrees equidistant [extensively], I draw a [recticurvilinear] triangle AB (see Fig. 6.10) which is a quadrant of a circle, the base of which is divided, for example, into eight equal parts by erect perpendicular lines at the points of division, which lines will measure the altitude of the surface upon any of its points you wish.... Then I assign increments of those lines equidistant among themselves, which increments represent increments of equidistant degrees, and it is clear because as is the proportion of the first increment to the second, so is the second to the third, and as is the proportion of the first line to the second, so is the second to the third, and so for the others. And if the base is divided into more parts than eight or even into fewer, provided that the division is made into equal parts, our statement [concerning the relation of increments] would be the same. These for now I assume without any proof, since, they can become apparent enough to one who contemplates the figure. From the above the difference is apparent between a latitude uniformly difform and a latitude uniformly difformly difform. For although in each latitude of the two the same proportion is maintained among increments of degrees equidistant among themselves, yet in a latitude uniformly difform there is maintained a proportion of equality [i.e. one to one] such that as the first degree exceeds the second, so the second exceeds the third, as is clear in figure BC (See Fig. 6.10), while in figure AB the first does not exceed the second by as much as the second exceeds the third, but rather by less. Whence in figure BC the increments of degrees equidistant among themselves are equal to each other, so they keep the same proportion, namely a proportion of equality. In figure AB, however, the increments are not equal to each other; whence it is granted that among themselves they maintain the same proportion, not however a proportion of equality,

[Jacobus de Sancto Martino, *Latitudes of Forms* 395]

but of inequality [where numerator and denominator differ]; and if it is asked what that proportion is, I say that it is a proportion of one and one-half to one, which for now I assume without proof....

10. About the subject of the last two propositions, several things are to be noted.... The third thing to be noted is that in such a figure (i.e., a segment of circle) its intension is terminated at the greatest degree of slowness and its remission begins from the greatest degree of slowness, namely in the middle point of the arc.... this is obvious in figures CD and DE (see Figure 6.10). The fourth thing to be noted is that in any semicircle whatever the intension of its latitude begins from the greatest degree of swiftness and is terminated at the greatest degree of slowness, namely, in the middle point of the arc. Its remission, to be sure, which begins in the middle point of the arc, begins from the greatest degree of slowness and is terminated at the greatest degree of swiftness, as is clear in figure CD....

COMMENTARY

The text of the *De latitudinibus* from which we cite these few remarks has recently been established by a former student of mine, Professor Thomas Smith. It is hoped that this text will be published by Mr. Smith and myself as a part of a joint volume on the medieval geometry of qualities. Mr. Smith shows that two names appear most often on the manuscripts of this work: Oresme and Jacob of Florence (or Jacobus de Sancto Martino). Both names appear in manuscripts dated in the last decade of the fourteenth century. Mr. Smith's conclusion on authorship: "It is possible that Maier also could be right that the original work was composed by Jacob of San Martino. If we were to assume that the original work bore the legend in its colophon to the effect that it was composed by Jacob of San Martino (of Florence? or Naples?) according to the doctrine of Oresme, we could explain the diverging tradition of attribution." The only specifically dated copies all date from the 1390's, although one codex in which the work appears (Thorn Codex R. 4º.2) may be dated 1359 (although I doubt this).

Passages 1–8 describe the figures already mentioned by Casali and Oresme, the rectangle for a uniform quality, the right triangle for a uniformly difform quality starting or ending with zero degree, the quadrangle for a uniformly difform quality neither beginning nor ending with zero. The only point of interest is that the author is using the term latitude to describe the whole quantity of the quality—the sum of all the intensions

—the whole figure. Oresme *(De configurationibus,* Part I, chap. 3, Doc. 6.1) already noted "that certain modern people call (but not correctly) the whole quality itself latitude" although most schoolmen (including Oresme) used it for differences of degrees or the intension itself at a particular point or instant (cf. the distinction made by John Dumbleton in his *Summa,* see note 15). It should be observed that, although he describes the various kinds of latitude, the author of this treatise is nowhere concerned with equating the various figures, one with the other. Hence it is not surprising that he does not use Oresme's equation of a right triangle and rectangle as a proof for the mean speed theorem. In fact, he does not mention the theorem. He generally specifies whether his latitudes are concerned with permanent or successive entities and recognizes that extension in "successives" is duration or time instead of the subject as it is in "permanents."

Proposition 30 (passage 9) presents us with one of the most interesting conclusions of the *De latitudinibus formarum.* It is concerned with a "latitude uniformly difformly difform." One would imagine that the English authors, particularly Swineshead, and perhaps also Oresme, would have conceived of this latitude as representing a uniform change of a rate of change, or as a movement such that the acceleration was changing uniformly. If we plotted velocity against time, the desired figure would be a segment of a parabola. However, the author of the *De latitudinibus* does not conceive of uniformly difformly difform as a uniform rate of change. But rather he defines it in the following way: in any two successive equal parts of extension (or time), the ratio of the excesses in intension (or velocity) is a constant. But he tells us that a quadrant of a circle is the figure representing a latitude uniformly difformly difform; and he seems to assume that the constant ratio of excesses of intension is $3/2$. Now in actuality the equation of a circle does not satisfy his general definition of a latitude uniformly difformly difform, for in fact his definition would only be satisfied by an equation which related intension and extension exponentially. Furthermore, a circle could not satisfy the condition that in equal parts of extension the excesses of intension are related as $3/2$.

I think we must conclude (1) that our author had little mathematical intuition, (2) that he thought a circle was representative of a "uniformly difformly difform latitude," and (3) that he thought a circle had the relationship which we have described. It is of interest to note that Galileo in his youth follows the same curious definition of this term as does this author (see commentary to Doc. 4.7).

Additional *notanda* are given here in passage 10. The third one indicates that in a latitude that has the form of the segment of a circle (up to and including the semi-circle) the rate of change of intension decreases to a minimum as we approach the maximum value of intension, i.e., as we approach maximum intension EF, the rate of change of the length of line AB approaches a minimum (see Fig. 6.11). Or to put it in more modern

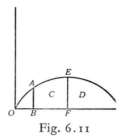

Fig. 6.11

terms, the slope of the curve approaches zero at point E. Similarly, as we start moving AB beyond EF and thus start to decrease AB, the rate of change of the decreasing AB starts from a minimum, i.e., the slope starts from zero. The fourth *notandum* notes that as we start from zero the rate of change of AB is at maximum. This again can be stated in modern terms by saying that at zero the slope is maximum.

*Tractatus de latitudinibus formarum**
(Editus a Jacobo de Sancto Martino secundum doctrinam Oresme?)

[1] Quia formarum latitudines multipliciter variantur que multiplicitas difficulter discernitur, nisi ad figuras geometricas consideratio
5   referatur....

---

* Text, prepared from extant manuscripts by Thomas Smith, is to be published upon the completion of a general volume on the medieval geometry of qualities by Mr. Smith and myself. The following manuscripts are known: Columbia Plimpton 171, ff. 4v–6v (*A*); Oxford, Bodl. Canon. Misc. 177, ff. 14r–16v (*B*); Paris, Bibl. de l'Arsenal 522, ff. 29r–33r (*P*); Vat. Chigi lat. F. IV. 66, ff. 13r–16r (*R*); Columbia Plimpton 187, ff. 144r–150r (*C*); Vienna, Nat.-bibl., cod. lat. 4953, ff. 1r–17v (*Vi*); Venice, Bibl. Naz. San Marco Lat., VI, 155, ff. 83r–87v (*V*); Thorn R. 4°. 2, pp. 198–206; Oxford, Bodl. Canon. Misc. 393, ff. 83r–87v; Oxford, Bodl. Canon. Misc. 506, ff. 458r–462v; Venice, Bibl. Naz. San Marco,

---

3–4 multiplicitas difficulter *RABV*
  multiplicitas difficile *PCVi*
  multiplices varietates difficillime *XYZ*
4 discernitur: distinguitur *A* discernuntur *XYZ*

4–5 consideratio referatur *RACViY*
  consideretur *P* consideratur *V*
  refferatur *B* quodammodo referuntur *XZ*

[2] 19ᵃ propositio. Omnis latitudo uniformis per totum ymaginanda est per quadrangulum rectangulum. Hoc probatur quia cum omnis latitudo uniformis sit eiusdem gradus per totum, ymaginanda est per figuram que sit eiusdem latitudinis per totum....

[3] 21ᵃ propositio. Omnis latitudo uniformiter difformis incipiens a non gradu ymaginanda est per triangulum rectilineum incipientem ab angulo acuto, terminatum ad angulum rectum....

[4] 22ᵃ propositio. Omnis latitudo uniformiter difformis incipiens a certo gradu et terminata ad non gradum ymaginanda est per triangulum incipientem ab angulo et terminatum ad angulum acutum....

[5] 23ᵃ propositio. Omnis latitudo uniformiter difformis incipiens a certo gradu et terminata ad certum gradum ymaginanda est per quadrangulum cuius duo anguli super basim sunt recti. Reliquorum vero alter acutus, alter est obtusus....

[6] 24ᵃ propositio. Nulla latitudo incipiens a non gradu et terminata ad non gradum est uniformis aut uniformiter difformis; licet possit habere partes uniformes vel uniformiter difformes....

[7] 26ᵃ propositio. Nulla latitudo secundum se totam difformiter difformis est per figuram rectilineam ymaginanda....

[8] 27ᵃ propositio. Omnis latitudo secundum se totam difformiter difformis ymaginanda est per figuram cuius altitudo terminatur per lineam curvam vel per lineas curvas....

Lat. VI, 62, ff. 139r–142v; Venice, Bibl. Naz. Lat. VI, 149, ff. 24r–27v; Munich, Bay. Staatsbibl., Cod. Lat. 4377, ff. 138r–141v; Munich, Bay. Staatsbibl., 26889, ff. 201r–206r. Vat. lat. 4829, ff. 124r–131v. These are in addition to the following editions: Padua, 1482; Padua, 1486 (*X*); Venice, 1505 (*Y*); Vienna, 1515 (*Z*). M. Curtze, "Eine Studienreise," *Centralblatt für Bibliothekswesen*, XVI. Jahrg., 6. u. 7. Heft. (1899), p. 263 (full article, pp. 257–306) mentions other manuscripts: Erfurt Ampl. Q. 384. Cod. Leips. Univ. 1480, Bamburgensis H. J. V. 2, Munich, Bay. Staats. 18985 and 19850, Kremsmünster 89 and 134, Melk 206 D. 44, and Vienna, Nat.-bibl. 2433. The edition will take into account as many of the manuscripts as are available. Here I have given variant readings as reported in Mr. Smith's preliminary edition. As usual, I have used capitalization for the letters of figures, although in the manuscripts these letters are lower-case letters.

---

7 quadrangulum rectangulum *RPB* quadrangulum recti angulum *V* figuram quadrangularem *C* figuram 4 angulam rectilineam *Vi* quadrilateram rectilineam *A* quadrangularem rectiangulum sive per quadrangulum rectiangulum *Y*

9 per¹: per tallem *B* / eiusdem latitudinis *ABCViV* latitudinis eodem *R*

12 terminatum: et terminato *B*
13 22ᵃ: 23ᵃ *B*
14 ad: e *R*
16 23ᵃ: 24ᵃ *B*
18 *ante* super *add. C* sunt *(?)*
20 24ᵃ: 25ᵃ *B* / *post* latitudo *add. C* est
21 ad: a *B*
23 26ᵃ: 27ᵃ *B*

[*De latitudinibus formarum* 399]

[9] 30ª propositio. Omnis latitudo uniformiter difformiter difformis ymaginanda est per triangulum habentem super basim unum angulum rectum et rectilineum, reliquos vero duos acutos et curvilineos.... Ad videndum autem quo modo in talibus figuris servetur eadem proportio inter excessus graduum eque distantium, describo triangulum AB qui est quarta pars circuli, cuius basis dividatur, gratia exempli, in 8 partes equales erectis lineis perpendicularibus in punctis divisionum, que linee mensurabunt altitudinem superficiei quelibet super puncto suo.... Deinde signo excessus illarum linearum eque distantium inter se, qui excessus representant excessus graduum eque distantium, et patet quod qualis est proportio primi excessus ad secundum, talis est secundi ad tertium, et qualis est proportio prime linee ad secundam, talis est secunde ad tertiam, et sic de aliis. Et eodem modo esset dicendum si basis dividatur in plures partes quam 8 vel etiam in pauciores, dummodo divisio fiat in partes equales. Ista pro nunc suppono sine aliqua probatione quia satis

29 *post* per *add.* R figuram
29–30 super... rectum: basim angulum rectum rectum X basim angulum rectum Z
30 reliquos vero duos $RPBCVVi$ et duos reliquos vero A reliquos vero XYZ
31 Ad *om.* XYZ / videndum: intelligendum PVi / autem: est XYZ / modo quo modo XYZ / servetur $APVViXYZ$ servatur BC servaretur R
32 excessus: se P ascensus XYZ / descripbo R
33 AB $PBCVViXYZ$ A et B R *om.* A / quarta: tertia P / cuius: et Vi / cuius basis: basis que P / dividatur: dividitur XYZ
34 gratia exempli *tr.* XYZ *ante* dividitur / in 8 $AB(?)RV$ *sed post* equales *add.* A vel in 10 *sed:* in octo aut v. P in viii C in 6 vel 10 Vi in 8 vel in 9 Y in sex partes Z in 6 partes X / equales erectis: equales in rectis B equales certis Y existentes XZ / perpendicularibus: perpendiculares XZ
35 punctis: puncto XYZ
36 signo: asigno (?) A signetur XYZ / excessum R / illarum *om.* P istarum V
36–37 illarum linearum *tr.* XYZ

37 eque distantium *tr.* PVi *post* inter se
37-38 qui... distantium *om.* C
37 representant excessus $ABVi$ representabant excessus P representat excessum $RXYZ$ representant excessum V
38 qualis est $RBPCViVXYZ$ talis sit A
38–39 proportio primi excessus: excessus primi $PXYZ$
38–40 primi... proportio *om.* BC
39 talis: qualis A
40 prime: quarte P
41 aliis: singulis A / Et eodem modo esset dicendum $BRCV$ concedendum esset A et eodem modo est PVi et eodem modo esset XYZ / si basis dividatur: si basis dividetur A de basi qui dividitur XZ si pars dividatur aut P
41–42 plures partes *tr.* R
42 quam... etiam *om.* P vel Vi / 8... pauciores: angulus XZ / divisio fiat $RBCVXYZ$; *tr.* $APVi$
43 pro: est C *om.* P
43–44 Ista... intuenti: Secunda pars et suppositio patere possunt sine alio exemplo satis XYZ
43 suppono: supponuntur A / sine... probatione *om.* BVi / aliqua: alia V
43–44 satis... figura BCV possunt satis

possunt apparere intuenti in figura. Ex hiis apparet differentia inter latitudinem uniformiter difformem et latitudinem uniformiter difformiter difformem. Nam licet in utraque latitudine servetur eadem proportio inter excessus graduum inter se eque distantium, tamen in latitudine uniformiter difformi servatur proportio equalitatis, ita quod quantum primus gradus excedit secundum, tantum secundus excedit tertium, ut patet in figura *BC*, in figura autem *AB* non quantum primus excedit secundum tantum secundus excedit tertium, ymo minus. Unde in figura *BC* excessus graduum inter se eque distantium sunt inter se equales, ideo servant eandem proportionem, scilicet proportionem equalitatis. In figura autem *AB* excessus inter se non sunt equales, unde licet inter se servent eandem proportionem, non tamen proportionem equalitatis, sed inequalitatis, et si queratur que proportio est ista, dico quod est proportio sexquialtera, quod pro nunc sine probatione suppono....

[10] Circa materiam illarum duarum ultimarum propositionum

apparere intuenti figuras *A* satis patent in figure intuenti *R* satis patet per figuras intuenti *P* //////// patet per figuras intuenti *Vi qui om.* quia satis
44 hiis: istis *B* illo *XYZ* / apparet: patet *P*
45 uniformiter difformem *tr. XYZ post* uniformiter difformiter difformem / et: et inter *Vi*
46 licet: hoc *XYZ* / utraque latitudine: eadem *A* eadem latitudine *XYZ* / servatur *XYZ*
47 inter se *om. A tr. R post* eque distantium / tamen: nam *Vi* et *XYZ*
48 difformi: difformiter *A R(?) V* / equalitat s: equalis *A*
48–49 ita quod quantum: in quocumque *XZ*
49 tantum *om. RXYZ* tantus *P*
49–50 tantum... tertium: secundus tertium, et tertius quartum *XYZ*
50 tertium: primum *P* / in figura[1]: per figuram *A*
50–51 ut... tertium *om. B*
50–52 in figura[2]... BC *om. A*
50–54 in figura[2]... equalitatis *om. XYZ*
51 excedit tertium: tertium *R*
51–52 minus unde *om. B*
52 BC: AB *B* / graduum *om. P* / inter se *tr. PVi post* eque distantium
53 inter se *om. PVi*
53–54 ideo... equales *om. AB* / eandem... equalitatis: inter se eandem proportionem equalitatis *C*
53 scilicet: secundum *P*
54 autem *om. P* / *post* excessus *add. XYZ* graduum / inter se *tr. VXYZ post* non sunt
55 licet *om. AC* / inter se *om. AXYZ* / se *om. Vi* / servent: servant *XZ* / eandem *om. Vi* / *post* tamen *add. Y* servet *et add. XZ* servant / proportionem *om. P*
56 sed inequalitatis *om. A* / sed... et: unde *XYZ* / queratur: queris *R* queritur *V*
56–57 que... ista: que sit iste proportio *Vi*
57 ista: ipsa *XYZ* / dico: dicitur *XYZ* / est proportio *om. Vi* est *A* / quod[2]: quam *XYZ* / pro *om. PB*
57–58 probatione: proportione *P* et *R correxit ex* propositione *aut* proportione
58 suppono: sub pono *R* presuppono *XYZ*
59 duarum: regularum *C*

60 plura notanda sunt.... Tertio notandum quod in qualibet tali figura sua intensio terminatur ad summum gradum tarditatis et sua remissio incipit a summo gradu tarditatis, videlicet, in medio puncto arcus.... patet in figuris CD et DE. Quarto notandum quod in quolibet semi-
65 circulo incipit intensio latitudinis sue a summo gradu velocitatis, et terminatur ad summum gradum tarditatis, scilicet in medio puncto arcus. Remissio vero eius, que incipit in medio puncto arcus, incipit a summo gradu tarditatis et terminatur ad summum gradum velocitatis, ut patet in figura CD.

60 notanda sunt: sunt manifeste A /
    Tertio: tertium *mg. B*
61 sua: summa PVi / tarditatis caliditatis Vi
61–62 et... videlicet: usque B
63 patet... DE *om. B* / Quarto: 4ᵐ

mg. B
64 *post* velocitatis *add. Vi* sue
64–65 et terminatur *iteravit R*
65 in $RBViV$ a APC
66 arcus¹ *om. B*
68 ut... CD *om. B*

Document 6.4

Blasius of Parma, *Questions on the Treatise on the Latitudes of Forms*

1. QUESTION III. Whether every uniformly difform latitude corresponds to its mean.... [This] is obvious from all those who commonly speak on this matter, as well as from rational exposition. It is argued as follows: I posit that A decelerates *(remittat motum suum)* uniformly from C degree [of velocity]—e.g., 4—to zero degree during half of the hour, while B accelerates *(intendat motum suum)* uniformly in the same half hour from exactly C degree [of velocity] to a degree which is twice C. With this posited, it follows that A and B [together] will traverse exactly as much [space], as they would traverse if they were moved continually with C degree [of velocity]. This is obvious, for just as much as B acquires of velocity increment *(latitudo motus)* in accelerating, that much exactly of velocity increment does A lose in decelerating. Therefore, B by its acceleration will traverse as much more [space than if moved uniformly with C velocity] as A by its deceleration traverses less [space than if moved

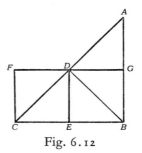

Fig. 6.12

uniformly with C velocity]. Therefore, exactly as much [space] will be traversed by A and B [together] as if they were moved continually with C degree of velocity [*gradus velocitatis*].

2. Furthermore, if C were a body uniformly difformly hot, whose latitude, for example, would be represented by triangle ABC (see Fig. 6.12),

[Blasius of Parma, *On Latitudes of Forms* 403]

and in whose more intense extreme were applied one corrupting [agent] and in whose more remiss extreme its contrary, in the same proportion, so that the corrupting [agent] would corrupt as much of latitude ABC in the more intense extreme as its contrary [agent] would introduce in the more remiss extreme, then it is evident that at the end of the action, latitude ABC would be uniform, and it would be of precisely the same [quantity] of intension as it was before the action of these [agents]....

3. ...I posit some suppositions: 1. Latitudes of forms are represented (*presentantur*) by geometrical figures. The author premises this supposition. 2. The line which divides two sides of any triangle into equal parts is one-half the third side. This is obvious from the fourth [proposition] of the sixth [book] of Euclid. Let there be a triangle ABC (see Fig. 6.12), and the line cutting two of the sides into equal parts we let be DE. I say that the line DE is one-half the line AB. For $BC/BE = AC/AD = AB/DE$, and $BC = 2BE$. Therefore $AB = 2DE$, and $DE = 1/2\ AB$. 3. Any parallelogram with one side equal to the mean degree and the other equal to the line of extension (*extensio*) is equal to the triangle by which a latitude uniformly difform is representable to us and whose mean degree line is the small side of the given parallelogram. This is apparent, [for], with triangle ABC given, whose mean line is DE, a parallelogram is formed of the aforementioned lines, and it is $BGCF$. This supposition is evident by the facts of geometry....

The second conclusion: every uniformly difform latitude, uniformly difformly acquired, whether it is a latitude of something with a "permanent" *esse* [like a quality] or is a latitude of something with a "successive" *esse* [like velocity], corresponds to its mean degree. This is clear by the first and third suppositions.

COMMENTARY

1. The first "proof" or explanation of the mean speed theorem given by Blasius is obviously very much like those found in the treatises of Heytesbury (or rather the author of the *Probationes* whom we have assumed to be Heytesbury) and Swineshead. I think there is little doubt that Blasius drew his proof from the English tradition. It is of the Merton *genre* that emphasizes the symmetrical gain and loss of velocity and space traversed in comparing uniformly accelerating and decelerating bodies with one moving uniformly at the mean velocity.

2. The second example, which concerns the action of two contrary agents which act so as to reduce a uniformly difform quality to uniformity,

is drawn directly from Giovanni di Casali's *De velocitate motus alterationis* (see Latin text of Doc. 6.2, pp. 389–90, lines 66–99), except that Casali converts a uniform quality into one that is uniformly difform by the continuous application of two contrary agents; in short Blasius' procedure is the reverse of Casali's. But it is clear that Blasius has used Casali's treatise as a source.

3. The ultimate source of Blasius' geometric proof of the mean speed theorem is either Casali's treatise or, more probably, the *De configurationibus qualitatum* of Oresme. Obviously, Blasius did not get the proof from the *De latitudinibus formarum*, since that treatise has no statement or proof of the mean speed theorem. Like the proof in the *De configurationibus* (but unlike that of Casali), Blasius appears to represent his intensity lines by vertical rather than horizontal perpendiculars. Needless to say, however, Blasius uses "latitude" in the sense labeled as "incorrect" by Oresme, i.e., as a name for the whole quantity of the quality or for the quantity of velocity. He obviously gets this usage from the *De latitudinibus formarum*, which served as the point of departure for his *Questiones*.

*Questiones super tractatu de latitudinibus formarum Blasii de Parma**

Questio III. Utrum quelibet latitudo uniformiter difformis correspondeat suo gradui medio.... patet per omnes communiter loquentes in hac materia, et etiam ratione. Arguitur sic: et volo quod A remittat motum suum uniformiter a C gradu, ut 4, in medietate huius hore, usque ad non gradum; et B intendat motum suum uniformiter, in eadem medietate illius, a C gradu precise ad gradum duplum ad C. Quo posito, sequitur quod A et B precise tantum spa-

---

* Edition of Venice, 1505 (*Ed;* no pag. but ff. 30r–32r; this quest. ff. 31r–32r). Cf. MS Oxford, Bodl. Canon. Misc. 177, ff. 97v–113v (*B;* old pag.; new pag. ff. 110v–113v; this quest. old pag. 99r–100v; new pag. 112r–113v); and cf. MS Venice, Bibl. Naz. San Marco Lat. VI, 155, ff. 88r–92 (*V*; this quest. ff. 90r–92r). Lower-case letters for figures again changed to capital letters. Additional manuscripts not used here are Bodl. Canon. Misc. 181, 15 c., ff. 64–66, Milan, Bibl. Ambros. F. 145, sup., 15 c., ff. 1r–5r; Bodl. Canon Misc. 393, 15c, 83–87, Vat. lat. 4829, 16c. Other editions are Padua, 1482 and 1486.

---

3 Questio III *Ed. om. V* quarto queritur *B* / uniformiter difformis *VEd om. B*
5 in *VB* de *Ed* / et etiam *Ed. om. V* et *B*
7–8 uniformiter *BV om. Ed*
8 medietate *VEd* hora usque *B* / illius *V om. BEd*
9 B *correxi ex* C *in MSS et Ed*

cium pertransibunt, quantum pertransirent si continue moverentur C gradu. Quod patet, nam quantumcunque B acquirit de latitudine motus per suam intensionem, tantum precise deperdet A de latitudine per remissionem sui motus. Ergo quanto magis per intensionem sui motus B pertransibit, tanto minus A per remissionem sui motus pertransibit. Ergo, tantum precise erit pertransitum ab A et B ac si continue moverentur C gradu velocitatis.

Preterea si C esset unum corpus uniformiter difformiter calidum cuius latitudo, gratia exempli, presentaretur per triangulum ABC, (*vide* Fig. 6.12) et in extremo eius intensiori applicaretur unum corrumpens et in extremo remissiori suum contrarium secundum eandem proportionem, ita quod quantum corrumpens corrumperet de latitudine ABC in extremo intensiori, tantum suum contrarium in extremo remissiori introduceret de latitudine, tunc patet quod in fine actionis latitudo ABC esset uniformis, et precise esset tante intensionis quante erat ante actionem istorum.

In hac questione erunt quatuor articuli. In quorum primo evidentie premittende sunt, in secundo distinctiones, in tertio conclusiones de quesito, in quarto difficultates. Quantum ad primum nota quod latitudo illa est uniformiter difformis cuius medius gradus per tantam latitudinem excedit non gradum sicut per quantam latitudinem ipse medius gradus exceditur ab intensiori gradu eiusdem latitudinis. Nota secundo quod non idem est sermo positus de latitudine uniformiter difformi motus localis et caliditatis, quoniam latitudo motus localis non habet esse permansivum, sed bene latitudo caliditatis. Et pro isto

---

10 pertransibunt *BV* transibunt *Ed* / pertransirent *BV om. Ed*
11 B *VEd* A B
12 intensionem *BEd* intensiorem *V* / deperdet *BV* deperdit *Ed*
14 B *tr. V* ante per / tanto *Ed* tantum *BV*
16 continue *VEd om.* B
17 Preterea *BV* Probatur sic *Ed* / difformiter *VEd* distan (!) B
18 presentaretur *BV* representaretur *Ed*
19 in *Ed om. BV*
20 suum *BV* etiam suum *Ed* / secundum *BV* applicetur secundum *Ed*
22 intensiori *BV* remissiori *Ed*
22–23 suum... latitudine *BV* introduceretur de latitudine in extremo intensiori *Ed*
26 In quorum primo *Ed* primo *BV*
27 premitte sunt *Ed* premittendo *BV* / in[1,2] *om. BV* / conclusiones *BEd* questiones *V*
28 in *om. BV*
29 illa B *om. Ed* illius *V* / tantam *VEd* totam B
31 eiusdem *BEd om. V* / *post* latitudinis *add. Ed* ut patuit in prohabita questione
31–32 Nota secundo *V tr. Ed* Nota primo B
32 idem *BV om. Ed* / est sermo *VEd tr.* B
33 et *BV* vel *Ed*
34 Et *V om.* B *Ed* / isto *BV* illo *Ed*

secundo notabili nota tertio qoud alius est sermo de latitudine motus quantum ad eius esse permansivum et quantum ad eius esse successivum. Nota quarto quod latitudo uniformiter difformis, quo ad eius esse successivum, potest dupliciter acquiri. Potest enim acquiri uniformiter, et etiam potest acquiri difformiter. Uniformiter ad istum sensum: quod si in hora debeat A latitudo uniformiter difformis acquiri, ita quod medietas acquiratur in medietate hore et alia in alia. Et secundum hoc erunt alie et alie conclusiones, et hoc de primo.

Quantum ad secundum, pono aliquas suppositiones. Prima: latitudines formarum per figuras geometricas presentantur. Hanc suppositionem auctor premittit. Secunda suppositio: cuiuslibet trianguli linea duo latera secans per equalia est subdupla ad tertium latus. Patet per quartam sexti Euclidis. Quoniam sit triangulus ABC (*vide* Fig. 6.12) et linea secans per equalia duo latera, DE, dico quod linea DE est subdupla ad lineam AB. Unde sicut BC ad BE et AC ad AD, sic AB ad ED. Sed BC est duplum ad BE. Ergo AB est duplum ad DE et per consequens linea ED est subdupla ad lineam AB.

Tertia suppositio: quodlibet paralegramon (!) ex linea medii gradus et linea extensionis constitutum est equale triangulo per quem nobis presentatur latitudo uniformiter difformis, cuius linea medii gradus est latus minus dati paralegrami. Hec patet: dato triangulo ABC et linea huius medii gradus DE, tunc fiat paralegramon constitutum ex predictis lineis et sit $BGCF$. Patet suppositio cui nota est geometria. Quarta suppositio: triangulus ABC datus est quadruplus ad triangulum eius partialem. Et ut hoc pateat: sit, gratia exempli,

---

35 nota *BEd* noto *V*
36 et *BV* et alius *Ed*
38 Potest...acquiri *BV* dupliciter acquiritur *Ed*
39 et etiam *Ed* et *B* etiam *V*
40 difformis *BV* difformiter *Ed*
41 ita quod *Ed* que *V* quod *B* / hore *Ed om. BV*
42 *post* primo *add. Ed* articulo
43 Prima *BV* Prima sit hec *Ed*
44 presentantur *BV* representantur *Ed*
45 premittit *V* premittit in littera *Ed* permittit *B*
46 secans *tr. V ante* est
48 DE *BV* sit DE *Ed*
49 sicut *B* sicut se habet *Ed* sic *V* / et *BV* ita se habet *Ed*
50 ED *BV* DE *Ed*
50–51 Sed... DE *VEd om. B*
51 ED *BV* DE sive *Ed*
52 suppositio *BV* suppositio est *Ed* / paralegramon *BV* paralogramum *Ed*
53 est *BEd om. V*
54 presentatur *BV* representatur *Ed*
56 huius *BV om. Ed*
57 et sit *BV om. Ed* / BGCF *correxi ex* BGCG *in MSS;* BGEG *Ed* / cui nota est geometria *BV* notata in geometria *Ed*
59 Et *V om. BEd* / pateat *VEd* patefiat *B* / gratia exempli *Ed om. BV*

triangulus *ABC* et linea *ED* que dividat per equalia duo latera trianguli que latera sint *AC* et *BC*. Dico tunc quod triangulus *ABC* est quadruplus ad triangulum eius partialem qui est *DEC*. Quod patet resolvendo quadrilaterum *ABDE* in tres triangulos isto modo: primo ducendo a puncto *D* usque ad lineam *AB* eque distantem linee *BE* et sit hec linea *DG*. Tunc triangulus *AGD* equalis est triangulo *DEC*. Deinde ducatur a puncto *D* ad punctum *B* linea *BD* et habes alios duos triangulos quorum quilibet cuilibet est equalis. Modo patet quod totus triangulus *ABC* resolutus est in quattuor triangulos equales, et per consequens totus triangulus est quadruplus ad quemlibet illorum et per consequens quadruplus ad triangulum *DEC*, et hoc fuit declarandum de secundo.

Quantum ad tertium pono conclusiones. Prima: non omnis latitudo uniformiter difformis quantum ad eius esse successivum correpondet suo gradui medio. Probatur per primam rationem. Unde sit *A* unum alterabile quod in hora acquirat sibi latitudinem caliditatis uniformiter difformem, non tamen acquirat eam uniformiter difformiter, sed bene difformiter. Sic scilicet quod in prima quarta hore huius alteretur a non gradu caliditatis usque ad 4 et in reliquis tribus quartis a 4 usque ad 8; tunc, si latitudo acquisita in prima quarta huius hore correspondeat suo gradui medio, *A* erit alteratum ut duo et si latitudo acquisita in reliquis tribus quartis correspondeat etc., tunc *A* erit alteratum ut 6 et per consequens in tota hora erit alteratum ut 8. Modo clarum est, quod si *A* fuisset alteratum ad caliditatem per hanc horam uniformiter ut 8, ergo maiorem latitudinem caliditatis acquisivisset *A* quam modo sit alteratum. Quare etc. et sic patet quod non omnis latitudo uni-

60 que *Ed om. BV* / dividat per equalia *BEd* per equalia dividat *V*
61 latera *BV om. Ed*
62 est quadruplus *BEd tr. V* / Quod *BEd om. V*
63 ABDE *BV* ABDC *Ed* / triangulos *BV* angulos *Ed* / isto *BV* illo *Ed*
64 primo ducendo *tr. B* / usque ad *Ed* versus *BV* / AB *VEd om. B*
65 et... DG *BV om. Ed* / equalis est *BV* erit equalis *Ed*
66 habes *V* sic habes *B* habebis *Ed*
67 cuilbet *BEd* cuiuslibet *V*
68 ABC *BV om Ed* / resolutus *VEd* resolutum *B*
70 *post* consequens *add. Ed* est
72 *post* prima *add. Ed* sit hec
74 Probatur per *BV* Patet propter *Ed* / Unde *BV* nam *Ed*
77 hore huius *BEd tr. V*
78 usque *VEd* ubique (?) *B* / a *VEd* ad *B*
79 huius hore *V om. BEd*
80 suo gradui medio *BV* etc. tunc *Ed*
81 reliquis *BV* aliis *Ed* / tribus *VEd* duabus sive tribus *B* / correspondeat *VEd* correspondet *B*
82 et... 8 *tr. Ed post* duo *in linea 80*
84 ergo *BEd om. V* / A *BV om. Ed*
85 non *VEd om. B*

formiter difformis quo ad eius esse successivum correpondet suo gradui medio. Secunda conclusio: omnis latitudo uniformiter difformis uniformiter difformiter acquisita tam quo ad eius esse successivum quam quo ad eius esse permanens correspondet suo gradui medio.
90 Patet per tertiam suppositionem cum auxilio prime. Tertia conclusio: cuiuslibet latitudinis uniformiter difformis incipientis a non gradu vel terminate ad non gradum, gradus medius est precise subduplus ad gradum summum. Patet per secundam suppositionem. Unde dicebatur quod linea DE est precise subdupla ad lineam AB. Et constat
95 quod ille due linee sunt duarum intensionum, quarum una est linea intensionis medii gradus alia est linea intensionis intensioris gradus. Quarta conclusio: nullius latitudinis uniformiter difformis incipientis a certo gradu et terminate ad certum gradum gradus medius est precise subduplus ad gradum summum. Patet per secundam supposi-
100 tionem: linea AB est precise dupla ad lineam DE, ergo est minor quam dupla ad quamlibet maiorem; sed quelibet linea cadens inter AB et ED est maior quam sit linea DE, ergo conclusio vera. Tenet consequentia, quia cuiuslibet latitudinis incipientis a certo gradu et terminate ad certum gradum, gradus medius cadit ultra lineam medii
105 gradus latitudinis terminate ad non gradum et sic patet conclusio. . . .

86 correspondet *BV* correspondeat *Ed*
92 gradus *BEd om. V*
93–94 dicebatur *VEd* patet *B*
94 precise *Ed om. BV*
96 intensioris gradus *BV* intensiorum graduum *Ed*
99 subduplus *BEd* duplus *V* / gradum *BV om. Ed*
102 AB *BEd* A et B *V* / ED *BV* CD *Ed*
103 et *BEd om. V*
104 cadit *BV* terminatur *Ed*
105 terminate *BV* terminante *Ed*

Document 6.5

Galileo Galilei, *The Two New Sciences**

THIRD DAY, THEOREM I, PROPOSITION 1. The time in which a certain space is traversed by a moving body uniformly accelerated from rest is equal to the time in which the same space would be traversed by the same body travelling with a uniform speed *(motu aequabili)*, whose degree of velocity *(velocitatis gradus)* is one-half of the maximum, final *(summum et ultimum)* degree of velocity of the original uniformly accelerated motion.

Let there be represented *(repraesentetur)* by extension *(extensio)* AB the time in which the space CD is traversed by a moving body accelerated from rest at point C (see Fig. 6.13). The maximum and last degree of

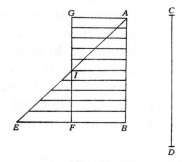

Fig. 6.13

all the degrees of velocity in the instants of time AB we let be represented by EB, constructed on AB. And, with AE drawn, all of the lines drawn from the individual points on line AB and parallel to BE will represent the increases of the degree of velocity after instant A. Then I bisect BE at F. A parallelogram $AGFB$ will be formed of the parallel lines FG, BA, and AG, BF. It will be equal to the triangle AEB, its side GF dividing AE into equal parts at I. For if the parallel lines of triangle AEB are extended up to IG, the aggregate of all the parallel lines contained in the quadrilateral will be equal to the aggregate of those contained in

* *Le Opere*, Ed. Naz., Vol. *8* (Florence, 1898), 208–12.

triangle *AEB*; for those in triangle *IEF* are equal to those contained in triangle *GIA*, while those which are in trapezium *AIFB* are common. Since each and every point on line *AB* corresponds to each and every instant of time *AB* and since the parallel lines drawn from these points and included in triangle *AEB* represent the growing degrees of the increasing velocity, while the parallel lines contained within the rectangle represent in the same way just as many degrees of nonincreasing but uniform velocity, it appears that there are assumed to be just as many moments of velocity *(momenta velocitatis)* in the accelerated motion represented by the growing parallel lines of triangle *AEB* as there are in the uniform motion represented by the parallel lines of *GB*. For the deficiency of velocity moments in the first half of the accelerated motion—the deficient moments being represented by the parallel lines of triangle *AGI*—is compensated for by the moments represented by the parallel lines of triangle *IEF*. It is obvious therefore that equal spaces will be traversed in the same time by two moving bodies, one of which is moved with a motion uniformly accelerated from rest, while the other is moved with a uniform motion having a moment half of the moment of the maximum velocity of the accelerated motion. Q.E.D.

Theorem II, proposition 2. If any moving body descends from rest with a uniformly accelerated motion, the spaces traversed in any times at all by that body are related to each other in the duplicate ratio of these same times, that is to say, as the squares of these times.

Let it be understood that the flow *(fluxus)* of time beginning from some first instant *A* (see Fig. 6.14) is to be represented by extension *(extensio)* *AB*, in which are taken any two time intervals *AD* and *AE*. And let *HI* be the line in which the moving body—beginning its motion from point *H*—descends with uniform acceleration. Let *HL* be the space traversed in the first time period *AD*, while *HM* we let be the space through which it will have descended in time *AE*. [With all these things posited,] I say thht the space *MH* to the space *HL* is in the duplicate ratio that time *EA* has to time *AD*; or let us say that the spaces *MH* and *HL* are in the same ratio as are the squares of *EA* and *AD*.

Let there be posited line *AC* making any angle at all with the line *AB*, while from points *D* and *E* draw parallel lines *DO* and *EP*; of these lines *DO* will represent the maximum degree of velocity acquired in instant *D* of the time interval *AD*, and *PE* the maximum degree of velocity acquired in instant *E* of time interval *AE*. Now it has been demonstrated above [in Theorem I] that, in regard to the spaces traversed,

[Galileo, *The Two New Sciences*

they are equal when one is traversed by a body moving with uniform acceleration from rest and the other is traversed in the same time by a body moving with a uniform motion whose velocity is one-half the

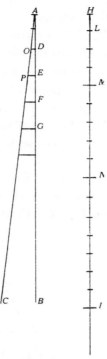

Fig. 6.14

maximum velocity acquired in the accelerated motion. It is therefore clear that the spaces MH and LH would be the same as those traversed in time intervals EA and DA by uniform motions whose velocities were one-half of PE and one-half of OD. Accordingly, if it is demonstrated that spaces MH and LH are in the duplicate ratio of the times EA and DA, our proposition will be proved. But in the fourth proposition of the first book [on uniform motion] it was demonstrated that the spaces traversed by two bodies moving with uniform motions are to each other as the product of the ratio of the velocities and the ratio of the times. Now in this case the ratio of the velocities is the same as the ratio of the times (for $1/2\ PE$ is to $1/2\ OD$—or PE is to OD—as AE is to AD); therefore the ratio of the spaces traversed is as the square of the ratio of the times. Q.E.D.

It is also evident from this that the same ratio of spaces is as the square of the ratio of the maximum degrees of velocity, namely lines PE and OD, since PE is to OD as EA is to DA.

Corollary I. Hence it is manifest that if we take any number of equal and consecutive time intervals starting from the first instant or the beginning of the motion, as for example AD, DE, EF, and FG during which spaces HL, LM, MN, and NI are traversed, these spaces will be related to each other as the odd numbers beginning with unity, i.e., 1, 3, 5, 7; for such is the ratio of the differences of the squares of the [velocity] lines—which lines are uniformly increasing by an amount equivalent to the smallest of these lines [i.e., to the line representing the velocity at the end of the first time interval]; or we may say [that this is the ratio of the differences] of the squares of the consecutive numbers beginning with unity.

SALV. Please suspend your lecture for a moment while I speculate on a certain idea that just now occurred to me. For its explanation I produce a small drawing (see Fig. 6.15) in order that it may be clearer

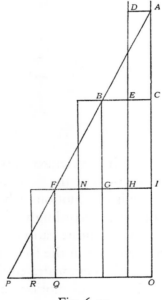

Fig. 6.15

to me and to you. Let the line AI represent the duration of time [measured] from the first instant A. Then after applying the line AF to A—making any angle you wish—and joining the termini I and F, I divide the time AI in half at C and draw CB parallel to IF. Next I consider CB as the maximum degree of velocity, [it being noted that] the velocity increases

[Galileo, *The Two New Sciences*

from rest at the first instant A in the same way that the lines parallel to BC and included in triangle ABC go on increasing—which increase [in velocity] is proportional to the increase in time. I then admit without dispute, in view of the preceding discussion, that the space traversed by the falling body whose velocity is increasing in the aforesaid way would be equal to the space which the same moving body would traverse if it were moved in the same time AC with a uniform motion whose degree of velocity was equal to EC, the half of BC. Furthermore, if we imagine that the moving body which has descended with accelerated motion is at instant C and has a degree of velocity BC, it is evident that, if it continued to move with the same degree of velocity BC without being accelerated further, it would traverse in the following interval of time CI a space twice that which it traversed in the equal time interval AC with a degree of uniform velocity EC, the half of degree BC. But because the moving body descends with velocity which is continuously and uniformly accelerated during all the equal time intervals, it will add to degree CB in the next time interval CI those moments of velocity that increase according to the parallel lines of triangle BFG, which triangle is equal to triangle ABC. Thus if we add to the degree of velocity GI one half of the degree FG, which is the maximum of those degrees acquired in the accelerated motion and determined by the parallel lines of triangle BFG, we shall have the degree of velocity IN. And this is the degree with which the body would be moved in uniform motion during time interval CI [if it were to traverse the same space as it does when accelerating]. Since the degree IN is triple the degree EC, it follows that the space traversed in the second time interval CI must be thrice that traversed in the first time interval CA. And if we imagine that another equal time interval IO is added to AI and the triangle extended to APO, it is evident that if the motion continued through the whole time interval IO at the degree of velocity IF acquired in the accelerated motion during time AI, the space traversed in time IO would be quadruple that traversed in the equal first time interval AC, because such a degree IF is quadruple EC. But if we continue the increase of uniform acceleration represented by triangle FPQ similar to that represented by triangle ABC, which acceleration when reduced to uniform motion adds [to IF] a degree equivalent to EC, and if we add [to IF such an additional degree] QR equal to EC, then the whole uniform velocity [equivalent to the accelerated motion] in time IO is five times that of the uniform [equivalent] velocity of the first time interval AC. And therefore the space traversed is five times that traversed in the first time interval AC.

It is thus evident by this simple calculation that [1] the spaces traversed in equal intervals of time by a moving body, which, starting from rest, goes on acquiring velocity in conformity with the increase of time, are related to each other as the odd numbers beginning with unity, 1, 3, 5; and [2] with the spaces traversed taken conjointly, that which is traversed in double time will be quadruple that traversed in half [of that double time], that traversed in triple time will be nine times as much [as that traversed in the first interval], and in general the spaces traversed are in the duplicate ratio of the times, i.e., as the squares of these times.

COMMENTARY

Again I have made the translation as literal as possible in order to illustrate the fact that Galileo is still using much of the same terminology as the fourteenth-century kinematicists. Like Blasius, Galileo uses the rectangle and triangle to "represent"* the uniform motion and the uniform acceleration. Time is still called by Galileo an "extension." Velocity intensities are still "degrees of velocity," the maximum velocity is the "summus" degree, and so on. The assumption of the areas as an aggregate of parallel lines which is clearly evident in all parts of the Oresme treatise is equally apparent in Galileo's proof. The inversion of the coördinates by Galileo so that the ordinate represents time and the abscissa represents velocity is similar to the procedure followed by Casali rather than to that of Blasius and Oresme. As will be evident in Chapter 11, the mean speed theorem was widely known in the sixteenth century and in fact appeared at least seventeen times in print. Galileo may have learned of its geometric proof from Casali, or from Blasius, or even from the 1494 edition of Heytesbury's *Regule solvendi sophismata* (see Plate 5).

The fortunes of the mean speed theorem with its geometric proof in the seventeenth century can be followed in the excellent treatment of A. Koyré in his *Études Galiléennes* (Vol. 2, Paris, 1939). I note also that Galileo had already virtually given the mean speed theorem and used the conventional triangular proof in his *Dialogue on the Two Great World Systems*, where in "The Second Day" he has Salviati say (translation of T. Salusbury as revised by G. de Santillana [Chicago, 1953], pp. 244–46; cf. *Le Opere*, Ed. Naz., Vol. 1, 255–56).

* Actually it is in the edition of Venice, 1505, of Blasius' *Questiones super tractatu de latitudinibus formarum* (see Doc. 6.4, Latin text, p. 406, line 44) that we find the verb "representare" used; in the two manuscripts used to prepare Document 6.4, the verb is "presentare." Needless to say, Galileo would probably have used an edition rather than a manuscript if he read the tract of Blasius.

[Galileo, *The Two New Sciences*

In the accelerate motion, the augmentation being continual, you cannot divide the degrees of velocity, which continually increase, into any determinate number, because, changing every moment, they are evermore infinite. Therefore, we shall be better able to exemplify our intention by describing a triangle, *ABC* (see Fig. 6.16). We take in the side *AC* as many equal parts as we please, *AD*, *DE*, *EF*, *FG*, and draw by the points *D*, *E*, *F*, and *G* straight lines parallel to the base *BC*. Now let us imagine the parts marked in the line *AC* to be equal times, and let the parallels drawn by the points *D*, *E*, *F*, and *G* represent to us the degrees

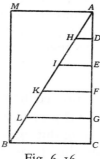

Fig. 6.16

of velocity accelerated and increasing equally in equal times; and let the point *A* be the state of rest, departing from which the body has, for example, in the time *AD* acquired the degree of velocity *DH*; in the second time we will suppose that it has increased the velocity from *DH* to *EI*, and so on, in the succeeding times, according to the increase of the lines *FK*, *GL*, etc. But because the acceleration is made continually from moment to moment, and not disjunctly from one certain part of time to another, the point *A* being put for the lowest moment of velocity, that is, for the state of rest, and *AD* for the first instant of time following, it is manifest that before acquiring the degree of velocity *DH*, made in the time *AD*, the body must have passed through infinite other lesser and lesser degrees gained in the infinite instants that are in the time *DA*, answering the infinite points that are in the line *DA*. Therefore, to represent to us the infinite degrees of velocity that precede the degree *DH*, it is necessary to imagine lines successively lesser and lesser, which are supposed to be drawn by the infinite points of the line *DA* and parallel to *DH*. These infinite lines represent to us the surface of the triangle *AHD*. Thus we may imagine any distance passed by the body, with a motion which begins at rest, and accelerates uniformly, to have spent and made use of infinite degrees of velocity, increasing according to the infinite lines that, beginning from the point *A*, are supposed to be drawn parallel to the line *HD*, and to the rest *IE*, *KF*, *LG*, and *BC*, the motion continuing as far as one will.

Now let us complete the whole parallelogram *AMBC*, and let us prolong as far as to the side *BM*, not only the parallels marked in the triangle, but those infinite others imagined to be drawn from all the points of the side *AC*; and like as *BC* was the greatest of those infinite parallels of the triangle, representing to us the greatest degree of velocity acquired by the movable in the accelerate motion, and the whole surface of the said triangle was the mass and sum of the whole velocity, with which in the time *AC* it passed such a certain space: so the parallelogram is now a mass and aggregate of a like number of degrees of velocity, but each equal to the greatest *BC*. This mass of velocities will be double the mass of the increasing velocities in the triangle, even as the said parallelogram is double to the triangle, and therefore if the body that, falling, did make use of the accelerated degrees of velocity answering to the triangle *ABC*, has passed in such a time such a distance, it is very reasonable and probable that, making use of the uniform velocities answering to the parallelogram, it shall pass with an even (i.e., uniform) motion in the same time a distance double that passed by the accelerate motion.

It is evident that Galileo has proved here that a body moving with a uniform movement at the final velocity of a uniformly accelerated motion beginning at rest—and continuing during the same period as the uniform acceleration—traverses double the space traversed in the course of the uniform acceleration. Notice that Galileo has assumed, and in fact specified, an infinite number of velocity degrees or parallel lines representing them. Notice further that the whole surface is designated as "the mass and sum of the whole velocity *(la massa e la summa di tutta la velocità)*." This identification of the space traversed with the total velocity was also made in the Middle Ages, as I have indicated above in Chapter 4, where I discussed the medieval distinction of the quality and quantity of velocity.

Document 6.6

The Journal of Isaac Beeckman
"On a Stone Falling in a Vacuum"*

(For dates 23 November—26 December 1618.)
WITH this agreed, things are moved downward toward the center of the Earth, the intermediate space being a vacuum, [in the following manner:] In the first moment so much space is traversed as can be by the attraction of the Earth *(per Terrae tractionem)*. With this motion continuing, in the second moment a new motion of attraction is added, so that double the space is gone through in the second moment. In the third moment, the doubled space remains, to which is added a third space resulting from the attraction of the earth, so that in the one [i.e., the third] moment a space triple the first space is gone through....

(p. 262) Since these moments are indivisibles *(individua)*, you will have a space such as ADE (see Fig. 6.17) through which something falls in

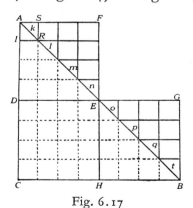

Fig. 6.17

one hour. The space through which it falls in two hours squares *(duplicat)* the proportion of time, i.e.:

---

* Translated from the Latin text in C. de Waard's edition of the *Journal tenu par Isaac Beeckman de 1604 à 1634*, Vol. I (The Hague, 1939), 260–62.

$$\frac{ADE}{ACB} = \left(\frac{AD}{AC}\right)^2.$$

For let the moment *(momentum)* of space through which something falls in one hour be of some magnitude, evidently $ADEF$. In two hours it will go through three such quantities, namely $AFEGBHCD$. But $AFED$ consists of ADE together with AFE and $AFEGBHCD$ consists of ACB together with AFE and EGB, i.e., with double AFE. So if the moment [of space] be $AIRS$, the proportion of space to space is as ADE plus k, l, m, n to ACB plus k, l, m, n, o, p, q, t, i.e., double k, l, m, n. But k, l, m, n is much less than AFE. Since, therefore, the proportion of space traversed to space traversed will consist of the proportion of triangle to triangle plus some equal additions to each terminus, and since [the sum of] these equal additions continually becomes smaller as the [individual] moments of space [considered] are smaller, it follows [that the sum of] these additional moments would be of no quantity *(nullius quantitatis)* when a moment of no quantity is taken. Moreover, such is the moment of space through which a thing falls. It remains, therefore, that the space through which something falls in one hour is related to the space through which it falls in two hours as the triangle ADE is to the triangle ACB. [Q.E.D.]

COMMENTARY

While Beeckman is still basically giving us the medieval triangle, unlike any previous author, including Galileo, he is taking into account the actual infinitesimal demonstration of the applicability of the triangle to the problem. He is thus closer to a correct demonstration by integral calculus than any of the other authors. His is certainly superior to Descartes' efforts to use the triangle. For this whole problem the reader is urged to read A. Koyré, *Études Galiléennes*, Vol. 2, 25–54.

Part III

MEDIEVAL DYNAMICS

Chapter 7

Aristotelian Mechanics and Bradwardine's Dynamic Law of Movement

WE have already remarked on the distinction between dynamics and kinematics which arose at Oxford in the first half of the fourteenth century.

But we have not yet discussed the dynamic law describing velocity in terms of an exponential relationship with the ratio of motive force and resistance that Thomas Bradwardine made so popular. In actuality, it would have been more proper from a chronological standpoint to describe this dynamic law before the Merton kinematic theories, for the Merton schoolmen ordinarily treated dynamics before kinematics. And it was, as we recall, in treating an objection to his dynamic theory that Bradwardine introduced the distinction between "qualitative" (instantaneous or intensive) velocity and "quantitative" velocity (the total velocity over some period of time measured by the distance traversed during that period of time).

Before describing Aristotle's "law" together with Bradwardine's critique, it will be useful to summarize the basic principles of Aristotelian physics generally accepted in the Middle Ages.[1]

Aristotle's physical picture distinguishes between celestial and terrestrial areas of activity. Looking for the moment at the terrestrial world, we are aware first of the doctrine of matter and form. There is a substratum, a prime matter, the independent existence of which is only potential but the

---

[1] The next six paragraphs are substantially taken from my *Greek Science in* *Antiquity* (New York, 1955), pp. 64-67.

actual existence of which is always in conjunction with form. Something is said to belong to classes as a result of its form, but it is said to be individual as the result of matter. In short, matter is the principle of individuation. Form gives something its essential character. It is by its form that we recognize it for what it is. But when we examine more deeply the principles of things, we are struck by four types of causation—one might say by four "factors" that are involved in things that exist in the terrestrial world. The first is the "material" factor, the second is the "formal," the third is the "efficient," and the fourth is the "final" or purposeful cause.

The four causes can be explained by analogy with something artificially produced. A bed is a bed because it is made of wood (the material cause), in a given shape (its formal cause), by a carpenter (its efficient cause), for the purpose of providing slumber (final cause).

As we look at the fundamental manifestations of matter and form in the terrestrial world, we detect or infer (for they are never in a pure state) the existence of four basic elements—earth, water, air, and fire—each having a distinctive pair of qualities, earth is cold and dry; water, cold and wet; air, hot and wet; fire, hot and dry. These elements tend to arrange themselves concentrically about the center of the world: the earth is a sphere at the center, and the water, air, and fire are successive shells about the core of the earth. There is constant change as the elements are transformed into one another when the substratum is differently formed, or as the elements compound themselves differently with one another. In the elementary changes, the substratum is common to the four elements. The commentators later suggest that there is some prime form joined with prime matter (and as potential as the prime matter) which constitutes corporeity, also common to all of these elements, and which renders matter susceptible to receiving three dimensions. Now, as we have suggested, the elements tend to seek a static arrangement according to their places about the center of the world, but because of the continuous change in elements and compounds brought about ultimately under the influence of the movements of celestial bodies, in actuality change and movement are the order of nature. The study of nature becomes the study of movement.

Movement in its broadest sense is said by Aristotle in some passages to include not only locomotion—i.e., change of place—but also qualitative alteration, quantitative augmentation and diminution, and even on occasion generation (coming into being) and corruption (passing away). In an effort to find a definition to include all these categories, he seizes upon a description of movement as the actualizing[2] of something that exists in potentiality. For example, something white undergoes qualitative alteration and becomes black. When actually white, it nevertheless is potentially its contrary, black. This particular qualitative movement is, then, the actualizing of the potential blackness. Such a definition seems to be an ambiguous locution, for the nature of *potential existence* is most elusive. Essential to this concept of movement was the doctrine of the existence of fundamental sets or pairs of contraries, with movement being the passage from one contrary of the pair to the other. This approach to movement as the actualizing of the potential seems to have been advanced to get around the basic criticism of the Eleatic philosophers of the preceding century, that change and movement do not exist, for if they did, something existent would follow from something nonexistent, or in short the nonexistent must exist, a fundamental contradiction. Aristotle is saying rather that movement and change are not examples of the existent arising from the nonexistent but only of changes of the mode of existence, changes from potential existence to actual existence.

Local movement of the elements and their compounds—i.e., terrestrial local movement—is governed by the doctrine of natural place. In the concentric arrangement of the elements, each element has its natural place, the earth at the center, the water adjacent to it, and the air and fire in successive places. Now if an element is removed from its natural place, it tends to return to that place by a straight-line movement. Hence if we pick up a stone and let go of it, since it is predominantly earth it falls downward in an

---

[2] Aristotle defines motion in two different ways: (1) "Motion is the actuality (ἐντελέχεια) of that which exists in potentiality in so far as it is in potentiality." (*Physics*, III.1, 201a. 10–11; *Metaphysics*, XI.9, 1065b.16), and (2) "Motion is the actuality of that which is movable in so far as it is movable." (*Physics*, III.2, 202a.7–8; *Metaphysics*, IX.9, 1065b.22–23). The word ἐνέργεια is used as equivalent to ἐντελέχεια in the definition of motion in *Physics*, III.2, 201b.31, and in the *Metaphysics*, IX.9, 1065b.22–23. The object of both of these definitions is, as Wolfson points out in his *Crescas' Critique of Aristotle* (Cambridge, Mass., 1929), p. 75, "to establish the nature of motion as something which is neither pure potentiality nor a complete actuality but a potentiality in the process of realization." Crescas adds a third definition: "Motion is a change or transition from potentiality to actuality." (*Ibid.*, pp. 75, 233; cf. the discussion of Averroës and others, pp. 253–30.) While the word ἐντελέχεια (*actus* in Latin) does not have a strictly verbal connotation and gives a sense of completion that appears to contradict the "incompleteness" that is the essence of motion, it is clear that the general purport of Aristotle's discussion is to describe motion as the "actualizing" (or, better, the "process of actualization") of that which is in potentiality. The word ἐνέργεια perhaps renders this idea better. For the latter term, see Wolfson's *Crescas' Critique*, p. 526, and my note 7 on p. 510 below.

attempt to return to its natural place near the center of the universe. The same is true with water; rain generated in the midst of the air tends to seek its natural place and so falls. Air and fire, on the other hand, tend upward, for their natural place is above us. The movement resulting from the tendency of elements to return to their natural places is called *natural movement*. Contrariwise, if we remove a body from its natural place—e.g., lift a stone—we have acted against the natural tendency of the body, producing thereby unnatural or *violent movement*.

It can be seen that local movement is intimately tied to qualitative changes or alteration, for if a transformation of elements, one into another, takes place and the new element is formed out of its natural place, then the tendency of that newly formed element will be to seek its natural place, thus producing movement. Similarly, if the proportion of elements in a mixture or a compound is altered sufficiently, movement can result. Thus something that is predominantly earth may by exposure to heat or some other influence change to a mixture predominantly of fire, and the result is the motion of the new mixture upward. And since both elementary and mixture changes are going on unceasingly, the terrestrial world is essentially in motion. The laws of movement that Aristotle deduces from gross experience, laws that necessitate the continual presence of mover, or at least a source of motion, and moved, force and resistance, for the continuance of movement, we shall deal with later.

We have spoken briefly of the terrestrial, or sublunar, world according to Aristotle. Distinct from it, but surrounding it and to be sure ultimately influencing its changes by its own movements, is the finite celestial world, including the moon, the sun, the planets, and the stars. Here the laws of movement and the nature of the constituent element differ from those in the sublunar world. Celestial bodies are formed of a fifth element, ether (the later quintessence); and ether is unchanging and eternal so far as its qualitative nature is concerned. This celestial fifth element possesses only the tendency for uniform, continuous, circular movement. Thus circular movement is natural to celestial bodies and actual movement is presumably engendered by souls or intelligences. Unlike sublunar rectilinear movement, the circular movement of the celestial bodies does not require the continued presence and substantial contact of both force and resistance.

The finer details of the system were, of course, much debated in the Middle Ages, and those debates are not without importance, as Anneliese Maier's works have so excellently shown.[3] We are here concentrating

---

[3] See the various works of A. Maier listed in the bibliography.

mainly on a few critical points of the system. However, in order that the reader may have at least one continuous and comprehensive summary of medieval Aristotelian physics, I have included as Document 7.1 a general late medieval description of movement, force, resistance, elements, and the like, which was commonly, if not universally, accepted in the fourteenth century. This short summary of a modified Aristotelian view was in fact put as a preface to the Venice edition of Bradwardine's *Proportions* in 1505. It is in all probability not by Bradwardine but dates perhaps from later in the century or even the next century when the "latitudes" vocabulary had become current.

But now let us focus on Aristotle's views on the quantitative laws which describe the force and resistance factors in determining distances traversed in periods of time.

The first point to recognize is that for Aristotle motion is a process arising from the continuous action of a source of motion or "motor" and a "thing moved." The source of motion or motor is a force[4]—either internal as in natural motion[5] or external as in unnatural motion—which

---

[4] In the passages we are about to discuss, Aristotle uses two principal words for "force": (1) δύναμις, which, as our succeeding quotations show, is generally rendered as *potentia* in the medieval translation from the Greek attributed to Moerbeke; it is often rendered as "power" in English, but sometimes as "force"; and (2) ἰσχύς, generally *virtus* in Moerbeke's translation, but on occasion also rendered as *potentia*; in other medieval translations we also have *vis* and *fortitudo*; in English the common translation is simply "force." In the passages where Aristotle deals with the dynamic rules, there is no clear distinction made between the two terms, although in other passages, we can see a distinction, with δύναμις carrying the idea of "capability" or "potency" and ἰσχύς as the actual strength of some forceful exertion. The Arabs tend to use *qūwat* to render both words in the few passages I have examined. Other Greek words used in dynamics by Aristotle and the commentators are: (1) ῥοπή (*inclinatio, impulsus,* "impulse"), (2) βία (*virtus violenta, violentia,* "muscular force" or "violence"), and (3) ὁρμή (*impetus,* "shock" or "impact").

[5] Some scholars, like Father J. A. Weisheipl, O.P., who particularly follow St. Thomas' interpretation, insist that we cannot call the source of motion in natural motion a motor. Weisheipl says "...it is clear that Aristotle did not explain natural motion by the constant exerted efficiency of a mover, as is often thought.... For Aristotle as well as for St. Thomas the form is *not the mover*, but the source of necessary and spontaneous movement. Avicenna in his *Sufficientia* and Algazel's paraphrase, *Maqâcid el-falâcifa*, propound the theory that in natural movement the form is the mover of the body which it informs. In a certain sense Averroes follows this opinion." *De natura et gravitatione* (River Forest, Ill., 1955), pp. 26–28. It does not appear so clear to me, however, that Aristotle would deny that when a body is away from its natural place, its form is a motive principle or intrinsic motor in the broad sense of these terms. The passages cited by Weisheipl do not prove that there is no *intrinsic* motor but only that there is no *external* motor. Notice that the author in Document 7.1 identifies "form" and "intrinsic motor."

during motion must be in contact with the thing moved. Thus he tells us in the beginning of Book VII of the *Physics*,[6] "Everything which is moved is necessarily moved by something. If it has not the source of motion within itself, then it is clear it is moved by another...." Or again, further on in the same book:[7]

Everything that is in local motion is being moved by itself or by something else. In the former case it is obvious that the moving principle and that which it moves are immediately in touch with each other, for since the source of movement is inherently contained in what it moves, there can be nothing intermediate between them. Whereas things that are moved by an external agent, are so moved either by pulling or pushing or carrying or turning.... But in any case the agent cannot move anything from itself to somewhere else or from somewhere else to itself, unless it is in contact with it. So it is clear that there is no intermediary between the mover and the moved in the case of local movement.

Motion then requires some source of motion and a thing moved in continuous contact.[8] More specifically it requires *force* and *resistance*. But

---

[6] *Physics*, VII.1.241b.34–36 (edition of W. D. Ross [Oxford, 1936]; lines 24–26 in older editions). The so-called new translation from the Greek, perhaps done by Moerbeke, reads (in the edition accompanying the commentaries of Averroës, Venice, 1495, f. 111r): "(Textus commenti 1) Omne quod movetur, necesse est ab aliquo moveri. Si quidem igitur in seipso non habet principium motus, manifestum est quod ab altero movetur, aliud enim erit movens."

[7] *Physics*, VII.2.243a.12–17, 244a.14–244b.2. The translation here given is from P. H. Wicksteed and F. M. Cornford, *Aristotle, The Physics*, Vol. 2 (Cambridge, Mass., London, 1934), 217, 219, 223. In the medieval translation cited in note 6, the text runs (f. 113r): "(Textus commenti 10)...Primum [quidem] igitur de loci mutatione dicamus. Hic enim primus motuum est. Omne igitur quod fertur, aut ipsum a seipso movetur, aut ab altero. Si quidem [igitur a] seipso, manifestum est quod in seipso movente existente simul movens et quod movetur erit, et nullum illorum medium. Quod autem ab alio movetur, quadrifarie dicitur. Qui enim sunt ab altero motus, quattuor sunt: pulsio, tractio, vectio, vertigo.... Manifestum igitur, quoniam quod movetur et movens simul sunt, et nullum ipsorum medium est." (The section just preceding the last section is quite different in this medieval translation from the Greek text now accepted. But since it does not bear on the point under discussion, I have here omitted it. The words in brackets have been added from the copy of this translation appearing with Walter of Burley's *Super libros Aristotelis de physica auscultatione...commentaria* [Venice, 1589], cc. 860–61.) I have altered the punctuation somewhat.

[8] Needless to say, there is no possibility of an inertial concept in a system that demands continuing force wherever there is motion. However, Aristotle does present such a concept as a kind of obvious absurdity when arguing against the existence of a void. He says (*Physics*, IV.8, 215a.19–22), speaking of a body moving in a void, "further, no one could give any reason why, having been set in motion, it should stop anywhere: for why here rather than there? Hence either it will remain at rest, or it must continue to move *ad infinitum* unless something stronger impedes it."

how are the force and resistance that determine the space traversed in a given time related to the distance and the time? Chapter 5 of Book VII gives us a partial answer for unnatural motion:[9]

If now A be the movent, B the thing moved, C the distance traversed, and D the time taken, then, in the same time [D], an equal force A will move $1/2$ B the distance 2 C and, in $1/2$ D, A will move $1/2$ B the distance C; for so the due proportion will be maintained. And if the same force [A] moves the same weight [B] a certain distance [C] in such-and-such a time [D] and half the distance in equal time, then half the force will move half the weight an equal distance in an equal time (i.e., $1/2$ A will move $1/2$ B a distance C in time D). Thus let E be half the force A and F half the weight B; the force and the weight are then similarly related, that is, in the same proportion, in the two cases, so that [the respective forces] will cause the respective weights to move an equal distance in an equal time. But if E moves F the distance C in the time D, it does not necessarily follow that E will in the same time move 2 $F(=B)$ the distance $1/2$ C; that is to say, if A moves B the distance C in the time D, it does not follow that $1/2$ A or F will move B, in time D, or in any part of it, some part of the distance C such as bears the same ratio to the whole of C that E bears to A (Aristotle, by a slip, says "A to E"); in fact it may happen that E cannot move B at all. For it does not

---

[9] *Physics*, VII.5.249b.30–250a.28. The English translation is that of T. L. Heath, *Mathematics in Aristotle* (Oxford, 1949), pp. 142–43, as are the additions and explanations, which, however, have been indicated by the system I have used elsewhere in this volume. The medieval translation runs (*edit. cit.* in note 6, ff. 120v–21v; and in Burley's text, cc. 919–25): "(Textus commenti 35)...Si igitur *a*, quod est movens, *b* autem, quod movetur, quantacunque autem mota longitudo est *c*, in quantocunque est tempore in quo est *d*. In equali ergo (igitur?) tempore equalis potentia in quo est *a* medietatem ipsius *b* per duplicationem ipsius *c* movebit; per ipsum autem *c* (! *e*) in medietate ipsius *d*, sic enim erit analogia. (Textus commenti 36) Et si eadem potentia *b* (! *delete*) [idem] in hoc tempore tantumdem movebit et medietatem in medietate movebit. Et media virtus medium movebit in equali tempore [per equale], ut si ipsius [potentie] *a* sit medietas, que est ipsum *e*, et ipsius *b*, *z* sit medium, similiter igitur se habebit secundum analogiam virtus ad grave. Quare equale et in equali tempore movebit. (Textus commenti 37) Et si *e* ipsum *z* movet in ipso *d* secundum *c*, non necessarium est in equali tempore *e* duplum ipso *z* movere secundum medietatem ipsius *c*. Si vero *a* ipsum *b* movebit in ipso *d*, quantum est ipsum *c* medietas ipsius *a* que est in quo est *e* ipsum *b* non movebit in tempore in quo est *d*, neque in aliquo [ipsius] *d* per aliam partem que est ipsius *c*, secundum quod est analogia ad totum *c*, sicut est *a* ad *e*; omnino enim si contingit non movebit nullum. Si enim tota virtus totam movet, medietas non movebit neque tantam neque in quolibet. Unus enim moveret navem, si quidem navem trahentium deciditur potentia in numerum et in longitudinem, quam omnes moverunt.... (Textus commenti 38) Si vero duo et utrumque horum autem utrumque movet tantum in tanto, et potentie composite compositum ex gravibus equali movebunt longitudine, et in equali tempore, simile namque est secundum analogum numerum." (Note: this translation agrees fairly closely with the so-called old Greek translation appearing in MS Paris, BN lat. 16141, first column, ff. 169v–170v.)

follow that, if the whole of a given force has moved [a thing] such-and-such a distance, the half of the force will move it any distance in any length of time whatever; were it so, we should have to suppose that one man could by himself move a vessel on the water, that is to say, if you have only to divide the total power of the hauliers and the distance they collectively moved the vessel by the number [of the gang].... But if we have two forces acting on two weights, and each moves one of the two weights such-and-such a distance in such-and-such a time, then the forces added together will move the weight made up of the two [together] an equal distance in an equal time.

In brief, the import of this whole statement is that with time and force constant, the distance traversed is inversely proportional to the resistance, or with distance and force constant, the time is directly proportional to the resistance—with the proviso that the force is sufficient to cause movement. Similar statements are made in the *De caelo*:[10] "there being a certain force which moves [the body], a smaller and lighter body will be moved a greater distance [in the same time] by the same force," and, "the speed of the lesser body will be to that of the greater as the greater body is to the lesser." The first of these statements appears to assume, as in the initial passage quoted, that, with time and force constant, the distance traversed varies inversely as the resistance, while the second passage considers space and time together and indicates that speed is inversely proportional to resistance. It is assumed further in unnatural motion (although it is not specifically stated so in these passages) that, so long as the force is great enough to produce motion, and with time and resistance constant, the distance traversed is directly proportional to the magnitude of the moving

---

[10] *De caelo*, III.2.301b.4–5, 11–13 (edition of D. J. Allan [Oxford, 1936]). The English translation is of Heath, *Mathematics in Aristotle*, pp. 143–44. A manuscript with one of the medieval versions of the *De caelo* was not available to me at the time I composed these notes. The translation accompanying the text of Thomas Aquinas, *In Aristotelis libros de caelo et mundo expositio* (Rome, 1952), p. 302, III.2, 442 (27) has the following: "...Quoniam enim virtus quaedam movens, minus et levius ab eadem virtute plus movebitur.... Velocitatem enim habebit quod minoris ad hoc quod maioris, ut maius corpus ad minus." The translation from the Greek in the Junta edition of the *De caelo* (Venice, 1574), f. 195v, reads: "(Textus commenti 27)...Cum enim potentia quaedam sit ea, quae movet, quod autem minus est, ac levius, ab eadem potentia plus moveatur....nam velocitas minoris sese habebit ad eam, quae est maioris, ut maius corpus ad minus." The Latin translation accompanying the text of Averroës (presumably by Michael Scot) in the same edition runs: "Quoniam, cum omne corpus motum non moveatur, nisi ab aliqua potentia, necesse est ut corpus minus, et levius, moveatur per spatium maius, quam corpus maius, et gravius, ex illa eadem potentia....erit igitur proportio velocitatis motus corporis minoris ad maius corpus, sicut proportio velocitatis maioris corporis ad corpus minus."

force.[11] Although these various statements do represent the beginnings of a quantified dynamics, it is evident that no precise metric statements are involved, particularly as to the measure of force.[12] The motive force in these unnatural movements is evidently constituted in the energy or activity of the motor, be it the physical-muscular energy of an animate being or the stored energy of a bow under tension, or something similar. The primary resistance in unnatural motion is weight, plus friction where this is significant.

Now before restating the substance of these passages in terms of simple proportions, we should add some significant passages regarding natural motion, which, with certain alterations as to what constitute force and resistance, can also be encompassed by the same basic proportionality statements. Several passages tell us that speed is directly proportional to weight, and weight is, of course, the inherent internal force of natural motion. We read in the *De caelo*:[13] "If such-and-such a weight moves

---

[11] This seems to be the substance of Hero of Alexandria's statement quoted below (see p. 433).

[12] In this regard the remarks of M. R. Cohen and I. E. Drabkin, *A Source Book in Greek Science* (New York, 1948), p. 203, are pertinent: "[Aristotle's] achievement consists [in these passages] in applying mathematics to physical phenomena, and in making certain abstractions which such treatment requires, e.g., in neglecting, as irrelevant, differences in the bodies moved other than weight (and, by implication, shape), in considering the medium perfectly homogenous, which it never is in nature, and defining force quantitatively in terms of the effect produced. But in reaching the result that, in 'forced' motion, the distance would vary directly as the product of force and time, and inversely as the weight, Aristotle fails to carry abstraction and analysis far enough. The effect of friction and the resistance of medium are included in the factor of weight. The basic case of a body moving without friction in a medium devoid of resistance, under the influence of a single constant force, is not considered."

[13] *De caelo*, I.6.273b.30–274a.2. Again the English translation is that of Heath, *Mathematics in Aristotle*, pp. 165–66. See the translation with Aquinas' *Expositio* (*edit. cit.* in note 10), p. 55, 73 (51): "Si enim tanta gravitas tantam in hoc tempore movet, tanta et adhuc in minori. (74) Et analogiam quam gravitates habent, tempora e converso habebunt: puta si media gravitas in hoc, duplum in medietate eius." The Junta Latin translation (*edit. cit.* in note 10, f. 37r) runs: "(Textus commenti 51)...Si enim tanta gravitas per tantum in hoc tempore movetur, tanta et adhuc in minori; et eam analogiam, quam gravitates habent, tempora econtrario habebunt, ut, si dimidia gravitas in hoc, dupla in dimidio huius." The greater freedom of the Latin translation accompanying the Averroës text is illustrated by the translation of the same passage, accompanying the above translation from the Greek: "...si quantitas huius gravitatis movetur per hoc spatium in hoc tempore, movebitur alia quantitas gravitatis in minori tempore, et etiam in minimo tempore. dicamus modo quod coniunctio temporum adinvicem est secundum coniunctionem gravitatis adinvicem. sed coniunctio tamen temporum differt a coniunctione gravitatis. si enim medietas gravitatis moveretur per aliquod spatium in aliquo tempore, movebitur duplum illius gravitatis finitae per illud spatium

such-and-such a distance in such-and-such time, such a weight with more added will move the same distance in a less time, and the times will have a ratio to one another which is the inverse of that of the weights; for example, if the half of the weight moves the distance in a given time, its double (i.e., the whole weight) will take half the time." The same thing is stated in the *Physics*:[14] "We observe that things which have a greater bias ('ροπήν, i.e., impulse) of weight or lightness, if conditions are in other respects the same, move more quickly over an equal distance, and that in the ratio which the magnitudes have to one another." With the simple and direct relation between weight and speed assumed, it is not surprising that, in the case of bodies of different sizes but the same material, the speed is then proportional to the volume.[15] Further, when we have the same volume of two bodies of different density, the "lighter" in a relative sense is that one "whose natural velocity in a downward direction is exceeded by that of the other."[16] Thus we have a dynamic definition of density which is elaborated upon later in a treatise known as the *Liber de ponderoso et levi*, a treatise of some influence in the medieval period. We shall quote its essential definitions and propositions shortly.

Further passages in the works of Aristotle hold that in natural motion, when the weight is the same, the speed is inversely proportional to the resistance, as in the case of unnatural motion. But for natural motion the resistance is the density of the medium. Thus we read in the *Physics*:[17]

per medietatem illius temporis."

[14] *Physics*, IV.8.216a.11–16. The English translation is that of Heath, *Mathematics in Aristotle*, p. 119. The Latin translation (*edit. cit.* in note 6, f. 59r; Burley edition, c. 480) runs: "(Textus commenti 74)... Videmus enim maiorem impetum* habentia, aut gravitatis aut levitatis, sic quidem alia similiter se habent figuris, citius lata per equale spacium finitum, et secundum rationem quam habent magnitudines adinvicem." (*The Burley edition has *inclinationem*; MS Paris, BN lat. 16141, f. 87r, has respectively, *motum*, *ponderosa*, and *ponderis* for the old translation from the Greek, Scot's translation, and Gerard of Cremona's translation, both of the latter translations from the Arabic.)

[15] *De caelo*, IV.2.309b.12–15.

[16] *Ibid.*, IV.1.308a.29–33.

[17] *Physics*, IV.8.215a.24–215b.10. English translation of Heath, *Mathematics in Aristotle*, p. 116. The medieval Latin text (*edit. cit.* in note 6, f. 57 r; Burley edition, cc. 461–53) runs: "(Textus commenti 71)... Videmus enim idem grave et leve corpus velocius ferri propter duas causas, aut in differendo per quid [fertur], ut per aquam aut per terram aut [per] aerem, aut in differendo quod fertur, si alia sint eadem, propter excellentiam gravitatis aut levitatis. Hoc quidem igitur per quod fertur causa est, que (quia?) impedit maxima (maxime?) quidem quod contra fertur, postea autem et manens. Magis autem quod non facile dividitur, huiusmodi autem pinguius est. Id autem in quo est a movebitur per b, quod autem in quo est c tempus per ipsum autem d, cum sit subtilius, in quo e tempus, si equalis longitudo est ipsi b que est ipsius d, secundum ana ogiam impedientis corporis. Sit enim b quidem aqua, d vero aer. Quanto ergo

We see the same weight moving faster owing to two causes, (1) that there is a difference in the medium through which it moves—e.g. it may be water or earth or air—or (2) that, other conditions remaining the same, there is a difference in the moving body due to excess of weight or lightness. Now the medium through which a thing moves causes a difference in speed because it impedes the moving thing, most of all if it (the medium) is moving in the opposite direction but to a certain extent also when it is at rest; the resistance is greater if the medium is not easily divided, that is to say, is more dense. Thus A will move through B in the time C, and through D, a thinner medium in the time E, and if the length of B is equal to D, the times will be in the proportion of [the resistance] of the impeding body. Let B be water and D air; then in proportion as air is thinner and more incorporeal than water, so much faster will A move through D than through B. Let then the speed be to the speed in the same ratio as that in which air differs from water [in thinness]. Then, if, for example, air [D] is twice as thin [as water, B], A will traverse B in twice the time it takes to traverse D—that is, the time C will be twice the time E.

Needless to say, the same remarks which were made about the lack of metric precision in regard to force in unnatural motion apply equally to the density of the medium in the formulations for natural motion.

We are now prepared to summarize the various passages in the following simple proportionality statements:[18]

(1) $S_1 : S_2 :: F_1 : F_2$ when $T_1 = T_2$ and $R_1 = R_2$, and assuming that movement occurs. S_1, S_2 are distances. T_1, T_2 are times. In "forced" movement F_1, F_2 are external forces; R_1, R_2 are weights. But in "natural" movements (like that of falling bodies) F_1, F_2 are weights and R_1, R_2 are the densities of the media the weights fall through.

(2) $T_2 : T_1 :: F_1 : F_2$ when $S_1 = S_2$, $R_1 = R_2$, and movement occurs.
(3) $S_1 : S_2 :: R_2 : R_1$ when $F_1 = F_2$, $T_1 = T_2$, and movement occurs.
(4) $T_1 : T_2 :: R_1 : R_2$ when $S_1 = S_2$, $F_1 = F_2$, and movement takes place.

One could, for the sake of economy, use a single modern formula to

(igitur?) subtilius aer aqua et incorporalius, tanto citius *a* per *d* movetur quam per *b*. Habet ergo (igitur?) eandem rationem secundum quam vere distat aer ab aqua velocitas ad velocitatem. Quare si dupliciter subtile est, in duplici tempore quod est ipsum *b* transibit *a* quam *d*. Et erit in quo est *c* tempus duplex eo in quo est *e*, et semper iam quantocunque sit incorporalius, et minus impeditivum et bene divisibilius per quod fertur, citius movebitur." As before, the words in brackets are accepted additions from the Burley edition; the words in parentheses are alternate and preferable readings from the Burley edition.

[18] Cf. I. E. Drabkin, "Notes on the Laws of Motion in Aristotle," *American Journal of Philology*, Vol. 59 (1938), 60–84, and Heath, *Mathematics in Aristotle*, pp. 142–46.

express all of these cases, although, of course, Aristotle does not do this:

(5) $V \propto S/T \propto F/R$ with V as the speed.[19]

Thus the basic dynamic formula which we might deduce from the scattered statements of Aristotle is that speed is proportional to the ratio of the motive force to the resistance, provided that the force is sufficiently great to overcome resistance and produce movement. Now suppose that we had a natural movement in a vacuum. The density of the medium would obviously be zero and thus the movement would take place instantaneously (or in the modern formula above, V would go to infinity as R goes to zero).[20] Since an instantaneous movement appears to Aristotle to lead to contradictions, it is unthinkable, and hence a vacuum does not exist.

It is difficult to know whether Aristotle's views had much currency before their criticism by the commentators of late antiquity. The most extensive summary of Aristotle's laws of motion is found in the treatise *De ponderoso et levi* which we mentioned above. It was attributed to Euclid and thus might be dated before the commentators. However, for the sake of convenience we shall postpone its treatment until after we have examined the views of Philoponus. Still, we ought to observe in passing that

---

[19] It can be argued that the functional expression $V \propto F/R$ does *not* faithfully represent the Aristotelian rules of motion for two reasons: (1) For Aristotle velocity is not a separate magnitude defined as the ratio of two unlike quantities, space and time, as the formula holds (see Chapter 3 above). (2) The statement of such a function implies continuous divisibility of force with a consequential continuous divisibility of velocity. But Aristotle specifically objected (as we have already noted) that force cannot be continually divided and still produce motion. Hence the formula does not apply in cases where F is less than or equal to R. It was because of Aristotle's objection that Bradwardine later interpreted Aristotle as holding that velocity increases arithmetically as the ratio of force to resistance increases geometrically (see p. 438 below). An interesting interpretation of Aristotle's view of motion as discontinuous and the historical importance of that view for the rise of psychophysics among the Islamic philosophers has been presented by Léon Gauthier, *Antécédents gréco-arabes de la psycho-physique* (Beyrouth, 1939).

[20] Aristotle, talking in terms of proportionality statements, expresses the substance of the argument as follows (*Physics*, IV.8.251b.12–22; Heath, *Mathematics in Aristotle*, pp. 116–17): "But there is *no* ratio in which the void is exceeded [in density] by body, just as *nothing* has no ratio to a number... Similarly the void cannot bear any ratio to the full; neither therefore can motion through the one bear a ratio to motion through the other; but if the thing moves through the thinnest medium such-and-such a distance in such-and-such a time, its motion through the void would exceed in speed any ratio whatever." The Latin translation (*edit. cit.* in note 6, ff. 57r, 58v) reads: "(Textus commenti 71) Vacuum autem nullam habet rationem qua excellatur a corpore, sicut neque nihil ad numerum.... Similiter autem et vacuum ad plenum nihil possible est habere rationem; ergo neque ad motum.... (Textus commenti 72) Sed si per subtilissimum in tanto tali quod fertur, per vacuum omnem exuperat rationem."

Hero of Alexandria (first century A.D.) in his *Mechanics* reflects Aristotelian ideas by his assertion of the direct relationship between force and ease of movement (i.e., speed?) in both natural and unnatural motion.[21] "Why do greater weights fall to the ground in shorter time than lighter ones? It is because, just as when the *external (min khārij)* motive force [applied] to them is greater they move more easily, so when their *internal (fī nafsihā)* force is greater they move more easily. And in natural motion the force and attraction *(al-jadhbu)* [downward] in a larger weight is greater than in a smaller weight."

The Aristotelian rules and their implications were submitted to severe criticism in the sixth century by the Neo-Platonic (and later Christian) commentator on Aristotle, John Philoponus, in his commentary on Aristotle's *Physics*. Anxious to prove that movement in a vacuum can take place, John criticized the basic dynamic laws of Aristotle in the following passage:[22]

Weight, then, is the efficient cause of downward motion, as Aristotle himself asserts. This being so, given a distance to be traversed, I mean through a void where there is nothing to impede motion, and given that the efficient cause of the motion differs, the resultant motions will inevitably be at different speeds, even through a void.... Clearly, then, it is the natural weights of bodies, one having a greater and another a lesser downward tendency, that cause differences in motion. For that which has a greater downward tendency divides a medium better. Now air is more effectively divided by a heavier body. To what other cause shall we ascribe this fact than that which has greater weight has, by its own nature, a greater downward tendency, even if the motion is not through a plenum?....

And so, if a body cuts through a medium better by reason of its greater downward tendency, then, even if there is nothing to be cut, the body will none the less retain its greater downward tendency.... And if bodies possess a greater or a lesser downward tendency in and of themselves, clearly they will possess this difference in themselves even if they move through a void. The same space will consequently be traversed by the heavier body in shorter time and by the lighter body in longer time, even though the space be void. The result will be due not to greater or lesser interference with the motion but to the greater or lesser downward tendency, in proportion to the natural weight of the bodies in question....

For if a body moves the distance of a stade through air, and the body is not at the beginning and at the end of the stade at one and the same instant, a definite time will be required, dependent on the particular nature of the body in question, for it to travel from the beginning of the course to the end (for, as I have indicated,

[21] *Mechanics*, II. 34d (edition of L. Nix [Leipzig, 1900]).

[22] Cohen and Drabkin, *Source Book*, pp. 217–19, translating Philoponus' commentary *(Commentaria in Aristotelem graeca,* Vol. *17* [Berlin, 1888], p. 678, line 24 to p. 682, line 11).

the body is not at both extremities at the same instant), and this would be true even if the space traversed were a void. But a certain *additional time* is required because of the interference of the medium. For the pressure of the medium and the necessity of cutting through it make motion through it more difficult.

Consequently, the thinner we conceive the air to be through which a motion takes place, the less will be the *additional time* consumed in dividing the air....

If a stone move the distance of a stade through a void, there will necessarily be a time, let us say an hour, which the body will consume in moving the given distance. But if we suppose this distance of a stade filled with water, no longer will the motion be accomplished in one hour, but a certain additional time will be necessary because of the resistance of the medium. Suppose that for the division of the water another hour is required, so that the same weight covers the distance through a void in one hour and through water in two. Now if you thin out the water, changing it into air, and if air is half as dense as water, the time which the body had consumed in dividing the water will be proportionately reduced. In the case of water the additional time was an hour. Therefore the body will move the same distance through air in an hour and a half. If, again, you make the air half as dense, the motion will be accomplished in an hour and a quarter. And if you continue indefinitely to rarefy the medium, you will decrease indefinitely the time required for the division of the medium, for example, the additional hour required in the case of water. But you will never completely eliminate this additional time, for time is indefinitely divisible.

For Philoponus, then, the fundamental and original determiner of movement is motive force. If there were no resistance, that is, if bodies were moving in a vacuum, a certain force would implant in a certain body a given movement, and the body would move a certain distance in a certain time (the original time): $S/T_o \propto F$.

John says that in addition to the original time there is an added time due entirely to the resistance. This added time is directly proportional to the density of the medium. Thus John is saying that $T_f = T_o + T_a$ and $T_a \propto R$ where T_f is the total time taken to move through S when there is resistance R, T_o is the time to move through S where there is no resistance, and T_a is the additional time due exclusively to the resistance. By this doctrine of original time, motion in a vacuum was saved.[23] Later

---

[23] Cosmas Indicopleustes, a contemporary of Philoponus, whose Christian cosmology is absurdly primitive, was influenced by the Aristotelian dynamics rather than that of Philoponus, for he tells us in his *Christian Topography*: "Imagine an area one hundred cubits deep, filled with one of the heavier elements, like water, for example. If a man grasps a stone in his hand and casts it [into the water], how long will it take the stone to reach the bottom? Assume that it will take four hours. Then if we suppose the area to be filled with a lighter element, like air, how much time

authors were also to interpret this view as saying essentially that speed is proportional to the *difference* of force and resistance (rather than to the *ratio* of force to resistance as Aristotle was interpreted as saying).

One other interesting interpretation of Aristotle's law that was to be joined with Philoponus' criticism by early modern writers like Benedetti and Galileo, but which some also believe to have originated in antiquity, was that the speed of fall in the case of natural movement is proportional to specific gravity rather than to gross weight. Some say (although I think wrongly) that Philoponus, in his celebrated passage criticizing Aristotle's law, implies this by an appeal to the experience of dropping unequal weights which are asserted to fall in about the same time. (See Chapter 9 below, where the substance of the passage is quoted.) It has also been suggested that such is the idea in the above-mentioned *Liber de ponderoso et levi* attributed to Euclid.

For the examination of this claim let us quote the pertinent parts of the *De ponderoso* asserting the relationship between force and speed:[24]

(Postulates)

(1) Bodies equal in volume are those which fill equal places.

(2) And those which fill unequal places are said to be of different volume.

(3) And what are said to be large, among bodies, are said to be capacious, among places.

(4) Bodies are equal in force, whose motions through equal places, in the same air or the same water, are in equal times.

(5) And those which traverse equal places in different times, are said to be different in force.

(6) And that which is the greater in its force, is the lesser in its time.

(7) Bodies are of the same kind which, if of equal volume, are of equal force.

(8) When bodies which are equal in volume are different in force with respect to the same air or water, they are different in kind.

(9) And the denser body is the more powerful.

---

will it take the stone to reach the bottom? Clearly less time, say two hours. If we assume a still lighter substance, we can say the stone will touch the bottom in one hour. In an even lighter medium, the stone will arrive in half an hour. The finer the medium, the less time will be required, until the medium has been so far reduced as to become incorporeal. In that case, necessarily, time will disappear also. Thus, a weight dropped into a medium that is without substance and completely incorporeal will reach the bottom without passage of time and remain there." M. Anastos, "Aristotle and Cosmas Indicopleustes on the Void," *Prosphora eis Stilpōna P. Kuriakidēn* (Thessalonika, 1953), pp. 35–50 (this passage, pp. 40–41).

[24] E. A. Moody and M. Clagett, *The Medieval Science of Weights* (Madison, 1952), pp. 40–47.

(Theorems)

I. Of bodies which traverse unequal places in equal times, that which traverses the greater place is of the greater force....

II. If, of two bodies of the same kind, one is a multiple in size of the other, then its force will be similarly related to the force of the other....

III. Of bodies of the same kind, the volumes and forces are proportional....

V. When the volumes and the forces of several bodies are in the same proportion, the bodies are of the same kind....

Now it seems clear to me that all that this author is saying is that, when the bodies are equal in volume, the one that is denser exerts the greater force (because it is of greater weight), and thus its velocity is greater in the same medium. He is not saying, as Moody seems to hold,[25] that because a larger weight would encounter a greater volume of the medium, the resulting velocity is the same as long as the different weights are of the same specific weight. Our author might have held such a view, but, as Dijksterhuis has pointed out, there is actually no specific consideration of the resistance of the medium in this treatise. The significance of this treatise historically is, as we have seen in Chapter 2, that it provided the chief source for a dynamic definition of specific weight.

But let us return to the main line of development. The views of Philoponus, and in particular his apparent rejection of the simple force and resistance ratio of Aristotle, were repeated by the Spanish Moslem Ibn Bājja (Avempace, d. 1138). This we know principally from their refutation by Ibn Rushd (Averroës, d. 1198) in his famous commentary on the *Physics* of Aristotle.[26] For Avempace as for Philoponus, the principal determiner

---

[25] *Ibid.*, p. 24. "Thus, if two bodies of the same density, but of unequal size or volume, fall in the same medium, they will fall at the same speed." See the review of Moody and Clagett, *Medieval Science of Weights* by E. J. Dijksterhuis, *Archives internationales d'histoire des sciences*, Vol. 6 (1953), 505, where he criticizes Moody's position.

[26] See the excellent article by E. A. Moody, "Galileo and Avempace: The Dynamics of the Leaning Tower Experiment," *Journal of the History of Ideas*, Vol. 12 (1951), 163–93, 375–422. Some of the pertinent remarks of Averroës concerning the theory of Avempace, showing its similarity to the position of Philoponus, can be cited (edition of Venice, 1495, ff. 57v–58r): "(Textus commenti 71)...Avempace autem bene movit hic questionem.... Et hec sunt verba eius in septimo sui libri, dicit: 'et ista resistentia que est inter plenum et corpus quod movetur in eo est illa, inter quam et potentiam vacui fecit Aristoteles proportionem in quarto, et non est sicut existimatur de eius opinione. Proportio enim aque ad aerem in spissitudine non est sicut proportio motus lapidis in aqua ad motum eius in aere. Sed proportio potentie continuitatis aque ad potentiam continuitatis aeris, est sicut proportio tarditatis accidentis rei mote ex illo in quo

of difference in speed is difference in motive power. The retardation caused by a medium appears to be a factor to be *subtracted* from the natural motion which the body would have in a vacuum.

With the translation of the *Physics* and the commentary of Averroës into Latin, both the original "law" as deducible from Aristotle's statements and the criticisms of Philoponus became available to Latin schoolmen. Under the influence of Averroës' treatment, there was a great deal of discussion concerning the determining factors in movement, and this was important for the ultimate distinction by Ockham of the kinematic and dynamic aspects of motion (see Chapter 4). But the *mathematical* exposition of the law inherited from the Arabs did not undergo important modification until Thomas Bradwardine's *Treatise on Proportions*, dated 1328.[27]

As both mathematician and natural philosopher, Bradwardine attempted to reconcile what he believed to be two facts: (1) No motion takes place when force is equal to, or less than, resistance[28] (a fact of experience, also implied by Aristotle's discussion). (2) The factors of force and resistance as determiners of speed are in some kind of ratio[29] rather than relatable

movetur, verbi gratia, aqua, ad tarditatem accidentem ei, quando movetur in aere....'
(58r) Et si hoc, quod dixit, concedatur, tunc demonstratio Aristotelis erit falsa, quoniam, si porportio subtilitatis medii ad subtilitatem alterius medii est sicut proportio tarditatis accidentis moto in altero eorum ad tarditatem accidentem ei in alio, non sicut proportio ipsius motus, non sequitur ut illud quod movetur in vacuo, moveatur in instanti. Quoniam tunc non aufertur ab eo nisi tarditas, que accidit ei propter medium, et remanet ei motus naturalis et omnis motus est in tempore, ergo illud quod movetur in vacuo, movetur in tempore necessario, et motu divisibili, et nullum sequitur impossibile. Hec igitur est questio Avempace.... Et causa istius erroris fuit existimare, quod tarditas et velocitas sunt motus additi motui et diminuti ab eo, sicut linea que additur linee, aut diminuitur ab ea. Cum igitur existimatur, quod velocitas est additio super motum naturalem, et tarditas diminutio, et causa istius additionis et diminutionis fuerit resistentia, sequitur ex hoc, ut proportio resistentis ad resistens sit sicut proportio tarditatis ad tarditatem, non sicut proportio motus ad motum...."

[27] See the edition of H. Lamar Crosby, Jr., *Thomas of Bradwardine's Tractatus de proportionibus* (Madison, 1954). The introduction misrepresents symbolically a number of Bradwardine's statements, as E. Grant will show in his edition of the works on proportionality of Nicole Oresme.

[28] This is what is meant by Bradwardine's crucial eighth conclusion (edition of Crosby, p. 114): "Ex nulla proportione aequalitatis vel minoris inaequalitatis motoris ad motum sequitur motus." Oresme succinctly states the essence of this idea when he uses it to criticize the conventional proportionality statement of Aristotle (see Doc. 7.2, passage 2): "...from this statement [of simple proportionality] it would follow that a power could move a resistance equal to itself and that any power, however small, could move any resistance, however large.... This is illogical and impossible."

[29] In his criticism of the first theory (Philoponus' theory) that the proportion of speeds follows the excess of the power

by a simple arithmetical difference. Aristotle is his authority for this second fact. Now it is clear that if we assume the mathematical relationship of $V \propto F/R$ which satisfies the second fact, then the first fact is contradicted. For, putting it in modern symbolic terms, if $F = R$, then V is not equal to zero as fact number (1) asserts but rather is equal to some value k greater than zero. The same thing results if we make F less than R; namely, V is not zero but some value k greater than zero.[30] So Bradwardine rejects any law of simple proportionality. For less cogent reasons he also rejects, among other positions, the possibility of a simple arithmetic relationship of force and resistance such as might be deduced from Philoponus' type of criticism, namely a relationship like $V \propto (F - R)$.[31] The problem then for Bradwardine was to find some proportionality which would leave F and R as a ratio but which would satisfy the fact that when they are equal no motion takes place. His answer was that "the proportion of velocities in motions follows the proportion of the power of the motor to the power of the thing moved."[32] At first glance this would appear to be merely a restatement summarizing Aristotle's proportionality statements. However, the succeeding conclusions show that his intention was to hold that the velocity increases arithmetically as the ratio of force to resistance increases geometrically, i.e., that the following relationship should be understood:[33]

$$F_2/R_2 = (F_1/R_1)^n \text{ where } (F_1/R_1) > 1 \text{ and } n = V_2/V_1.$$

of the moved over that of the resistance, Bradwardine makes it abundantly clear that the force and resistance are related as a ratio. (See the whole discussion, *edit. cit.*, pp. 86–93, 110).

[30] This type of criticism is certainly implied by Bradwardine (*edit. cit.*, p. 98): "Est autem ista positio ex mendacio arguenda, quia aliqua potentia motiva localiter potest movere aliquod mobile aliqua tarditate, et potest movere dupla tarditate. Ergo (per istam positionem) potest movere duplum mobile. Et potest movere quadrupla tarditate; igitur quadruplum mobile, et sic in infinitum. Igitur quaelibet potentia motiva localiter esset infinita.... Tertio est ista positio super mendacio arguenda, quoniam experimentum sensibile docet huius positionis contrarium. Videmus enim quod, uno homine movente aliquod ponderosum (quod potest vix solus movere motu valde tardo) si alius sibi adiungatur, illi duo movent illud multo plus quam in duplo velocius." Oresme's criticism in Document 7.2 is most succinct in presenting in medieval proportionality terms the argument I have here given.

[31] Bradwardine opposes four different theories of the dynamic measure of motion (*edit. cit.*, pp. 86–110). These theories are summarized in our Document 7.3, passage 4 and the appropriate notes of criticism.

[32] Bradwardine outlines his theory in Chapter III of his *Tractatus de proportionibus* (*edit. cit.*, pp. 110–24). His general statement of the theory (p. 110): "Scientia autem veritatis ponit quintam opinionem, dicentem quod proportio velocitatum in motibus sequitur proportionem potentiae motus ad potentiam rei motae." It is also presented succinctly in Document 7.3, end of passage 4 and all of passage 5.

[33] See notes 63–70.

As Maier has pointed out,[34] in modern terms we would say (but not without anachronism in notation) that this is an exponential relationship, statable as follows:[35]

$$V = \log_a (F/R'; \text{ where } a = F_1/R_1.$$

It can be seen that Bradwardine has at least partially solved his problem, for if F is equal to R, $\log_a 1 = 0$, since the log of one to any base is zero, and thus V is zero. As he puts it, no motion arises from a proportion of equality (see note 28). In the case where F_1 is less than R_1, he states categorically that motion does not arise. It is clear that Bradwardine's formula was not submitted to any crucial experimental tests and that the empirical evidences adduced for it are not decisive. So far as representation of experience is concerned, Philoponus' criticism interpreted by the formulation $V \propto (F - R)$ is much better. And it was in fact this formulation which Galileo accepted in his youth.[36]

One other important aspect of Bradwardine's law should be noticed.

---

[34] A. Maier, *Die Vorläufer Galileis im 14. Jahrhundert* (Rome, 1949) p. 92.

[35] Bradwardine's formula relating an arithmetical increase in velocity to a geometrical increase of the ratio of force to resistance was foreshadowed to some extent by the theory of the Arabic philosopher al-Kindī (d. after 870), relating an arithmetic ordering of degrees of intensity with a geometrical increase of the determinate qualitative powers. Thus al-Kindī assumed that a medicament or compound is "in equality" (i.e., will not produce a sensible hot or cold effect on a patient) when its frigidity (i.e., power of frigidity) is equal to its calidity (i.e., power of calidity). It is hot in the first degree (or will produce on a body or patient an effect of "hotness in the first degree") when its calidity is twice its frigidity. Further, it is hot in the second degree when its calidity is four times its frigidity; it is hot in the third degree when its calidity is eight times its frigidity; and finally it is hot in the fourth degree when its calidity is sixteen times its frigidity. See his *De investigandis compositarum medicinarum gradibus*, with Peter of Abano, *Supplementum in secundum librum compendii secretorum medicinae Joannis Mesues* (Venice, 1581), ff. 269r–273r, and particularly, f. 271r, c.1. See also M. Clagett, *Giovanni Marliani and Late Medieval Physics* (New York, 1941), pp. 34–35. Cf. the Arabic text and expert analysis of L. Gauthier, *Antécédents gréco-arabes de la psychophysique* (Beyrouth, 1939). Since this treatise of al-Kindī was translated by Gerard of Cremona, it was no doubt available to Bradwardine. Furthermore, Bradwardine was one of the first to make the analogy between velocity as intensity of motion and the intensity of permanent qualities. It is not then much of a step to go from (1) an exponential relationship between qualitative degrees of intensity and qualitative powers to (2) an exponential relationship between degrees of motion and the ratio of motive and resistive powers. It is certainly within the realm of probability that Bradwardine took precisely this step. Al-Kindī's theory was opposed by Averroës in his *Colliget*, edition in the *Aristotelis opera omnia...cum commentariis Averrois*, Vol. *10* (Venice, 1574), ff. 130r–132r; Cf. Gauthier, *op. cit.*, pp. 72–98.

[36] Moody, *op. cit.* in footnote 26 above, pp. 172–73, Cf. Chapter 11, footnote 130.

His law was applied in at least one instance to instantaneous speed (or "qualitative" velocity as Bradwardine called it) rather than to average or total velocity over some given duration of time, as Aristotle's law had been. We have already stressed this distinction and given the pertinent passages in Chapter 4.

For the actual presentation of the law, we have chosen two documents less complicated and detailed than the original treatise of Bradwardine. The first of these (Doc. 7.2) is a succinct and lucid passage from Nicole Oresme's French commentary on the *De caelo* of Aristotle, written almost half a century later. In a few well chosen sentences, Oresme presents the Bradwardine "correction" of Aristotle's law of simple proportionality. The second document (Doc. 7.3) consists of a small tract abbreviating the dynamic sections of Bradwardine's treatise. I have included almost the whole tract in order to illustrate the Euclidian vocabulary of proportions as applied to motion. Early manuscripts of this treatise indicate that it was composed from the longer treatise about the middle of the fourteenth century. And although it no doubt precedes the Oresme treatise in date of composition, I have allowed it to follow Oresme's treatment because of the simplicity and lucidity of the latter.

The popularity of Bradwardine's treatise, both in the original form and in the abbreviated tract given here, is attested by the great number of manuscripts of these treatises still extant, as well as by the numerous reworkings of his ideas.[37] Thus, for example, Bradwardine's dynamic formulation made an immediate impression at Oxford among his junior contemporaries. Of these, Richard Swineshead[38] and John Dumble-

---

[37] Crosby in his edition of the *Tractatus de proportionibus* (pp. 59–61) lists some thirty manuscripts of Bradwardine's work. For a list of some of the manuscripts of the *Abbreviatus*, see the Bibliography at the end of our work. In addition to the reworkings mentioned here in this chapter, a number are given in Chapter 11.

[38] Swineshead's *Liber calculationum* is much concerned with Bradwardine's dynamics, particularly in the chapter known as *Regule de motu locali*, which begins with the Bradwardine formulation immediately (MS Vat. lat. 3064, f. 98r.): "Hic incipiunt quedam regule de motu locali supponendo motum attendi penes proportionem geometricam. Prima regula est hec: ubicunque aliqua potentia crescat respectu resistentie non variate tantam proportionem acquiret respectu illius resistentie per quantam ipsa fiet maior...." (Cf. MS Pavia, Bibl. Univ. Aldini 314, f. 76r, c.1.) In this treatise a detailed attempt is made to relate the various kinematic theorems to sundry variations in forces and resistances. The two short tracts on local motion contained at the end of *Liber calculationum* in Cambridge, Gonv. and Caius, 499/268, ff. 212r–213r, 213r–215r, and which appear instead of the *Regule de motu locali*, also

Plate 8: A page from a fourteenth-century manuscript of *De latitudinibus formarum*. MS Columbia University Libraries, Plimpton 171, f. 5r. Copied at Bologna in 1395.

Plate 9: A page from a late thirteenth-century manuscript giving the text of the *Liber de ponderoso et levi* attributed to Euclid. MS Florence, Bibl. Naz., Conv. Soppr. J. I. 32, f. 47r.

ton[39] gave detailed elaborations going much beyond the original treatment of Bradwardine. It also spread with rapidity to France. John Buridan, composing his *Physics* not much later than 1340 (if not earlier), clearly was influenced by Bradwardine's treatment.[40] Similarly Buridan's students Nicole treat the Bradwardine rules. The *Questiones super physicam Aristotelis Ricardi* (Venice, Bibl. Naz. San Marco Lat. VI, 72, ff. 81r–112r, 163r–163v) could be by Swineshead. At least it ought to be compared in detail with the above mentioned treatises. It certainly seems less subtle than the *Liber calculationum*. Apropos of the proportionality problem, we read on f. 82r: "Questio est ista, utrum in omni motu potentia motoris excedit potentiam rei moti." Note: this author mentions Heytesbury ("reverendum magistrum G. Hesberi," f. 163v). It is noteworthy that he distinguishes the different versions of proposition 1 of the *De ponderibus* treatises, attributing the reading found in the *De ratione ponderis* to Alchindus and that found in the *Elementa de ponderibus* to Jordanus (f. 88v).

[39] John Dumbleton, *Summa de logicis et naturalibus*, Part III, chap. 7, takes up the Bradwardine dynamics in detail. It begins (MS Cambridge, Peterhouse 272, f. 24v, c. 1): "Sequitur dicere qualiter motus sequitur proportionem.... hoc sic patet [quod] si omnis motus velocius fit secundum quod motor dominatur super resistentiam, set cum motus non sequitur excessum potentie motoris super motum nec proportionem motorum adinvicem, ergo diversitas motuum sequitur proportionem proportionum." It became conventional, at least from the time of Dumbleton and Buridan (see footnote 40), to express the Bradwardine formula by saying that velocity follows "the proportion of proportions." Moody, who studied this section of Dumbleton's work ("Rise of Mechanism in 14th Century Natural Philosophy," unpublished mimeographed pamphlet, p. 38) says: "This dynamic analysis is a development of the theory of Bradwardine's *Tractatus de proportionibus velocitatum...*, but whereas Bradwardine had considered only the differences in proportion of motive power to resistance, determined by different rates of uniform displacement, Dumbleton applies the ideas of intension and remission to a consideration of the relation between changes in the ratio of motive power to resistance, to rates of acceleration or deceleration, in function of time differences." On the other hand, Crosby's judgment as to the implications of Dumbleton's analysis would appear to me to be incorrect. He says in his edition of Bradwardine's treatise (p. 52): "Dumbleton's theory of the manner in which an 'intensive' magnitude may be said to increase then leads inevitably, not only to his conclusions regarding the relation of distance to time in a uniformly accelerated motion, but to the realization that a constant proportion of force to resistance must produce just such a motion." Cf. A. Maier's review of Crosby's volume in *Isis*, Vol. 48 (1957), 85–87.

[40] John Buridan, *Super octo phisicorum libros Aristotelis* (Paris, 1509), Bk. VII, quest. 7, ff. 107r–108r: "Queritur septimo circa ultimum capitulum huius septimi in quo Aristoteles ponit multas regulas de comparationibus motuum secundum habitudinem ad motores. Et erit hec questio de primis duabus regulis, videlicet, utrum hec due regule sunt vere, scilicet, iste: [1] Si aliqua virtus movet aliquod mobile per aliquod spacium in aliquo tempore, eadem vel equalis virtus movebit medietatem illius mobilis per duplex spacium in equali tempore. [2] Secunda regula est, si aliqua virtus movet aliquod mobile per aliquod spacium in aliquo tempore, eadem totalis virtus movebit medietatem illius (107v) mobilis per equale spacium in dimidio tempore... Prima [conclusio] est quod illa regula non est universaliter vera, quia, si illa esset vera, sequitur quod etiam illa alia regula esset vera: si aliqua virtus

Oresme and Albert of Saxony treated the Bradwardine formulation on more than one occasion.[41] Bradwardine's ideas also appeared in Italy within about

movet aliquod mobile per aliquod spacium in aliquo tempore, eadem vel equalis virtus movet duplex mobile per dimidium spacium in equali tempore. Sed illa regula est falsa... quia tantum mobile vel tanta resistentia mota a tua virtute posset signari, quod tu non posses movere duplam per aliquod spacium nec in aliquo tempore, et causa huius est quia virtus motoris debet excedere virtutem resistentie et forte non excederet si resistentia duplaretur. Verbi gratia, tria movent duo quia excedunt ea, sed duplatis duobus erunt quattuor et ita non excedunt illa quattuor; ideo nullo modo movebunt.... Ex istis sequitur tertia conclusio, scilicet, quod proportio velocitatum motuum non attenditur penes proportionem resistentiarum, manente eodem vel equali motore.... Et tunc sequitur quinta conclusio, scilicet, quod proportio velocitatum motuum non debet attendi penes proportionem moventium, stante eadem vel equali resistentia.... Deinde etiam nos supponimus quod movens debet fortius quam resistentia, ita quod moventis ad mobile debet esse proportio maioris inequalitatis. Et tunc ponitur sexta conclusio, quod proportio velocitatum motuum non debet attendi secundum proportionem excessuum quibus moventia excedunt mobilia.... (108r) Septima conclusio infertur, quod velocitas attendi debet penes proportionem maioris inequalitatis moventis ad resistentiam, ita quod quanto erit maior proportio maioris inequalitatis moventis ad resistentiam, tanto erit motus velocior, et ita proportio velocitatum motuum attendenda est penes proportionem proportionum moventium ad suas resistentias, ita quod in qua proportione proportio istius a ad b est maior quam proportio ipsius c ad d in eadem proportione a movet ipsum b velocius quam c, d.... Pono octavam conclusionem, scilicet, quod iste due regule Aristotelis, de quibus querebatur, habeant veritatem in quibuscunque proportio motoris ad resistentiam est dupla, sed in aliis non sunt vere.... sed secunda pars conclusionis declaratur tam ex predictis quam exemplariter; quia, si sex movent duo, proportio est tripla; et si auferatur medietas duorum, tunc sex movent unum et est proportio sextupla et non movebunt dupla velocitate, quia proportio sextupla non est dupla ad triplam, non enim continet duas triplas, sed continet solum unam triplam cum una dupla, quia sex primo continent bene unam triplam, que est trium ad unum, sed ultra non continet nisi unam duplam, que est sex ad illa tria; dupla autem ad triplam esset proportio novem ad unum, quia contineret duas triplas, scilicet, unam trium ad unum, et aliam novem ad illa tria, et sic dupla ad unum esset tripla ad triplam. Ideo ergo non sex, sed novem, moveret unum dupliciter quam si (! tria?) unum."

[41] In addition to the section of the *Du ciel* given as Document 7.2, Oresme earlier treated the Bradwardine dynamics in the fourth chapter of his *Proportiones proportionum* (see the Bibliography for manuscripts of this work). Also attributed to Oresme is a complete tract on the velocity problem entitled *De proportionibus velocitatum in motibus* (MS Paris, Bibl. de l'Arsenal 522, ff. 126r–168v). Its form is obviously taken from the English treatises. It has a mathematical introduction on proportions (ff. 126r–130v), a selection elaborating the Bradwardine dynamics as in the manner of Swineshead and Dumbleton (ff. 130v–145r), and a section with the kinematic treatment of motion (ff. 145r–150v). In this section the author holds against the English that in rectilinear motion, the speed of a moving body is measured by the linear space described by the "mean point" *(medio puncto)* rather than by the fastest point (see the fifth conclusion, f. 145v, and the sixth conclusion, f. 146r). However, in circular motion he holds to the Bradwardine conclusion that the measure is the space described by the fastest point (f. 146v). Uniform motion

[Bradwardine's Dynamic Law of Movement]

a generation after their original exposition.[42] To illustrate this, we have included as our final document (Doc. 7.4) a *Questio* of one Francischus de Ferraria, composed at Padua in 1352. Francischus' question takes its point of departure from Bradwardine's treatise. Bradwardine he simply refers to as the *Magister de proportionibus*. It will be noticed that I have included some passages that illustrate that not only the English dynamics but the kinematics as well had already spread to Italy by 1352.

Bradwardine's law, then, was widely accepted in the last centuries of the Middle Ages, although an occasional dissenting voice was heard, e.g., that of Blasius of Parma at the end of the fourteenth century, Giovanni Marliani in the middle of the fifteenth century,[43] and A. Achillini and B. Vittori[44] in the beginning of the sixteenth century. Quite different refutations of the older Aristotelian rules were made by Giovanni Battista Benedetti[45] and Galileo Galilei in the later sixteenth century.[46]

both as to subject and to time is defined in the Merton manner (f. 147v), as is uniformly difform motion as to subject and to time (f. 147v). The mean speed theorem as applied to both subject and time is given with proofs (f. 147v): "Tertio nota quod velocitatem motus localis uniformiter difformem quo ad tempus correspondere gradui medio nihil aliud est quam mobile quod sic uniformiter difformiter quo ad tempus movetur tantum spacium precise in dato tempore pertransire quantum ipsum precise pertransiret si ipsum continue per idem tempus uniformiter moveretur gradu medio illius velocitatis uniformiter difformis. Sed quod latitudinem motus localis uniformiter (148r) difformem quo ad subiectum gradui medio quo ad subiectum correspondere nihil aliud est quam mobile quod sic uniformiter difformiter quo ad subiectum movetur tantum spacium precise transiret in dato tempore quantum pertransiret si quilibet punctus eius moveretur continue per idem tempus uniformiter gradu velocitatis quo movetur punctus eius qui medio gradu quo ad subiectum movetur...." Other kinematic theorems drawn directly from Heytesbury or Swineshead are given, e.g., (f. 149r), "omnis latitudo motus localis uniformiter difformis, non uniformiter sed difformiter quo ad tempus quia si (?) continue velocius et velocius acquisita, remissiori gradui quam sit gradus medius illius tocius latitudinis correspondet." The treatise concludes with a section on the velocity of augmentation (ff. 150v–155r) and one on the velocity of alteration (ff. 155r–168v). Oresme's contemporary at Paris, Albert of Saxony, treated the Bradwardine dynamics in a more modest fashion twice (and both treatments are practically identical), once in his *Tractatus de proportionibus* (edition of Venice, 1496, and many manuscripts) and in his *Questiones in octo libros phisicorum* (Paris, 1518).

[42] Maier, *Die Vorläufer*, pp. 104–7.

[43] M. Clagett, *Giovanni Marliani and Late Medieval Physics* (New York, 1941), chapter 6.

[44] *Ibid.*, p. 144. The pertinent work of Alessandro Achillini is his *De proportione motuum (Opera omnia*, [Venice, 1545]); that of Benedetto Vittorio is his *Commentaria in tractum proportionum Alberti de Saxonia* (Bologna, 1506).

[45] G. B. Benedetti, *Demonstratio proportionum motuum localium contra Aristotelem et omnes philosophos* (Venice, 1554). See also his *Diversarum speculationum mathematicarum et physicarum liber* (Turin, 1585), pp. 168–81. Passages from these works are quoted below in Chapter 11, notes 121 and 122.

[46] Galileo Galilei, *De motu* (*Le Opere*,

We may note, in conclusion, that unsuccessful as Bradwardine's law was in representing dynamical experience, it was, nevertheless, highly influential in its kinematic by-products at Merton College, as we have already shown in earlier chapters.

Ed. Naz., Vol. *1* [Florence, 1890], 262–84, 294–96). See passage quoted in Chapter 11, footnote 130.

Document 7.1

An Anonymous Treatment of Peripatetic Dynamics

1. NOTE that there are six species of motion, viz., augmentation, diminution, change of place, generation, corruption, and alteration. Since local motion is the most common among all of these motions, one should speak of it first.

It ought to be known that local motion (i.e., change of place) is spoken of in many ways: some is natural local motion, some violent, some [motion] of ascent, some [motion] of descent, some uniform, some difform, some uniformly difform, some difformly difform, some uniform as to parts and difform as to time, some uniform as to time and difform as to parts.

We should see in the first place what motion is. It ought to be known that motion is the act of being in potency (i.e., the actualizing of the potential), or motion is the perfection of the thing moved, or motion is the accident extended in the subject, or motion is the acquisition part by part of that perfection toward which the thing moved tends.

2. Natural local motion is motion arising from intrinsic form, as the motion of a heavy body from [its] gravity, which gravity indeed is the intrinsic form of the heavy body. This is like earth descending naturally because of its gravity... or just as any light body is naturally mobile [upward] by lightness, which lightness is indeed the intrinsic form of the light body. It is by this form that it is a naturally mobile body.

And the intrinsic form of a moving body is called the intrinsic motor of the moving body. From this it is evident that natural motion by form and by intrinsic motor are identical, since the intrinsic form and the intrinsic motor by which a moving body is moved are the same thing.

Finally it ought to be known that any element is naturally a moving body. For this it ought to be noted that there are four elements, namely, fire, air, water, and earth. Each of these has its natural place toward which it is naturally moved if it is at a distance from that [place]. Of these elements one is in the highest place and one in the lowest. The highest place of the elements is the place of fire. Fire is naturally mobile toward this place if it is distant from that place, as the result of its intrinsic form or motor (which are the same). And the intrinsic motor is the intrinsic form by means of which fire is moved naturally to its natural place. This intrinsic form in fire is lightness *(levitas)*. This lightness, then, is the instrument by means of which the fire becomes a moving body.

Earth is the lowest of the elements. Its natural place is the lowest place toward which earth is naturally mobile, if it is at a distance from that place. This [motion towards its natural place] is from its intrinsic form, and the intrinsic form of earth by which it is naturally mobile is heaviness *(gravitas)*. The intermediate places between the highest place *(locus supremus)* and the lowest place *(locus infimus)* are the natural places of air and water, which are the intermediate elements. And the natural place of water is situated immediately above the place of earth. Immediately above the [natural] place of water is situated that of air. Just as in the case of fire and earth, water and air are naturally mobile toward their natural places by means of their intrinsic forms.

Just as there are four elements, so there are four prime qualities following the elements. These four are hotness (calidity), wetness (humidity), coldness (frigidity), and dryness (siccity). All qualities other than these four result from these four and are called secondary qualities. Of this type are the following: whiteness, blackness, sweetness, bitterness, heaviness and lightness, and [qualities of] this kind. Any one of these is a secondary quality, and it results from a certain proportion of the prime qualities. The prime qualities are elementary forms. Of the elementary forms certain are accidental and certain substantial. The substantial form of fire is lightness, as some say, or ignitability *(igneitas)*, as others say. The accidental forms of fire are calidity and dryness, each of which is maximum *(summa)* if the fire is at a maximum or remitted (decreased) if the fire is remitted. Calidity is the dominant quality in fire, there being in every element two qualities, one of which is dominant over the other.

The accidental forms of air are humidity and calidity, of which humidity is dominant, and each of which is maximum if the air is at a maximum. The accidental forms of water are humidity and frigidity, of which frigidity is dominant, and each of which is maximum if the water is at a maximum. The accidental forms of earth are frigidity and dryness, of which dryness is dominant.

And note briefly that each of the elements has three forms, two of which are accidental and one substantial. The accidental elementary forms are prime qualities, viz.: calidity, humidity, frigidity, dryness. By means of these [accidental] forms the elements are active and passive in turn. But by means of the substantial forms they are not active and passive elements, but only moving bodies *(mobilia)*, as by means of heaviness or lightness, which are the substantial forms of the elements.

3. Note further that local movement, which is change of place, occurs when anything is moved from one place to another place. The place from which a body is moved is called the *terminus a quo*, because the body is moved from that place; while the place toward which motion takes place is called the *terminus ad quem*. To that place a body is moved, because every motion takes place from a *terminus a quo* to a *terminus ad quem*.

A moving body is moved toward a place either naturally or violently. If [it is moved] naturally, then the place is the natural place of that body in which it seeks to rest. Thus if [some] earth should be distant from its natural place, which is the center of the universe, it would be moved naturally by means of its form, which is gravity, until it should have the same center as the center [of the universe]; i.e., if the earth is pure and [if] the earth is said to be concentric with the center of the universe when the center of the earth coincides with the center of the universe. The center of such earth is the point so uniting the halves of the earth, that there is just as much earth above the point as below it. The center of the universe is designated as a single point because it is equally distant [from all points on the] circumference of the final sphere, i.e., the heavens. That point is the middle point of the whole universe toward which all earth is mobile if nothing impedes it. [On the other hand], a naturally inclining moving body is moved violently when it is moved against the natural inclination, i.e., away from the natural place, as earth moving up or fire down. And the violence which makes the moving body move violently is called the extrinsic motor, as a propellent or thrower who puts a stone into movement violently or an archer who makes the arrow fly, and other similar examples. It is evident from this consideration that every movement of propulsion is violent movement.

It is to be noted that every moving body is either animate or inanimate. An animate body is moved in many ways—up, down, forward, backward, and by a progressive movement, just as animals who are capable of progression *(gradativa)* are moved by a progressive movement. Such movement arises from the soul *(anima)* or from another form which is in the soul, but not from a particular elementary form. So an animal moves up in a leaping motion through the action of his soul or by another particular form [of his soul], but not by an elementary form. However, if such an animal were up in the air and should fall, the movement of descent of the animal would arise from a quality which is an elementary form. He descends then not by his own volition as he ascended [but] through an intrinsic motor.

4. It is noteworthy that there are two kinds of intension, quantitative and qualitative. Quantitative intension takes place by adding quantity to the maximum degree of that quantity *(gradus summus illius quantitatis)*. Qualitative intension can take place through rarefaction or condensation, e.g., if there were a body one half of which was at a maximum hotness *(summe calida)* and in the other part the heat was remitted. Then if the more remiss (i.e., less hot) half were condensed, while the other part remained as before, the whole is understood to be hotter because the ratio of the more intense to more remiss part will be continually in greater proportion....

It should be noted that qualitative intension takes place by the remission of its contrary. This is not so for quantitative intension, since however great is substance or quantity there is nothing contrary to either substance or quantity. It also follows that something which has become "qualified" *(qualificatum*, i.e., has been made to have a certain quality throughout) is not intended by the remission of its contrary.

If it is said that there is something contrary to substance because fire is contrary to water, it is answered that they are contraries not according to substance but according to qualities.

And it is to be noted that if something "qualified" *(qualificatum)*, such as a hot body or a cold body, suffers remission [in that quality], then it can be intended (i.e., increased again in that quality) by the remission (decrease) of some [other contrary] quality. For example, if a hot body has been remitted, it is necessary that there is a mixture of elements, i.e., of fire and others, because, if the mixture is hot, the hot element must be dominant in the mixture. But since there are no hot elements but fire and air [and fire is the only element where it is dominant], then it follows in a mixed hot body that fire is dominant. And since no hot mixture which is remitted is composed of anything but contraries, it follows, therefore, that a [mixed] hot body is composed of contrary elements and so it follows that a remitted hot body is hotter by the corruption of the contrary element.

And just as space is that which is to be acquired by local motion, so latitude is that which is acquired by motion of alteration. And in the same way that a moving body which describes more space in an equal time is moved more swiftly as to space, so that which by alteration acquires more latitude in a certain time is altered more swiftly. And in the same way that velocity in local motion becomes sometimes more intense and sometimes more remiss by means of the variation of the proportion from which such velocity arises, so similarly the velocity of the movement of alteration is varied by a variation of the proportion from which velocity in alteration arises.

It is remarked that every alteration is intension or remission, and because every intension or remission is in quality, it follows that every alteration is in quality, and alteration in quality is action in quality. Every qualitative action takes place by reason of a contrariety between qualities. It follows that alteration is between contrary qualities. And alteration in qualities does not take place from a proportion of equality or lesser inequality. For from such a proportion action does not arise. It follows, therefore, that it takes place from a proportion of greater inequality. Hence, if there is an action between qualities, one must be dominant above the other, and the dominant quality will be active *(activa)*, while the other is passive *(passiva)*. Then the thing so qualified to be dominant is called "the agent" *(agens)*; and that which suffers (the action) of the agent is called "the patient" *(passum)*. The patient receives action from the agent. The agent resists the patient....

Some qualities are uniform. Some are difform. A uniform quality is for example uniform calidity or frigidity, etc. There is uniform calidity when any part of the quality is equally intense as any other part. A difform quality exists when one part is more intense than another. A quality uniformly difform is, or is called, that latitude itself of calidity. There is a uniformly difform calidity when, any two immediate parts having been taken, the most intense calidity which is not in the more intense part is the most remiss calidity which is not in the remiss part. "The most intense degree which is not in the more intense part" ought to be expounded as follows: That degree is not in the more intense part, and in the more intense part there is a degree more remiss than any degree more intense than the accepted degree. "The most remiss degree which is not in the more remiss part" ought to be expounded as follows: That degree is not in the more remiss part, and in the remiss part there is a degree more intense than any degree more remiss than the accepted degree.

It should be noted that whenever some part of some patient is altered, the whole is also said to be altered.

And it is to be noted that latitude of quality is called the distance between distant *(distantes)* degrees; and just as there are infinite points in every line and infinite instants in every [period] of time, so there are infinite degrees in every latitude. And just as a mean line *(linea media)* falls between two points of some line, so between all the degrees of a certain latitude there is a mean latitude. And just as two points in the same line are not immediate, in the same way two degrees in the [same] latitude are not immediate.

It should be noted that every degree is uniform and none difform. And the degree of a quality is nothing else but the denomination of a quality in such a degree; for example, a remitted degree of calidity is a certain distance from the maximum *(summus)* degree of calidity. That degree which is more distant from the maximum degree is more remiss than that degree which is less distant from the maximum degree. And in so far as one degree of one latitude is more distant from another degree of the same latitude, so it is either more intense or more remiss [than that degree].

Whence a rule is noted: Such as is the proportion of the distance of some degree from zero degree, to the distance of another degree from zero degree, so is the proportion of the one degree to the other degree. For example, if the proportion of latitude between A degree and zero degree is double the [proportion of the] latitude between B degree and zero, it follows that A is double B, and conversely. And it is said that A degree is not spoken of as twice as intense as B because it (A) is mixed with its contrary [in an amount] only half as much as [is] B. Nor is it [said to be doubly intense] because it is half as distant from the maximum degree. But rather [it is spoken of as being doubly more intense than B] because it is twice as distant from the zero degree of its latitude.

[We can argue negatively saying that] one degree would be doubly more intense than another because it is doubly less mixed with its contrary [in the following way:] Beneath the maximum there is some degree which is a certain amount mixed with its contrary, and there is [also] some degree which is doubly more [mixed with its contrary], and so on to infinity; thus in infinity it follows that there would be some degree beneath the maximum which is infinitely remiss. And since A degree is more intense than some [one] of this series, it follows that A degree is infinitely intense, which is false and impossible. Therefore, etc.

5. From this it follows that all local motion is successive, and every succession in motion is caused by resistance. It follows that resistance is required for all local motion, and some resistance is intrinsic, some extrinsic, in a moving body. And intrinsic resistance of a moving body is

resistance which is in the moving body. For example, if a mixed heavy body is compounded of fire and earth with earth the dominant element, then that mixed body is mobile because of [the earth's] gravity, and it is moved more slowly because of the levity coextensive with it than it would be if it did not have levity. It follows, therefore, that the levity is the intrinsic resistance of the moving body. Then if the levity were continually decreased, the mixed body would continually accelerate its movement. And thus it follows that diminution of resistance is the cause of acceleration *(intensio motus)*. And increase of resistance is the cause of deceleration *(remissio motus)*.

The extrinsic resistance to a moving body is the resistance of the medium in which that moving body moves. In itself a medium is rarer and denser, i.e., fine and gross. A rarer medium resists less than a denser medium....

And note that without resistance, motion does not take place, and the absence of resistance is the cause of why a simple body is not moved in a vacuum. Because a simple body does not have intrinsic resistance, and there is no extrinsic resistance in a vacuum, and [further] there is no resistance except intrinsic or extrinsic [resistance], it follows that a simple body does not have resistance, and consequently is not moved in a vacuum. Because without resistance motion does not take place—except one that is infinitely fast—it follows that a pure, simple body is not mobile in a vacuum unless it should be mobile infinitely fast; because if this were the case, it would traverse space immediately, which is impossible....

A moving body can be moved locally *per se* or *per accidens*, *per se* when it changes place by its own motion, *per accidens* when it changes place as a result of the action of another motion....

6. Some local motion is uniform, some difform. Uniform local motion is movement arising from a single proportion, such as a double or triple proportion. Or uniform motion is that which is equally intense with respect to all of its parts. Difform local motion is motion arising from diverse proportions, one of which is less than another, or difform local motion is one in which certain parts are more intense and certain more remiss.

Difform motion is divided into uniformly difform and difformly difform motion. Motion uniformly difform is a latitude arising from a latitude of uniform proportion so that any degree of that latitude of motion uniformly difform is the most remiss [degree] which is not below that degree (i.e., any degree is less than all the succeeding degrees above it up to the maximum).... Or any degree at all is the most intense degree [of all the degrees of that part of the latitude] which is not above the assumed degree (i.e., any degree is greater than all the degrees below it)....

Motion difformly difform is motion in which one part is more intense than another or in which one half is moved at a certain degree [of velocity] and the other half at a degree double to the first. Then such a body is moved difformly and yet not uniformly difformly. Hence [we say] difformly difformly.

Motion uniform as to part occurs when any part at all is moved equally swiftly as any other part. Motion is difform as to time when more is traversed by that movement in one part of the time than in another part of the time equal to it.

Motion uniform as to parts and difform as to time is uniform *(sic)* movement which is continually increased in such a fashion that all of its parts will remain equally intense. Then this motion will be continually uniform as to parts because all parts will continually remain equally intense, and it is difform as to time because always more would be traversed by that movement in one part of the time than in another part equal to it.

Motion difform as to parts and uniform as to time is motion of the sort that occurs in the movement of a radius which continually rotates uniformly. It is at rest in the end point (center), and its most swiftly moving point will continually be moved uniformly and at a single degree [of velocity]. The whole radius ought to be said to be moved in that [same] motion in which is moved the most swiftly moving point. So it follows that the whole is moved uniformly, and continually so, as to time because the swiftest moving point will traverse just as much in one [part of the] time as in another [part] equal to it, but it is moved difformly as to parts because any part of the radius terminated below the swiftest moving point is moved more slowly than another part terminated at the swiftest moving point. Therefore one part is moved more swiftly than another. Consequently the motion is difform as to parts....

Some local motion is straight line *(rectus)*, some circular. Circular motion occurs when something is moved by describing a circle. Straight-line motion occurs when something is moved rectilinearly by describing a straight line. Retrograde [motion] occurs when something is moved in rectilinear motion from one position to another and then returns to the position from which it started.

And note that for every motion three things are required: a moving body, something acquired by the movement, and time. And such condi-

[*Peripatetic Dynamics* 453]

tions have been manifested in three predicaments, i.e., quality, quantity, and place *(ubi)*—properly latitude—and motion in quality is called alteration, [motion] in quantity [is called] augmentation and diminution, and [motion] in place is called change of place. Let these things suffice.

COMMENTARY

1. Although this piece accompanies Bradwardine's *Proportiones* in the edition of Venice, 1505, as an introduction, its absence from the manuscripts of the *Proportiones* makes it highly unlikely that it was composed by Bradwardine as such an introduction. It is difficult to pin it down chronologically on the basis of internal evidence. One can only be confident that it dates from after the time of the Merton kinematic distinctions, since it shows the influence of those distinctions, e.g., uniformly difform or difformly difform motions, etc. Notice in the opening passage that the author is virtually equating motion with change as he details the various kinds of motions. Notice further the Aristotelian definition of motion as the "act of a being in potency," i.e., the actualizing of the potential.

2. The second passage describes the Aristotelian theory of the four elements, the four prime qualities, the secondary qualities. Distinction is made between accidental and substantial qualities and forms.

3. Then follows a definition of local motion as change of place and of the expressions *terminus a quo* and *terminus ad quem*. Natural and violent movements are distinguished in the Aristotelian manner, as are animate and inanimate movements.

4. Intension is discussed in something like the Mertonian manner, a distinction being drawn between qualitative intension and quantitative intension. Intension and remission of qualities is then elaborated. One of the most interesting statements in this passage occurs where latitude is described as the "distance between separate degrees," and "just as there are infinite points in every line and infinite instants in every [period] of time, so there are infinite degrees in every latitude." It is not surprising then that I was tempted in Chapter 4 to identify degree with a magnitude of instantaneous velocity.

5. We are then reminded of the necessity of resistance to movement, either as intrinsic or extrinsic resistance. The diminution of resistance brings about acceleration, while its increase brings about deceleration. In a vacuum a body would either not move or would move with an infinite velocity since there would be no resistance.

6. The usual Mertonian distinctions between uniform and difform

movements are presented. The reader will do well to compare this section with the documents appended to Chapter 4, particularly Documents 4.3–4.5.

[Sumulus de motu incerti auctoris*]

(9r) Nota quod sex sunt species motus, scilicet, augmentatio, diminutio, loci mutatio, generatio, et corruptio, et alteratio. Et quia inter omnes motus, motus localis est communissimus, ideo primo
5 dicendum est de motu locali. Et sciendum est quod motus localis multipliciter dicitur. Quidam est motus localis naturalis. Quidam violentus, et quidam ascensus, et quidam descensus, quidam uniformis, quidam difformis, quidam uniformiter difformis (*text:* uniformis), quidam difformiter difformis, quidam uniformis quo ad partes
10 et difformis quo ad tempus, et quidam uniformis quo ad tempus et difformis quo ad partes.

In principio videndum est quid sit motus.

Et sciendum est quod motus est actus entis in potentia, vel motus est perfectio rei mote, vel motus est accidens extensum in subiecto, vel
15 motus est acquisitio partis post partem illius perfectionis ad quam vadit res mota.

Motus localis naturalis est motus proveniens a forma intrinseca, sicut motus gravis a gravitate, que quidem gravitas est forma intrinseca gravis, sicut terra naturaliter a gravitate descendit. Et sicut
20 quodlibet grave a gravitate naturaliter movetur, sic et quodlibet leve a levitate naturaliter est mobile, que quidem levitas est forma intrinseca levis et a qua forma ipsum est naturaliter mobile et vocatur forma intrinseca mobilis motor intrinsecus mobilis.

Per hoc patet, quod idem est movere naturaliter et a forma et a
25 motore intrinseco, et motor intrinsecus nam (et?) forma intrinseca a qua mobile movetur idem sunt.

Ulterius est sciendum quod quodlibet elementum naturaliter est mobile. Pro quo est notandum quod quatuor sunt elementa, scilicet, ignis, aer, aqua, et terra, quorum quodlibet habet locum suum natura-
30 lem ad quem est naturaliter mobile, si ipsum ab isto distet. Istorum elementorum quoddam est supremum, quoddam infimum. Locus supremus elementorum est locus ignis, ad quem locum ignis est naturaliter mobilis, si ab illo loco distet, a forma intrinseca seu a

---

* Printed as an introduction to the *Proportiones* of Thomas Bradwardine (Venice, 1505), ff. 9r–10r. I have altered punctuation and capitalization somewhat.

[*Summulus de motu*

motore intrinseco, quid idem est. Et est motor intrinsecus forma
intrinseca, a qua ignis est naturaliter mobilis ad suum locum naturalem, que quidem forma in igne est levitas, que levitas est instrumentum mediante quo ignis est mobilis.

Elementum infimum est terra, cuius locus naturalis est locus infimus, ad quem locum terra est naturaliter mobilis, si ab illo loco distet, et hoc a forma intrinseca, et est forma intrinseca terre gravitas a qua terra est naturaliter mobilis. Loca intermedia inter locum supremum et locum infimum sunt loca naturalia aque et aeris, que sunt intermedia, et est locus naturalis aque locus immediate supra locum terre situatus, supra quem locum immediate ponitur locus aeris. Et sicut terra et ignis ad sua loca naturalia a suis formis intrinsecis sunt mobiles, eodem modo aer et aqua ad sua loca naturalia a suis formis intrinsecis sunt mobiles.

Et sicut quatuor sunt elementa, sic sunt quatuor qualitates prime, que sunt qualitates consequentes elementa, et sunt iste quatuor: caliditas, humiditas, frigiditas, et siccitas. Omnes alie qualitates ab istis quatuor resultant ex istis quatuor, et vocantur qualitates secunde, cuiusmodi sunt: albedo, nigredo, dulcedo, et amaritudo, gravitas, levitas, et huiusmodi; quarum quelibet est qualitas secunda et quarum quelibet resultat ex certa proportione qualitatum primarum. Qualitates prime sunt forme elementares, et sunt forme elementares quedam accidentales, quedam substantiales. Forma substantialis ignis est levitas, ut dicunt quidam, sive igneitas, ut dicunt [alteri]. Forme accidentales ignis sunt caliditas et siccitas, quarum utraque est summa si ignis est summus, et altera remissa si ignis sit remissus. Et est caliditas qualitas dominans in igne, quia in quolibet elemento sunt due qualitates, quarum una super alteram est dominans. Forme accidentales aeris sunt humiditas et caliditas, quarum humiditas est dominans, et utraque est summa si aer sit summus. Forme accidentales aque sunt humiditas et frigiditas, quarum frigiditas est dominans, et utraque est summa si acqua est summa. Forme accidentales terre sunt frigiditas et siccitas, quarum siccitas est dominans.

Et nota breviter quod quodlibet elementum habet tres formas, quarum due sunt accidentales et una substantialis et sunt forme accidentales elementares qualitates prime, videlicet, caliditatis, humiditatis, frigiditatis, et siccitatis, et mediantibus illis formis sunt elementa activa et passiva adinvicem. Sed mediantibus formis substantialibus non sunt elementa activa et passiva, sed solum mobilia, sicut a gravitate et levitate que sunt forme substantiales elementorum.

Et nota quod motus localis qui est mutatio loci est quando aliquis movetur ab uno loco ad alium locum, et vocatur locus a quo mobile movetur terminus a quo, quia ab illo loco fit motus, et locus ad quem fit motus, terminus ad quem, ad illum locum mobile movetur, quia omnis motus fit a termino a quo ad terminum ad quem.

Et movetur mobile ad locum vel naturaliter vel violenter. Si naturaliter, tunc locus est naturalis illius mobilis in quo loco mobile appetit quiescere, sicut terra, si distaret a loco suo naturali, qui est centrum mundi, naturaliter moveretur a sua forma, que est gravitas, quousque foret concentrica cum centro, et hoc si sit terra pura, et dicitur terra esse concentrica cum centro mundi, quando centrum illius terre est simul cum centro mundi.

Et vocatur centrum talis terre, cuius punctus copulans medietates terre adinvicem, sicut quod tantum ex illa terra sit supra illum punctum sicut sub illo puncto.

Et centrum totius mundi vocatur unus punctus, quia ex omni parte equaliter distat a circumferentia supremi orbis, qui est celum, et ille punctus est medius punctus totius terre ad quem punctum est omnis terra naturaliter mobilis, si non impediatur. Et sicut naturaliter inclinans violenter movetur mobile quando movetur contra inclinationem naturalem, sicut a suo loco naturali, sicut terra versus sursum, ignis verus deorsum, et hoc ab aliquo violentante ipsum, et vocatur illud quod facit mobile moveri violenter motor extrinsecus, sicut pellens lapidem vel iaciens facit lapidem moveri violenter et sagittans sagittam, et alia similia.

Et patet ex hoc quod omnis motus qui sit per pulsum est motus violentus.

Et notandum est quod omne mobile vel est animatum vel inanimatum. Mobile animatum multipliciter movetur, scilicet, sursum et deorsum, ante et retro, et motu progressivo, sicut animalia que sunt gradativa moventur illo motu progressivo, et talis motus provenit ab anima vel ab alia forma que est in anima, et non a propria forma elementari, sicut animal est motum versus sursum motu saltationis qui provenit ab anima vel ab alia propria forma et non a forma elementari. Sed si tale animal foret sursum in aere et descenderet, talis motus descensus animalis proveniret a qualitate que est forma elementaris. Et hoc si non sit animal volatile, quia tale ascendit, et descendit per motorem intrinsecum.

[*Summulus de motu* 457]

Notandum quod duplex est intensio, scilicet, intensio quantitativa et qualitativa. Intensio quantitativa est per accessum quantitatis ad gradum summum illius quantitatis. Intensio qualitativa potest fieri per rarefactionem vel condensationem. Ut verbi gratia, si foret unum corpus, cuius una medietas foret summe calida et alia medietas remisse calida. Si medietas remissior condensaretur, stante alia medietate sicut prius, totum intelligitur calidius, quia pars intensior ad partem remissiorem continue se habebit in minori proportione. Et consimiliter per condensationem partis intensioris posset remitti, et si nulla qualitas in eo intenderetur vel remitteretur, tunc non sequitur: hoc intendetur, ergo aliqua qualitas in eo intendetur. Nec sequitur: hoc remittetur, ergo qualitas in eo remittetur.

Et notandum est quod intensio qualitativa semper fit per remissionem sui contrarii; sed non sic est de intensione (9v) qualitativa (! quantitativum?), ut probatum est, quia quantum est substantia vel quantitas, sed neque substantie neque quantitati est aliquid contrarium. Sequitur quod qualificatum cuiusmodi est calidum, frigidum, album, vel nigrum non intenditur per remissionem sui contrarii.

Si dicatur quod substantie est aliquid contrarium, quia ignis aque contrariatur, pro illo dicitur quod non secundum substantiam sed qualitates in eis contrariantur.

Et notandum quod qualificatum, cuiusmodi est calidum, frigidum, si sit remissum potest intendi per remissionem alicuius qualitatis. Verbi gratia, si sit calidum remissum, oportet quod sit mixtum ex elementis, videlicet, ex igne et aliis, et quia si mixtum sit calidum, oportet quod elementum calidum dominetur in tali mixto, et cum nullum sit elementum calidum nisi ignis vel aer, sequitur quod in mixto calido dominetur ignis; et cum nullum mixtum calidum remissum componatur nisi ex contrariis, sequitur ergo quod calidum componitur ex elementis contrariis et sic sequitur quod calidum remissum sit calidius per corruptionem elementi contrarii. Et sicut acquirendum per motum localem est spacium, sic quod acquiritur per motum alterationis est latitudo. Et eodem modo sicut mobile localiter velocius movetur, quid plus in equali tempore pertransit de spacio, eodem modo quid maiorem latitudinem in aliquo certo tempore acquirit per alterationem velocius alteratur. Et eodem modo sicut velocitas in motu locali aliquando fit intensior et aliquando remissior per variationem proportionis a qua fit talis velocitas, eodem modo velocitas motus alterationis variatur per variationem proportionis a qua provenit velocitas in alteratione.

Et notandum quod omnis alteratio est intensio vel remissio, et quia omnis intensio vel remissio est in qualitate, sequitur quod omnis alteratio est in qualitate, et alteratio in qualitate est actio in qualitate et omnis actio qualitatum fit ratione contrarietatis inter qualitates. Sequitur quod alteratio est inter qualitates contrarias. Et alteratio inter qualitates non fit a proportione [e]qualitatis neque minoris inequalitatis, quia a tali non provenit actio. Sequitur ergo quod fit a proportione maioris inequalitatis. Oportet ergo quod si est actio inter contraria, quod una illarum qualitatum super aliam denominetur, et qualitas dominans erit activa, alia vero passiva; tunc illud qualificatum mediante qualitate dominante dicitur agens et quid patitur ab agente dicitur passum, et recipit actionem ab agente, et resistit agens passo. Per augmentationem potentie agentis, stante resistentia passi nec aucta nec diminuta, velocitas alterationis intenditur. Et per augmentationem potentie agentis sicut per augmentationem resistentie passi tardatur motus alterationis. Et dicitur alteratio motus mediante quo res sit altera[ta].

Qualitatum quedam est uniformis, quedam difformis. Qualitas uniformis est ut caliditas uniformis vel frigiditas et sic de aliis. Et est caliditas uniformis quando illius caliditatis quelibet pars cum qualibet parte est [eque] intensa. Qualitas difformis est quando una pars alia parte est intensior.

Qualitas uniformiter difformis est, vel vocatur, illa latitudo caliditatis; et est uniformiter difformis caliditas, cuius quibuscunque duabus partibus captis immediatis, intensissima caliditas que non est in parte intensiori est remississima que non est in parte remissiori. Gradus intensissimus qui non est in parte intensiori debet sic exponi: Ille gradus non est in parte intensiori et quocunque gradu intensiori accepto gradus remissior est in parte intensiori. Gradus remississimus qui non est in parte remissiori debet sic exponi: Ille gradus non est in parte remissiori et quocunque gradu remissiori accepto gradus intensior est in parte remissiori.

Et notandum quod quandocunque aliqua pars alicuius passi est alterata, totum dicitur et[iam] alteratum.

Et notandum quod latitudo qualitatis vocatur illa distantia que est inter gradus distantes, et sicut in omni linea sunt infinita puncta et in omni tempore infinita instantia, sic in omni sunt infiniti gradus. Et sicut inter duo puncta alicuius linee cadit linea media, sic inter omnes gradus certe latitudinis latitudo est media. Et sicut duo puncta in eadem linea non sunt immediata, eodem modo nec duo gradus in latitudine.

Et notandum quod omnis gradus est uniformis et nullus difformis, et est gradus qualitatis nihil aliud quam denominatum qualitatis in tali gradu. Verbi gratia, gradus remissus caliditatis est caliditas in certa distantia a summo gradu caliditatis, et est ille gradus qui per plus distat a gradu summo remissior quam ille qui per minus distat a gradu summo, et inquantum unus gradus unius latitudinis ab alio gradu eiusdem latitudinis per plus distat, tanto est intensior et remissior.

Unde notandum est pro regula quod qualis est proportio distantie alicuius gradus ad non gradum ad distantiam alterius gradus ad non gradum, consimilis erit proportio unius gradus ad alium gradum; ut, verbi gratia, si sit proportio dupla latitudinis inter A gradum et non gradum ad latitudinem inter B gradum et non gradum, sequitur quod A sit duplum ad B, et econtra. Et dicitur quod A gradus non dicitur in duplo intensior quia in duplo minus admiscetur cum suo contrario quam B, nec quia in duplo minus distat a gradu summo, sed dicitur in duplo intensior quia in duplo distat a non gradu sue latitudinis. Si unus gradus foret alio in duplo intensior quia in duplo minus admiscetur cum suo contrario, cum citra gradum summum sit aliquis gradus aliqualiter admixtus cum suo contrario, et aliquis in duplo magis, et sic in infinitum, sequitur quod citra gradum summum foret aliquis gradus infinite remissus, et cum gradus A sit intensior quam aliquis illorum, sequitur quod gradus A sit infinitus intensive, quid est falsum et impossibile. Ergo et cetera.

Ex quo sequitur quod omnis motus localis est successivus et omnis successio in motu causatur ex resistentia. Sequitur quod ad omnem motum localem requiritur resistentia, et quedam est resistentia intrinseca, quedam extrinseca, mobilis. Et resistentia intrinseca mobilis est resistentia que est in mobili. Verbi gratia, si foret unum mixtum grave compositum ex igne et terra, sic quod terra foret elementum dominans, tunc illud mixtum est mobile a gravitate et tardius est mobile propter levitatem secum coextensam quam si non haberet levitatem. Sequitur ergo quod levitas est resistentia intrinseca mobilis; et tunc si illa levitas continue minoretur, mixtum illud continue velocitaret motum suum. Et sic sequitur quod diminutio resistentie est causa intensionis motus et maioratio resistentie est causa remissionis motus. Resistentia extrinseca mobilis est resistentia medii in quo movetur illud mobile. Per se medium est rarius et densius, i.e., subtile et grossum. Medium rarius minus resistit quam medium densius, subtilius minus quam grossius. Aereum subtilius est quam aqueum, et per consequens minus resistens. Et sic mobile quod movetur in medio aereo minus habet de resistentia extrinseca quam quod movetur in aqueo.

Et nota quod sine resistentia non fit motus, et defectus resistentie est causa quare simplex non movetur in vacuo. Quia simplex non habet resistentiam intrinsecam, nec in vacuo est resistentia extrinseca, et nulla est resistentia nisi intrinseca vel extrinseca, sequitur quod simplex in vacuo non habet resistentiam, et per consequens non movetur in vacuo. Quia sine resistentia non fit motus nisi fit infinite velox, sequitur quod simplex purum in vacuo non est mobile, nisi foret infinite velociter mobile, (10r) et simplex non est mobile in infinitum velociter; quia si sic, subito pertransiret spacium, quid est impossibile. Sed in infinitum velociter quodlibet simplex est mobile, et nullum simplex est mobile in infinitum velociter. Et differentia inter illa duo est hoc. Simplex est mobile in infinitum velociter debet sic exponi: Simplex est mobile aliqua velocitate que est infinita, et illud est falsum, quia non contingit dare velocitatem infinitam. Sed in infinitum velociter simplex est mobile debet sic exponi: Aliqualiter velociter simplex est mobile, et in duplo velocius simplex est mobile, et sic in infinitum. Et istud est verum, quia simplex si poneretur in medio ubi non haberet resistentiam nisi ex medio posset per subtiliationem medii in infinitum velocitare motum, quia medium in infinitum potest subtiliari.

Mobile localiter potest moveri per se vel per accidens, per se quando per motum proprium mutat locum, per accidens quando per motum alienum mutat locum, sicut clavus in navi movetur ad motum navis.

Motus localis quidam est uniformis, quidam difformis. Motus localis uniformis est motus proveniens ab unica proportione, sicut a proportione dupla vel tripla. Vel motus uniformis est qui secundum omnes suas partes est eque intensus.

Motus localis difformis est motus proveniens ex diversis proportionibus, quarum una alia est minor. Vel motus localis difformis est motus, cuius quedam partes sunt intensiores, quedam remissiores.

Motus difformis dividitur, quidam est uniformiter difformis, qui-

[*Summulus de motu* 461]

dam difformiter difformis. Motus uniformiter difformis est latitudo proveniens a latitudine uniformis proportionis, ita quod illius latitudinis motus uniformiter difformis, quilibet gradus sit remississimus qui non est sub illo gradu, et debet exponi sic: Ille gradus non est sub illo et quocunque remissiori gradu intensior est sub illo gradu. Vel intensissimus qui non est supra illum gradum et quocunque gradu intensiori adhuc remissior est supra illum.

Motus difformiter difformis est talis motus, cuius una pars alia parte est intensior, vel cuius una medietas movetur aliquo certo gradu et alia medietas gradu duplo ad illum gradum; tunc illud movetur difformiter, et non uniformiter difformiter, ergo difformiter difformiter.

Motus uniformis quo ad partem est quando quelibet pars movetur eque velociter cum qualibet parte.

Motus ille est difformis quo ad tempus quando illo motu plus contingit pertransiri in una parte temporis quam in alia sibi equali.

Motus uniformis quo ad partes et difformis quo ad tempus est motus uniformis qui continue intendetur sic quod continue manebunt omnes partes eque intense, tunc ille motus continue erit uniformis quo ad partes, quia continue manebunt omnes partes eque intense, et difformis quo ad tempus, quia continue plus pertransiretur illo motu in una parte temporis quam in alia parte sibi equali.

Motus difformis quo ad partes et uniformis quo ad tempus est talis motus qui est motus semidiametri qui continue circumvoluitur uniformiter, quiescens in finali puncto, eius punctus velocissime motus continue movebitur uniformiter et unico gradu, et totum debet dici moveri illo motu quo movetur punctus velocissime motus et sic sequitur quod totum movetur uniformiter et continue uniformiter quo ad tempus, quia punctus velocissime motus tantum pertransibit in uno tempore sicut in alio sibi equali, et difformiter quo ad partes, quia quelibet pars terminata citra punctum velocissime motum tardius movetur quam alia pars terminata ad punctum velocissime motum; ergo una pars alia parte velocius movetur, et per consequens difformiter, et quo ad partes; et quodlibet mobile localiter, ita velociter movetur sicut punctus velocissime motus illius mobilis uniformiter quo ad tempus. Diameter vocatur linea transiens ab uno puncto in circumferentia ad punctum in circumferentia sibi oppositum. Et medietas illius linee vocatur semidiameter et terminatur ad centrum circuli, et vocatur centrum circuli talis punctus a quo linee protracte ad circumferentiam sunt equales, et est circulus figura rotunda.

310 Et est motus localis quidam rectus, quidam circularis. Motus circularis est quando aliquid movetur describendo circulum. Motus rectus quando aliquid movetur recte linealiter et describendo lineam.

Retrogradus quando aliquid movetur motu recto ab aliquo situ ad alium et redit ad situm e quo incipit.

315 Et nota quod ad omnem motum tria requiruntur: mobile, et acquisitum per motum, et tempus, et tantum in tribus predicamentis est contentus, scilicet, qualitate, quantitate, et ubi, proprie latitudo; et motus in qualitate vocatur alteratio, et in quantitate augmentatio et diminutio, et in ubi loci mutatio, et hec sufficiant.

Document 7.2

Nicole Oresme, *On the Book of the Heavens and the World of Aristotle**

Book I, question 12

TEXT (from Aristotle): Again, the proportion of the larger power is to the smaller power as the proportion of the time in which the smaller mover moves is to the time in which the larger power moves. Thus, if a weight is moved through a space in an hour, then a weight which is twice as heavy will describe the same space in one-half of an hour.

1. GLOSS: He (Aristotle) wishes to say that just as the motive power is greater, so it moves in less time and more quickly, if other things are equal. He says the same thing in the seventh book of the *Physics* (VII.5, 249b–250a).

2. But, saving his reverence, it is not well stated, because from this statement it would follow that a power could move a resistance equal to itself and that any power, however small, could move any resistance, however large. I shall demonstrate and prove this, it having been posited that a power moves a resistance with a certain swiftness *(isneleté)*. I make the supposition that it is possible for a power to be just that much less than the original power that it can move this same resistance by a speed which is exactly one-half the posited velocity. And I suppose that there can be another power which can move it with a speed one-fourth of the posited velocity, and another with a speed of one-eighth that velocity and so on. And according to Aristotle here and in the seventh book of the *Physics*, the second power will be one-half of the first, the third will be one-fourth the first, and so on. Thus any power, however small, would move the resistance some degree more slowly than the large power. So, for example, if a power is as 8 and the resistance is as 4, and the movement

---

* Translated from the Old French edition of A. D. Menut and A. J. Denomy in *Mediaeval Studies*, Vol. *3* (1941), 216–17.

takes place in one day, then according to Aristotle the power which would produce such a movement in two days exactly would be 4, and hence it would be equal to the resistance. Further, that power which would produce this movement in four days would be 2, and so it would be less than the resistance. A similar situation obtains if we proceed further. This is illogical *(inconvenient)* and impossible.

3. In this matter the supposition should be, not that the increase or decrease of the swiftness follows the increase or decrease of the motive power proportionally, but that it follows the increase or decrease of the proportion which the motive power has to the resistance. So if a power is as 3 and a resistance is as 1, it is a triple proportion. If the power were as 6, it would not move the same resistance with a velocity double the original one, but the power which is as 9 would accomplish this doubling, because the proportion of 9 to 1 is "double" *(doble)* the proportion of 3 to 1. Also if 18 could move 8 in one day, the power which would produce this movement in 2 days will not be 9, but will be 12, and that which would produce it in 4 days is something which is greater than eight in a proportion which is not in numbers (i.e., is not a rational number), but which is called the square root of one and one-half *(medietas sesquialtere)*.

COMMENTARY

One of the ambiguities of the fourteenth century vocabulary on proportions is clearly illustrated in this passage. When the schoolmen spoke of one proportion being "double" another, they meant that it was the "square" of that proportion. The words "triple," "quadruple," etc., when applied to the comparison of proportions meant "cubed," "raised to the fourth power," etc. On the other hand, when the expression a "double proportion" was used, it meant a proportion of 2 : 1. Similarly a "triple proportion" was 3 : 1, etc.

The examples given by Oresme can be illustrated by using the formula

$$F_2 = R_2 \left(\frac{F_1}{R_1}\right)^n \text{ and } n = \frac{V_2}{V_1}.$$

In the first example, Oresme seeks F_2 when $R_2 = R_1 = 1$, $F_1 = 3$, and $n = 2$; thus $F_2 = 1 \, (3/1)^2 = 9$. In the second example $R_2 = R_1 = 8$, $F_1 = 18$, $n = \frac{1}{2}$.

Thus $F_2 = 8 \, (18/8)^{\frac{1}{2}} = 12$.

Finally in the third example $F_2 = 8 \, (9/4)^{\frac{1}{2}} = 8(3/2)^{\frac{1}{2}}$.

Document 7.3

A Brief Tract on Proportions Abridged from the Book on Proportions of Thomas Bradwardine, the Englishman

1. EVERY proportion is spoken of "commonly" or it is spoken of "properly." Commonly speaking it is the relationship *(habitudo)* of two things mutually compared. Properly speaking it is the mutual relationship of two things of the same kind. Proportion properly speaking is spoken of in two ways: rational and irrational. A rational [proportion] is one which is immediately denominated by some number. An example is that of a double [proportion which is immediately denominated] by two. An irrational proportion is one which is not denominated immediately by some number. An example [of this kind of proportion] is the square root of a double proportion *(medietas proportionis duple)*.

Communicant, commensurable, or rational quantities are those which have one common measure which measures exactly any one of the quantities you please. For example, in the case of the quantities two feet and three feet, each is measured by one foot.

Noncommunicant, incommensurable, or irrational quantities are those for which there is no common measure which measures exactly any one of them you please. An example of such quantities occurs in the diagonal *(dyameter)* and side of a square. The diagonal of a square is the line of a square protracted from one angle to another angle opposite it. A square is a figure having all sides equal.

Some proportions are proportions of equality, others proportions of inequality. A proportion of equality is the ratio of two equal quantities, such as of 4 to 4. A proportion of inequality is the ratio of two unequal quantities, such as 4 to 2. A proportion of inequality is spoken of in two ways: a proportion of greater inequality and a proportion of lesser inequality. A proportion of greater inequality occurs when the larger number is compared to the smaller, as 4 to 2. A proportion of lesser inequality occurs when the lesser quantity is compared to the greater, as 2 to 4.

A proportion of greater inequality is spoken of in five ways: multiple (*multiplex*), superparticular (*superparticularis*), superpartient (*superparciens*), multiple superparticular, and multiple superpartient.

A multiple proportion is a ratio of a larger to a lesser quantity in which the larger contains the lesser a certain number of times, for example, 6 to 2. A superparticular proportion occurs when the larger number contains the lesser number [once] and in addition contains some aliquot part of the latter, for example, 6 to 4. A superpartient proportion is formed when the greater contains the lesser once and in addition contains some aliquot parts from which no single aliquot part of the lesser number or quantity is made, for example, 8 to 5. A multiple superparticular proportion occurs when the greater number contains the smaller [a number of times] and in addition some aliquot part of the lesser number, for example, 14 to 4. Moreover, an aliquot part is one which taken some number of times will produce its whole. A nonaliquot part is one which taken any number of times will not exactly produce its whole. A multiple superpartient occurs when the greater number contains the lesser [a number of times] and in addition contains some aliquot parts which taken together do not make a single aliquot part of the lesser quantity, for example 16 to 6.

For the denomination of proportions it should be known that a double proportion exists when the antecedent number contains the consequent number twice, a triple when it contains it thrice, and so to infinity. The antecedent is the first term of the proportion and the consequent the second.

A sesquialterate proportion exists when the greater number contains the lesser number once and one-half, for example, 3 to 2. A double sesquialterate proportion exists when tye greater number contains the lesser twice and one half, for example, 5 to 2. And so on ascending toward infinity.

A sesquitertiate proportion exists when the greater number contains the lesser once and one-third, for example, 4 to 3. A double sesquitertiate exists when the greater contains the lesser twice and one-third, for example 7 to 3. And so on in ascending to infinity. All of these proportions are of the superparticular species.

But finally for the denomination of proportions of the superpartient species, it should be observed how many times the greater contains the lesser. If once and two such parts, like 5 to 3, then the proportion is a superbipartient proportion. If once and three such parts, then it is a supertripartient proportion, of which kind is the proportion 7 to 4. After it has been seen how many times the greater quantity contains the lesser and in addition how many extra aliquot parts there are, which when taken together do not make a single aliquot part of the lesser number, one ought to observe how much any one of these parts is with respect to the lesser number, whether a third or a fourth, and similarly with other numbers. And according to that [determination], the last part of the denomination is fashioned. Let us take, for example, the proportion of 5 to 3, which has this last part of the denomination. Clearly it is superbipartient because it contains two additional parts out of which no single aliquot part with respect to the lesser number can be formed. But it has as this last part of the denomination, "thirds," because either of the additional parts is a third with respect to the lesser number. Hence, the proportion of 5 to 3 is a superbipartient thirds. If the greater contains the lesser number once and three such parts in addition, out of which parts no single aliquot part with respect to the lesser number can be formed, and any of which parts is a fourth of the lesser number, then the proportion is spoken of as supertripartient fourths. The proportion of 7 to 4 is one of this kind. And thus in ascending to infinity. If the greater number contains the lesser number twice and two such parts in addition, from which parts no single aliquot part of the lesser number can be formed, and each of which parts is one third of the lesser number, then it is spoken of as double superbipartient thirds. The proportion of 8 to 3 is one of this kind.

What difference is there between proportion and proportionality? Proportion is spoken of in two ways, as was said earlier. Proportionality, however, is the order of proportions. It is spoken of in three ways: arithmetic, geometric, and harmonic proportionality. Arithmetic proportionality exists when there are many terms and the excess of the first beyond the second is equal to the excess of the second over the third, and in the same way with the succeeding terms. An example of such proportionality is 6, 5, 4. Geometric proportionality exists when there are many terms and the proportion of the first to the second term is equal to the proportion of the second to the third term, and in the same way with the succeeding terms. Example: 8, 4, 2, so that a term is always double the succeeding term. Harmonic proportionality exists when there are three terms and the proportion of the first term to the third is equal to the proportion of the excess of the first over the second term to the excess of the second over the third term. Example: 6, 4, 3. There exist interchangeable proportionals when the antecedent of one is to the antecedent of the other as the consequent of one is to the consequent of the other. Example: 8, 4, 2, 1. Equal proportionality exists when there are two [sets of] numbers, and in each [set] there are three terms and the first term of the first set is to the second term of the first set as the first term of the second set is to the second term of the second set, and the first term of the first set is to the third term of the first set as the first term of the second set is to the third term of that set. Example: 8, 4, 2 and 4, 2, 1.

2. There follow certain suppositions:

[First supposition:] All proportions are equal whose denominations are equal.

The second supposition is this: If there are three proportional terms and the first is greater than the second and the second greater than the third, then the proportion of the first to the last is composed of the proportion of the first to the second and of the second to the third. [47]

Third supposition: When there are many terms and the first term is more than the second and the second is more than the third, and similarly with respect to the other terms up to the last proportional term, then the proportion of the first to the last is composed of the proportion of the first to the second and that of the second to the third, and thus for the other terms. [48]

Fourth supposition: When equals are compared to some third quantity, the proportion between one of the equals and the third quantity is the same as the proportion between the remaining equal and the third quantity.

Fifth supposition: If two unequal quantities are compared to some third quantity, the proportion of the greater of these to that third quantity will be greater than the proportion of the lesser of them to the third quantity. Conversely, the proportion of that third quantity to the lesser quantity will be greater than that of the third quantity to the greater quantity.

Sixth [supposition]: When anything is composed of two equals, that composite will be double either of the equals; if it were composed of three equals, it will be triple any of them, etc. [49]

---

[47] If $a/b = b/c$ and $a > b > c$, then $a/c = (a/b)(b/c)$. This and the succeeding notes attempt to give modern representations of the rhetorical statements of Bradwardine.

[48] If $a/b = b/c = c/d \ldots y/z$ and $a > b > c > d \ldots y > z$, then $a/z = (a/b)(b/c)(c/d) \ldots (y/z)$, where y and z are any last tems in a series of continually proportional terms.

[49] The term "composite" throughout this treatise means "composed by proportions." Thus "double" when used with composites and proportions usually means

[*Brief Tract on Bradwardine's Proportions* 469]

Seventh supposition: When something is composed of two unequals, that something is more than double the lesser and less than double the greater.[50] For example, three feet, the composite of two feet and one foot, is more than double one foot and less than double two feet. The first five suppositions are proved in the fifth book of the *Elements* of Euclid. The two last are obvious by reasoning.

3. There follow certain conclusions:

The first conclusion to be proved from these suppositions is this: If there is a proportion of greater inequality of the first to the second term and of the second to the third (i.e., in continually proportional terms), then the proportion of the first to the third will be a proportion double (i.e., the square of) the proportion of the first to the second and of the second to the third.[51] This [conclusion] I prove thus: I take three terms and argue in this manner: Between 8 and 4 there is a double proportion, and between 4 and 2 there will also be a double proportion. Hence the proportions are equal. The consequence holds from the first supposition: "All proportions etc." And then I argue by the second supposition: "When three terms etc." But here in this conclusion there are three terms. Hence the proportion of the first to the last is composed of them, but the proportion between the first and second is equal to that between the second and third. Consequently, the proportion of the first to the last [term] is composed of two equals. Hence it is double either of them. The consequence holds from the sixth supposition.

The second conclusion is this: If there are four terms proportional in geometric proportionality, the proportion of the first to the last is triple (i.e., the cube of) the proportion of the first to the second, and in the same way concerning [the proportions between] the others (i.e., the succeeding terms).[52] I prove this thus: I take these four terms, 8, 4, 2, and 1. Then

---

"squared." This supposition originally simply meant that if $c = a + b$ and $a = b$, then $c = 2a = 2b$. Using composite in case of proportions, this supposition means that, if $a/c = (a/b)(b/c)$ and $a/b = b/c$, then $a/c = (a/b)^2$ and $a/c = (b/c)^2$.

[50] Similarly this originally meant simply: If $c = a + b$ and $a > b$, then $c > 2b$ and $c < 2a$. But applied to proportions this supposition, assigning the word "double" to mean "squared," becomes:
If $a/c = (a/b)(b/c)$ and $a/b > b/c$, then $a/c > (b/c)^2$ and $a/c < (a/b)^2$.

[51] Here, as I suggested in note 49 above, the interpretation of a "double" as a "squared" appears. Thus, if $a/b = b/c$, then $a/c = (a/b)^2 = (b/c)^2$. Notice that the example he gives involves $a/b = b/c = 2$. This is chosen because of the coincidence of $a/c = (a/b)(b/c) = (a/b) + (b/c)$ or
$$a/c = (a/b)^2 = (b/c)^2 = 2a/b = 2b/c,$$
when $a/b = b/c = 2$.

[52] If $a/b = b/c = c/d$, then
$$a/d = (a/b)^3 = (b/c)^3 = (c/d)^3,$$
assuming "triple" is "cubed."

in this example *(hic)* there are four terms and the first is more than the second and the second is more than the third, and similarly concerning the others (i.e., the succeeding terms). Hence the proportion of the first to the last is composed of the proportion of the first to the second, the second to the third, and similarly of the others. The consequence holds from the third supposition. But the proportions here between the first and the second, the second and the third, and similarly with regard to the others, are equal. Hence, the proportion of the first to the last is composed of three equals. Hence it is triple (i.e., the cube of) any one of them. The consequence holds from the sixth supposition. And if there are five terms, the proportion of the first to the last is quadruple to any one of the proportions. If there are six terms, it will quintuple any one of them, and thus to infinity.

The third conclusion is this: If the first term is more than double the second, and the second term is exactly double the third, the proportion of the first to the third will be less than double (i.e., less than the square of) the proportion of the first to the second.[53] This conclusion I prove thus: I take three terms, 6, 2, 1. Then I argue by the second supposition: "If there are three proportional terms, etc." But in this case here there are three such terms. Therefore, the proportion of the first to the last is composed out of them. But the proportion of the first to the second is more than that of the second to the third. Hence the proportion of the first to the last is composed of two unequals. Hence it will be more than double the lesser proportion and less than double the larger proportion. The consequence holds by the seventh supposition. But the proportion of the first to the second is more than that of the second to the third. Consequently, the proportion of the first to the last will be less than double the proportion of the first to the second.

The fourth conclusion: If the first term is double the second and the second is more than double the third, the proportion of the first to the last will be less than double the proportion of the second to the third....[54] [The proof given is similar to that supporting the third conclusion.]

The fifth conclusion is this: If the first term is less than double the second, and the second is precisely double the third, then the proportion of the first term to the last will be more than double the proportion of the first to the second.[55] I prove this conclusion thus: I take three terms, 6, 4, and 2. Then I argue by the second supposition: "If there are three pro-

[53] If $a/b > 2$ and $b/c = 2$, then $a/c < (a/b)^2$.
[54] If $a/b = 2$ and $b/c > 2$, then $a/c < (b/c)^2$.
[55] If $a/b < 2$, $b/c = 2$, then $a/c > (a/b)^2$.

[*Brief Tract on Bradwardine's Proportions* 471]

portional terms etc." But in this case there are three such terms. Therefore, the proportion of the first to the last will be composed of these. But the proportion of the second to the third will be more than that of the first to the second. Therefore, the proportion of the first to the last is composed of two unequal proportions. Hence, it is more than double the lesser proportion and less than double the greater proportion. But the proportion of the second to the third will be more than that of the first to the second. Consequently, the proportion of the first to the last will be more than double the proportion of the first to the second.

The sixth conclusion: If the first term is double the second and the second is less than double the third, the proportion of the first to the last will be more than double the proportion of the second to the third ....[56] [The proof of this conclusion is similar to that supporting the fifth conclusion.]

The seventh conclusion: There is no proportion greater or less than a proportion of equality because [one] proportion of equality is neither more nor less than [another] proportion of equality, all proportions of equality being equal and hence no one being greater than another.[57] Nor is a proportion of greater inequality greater than a proportion of equality, because if so, then a proportion of equality could be increased [geometrically] until it was equal to it (i.e., to a proportion of greater inequality). The consequent is false because however much a proportion of equality is increased [geometrically], it will always remain a proportion of equality. Consequently, it will be neither more nor less than it now is. Similarly, if a proportion of greater inequality is less than a proportion of equality, then a proportion of equality can be diminished until it is equal to it (i.e., to a proportion of greater inequality). But the consequent is false, as above. It should be argued in the same way with respect to a proportion of lesser inequality compared to a proportion of equality.

But against this conclusion it is argued thus: Let three terms be taken: 6, 4, 6. Then, by the second supposition, the proportion of the first to the last is composed out of the proportion of the first to the second and of the second to the third, and every composite is greater than any part com-

---

[56] If $a/b = 2$ and $b/c < 2$, then $a/c > (b/c)^2$.

[57] The purport of this conclusion is simply that, if we assume a geometric proportionality which we can represent: $a/z = (a/b)^n$, then if $a/b = 1$ (or is a proportion of equality, as the schoolmen said), then a/z must always equal 1, regardless of what value n is, i.e., how many times a/b is taken. One can say that, in the sense that a/b can never geometrically produce a proportion greater than 1, a/z as a proportion of greater equality could never be produced by a/b and thus by "composition" is not comparable to a/b.

posing it. Hence, the proportion of the first to the last is greater than the proportion of the first to the second or of the second to the third; but the proportion of the first to the last is a proportion of equality, and the proportion of the first to the second is a proportion of greater inequality, and the proportion of the second to the third is a proportion of lesser inequality. Hence, the proportion of equality is greater than a proportion of greater inequality, and similarly than a proportion of lesser inequality.

This argument is answered by noting that the second supposition is understood only in the way it is described in that treatise [of Bradwardine].[58] The conclusion is also objected to in another way as follows:

I take these three terms, 6, 4, 4. Then by the fifth supposition the proportion between 6 and 4 will be greater than between 4 and 4 and the proportion of 6 to 4 is a proportion of greater inequality and 4 to 4 is one of equality. Hence, a proportion of greater inequality will be more than a proportion of equality. In answer to this argument, it is said that the fifth supposition has to be understood for two unequal [proportions] compared with a third in the same kind of a proportion....

4. Some erroneous opinions.

There are three opinions especially to be disproved. The first of these posits that the velocity in movements follows the excess of the motive power over the power of the thing moved.[59] Hence the meaning *(sententia)* of this opinion is this: If there are two powers and two resistances, then by the amount more that the first power exceeds its resistance than the second power exceeds its resistance, by that same amount will the first power move more swiftly with its resistance than will the second [power] with its. Example: If a is a power as 8, b its resistance as 4, c another power such as 2, and d its resistance as 1, then by this opinion a will be moved with b 4 times as fast as c with d, because the excess of a over b is quadruple the excess of c over d.

Against this opinion it is argued thus: It is against Aristotle's opinion in the seventh [book] of the *Physics* and also in many other places where he posits that the velocity of movement follows a proportion. But it is answered

---

[58] The example taken in this argument is 6, 4, 6. It does not fulfill the requirement of the second supposition of continually proportional terms with the first term greater than the second, and the second greater than the third. Hence the objection must be set aside. Nor does the objection which follows fulfill the conditions of the fifth supposition; hence it too can be set aside.

[59] The first opinion can be represented as follows: $\dfrac{V_2}{V_1} = \dfrac{F_2 - R_2}{F_1 - R_1}$, where V_2 and V_1 are velocities, F_2, and F_1 powers, and R_2 R_1 resistances.

[*Brief Tract on Bradwardine's Proportions* 473]

perhaps that Aristotle understands by proportion arithmetical proportion. Against this gloss, it is argued thus: In the fifth [book] of the *Elements* of Euclid this is the first conclusion: If there is a comparison of some quantities to just as many others, and they are equally more, equally less, or similar, as any one of them compares to its corresponding quantity, then also the whole totaled out of these quantities has itself in the same way to the total of the corresponding quantities, accepted in like manner. Whence the judgment of the conclusion is this: If there are many quantities compared throughout and any one of them at all is to the term it is compared with in the same proportion as any other to its comparable term, then the total aggregated of all the antecedents is in the same proportion to the aggregate of all the consequences as the first antecedent is to its consequence. This, however, is not true of arithmetical proportion.

But against this opinion it is argued *per experimentum*, because if a fly were carrying a single small weight and a strong man were carrying just as much weight, it stands that the fly with its weight is moved with much greater velocity than that with which the man is moved with his weight. Yet the power of the man exceeds his resistance (the weight) by a much greater excess than the power of that fly exceeds his [resistance]. Hence that opinion is false, etc.

The second opinion posits that movement follows the proportion of the excess of the powers over their resistances. The meaning of this opinion is this: If there are two powers and two resistances, and the excess of the first power has itself in greater proportion to its resistance than the excess of the second power to its resistance, then the first power will be moved by greater velocity with its resistance than the second with its. [60] Example: Let a be a power as 5, and let b be its resistance as 1; and let c be a power as 3, d its resistance as 1. Then, by that opinion, a will be moved twice as fast with b as c with d, because the proportion between the excess by which a exceeds b and b is twice as large as the proportion between the excess by which c exceeds d and d, since the first is a quadruple and the second is a double proportion, as is obvious to the attentive.

Against that opinion it is argued by absurdity as follows: Some power will be moved with some resistance and [yet] nothing can be moved more swiftly or more slowly than it. I prove this thus: I posit that a is a power

---

[60] The second opinion can be represented as follows:

$$\frac{V_2}{V_1} = \frac{\dfrac{F_2 - R_2}{R_2}}{\dfrac{F_1 - R_1}{R_1}} = \frac{(F_2 - R_2) R_1}{(F_1 - R_1) R_2}.$$

as 8, b its resistance as 4. Then the excess of a power over its resistance b is 4. Its resistance is also 4. Therefore, the excess of no power can have itself in less proportion nor in greater proportion than that excess has to its resistance [the proportion of that excess 4 to its resistance 4 being a proportion of equality]. The consequence holds by the seventh conclusion, the one *in re* a proportion of equality. Hence by this position nothing can be moved more swiftly nor more slowly. Yet that this conclusion is false is obvious, because however large that power is, another such power can be moved with greater resistance. Therefore, it can be moved more slowly with less resistance, and consequently faster. Therefore, the conclusion is false.

The third opinion posits that velocity in movements with the motor remaining the same or equal follows the proportion of the patient, and with the patients remaining the same or equal follows the proportion of the motor.[61] This opinion has two members. Whence the meaning *(sententia)* of the first part is this: If there is a power sufficient to be moved with two unequal resistances, just as one resistance is less than the other so it is sufficient that that power is moved with less resistance more quickly than it is with greater resistance. Example: if a is one power as 8, b its resistance as 4, c another resistance as 2, then by this opinion a will be moved twice as quickly with c resistance than with b, because c resistance is doubly less than b resistance.

But the meaning of the second part is this: If two unequal powers suffice to be moved with a single resistance; just as the one power is greater than the other, so it suffices that it is moved more quickly with that resistance than the lesser power. Example: if a is a power as 8, b a power as 4, c a resistance as 2, then just as a power is greater than b, so a power is sufficient to be moved more quickly with c resistance than b power with c.

That this opinion is insufficient I prove thus: It does not teach with what the velocity of a movement is measured where there are diverse powers and diverse resistances, but only where there are two powers and one resistance or two resistances and one power. Against the first part of the

---

[61] This appears to be the traditional Aristotelian law:
1st part: $V_2/V_1 = F_2/F_1$, with R_1 and R_2 equal.
2nd part: $V_2/V_1 = R_1/R_2$, with F_1 and F_2 equal.

Taken together (although they are not specifically done so in this version) these could be represented:
$$\frac{V_2}{V_1} = \frac{F_2 R_1}{F_1 R_2}.$$

opinion it is argued by the following impossible conclusion:[62] It suffices that some power is moved with a certain resistance; and yet with the given power remaining the same, no resistance can be found with which that is moved twice as slowly as before. That the conclusion is impossible is obvious enough because however great will be that power, if one resistance grows to equality with its power and the power remains the same, it will be moved with a speed approaching infinite slowness with that power. Hence it is false that it does not move [at least] doubly slower. Therefore, the conclusion is false. And that the conclusion follows out of the first part of that opinion I prove thus: If a is a power such as 8, with b its resistance as 4, then I prove that a cannot be moved doubly slowly with some resistance other than b; because if so, let c be such a resistance. Then either c is equal to a or less or more. It is not equal or more than a because then a would not be sufficient to be moved with c by that conclusion, since movement cannot arise from a proportion of equality nor from one of lesser inequality. If, however, c is less than a and greater than b, then I argue thus: a is sufficient to be moved twice as slowly with c as with b, hence doubly more swiftly with b than with c. The consequent is false because b is not doubly less than c, and similarly a is precisely double b, and c is greater than b. Therefore, a is not sufficient to be moved twice as slowly with c as with b. The consequence holds through the first part of that opinion. Similarly, it is argued concerning any resistance at all. Hence follows the conclusion with respect to the first part of the opinion.

Against the second part of the same opinion it is argued *per experimentum*: if there were one man carrying one large weight which he was just able to carry and if another man equally strong should help him, then the two together would move that weight more than doubly fast than one of them would move it by himself. Yet the power composed of these two is nothing but twice as great as the power of one of them alone. Hence, the second part of this opinion is false.

A fourth opinion posits that velocity of movement does not follow a proportion because there is no proportion between power and resistance. But all the arguments for this opinion proceed only from a proportion properly speaking. Hence I shall not argue against it at present.

The fifth opinion, wihch is the true one, posits that velocity of move-

---

[62] The whole force of the succeeding criticism is to show that, if this third position is followed, we arrive at the contradiction of having a velocity even where there is a proportion of equality or lesser inequality, i.e., when the power is equal to the resistance or less than it.

ment follows the geometric proportion of the power of the motor above the power of the thing moved.[63] Whence the meaning of that opinion is this: If there are two powers and two resistances and the proportion between the first power and its resistance is greater than the proportion between the second power and its resistance, the first power will be moved more rapidly with its resistance than the second with its, just as one proportion is greater than another. Example: Let a be a power as 8, b its resistance as 2, and c be a power as 6, d its resistance as 3. Let a be moved with b and c with d. Then a will be moved doubly fast with b than c with d, because the proportion between a and b is doubly greater than that between c and d. This opinion is obvious by Aristotle in the seventh [book] of the *Physics*, where he says that the velocity of motion follows the proportion of the power of the motor above its resistance.[64]

5. According to that opinion, the following conclusions are to be conceded:

If one power has itself in a double proportion to its resistance, when the power is duplicated with the resistance remaining the same, the movement will be doubled.[65] I prove this thus: Let a be a power as 4, b its resistance as 2, c another power as 8. Then I argue thus by the first conclusion: If there is a proportion of greater inequality of first to second and second to third, etc. But here are three such terms and the proportion of the first to the second is equal to that of the second to the third. Hence the proportion of the first to the last will be double to any one of them. But velocity of movement follows geometric proportion, etc. Hence c will move twice as quickly with b as a with b. But if a were doubled, then it would be sufficient that a be moved with b at the same velocity as now is sufficient for c. Hence, the conclusion is true.

The second conclusion is this: If any power has itself in a double pro-

---

[63] The position followed by Bradwardine can be represented as follows: $V_2/V_1 = \log_a (F_2/R_2)$ where $a = F_1/R_1$ and $F_1/R_1 > 1$. Or we can write this: $F_2/R_2 = (F_1/R_1)^n$ where $n = V_2/V_1$.

[64] Now it seems clear that Aristotle follows the third opinion already rejected. But in order to save Aristotle and bring the master in line with his own position, Bradwardine picks an example where the two positions are equivalent. This is when $F_2/R_2 = 2(F_1/R_1)$. Thus, using the figures in the example, $F_2 = 8$, $R_2 = 2$, $F_1 = 6$, $R_1 = 3$, then according to both the third and his own position $V_2/V_1 = 2$.

[65] This can be shown by using the second form of the formula developed in note 63.

(1) Assume $F_2/R_2 = (F_1/R_1)^n$, where $n = V_2/V_1$.

(2) We are told that $F_2 = 2F_1$, $R_2 = R_1$, $F_1 = 2R_1$.

(3) Then substituting (2) in (1): $4R_1/R_1 = (2R_1/R_1)^n$, or $4 = 2^n$, and $n = \log_2 4 = 2 = V_2/V_1$. Q.E.D.

[Brief Tract on Bradwardine's Proportions

portion to its resistance, it suffices that that power will be moved with half that resistance twice as rapidly as it will be with the total resistance.[66] I prove this thus: Let a be a power such as 8, b its resistance such as 4, and c another resistance as 2. Then I argue by the first conclusion thus: "If the proportion of the first to the second is a proportion of greater inequality, and so is the proportion of the second to the third, etc." But here it is so. Hence the proportion of the first to the last will be double either of these [proportions]. But the motion follows the proportion, etc. Hence it suffices that a will be moved with c twice as rapidly as with b, but c is one half b. Hence the conclusion is true.

The third conclusion is this: If one power has itself in more than double proportion to its resistance, with the power duplicated and the resistance remaining the same, the movement will not be doubled.[67] I prove this thus: Let a be a power as 6, b a resistance as 2, c a power as 12. Then I argue by the fourth conclusion: "If the first is double the second and the second is more than double the third, etc." But here it is so. Hence the proportion of the first to the last is less than double the proportion of the second to the third. But the motion follows the proportion. Therefore c will not be moved with b twice as rapidly as a with b. But if a were doubled, then it would suffice that a is moved with b just as it now suffices that c is moved with b. Hence the conclusion is true.

The fourth conclusion: If one power has itself in a proportion more than double its resistance, it suffices that that power is not moved twice as rapidly with half that resistance as with the whole.[68] I prove this thus: Let a be one power as 6, b a resistance as 2, and c a resistance as 1. Then I argue by the third conclusion: "If the first is more than double the second and the second is exactly double the third, etc." But here it is so. Hence the proportion of the first to the last is less than double the proportion of the first to the second. But motion follows the proportion. Hence a is not moved with c twice as rapidly as with b, and c is one-half of b. Therefore, the conclusion is true.

The fifth conclusion is this: If a power has itself in less than double

---

[66] Similarly as in note 65, with
(1) $F_2 = F_1$, $R_2 = R_1/2$, $F_1 = 2R_1$.
(2) Then $4R_1/R_1 = (2R_1/R_1)^n$ and $n = 2 = V_2/V_1$. Q.E.D.

[67] Using same formulations as in note 65, with
(1) $F_1 = (2+e) \cdot R_1$, $R_2 = R_1$, $F_2 = 2F_1$.
(2) Then $2(2+e) = (2+e)^n$ or $2 = (2+e)^{n-1}$.
(3) Now n has the largest possible value when e is zero, and this value is $n = 2$.
(4) Therefore, for all values of e greater than zero, $n < 2$, or since $n = V_2/V_1$, $V_2/V_1 < 2$. Q.E.D.

[68] The same ultimate equation is formed as in note 67: $2 = (2+e)^{n-1}$. Thus the same reasoning can be used as in note 67.

proportion to its resistance, the power doubled will move [the same resistance] more than twice as rapidly.[69] I prove this thus: Let a be a power as 6, b a resistance as 4 and c another power as 12. Then I argue by the sixth conclusion: "If the first is double the second, and the second less than double the third, the proportion of the first to the third, etc." But here is it so. Hence the proportion of the first to the last will be more than double the proportion of the second to the third. But motion follows the proportion. Therefore, c will be moved with b more than twice as quickly as a with b. But if it were doubled, then it would suffice that a be moved with b in such a velocity as now it is sufficient for c to be moved by b. Therefore, the conclusion is true.

The sixth conclusion is this: If any power has itself in less than double proportion to its resistance,[70] it suffices that that power is moved more than doubly fast with half of that resistance than with the whole. I prove this thus: Let a be a power as 6, b a resistance as 4, c another resistance as 2. Then I argue by the fifth conclusion: "If the first is less than double the second, and the second is precisely double the third, etc." But here it is so, as is obvious in the example. Hence the proportion of the first to the last will be more than double the proportion of the first to the second. But motion follows the proportion. Hence it is sufficient that a is moved more than twice as quickly with c as with b, and c is one-half b. Therefore, the conclusion is true.

Against this [fifth] opinion it is argued as follows: From that opinion follow three contradictions *(inconvenientia)*. The first is this, that unequal movements arise from equal proportions. The second is that the same velocity arises from a lesser proportion as from a greater proportion. The third is that from a lesser proportion, a greater velocity arises than from a greater proportion. But all of these are false. Hence the fifth opinion is false. Moreover the falsity of these three is obvious from that opinion itself.

The first contradiction is proved: Let a and b be two certain unequal powers. Let the power of a be as 8, the power of b as 4. Let c and d be two other uniform things proportional to a and b, and the resistance of c

---

[69] Using the formulation of note 65, this would reduce to:
(1) $2 = (2 - e)^{n-1}$.
(2) Now here the minimum value of n exists when e is zero and that minimum value is $n = 2$.
(3) Therefore, when e is any value greater than zero, n must have a value greater than 2, and thus $V_2/V_1 > 2$. Q.E.D.

[70] This reduces to the same formulation as in note 69, namely: $2 = (2 - e)^{n-1}$. Thus the same conclusion follows, i.e., $V_2/V_1 > 2$.

[*Brief Tract on Bradwardine's Proportions* 479]

be of a foot in size, d a half foot. Let the resistance c be as 2 and the resistance d be as 1. And it is posited that a traverses c, and b, d. Then I enquire whether in equal time a will traverse c as b, d, or not. This is obvious by that opinion, since a is in equal proportion to c as b is to d, by the case [posited]. But motion follows the proportion. Therefore a will traverse c in the same time as b traverses d. And if so, the first conclusion follows. I prove this thus: a will describe more than b, and they were moved from equal proportions. Hence the conclusion follows. That the second follows, is proved thus: The resistance of c would be increased until a is sufficient to traverse one-half of c in the time in which b suffices to traverse the whole d. Then it is argued thus: Now the proportion of a to c is equal to the proportion of b to d, and in the end the proportion of a to c will be less than it is now, because c resistance will be greater than it now is, as the case posits. And then the proportion of b to d will be as in the beginning, and the movement of a will be equally swift as the movement of b, since they will traverse equal spaces in equal time. Hence from a lesser proportion arises an equally fast movement as from a large proportion.

That the third contradiction follows is proved thus: Again let the resistance of c be decreased, but not to the point of equalling the resistance in the first case. Then the proportion of a to c will still be less than in the first case. In the first case the proportion of a to c was as that of b to d. Hence in this case a to c will be less than b to d, and the movement of a will be swifter in this case than the movement of b, because c now will traverse more in this case, and yet b is as in the second case. But a was sufficient to traverse just as much as b in an equal time. Hence a in this case is sufficient to traverse more than b. Hence the movement of a is swifter than b, and a will be moved in c from a lesser proportion than b in d. Hence a swifter movement arises from a lesser proportion than from a greater one. So the third contradiction follows.

It is answered that in the time in which b will traverse d, a will traverse one-half of c. Then to the argument—a has itself to c in the same proportion as b to d, hence a suffices to traverse c in the same time as b does d—the consequence is denied. For something can have itself in equal proportion in two ways, either as to quality or as to quantity, i.e., in the manner of equally fast in those media or in traversing them in an equal time. By the first method: the proportion of a to c is equal to the proportion of b to d. But [this is] not [so] by the second method. Therefore, it does not follow that a is sufficient to traverse c in the same time as b traversed d. Therefore, etc. Whence note that no motion can arise from a proportion of lesser inequality nor from a proportion of equality, but every movement arises from a proportion of greater inequality. Thus, however much heavier a mixture is than a simple [element], it can be moved in the same medium with a greater velocity, a lesser velocity and a velocity equal to that of the simple. I prove this thus: Let a be a mixed heavy body whose gravity is as 8 and its lightness is as 2, and b a simple heavy body whose gravity is as 4. Let c be a medium where resistance is as 2. Let a and b begin to be moved in c; a begins to be moved in c with the same velocity as b in c, and in equal [time], because the power of a is as 8, its levity, i.e., its intrinsic resistance, is as 2, its extrinsic resistance is as 2. Therefore, the whole of its resistance is as 4. But between 8 and 4 there is a double proportion. Hence a begins to be moved from a double proportion, and the power of b is as 4, and one resistance of it is as 2. But between 4 and 2 there is a double proportion. Hence a and b begin to be moved equally and equally swiftly. And I prove that a can be moved with b less swiftly than b with c, because if the resistance of c is diminished until it is as 1, then the power will be as 8, its intrinsic resistance as 2, its extrinsic resistance as 1. Hence the total resistance will be as 3. But 8 to 3 is a proportion double superbipartient thirds. Now the power of b will be as 4, its resistance as 1. Therefore, b will be moved from a quadruple proportion, but a will not be moved with any such velocity. Hence a is moved with less velocity than b. Also I prove that a can be moved from a greater velocity in c than b in c, because if the resistance of c grows until it is as 3, then the power of a will be as 8, its intrinsic resistance as 2, and its extrinsic resistance as 3. Hence the total resistance will be as 5. But between 8 and 5 there is a proportion supertripartient fifths. But the power of b is as 4, and its whole resistance is as 3. But between 4 and 3 there is a sesquitertiate proportion, which is less than the proportion by which a will be moved in c. Therefore, the whole conclusion is true, etc. The end.

COMMENTARY

I have already made detailed comments in the footnotes to the translation. However, I might point out again to the reader that the Euclidean theory of proportions that is represented here was the mathematical language of physical laws and descriptions in antiquity and the Middle Ages. It was initially given in Book V of the *Elements* but it was much commented upon and extended in mathematical and physical writings of the fourteenth century. The most notable extension of the theory of proportions in the fourteenth century was that made by Nicole Oresme in two treatises, the *Algorismus proportionum* and the *De proportionibus proportionum*, where we not only see the rules of manipulating with integer and fractional exponents laid out but some remarkable discussion of irrational exponents, unsurpassed until early modern times. New and critical editions of these works are being prepared by a former student of mine, Edward Grant.

As I suggested in the body of the chapter, this *Abbreviatus* of the dynamic section of Bradwardine's *Tractatus de proportionibus* (Chapters I–III) was composed by some unknown figure about the middle of the fourteenth century. That it was fairly popular is demonstrated by the number of manuscripts of it (see Bibliography).

*Tractatus brevis proportionum: abbreviatus ex libro de proportionibus D. Thome Braguardini Anglici**

Omnis proportio vel est communiter dicta, vel proprie dicta. Communiter dicta est duarum rerum comparatarum adinvicem habi-
5 tudo. Proprie dicta est duarum rerum eiusdem generis adinvicem habitudo. Proportio proprie dicta dicitur duobus modis, scilicet, rationalis et irrationalis. Rationalis est que inmediate denominatur ab aliquo numero, ut dupla a duobus. Irrationalis est que non inmediate denominatur ab aliquo numero, ut medietas proportionis duple.
10 Quantitates communicantes commensurabiles seu rationales sunt quarum est una mensura communis quamlibet istarum precise mensurans, sicut bipedalis, tripedalis, quarum utraque pedali mensuratur. Quantitates non communicantes incommensurabiles seu irrationales dicuntur, quibus non est una ratio communis quamlibet istarum precise
15 mensurans, sicut dyameter et costa quadrati. Unde dyameter quadrati est linea quadrati protracta de uno angulo ad alium angulum sibi oppositum. Quadratum est una figura cuius omnes coste sunt equales.

Quedam est proportio equalitatis, et quedam inequalitatis. Proportio equalitatis est duarum quantitatum equalium adinvicem habitudo,
20 ut 4 ad 4. Proportio inequalitatis est duarum quantitatum inequalium adinvicem habitudo, ut 4 ad 2.

---

* Edition of Vienna, 1515, ff. 17v–27r. I have altered the punctuation slightly to bring it in line with modern usage, I have expanded the abbreviations, and freely altered capital and lower-case letters. There is some variety in the spelling, e.g., "tertium" and "tercium," "proportio" and "proporcio." I have left these inconsistencies as in the edition. For a number of manuscripts of this and similar *Abbreviations*, see this work in the Bibliography (under Bradwardine).

Proportio inequalitatis dicitur duobus modis: quedam maioris inequalitatis, quedam minoris inequalitatis. Proportio maioris inequalitatis est quando maius comparatur ad minus, ut 4 ad 2. Proportio minoris inequalitatis est quando minus comparatur ad maius, ut 2 ad 4.

Proportio maioris inequalitatis dicitur quinque modis: scilicet, multiplex, superparticularis, superparciens, multiplex superparticularis, multiplex superparciens. Multiplex vero proportio est habitudo maioris quantitatis ad minorem ipsam multotiens continentis, ut 6 ad 2. Superparticularis est quando maius continet minus et aliquam partem aliquottam ultra, ut 6 ad 4. Superparciens est quando maius continet minus et aliquas partes aliquottas ultra ex quibus non fit una pars aliquotta respectu minoris numeri vel quantitatis, sicut 8 ad 5. Multiplex superparticularis est quando maius multotiens continet minus et aliquam eius partem aliquottam ultra, ut 14 ad 4. Pars autem aliquotta est que aliquociens sumpta reddit equaliter suum totum. Pars non aliquotta est que aliquociens sumpta non reddit equaliter suum totum. Multiplex superparciens est quando maius multociens continet minus et aliquas partes aliquottas ultra ex quibus non fit una pars aliquotta respectu minoris quantitatis, ut 16 ad 6.

Pro denominatione proportionum est sciendum quod proportio dupla est quando antecedens bis continet consequens. Tripla quando ter, et sic in infinitum. Antecedens est primus terminus proportionis et consequens secundus. Sesquialtera proportio est quando maius semel continet minus et eius medietatem ultra, ut 3 ad 2. Dupla sesquialtera est quando maius bis continet minus et eius medietatem ultra, ut 5 ad 2, et sic in infinitum ascendendo. Sesquitertia est quando maius semel continet minus et eius tertiam partem ultra, ut 4 ad 3. Dupla sesquitercia est quando maius bis continet minus et eius terciam partem ultra, ut 7 ad 3, et sic in infinitum ascendendo; et omnes ille proportiones sunt in specie superparticularis.

Sed ulterius pro denominatione proportionum in specie proportionis superparcientis est videndum quociens maius continet minus. Si enim semel et tales duas partes, ut 5 ad 3, est proportio superbiparciens. Si vero semel et tales tres partes, tunc est proportio supertriparciens, qualis est proportio 7 ad 4. Viso autem quotiens maius continet minus et quot tales partes ultra ex quibus non fit una pars aliquotta, videndum est quanta sit quelibet illarum partium respectu minoris numeri, numquid 3 vel 4, et sic de aliis. Et iuxta illud constituitur particula ultima denominationis illius proportionis, ut capta proportione inter 5 et 3 istam particulam denominationis, scilicet superbiparciens, habet, quia continet duas partes ultra ex quibus non fit una pars aliquotta respectu minoris numeri. Sed istam particulam denominationis, scilicet tertias, habet, quia queque duarum partium ultra contentarum est tripla respectu minoris numeri. Ideo proportio 5 ad 3 est superbiparciens tertias. Si maius semel continet minus et tales tres partes ultra ex quibus non fit una pars aliquotta respectu minoris numeri quarum queque est quarta respectu minoris, tunc dicitur proportio supertriparciens quartas, qualis est proportio 7 ad 4, et sic ascendendo in infinitum. Si maius bis continet minus et tales duas partes ultra ex quibus non fit pars aliquotta respectu minoris numeri, tunc dicitur proportio dupla superbiparciens tertias, qualis est proportio 8 ad 3.

Que est differentia inter proportionem et proportionalitatem? Unde proportio dicitur duobus modis, ut dictum est prius. Proportionalitas autem est ordo proportionum et dicitur tribus modis, scilicet, proportionalitas arismetrica, geometrica, et armonica. Arismetrica est quando sunt multi termini et equalis est excessus inter primum et secundum sicut inter secundum et tertium, et sic de aliis. Exemplum: ut 6, 5, 4. Geometrica est quando sunt multi termini et equalis est proportio inter primum et secundum sicut inter secundum et tertium, et sic de aliis. Exemplum: ut 8, 4, 2, ita quod semper dupletur numerus sequens. Armonica est quando sunt tres termini et equalis est proportio inter primum et tertium sicut excessus inter primum et secundum habet se ad excessum inter secundum et tertium. Exemplum: sicut 6, 4, 3. Permutatim proportionabilia sunt quando sicut se habet antecedens unius ad antecedens alterius sic se habet consequens unius ad consequens alterius, ut 8, 4, 2, 1. Equalis proporcionalitas est quando sunt duo numeri et in utroque sunt tres termini et sicut se habet primus terminus primi numeri ad suum secundum sic se habet primus terminus secundi numeri ad suum secundum. Et sicut se habet primus terminus primi numeri ad suum tertium sic se habet primus terminus secundi numeri ad suum tercium. Exemplum: ut 8, 4, 2 et 4, 2, 1 etc.

Sequuntur quedam suppositiones.

Omnes proportiones sunt equales quarum denominationes sunt equales.

Secunda suppositio est ista: Si sunt tres termini proporcionales et primus terminus est maior secundo et secundus maior tercio, tunc proportio primi ad ultimum componitur ex proportione primi ad secundum et secundi ad tercium.

Tercia suppositio: Quando sunt multi termini et primus terminus est maior secundo et secundus maior tercio, et sic de aliis usque ad extremum proporcionale, tunc proporcio primi ad ultimum componitur ex proporcione primi ad secundum et secundi ad tercium, et sic de aliis.

Quarta suppositio: Quando aliqua equalia comparantur ad aliquod tercium, eadem erit proportio inter unum illorum ad illud tercium sicut inter reliquum illorum et inter illud tertium.

Quinta suppositio: Si due quantitates inequales comparantur ad aliquod tertium, maior erit proportio maioris istarum ad illud tercium quam minoris istarum ad illud tercium; et econverso maior erit proportio illius tercii ad minorem quantitatem quam istius tercii ad maiorem quantitatem.

Sexta: Quando aliquod est compositum ex duobus equalibus, illud compositum erit duplum ad quodlibet illorum; si fuerit compositum ex tribus equalibus, erit triplum ad quodlibet illorum.

Septima suppositio: Quando aliquod est compositum ex duobus inequalibus, illud est plus quam duplum ad minus et minus quam duplum ad maius, ut tripedale compositum ex pedali et bipedali est plus quam duplum ad pedale et minus quam duplum ad bipedale. Quinque prime suppositiones sunt probate quinto Elementorum Euclidis. Et due ultime satis patent per rationem.

Sequuntur quedam conclusiones.

Prima conclusio probanda ex illis suppositionibus est ista: Si fuerit proportio maioris inequalitatis primi ad secundum et secundi ad tercium, tunc proportio primi ad tercium erit proportio dupla ad proportionem primi ad secundum et secundi ad tercium. Hoc probo sic: Nam capiendo illos tres treminos: 8, 4, 2, tunc arguo sic: Inter 8 et 4 est proportio dupla et tunc 4 ad 2 etiam erit proportio dupla. Ergo proportiones sunt equales. Tenet consequentia per primam suppositionem: Omnes proportiones etc. Et tunc arguo per secundam suppositionem: Quando tres termini etc. Sed hic sunt tales tres termini; igitur proportio primi ad ultimum componitur ex istis. Sed equalis est proportio inter primum et secundum sicut inter secundum et tercium. Ergo proportio primi ad ultimum componitur ex duobus equalibus. Ergo est dupla ad quamlibet illarum. Consequentia tenet per sextam suppositionem.

[*Tractatus brevis proportionum* 485]

Secunda conclusio est ista: Si fuerint 4 termini proportionales proportionalitate geometrica, proportio primi ad ultimum erit tripla ad proportionem primi ad secundum, et sic de aliis. Quod probo sic. Et capio istos quatuor terminos 8, 4, 2, 1. Tunc hic sunt quatuor termini et primus est maior secundo et secundus maior tertio, et sic de aliis. Ergo proportio primi ad ultimum componitur ex proportione primi ad secundum et secundi ad tertium, et sic de aliis. Tenet consequentia per terciam suppositionem. Sed equalis est proportio inter primum et secundum et secundi ad tercium, et sic de aliis. Ergo proportio primi ad ultimum componitur ex tribus equalibus. Ergo est tripla ad quamlibet earum. Tenet consequentia per sextam suppositionem. Et si sunt quinque termini, proportio primi ad ultimum est quadrupla ad quamlibet istarum. Si sex termini, quintupla; et sic in infinitum.

Tercia conclusio est illa: Si fuerit primum maius quam duplum secundi, fueritque secundum equaliter duplum tercii, proportio primi ad tertium erit minor quam dupla ad proportionem primi ad secundum. Quod probo sic: Capio illos tres terminos, 6, 2, 1. Tunc arguo per secundam suppositionem: Si sunt tres termini etc. Sed hic sunt tales termini. Igitur proportio primi ad ultimum componitur ex istis. Sed maior erit proportio primi ad secundum quam secundi ad tercium. Ergo proportio primi ad ultimum componitur ex duabus inequalibus. Ergo erit plus quam dupla ad minus et minus quam dupla ad maius. Consequentia tenet per septimam suppositionem. Sed maior erit proportio primi ad secundum quam secundi ad tertium. Ergo proportio primi ad ultimum erit minor quam dupla ad proportionem primi ad secundum.

Quarta conclusio: Si fuerit primum duplum secundi et fuerit secundum maius quam duplum tertii, proportio primi ad ultimum erit minor quam dupla ad proportionem secundi ad tertium. Quod probatur sic: Et capio istos tres terminos, 6, 3, 1. Tunc arguo per secundam suppositionem: Si sunt tres termini etc. Sed hic sunt tales tres termini. Ergo proportio primi ad tertium componitur ex proportione primi ad secundum et secundi ad tercium. Sed maior erit proportio secundi ad tercium quam primi ad secundum. Ergo proportio primi ad ultimum erit minor quam dupla ad proportionem secundi ad tertium.

Quinta conclusio est illa: Si fuerit primum minus quam duplum secundi, fueritque secundum equaliter duplum tercii, proportio primi ad ultimum erit maior quam duplum ad proportionem primi ad secundum. Quod probo sic: Et capio illos tres terminos 6, 4, 2. Tunc arguo per secundam suppositionem: Si sunt tres termini etc. Sed hic sunt tales tres termini. Ergo proportio primi ad ultimum componitur ex istis. Sed maior erit proportio secundi ad tercium quam primi ad secundum. Ergo proportio primi ad ultimum componitur ex duabus inequalibus. Ergo est plus quam dupla ad minus et minus quam dupla ad maius. Sed maior erit proportio secundi ad tercium quam primi ad secundum. Ergo proportio primi ad ultimum erit maior quam dupla ad proportionem primi ad secundum.

Sexta conclusio: Si fuerit primum duplum secundi, fueritque secundum minus quam duplum tercii, proportio primi ad ultimum erit maior quam dupla ad proportionem secundi ad tercium. Quod probo sic: Et capio istos tres terminos 8, 4, 3. Tunc arguo per secundam suppositionem: Si sunt tres termini etc. Sed hic sunt tales tres termini. Ergo proportio primi ad ultimum componitur ex istis. Sed maior erit proportio primi ad secundum quam secundi ad tercium. Ergo proportio primi ad ultimum componitur ex duabus inequalibus. Ergo est plus quam dupla ad minus et minus quam dupla ad maius. Sed maior est proportio primi ad secundum quam secundi ad tercium. Ergo proportio primi ad ultimum est maior quam dupla ad proportionem secundi ad tercium.

Septima conclusio: Quod proportione equalitatis nulla est proportio maior vel minor, quia proportio equalitatis non est maior vel minor proportione equalitatis, quia omnes proportiones equalitatis sunt equales, ergo nulla est alia maior. Nec proportio maioris inequalitatis est maior proportione equalitatis; quia si sic, ergo proportio equalitatis potest augeri quousque fuerit equalis isti. Consequens est falsum, quia quantumcunque augetur semper erit proportio equalitatis, et per consequens nec maior nec minor quam nunc est. Similiter si proportio maioris inequalitatis sit minor proportione equalitatis, ergo proportio equalitatis potest diminui quousque fuerit equalis isti. Consequens est falsum ut prius. Et similiter est arguendum de proportione minoris inequalitatis et de proportione equalitatis. Sed contra istam conclusionem arguitur sic: Capiantur isti tres termini 6, 4, 6. Tunc per secundam suppositionem proportio primi ad ultimum componitur ex proportione primi ad secundum et secundi ad tercium, et omne compositum quacunque parte ipsum componente est maius. Ergo proportio primi ad ultimum est maior quam proportio primi ad secundum vel secundi ad tertium. Sed proportio primi ad ultimum est proportio equalitatis et proportio primi ad secundum est proportio maioris inequalitatis et proportio secundi ad tercium est proportio minoris inequalitatis. Ergo proportio equalitatis est maior proportione maioris inequalitatis, et similiter proportione minoris inequalitatis.

Pro illo dicitur quod secunda suppositio sic intelligitur ut in illo tractatu describitur. Sed aliter obicitur contra conclusionem sic: Et capio illos tres terminos 6, 4, 4. Tunc per quintam suppositionem maior erit proportio inter 6 et 4 quam inter 4 et 4, et proportio 6 ad 4 est proportio maioris inequalitatis et 4 ad 4 est equalitatis. Ergo proportio maioris inequalitatis erit maior proportione equalitate.

Pro illo dicitur quod quinta suppositio habet intelligi de duobus inequalibus comparatis ad illum tercium in eodem genere proportionis. Aliter obicitur contra eandem conclusionem sic: Capio tres certos terminos 6, 6, 6. Tunc per primam conclusionem adiuncta secunda suppositione proportio primi ad ultimum est dupla ad quamlibet illarum. Ergo una proportio equalitatis est maior alia. Pro illo dicitur quod prima conclusio loquitur ubi est proportio maioris inequalitatis primi ad secundum et secundi ad tertium. Et intellectus secunde suppositionis patet ex predictis. Et sic cessat omnis obiectio.

De quibusdam opinionibus erroneis.

Tres sunt opiniones specialiter improbande. Quarum prima ponit velocitatem in motibus sequi excessum potentie motive super potentiam rei mote. Et ideo sentencia illius opinionis mote est hec: Si sunt due potentie et due resistentie, et tunc per maiorem excessum excedat prima potentia suam resistentiam quam secunda suam, tunc in maiori velocitate movebit prima potentia cum sua resitentia quam secunda cum sua. Exemplum: si a sit una potentia ut 8, b sua resistentia ut 4, et sit c una alia potentia sicut 2 et d sua resistentia ut 1, tunc per opinionem in quadruplo maiori velocitate movebitur a cum b quam c cum d, quia in quadruplo est maior excessus a super b quam c super d. Contra illam opinionem arguitur sic: Ipsa est contra Aristotelem in septimo Phisicorum et etiam in multis aliis locis ubi ponit velocitatem motus sequi proportionem. Sed forte dicitur quod Aristoteles per proportionem intelligit proportionem arithmetricam. Contra illam glosam arguitur sic: Nam quinto Elementorum Euclidis est prima conclusio hec: Si fuerit aliquarum quantitatum comparatio ad earum totidem, aut eque maiores aut eque minores aut consimiles, necesse est quemadmodum una illarum ad sui comparem quantitatem, totum quoque ex his aggregatum ad omnes illarum pariter acceptas similiter se habere. Unde sententia conclusionis est hec: Si fuerint multe quantitates comparate ad tot et in equali proportione se habet quelibet illarum ad suum comparatum sicut alia ad suum comparatum, tunc aggregatum ex omnibus antecedentibus se habet in eadem proportione ad aggregatum ex omnibus consequentibus sicut primum antecedens se habet ad suum consequens, quod tamen non est verum de proportione arithmetica.

Sed contra illam opinionem per experimentum arguitur, quia si esset una musca portans unum parvum pondus et unus fortis homo portans tantum pondus, tunc multo maiori velocitate stat muscam moveri cum suo pondere quam est velocitas cum qua movetur homo cum suo pondere, et tamen per multum maiorem excessum excedit potentia illius hominis suam resistentiam quam potentiam illius musce suam. Ergo opinio illa falsa etc.

Secunda opinio ponit velocitatem motus sequi proportionem excessus potentiarum super suas resistentias. Sententia illius opinionis est hec: Si sunt due potentie et due resistentie, et in maiori proportione se habet excessus prime potentie ad suam resistentiam quam excessus secunde potentie ad suam resistentiam, tunc maiori velocitate movebitur prima potentia cum sua resistentia quam secunda cum sua. Exemplum: Sit a una potentia ut 5, et sit b sua resistentia sicut 1; et sit c potentia ut tria, d sua resistentia ut 1. Tunc per illam opinionem movebitur in duplo velocius a cum b quam c cum d, quia in duplo est maior proportio inter excessum per quem a excedit b ad b quam est proportio inter excessum per quem c excedit d ad d, quia prima est quadrupla, secunda est dupla, ut patet advertenti. Contra illam opinionem arguitur per illam conclusionem impossibilem: Aliqua potentia movebitur cum aliqua resistentia et nihil potest moveri velocius et tardius ea. Quod probo sic: Et pono quod a sit una potentia ut 8, sit b sua resistentia sicut 4. Tunc excessus inter a potentiam et suam resistentiam est 4, et restentia etiam ut 4. Ergo nullius potentie excessus potest se habere in minori proportione nec in maiori ad suam resistentiam quam se habet ille excessus ad suam. Consequentia tenet per septimam conclusionem, scilicet proportione equalitatis. Ergo per positionem nihil potest moveri velocius vel tardius. Quod tamen illa conclusio sit falsa satis patet, quia quantumcunque fuerit illa potentia magna alia potentia tanta potest moveri cum maiori resistentia; ergo et tardius cum minori resistentia; et per consequens velocius. Ergo conclusio falsa.

Tertia opinio ponit quod velocitas in motibus, manente eodem motore vel equali, sequitur proportionem passoris, et manente eodem passo vel equali, sequitur proportionem motoris. Et illa opinio est bimembris. Unde sententia prime partis est illa: Si fuerit una potentia sufficiens moveri cum duabus resistentiis inequalibus, sicut resistentia una est alia minor ita velocius sufficit illam potentiam moveri cum resistentia minori quam cum resistentia maiori. Exemplum, ut si a sit una potentia sicut 8, b sua resistentia sicut 4, c una alia resistentia ut 2. Tunc per opinionem in duplo velocius movebitur a cum c resistentia quam cum b, quia in duplo minor est c resistentia quam b resistentia. Sed sententia secunde partis est hec: Si due potentie inequales sufficiant moveri cum una resistentia, sicut potentia una est alia maior ita velocius sufficit illam moveri cum illa resistentia quam potentiam minorem. Exemplum, si a sit una potentia ut 8, b una potentia sicut 4, c una resistentia ut 2, tunc sicut a potentia est maior b ita velocius sufficit a potentiam moveri cum c resistentia quam b cum c. Quod autem illa opinio sit insufficiens, hoc probo sic: Quia non docet penes quid velocitas motus habet attendi ubi sunt diversi potentie et diverse resistentie, nisi solum ubi sunt duo potentie et una resistentia vel due resistentie et una potentia. Contra primam partem opinionis arguitur per illam conclusionem impossibilem: Aliquam potentiam sufficit moveri cum certa resistentia et tamen cum nulla resistentia sufficit moveri in duplo tardius, stante potentia certa data. Quod conclusio sit impossibilis satis patet, quia quantacunque fuerit illa potentia, si una resistentia crescat ad equalitatem sue potentie stante potentia, in infinitum tarde movebitur cum illa. Ergo falsum est quod non in duplo tardius. Ergo conclusio falsa. Et quod conclusio sequatur ex prima parte illius opinionis, probo sic: Sit a una potentia sicut 8, b sua resistentia ut 4. Tunc probo quod a non potest moveri in duplo tardius cum aliqua resistentia alia quam cum b, quia si sic, sit c talis resistentia. Tunc vel c est equalis a, vel minor vel maior. Non equalis a nec maior, quia tunc a non sufficeret moveri cum c per illam conclusionem, cum nec a proportione equalitatis nec a proportione minoris inequalitatis provenire potest motus. Si tamen c sit minor a et maior b, tunc arguo sic: In duplo tardius sufficit a moveri cum c quam cum b; ergo in duplo velocius cum b quam cum c. Consequens est falsum, quia b non est in duplo minus c et similiter a est precise duplum ad b et c maius b. Ergo a non sufficit moveri in duplo tardius cum c quam cum b. Consequentia tenet per primam partem illius opinionis. Et similiter arguitur de quacunque resistentia. Sequitur ergo conclusio ex prima parte illius opinionis.

Contra secundam partem eiusdem arguitur per experimentum, quod (quia?) si esset unus homo portans unum magnum pondus quod ipse vix poterit portare, et unus alter homo eque fortis iuvaret ipsum, tunc isti duo plus quam in duplo velocius moverent illud pondus quam unus illorum per se moveret. Et potentia composita ex illis duobus non est nisi in duplo maior quam potentia unius illorum per se. Ergo secunda pars illius opinionis est falsa.

Quarta opinio ponit quod velocitas motus non sequatur proportionem, quia nulla est proportio inter potentiam et resistentiam. Sed omnia argumenta pro illa opinione solum procedunt de proportione proprie dicta. Ideo contra eam non arguo ad presens.

Quinta opinio, que est vera, ponit quod velocitas motus sequitur proportionem geometricam potentie motoris super potentiam rei mote. Unde sententia illius opinionis est hec: Si sint due potentie et due resistentie et maior sit proportio inter primam potentiam et suam resistentiam quam inter secundam et suam resistentiam, velocius movebitur prima potentia cum sua resistentia quam secunda cum sua, sicut una proportio est maior alia. Exemplum: Sit a una potentia sicut 8, b sua resistentia sicut 2, et sic c una alia potentia ut 6, d sua resistentia ut 3. Moveatur a cum b et c cum d. Tunc in duplo velocius movebitur a cum b quam c cum d, quia in duplo maior est proportio inter a et b quam inter c et d. Ista opinio patet per Aristotelem septimo Physicorum, ubi dicit quod velocitas motus sequitur proportionem potentie motoris super resistentiam. Iuxta illam opinionem sunt ille conclusiones concedende.

Prima: Si sit una potentia se habens in proportione dupla ad suam resistentiam, duplata potentia, stante resistentia, duplabitur et motus. Quod probo sic: Sit a una potentia sicut 4, b resistentia sicut 2, c una alia potentia sicut 8. Tunc arguo sic per primam conclusionem: Si fuerit proportio maioris inequalitatis primi ad secundum et secundi ad tercium, etc. Sed hic sint tales tres termini, et equalis est proportio primi ad secundum et secundi ad tercium. Ergo proporcio primi ad ultimum erit dupla ad quamlibet illarum. Sed velocitas motus sequitur proportionem etc. Ergo in duplo velocius movebitur c cum b quam a cum b. Sed si a duplaretur, tunc tanta velocitate sufficeret a moveri cum b quanta nunc sufficit c. Ergo conclusio vera.

Secunda conclusio est hec: Si aliqua potentia se habet in dupla proportione ad suam resistentiam, in duplo velocius sufficit illam potentiam moveri cum medietate illius resistentie quam cum tota resistentia. Quod probo sic: Sit *a* una potentia sicut 8, *b* sua resistentia ut 4, et *c* una alia resistentia sicut 2. Tunc arguo per primam conclusionem sic: Si fuerit proportio maioris inequalitatis primi ad secundum et secundi ad tercium etc. Sed hic est sic. Ergo proportio primi ad ultimum erit dupla ad quamlibet illarum. Sed motus sequitur proportionem etc. Ergo in duplo velocius sufficit *a* moveri cum *c* quam cum *b*. Sed *c* est medietas *b*. Ergo conclusio vera.

Tercia conclusio est ista: Si fuerit una potentia habens se in maiori proportione quam dupla ad suam resistentiam, duplata potentia, stante resistentia, non duplabitur motus. Quod sic probo: Sit *a* una potentia sicut 6, *b* una resistentia sicut 2, *c* una potentia sicut 12. Tunc arguo per quartam conclusionem: Si fuerit primum duplum secundi, fueritque secundum maius quam duplum tercii etc. Sed sic est hic. Ergo proporcio primi ad ultimum est minor quam dupla ad proportionem secundi ad tercium. Sed motus sequitur proportionem. Ergo non in duplo velocius *c* movebitur cum *b* quam *a* cum b. Sed si *a* duplaretur, tunc tanta velocitate sufficeret *a* moveri cum *b* sicut nunc sufficit *c* moveri cum *b*. Ergo conclusio vera est.

Quarta conclusio: Si fuerit una potentia se habens in maiori proportione quam dupla ad suam resistentiam non in duplo velocius sufficit illa moveri cum medietate illius resistentie quam cum toto. Quod probo sic: Sit *a* una potentia sicut 6, *b* una resistentia sicut 2, *c* una resistentia sicut 1. Tunc arguo per terciam conclusionem: Si fuerit primum maius quam duplum secundi, fueritque secundum equaliter duplum tercii etc. Sed sic est hic. Ergo proportio primi ad ultimum et (! est) minor quam dupla ad proportionem primi ad secundum. Sed motus sequitur proportionem. Ergo non in duplo *a* velocius movetur cum *c* quam cum *b*, et *c* est medietas *b*. Ergo conclusio vera.

Quinta conclusio est illa: Si fuerit una potentia se habens in minori proportione quam dupla ad suam resistentiam, duplata potentia movebitur plus quam in duplo velocius. Quod sic probo: Sit *a* una potentia sicut 6, *b* una resistentia sicut 4, *c* una alia potentia sicut 12. Tunc arguo per sextam conclusionem: Si fuerit primum duplum secundi, fueritque secundum minus quam duplum tercii, proportio primi ad tercium etc. Sed sic est hic. Ergo proportio primi ad ultimum erit maior quam dupla ad proportionem secundi ad tertium. Sed motus sequitur proportionem. Ergo plus quam in duplo velocius movebitur c cum b quam a cum b. Sed si duplaretur, tunc tanta velocitate sufficeret a moveri cum b sicut nunc sufficit c cum b. Ergo conclusio vera.

Sexta conclusio est hec: Si aliqua potentia se habet in minori proportione quam dupla ad suam resistentiam, plus quam in duplo velocius sufficit illam potentiam moveri cum medietate illius resistentie quam cum tota. Quod probo sic: Sit a una potentia sicut 6, b una resistentia sicut 4, c una alia resistentia sicut 2. Tunc arguo per quintam conclusionem: Si fuerit primum minus quam duplum secundi, fueritque secundum equaliter duplum tertii etc. Sed sic hic est, ut patet exemplo. Ergo proportio primi ad ultimum erit maior quam dupla ad proportionem primi ad secundum. Sed motus sequitur proportionem. Ergo plus quam in duplo velocius sufficit a moveri cum c quam cum b, et c est una medietas b. Ergo conclusio vera.

Contra istam opinionem arguitur sic: Ex illa opinione sequuntur tria inconvenientia. Primum est hoc, quod a proportionibus equalibus proveniunt motus inequales. Secundum, quod a minori proportione provenit tanta velocitas sicut a maiori. Tercium est, quod a minori proportione provenit maior velocitas quam a maiori. Sed omnia ista sunt falsa. Ergo opinio quinta falsa est. Falsitas autem illorum trium per istam opinionemmet patet. Et primum inconveniens probatur: Sint a et b due certe potentie inequales. Sit potentia a ut 8, potentia ipsius b ut 4. Sint c, d due alie res uniformes proportionales ipsis a, b, et resistentia c pedalis quantitatis, d semipedalis. Sit resistentia c sicut 2 et resistentia d sicut 1. Et ponatur quod a pertranseat c et b, d. Tunc quero numquid in equali tempore a pertransibit ipsum c sicut b, d vel non, quod sic patet per illam opinionem, cum in equali proportione se habet a ad c sicut b ad d per casum. Sed motus sequitur proportionem. Igitur in equali tempore a pertransibit c sicut b, d. Et si sic, sequitur prima conclusio. Quod probo sic: a plus pertransibit quam b et a proportionibus equalibus moveba[n]tur. Igitur sequitur conclusio. Quod secundum sequatur, probatur sic: Maioretur resistentia c quousque a sufficit pertransire medietatem c in tempore in quo b sufficit pertransire totum d. Tunc arguitur sic: Nunc est equalis proportio a ad c sicut b ad d, et in fine erit minor proportio a ad c quam nunc est, quia c resistentia erit maior quam nunc, ut ponit casus. Et tunc erit tanta proportio b ad d sicut in principio et motus a erit eque velox cum motu b, quia pertransibunt equalia spacia in tempore equali. Igitur a proportione minori provenit motus eque velox sicut a proportione maiori.

Quod tertia conclusio sequatur, probatur sic. Minuetur iterum resistentia *c*, non tamen ad equalitatem resistentie in primo casu. Tunc adhuc minor erit proportio *a* ad *c* quam in primo casu, et in primo casu fuit tanta proportio *a* ad *c* sicut *b* ad *d*. Ergo in illo casu erit minor *a* ad *c* quam *b* ad *d* et motus *a* erit velocior in illo casu quam motus *b*, quia *a* nunc plus pertransibit quam in illo casu, et *b* tamen sicut in secundo casu. Sed *a* tantum sufficiebat pertransire sicut *b* in equali tempore. Ergo *a* in illo casu sufficit plus pertransire quam *b*. Ergo motus *a* est velocior quam *b* et *a* movebitur in *c* a proportione minori quam *b* in *d*. Ergo a proportione minori provenit motus velocior quam a maiori. Igitur sequitur tertia conclusio.

Pro illo dicitur quod in tempore in quo *b* pertransibit *d*, *a* pertransibit medietatem *c*. Et tunc ad argumentum, *a* habet se in equali proportione ad *c* sicut *b* ad *d*, ergo in equali tempore sufficit *a* pertransire *c* sicut *b*, *d*, negetur consequentia. Quia aliqua se habere in equali proportione ad sua media est dupliciter, aut quo ad qualitatem aut quo ad quantitatem, id est, ad modum in illis modis (! mediis?) eque velociter vel pertranseundo ea in tempore equali. Primo modo est equalis proportio *a* ad *c* sicut *b* ad *d* et non secundo modo. Igitur non sequitur quod *a* in equali tempore sufficit pertransire *c* sicut *b*, *d*; igitur etc. Unde nota quod nullus motus potest provenire a proportione minoris inequalitatis, nec a proportione equalitatis. Sed omnis motus provenit a proportione maioris inequalitatis. Unde quantumcunque fuerit unum mixtum gravis simplici potest in eodem medio maiori velocitate, minori velocitate, et equali velocitate moveri cum illo simplici. Quod probo sic: Et sit *a* unum grave mixtum cuius gravitas sit 8 et sua levitas sicut 2, et *b* unum grave simplex cuius gravitas sit sicut 4. Sic *c* unum medium cuius resistentia sit sicut 2. Incipiant *a*, *b* moveri in *c*; tanta velocitate incipit *a* moveri in *c*, sicut *b* in *c*, et equali, quia potentia *a* est sicut 8, sua levitas, id est, resistentia intrinseca sicut 2, resistentia eius extrinseca sicut 2. Ergo tota eius resistentia est sicut 4. Sed inter 8 et 4 est dupla proportio. Ergo *a* incipit moveri a proportione dupla. Et potentia *b* est sicut 4 et una resistentia eius sicut 2. Sed inter 4 et 2 est proportio dupla. Ergo *a* et *b* eque velociter et equaliter incipiunt moveri. Et probo quod minori velocitate potest *a* moveri cum *b* quam *b* cum *c*, quia diminuatur resistentia *c* usque fuerit sicut 1. Tunc potentia erit sicut 8, sua intrinseca resistentia sicut 2, extrinseca sicut 1. Ergo tota resistentia sua erit sicut 3. Sed inter 8 et 3 est proportio dupla superbipartiens tertias. Sed potentia *b* erit sicut 4, eius resistentia sicut 1. Ergo *b* movebitur a proportione quadrupla. Sed *a* non tanta. Ergo minori velocitate movebitur quam *b*. Etiam probo quod a maiori velocitate potest moveri *a* in *c* quam *b* in *c*, quia crescat resistentia *c* quousque fuerit ut 3. Tunc potentia *a* erit sicut 8, sua resistentia intrinseca sicut 2, et extrinseca sicut 3. Ergo tota sua resistentia erit sicut 5. Sed inter 8 et 5 est proportio supertriparciens quintas. Sed potentia *b* est sicut 4, et tota sua resistentia est sicut 3. Sed inter 4 et 3 est proportio sesquitertia, que est minor quam proportio a qua *a* movebitur in *c*. Ergo tota conclusio vera etc. Finis.

Document 7.4

Francischus de Ferraria
*On the Proportions of Motions**

1. A QUESTION determined by me, Francischus de Ferraria, at Padua at the request of certain scholars in the year of the Lord MCCCLII on the tenth day of December. ... at the request of certain people I have proposed to write again [on this subject] in the form of a *questio*, to be better able to add something to the incomplete parts, and also to cut out superfluities in the prolix parts.... and we propose its title as follows: Whether velocity and slowness in motion is to be measured by *(sit atendenda!)* the proportion of the motive powers to the resisting powers, which is nothing else than to inquire as to what is the cause that one agent moves its resistance in any motion with a certain degree of velocity *(tanto gradu velocitatis)* and no more and no less....

2. Regarding the answer to our question there have been five famous opinions. The first opinion is that velocity and slowness in motion is measured by the excess of the motive power beyond the power of the thing moved, so that however much the agent exceeds its resistance, by that much it moves more swiftly, and however much less, by that amount it moves more slowly....

Moreover this opinion is false and erroneous... for it follows [from it] that a power of six would act more swiftly on a resistance of four than a power of two would act on a resistance of one. The consequent is fallacious, since a power of two has a greater proportion to a resistance of one than six has to four, the one being a double proportion, the other a sesquialterate proportion.... In the third place it follows [from it] that if two motors separately move two separate moving bodies through a certain space in a certain time, the two motors joined together will not move the two mobiles joined through the same space. The consequent is similarly

---

* For the sake of economy, a few sentences appearing, in the Latin text have been omitted in the translation.

against the rules of Aristotle [given] in the end of the seventh [book] of the *Physics*....

3. There is then a second opinion which poses that velocity and slowness in motion is measured by the proportion of the excess of the power of the motor over the power of the thing moved.... This opinion is false and is to be thoroughly refuted like the preceding one and by means of the same arguments and ideas.... The third opinion is that velocity and slowness in motion are measured by the proportion of the patients with the agent equal or the same, and by the proportion of the agents with the patient equal or the same.... This opinion is false and insufficient.... This opinion is false, for from it follows this fallacy, namely, that some motive power could move any mobile at all, and that some mobile could be moved by any motive power at all.... The fourth general opinion is that velocity and slowness in motion do not follow an excess or a proportion of motive power to resistance but rather follow a certain mastery *(dominium)* or natural habitude of the motor for motion.... This is erroneous, is very weakly urged, and is based on a weak foundation.... This opinion, abusive as it is, is to be thoroughly refuted. For if there would not be a proportion between motive powers and resistances because they (powers and resistances) are not quantitative [as this theory holds], in the same way there would be no proportion between sounds, and consequently the whole modulation of music would be lost.... It is also false because this opinion has asserted that excess ought to be in quantity only, while it is [actually] also in force, since it is divisible intensively as well as extensively.... The fifth is the opinion posing that velocity and slowness in motion are measured by the facility and difficulty of action of the agent on the patient.... But this opinion is false as to nature *(quo ad quid)* and moreover insufficient as to nature. For it is false in that it poses that velocity and slowness do not follow a proportion, which, however, it will be clear below, is false....

4. It remains to give over the third principal part [of our work] to posing a truer opinion on our question and to making the matter of proportions clear with some additional material.... The first conclusion is that velocity and slowness in motion are measured by the proportion of agents to patients. The second conclusion is that that proportion by which velocity and slowness are measured and diversified is a proportion of greater inequality, so that no single motion can arise from a proportion of equality. The third conclusion is that not every proportion of greater inequality is sufficient to produce motion or action.

[Francischus de Ferraria, *On Proportions*   497]

5. ... Moreover it is to be noted in the fuller clarification of this [first] conclusion that this conclusion ought to be universally in terms of a geometric proportion, for if it were understood arithmetically, then it would only be saying that velocity and slowness are measured by an excess, which was the first opinion refuted above. Hence when variation of velocity and slowness is argued from variation of proportion, we ought to examine carefully whether it is a geometric or arithmetic proportion that is varied....

6. The second conclusion is made clear: If from a proportion of equality or lesser inequality motion could arise, it would follow that this motion would be neither swifter nor slower than some [other] motion. But such a consequent is false, for there can be given a motion twice as fast or three times as fast as any given motion.... That no proportion of greater inequality is more or less than a proportion of equality, is proved, for if it were [so that a proportion of greater equality could be more or less than a proportion of equality], then a proportion of equality taken some number of times would equally produce a proportion of greater inequality, which is false, because it would always produce less....

7. I shall clarify the third conclusion, and I understand this as regards motion of alteration, for, as regards local motion, while it may be doubtful, I do not now believe it to be true. This conclusion is obvious, it having been assumed that *a* is uniformly nonuniform *(uniformiter difformiter)* hot, terminated exclusively in the more intense extreme at a maximum degree *(gradum summum)* and in the more remiss extreme at zero degree *(non gradum)*. I then accept two immediately adjacent parts of this body—one terminated exclusively in its more remiss extreme at the mean degree *(medium gradum)* of the whole latitude and the other in a similar manner. Of these let the more intense be called *a*, the more remiss *b*, and *c* the mean degree. Then I argue thus: *a* has a domination or proportion of greater inequality and excess over *b*, and yet *a* cannot act on *b*. Therefore, the said conclusion is true. The consequence is clearly valid, but the antecedent I [now] clarify—for if *a* were to act on *b*, it either acts with *c* degree or with a [degree] more intense or more remiss. It is not to be given that *c* degree acts on *b*, because, the body being uniformly difformly hot and *a* and *b* being immediate, *c* is the most intense [degree] not in *a* and the most remiss [degree] not in *b*. Thus it is valid that *c* is not in *a* and anything more intense [than it] is in *a*; similarly *c* is not in *b* and anything more remiss than it is in *b*, and nothing which is not in *b* is as remiss as it. Hence *c* is the most intense degree not in *a* and the most remiss degree not in *b*, as

is obvious by the common exposition of these terms. If it is said that *a* acts with a more remiss degree [than *c*], [I argue] against this by the same reasoning, since it does not have a more remiss degree. Hence it does not induce a more remiss degree. Nor similarly does it act with a degree more intense.... That this conclusion is not verified in regard to local motion is obvious by the authority of Jordanus in the *De ponderibus*, to the effect that any weight at all on a balance can lift a weight less than it....

8. The third corollary is that some motion can *in infinitum* be faster or slower than any given motion.... From the third conclusion I deduce the corollary, that no uniformly difform thing can be altered intrinsically. The reason is that the parts immediately adjacent are not distant by any latitude (i.e., increment of degrees) but only by a degree, and a degree is posited to be an indivisible....

COMMENTARY

There can be no doubt about the date of this treatise given here at the beginning, since the date is repeated on another page of the unique manuscript (MS Bodl. Canon. Misc. 226, 64r) which gives us the title alone. "Questio de proportionibus motuum determinata per magistrum Francischum de Ferraria in studio paduansis. Anno domini M⁰ IIĪ LII die X^a mensis decenbris. Amen." On still another page—not a part of the treatise (63v), we find written on one end of the sheet, "Questio de proportionibus determinata," and inverted on the other end, the date "M⁰ IIĪ XLVII, Indictione XV." I am sure that the part of the manuscript including this *questio* is an autograph of Francischus, for the scribe has doodled at random on folios 63r, 63v, 64r, repeating his name in every case, and on f. 63v in the same hand as the text we read: "Iste quaternus est mei francisci de ferraria filii cuiusdam domini Barthy (Bartholomei? *or* Bartholi?)." It is of interest to note that biographical details regarding Francischus are almost non-existent, although A. Gloria in his *Monumenti della Università di Padova* (Vol. *1* [Padua, 1888], 508) gives us the following brief reference which may relate to our Francischus in spite of the difference in the father's name given: "Francesco da Ferrara, figlio di Benvenuto, nel Gennaio 1378 fu presenta nella residenza del Podestà a una cessione fatta dal professore Giovanni Francesco da Monselice al professore Pietro Corialti da Tossignano."

1–3. Writing a *questio*, Francischus naturally starts off with "principal reasons," which we have omitted here. We have also omitted his detailed

[Francischus de Ferraria, *On Proportions* 499]

presentations and refutations of the various opinions regarding the proportions of motions. He takes them primarily from Bradwardine, although he does have some original arguments. But for a discussion of the similar opinions of Bradwardine, the reader is referred to the commentary to Document 7.3.

4–6. The presentation of Bradwardine's law is very much in the spirit of Bradwardine's discussion. I have given only enough of these passages to show how closely Francischus is following Bradwardine.

7–8. The extracts presented in this passage are most interesting since they do not suggest anything in Bradwardine's treatise, but rather the treatises of his Merton successors. Notice that the Merton terminology is present: *uniformiter difformiter*, *gradus summus*, etc. This whole passage resembles Giovanni di Casali's work (of 1346?)—and, of course, Casali borrowed directly from the Merton masters. The definition of uniformly difform implied in the argument in passage 7 is very much like that given in Document 7.1, passage 4, particularly the expression "*c* is the most intense [degree] not in *a* and the most remiss [degree] not in *b*" when *a* and *b* are any immediate parts of something uniformly difform and *c* separates them. The distinction between a *gradus* as a number representing intensity, but measuring only an indivisible of extension, and *latitude* as an increment of degrees or qualitative "distance" between different *gradus* is nicely brought out in passage number 8 (cf. again Doc. 7.1, passage 4).

*Questio de proportionibus motum Francischi de Ferraria**

[1] (58r) Questio determinata per me Francischum de Ferraria Padue ad preces quorundam scolarium anno domini MCCCLII die decima mensis Decenbris. Deo gratias.

Quoniam materia de proportionibus agentium ad passa et moventium et resistentiarum ad invicem in velocitate et tarditate motuum eorundem, de earundemque causis que multum naturali phylosophie existunt utilissima, multum latitat hodie philosophos naturales modernos, sitque scientia multum ardua et dificilis quod modernorum doctorum et naturalium sapientum insinuat discolia que signum dificultatis in rebus existit eficax ac perspicuum, fueritque latitanter ac obscure aliqualiter in forma tractatuum a quampluribus valentissimis pertractata, tamen ut intellectus studentium in ea clarius elucescant ad

---

* From the unique manuscript, Oxford, Bodl. Canon. Misc. 226, ff. 58r–63r. Note the frequent use of single consonants where double consonants are more usual, e.g., *atenditur*, *dificilis*, and *suficiens*.

sepissimas preces quorundam commotus proposui in forma questionis rescribere, diminutis adere, aditis quoque superflua iuxta posse melius resecare, quam sub tituli questionis forma et titulo proponemus ut sic: Utrum velocitas et tarditas in motu sit atendenda penes proportionem potentiarum moventium ad potentias resistentes; quod nichil est aliud nisi querere, que est causa quod unum agens movet suam resistentiam in quocunque motu sit tanto gradu velocitatis et non maiori nec minori et quare unum agens certo gradu velocitatis movet velocius vel tardius suam resistentiam quam aliud suam; utrum videlicet hec sint quia unum agens tantam proportionem habet ad suam resistentiam et non maiorem nec minorem et quia unum agens habet maiorem proportionem ad suum passum quam aliud ad suum, ideo unum velocius movet alio....

[2] (59v) Circa quesitum enim questionis nostre quinque sunt opiniones famose. Prima opinio est quod velocitas et tarditas in motu atenditur penes excessum potentie moventis supra potentiam rei mote, ita quod quanto agens excedit suam resistentiam tanto velocius movet et quanto minus tanto tardius.... (60r) Est autem illa opinio falsa et erronea, cuius defectum et error in eius reprobatione apparet. Ex hac enim quampluria absurda in physica et inconvenientia sequuntur. Quod ostendo, primo enim sequitur quod potentia existens sicut 6 velocius* ageret in resistentiam sicut 4 quam potentia sicut 2 in resistentiam sicut 1. Consequens est inconveniens, nam potentia sicut 2 habet maiorem proportionem ad resistentiam que est 1 quam 6 ad 4, nam illa est proportio dupla, ista autem sexquialtera.... Tertio sequitur quod si duo motores separati movent duo mobilia separata per aliquod spacium in aliquo tempore, illi duo motores coniuncti non movebunt illa duo mobilia coniuncta per idem spacium in equali tempore. Consequens similiter est contra regulas Aristotelis in fine septimi physicorum. Et patet, quia secundum eandem proportionem se habet motor compositus ad motum compositum et motor simplex ad motum simplicem....

[3] Est igitur secunda opinio que ponit quod velocitas et tarditas in motu atenditur penes proportionem excessus potentie motoris supra potentiam rei mote.... Est autem illa opinio falsa et multo reprobanda ut precendens eisdem rationibus et motivis.... Tertia opinio est quod velocitas et tarditas in motu attenditur penes proportionem passorum, manente eodem agente vel equali, et penes propor-

_____
* *corr. ex* ita velociter

tionem agentium, manente eodem passo vel equali.... Est igitur hec opinio falsa et insuficiens.... hec opinio falsa, nam ex ea sequitur hoc inconveniens, quod aliqua patentia motiva quodlibet mobile movere possit, et quod aliquod mobile a qualibet potentia motiva moveri possit.... (60v) Quarta opinio generalis est quod velocitas et tarditas in motu non sequitur excessum nec proportionem potentie motive ad resistentiam, sed sequitur quoddam dominium et habitudinem naturalem motoris ad motum.... Hec autem sicut erronea et multo(?) viliter movetur et debili fundamento fundatur.... Ista autem opinio sicut abusiva est multo reprobanda, quia si inter potentias motivas et resistentias non esset proportio, quia non sunt quantitative, tunc nec eodem modo inter voces. Et per consequens totius musice modulatio deperderetur, nam tonus in sexquioctava proportione consistit, diatertson in sexquitertia [proportione], diapente in sexquialtera [proportione].... Item falsum est quod illa opinio asservit quod excessus solus debeatur quantitati, quia etiam virtuali, cum sit divisibilis saltim intensive et etiam extensive.... Et cum dicitur omne excedens dividitur in excellentiam et in quod exceditur, dico quod sicut proportio seu excessus est duplex, sic excedens dividitur dupliciter. Si enim sit excessus seu proportio proprie, quia quantum proprie dividitur, quia talis est vere divisibilis et vere quanta. Si autem sit excessus seu proportio communiter sumpta, non dividitur proprie sed communiter, quod nihil aliud est nisi remitti. Et tunc erit sensus quod omne excedens potest remitti ad extremitatem eius quod exceditur, sic quod virtualiter in excedente continetur tota latitudo qua excellit et que exceditur.... Quinta est opinio ponens quod velocitas et tarditas in motu atenduntur penes facilitatem et difficultatem agendi agentis in passum.... Sed illa opinio quo ad quid est falsa, quo autem ad quid insuficiens. Est enim falsa in hoc quod ponit velocitatem et tarditatem non sequi proportionem, quod tamen patebit inferius esse falsum....

[4] (61r) Expedito igitur tertio principali, positisque aliorum opinionibus eisque iuxta auctorum imaginationem et mentem iuxta posse multius reprobatis, breviter restat ad tertium accedere principale ad ponendum unam opinionem in nostro proposito veriorem, et ipsam cum anexis materiam de proportionibus declarantem. Sub trium conclusionum numero reedigam breviter. Prima igitur conclusio est quod velocitas et tarditas in motu taliter atenduntur penes proportionem agentium ad passa. Secuna conclusio quod illa proportio penes quam atenditur et diversificatur velocitas et tarditas motuum est proportio

maioris inequalitatis, sic quod non a proportione equalitatis possit ullatenus singulus motus. Tertia conclusio quod non quelibet proportio maioris inequalitatis est sufficiens ad causandum motum nec actionem....

[5] Ad huius tamen conclusionis evidentiam pleniorem est notandum quod hec conclusio universaliter debet [intelligi] de proportione geometrica; si enim de arismetica intelligeretur, tunc nihil esset aliud dicere nisi ponere velocitatem et tarditatem atendi penes excessum, que fuit prima opinio superius reprobata. Et ideo quando ex variatione proportionis arguitur et concluditur variatio velocitatis et tarditatis debemus diligenter inspicere an varietur proportio geometrica an arismetica....

[6] Secunda conclusio declaratur: Si a proportione equalitatis vel minoris inequalitatis possit provenire motus, sequeretur quod iste motus non esset velocior nec tardior aliquo motu, consequens est falsum, quia omni motu dato potest dari motus in duplo velocior, et in triplo, ut inferius ostendetur.... Quod autem nulla proportio maioris inequalitatis sit maior vel minor proportione equalitatis probatur, quia si foret, tunc proportio equalitatis aliquotiens sumpta rederet equaliter proportionem maioris inequalitatis, quod tamen est falsum, quia semper redit minorem....

[7] (61v) Tertiam autem conclusionem declarabo et hoc intelligo de motu alterationis, quia de motu locali pro nunc licet dubitum habeat, non credo veritatem habere. Hec conclusio patet, sumpto quod a sit uniformiter difformiter calidum terminatum in extremo intensiori ad gradum summum exclusive et in extremo remissiori ad non gradum. Deinde accipio duas partes immediatas huius corporis, partes quarum una sit versus extremum intensius terminata in extremo remissiori ad medium gradum totius latitudinis exclusive et aliam similiter, quarum intensior vocetur a, remissior b, c autem sit gradus medius. Tunc arguo sic: a dominationem seu proportionem maioris inequalitatis et excessus habet supra b; et tamen a non potest agere in b. Igitur conclusio dicta vera est. Consequentia valet clarissime, sed antecedens declaro, quia si a ageret in b, aut igitur c gradu* aut intensiori** aut remissiori.*** Non est dandum quod c gradum agat in b, quia ipsum non habet, quia istud corpus sit uniformiter difformiter calidum et a

* gradum *in MS*
** intensiorem *in MS*
*** remissiorem *in MS*

et b sint immediate, c est intensissimus* qui non est in a et remissimus** qui non est in b. Sic valet quod c non est in a et quilibet intensior est in a. Similiter c non est in b et quilibet remissior est in b et nullus qui non est in b est ita remissus sicut iste. Igitur c est intensissimus* qui non est in a et remissimus** qui non est in b, ut patet per istorum terminorum expositionem communem. Si dicatur quod a agit gradu*** remissiori,† contra [arguo] per idem, quia nullum remissiorem habet, igitur nullum remissiorem inducit; nec similiter agit gradu*** intensiori,†† quia non potest inducere gradum intensiorem quin prius inducat remissiorem.... Quod autem hec conclusio de motu locali non verificetur patet per auctoritatem Jordani de ponderibus, scilicet, quod quelibet pondus in equilibera leveat pondus minus eo....

[8] Tertium corollarium est quod omni motu dato in infinitum potest motus velocior et tardior dari.... Ex tertia autem conclusione talem elicio corollarium, quod nullum uniformiter difforme potest ab intrinseco alterari, et rationabilis causa est quia partes ille sic immediate ad invicem non distant per latitudinem sed solum per gradum, qui gradus ponitur consistere in indivisibili....

* intensius *in MS*
** remissius *in MS*
*** gradum *in MS*
† remissiorem *in MS*
†† intensiorum *in MS*

Chapter 8

John Buridan and the Impetus Theory of Projectile Motion

ONE of the main points of criticism of Aristotelian dynamics in the fourteenth century centered in Aristotle's explanation of projectile motion. Obviously such motion was difficult to explain, for the Aristotelian view of violent motion demanded the *continuing* presence of a motor in contact with the thing moved. Aristotle supposed that in some way the original force communicated to the medium or air the power to continue the projectile's motion. There are two principal passages which outline possible explanations. The first occurs in the fourth book of the *Physics*, but it merely mentions without elaboration the possible role of the air:[1] "Again, as it is, things *thrown* continue to move, though that which impelled them is no longer in contact with them, either because of "mutual replacement" (ἀντιπερίστασις) as some say, or because the air which has been thrust forward thrusts them with a movement quicker than the motion by which the object thrown is [naturally] carried to its proper place." However, the second passage from the eighth book of the *Physics* tells us something more about the detailed mechanics of these theories:[2]

[1] *Physics*, IV.8.215a.14–17 (edition of W.D. Ross [Oxford, 1936]). The English translation is that of T. L. Heath, *Mathematics in Aristotle* (Oxford, 1949), p. 115. The new medieval Latin translation of Moerbeke runs (edition of Venice, 1495, f. 56v): "(Textus commenti 68)... Amplius nunc quidem moventur proiecta proiectore non tangente, aut propter repercussionem, sicut quidam dicunt, aut ex eo quod pellit pulsus aer velociori motu illius quod pellitur motu secundum quem fertur in proprium locum." Note that *antiperistasis*, which is here translated *repercussionem* in this translation, is rendered in the translation from the Arabic accompanying the commentary of Averroës (in this edition), *successionem et motionem*. The translation from the Greek, called the old translation, has (MS Paris, BN lat. 16141, f. 84v, c. 1) *repercussionem*.

[2] *Physics*, VIII.10.266b.27–267a.20. The English translation is again that of Heath, *Mathematics in Aristotle*, pp. 155–56. The Latin translation of Moerbeke has the following (*edit. cit.*, ff. 158v–159r): "(Textus

With regard to things which are moved in space, it will be well, first of all, to dispose of a certain difficulty. It if is true that everything that is moved, except things which are self-moved, is moved by something, how comes it that some things are moved continuously, though that which has caused them to move is no longer in contact with them, as, for instance, things *thrown?* If that which has caused their motion sets something else in motion too, say the air, which when set in motion also causes motion, it is no less impossible that it (the air) should be in motion when the original movent is no longer in contact with it or moving it; all the things [that are moved] must be moved, and must have ceased to be moved, at the same time, [their motion ceasing] whenever the first movent ceases to operate, even if, like the magnet, it communicates to what it has moved the power of causing motion. We must, therefore, hold that the original movent gives the power of causing motion to air, or water, or anything else which is naturally adapted for being a movent as well as for being moved. But this thing does not simultaneously cease to be a movent and to be moved; it ceases to be moved at the same moment as the movent acting upon it ceases to move it, but it is still a movent. Hence it moves something else contiguous to it; and the same applies to that again. The motion comes to cease whenever the power of causing motion communicated to the next member of the series becomes less and less continually; and it finally ceases when the member of the series immediately preceding no longer makes the next a movent but only causes it to be moved. The motion of the last two, of the one as movent and of the other as moved, must cease simultaneously, and with this the whole motion. Now this kind of motion takes

---

commenti 82) De his autem que feruntur bene se habet dubitare quandam dubitationem primum. Si enim quod movetur omne movetur ab aliquo, quecunque non ipsa seipsa movent quomodo moventur aliqua continue non contingente movente, ut proiecta. Si autem simul movet et aliud aliquid movens, ut aerem, qui motus movet, similiter impossibile est primo non contingente nec movente moveri. Sed simul omnia moveri et quiescere cum primum movens quiescet, et si facit sicut lapis ut movere quod ferrum movet. Necesse hoc quidem dicere quod primum movens facit possibile movere aut aerem huiusmodi aut aquam aut aliud aliquod tale, quod aptum natum est movere et moveri. Sed non simul pausat movens et quod movetur. Sed quod movetur quidem simul cum movens quiescat; movens autem adhuc est; unde movetur cum alio habitum et in hoc eadem ratio est. Pausat enim cum minor potentia in movente sit habito. Penitus autem quiescit cum non amplius faciat primum movens, sed quod movetur solum. Hec autem necesse est simul pausare movens quidem autem et quod movetur et totum motum. Hic quidem igitur in contingentibus, aliquando quidem moveri, aliquando autem quiescere fit motus et non continuus, sed videtur. Aut enim consequenter est eorum que sunt aut tangentium est; non enim unum est movens sed habita ad invicem sunt. Unde et in aere et in aqua huiusmodi modus, quem dicunt quidam antiperistasim. Impossibile autem aliter est opposita solvere nisi dicto modo. Antiperistasis autem simul omnia moveri facit et movere; quare et quiescunt. Nunc autem unum aliquod quod movetur continue a quolibet, non enim ab eodem." Here it should be noted the translator has used *antiperistasis*.

place in things that can be at one time in motion and at another time at rest, and the motion is not continuous, but only appears to be. For it is a motion of [a series of] things which are either successive or in contact, since the motion is not one but a series of movements contiguous to one another. This is why such motion occurs in air and water, and some thinkers call it "mutual replacement." But the difficulties involved cannot be disposed of in any way other than that which we set out. The theory of "mutual replacement" makes the whole series of things move and cause motion simultaneously, so that they must also all cease to move at the same time; whereas the appearance presented to us is that of some one thing moving continuously. What then keeps it in motion, seeing that it cannot be the same movement [all the time]?

Aristotle, then, seems to be describing two theories involving the air as the continuing motor: (1) the air accomplishes the continuance of the projectile motion by some kind of mutual replacement of the air by the projectile so that the air comes around behind to act as a movent, the so-called theory of *antiperistasis*, or (2) the air not only is directly moved by the original motor but (due to the air's special nature) *it simultaneously receives the power or force*[3] *to act as a movent;* then by this power it moves the air next to it and the projectile as well; the air next to the original air is not only moved, but likewise receives the power of acting as a movent; and so the projectile is successively pushed in the direction which the original motor intended. It would appear from the passage quoted that Aristotle rejects the first of these views, i.e., the theory of *antiperistasis* or mutual replacement, a theory which he probably took over from Plato, who employed it to explain respiration, the action of a medical cupping glass, as well as projection (*Timaeus*, 79–80). Incidentally, that theory is elaborated somewhat by Simplicius, who tells us in his commentary to this passage,[4]

---

[3] For a discussion of the use of the terms "power" and "force," see note 4 in the preceding chapter, p. 425.

[4] Simplicius, *In Aristotelis physicorum libros...commentaria*, edition of H. Diels in *Commentaria in Aristotelem graeca*, Vol. *10* (Berlin, 1895), 1350, lines 31–36. For English translation, see Heath, *Mathematics in Aristotle*, p. 157. Compare Simplicius' account with the passage from the *Timaeus* of Plato cited in the text and which I give in the Jowett translation (*Dialogues of Plato*, Vol. *2* [New York, 1871], 79–80, pp. 570–71): "Let us further consider the phenomena of respiration, and inquire what are the real causes of it. They are as follows: Seeing that there is no such thing as a vacuum into which any of those things which are moved can enter, and the breath is carried from us into the external air, the next point is, as will be clear to everyone, that it does not go into a vacant place, but pushes its neighbor out of its place, and that which is thrust out again thrusts out its neighbor; and in this way of necessity everything at last comes round to that place from whence the breath came forth, and enters in there, and follows with the breath, and fills up the place; and this goes on like the circular

that in *antiperistasis*, "as one body is extruded by another, there is interchange of places, and the extruder takes the place of the extruded, that again extrudes the next, the next the succeeding one (if there are more than one), until the last is in the place of the first extruder." It should be observed that in the Middle Ages the theory of *antiperistasis* or mutual replacement was elaborated further, as for example by John Buridan (in Doc. 8.2) who describes it thus: "The first one (theory), which he calls 'antiperistasis,' holds that the projectile swiftly leaves the place in which it was, and nature, not permitting a vacuum, rapidly sends air in behind to fill up the vacuum. The air moved swiftly in this way, and impinging upon the projectile, impels it along further."

Simplicius' contemporary of the sixth century, John Philoponus, in his commentary on the *Physics*, rejects by an appeal to experience and reason both explanations involving the air as the continuator of projectile motion:[5]

> Let us suppose that *antiperistasis* takes place according to the first method indicated above, namely, that the air pushed forward by the arrow gets to the rear of the arrow and thus pushes it from behind. On that assumption, one would be hard put to it to say what it is (since there seems to be no counter force) that causes the air, once it has been pushed forward, to move back, that is along the sides of the arrow, and, after it reaches the rear of the arrow, to turn around once more and push the arrow forward. For, on this theory, the air in question must perform three distinct motions: it must be pushed forward by the arrow, then move back, and finally turn and proceed forward once more. Yet air is easily moved, and once set in motion travels a considerable distance. How, then, can the air, pushed by the arrow, fail to move in the direction of the impressed impulse, but instead, turning about, as by some command, retrace its course? Furthermore, how can this air, in so turning about, avoid being scattered into space, but instead impinge precisely on the notched end of the arrow and again push the arrow on and adhere to it? Such a view is quite incredible and borders rather on the fantastic....

Now there is a second argument which holds that the air which is pushed in the first instance (i.e., when the arrow is first discharged) receives an impetus to motion of a wheel, because there can be no such thing as a vacuum.... The phenomena of medical cupping-glasses and of the swallowing of drink and of the hurling of bodies, whether discharged in the air or moving along the ground, are to be explained on a similar principle...."

[5] John Philoponus, *In Aristotelis physicorum libros commentaria*, edition of H. Vitelli in *Commentaria in Aristotelem graeca*, Vol. *17* (Berlin, 1888), 639, line 22; 640, line 5; 641, lines 7–26. The English translation is that of I. E. Drabkin in Cohen and Drabkin, *A Source Book in Greek Science* (New York, 1948), pp. 221–22.

motion, and moves with a more rapid motion than the natural [downward] motion of the missile, thus pushing the missile on while remaining always in contact with it until the motive force (δύναμις) originally impressed on this portion of air is dissipated. This explanation, though apparently more plausible, is really no different from the first explanation by *antiperistasis*, and the following refutation will apply also to the explanation by *antiperistasis*.

In the first place we must address the following question to those who hold the views indicated: "When one projects a stone by force (βία), is it by pushing the air behind the stone that one compels the latter to move in a direction contrary to its natural direction? Or does the thrower impart a motive force to the stone, too?" Now if he does not impart any such force to the stone, but moves the stone merely by pushing the air, and if the bowstring moves the arrow in the same way, of what advantage is it for the stone to be in contact with the hand, or for the bowstring to be in contact with the notched end of the arrow?

For it would be possible, without such contact, to place the arrow at the top of a stick, as it were on a thin line, and to place the stone in a similar way, and then, with countless machines, to set a large quantity of air in motion behind these bodies. Now it is evident that the greater the amount of air moved and the greater the force (βία) with which it is moved the more should this air push the arrow or stone, and the further should it hurl them. But the fact is that even if you place the arrow or stone upon a line or point quite devoid of thickness and set in motion all the air behind the projectile with all possible force, the projectile will not be moved the distance of a single cubit.

If the motion of the air does not move the projectile but rather resists the movement of the projectile, what is it that is causing the continued motion of the projectile? Newtonian physics rejecting the necessity of a continuing force answers in terms of inertia. Philoponus' explanation is that an incorporeal kinetic force (κινητικήν τινα δύναμιν ἀσώματον) has been impressed (ἐνδίδοσθαι) in the body (not the medium) and this impressed force continues the movement of the body until it is spent by the resistance to movement presented by the weight of the body and perhaps the resistance of the air.[6]

From these considerations and from many others we may see how impossible it is for forced motion to be caused in the way indicated. *Rather is it necessary to assume that some incorporeal motive force is imparted by the projector to the projectile*,* and that the air set in motion contributes either nothing at all or else very little to this motion of the projectile. If, then, forced motion is produced as I have suggested, it is quite evident that if one imparts motion "contrary to nature" or forced

---

[6] *Ibid.*, Vitelli edition, p. 642, lines 3–9; translation of Drabkin, *Source Book*, p. 223.

* The italics are those of the translator.

motion to an arrow or a stone the same degree of motion will be produced much more readily in a void than in a plenum. And there will be no need of any agency external to the projector....

This motive force or kinetic power is also identified with incorporeal kinetic energy[7] (ἐνέργεια τις ἀσώματος κινητική).[8] This energy is compared with the incorporeal energies emanating to the eyes from objects seen.

The views of Philoponus appear to have been of considerable influence among Islamic authors in a number of different areas of mechanics, and particularly in discussions of dynamics. Thus the famous physician and philosopher Ibn Sīnā (Avicenna d. 1037) gives a discussion of projectile motion in his *Book of the Healing of the Soul* (which is a commentary on the works of Aristotle) that shows the influence of Philoponus' prior discussion:[9]

As for the case where there is [violent motion with the] separation of the moved [from the motor] like the projectile or that which is rolled, the scientists disagree in their opinions. There are some who hold that the cause lies in the tendency

---

[7] I do not mean to imply a modern physical definition of "energy" when I use this term here, but rather I use "energy" in the sense of "activity." The Greek word is sometimes used interchangeably with ἐντελέχεια. Strictly, the latter is to be rendered as "completeness" or "actuality," while the former is better translated as "activity" or "actualization." But, as H. A. Wolfson points out in his *Crescas' Critique of Aristotle* (Cambridge, Mass., 1929), p. 526, "Aristotle commonly uses these terms without distinction." (See *Metaphysics*, IX.9, 1065b.22–23.) In a later book, *The Philosophy of the Church Fathers* (Cambridge, Mass., 1956), pp. 465–67, Wolfson explores the different usages of ἐνέργεια, which he commonly renders as "operation." The sense that appears to bear on the passage under consideration here is that of "operations of qualities," particularly where they apply to "states of qualities which are not permanent but arise from certain temporary causes."

[8] Philoponus, Vitelli edition, p. 642, lines 11–14.

[9] Ibn Sīnā, *Kitāb al-Shifā* (Arabic text), Vol. *1* (Teheran, 1885), 154–55; cf. p. 605. S. Pines, "Études sur Awḥad al-Zaman Abu 'l-Barakât al-Baghdâdî," *Revue des études juives*, New Series, Vol. *3* (1938), 3–64, and Vol. *4* (1938), 1–33, was the first scholar to study the history of the *mail* theory and my account, particularly for the ideas of Abū 'l-Barakāt where no Arabic text was available to me, is largely drawn from these articles. While Pines did not translate the whole passage, which I have here translated, he did translate parts of it, in addition to other parts which I did not translate (see particularly Vol. *3*, 33–50). Pines found an even earlier account of the influence of Philoponus in some glosses of the late tenth century, perhaps by Yaḥya ibn Adī (d. 973–74). See S. Pines, "Un précurseur Bagdadien de la théorie de l'impetus," *Isis*, Vol. *44* (1953), 247–51. In this account the author, like Philoponus, refutes the two Aristotelian theories employing the air as the continuing agent of motion. He then gives us the theory of impressed force (p. 249): "It is necessary to say that a power *(qūwa)* which is incorporeal comes from the projector into the projectile; it makes it (the projectile) arrive at the terminus [of its motion]; then it (the force) is dissipated."

of the air which has been pushed to get behind the projectile and to unite there with a force which presses against that which is in front of it. There are others who say that the pusher pushes the air and the projectile together, but the air is more receptive to pushing and so it is pushed more swiftly and thus pulls that which has been placed in it. And there are those who hold that the cause is in that force which the moved acquires from the mover and which persists in it for a period until it is abolished by the opposing friction of that [medium] which touches it and is displaced by it. And just as the force is weakened in the projectile, so the natural inclination *(mail)* and the action of friction becomes dominant over it, and thus the force is abolished and consequently the projectile passes in the direction of its natural inclination (i.e., it falls to the earth). The supporters of the viewpoint of the movement of the air [as the cause of the continuance of projectile motion] have said that it is not remarkable that there is movement of the air which attains such power as to carry rocks or large bodies, for so it is that a [loud] noise sometimes crushes the top of a hill, and that there are hills whose foundations have crumbled as the result of noise. Also thunder demolishes firmly built buildings, turns topsy-turvy the summits of hills and cleaves hard rocks.... [But] how is it possible that we can say that the air which has returned behind the projectile has united and pressed forward that which is in front of it? And what has caused the movement which is forward in direction to unite and push what is behind it? [Or] how can we say that the mover conveyed a force to the moved? For the force is not "natural," nor "psychic," nor "accidental."...And if the mover conferred force, then it would be more strongly active in the beginning of its duration. But in actual fact it is found that it is more strongly active in the middle of the movement. If the cause of [projectile] movement were the air carrying the projectile, then we would have a cause for that [acceleration in the middle of the trajectory]. It is that the air is rarefied by the movement and its speed is increased and thereby that of the motion [of the projectile].... This cause is not found [to be true]. And some people have spoken for the doctrine of "engendering." They say that it is of the nature of movement that [another] movement is engendered after it; and it is the nature of "propension" *('i 'timād)* that [another] propension is engendered after it. They affirm that motion ceases momentarily; then a period of rest follows it; and then after this another movement is engendered out of the propension [for movement]. This theory is distasteful. For certainly the "engendered" is not classified as a *ḥadith* ("something happening for the first time") following upon "that which never was," and every *ḥadith* is after "that which never was," for so a "first happening" is the cause of a "first happened." If this [engendering] were the cause, it would follow necessarily that the first movement is coexistent with the second. Also if this were [the cause], there would be lacking—although it is necessary—a lasting cause for the movement.... *But when we verified the matter we found the most valid opinion to be that of those who hold that the moved receives an inclination* (mail) *from the mover. The inclination is that which is perceived by the senses to be resisting a forceful effort to bring natural motion to rest or to change one violent motion into another.**

While this and other passages are no doubt influenced by the discussion given by Philoponus, it ought to be observed that, at least in this passage, Avicenna is supporting a theory that is somewhat different from the theory of impressed force affirmed by his predecessor. In the first place, Avicenna reports on four opinions in regard to the continuance of projectile motion before presenting the one which he supports. The first two theories are clearly the two theories involving the air as the continuing mover, theories noted by Aristotle. A third theory is that which holds that the projector communicates to the projectile a force or power that accomplishes the continuance of the movement and lasts until overcome by the natural inclination of the projectile downward and the friction of the medium. A fourth theory described by Avicenna—the so-called "engendering" theory—considers movement as discontinuous or atomic, where moments of movement are interrupted by moments of rest and the continuance of movement of the projectile from one moment to the next is by means of a propensity for movement that remains after each rest to inaugurate once more another movement. Pines has found that this theory had considerable currency among the Islamic authors and that some authors even equated the *'i 'timād* theory with the so-called *mail* theory accepted by Avicenna.[10]

Now in the *mail* theory, which is the fifth reported by Avicenna, projectile motion is said to continue as the result of an inclination *(mail)* transferred by the force of the original projector to the projectile. Notice that he says that this inclination is that which is perceived as resisting immobilization. This, of course, sounds not unlike the *vis inertiae* of Newtonian physics, at least so far as movement is concerned.[11] Avicenna at least theoretically distinguished the *mail* from the moving force *(qūwat muharrikat)*, for a *mail* is the instrument of force. Every force communicates its action by means of a *mail*; the *mail* thus can continue to persist after the initial exertion of a force and even after the completion of the movement.[12] The *mail* is something permanent (but destructible), and something nonsuccessive. Avicenna distinguished three categories of *mail*: (1) *mail nafsānī*—psychic *mail*, (2) *mail ṭabī'ī*—natural *mail*, and (3) *mail qasrī*—

---

* The italics are mine.

[10] Pines, "Études," Vol. *3*, 45–48.

[11] Thus in a modern description of inertia we are apt to see it defined as "a general property of matter of which we become conscious through our muscle senses whenever by our muscles we change the motion of matter."

[12] Pines, "Études," Vol. *3*, 49–52. Pines here discusses the ontology of *mail* which he prefers to transliterate *mayl*.

unnatural or violent *mail*. Categories (2) and (3) interest us. The natural motion of falling bodies is accomplished by the "natural *mail*." Hence gravity acts through the natural *mail*. It is the *mail qasrī* however, which is imparted to a projectile and continues its motion. Avicenna sometimes used the expression "impressed force" (*qūwat mustafādat*) in the same context as *mail qasrī*, thus obscuring his distinction between "inclination" and "force." The action of the *mail* differs according to the weight of the body to which it is communicated. And thus it would seem that Avicenna takes the first step towards quantification of the *mail* or impressed force. One most interesting observation, repeated by John Buridan in his fourteenth-century exposition of the impetus theory, held that if there were no resistance to the *mail*, it would last indefinitely, and thus so would the movement it generates. Such would be the case of a body moving in a void. "If the violent movement of the projectile is produced by a force operating in the void, it ought to persist, without annihilation or any kind of interruption."[13] But since there does not appear in nature any case of everlasting movement, a void does not exist. This argument had already been used by Aristotle—but of course without application to a theory of impressed *mail* or force.[14] It is worth noting that one of the most important Islamic doctors succeeding Avicenna, Abū 'l-Barakāt (died about 1164) modified the nature of the *mail qasrī* to meet the argument of Avicenna. The *mail qasrī* for Abū 'l-Barakāt was self-expending rather than of a permanent nature. Thus there could be a violent motion in a void. The violent motion would come to an end without external resistance as the result of the self-exhaustion of the *mail qasrī*. Thus for projectile motion in a medium, Abū 'l-Barakāt would recognize three factors tending to bring the movement to an end:[15] (1) air resistance, (2) the natural gravity, and (3) self-exhaustion due to the progressive increase of the period of time away from the source of movement.

One other crucial point of difference between the concepts of *mail* in the works of the two Islamic authors was in the nature of its residence in a body. Avicenna believed that only a single *mail* could reside in a

---

[13] *Ibid.*, Vol. *3*, *55*; *Kitāb al-Shifā*, Vol. *1*, 61. Avicenna goes on to say: "... when a force exists in a body, it must necessarily either persist or be annihilated. But if it persists, the movement will persist in a permanent fashion...." There is no mention of *impressed* force or *mail* in this passage, but in view of Avicenna's avowed explanation of projectile motion, we are justified in assuming that it is such an impressed force which is understood in this passage.

[14] *Physics*, IV.8.215a.19. See Chapter 7 above, p. 456, note 8.

[15] Pines, "Études," Vol. *4*, *3*.

body at one time.[16] Thus in the case of projectile movement, the propelling force communicated a *mail qasrī* to the body producing its movement. The gravity of the body would tend to produce a natural *mail* in the body, but it would not be able to do so until the *mail qasrī* was completely destroyed and the projectile had been brought to momentary rest (the *quies media* of the Latin authors). After such a rest the gravity introduces a natural *mail* which would then produce the downward movement. Now Abū 'l-Barakāt, opposing the idea of *quies media*, proposed that contrary inclinations can exist simultaneously in a body, as is obvious from the experience of people tugging on a ring from opposite directions.[17] In the case of projectiles, as long as the body is moving upward, the *mail qasrī* is dominant, the natural *mail* acting as resistance. But as the *mail qasrī* is dissipated, the natural *mail*, being constant, finally becomes dominant, and the body starts to move downward, the remnant of the *mail qasrī* offering continually weaker and weaker resistance to the natural *mail*. As the natural *mail* becomes more and more dominant, and the resisting *mail qasrī* becomes weaker and weaker, the falling body accelerates.[18] This explanation is very much like a theory reported by the Greek commentator Simplicius as the theory of Hipparchus. We shall examine it more closely in the next chapter. The further important addition made by Abū 'l-Barakāt to the theory of falling bodies in terms of the continuous augmentation of natural inclinations, we shall also examine in the next chapter. Other Islamic authors treat of the *mail* theory, but we have given sufficient examples here.[19]

One of the crucial questions is how influential the Philoponus-Islamic views were in the development of the Latin version of the theory of the impressed force—the so-called impetus theory. The important passages from Avicenna's treatment do not appear to have been translated in full into Latin.[20] Furthermore, the short, ambiguous allusion to the theory found in the medieval Latin version of Alpetragius' (al-Biṭrūjī, *fl.* 1200) popular astronomical work is of no apparent significance in the Latin

[16] *Ibid.*, Vol. *3*, 63.
[17] *Ibid.*, Vol. *4*, 7.
[18] *Ibid.*, Vol. *4*, 9.
[19] *Ibid.*, Vol. *3*, 53, n. 210; Vol. *4*, 12–15 for mention of the views of Fakhr al-Dīn, Naṣīr al-Dīn al-Ṭūsī, Laukwarī, and others. See also Pines, "Quelques tendences antipéripatéticiennes de la pensée islamique," *Thales*, Les Années 1937–39 (1940), pp. 210–19.
[20] Pines, "Études," Vol. *4*, 17–20. The main description of the *mail* theory in the *Kitāb al-Shifā* is missing in the abbreviated Latin translation known as the *Liber de sufficientia*, although the passage cited in footnote 13 is in the translation. But that passage, as we remarked, does not specifically mention *mail* or impressed force. Cf. A. Maier, *Zwei Grundprobleme der scholastischen Naturphilosophie*, 2d ed. (Rome, 1951), pp. 129–33.

development of the theory.[21] There was also no apparent influence from Philoponus' commentary itself, since it does not appear to have been translated in full, although, as I have indicated before, at least a fragment of it was translated. In spite of the difficulty of finding any direct contacts with the earlier theories of impressed force and inclination, we know that a number of schoolmen of the thirteenth century considered and rejected theories that held the continuation of projectile motion to arise from a continuation of the force of the projector in the projectile rather than in the medium.[22] Thus Roger Bacon in his *Questiones* on Aristotle's *Physics* rejects the continuation or influence (i.e., inflowing) of the power of the projector in the projectile, because in such a case there would be no substantial contact of mover and moved, which is necessary in motion.[23]

[21] It was Pierre Duhem in his *Études sur Léonard de Vinci*, Vol. *2* (Paris, 1909), 191, who first suggested that the impetus theory passed to the West in the *De motibus celorum* of al-Bṛitūjī (Alpetragius). But the passage in the form quoted by Duhem, which clearly describes the impetus theory, comes from the Latin translation of the Hebrew text, which was not known until much too late to influence the development of the impetus theory in the West. In the earlier translation from the Arabic, made by Michael Scot, the passage is so abbreviated that it could hardly have been the source of the Western impetus theory. See the edition of this translation made by F. J. Carmody (Berkeley, Calif., 1952), chapter 7, pp. 93–94: "Et quia diminuitur virtus [celi] proveniens ex superiori paulatim, sicut in lapide proiecto aut sagitta emissa—que paulatim diminuitur quousque desinat esse, et tunc quiescit sagitta; et sic in ista virtute celi quousque perveniatur ad terram que quiescit naturaliter...; tunc ista corpora non sunt nisi sicut sagitta aut lapis habens virtutem decisam a virtute motoris." For the refutation of Duhem's idea, see Maier, *Zwei Grundprobleme*, pp., 127–29.

[22] The standard treatment of the Latin impetus theory is the excellent work of Maier already cited in footnote 20. She revises and greatly extends the pioneer studies of P. Duhem, who was the first to discover the existence of the impetus theory in the medieval scholastic literature (see the *Études* cited in footnote 21). Maier's most recent summary and evaluation of the theory is to be found in her "Die naturphilosophische Bedeutung der scholastische Impetustheorie," *Scholastik*, *30* Jahrg., Heft *3*, (1955) 321–43. The use of *impetus* by Thierry of Chartres in the twelfth century seems to be in the sense of the propulsive power that a projectile will have as the result of being in motion, and this arises from the thrust of projection which depends in turn on how firmly based the projecting agent is on something solid. Thierry of Chartres, *De Opere sex dierum* (edited by M. Hauréau, *Notices et extraits de manuscrits de la Bibliothèque Nationale*, Vol. *32*, 2nd part [1888], 177): "Cum lapis ejicitur, ex projicientis innixu circa aliquod solidum impetus projecti contingit; unde quanto firmius se infigit tanto jactus projicientis est impetuosior." ("When a stone is thrown, the impetus of the projectile arises from the pressing of the projector on something solid; whence by the amount that it presses more firmly, so the projection of the projector is more impetuous.") Cf. E. Gilson, *History of Christian Philosophy in the Middle Ages* (New York, 1955), p. 147.

[23] *Questiones supra libros octo physicorum Aristotelis*, in *Opera hactenus inedita Rogerii Baconi*, fasc. XIII, edition of F. M. Delorme, O.F.M., with the collaboration of R. Steele (Oxford, 1935), Bk. VII, p. 338:

Thomas Aquinas' rejection of the continuance of projectile motion by an impressed force in the projectile is somewhat more extensive:[24]

He (Aristotle) says first, therefore, that the force of the violent motor uses the air as a certain instrument for both, i.e., for motion upward and motion downward. Air, moreover, is both innately light and heavy.... Thus air, according as it is light, will complete the violent motion upward—but only as it is being moved, for the [original] source of such motion is the power *(potentia)* of the violent motor—while it (the air) completes motion downward, according as it is heavy. For the force of the violent motor by means of a certain impression *(impressio)* transfers motion to both, i.e., to the air moved upward and to the air moved downward, or to both the air and the heavy body, as, for example, a stone. However, it

"Et hec excitata movet ipsum grave vel leve et est ei presens secundum substantiam, et sic partes aeris divise reinclinantes et pellentes ipsum grave vel leve continue sunt ei conjuncte secundum substantiam, et ita omne movens conjunctum est moto secundum substantiam, loquendo de hoc movente et moto naturaliter, et non sufficit quod secundum influentiam virtutis. Ergo in projectione non sufficit continuatio virtutis proicientis in projectum, set oportet quod sit conjunctum secundum subtantiam semper. Et hoc concedendum est. Dicendum ad primum quod proiciens proximum simul est cum projecto secundum substantiam; et sunt ibi, ut dictum est, plura moventia continue, partes scilicet aeris reinclinantes ita quod proximum movens semper [sit] simul secundum substantiam, et non oportet quod primum proiciens continuet suam virtutem...."

[24] Thomas Aquinas, *In libros Aristotelis de caelo et mundo expositio*, Bk. III.2, lect. 7 (in *Opera omnia*, Vol. *3* (Rome, 1886), 252 (cf. the recent edition [Rome, 1952], 305, c. 1): "Dicit ergo primo quod virtus motoris violenti utitur aere tanquam quodam instrumento *ad ambo*, idest ad motum sursum et ad motum deorsum. Aer autem natus est esse levis et gravis.... Sic igitur aer, secundum quod est levis, perficiet motum violentum qui est sursum (ita tamen prout movetur, et fuerit principium talis motionis potentia violenti motoris): motum autem qui est deorsum perficit secundum quod est gravis. Virtus enim violenti motoris, per modum cuiusdam impressionis, tradit motum *utrique*, idest vel aeri sursum moto et deorsum moto, vel etiam aeri et corpori gravi, puta lapidi. Non est autem intelligendum quod virtus violenti motoris imprimat lapidi qui per violentiam movetur, aliquam virtutem per quam moveatur, sicut virtus generantis imprimit genito formam, quam consequitur motus naturalis: nam sic motus violentus esset a principio intrinseco, quod est contra rationem motus violenti. Sequeretur etiam quod lapis, ex hoc ipso quod movetur localiter per violentiam, alteraretur: quod est contra sensum. Imprimit ergo motor violentus lapidi solum motum: quod quidem fit dum tangit ipsum. Sed quia aer est susceptibilior talis impressionis, tum quia est subtilior, tum quia est quodammodo levis, velocius movetur per impressionem violenti motoris, quam lapis: et sic, desistente violento motore, aer ab eo motus ulterius propellit lapidem, et etiam aerem coniuctum; qui etiam movet lapidem ulterius, et hoc fit quousque durat impressio primi motoris violenti, ut dicitur in VIII *Physic*. Et inde est quod, quamvis motor violentus non sequatur ipsum mobile quod per violentiam fertur, puta lapidem, ut praesentialiter ipsum moveat, tamen movet per impressionem aeris: si enim non esset tale corpus quale est aer, non esset motus violentus. Ex quo patet quod aer est instrumentum motus violenti necessarium...."

ought not to be thought that the force of the violent motor impresses in the stone which is moved by violence some force *(virtus)* by means of which it is moved, as the force of a generating agent impresses in that which is generated the form which natural motion follows. For [if] so, violent motion would arise from an intrinsic source, which is contrary to the nature *(ratio)* of violent motion. It would also follow that a stone would be altered by being violently moved in local motion, which is contrary to sense. Therefore, the violent motor impresses in the stone only motion and only so long as it touches it. But because the air is more susceptible to such an impression—both because it is finer in structure *(subtilior)* and because it is "light" in a certain way—it is moved more quickly by the impression [it receives] from the violent motor than is the stone. And so, when the violent motor has ceased [its action], the air moved by it propels the stone further, and [it] also [propels] the air with which it is in contact, and this [latter air] moves the stone [still] further, and this happens so long as the impression of the first violent motor lasts, just as is stated in [Book] VIII of the *Physics*. And thence it is that, although the violent motor does not follow the mobile which is borne along by violence, as, for example, a stone, so that it moves it by being present, yet it moves [it] by the impression in the air. For if there were not such a body as air, there would not be violent motion. From which it is evident that air is the necessary instrument of violent motion....

Thus Thomas Aquinas rejects an impressed force in the projectile on the basis (1) that this would put the source of movement within the projectile, and thus violent motion would arise from an intrinsic source, which seems contrary to the nature of violent motion, and (2) that such a theory would mean that the projectile suffers alteration in being moved by violence, and this evidently does not happen. But unlike the projectile (which can receive only motion from the projector, and only so long as the projector is in contact with it), the air receives not only motion from the projector, but it can receive as well an "impression" of the force of the projector, by which impression the adjacent air is moved and it moves the projectile. The air can receive this impression, while the projectile cannot, because of its fine structure and because it is a light body in certain respects.

In addition to the theory of impressed force rejected by Aquinas, we find described another theory by Peter John Olivi (d. 1298), a theory which he probably did not himself hold. This theory, outlined in two slightly different forms in his *Questions on the Second Book of the Sentences*, holds that a "similitude" or "impression" of the motive force is introduced into the moving body (rather than in the medium) and causes there successively an "inclination of the moving body toward the terminus of motion;"[25]

---

[25] The first analysis of Olivi's treatment of the problem of violent motion was

An opponent of this theory might object as follows:[26] "Since the motive force is present in the moving body and is sufficient for producing motion as well as for causing the influx of its species, it seems that it could just as well *per se* and immediately cause motion as to do so by means of an influx. It is, moreover, superfluous to posit two or more [causes] where one is sufficient." But the supporters of the theory of the influx of a similitude or impression of the motive power in the moving body answer as follows:[27] "Because the force of the motor is not intrinsic in this way in the mobile, nor is it an actual or formal applying of the mobile to the terminus of motion—as is the impression which has flowed out from the motor into the moving body—therefore, the force of the motor is not sufficient for causing motion without an influx of its impression."

given by B. Jansen, "Olivi der älteste scholastische Vertreter des heutigen Bewegungsbegriffs," *Philosophisches Jahrbuch der Görresgesellschaft*, Vol. *33* (1920), 137–52. For the passage here referred to, see p. 145; cf. his edition, *Quaestiones in secundum librum sententiarum*, Vol. *1* (Quaracchi, 1922), quest. 29, p. 499. "Alii vero volunt quod primo fiat aliqua similitudo seu impressio a motore in mobili, ex qua impressione causetur in eodem successive inclinatio mobilis ad terminum motus, ad hanc autem secundum eos sequitur motus immediate. Tertii sunt qui cum mediis concordant in omnibus, quando contingit motum durare in absentia motoris; quando vero motus non potest esse nisi praesente motore, tunc dicunt quod motus sequitur immediate ad primam impressionem influxam a motore, ita quod non oportet aliquam aliam inclinationem interponi." In question 31 (*edit. cit.*, p. 563; Jansen, "Olivi," p. 144) a comparison is drawn between a *vis formativa* and the *impulsus* or *inclinationes* given to projectiles by projectors: "Nam vis formativa non agit nisi sicut virtus instrumentalis alicuius principalis agentis, sicut suo modo impulsus seu inclinationes datae projectis a projectoribus movent ipsa projecta etiam in absentia proicientium."

[26] Jansen, "Olivi," p. 146: "Cum enim virtus eius motiva sit praesens mobili et sufficiens ad movendum aeque bene sicut et ad suam speciem influendam, videtur quod aeque bene poterit per se et immediate causare motum sicut et per mediam influentiam. Superfluum autem est ponere duo vel plura, ubi sufficit unum."

[27] Olivi, *In secundum librum sententiarum*, *edit. cit.*, p. 503; Jansen, "Olivi," p. 146: "...quia virtus motoris non est sic intrinseca mobili nec est actualis et formalis applicatio mobilis ad terminum motus, sicut est impressio influxa a virtute motoris in mobile, idcirco virtus motoris in mobile non sufficit ad causandum motum absque influxu suae impressionis." A second objection against this theory is that from it would follow an absurdity, namely the identification of the *influentia* and the *motus mobilis*: "Si ergo ista influentia causat motum—alias enim non necessario sequeretur ad eam—: ergo ipsa movebit mobile. Sed omne tale est vere motus; ergo ipsa erit verus motus mobilis; quod absurdum videtur." The answer *(loc. cit.)* given by the defenders of the theory follows: "Causare enim motum non est semper idem quod movere proprie sumptum; nam levitas ignis non dicitur proprie movere ignem in sursum, quamvis causet eius motum, nec inclinatio data lapidi a proiectore dicitur proicere vel movere lapidem, quamvis causet eius motum. Motus enim proprie non dicitur nisi ille qui influit impressionem in mobile, per quam ipsum movet."

It is evident that the proponents of this theory which holds the force of the motor causes a similitude, impression, or species of itself to flow into the moving body, thereby causing its continuing motion, have borrowed from the more general emanation theory of the multiplication and influence of species that Bacon had described.[28] Furthermore, Maier has found another work of Olivi in which he combats the above idea that local motion can be continued by a species or similitude of the motive force and in which he anticipates Ockham's entirely different idea that, since motion is not a separate entity but rather is a mode of relationship, projectile motion, as any local motion, does not require a continuing cause.[29] But, as Maier has pointed out, this view of motion held by Olivi is not the modern inertial principle, since he specifically refutes the idea that such a mode of relationship could last indefinitely.

Before turning to the fourteenth century, it should be noted that Moody has suggested a possible connection between the use of *impulsus* in the fourth book of the *De ratione ponderis* attributed to Jordanus and the *impetus* theory developed by Buridan in the next century.[30] However, since it is a mechanical work not interested in the ontology of the mechanical terms and ideas used, it is most difficult to decide what is meant by *impulsus*.

Now in the fourteenth century several different trends in the solution of the problem of projectile motion can be noted: (1) The first trend centers in the elaboration and support for a theory holding that the original mover leaves behind a force in the projectile which is the principal continuer of the motion of the projectile. This is an impressed force of the temporary, self-expending variety. The chief exponent of this in the early part of the century is Franciscus de Marchia. (2) The second lies in a vigorous nominalist refutation of this doctrine as well as of the opinion of Aristotle holding for the reception of a power by the medium. This refutation, whose principal exponent is William Ockham, reduces motion (including projectile motion) merely to the body being at successive positions, and sees no need for the positing of a further cause in projectile motion. (3) A third opinion—that of John Buridan—alters the temporary impressed force, making it a permanent quality called *impetus* and giving it rudimentary quantification in terms of the prime matter of the projectile and the

[28] Maier, *Zwei Grundprobleme*, pp. 142–153.
[29] Maier, "Die naturphilosophische Bedeutung," p. 326.
[30] E. A. Moody and M. Clagett, *Medieval Science of Weights* (Madison, 1952), pp. 409, 412. Cf. the review by E. J. Dijksterhuis in *Archives internationales d'histoire des sciences*, Vol. *6* (1953), 505.

velocity imparted to the projectile. (4) The last general trend is characterized by the alteration of *impetus* so that it becomes once more temporary or nonpermanent, with the additional confusion as to whether it tends to produce uniform motion or acceleration. Buridan's student Oresme can be singled out as important in the development of this trend.

Let us examine these trends in somewhat greater detail. The first trend we have illustrated by Document 8.1, which is a *reportacio* of a selection from the *Sentence Commentary* of Franciscus de Marchia and is dated 1323 (although Franciscus lectured on the *Sentences* of Peter Lombard in 1319–20).[31] As the reader examines Document 8.1, he ought to note the following important points made by Franciscus de Marchia: (1) The continuing motion of the projectile is caused primarily by a "force left behind" *(virtus derelicta)* in the projectile. A similar force left behind in the medium is of assistance in the motion. Thus Franciscus visualizes a simple emendation of the Aristotelian theory, transferring the principal impressed power to the projectile (see passages 8–12). (2) The *virtus derelicta* is not simply permanent, as is the calidity of fire, but rather is temporary—"for a time," as in the manner of the heat introduced into some body by fire (see passage 11). Hence Marchia's *virtus derelicta* is temporary as was the *mail* of Abū 'l-Barakāt. (3) The assumption of a *virtus derelicta* in the projectile as the principal continuator of the motion is preferable as an explanation to a force left in the medium for two reasons: first, because the principle of economy is better served; and, second, because the appearances or phenomena are better accounted for (see passage 11). (4) Granting the *virtus derelicta* theory, one could explain the motion of heavens by saying that "intelligences" impressed similar forces in the heavens, which would continue their motion "for a time" (see passage 12). It is obvious that an indefinitely lasting force is not here implied.

Other schoolmen contemporary with Franciscus de Marchia considered or reported on the impressed force theory, as Maier has shown.[32]

Radically different from the solution of Franciscus de Marchia was that of William Ockham. While apparently at first holding to a kind of "action at a distance" explanation of projectile motion, his mature consideration of the problem appears to have been determined by his general terminalistic analysis of motion. For he denied the existence of motion as an entity separate from the moving body, indicating rather that it was a term standing for a series of statements that the moving body is now here, now

[31] Maier, *Zwei Grundprobleme*, pp. 164–165.   [32] Ibid., pp. 197–200.

here, etc. (see Chapter 10, pp. 589–90). In the case of projectile motion, Ockham argued vigorously against the Aristotelian position and its modification, both of which held the necessity of a continuing force; he declared rather that the motion of the projectile is simply *secundum se*.[33] Again it should be observed that we do not have in this terminalistic analysis the modern inertial doctrine, for there is no attempt to assert the *indefinite* continuance of this relational mode of the moving body as a state or a condition.

Of the various successors of Franciscus de Marchia holding some form of an impressed force theory, we must above all single out John Buridan, although how far he was influenced by Marchia is not easy to say. Buridan was the principal exponent of the "new" impetus dynamics at the University of Paris. He is variously reported at the University of Paris during the period 1328–58.

John was in all probability born at Béthune in the diocese of Arras, probably close to 1300.[34] He is first mentioned in a university document of February, 1328, as rector of the university. In the next year, another documentary reference to him calls him a celebrated philosopher *(celeber philosophus)*. In 1340 he was again rector of the university, and in 1342 he is mentioned as receiving a benefice in Arras, "while lecturing at Paris

---

[33] William Ockham, *Questiones super quattuor libros sententiarum* (Lyons, 1495), Bk. II, quest. 26, part m: "Item notandum quod in motu proiectionis est magna difficultas de principio motivo et effectivo illius motus, quia non potest esse proiiciens, quia potest corrumpi existente motu. Nec aer, quia potest moveri motu contrario sicut si sagitta obviaret lapidi, nec virtus in lapide, quia quero a quo causatur illa virtus, non a proiiciente, quia agens naturale equaliter approximatum passivo equaliter causat semper effectum; sed proiiciens quantum ad omne absolutum et respectivum in eo potest equaliter approximari lapidi et non movere sicut quando movet potest, enim manus mea tarde moveri et approximari alicui corpori, et tunc non movebit ipsum localiter et potest velociter et eum impetu moveri, et tunc approximatur eo modo sicut prius, et tunc causabit motum et prius non ergo ista virtus quam tu ponis non potest causari ab aliquo absoluto vel respectivo in proiiciente nec a motu locali ipsius proiicientis, quia motus localis nihil facit ad effectum nisi approximare activa passivis sicut sepe prius dictum est. Sed omne positum in proiiciente equaliter approximatur proiecto per motum tardum sicut per velocem. Ideo dico quod ipsum movens in tali motu post separationem mobilis a primo proiiciente est ipsum motum secundum se et non per aliquam virtutem absolutam in eo respectivam, ita quod hoc movens et motum est penitus indistinctum...." Punctuation altered. Cf. P. Boehner, *Ockham, Philosophical Writings* (Edinburgh, 1957), pp. 139–41.

[34] The pertinent facts of Buridan's life are summarized by E. A. Moody in his edition of Buridan's *Quaestiones super libris quattuor de caelo et mundo* (Cambridge, Mass., 1942), pp. xi–xiv. But Edmond Faral, "Jean Buridan, maître èn arts de l'Université de Paris," *Histoire littéraire de la France*, Vol. 28, 2e partie, is the most important biographical source for Buridan.

on the books of natural, metaphysical, and moral philosophy." Continued scattered references to him are made until a document of 1358, in which he appears as a signatory along with his almost equally celebrated successor, Albert of Saxony. His most important works, so far as mechanics is concerned, were his *Questions* on Aristotle's *De caelo*, and his three different treatments of the *Physics*. We include an important question from his third version of the *Physics* in this chapter (see Doc. 8.2). For no sufficient reason, his exposition of the impetus theory in his works on the *De caelo* and the *Physics* usually is dated about 1340 or perhaps even earlier. However, it should be pointed out that there exists a manuscript at the Vatican (Vat. lat. 2163, f. 1r) of his final version of the *Questions on the Physics*, which seems to indicate 1357 as the date of composition of at least the table of questions treated, if not the questions themselves.[35] A late reference to Buridan in 1366 uncovered by Michalski is no doubt erroneous.[36] It has been suggested that Buridan died from plague in 1358, but there is no real evidence to support this view.

In the same loose sense that Bradwardine was the "founder" of a school of mechanicians at Merton College, so Buridan was the "founder" of such a school at Paris. But Buridan's principal interest so far as mechanics was concerned was in dynamics. And his exposition of the theory of the impressed force (called by him *impetus*) was the starting point of similar discussions of his principal successors at Paris in the second half of the century, namely Nicole Oresme, Albert of Saxony, and Marsilius of Inghen. The complete account of Buridan's theory is given in Document 8.2.

The main points to notice in Buridan's exposition of the impetus theory are the following:

1. *The impetus imparted by the projector to the projectile varies, on the one hand, as the velocity of the projectile (initially and immediately introduced), and as the quantity of matter of the body in movement on the other hand:*

And by the amount the motor moves that moving body more swiftly, by the same amount it will impress in it a stronger impetus.... Hence by the amount more there is of matter, by that amount can the body receive more of that impetus and more intensely. Now in a dense and heavy body, other things being equal,

---

[35] MS Vat. lat. 2163, f. 1r. Cf. Vienna, Nat.-bibl. cod. 5424, f. 1r, where the date is obscure but also looks like 1357.

[36] K. Michalski, "Les courants philosophiques à Oxford et à Paris pendant le XIVe siècle," *Bulletin international de l'Académie polonaise des sciences et des lettres. Class d'histoire et de philosophie et de philologie. Les Années 1919, 1920* (Cracovie, 1922), p. 82, cites manuscripts dated 1372 and 1387; but these are no doubt copy rather than composition dates.

there is more of prime matter than in a rare and light one.... And so if light wood and heavy iron of the same volume and of the same shape are moved equally fast by a projector, the iron will be moved farther because there is impressed in it a more intense impetus, which is not so quickly corrupted as the lesser impetus... (see passages 4–5).

Thus Buridan gives a quasi-quantitative definition of *impetus* at the time of its imposition. We must say "quasi" because there is no formal discussion of its mathematical description. The exact nature of the impetus as conceived by Buridan is difficult to pin down. He spoke of it as motive force and as the reason for the continued movement, for he says:

Thus we can and ought to say that in the stone or other projectile there is impressed something which is the motive force of that projectile.... the motor in moving a moving body impresses in it a certain impetus or a certain motive force of the moving body, [which impetus acts] in the direction toward which the mover was moving the moving body, either up or down, or laterally, or circularly.... It is by that impetus that the stone is moved after the projector ceases to move. But that impetus is continually decreased by the resisting air and by the gravity of the stone, which inclines it in a direction contrary to that in which the impetus was naturally predisposed to move it (see passage 4).

Yet it is clearly not a force of the same species as the original projecting force. This is borne out by a careful reading of the various passages in Document 8.2 below. It appears that while the impetus is unquestionably a force, it nevertheless seems close to being the effectiveness which the original force has on a particular body, an effectiveness measurable in terms of the velocity immediately supplied to the body and the quantity of matter in the body. It is certainly doubtful whether Buridan meant to relate impetus and the immediate velocity imparted to a body exponentially by Bradwardine's law, although Buridan knew and used that law.[37] One cannot help but compare Buridan's impetus with Galileo's *impeto* (see Chapter 11, note 131) and Newton's quantity of motion (momentum), even though on the face of it they are ontologically different from impetus considered as a kind of force. But while the affirmed ontology of impetus would seem to differentiate it from the later concepts, yet the terms of its *measure* as presented by Buridan make an analogue with momentum, i.e., this quality which is motive force for Buridan turns out to be described in dimensions analogous to those of Newton's momentum.

2. Like Avicenna, but unlike Abū 'l-Barakāt and Franciscus de Marchia,

---

[37] See above, Chapter 7, note 40. Cf. Maier, *Die Vorläufer Galileis im 14. Jahrhundert* (Rome, 1949), pp. 145–46.

Buridan conceived his impetus as a permanent quality (although of course destructible by contrary agents), and as such it is not self-expending merely as the result of separation from the principal motivating force, but must be overcome by the resistance of the air and the contrary inclination of the body. "The third conclusion is that impetus is a thing of permanent nature distinct from the local motion in which the projectile is moved.... the impetus is a quality naturally present and predisposed for moving a body in which it is impressed, just as it is said that a quality impressed in iron by a magnet moves the iron to the magnet.... it is remitted, corrupted, or impeded by resistance, or a contrary inclination" (see passage 9). And we are definitely told in another place that impetus would last indefinitely were it not for these contrary agents (and of course they are always present in the terrestrial area), for he says that, "the impetus would last indefinitely *(in infinitum)* if it were not diminished by a contrary resistance or by an inclination to a contrary motion...."[38] And in still another place,[39] while suggesting that the impetus theory explains the continued motion of the smith's wheel or mill after one has stopped actively turning it, he says, "And perhaps if the mill would last forever *(semper duraret)* without some diminution or alteration of it, and there were no resistance corrupting the impetus, the mill would be moved perpetually *(perpetue)* by that impetus." Being a permanent quality, it is for Buridan distinct from the motion itself.

3. The characteristic of permanence which Buridan assigned to his impetus made it plausible for him to explain the everlasting movement of the heavens by the imposition of impetus by God at the time of the world's creation: "... it does not appear necessary to posit intelligences of this kind, because it could be answered that God, when He created the world,

[38] John Buridan, *Questiones in metaphysicam Aristotelis*, Bk. XII, quest, 9, through E. Borchert, *Die Lehre von der Bewegung bei Nicolaus Oresme*, (Beiträge zur Geschichte der Philosophie und Theologie des Mittelalters, Bd. 3, Heft 3 [Münster, 1934]), p. 43: "... multi ponunt quod proiectum post exitum a projiciente movetur ab impetu dato a projiciente et movetur quamdiu durat impetus fortior quam resistentia; et in infinitum duraret impetus nisi diminueretur et corrumperetur a resistente contrario vel ab inclinante ad contrarium motum: et in motibus celestibus nullum est resistens contrarium...."

[39] John Buridan, *Quaestiones super libris quattuor de caelo et mundo*, edition of Moody, Bk. II, quest. 12, p. 180: "Et experimentum habetis, quod si mola fabri magna et valde gravis velociter movetur a te, motu reversionis, et cessares eam movere, adhuc ab ipso impetu acquisito ipsa diu moveretur; imo tu non posses eam statim quietare, sed propter resistentiam ex gravitate illius molae, ille impetus continue diminueretur donec mola cessaret; et forte si mola semper duraret sine aliqua eius diminutione vel alteratione, et non esset aliqua resistentia corrumpens impetum, mola ab illo impetu perpetue moveretur."

moved each of the celestial orbs as He pleased, and in moving them impressed in them impetuses which moved them without his having to move them any more .... And these impetuses which he impressed in the celestial bodies were not decreased or corrupted afterwards because there was no inclination of the celestial bodies for other movements. Nor was there resistance which would be corruptive or repressive of that impetus." The use of impetus to explain the continuing movement of the heavens is the closest that Buridan comes to the inertial idea of Newton's mechanics. It can scarcely be doubted that impetus is analogous to the later inertia, regardless of ontological differences.

One serious defect of this medieval theory is that there was no sure distinction between rectilinear and circular impetus. It was equally possible to impose rectilinear or circular impetus. We must await the sixteenth century for a clarification of the directional aspects of impetus. One point should be made clear, and that is that the suggestion of the use of impetus to account for the continuing movement of the heavens more economically than by the use of intelligences, although it was made by Buridan in more than one place, is still a rather incidental suggestion. It is probably an exaggeration to say that Buridan by this doctrine was seeking to apply a single mechanics to terrestrial and celestial phenomena. It is evident throughout his writings that he accepts the basic Aristotelian dichotomy described above in Chapter 7; and in fact he asserts in these very passages that there is no resistance in the heavens, while of course there is resistance in the terrestrial area. This immediately separates the two areas.

4. Like Abū 'l-Barakāt and his successors in Islam, Buridan also applied the impetus theory to the problem of the acceleration of falling bodies (see passage 6). The continued acceleration, he believed, is produced because the gravity of the body is continually impressing more and more impetus in the body. The continually growing impetus produces the continually growing velocity. We shall discuss this theory at greater length in the next chapter, and we shall discuss its implication as impetus is changed from cause to effect in early modern times. We also shall see how Nicole Oresme, the principal successor of Buridan at Paris, altered the concept of impetus. And finally in Chapter 11 we shall give a brief survey of the history of this interesting idea from its propagation in the fourteenth century through its spread to European universities in the later Middle Ages and early modern times.

Document 8.1

Franciscus de Marchia
On the Sentences of Peter Lombard
(A *Reportacio* of the Fourth Book)*

1 ... THE first conclusion [concerning the violent motion of a stone upward] is that violent motion of this kind, i.e., of a heavy body upward, does not arise from the hand as movent. I prove this first [conclusion] [1] because when the hand has ceased to move the heavy body, it (the body) still continues its motion a certain amount, [2] because, even if we let the hand be destroyed or move something else with an opposite motion, i.e., downward, nevertheless the heavy body still continues the motion begun by it (the hand).

2. The second conclusion is that this motion does not effectively arise from the natural form of this heavy body which is moved. This is proved, since every motion arising from an intrinsic source or natural form of the moving body is natural; yet this motion is not natural but violent; therefore, etc. . . . .

3. The third conclusion is that motion of this kind does effectively arise from the natural form of the medium—e.g., the water or air—in which the body is moved . . . .

4. The fourth conclusion is that it does not arise from a series *(ordo)* of parts of the medium as they successively move themselves. For perhaps some would say that motion of this kind is caused in this way, since the movent of the stone or any other heavy body initially moves the part of the air near it, and that [part] which is first moved then moves another part, and so on serially *(et sic per ordinem)*. And thence it is that, with the hand stopping, the motion does not cease, since, although with the stopping of the hand the first part of the air next to it stops, still all the other

---

* For the Latin text of the question here translated, see A. Maier, *Zwei Grund- probleme der scholastischen Naturphilosophie,* 2d ed. (Rome, 1951), pp. 166–80.

[parts] do not, and, therefore, the parts of the air which are moved carry along that heavy body. And so they say that, although this motion does not arise from the hand, nor from the form of the heavy body, nor even determinately by some part of the medium, still it arises from the whole medium indeterminately assumed. The Philosopher argues against this at the end of the eighth [book] of the *Physics* . . . .

5. The fifth conclusion is that it does not arise from itself. This is evident, since nothing can effectively arise from itself; therefore, etc.

6. The sixth conclusion is that it cannot arise from celestial form. This is evident, since celestial form is determinate for one motion.

(The seventh conclusion is lacking.)

8. The eighth and last affirmative conclusion, which follows out of the aforementioned, is that motion of this kind arises immediately from some force left behind *(virtus derelicta)* by means of an initial action of the first motor, for example, a hand. And this is the opinion of the Philosopher and also of the Commentator at the end of the eighth [book] of the *Physics*, comment 27 . . . .

9. But then there is a doubt as to where a force of this kind resides *(sit)* subjectively, i.e., whether in the heavy body which is moved or in the medium itself, and also [a doubt as to] what it (the force) is in itself formally.

In answer to this, the Philosopher seems to say at the end of the eighth [book] of the *Physics*, that a force of this kind is subjectively and formally in the medium, e.g., in air or in water, and not in the moving body. For the Philosopher imagines that air and water are of a quicker motion than a stone or heavy bodies of this kind which are moved in them. Therefore, the first part of the air next to the hand which impels some heavy body is moved more quickly than that heavy body is innately capable of moving downward and, therefore, that swifter motion of the air impedes the downward motion of the heavy body. And thence it is that a heavy body which has been impelled ascends until it arrives at some part of the air which is not moved more quickly than the heavy body is moved downward. Nay, the motion of the heavy body downward is swifter than that motion; and, therefore, the motion of that part of the medium does not impede the motion of the heavy body. Accordingly, that heavy body at that time immediately begins to descend. And thus the Philosopher says there, that a force of this kind, caused by the hand and continuing the motion, is received subjectively in the air. For the parts of the air are innately constituted for condensation and rarefaction. Whence the Philosopher, and the Commentator as well, imagine that just as when a stone is thrown into water and there are thereby formed and generated in the water certain circles, so similarly a stone thrown in air makes in the air certain invisible circles, the first of which—because it is moved more quickly than the stone would by itself descend—carries the stone along to the second circle, and the second to the third, [still] impeding its (the stone's) proper motion downward. And so the circles of this kind caused in the air carry the stone until it arrives at some [circle] whose motion is not swifter than the downward motion of the heavy body, and then the circles stop and the heavy body begins to descend.... And because these [circles] are related in succession *(consequenter se habentes)*, i.e., not continuous, hence the Philosopher concludes that violent motion of this kind is not continuous but related in succession....

10. The Philosopher in three ways proves this, i.e., that forces of this kind are received in the medium. [1] Contraries are the causes of contraries; but it is apparent to the senses that when a not-very-heavy body is thrown upward and the wind is blowing in the opposite direction, sometimes it (the body) returns to the projector, which situation cannot arise from anything but the contrary motion of the air impelling it downward. Therefore, by that same reason, its motion arises from a contrary motion of the air, evidently by a force received in the air which impels it upward; therefore, etc. [2] Every motion which arises from an intrinsic source is natural. For nature is the source *(principium)* of motion and rest, according to the Philosopher in the second [book] of the *Physics*. But the motion of a stone upward is not natural but violent. However, if it were to arise from some force received in the stone, it would be natural because it would arise from an intrinsic source in the moving body; therefore, etc.... [3] Further, every mobile which receives some force from the motor becomes accustomed *(assuescit)* to it or to its act, as is evident in the case of the hand of the writer. But a stone which is thrown never becomes accustomed to motion upward however often it is thrown; therefore, it receives no force from the motor.

11. But against this I argue and show that a force of this kind is rather in the stone or in any other heavy body than in the medium.... But whatever is so regarding the subject of this force, [we can be sure] at least [that] a force of this kind which continues the motion already begun is to be posited, either in the medium, or, which I believe more, in the moved body. Whence it is to be known that the force moving some heavy body upward is twofold: [1] one which begins the motion or determines the

heavy body for some motion—and this force is the force of the hand; [2] another force which comes after the motion has begun and continues it —and this is caused or left behind by the first [force] with the object of producing motion. For unless some force other than the first one is posited, it is impossible to give a cause for the succeeding motion, as was deduced above. And this force, regardless of which subject it is posited as being in, continues and follows the motion according to the proportion and mode determined by the first [force]; and this force is a neutral force not having a contrary, since it follows motion according to any difference of position. And if one asks of what sort *(qualis)* is a force of this kind, it can be answered that it is not simply permanent nor simply fluent, but almost medial [between them], since it lasts (i.e., is permanent) for a certain time—just as calidity generated by fire does not have to be simply permanent as fire does—nor in addition is it simply fluent as is "heating" itself.... It seems preferable that a force of this kind resides *(sit)* in the body which is moved rather than in the medium, regardless of what the Philosopher and the Commentator have said on this matter. [This theory is preferred] because [1] it would be in vain that something should be done by many [causes] which can be done by few *(quia frustra fit per plura quod potest fieri per pauciora)*—now it does not appear necessary to posit something other than the moving body or the force received in it and the original motor as the effective cause[s] of motion; therefore, the medium is not [the cause]. [The theory is also preferred] because [2] in positing this, all phenomena *(apparentia)* are accounted for *(salvantur)* better and more easily.... [As for example the phenomenon] that a stone, or any other body which is thrown—such as a javelin—divides the medium through which it courses....

12. By accepting this theory, one can answer the arguments of the Philosopher. To the first, when it is said that sometimes the body which is thrown returns to the projector etc., I concede this when the motion of the air in the opposite direction exceeds the force of projection. However, when it does not exceed it, it (the body) does not return but is moved further, and this by means of the force left behind in the stone by the original force or initial mover, since just as the motion of the stone surpasses and exceeds the motion of the air, so the accession of the force left behind in the stone from the hand which impels it exceeds the accession of force caused in the air by that which moves it. For I do not deny that a force of this kind is received also in the medium.... From which it follows that, when a stone or some heavy body, or even a light body, is

moved in a medium, two motions concur there, evidently the motion of this stone which arises immediately from the force left behind in the stone and also the motion of the air, which likewise produces—although not immediately—the motion of the stone. For both air which is moved and the force of the stone caused in it by the impellent carry along the stone.... [The author then answers the remaining arguments of Aristotle, although we here omit them. A further discussion of "natural instrument" leads to the following:] From this it follows finally that with the intelligences ceasing to move the heavens, the heavens would still be moved, or revolve, for a time *(ad tempus)* by means of a force of this kind following and continuing the circular motion, as is evident in a potter's wheel which revolves for a time after the first mover has ceased to move [it].

COMMENTARY

Miss Maier, just preceding her text of this question, notes that this *reportacio* has a date of 1323 in the explicit (MS Vat. Chigi lat. B. VII, 113). The over-all question being treated—in which the problem of violent motion is presented as a similar question—is the following: Utrum in sacramentis sit aliqua virtus supernaturalis insistens sive eis formaliter inhaerens.

In the seven principal conclusions (passages 1–8), Marchia makes it clear, by excluding other possibilities, that the continuator of the motion is a residual force left behind by the primary or original force. He admits that there is some doubt as to where this residual force resides, i.e., as to which is its subject, the medium or the projectile. In outlining the theory of Aristotle and Averroës holding for the medium as the subject, he gives the detailed mechanics. It involves an analogy between the spreading of waves in water disturbed by a stone and a supposed similar series of waves produced by the throwing of a stone in air.[40] Somehow the stone is thought to be carried along by these waves successively. Marchia points out further that Aristotle held this kind of violent motion to be non-continuous.

[40] Compare the similar analogy given by Averroës in his *Commentarium in physicam* (Venice, 1495), Bk. VIII, commentum 82, f. 159r: "... et hoc quod auctor dicit manifeste videtur sensu in motu aque et lapide proiecto in illam. Videmus enim partem que sequitur lapidem moveri undique, deinde transfertur motus ab illa parte ad partem consequentem: deinde de illa ad aliam quousque cesset. Partes vero aque non simul moventur sicut simul moventur partes figurati corporis; et hoc patet in circulis qui in aqua fiunt quando lapides cadunt in ea.... (159v) Et universaliter motus sagitte in aerem similis est multum motui navis in unda que portat eam: qui existimatur esse unus, cum sit motus successivus per undationem que fit in aqua: et talis undatio fit in aere."

[Franciscus de Marchia, *On Peter Lombard*

In passage 10, Marchia notes the objection made by supporters of the theory posing the medium as the subject of the motive force: If the projectile is the subject of such a residual force, the force would be intrinsic in the projectile and thus we would have a case of natural motion; whereas in fact we have a case of violent motion.

In passage 11 we see that Franciscus conceives of the residual force in the projectile as being neither simply permanent nor simply fluent, but in a sense medial between them. It is only "permanent (i.e., lasting) for a time" *(permanens ad tempus)* which is of course quite different from Buridan's *impetus* (see Doc. 8.2, passage 9), which is presented as a permanent quality. As we noticed in the body of the chapter, Marchia preferred the theory of a residual force in the projectile rather than in the medium for reasons of economy and better accordance with experience.

In spite of his preference for the theory that the *virtus derelicta* in the projectile is the principal continuator of the motion of the projectile, he admits, in passage 12, that there is also a *virtus derelicta* in the medium, which force assists in the continuance of the motion. As we shall see in Chapter 11, many authors in the course of the fourteenth, fifteenth, and sixteenth centuries accepted a supplementary role for the air in causing the continuance of the projectile motion, no doubt with the object of "saving" Aristotle. It should also be noticed that, in the final statement we have included in passage 12, Franciscus suggests the possibility that intelligences could introduce a force of this kind in the heavenly bodies, which force could continue their motion for a time. To support this suggestion, he gives the analogy of the potter's wheel that continues to move after the original mover has ceased his action. Once more it should be remarked that no inertial analogy can be drawn here, as it can be in the case of similar statements by Buridan, for, unlike Buridan, Franciscus conceives of the force as lasting only "for a time" rather than continuing indefinitely.

Document 8.2

John Buridan, *Questions on the Eight Books of the Physics of Aristotle**

1. BOOK VIII, QUESTION 12. It is sought whether a projectile after leaving the hand of the projector is moved by the air, or by what it is moved.

It is argued that it is not moved by the air, because the air seems rather to resist, since it is necessary that it be divided. Furthermore, if you say that the projector in the beginning moved the projectile and the ambient air along with it, and then that air, having been moved, moves the projectile further to such and such a distance, the doubt will return as to by what the air is moved after the projector ceases to move. For there is just as much difficulty regarding this (the air) as there is regarding the stone which is thrown.

Aristotle takes the opposite position in the eighth[41] [book] of this work (the *Physics*) thus: "Projectiles are moved further after the projectors are no longer in contact with them, either by antiperistasis, as some say, or by the fact that the air having been pushed, pushes with a movement swifter than the movement of impulsion by which it (the body) is carried towards its own [natural] place." He determines the same thing in the seventh and eighth [books] of this work (the *Physics*) and in the third [book] of the *De caelo*.

2. This question I judge to be very difficult because Aristotle, as it seems to me, has not solved it well. For he touches on two opinions. The first one, which he calls "antiperistasis," holds that the projectile swiftly leaves the place in which it was, and nature, not permitting a vacuum, rapidly sends air in behind to fill up the vacuum. The air moved

---

* Translated from the edition of Paris, 1509, with the modification of A. Maier, *Zwei Grundprobleme der scholastischen Naturphilosophie*, 2d ed. (Rome, 1951), pp. 201–14.

[41] This is a statement from the fourth book rather than the eighth book; see footnote 1 above.

swiftly in this way and impinging upon the projectile impels it along further. This is repeated continually up to a certain distance.... But such a solution notwithstanding, it seems to me that this method of proceeding was without value because of many experiences *(experientie)*.

The first experience concerns the top *(trocus)* and the smith's mill (i.e. wheel—*mola fabri*) which are moved for a long time and yet do not leave their places. Hence, it is not necessary for the air to follow along to fill up the place of departure of a top of this kind and a smith's mill. So it cannot be said [that the top and the smith's mill are moved by the air] in this manner.

The second experience is this: A lance having a conical posterior as sharp as its anterior would be moved after projection just as swiftly as it would be without a sharp conical posterior. But surely the air following could not push a sharp end in this way, because the air would be easily divided by the sharpness.

The third experience is this: a ship drawn swiftly in the river even against the flow of the river, after the drawing has ceased, cannot be stopped quickly, but continues to move for a long time. And yet a sailor on deck does not feel any air from behind pushing him. He feels only the air from the front resisting [him]. Again, suppose that the said ship were loaded with grain or wood and a man were situated to the rear of the cargo. Then if the air were of such an impetus that it could push the ship along so strongly, the man would be pressed very violently between that cargo and the air following it. Experience shows this to be false. Or, at least, if the ship were loaded with grain or straw, the air following and pushing would fold over *(plico)* the stalks which were in the rear. This is all false.

3. Another opinion, which Aristotle seems to approve, is that the projector moves the air adjacent to the projectile [simultaneously] with the projectile and that air moved swiftly has the power of moving the projectile. He does not mean by this that the same air is moved from the place of projection to the place where the projectile stops, but rather that the air joined to the projector is moved by the projector and that air having been moved moves another part of the air next to it, and that [part] moves another (i.e., the next) up to a certain distance. Hence the first air moves the projectile into the second air, and the second [air moves it] into the third air, and so on. Aristotle says, therefore, that there is not one mover but many in turn. Hence he also concludes that the movement is not continuous but consists of succeeding or contiguous entities.

But this opinion and method certainly seems to me equally as impossible as the opinion and method of the preceding view. For this method cannot solve the problem of how the top or smith's mill is turned after the hand [which sets them into motion] has been removed. Because, if you cut off the air on all sides near the smith's mill by a cloth *(linteamine)*, the mill does not on this account stop but continues to move for a long time. Therefore it is not moved by the air.

Also a ship drawn swiftly is moved a long time after the haulers have stopped pulling it. The surrounding air does not move it, because if it were covered by a cloth and the cloth with the ambient air were withdrawn, the ship would not stop its motion on this account. And even if the ship were loaded with grain or straw and were moved by the ambient air, then that air ought to blow exterior stalks toward the front. But the contrary is evident, for the stalks are blown rather to the rear because of the resisting ambient air.

Again, the air, regardless of how fast it moves, is easily divisible. Hence it is not evident as to how it would sustain a stone of weight of one thousand pounds projected in a sling or in a machine.

Furthermore, you could, by pushing your hand, move the adjacent air, if there is nothing in your hand, just as fast or faster than if you were holding in your hand a stone which you wish to project. If, therefore, that air by reason of the velocity of its motion is of a great enough impetus to move the stone swiftly, it seems that if I were to impel air toward you equally as fast, the air ought to push you impetuously and with sensible strength. [Yet] we would not perceive this.

Also, it follows that you would throw a feather farther than a stone and something less heavy farther than something heavier, assuming equal magnitudes and shapes. Experience shows this to be false. The consequence is manifest, for the air having been moved ought to sustain or carry or move a feather more easily than something heavier....

4. Thus we can and ought to say that in the stone or other projectile there is impressed something which is the motive force *(virtus motiva)* of that projectile. And this is evidently better than falling back on the statement that the air continues to move that projectile. For the air appears rather to resist. Therefore, it seems to me that it ought to be said that the motor in moving a moving body impresses *(imprimit)* in it a certain impetus *(impetus)* or a certain motive force *(vis motiva)* of the moving body, [which impetus acts] in the direction toward which the mover was moving the moving body, either up or down, or laterally, or circularly.

[Buridan, *Questions on the Physics* 535]

And by the amount the motor moves that moving body more swiftly, by the same amount it will impress in it a stronger impetus.* It is by that impetus that the stone is moved after the projector ceases to move. But that impetus is continually decreased *(remittitur)* by the resisting air and by the gravity of the stone, which inclines it in a direction contrary to that in which the impetus was naturally predisposed to move it. Thus the movement of the stone continually becomes slower, and finally that impetus is so diminished or corrupted that the gravity of the stone wins out over it and moves the stone down to its natural place.

This method, it appears to me, ought to be supported because the other methods do not appear to be true and also because all the appearances *(apparentia)* are in harmony with this method.

5. For if anyone seeks why I project a stone farther than a feather, and iron or lead fitted to my hand farther than just as much wood, I answer that the cause of this is that the reception of all forms and natural dispositions is in matter and by reason of matter. *Hence by the amount more there is of matter, by that amount can the body receive more of that impetus and more intensely* (intensius). *Now in a dense and heavy body, other things being equal, there is more of prime matter than in a rare and light one. Hence a dense and heavy body receives more of that impetus and more intensely, just as iron can receive more calidity than wood or water of the same quantity.* Moreover, a feather receives such an impetus so weakly *(remisse)* that such an impetus is immediately destroyed by the resisting air. *And so also if light wood and heavy iron of the same volume and of the same shape are moved equally fast by a projector, the iron will be moved farther because there is impressed in it a more intense impetus, which is not so quickly corrupted as the lesser impetus would be corrupted. This also is the reason why it is more difficult to bring to rest a large smith's mill which is moving swiftly than a small one, evidently because in the large one, other things being equal, there is more impetus.* And for this reason you could throw a stone of one-half or one pound weight farther than you could a thousandth part of it. For the impetus in that thousandth part is so small that it is overcome immediately by the resisting air.

6. From this theory also appears the cause of why the natural motion of a heavy body downward is continually accelerated *(continue velocitatur)*. For from the beginning only the gravity was moving it. Therefore, it moved more slowly, but in moving it impressed in the heavy body an impetus. This impetus now [acting] together with its gravity moves it. Therefore, the motion becomes faster; and by the amount it is faster, so

---

* The italics here and elsewhere are, of course, mine.

the impetus becomes more intense. Therefore, the movement evidently becomes continually faster.

[The impetus then also explains why] one who wishes to jump a long distance drops back a way in order to run faster, so that by running he might acquire an impetus which would carry him a longer distance in the jump. Whence the person so running and jumping does not feel the air moving him, but [rather] feels the air in front strongly resisting him.

Also, since the Bible does not state that appropriate intelligences move the celestial bodies, it could be said that it does not appear necessary to posit intelligences of this kind, because it would be answered that God, when He created the world, moved each of the celestial orbs as He pleased, and in moving them He impressed in them impetuses which moved them without his having to move them any more except by the method of general influence whereby he concurs as a co-agent in all things which take place; "for thus on the seventh day He rested from all work which He had executed by committing to others the actions and the passions in turn." And these impetuses which He impressed in the celestial bodies were not decreased nor corrupted afterwards, because there was no inclination of the celestial bodies for other movements. Nor was there resistance which would be corruptive or repressive of that impetus. But this I do not say assertively, but [rather tentatively] so that I might seek from the theological masters what they might teach me in these matters as to how these things take place....

7. The first [conclusion] is that that impetus is not the very local motion in which the projectile is moved, because that impetus moves the projectile and the mover produces motion. Therefore, the impetus produces that motion, and the same thing cannot produce itself. Therefore, etc.

Also since every motion arises from a motor being present and existing simultaneously with that which is moved, if the impetus were the motion, it would be necessary to assign some other motor from which that motion would arise. And the principal difficulty would return. Hence there would be no gain in positing such an impetus. But others cavil when they say that the prior part of the motion which produces the projection produces another part of the motion which is related successively and that produces another part and so on up to the cessation of the whole movement. But this is not probable, because the "producing something" ought to exist when the something is made, but the prior part of the motion does not exist when the posterior part exists, as was elsewhere stated. Hence, neither does the prior exist when the posterior is made. This consequence is obvious from this reasoning. For it was said elsewhere that motion is nothing else than "the very being produced" *(ipsum fieri)* and the "very being corrupted" *(ipsum corumpi)*. Hence motion does not result when it *has been* produced *(factus est)* but when it *is being* produced *(fit)*.

8. The second conclusion is that that impetus is not a purely successive thing *(res)*, because motion is just such a thing and the definition of motion [as a successive thing] is fitting to it, as was stated elsewhere. And now it has just been affirmed that that impetus is not the local motion.

Also, since a purely successive thing is continually corrupted and produced, it continually demands a producer. But there cannot be assigned a producer of that impetus which would continue to be simultaneous with it.

9. The third conclusion is that that impetus is a thing of permanent nature *(res nature permanentis)*, distinct from the local motion in which the projectile is moved. This is evident from the two aforesaid conclusions and from the preceding [statements]. And it is probable *(verisimile)* that that impetus is a quality naturally present and predisposed for moving a body in which it is impressed, just as it is said that a quality impressed in iron by a magnet moves the iron to the magnet. And it also is probable that just as that quality (the impetus) is impressed in the moving body along with the motion by the motor; so with the motion it is remitted, corrupted, or impeded by resistance or a contrary inclination.

10. And in the same way that a luminant generating light generates light reflexively because of an obstacle, so that impetus because of an obstacle acts reflexively. It is true, however, that other causes aptly concur with that impetus for greater or longer reflection. For example, the ball which we bounce with the palm in falling to earth is reflected higher than a stone, although the stone falls more swiftly and more impetuously *(impetuosius)* to the earth. This is because many things are curvable or intracompressible by violence which are innately disposed to return swiftly and by themselves to their correct position or to the disposition natural to them. In thus returning, they can impetuously push or draw something conjunct to them, as is evident in the case of the bow *(arcus)*. Hence in this way the ball thrown to the hard ground is compressed into itself by the impetus of its motion; and immediately after striking, it returns swiftly to its sphericity by elevating itself upwards. From this elevation it acquires to itself an impetus which moves it upward a long distance.

Also, it is this way with a cither cord which, put under strong tension and percussion, remains a long time in a certain vibration *(tremulatio)* from which its sound continues a notable time. And this takes place as follows: As a result of striking [the chord] swiftly, it is bent violently in one direction, and so it returns swiftly toward its normal straight position. But on account of the impetus, it crosses beyond the normal straight position in the contrary direction and then again returns. It does this many times. For a similar reason a bell *(campana)*, after the ringer ceases to draw [the chord], is moved a long time, first in one direction, now in another. And it cannot be easily and quickly brought to rest.

This, then, is the exposition of the question. I would be delighted if someone would discover a more probable way of answering it. And this is the end.

COMMENTARY

The reader's attention is first called to the refutation in passages 2 and 3 of the two theories presented by Aristotle. The first point worth noting is that it is largely on the basis of experience that these two theories are shown to be inadequate. The *experientie* adduced against *antiperistasis*, i.e., the mechanical action of the air, are the following: (1) The spinning of a top or smith's wheel takes place without leaving its place of motion and the air can hardly be said to come behind the moving body to continue its motion. (2) The sharpening to a point of the posterior end of a lance does not thereby reduce its speed, as one would expect if this theory were correct. (3) In the course of the continuation of the movement of a ship in a river after the haulers have stopped pulling it, a sailor on deck does not feel the air pushing him from behind but rather feels it resisting him. Nor if behind some cargo would he be pushed against it; and similarly straws in the rear are not bent over. Similar "experiences" are brought against the second theory which held for a successive communication of motive power to the parts of the air.

In passage 4 Buridan states his acceptance of the theory which posited that the projectile motion is continued because the motor impresses in the projectile an *impetus* or motive force. He relates the intensity of impressed impetus to the velocity imparted by the original force to the projectile, i.e., the greater the velocity, the greater the impetus. Quite evidently this velocity is the speed of the projectile immediately after the original force of action has ceased. At the same time, he says that the impetus is made to decrease (i.e., is remitted and corrupted) in the same way that the motion is made to decrease, by the resistance and contrary inclination of the moving body. Perhaps Buridan might hold that the factors of impetus, resistance, and the continuing speed of the projectile are related as Bradwardine held generally for all cases of motion considered dynamically. However, he makes no such statement; and one might suppose from his arguments that, were there no resistance, not only would the *impetus* last indefinitely (as he states) but also that the movement maintained by the impetus would be both finite and uniform. However, such a conclusion would be difficult to fit in with the Aristotelian framework, which Buridan generally accepted. And hence our puzzlement as to just what kind of force this *impetus* really is.

Then Buridan in passage 5 relates the intensity of impetus imparted to the projectile with the quantity of prime matter in the projectile: This "quantity of prime matter" is a kind of analogue of the "mass' of early modern physics.

It is interesting to speculate on how it happened that Buridan seized upon velocity and quantity of matter as the two factors in determining the intensity of the *impetus*. It seems to be another case of the simultaneous consideration of extensive and intensive factors (see Chapter 4). The schoolmen had considered the effectiveness of a heat agent (i.e., its *potentia*) to be dependent on both the intensity of heat and the extent of the subject through which the heat is distributed, in short on a quantity of heat. Now *impetus* seems semantically to have the meaning of something like force of impact; its effectiveness, according to Buridan, would depend on how fast it is going (an intensive factor) and how much matter there was in motion (extensive factor). The analogy is apparent.[42]

Having outlined the measure of impetus in terms of the velocity im-

---

[42] One final point should be observed in regard to the measure of impetus. As I have just said, the concept of impetus no doubt has an intimate relationship with that of impulsion and impact, and the primitive use of the Latin term is in fact to represent these ideas of impulsion and impact. While there is no formal discussion of the quantification of impact in antiquity, we can remark from our quotation of Strato's views in Chapter 9 (see there p. 546), that he discussed the question of the greater impact of bodies falling from a greater height. Incidentally, from this discussion we see that he considers as possible factors producing greater impact, greater weight, a larger body (i.e., more matter?), and greater velocity. Galileo tells us somewhat the same thing (see Doc. 9.5, passage 1). For him the impulse of fall produces a force of impact. Impact, then, is an "effect brought about by the [weight and/or quantity of matter of?] the falling body together with the velocity acquired during the fall, an effect which will be greater and greater according to the height of fall, that is, according as the velocity of the falling body becomes greater" Notice that Galileo does not specify whether he means the weight or the quantity of matter as the other factor beyond velocity, but it is probably weight that he has in mind here. So for Galileo impact is probably measured by weight and velocity conjointly.

parted to the projectile and the quantity of matter in the projectile, Buridan then in passage 6 notes that the continuous impress of impetus in a falling body by gravity can account for its acceleration. This idea is taken up in greater length by Buridan in a question on the *De caelo*, which we have translated and discussed in the next chapter (see Doc. 9.1). The impetus theory also explains the fact that a broad jumper takes a long initial run before making his jump. Buridan incidentally adds that such a jumper never feels the air moving him along from behind but rather feels it resisting him. In this passage Buridan also presents his tentative hypothesis that the impetus theory could be used to account for the continuing motion of the heavenly bodies, thus eliminating the necessity of positing Intelligences as the movers. As we remarked in the body of the chapter, it is here that Buridan comes closest to producing an analogue with the later inertial idea.

While not discussing here the ontology of movement, Buridan in his first conclusion of this question opposes the identification of impetus and the motion of the projectile, and initially on the ground that, since impetus produces the continuing motion, something cannot produce itself. Buridan thus appears to accept the continuance of motion as a new effect for which a cause must be sought. And if the continuing motion is such an effect, it cannot be identical with its cause, namely, impetus. And so, while fundamentally in opposition to William Ockham in conceiving of the continuing motion as a new effect demanding a causal explanation, Buridan does not in this passage appear to be opposing specifically Ockham's view that the motion of a projectile is merely *secundum se*, for Buridan would appear to be arguing against those who would identify impetus with motion as an entity separate from the projectile.

In addition to rejecting the identification of impetus and motion, Buridan rejects the complementary view that impetus is a successive entity of the same kind as motion (see passage 8). Rather, impetus must be something permanent, distinct from the local motion of the projectile. However, it is corrupted and diminished in the same way as the motion, i.e., by resistance and contrary inclination.

Finally, according to Buridan (passage 10), the impetus theory gives a satisfactory explanation of the phenomena of rebound, the vibration of a bow string or the string of a musical instrument, and the pendular swing of a bell.

Chapter 9

The Free Fall of Bodies

HISTORICALLY speaking the two most important problems of dynamics were (1) the continuation of projectile motion and (2) the acceleration of freely falling bodies. In Chapter 8 we examined some of the results of the criticism of Aristotelian views concerning the first of these problems. We saw that the impetus mechanics as elaborated by John Buridan at Paris yielded the most satisfactory solution to that problem prior to the rise of modern mechanics.

Now the second problem can be resolved into two principal subsidiary questions: (*a*) What is the cause of acceleration? (*b*) How is the acceleration to be described kinematically? We have touched on the kinematic description of uniform acceleration in Chapter 5, but primarily from an abstract point of view without particular attention to the problem of freely falling bodies. In this chapter we are interested in seeing, then, what the schoolmen had to say about the problem of acceleration in the context of nature, i.e., as regards freely falling bodies. We shall see that the most significant answer given to subquestion (*a*) was given in terms of the impetus mechanics of Buridan, but at the same time that there was much confusion as to an explicit answer to (*b*). The questions (*a*) and (*b*) are of course by no means unconnected. For in the few passages that seem to imply the correct answer to (*b*), namely, that the velocity of fall increases directly as the time of fall, the answer follows from having considered the continuous introduction of successive *impetus* in the course of the *time* of fall. Thus, introduction of this impressed force is an act that must take place in time; and so the more time, the more impetus, and hence the greater velocity. On the other hand, when the medieval schoolmen did not think in terms of the time necessary for the introduction of impetus, but rather in terms exclusively of the measure of speed, they fell naturally into the error of thinking, the greater the distance of fall, the greater the speed. As we shall point out, many authors held both views simultaneously, apparently not realizing the contradiction involved. This was even true of Galileo as late as 1604 (see the commentary to Doc. 9.5). But let us address ourselves first to the answers given to (*a*) and (*b*) before the fourteenth century. We shall not go into the important larger question of the cause of the movement of the fall in general, a question which Miss Maier has carefully and penetratingly discussed.[1]

Aristotle was by no means clear in his views on either the cause or the measure of the acceleration of falling bodies. He appears to have believed that acceleration depends on the increasing proximity of the body to its natural place, i.e., the center of the world. This increasing proximity produces additional weight, which in accordance to his dynamical rules (see Chapter 7) would bring about the observed quickening. Yet, at the same time, in refuting infinite locomotion, Aristotle appears to imply two mutually incompatible ideas, namely that the speed of fall varies with the proximity to the center and that it also varies as the distance of fall:[2]

That locomotion cannot proceed to infinity is indicated by the fact that earth moves more quickly the nearer it is to the center, and fire the nearer it is to the upper region. If movement had been through an infinite distance, the speed would have been infinite too, and if the speed, then the weight and the lightness respectively. For as that which by virtue of speed is lower than another body would have owed its speed to its weight, so, if its increase of weight had been infinite, the

---

[1] A. Maier, *An der Grenze von Scholastik und Naturwissenschaft*, 2d ed. (Rome, 1952), pp. 413–82.

[2] *De caelo*, I.8.277a.27–277b.8 (edition of D. J. Allan [Oxford, 1936]). The English translation is that of T. L. Heath, *Mathematics in Aristotle* (Oxford, 1949), pp. 167–68. See translation from the Greek accompanying Aquinas' *In Aristotelis libros de caelo et mundo.... expositio* (Rome, 1952), p. 81, 119 (88): "Argumentum autem est eius quod non in infinitum ferri, et terram quidem, quanto utique propinquior fit medio, velocius ferri, ignem autem quanto utique ei quod est sursum. Si autem infinitum esset, infinita velocitas. Si autem velocitas, et gravitas et levitas..." p. 86, 122 (89): "neque velocius utique in fine ferretur, si vi et extrusione: omnia enim a vi inferente longius facta, tardius feruntur." I also include here the Latin translation of Michael Scot which accompanies the *Commentary* of Averroës and which was, of course, done from the Arabic (edition of Venice, 1574, ff. 75v–58r), textus commenti 88: "Et signum quod movens non movetur in infinitum, est terra. terra enim quantumcunque appropinquaverit medio, citius movetur: similiter ignis, quanto magis ad superius, tanto citius movetur. si igitur motus esset in infinitum, velocitas esset in infinitum. si igitur velocitas esset secundum hoc, gravitas, et levitas secundum hoc essent, scilicet in infiinitum.... (58r) Et etiam, si motus elementorum esset cum constrictione, non esset motus terrae, et ignis, cum appropinquaverint suis locis, velocior. omne enim quod a suo expellente removetur, tarde movetur, et dicemus quod ad locum, ex quo elementa moventur violente, moventur naturaliter."

[The Free Fall of Bodies

increase of its speed would also have been infinite .... Nor would their movement have been quicker towards the end if it had been due to compulsion or "squeezing-out" (i.e., relative displacement); for everything moves more slowly as it leaves the source of compulsion farther behind, and, if compulsion drives it from any place, it tends, if there is no compulsion, to that place.

Simplicius, the well-known Neo-Platonic commentator on Aristotle of the sixth century A.D., interprets Aristotle in this manner:[3] "Aristotle holds that as bodies approach the whole mass of their own element, they acquire a greater force therefrom and recover their form more perfectly; that thus it is by reason of an increase of weight that earth moves more swiftly when it is near the center."

In addition to this theory assigning greater perfection of form as the cause of increased weight and thus of acceleration, Simplicius immediately afterwards describes two other oft-repeated theories:

Hipparchus, on the other hand, in his work entitled *On Bodies Carried Down by Their Weight* declares that in the case of earth thrown upward it is the projecting force that is the cause of the upward motion, so long as the projecting force overpowers the downward tendency of the projectile, and that to the extent that this projecting force predominates, the object moves more swiftly upwards; then, as this force is diminished (1) the upward motion proceeds but no longer at the same rate, (2) the body moves downward under the influence of its own internal impulse, even though the original projecting force lingers in some measure, and (3) as this force continues to diminish the object moves downward more swiftly, and most swiftly when this force is entirely lost.

Now Hipparchus asserts that the same cause operates in the case of bodies let fall from above. For, he says, the force which held them back remains with them up to a certain point, and this is the restraining factor which accounts for the slower movement at the start of the fall. Alexander replies: "This may be true in the case of bodies moved by force or kept by force in the place opposite their natural place, but the argument no longer applies to bodies which on coming into being move in accordance with their own nature to their proper place."

On the subject of weight, too, Hipparchus contradicts Aristotle, for he [Hipparchus] holds that bodies are heavier the further removed they are from their natural places. This, too, fails to convince Alexander. "For," writes Alexander, "it is far more reasonable to suppose that when bodies change their natures, as when light bodies become heavy, they still retain something of their former nature

---

[3] Simplicius, *In Aristotelis de caelo libros commentaria*, edition of J. L. Heiberg in *Commentaria in Aristotelem graeca*, Vol. 7 (Berlin, 1894), p. 264, line 22–p. 267, line 6. The English translation of this and the immediately following quotation is by Drabkin, M. R. Cohen and I. E. Drabkin, *A Source Book in Greek Science* (New York, 1948), pp. 209-11.

when they are still at the very beginning of their downward fall and are just changing to that form by virtue of which they are carried downward, and that they become heavier as they go along, than to suppose that they still keep the force imparted to them by that which originally kept them up and prevented them from moving downward. Furthermore, if it is in the nature of the heavy to be below (for this is why its natural motion is toward that place), objects would be heaviest and would have assumed their proper form in this regard whenever they were below; and, since they have their perfection in downward movement, it would be reasonable to suppose that they receive an additional weight the nearer they come to that place. For if these bodies move downward more swiftly in proportion to their distance from above, it would be unreasonable to suppose that they exhibit this property in proportion as they are less heavy. For to hold such a view is to deny that these bodies move downward because of weight....

Now there are not a few who assert that bodies move downward more swiftly as they draw nearer their goal because objects higher up are supported by a greater quantity of air, objects lower down by a lesser quantity, and that heavier objects fall more swiftly because they divide the underlying air more easily. For just as in the case of bodies which sink in water the lighter they are the more does the water seem to hold them up and resist the downward motion, so it is fair to suppose that the same thing happens in air, and that the greater the amount of underlying air, the more do lighter objects seem buoyed up. Similarly, a greater amount of fire moves upward more swiftly since it divides the air above it more easily, and in proportion as the quantity of air above it is greater, that which moves upward through air moves more slowly. And though air may not be similar by nature to water, air does, since it is corporeal, impede the motion of objects passing through it. If this is the case, acceleration is due, not to addition of weight, but to diminution of the resistant medium....

"The reason given by Aristotle for acceleration in natural motion, namely, an addition of weight or lightness, is," says Alexander, "a sounder reason and more in accordance with nature. Aristotle would hold that acceleration is due to the fact that as the body approaches its natural place it attains its form in a purer degree, that is, if it is a heavy body, it becomes heavier, and if light, lighter."

Now in the first place I think it worthwhile to investigate how the acceleration of bodies as they approach their natural places (a fact said to be universally acknowledged) is to be explained. Again, if it is a case of addition of weight or lightness, it follows that a body weighed in air at the surface of the earth should appear heavier than if weighed in air from a high tower, or tree, or sheer precipice (the weigher stretching himself out over the edge). Now this seems impossible, unless, indeed, one were to say that in this case the difference in weight is imperceptible.

Some believe that Hipparchus' theory of a residual resistance referred to in the quotation above also implies a theory of impressed force as described in the preceding chapter. Whether this is true or not, the theory itself was influential. It was elaborated upon by the Arab author Abū 'l-Barakāt, and we shall return to it again shortly. It also attracted Galileo in his earlier writings.

The third theory of the cause of acceleration advanced by Simplicius is that bodies move more rapidly closer to the ground because there is less air resistance below. This opinion had some currency in later times, being one of the opinions rejected by Buridan (see Doc. 9.1 below). It is not surprising that Simplicius' views were influential on Buridan and other authors of the fourteenth century. For Simplicius' commentary on the *De caelo*, which includes the above-quoted passages, was rendered into Latin by William Moerbeke in 1271 and, probably, earlier by Robert Grosseteste.[4]

It is worth noting in connection with this quotation of Simplicius that he cites Alexander of Aphrodisias, a commentator on Aristotle in the third century A.D., to the effect that although acceleration is due to the fact that as a body approaches its natural place it attains its form in a purer degree, still bodies move down more swiftly in proportion to their distance of fall. In this theory Alexander was perhaps clarifying Aristotle, or following Strato, who, some six centuries earlier in his lost work *On Motion*, is reported by Simplicius to have asserted that as a body accelerates it completes the last part of its trajectory in the shortest time. We have already seen in Chapter 5 that Strato's discussion is equivalent to defining acceleration kinematically as "a movement such that equal spaces are traversed in succeeding periods of less time, e.g., to say, at continually greater speed." I think it will be useful at this point to quote the major part of the Simplicius passage in which Strato's views are given. Not only are we to remark the kinematic passage once more but also the experimental or empirical evidences given by Strato to confirm the very fact of acceleration:[5]

It may therefore not be out of place to set forth the indications [of acceleration] given by Strato the Physicist. For in his treatise *On Motion*, after asserting that a body so moving completes the last stage of its trajectory in the shortest time, he

---

[4] M. Grabmann, *Guglielmo di Moerbeke, O.P., Il traduttore delle opere di Aristotele* (Rome, 1946), pp. 129-31. See D. J. Allan, "Mediaeval Versions of Aristotle, *De caelo*, and of the Commentary of Simplicius," *Mediaeval and Renaissance Studies*, Vol. 2 (1950), 82-120, and particularly p. 104.

[5] Simplicius, *In Aristotelis physicorum libros...commentaria*, edition of H. Diels in *Commentaria in Aristotelem graeca*, Vol. 10 (Berlin, 1895), 916, lines 11-30. The English translation is that of Drabkin, *Source Book*, pp. 211-12.

adds: "In the case of bodies moving through the air under the influence of their weight this is clearly what happens. For if one observes water pouring down from a roof and falling from a considerable height, the flow at the top is seen to be continuous, but the water at the bottom falls to the ground in discontinuous parts. This would never happen unless the water traversed each successive space more swiftly." By "this" Strato means the breaking up of the object as it approaches the ground.

Strato also adduces another argument, as follows: "If one drops a stone or any other weight from a height of about an inch, the impact made on the ground will not be perceptible, but if one drops the object from a height of a hundred feet or more, the impact on the ground will be a powerful one. Now there is no other cause for this powerful impact. For the weight of the object is not greater, the object itself has not become greater, it does not strike a greater space of ground, nor is it impelled by a greater [external force]. It is merely a case of acceleration. And it is because of this acceleration that this phenomenon and many others take place."

The proof given, I think, indicates that an object raised but slightly from the earth is slow to move, for it is still, as it were, on the earth, but when an object moves toward its natural place from a distance considerably above the earth its power always keeps increasing as it approaches.

Now Strato does not appear to have been alone in attempting to describe acceleration by using empirical or experimental evidence. A contemporary of Simplicius, John Philoponus, in his commentary on Aristotle's *Physics* includes in his criticism of Aristotle's "law" of dynamics (see Chapter 7) a famous passage which appears to indicate that he had dropped bodies of different weight:[6]

But this is completely erroneous, and our view may be corroborated by actual observation more effectively than by any sort of verbal argument. *For if you let fall from the same height two weights of which one is many times as heavy as the other, you will see that the ratio of the times required for the motion does not depend on the ratio of the weights, but that the difference in time is a very small one.** And so, if the difference in the weights is not considerable, that is, if one is, let us say, double the other, there will be no difference, or else an imperceptible difference, in time, though the difference in weight is by no means negligible, with one body weighing twice as much as the other.

It is obvious, then, that neither Stevin nor Galileo was the first to per-

---

* The italics are those of the translator.
[6] John Philoponus, *In Aristotelis physicorum libros*, edition of H. Vitelli in *Commentaria in Aristotelem graeca*, Vol. *17* (Berlin, 1888), 683, lines 18-25. The English translation is that of Drabkin, *Source Book*, p. 220.

form such an experiment; nor in all likelihood was Philoponus. But Philoponus does give us the first record of such an experiment used to refute or confirm a dynamic law. It has sometimes been thought that Philoponus is using two different weights of the same material, and that he is thus proving that velocity of fall increases not with absolute weight, but with specific gravity, but no such specification is made by Philoponus.

We have already suggested in Chapter 8 the importance of the influence of Philoponus on Arabic authors, and we outlined there the principal contributions to the theory of *mail* and impressed force by Avicenna and Abū 'l-Barakāt. Furthermore, we indicated that the latter held a view that opposing inclinations could be resident in a projectile at the same time. Thus when a body begins to fall, there is still some violent *mail* left in it from its original trajection to a position not natural to it. This resident violent *mail* opposes the natural *mail* which operates to move the body downward; and although the violent *mail* is continually weakened, since it is not being replenished, still it serves, particularly at the beginning, to slow down the free fall of the body. But there is still another *continuing cause* of acceleration. So long as a body is away from its natural place, its gravity continues to impress natural *mail* in it. So that during the whole duration of the movement more and more natural *mail* is being introduced in the falling body, and thus as the *mail* increases, the speed increases. For a description of this second cause of acceleration let us turn to to Abū 'l-Barakāt himself.[7]

The productive cause of the violent *mail*, being resident in that which effects the violent (projectile) movement, is separated from the moving body and so does not produce in it successive inclinations (literally *mail* after *mail*) to replace the portion of the *mail* weakened by resistance. On the contrary, the source of the natural *mail* is found in the stone, [and thus] it supplies it with one *mail* after another.... The source of the natural *mail* is not separated [from the moving body], and so it continues to act until it causes the body to arrive at its natural place. *So long as the force (i.e., gravity) is acting outside of the natural place of the body, it produces successive inclinations in such a way that the force of the mail increases throughout the duration of the movement*....*

Now since increased force means increased speed, acceleration must result. At least one other Islamic author, Fakhr al-Dīn al-Rāzī (d. 1210), made this successive introduction of natural inclinations the exclusive

---

* The italics are mine.

[7] S. Pines, "Études sur Awhad al-Zamân Abu 'l-Barakât al-Baghdâdî," *Revue des études juives*, New Series, Vol. *4* (1938), 10–11, gives the appropriate Arabic passage with a French translation.

cause of acceleration.[8] As we shall see shortly, this is very much like the theory of John Buridan, produced apparently independently at Paris in the first half of the fourteenth century.

The significance of this theory is that it seems to suggest the germ of the basic idea of Newtonian mechanics, namely that a continually applied force produces acceleration. But, as we said in the previous chapter, it is still much confused, because this theory of Abū 'l-Barakāt (and also later that of Buridan) preserves the Aristotelian idea that all motion is a process arising from the joint action of motive force and resistance; and so the *mail* or *impetus* that is being continually introduced by gravity is not actually momentum or a quantity of a movement but rather is itself a kind of force or cause of movement. Thus the step that must be taken to transform the *mail* or *impetus* mechanics into inertial mechanics is to conceive of the *impetus*, not as a force or "cause" productive of movement, but as an "effect" which we can call momentum or quantity of movement.

As we pass to the Latin West, we find early references to free fall that occur in several theorems of Book IV of the thirteenth-century treatise on statics attributed to Jordanus, the *Liber de ratione ponderis*. In Chapter 5 (see p. 260) we quoted the most interesting of these theorems (R4.16) and came to the conclusion that for Jordanus acceleration is defined as a movement such that "in equal periods of time greater and greater space is traversed, i.e., in equal periods of time the speed is greater." And we also noticed that this is the acceleration that falling bodies possess. Thus in this theorem of Jordanus we find the root of the correct description of the acceleration of falling bodies, namely, that in such an acceleration the speed is directly proportional to the time of fall. There is of course no mention in this theorem of Jordanus of a *uniform* acceleration described in infinitesimal terms, nor is there made explicit the measure of the distance of fall in terms of the time of fall.

Another theorem of the *De ratione ponderis* (R4.06) also relates the increase of speed with the time of fall, but more ambiguously so. This theorem, furthermore, gives a curious view of the cause of acceleration, saying that as the body descends it sets up movement in the resisting air in such a way that it continually becomes less resistant, thereby permitting the gravity to be continually more effective and the body to accelerate.[9]

---

[8] *Ibid.*, p. 14.
[9] E. A. Moody and M. Clagett, *The Medieval Science of Weights* (Madison, 1952), pp. 216–17: "Res gravis, quo amplius descendit, eo fit descendendo velocior. In aere quidem verum magis, in aqua minus;

R4.06 The longer a heavy body falls, the faster it becomes in descending.

In air, indeed, more, and in water less. For air is related to all movements. When therefore a heavy body is falling, in its first movement it will draw along those parts of the medium which are behind it, and it will move the parts just beneath it; and these, set in motion, move the parts next to them, so that these in turn, being set in motion, offer less resistance to the gravity of the falling body. Hence it becomes heavier, and pushes the receding parts of the medium still more, so that they presently cease to be pushed, and even pull. And so it comes about that the gravity of the falling body is aided by the traction of those parts of the medium, and their movement in turn is aided by the body's gravity. Hence its velocity also is observed to be continuously multiplied.

In still another theorem of the *De ratione ponderis* (R4.08) we are told that "If . . . . it (a body) falls with its own natural motion, the more it is moved *(plus movetur,* longer in time? or farther in space?) the faster its motion becomes, and therefore so much the heavier; and it gives a greater impulsion, when in motion, than without being in motion, and the more it is moved the greater is the impulsion."[10] The "impulsion" of this passage is perhaps in some respects like the "impetus" later employed by John Buridan. Thus Jordanus seems to be saying in this passage: $I \propto V$, and either $V \propto t$ or $V \propto S$, where I is the impulsion, V is the speed, t is the time of fall, and S is the distance of fall. The reader is urged to compare this passage with Buridan's exposition of falling bodies (Doc. 9.1, passage 6).

Numerous authors of the thirteenth century discussed the variations of force or resistance that would be necessary to cause acceleration, as Duhem and Maier have shown.[11] With Roger Bacon we begin at least to have the idea that two forces are involved. One is the perpetual force due to natural weight, while the other is a force that becomes effective as we

habet se enim aer ad omnes motus. Res igitur gravis descendens, primo motu trahit posteriora et movet proxima inferiora; et ipsa mota, movent sequentia, ita ut illa mota gravitatem descendentis impediant minus; unde gravius efficitur, et cedentia amplius impellit, ita ut iam non impellantur sed etiam trahant. Sicque fit ut illius gravitas tractu illorum adiuvetur, et motus eorum gravitate ipsius augeatur; unde et velocitatem illius continue multiplicari constat."

[10] *Ibid.*: "Omne motum plus movet. Si quidem ex impulsu moveatur, certum est quod impellere habet. Si autem motu proprie descendat, quo plus movetur, velocius fit, et eo ponderosius; atque plus impellit motum quam sine motu, et quo plus movetur, eo amplius."

[11] P. Duhem has treated the problem in his *Études sur Léonard de Vinci,* Vol. *3* (Paris, 1913). A Maier takes up the various views of the cause and description of the acceleration of falling bodies in her *An der Grenze von Scholastik und Naturwissenschaft,* 2d ed. (Rome, 1952), pp. 183–218.

approach the natural place of the heavy body.[12] And with Aegidius Romanus we find emphasis put on the idea that the speed of fall increases as the distance from the beginning of the fall.[13] Walter Burley in the first half of the fourteenth century also believes the measure of the speed to be the distance of fall from its origin rather than proximity to natural place. He notes that some explain the acceleration by the continual acquisition of "accidental gravity."[14] While he does not appear to have meant "impetus" by his term "accidental gravity" but rather the older Aristotelian

---

[12] Duhem, *Études*, Vol. *3*, 74.

[13] Maier, *An der Grenze*, pp. 195–96.

[14] Walter of Burley, *Super Aristotelis libros de physica auscultatione...commentaria* (Venice, 1589), commenting on text 66 of Book VIII, cc. 1099–1100: "Unde numero quod motus naturalis incipit a quiete violenta, et motus violentus incipit a quiete naturali. Ergo motus naturalis quanto plus distat a quiete a qua incipit motus est velocior ex distantia eius a quiete a qua incipit moveri; sed in motu violento est econverso. Ista igitur propositio, scilicet, omnia quanto plus distant a quiete feruntur velocius; detur (?) intelligi de motu naturali.... Intelligendum quod illud quod communiter dicitur, scilicet, quod motus naturalis rectus intenditur in fine propter approximationem ad terminum motus, non est verum, quoniam grave non moveret velocius propter solam approximationem ad centrum; quia accipiantur duo eque gravia cum omnibus aliis paribus, ut scilicet quod sint eiusdem figurae, eiusdem magnitudinis, et sic de de aliis promoventibus ad velocitatem motus, si sint *a* et *b*, et ponatur *a* superius in aere in loco distante a terra per decem stadia, et sit locus *c*, et *b* ponatur superius in aere in loco distante a terra per unum stadium, et sit ille locus *d*, et moveatur *a* deorsum, et cum venerit *a* ad locum distantem a terra per spatium unius stadii, incipiat *b* descendere; et sit *e* instans in quo utrumque illorum, scilicet *a* et *b*, distant solum per spatium unius stadii, manifestum est quod post *e* instans *a* velocius descendet quam *b*, et tamen in *e* (*corr. ex* c) instanti sunt aeque approximata terre. Ergo maior approximatio (*corr. ex* approximato) ad locum naturalem non est causa velocitatis (! velocitationis?) motus naturalis, sed maior distantia a quiete violenta in termino a quo est causa velocitatis (! velocitationis?) motus naturalis.... (c. 1100) dicunt quidam quod grave in descendendo continue (*corr. ex* continentiae) acquirit novam gravitatem accidentalem; et quia fit continue gravius et gravius, ideo continue velocitatur motus eius.... Videtur tamen mihi quod, quia aer est cum gravibus gravis et cum levibus levis, ideo, cum grave descendit, semper maior et maior pars aeris praecedens ipsum movetur deorsum et etiam continue maior et maior pars aeris insequitur ipsum; ideo motus eius velocitatur, quia continue medium gravius et gravius praecedit ipsum cedens ei et medium continue gravius et gravius subsequitur ipsum continue fortius pellens ipsum; et ideo quanto a remotiori movetur, tanto velocius movetur, quia continue iuvatur motus eius magis et magis a medio tam a parte anteriori quam a parte posteriori." Punctuation altered. See the additional passage quoted by Maier, *An der Grenze*, p. 198, linking the theory of accidental gravity to the Commentator (Averroës). It is of interest to note that Burley in commenting on the first text of Book VII makes a reference to the *virtus derelicta* theory of projectile motion which we have treated in Chapter 8 (edition of Venice, 1589, c. 826, part B), "Ideo aliqui credunt quod proiectum primo proiiciente cessante movetur ab aliqua virtute derelicta in eo a primo proiiciente...."

idea of increasing weight, still it is of interest that Buridan and Albert of Saxony both identify this "accidental gravity" with "impetus," as is evident in Documents 9.1 (passage 7) and 9.2 (passage 2).

The mysterious author of the tract known as *Tractatus de sex inconvenientibus* reviews the principal theories of the cause of acceleration and concludes that "the acceleration *(velocitatio)* of a heavy body in its descent downward is the result of several causes, although one is more fundamental than the others.... I say that decrease of resistance is the principal cause."[15]

We are now in a position to consider Buridan's contribution to the acceleration problem. We are immediately struck by the similarity of Buridan's views with those of his Arabic predecessors, although there is no evidence of his having had access to their accounts. I would suspect rather that Buridan either developed further the implications of the passages in the *De ratione ponderis* of Jordanus, or he altered and clarified the opinions of those who held the cause of acceleration to lie in the acquisition of accidental gravity.

Now Buridan in his *questio* on this matter (Doc. 9.1) is primarily interested in the cause of acceleration. He marshals and rejects (largely on empirical grounds) a number of theories as to the cause of acceleration, theories drawn largely from Averroës, and possibly Simplicius. After rejecting the earlier theories, he then presents what he believes to be the correct theory, namely, that acceleration is explained by the continual impression of impetus by gravity:

...it follows that one must imagine that a heavy body not only acquires motion unto itself from its principal mover, i.e., its gravity, but that it also acquires unto itself a certain impetus with that motion. This impetus has the power of moving the heavy body in conjunction with the permanent natural gravity. And because that impetus is acquired in common with the motion, hence the swifter the motion is, the greater and stronger the impetus is. So, therefore, from the beginning the heavy body is moved by its natural gravity only; hence it is moved slowly. Afterwards it is moved by that same gravity and by the impetus acquired at the same time; consequently, it is moved more swiftly. And because the movement becomes swifter, therefore the impetus also becomes greater and stronger, and

---

[15] *Tractatus de sex inconvenientibus*, edition of Venice, 1505, quest. IIII, primus articulus (f. G4 r, c.2): "Utrum velocitatio motus sit ab aliqua certa causa.... (f. G5 r, c.2) nam velocitatio gravis versus deorsum in suo descensu est a pluribus causis, licet una sit principalior aliis. Unde dico cum magistro Adam de pipeltelle quod minoritas resistentie est causa principalis; et continuatio motus, propinquitas [ad locum naturalem], pulsus medii, gravitas accidentalis, inclinatio naturalis, quia et appetitus, sunt cause partiales."

thus the heavy body is moved by its natural gravity and by that greater impetus simultaneously and so it will again be moved faster; and thus it will always and continually be accelerated to the end (see Doc. 9.1, passage 6).

Thus Buridan appears to say that gravity acting (throughout the time? or distance?) of fall imposes continuously growing impetus in the body, and, with it, continually growing velocity. The question of whether Buridan believed the speed proportional to the time or distance of fall, or both at the same time, we shall discuss in the commentary to Document 9.1. But in any case, we must be careful not to impose on Buridan's views greater quantitative precision than they actually have.

Of Buridan's successors the most interesting and important one is Nicole Oresme, who altered Buridan's view of the nature of impetus. In his Latin commentary on Aristotle's *De caelo*, Oresme held that impetus arises from an initial acceleration and then acts to accelerate further the speed. "...I say that this is the cause of the acceleration *(velocitatio)* of a heavy body in the end: Because it is accelerated in the beginning, it acquires such an impetus and this impetus is a coassistor for producing movement. Thus with other things equal, the movement is faster."[16] Sometime later in his French commentary on the *De caelo*, he says much the same sort of thing (see Doc. 9.3). The acquisition of impetus (called in French *impetuosité*) comes from acceleration. As Miss Maier has pointed out, the question of how the initial acceleration takes place is not clearly answered by Oresme. As to the nature of Oresme's impetus and its generation, he says,[17] "it is a certain quality of the second species...; it is generated by the motor by means of motion, just as it would be said of heat, when the motion is the cause of heat.... It is corrupted by the retardation of motion because for its conservation, speed *(velocitas)* or acceleration *(velocitatio)* is required." Thus Oresme's impetus differed from Buridan's in two major respects. (1) It was no longer simply a function of velocity but apparently of acceleration as well. (2) It was no longer considered to be of permanent nature. Hence, it is not surprising that Oresme does not explain the movement of the heavens in terms of impetus.

---

[16] Maier, *Zwei Grundprobleme der scholastischen Naturphilosophie*, 2d ed. (Rome, 1951), p. 246, gives the text of Oresme's latin *De caelo*, Bk. II, quest., 13, not available to me: "Istis positis dico quod ista est causa velocitationis gravis in fine, quia ex eo quod velocitatur in principio acquirit talem impetum et iste impetus coadiuvat ad movendum. Ergo ceteris paribus est motus velocior."

[17] *Ibid.*, p. 248: "...et potest dici quod est quaedam qualitas de secunda specie... generatur a motore mediante motu sicut diceretur de calore cum motus est causa caloris.... corrumpitur per retardationem motus, quia ad conservationem eius requiritur velocitas vel velocitatio."

[The Free Fall of Bodies 553]

It should be remarked that Oresme later applied his doctrine of the impetus produced from acceleration to projectile motion. In such motion the propellant introduces an initial acceleration into a body which causes an impetus *(qualité motive novelle* or *force ou rèdeur)* which in turn moves the body after there is no longer any contact with the projector.[18] The impetus then by its action further accelerates the movement of the projectile until it is sufficiently weakened by resistance, at which time deceleration occurs.

One of the most interesting parts of Oresme's exposition concerning falling bodies is that where he suggests that if we pierced the earth so that a body could fall to its center, the body would acquire an *impetuosité* which would carry it beyond the center of the earth; and so, rather than coming to rest immediately at the center of the earth, it would oscillate about the center of earth at gradually decreasing distances until it finally came to rest (see Doc. 9.3).[19] This was a view also expressed by Oresme's contemporary, Albert of Saxony (see Doc. 9.2).

Perhaps even more interesting than Oresme's account of the impetus theory is his apparent acceptance of the correct description of the acceleration of falling bodies in his Latin commentary on the *De caelo* of Aristotle:[20]

---

[18] *Ibid.*, p. 253.

[19] Cf. Oresme's *Questiones de spera*, MS Florence, Bibl. Riccard. 117, f. 127r: "Utrum grave partiale, sicut est lapis, si non esset impeditum, moveretur ad centrum et ibi quiesceret naturaliter...(127v) conceditur quod si grave sicut lapis descendat ad centrum et non haberet impedimentum adhuc iret ultra centrum et iret et rediret faciendo multas reflexiones quousque tandem quiesceret, sicut directe quando pilla dimittitur cadere super terram et adhuc revertitur, quamvis esset in loco in quo potuisset quiescere et sicut etiam in equilibra quando unum brachium elevatur et dimittitur, iret vel totiens replectitur. Ad probationem conceditur quod quando aliquid est in loco suo naturali non debet ab illo moveri nisi sit aliquid violentans ipsum. Et si queratur quid est illud, potest dici quod tale movetur ab aere insequente propter vehementem impulsum, sicut etiam est in motu proiectorum. Aliter potest dici quod hoc est propter quandam virtutem acquisitam in motu naturali, que potest vocari impetus, et illa virtus est causa quare potest moveri velocius in fine quam in principio; et est que causa quare grave proiectum movetur post recessum a manu, quia ibi inducitur quedam virtus a proieciente."

[20] Maier, *An der Grenze*, p. 214. I again quote the Latin passage from Oresme's *De caelo*, Bk. II, quest. 7: "...dupliciter potest intendi aliquod continue velocitari. Uno modo sic, quod fiat additio velocitatis per partes aequales, vel aequivalenter, v.gr. quod in ista hora movetur aliqua velocitate et in secunda duplo velocius et in tertia triplo etc., et eodem modo de partibus proportionalibus horae. Et isto modo ex hoc sequeretur infinita velocitas, si procederetur in infinitum, quia omnis velocitas data excederetur per hunc modum. 2º modo potest imaginari additio velocitatis non per partes aequales, sed per continue proportionales et minores, vel aequivalenter, ut si modo velocitas esset unius gradus, deinde unius gradus cum dimidio, et postea unius gradus cum dimidio et quarta parte eius. Et isto modo nunquam excederetur velocitas dupla,

That something is "continually accelerated" can be understood in two ways. In one way thus: An addition of velocity takes place by equal parts, or equivalently. For example, in this hour it is moved with some velocity, and in the second twice as quickly, and in the third three times as quickly, etc. In the same way [such an addition can be made] in proportional parts of the time. And in this way infinite velocity would accordingly follow if one proceeded to infinity, because any given velocity would be exceeded in this way [of increase]. In the second way, addition of velocity can be imagined not by equal parts but by continually proportional and smaller parts, so that if now the velocity were of one degree, next it [would be] $1 + 1/2$ degrees, and then $1 + 1/2 + 1/4$ degrees, etc. By this way a double velocity (i.e., one of 2 degrees) would never be exceeded, even though one proceeded to infinity. Now as for the question at hand [of the acceleration of falling bodies], velocity in the motion of a heavy body increases in the first way and not the second.

It should be observed that this description is still somewhat ambiguous. Oresme tells us that velocity can continually increase divergently or convergently. In explaining divergent increase, he notes two examples. The first is when there is an arithmetical increase of velocity in equal periods of time. The second is when there is an arithmetical increase of velocity in continuously proportional parts of time, e.g., $t/2$, $t/4$, $t/8$, etc. The only example he gives of the second or convergent increase of velocity is that of a motion where velocity increases convergently in equal periods of time, i.e., V, $V + V/2$, $V + V/2 + V/4$, etc. Presumably, he would also assume that there could be a convergent increase of velocity in proportional parts of the time, although he does not give this case. Now it would appear that Oresme believed that, in the natural motion of a heavy body, the first method of arithmetical increase in equal periods of time was the proper description, although he might mean that the velocity increase was arithmetical with some other than equal divisions of the time period. It should be clear that Oresme has not, any more than Jordanus, considered in this passage the infinitesimal aspects of the description of uniform acceleration evident in the Merton treatment of "motion uniformly difform," although, of course, Oresme was quite familiar with that kinematic activity, as we have shown in Chapter 6 above.

Now Albert of Saxony, Oresme's contemporary at Paris, in addition to having been influenced by the impetus discussions of Buridan, and probably of Oresme, was clearly influenced by Oresme's discussion of the meas-

---

quamvis sic procederetur in infinitum. Modo ad propositum: velocitas in motu gravium fit primo modo et non secundo modo."

[The Free Fall of Bodies

ure of the acceleration of falling bodies. We have presented Albert's discussion as Document 9.2. After initially rejecting a number of possibilities, Albert incorrectly settles for the view so popular since the time of Strato, Alexander of Aphrodisias, and others, namely, that the speed increases in direct proportion to the distance of fall, and this increase is arithmetic rather than geometric. In this discussion he does not appear to discuss the possibility that speed increases arithmetically as the time of fall, as had Oresme. But elsewhere, like Buridan, he seems to imply that such is the case (see the commentary to Doc. 9.2). If this is so, he would appear not to have realized that this was contradictory to his belief that the velocity grows directly as the distance of fall. But Buridan and Albert are not alone in their confusion. For Leonardo da Vinci similarly confuses and identifies time of fall and distance of fall. In fact Leonardo gives us both views in one statement (see Doc. 9.4, passage 2). Furthermore, Galileo was just as confused in 1604 when he stated that the "proposition" $S \propto t^2$ could be derived from the "principle" $V \propto S$ (see commentary to Doc. 9.5).

So far as I know, the first statement of free fall with the infinitesimal implications of the Merton discussions explicitly applied is found in the *Questiones super octo libros Physicorum Aristotelis* published by the Spaniard Domingo de Soto in 1555, where he says:[21]

Movement uniformly nonuniform *(uniformiter difformis)* as to time is nonuniform in such a manner that if it is divided according to time (i.e., according to before and after), the middle point of any part at all exceeds [in velocity] the least velocity *(remissum extremum)* of that part by the same proportion that the mean is exceeded by the greatest velocity *(intensissimum)* [of that part]. This species of movement

[21] Edition of Salamanca, 1555, f. 92v, c.2 (edition of Venice, 1582, p. 339): "Motus uniformiter difformis quo ad tempus est motus ita difformis, ut si dividatur secundum tempus (scilicet secundum prius et posterius) cuiusque partis punctum medium illa proportione excedit remississimum extremum illius partis, qua exceditur ab intensissimo. Hoc motus species proprie accidit naturaliter motis et proiectis. Ubi enim moles ab alto cadit per medium uniforme, velocius movetur in fine, quam in principio. Proiectorum vero motus, remissior in fine, quam in principio; atque adeo primus uniformiter difformiter (*sic* but should be deleted) intenditur; secundum vero uniformiter difformiter (! but delete?) remittitur.... (93v, c.2; p. 343) Respondetur nihilominus, velocitatem motus uniformiter difformem quo ad tempus est aestimari penes gradum medium et ab eo denominari." Note that Domingo first gives natural motion, i.e., fall of bodies, as an example of motion uniformly difform, but then at the end he inadvertently errs by saying that it is "increased uniformly difformly." That this is merely a slip is evident from still another place where he defines natural motion as always increasing in swiftness (see Chapter 5 above, note 4). For other Merton extracts from Domingo's work, see Chapter 11 below.

belongs properly to things which are moved naturally and to projectiles. For when a body *(moles)* falls through a uniform medium, it is moved more quickly in the end than in the beginning. On the other hand, the movement of projectiles [upward] is less quick *(remissior)* in the end than in the beginning. And so the first is uniformly increased, while the second is uniformly decreased.

Soto then goes on to say, just as the Merton College kinematicists had said, that a movement uniformly accelerated is measured or denominated *(denominatur)* with respect to the space traversed in a given time by its mean velocity. The example (a thought experiment only) which Soto gives is of a body falling through an hour which accelerates uniformly from zero degree of velocity to a velocity of eight. Then it would traverse, he says, just as much space as another body moving uniformly through the hour with a speed of four. Soto, then, has applied the mean law of Merton College to falling bodies.

We have shown in Chapter 11 something of the widespread repetition in the later Middle Ages and early modern period of the impetus theory to explain both projectile and natural motion. At this point, I should like merely to contrast with the medieval discussions the treatment of falling bodies given by Galileo in his *Two New Sciences* of 1638. I have presented in Document 9.5 some excerpts to illustrate Galileo's treatment of the question. Notice his impatience with the *causal* question. He is interested only in the *kinematic* description of falling bodies. We already have given in Chapter 6 (Doc. 6.5) his proof of the mean speed theorem. Now he shows by means of the inclined plane experiment given below that a deduction from the mean speed theorem, namely $S \propto t^2$, applies to the fall of bodies. His procedure is this. Accept as a fundamental principle that uniform acceleration is defined as $V \propto t$. From this deduce that $S \propto t^2$. Reason that the case of free fall is like the case of balls rolling down inclined planes. Now show by rolling balls down an inclined plane that the relationship $S \propto t^2$ holds. Conclude, then, that it also holds for free fall.

Regardless of how well he performed his experiments and what data came out of those experiments, Galileo's treatment was certainly the starting point of modern investigations of the problem of the acceleration of falling bodies.

Document 9.1

John Buridan, *Questions on the Four Books on the Heavens and the World of Aristotle**

1. BOOK II, QUESTION 12. Whether natural motion ought to be swifter in the end than the beginning.... With respect to this question it ought to be said that it is a conclusion not to be doubted factually *(quia est)*, for, as it has been said, all people perceive that the motion of a heavy body downward is continually accelerated *(magis ac magis velocitatur)*, it having been posited that it falls through a uniform medium. For everybody perceives that by the amount that a stone descends over a greater distance and falls on a man, by that amount does it more seriously injure him.

2. But the great difficulty *(dubitatio)* in this question is why this [acceleration] is so. Concerning this matter there have been many different opinions. The Commentator (Averroës) in the second book [of his commentary on the *De caelo*] ventures some obscure statements on it, declaring that a heavy body approaching the end is moved more swiftly because of a great desire for the end and because of the heating action *(calefactionem)* of its motion. From these statements two opinions have sprouted.

3. The first opinion was that motion produces heat, as it is said in the second book of this [work, the *De caelo*], and, therefore, a heavy body descending swiftly through the air makes that air hot, and consequently it (the air) becomes rarefied. The air, thus rarefied, is more easily divisible and less resistant. Now, if the resistance is diminished, it is reasonable that the movement becomes swifter.

But this argument is insufficient. In the first place, because the air in the summer is noticeably hotter than in the winter, and yet the same stone falling an equal distance in the summer and in the winter is not moved with appreciably greater speed in the summer than in the winter; nor does it strike harder. Furthermore, the air does not become hot through move-

---

* Translated from the Latin edition of E. A. Moody, (Cambridge, Mass., 1942), pp. 176–81.

ment unless it is previously moved and divided. Therefore, since the air resists before there has been movement or division, the resistance is not diminished by its heating. Furthermore, a man moves his hand just as swiftly as a stone falls toward the beginning of its movement. This is apparent, because striking another person hurts him more than the falling stone, even if the stone is harder. And yet a man so moving his hand does not heat the air sensibly, since he would perceive that heating. Therefore, in the same way the stone, at least from the beginning of the case, does not thus sensibly heat the air to the extent that it ought to produce so manifest an acceleration *(velocitatio)* as is apparent at the end of the movement.

4. The other opinion which originated from the statements of the Commentator is this: Place is related to the thing placed as a final cause, as Aristotle implies and the Commentator explains in the fourth book of the *Physics*. And some say, in addition to this, that place is the cause moving the heavy body by a method of attraction, just as a magnet attracts iron. By whichever of these methods it takes place, it seems reasonable that the heavy body is moved more swiftly by the same amount that it is nearer to its natural place. This is because, if place is the moving cause, then it can move that body more strongly when the body is nearer to it, for an agent acts more strongly on something near to it than on something far away from it. And if place were nothing but the final cause which the heavy body seeks naturally and for the attainment of which the body is moved, then it seems reasonable that that natural appetite *(appetitus)* for that end is increased more from it as that end is nearer. And so it seems in every way reasonable that a heavy body is moved more swiftly by the amount that it is nearer to [its] downward place. But in descending continually it ought to be moved more and more swiftly.

But this opinion cannot stand up. In the first place, it is against Aristotle and against the Commentator in the first book of the *De caelo*, where they assert that, if there were several worlds, the earth of the other world would be moved to the middle of this world....

Furthermore, this opinion is against manifest experience, for you can lift the same stone near the earth just as easily as you can in a high place if that stone were there, for example, at the top of a tower. This would not be so if it had a stronger inclination toward the downward place when it was low than when it was high. It is responded that actually there is a greater inclination when the stone is low than when it is high, but it is not great enough for the senses to perceive. This response is not valid, because if that stone falls continually from the top of the tower to the earth, a double or triple velocity and a double or triple injury would be sensed near the earth than would be sensed higher up near the beginning of the movement. Hence, there is a double or triple cause of the velocity. And so it follows that that inclination which you posit not to be sensible or notable is not the cause of such an increase of velocity.

Again, let a stone begin to fall from a high place to the earth and another similar stone begin to fall from a low place to the earth. Then these stones, when they should be at a distance of one foot from the earth, ought to be moved equally fast and one ought not be swifter than the other if the greater velocity should arise only from nearness to [their] natural place, because they should be equally near to [their] natural place. Yet it is manifest to the senses that the body which should fall from the high point would be moved much more quickly than that which should fall from the low point, and it would kill a man while the other stone [falling from the low point] would not hurt him.

Again, if a stone falls from an exceedingly high place through a space of ten feet and then encountering there an obstacle comes to rest, and if a similar stone descends from a low point to the earth, also through a distance of ten feet, neither of these movements will appear to be any swifter than the other, even though one is nearer to the natural place of earth than the other.

I conclude, therefore, that the accelerated natural movements of heavy and light bodies do not arise from greater proximity to [their] natural place, but from something else that is either near or far, but which is varied by reason of the length of the motion *(ratione longitudinis motus)*. Nor is the case of the magnet and the iron similar, because if the iron is nearer to the magnet, it immediately will begin to be moved more swiftly than if it were farther away. But such is not the case with a heavy body in relation to its natural place.

5. The third opinion was that the more the heavy body descends, by so much less is there air beneath it, and the less air then can resist less. And if the resistance is decreased and the moving gravity remains the same, it follows that the heavy body ought to be moved more swiftly.

But this opinion falls into the same inconsistency as the preceding one, because, as was said before, if two bodies similar throughout begin to fall, one from an exceedingly high place and the other from a low place such as a distance of ten feet from the earth, those bodies in the beginning of their motion are moved equally fast, notwithstanding the fact that one of them has a great deal of air beneath it and the other has only a little.

Hence, throughout, the greater velocity does not arise from greater proximity to the earth or because the body has less air beneath it, but from the fact that that moving body is moved from a longer distance and through a longer space.

Again, it is not true that the less air in the aforementioned case resists less. This is because, when a stone is near the earth, there is still just as much air laterally as if it were farther from the earth. Hence, it is just as difficult for the divided air to give way and flee laterally [near the earth] as it was when the stone was farther from the earth. And, in addition, it is equally difficult or more difficult, when the stone is nearer the earth, for the air underneath to give way in a straight line, because the earth, which is more resistant than the air, is in the way. Hence, the imagined solution *(imaginatio)* is not valid.

6. With the [foregoing] methods of solving this question set aside, there remains, it seems to me, one necessary solution *(imaginatio)*. It is my supposition that the natural gravity of this stone remains always the same and similar before the movement, after the movement, and during the movement. Hence the stone is found to be equally heavy after the movement as it was before it. I suppose also that the resistance which arises from the medium remains the same or is similar, since, as I have said, it does not appear to me that the air lower and near to the earth should be less resistant than the superior air. Rather the superior air perhaps ought to be less resistant because it is more subtle. Third, I suppose that if the moving body is the same, the total mover is the same, and the resistance also is the same or similar, the movement will remain equally swift, since the proportion of mover to moving body and to the resistance will remain [the same]. Then I add that in the movement downward of the heavy body the movement does not remain equally fast but continually becomes swifter.

From these [suppositions] it is concluded that another moving force *(movens)* concurs in that movement beyond the natural gravity which was moving [the body] from the beginning and which remains always the same. Then finally I say that this other mover is not the place which attracts the heavy body as the magnet does the iron; nor is it some force *(virtus)* existing in the place and arising either from the heavens or from something else, because it would immediately follow that the same heavy body would begin to be moved more swiftly from a low place than from a high one, and we experience the contrary of this conclusion....

From these [reasons] it follows that one must imagine that a heavy body not only acquires motion unto itself from its principal mover, i.e., its gravity, but that it also acquires unto itself a certain impetus with that motion. This impetus has the power of moving the heavy body in conjunction with the permanent natural gravity. And because that impetus is acquired in common with motion, hence the swifter the motion is, the greater and stronger the impetus is. So, therefore, from the beginning the heavy body is moved by its natural gravity only; hence it is moved slowly. Afterwards it is moved by that same gravity and by the impetus acquired at the same time; consequently, it is moved more swiftly. And because the movement becomes swifter, therefore the impetus also becomes greater and stronger, and thus the heavy body is moved by its natural gravity and by that greater impetus simultaneously, and so it will again be moved faster; and thus it will always and continually be accelerated to the end. And just as the impetus is acquired in common with motion, so it is decreased or becomes deficient in common with the decrease and deficiency of the motion.

And you have an experiment [to support this position]: If you cause a large and very heavy smith's mill [i.e., a wheel] to rotate and you then cease to move it, it will still move a while longer by this impetus it has acquired. Nay, you cannot immediately bring it to rest, but on account of the resistance from the gravity of the mill, the impetus would be continually diminished until the mill would cease to move. And if the mill would last forever without some diminution or alteration of it, and there were no resistance corrupting the impetus, perhaps the mill would be moved perpetually by that impetus.

7. And thus one could imagine that it is unnecessary to posit intelligences as the movers of celestial bodies since the Holy Scriptures do not inform us that intelligences must be posited. For it could be said that when God created the celestial spheres, He began to move each of them as He wished, and they are still moved by the impetus which He gave to them because, there being no resistance, the impetus is neither corrupted nor diminished.

You should note that some people have called that impetus "accidental gravity" and they do so aptly, because names are for felicity of expression. Whence this [name] appears to be harmonious with Aristotle and the Commentator in the first [book] of this [work, the *De caelo*], where they say that gravity would be infinite if a heavy body were moved infinitely, because by the amount that it is moved more, by that same amount is it moved more swiftly; and by the amount that it is moved more swiftly, by that amount is the gravity greater. If this is true, therefore, it is necessary that a heavy body in moving acquires continually more gravity, and that gravity is not of the same constitution *(ratio)* or nature as the first natural gravity, because the first gravity remains always, even with the movement stopped, while the acquired gravity does not remain. All of these statements will appear more to be true and necessary when the violent movements of projectiles and other things are investigated....

COMMENTARY

This selection (passages 3–5) tells us of three common explanations of the cause of the acceleration of falling bodies in addition to the fourth, the impetus explanation which Buridan is supporting: (1) a heating of the medium which decreases its resistance and thus increases the velocity; (2) proximity to natural place which acts by some virtue or other (like that of the magnet) as a moving cause, this virtue being increased as the body comes closer (see the discussion of Simplicius and Aristotle in the introductory remarks of this chapter); (3) as the body falls there is continually less air beneath it acting as resistance; hence the velocity increases (see the discussion of Simplicius above);[22] (4) the impetus explanation, i.e., gravity continually introduces an impetus which acting as a supplementary increasing cause of movement, and acting with the gravity, produces the acceleration. Among the other explanations not mentioned by Buridan were two which centered around the supplementary action of the medium: (5) one which held that air stirred up by the movement is able to get behind the falling body and give it supplementary pushes (a theory taken over from Aristotle's explanation of the continuance of projectile motion, see Chapter 8);[23] and (6) the falling body not only draws the air behind, but in pushing the air beneath it, it sets it in motion, and this air sets other air in motion, and the drawing action of the air makes it less resistant and helps the gravity of the body (a theory found in the *Liber*

---

[22] Something of this sort is understood by the author of the *De sex inconvenientibus* (see footnote 15 and cf. Maier, *An der Grenze*, pp. 190–91). A somewhat different view was held by Durandus de St. Porciano, who held that "by the amount that the air is closer to the earth, by that amount is it less light and exerts itself less against the motion of the heavy body" ("motus naturalis sit intensior in fine quam in principio: causa est minor resistentia medii, supposita eadem inclinatione mobilis, quanto enim aer est terrae propinquior, tanto est minus levis et minus nititur contra motum gravis"). *Sent.*, Bk. II, dist. 14, quest. 1, quoted through Maier, *An der Grenze*, p. 190.

[23] It is this theory that is referred to by "pulsus medii" by the author of the *De sex inconvenientibus, edit. cit.* in footnote 15.

de ratione ponderis attributed to Jordanus and already quoted above in this chapter, note 9). One other opinion drawn from Simplicius (and supported later by Galileo in his early work *De motu*) held (7) that when a body starts to fall, it still has unnatural lightness which is a holdover from its previous violent movement, and this lightness slows up the body until it is dissipated by the action of the gravity.

Like Albert of Saxony later (see Doc. 9.2), Buridan seems to believe that the increasing velocity is a simple function of the increasing distance of fall (see the second paragraph of passage 5). But he does say in one place, you will notice, that the increasing velocity is dependent on the "length of movement" (passage 4, last paragraph), which is somewhat ambiguous but sounds as if he is talking of time. Furthermore, the adverb he uses in his exposition of the impetus theory (passage 6) and its application to falling bodies is temporal, e.g., he speaks of the velocity *continually (continue)* increasing. I believe actually he made no clear distinction between the mathematical difference involved in saying that the velocity increases directly as the distance of fall and saying that it increases directly as the time of fall. The reader should compare this discussion with the discussions of Nicole Oresme (note 20), Albert of Saxony (Doc. 9.2) and Galileo (Doc. 9.5).

The source of the continually increasing impetus acquired by the falling body clearly appears to be the continually acting gravity, and, furthermore, the increasing velocity caused by the gravity and continually impressed impetus is a measure of the increased impetus. Hence this whole passage seems to confirm his previous description of impetus as varying directly with the velocity it maintains.

The strongly empirical and observational character of Buridan's refutation of the "erroneous" opinions as well as of the exposition of his own theory should be noticed. For example (see passage 4), if the theory that held that the cause of acceleration lay in the proximity of the body to its natural place were tested by dropping a body from a high place and another body from a low place, according to the theory when these stones "should be at a distance of one foot from the earth, [they] ought to be moved equally fast and one ought not to be swifter than the other if the greater velocity should arise only from nearness to [their] natural place because they should be equally near to [their] natural place. Yet it is manifest to the senses that the body which should fall from the high point would be moved much more quickly than that which should fall from the low point, and it would kill a man while the other stone [falling from the low point] would not hurt him." [24] Notice then that, like Strato, Buridan takes as his practical measure of velocity of fall the force of impact, as also did Leonardo and Galileo later.

[24] Compare the argument given by Strato in the passage quoted earlier in this chapter (p. 546) and also that of Burley in footnote 14.

Document 9.2

Albert of Saxony, *Questions on the [Four] Books on the Heavens and the World of Aristotle*

1. BOOK II, QUESTION 4. Whether every natural movement is swifter in the end than in the beginning.... Fourth distinction: Motion can be understood to be increased intensively [i.e., its speed increased] in two ways: [1] in one way [divergently] by doubling, tripling, quadrupling, and so on; and [2] in another way [convergently] so that in the first place *(primo)* there is a certain speed, and then in the second place *(secundo)* there is added to it some degree of velocity, then in the third place there is added half of that degree of velocity, in the fourth place one-fourth that degree, and so on.

Then let this be the first conclusion: If the increase of speed took place in the first way [i.e., divergently] then it would become infinite....

Second conclusion: If speed were increased in the second way [i.e., convergently], it would not be necessary for it to become infinite. Rather it could in a given case be increased indefinitely *(in infinitum)* and yet never thereby become triple itself [since it could, for example, converge toward some speed less than triple or even double itself, say to some speed three-halves itself]....

Third conclusion: Natural motion in the end, i.e. toward the end rather than immediately at the end—when it begins and rushes toward the end—is increased in the second (? should be "first") way. This is obvious, for otherwise a heavy body an infinite distance away would not be moved with an infinite velocity before it would arrive at the center [of the world]. Aristotle says the opposite of this.

But it should be known that natural motion does not accelerate by double, triple, and so on in such a way that in the first proportional part of the hour it is a certain speed and in the second proportional part of the hour twice as fast, and so on. Nor also does it accelerate in such a way that in the first proportional part of the space traversed, for example, the first half of that space, it would be a certain velocity, and after the second proportional part of the space has been traversed it then would be a velocity twice as fast, and so on. For then it would follow that any natural motion at all, which would last through any time as small as you like or traverse any space as small as you wish, would attain before the end any degree at all of velocity. Now this is false.... Therefore in the third conclusion it is understood that the speed is increased by double, triple, etc. in such a fashion that when some space has been traversed by this [motion], it has a certain velocity, and when a double space has been traversed by it, it is twice as fast, and when a triple space has been traversed by it, it is three times as fast, and so on....

2. (32v) ... There is another opinion concerning the cause of the acceleration of a natural movement in its end, and I approve of this one. According to this opinion, it is to be imagined that the heavy body in its motion acquires unto itself beyond its natural gravity a certain impetus or accidental gravity, which helps the natural gravity to move the heavy body more quickly.... and according as that natural body is moved a longer and longer time *(movetur diutius et diutius)*, so accordingly more and more impetus is acquired unto itself, and so thereby is it continually moved more and more swiftly, unless, however, it would be impeded by an increase of its resistance greater than the impetus thus acquired.... For a heavy body in descent acquires unto itself such an impetus which is not immediately corrupted, and hence, when the body hits an obstacle, that impetus which is not yet corrupted inclines the body to motion, and not being able to move the body downward any more, it conversely moves it upward. It continues to do so until that impetus is corrupted. According to this [theory], it would be said also that if the earth were completely perforated, and through that hole a heavy body were descending quite rapidly toward the center, then when the center of gravity *(medium gravitatis)* of the descending body was at the center of the world, that body would be moved on still further [beyond the center] in the other direction, i.e., toward the heavens, because of the impetus in it not yet corrupted. And, in so ascending, when the impetus would be spent, it would conversely descend. And in such a descent it would again acquire unto itself a certain small impetus by which it would be moved again beyond the center. When this impetus was spent, it would descend again. And so it would be moved, oscillating *(titubando)* about the center until there no longer would be any such impetus in it, and then it would come to rest.

[Albert of Saxony, *On the Heavens and the World*

COMMENTARY

Here Albert takes up the question of what kind of an increase of speed takes place in the acceleration of natural bodies. Its dependence on the treatment of Oresme (see note 20) is obvious. The first point he is trying to make in the passage I have translated is the same as that of Oresme and is that there are two possibilities: Either velocity is increasing divergently as the series of cardinal numbers, or it is increasing in such a way that it is converging toward some final velocity, i.e., it is converging from the initial velocity V_o toward a final velocity V_f, where $V_f = V_o + 2a$, where a is the first given increment of speed; and furthermore any velocity (V_n) after V_o is given as follows: $V_n = V_o + a + a/2 + a/4 + \ldots + a/2^{n-1}$.

By an apparent error in the text, Albert seems to say first that the velocity is convergent, but then, as his explanation makes clear, the increase in speed is divergent, the increase in speed being arithmetical. At this point he does not state what variable this increasing velocity is a function of. Hence the passage is mathematically ambiguous. But in the next passage he attempts to clear up the ambiguity. He notes first that the arithmetical increase in speed *does not* take place in the succeeding *proportional* parts of time of fall ($t/2$, $t/4$, etc.), nor does it take place upon completion of the succeeding proportional parts of the distance traversed (e.g., $S/2$, $S/4$). Rather he thinks that the speed in free fall increases simply as the distance of fall ($V \propto S$). This erroneous opinion later also seduced Galileo before he found the correct one, namely that $V \propto t$. Now Albert does not anywhere really consider or mention the correct solution, although Oresme does so in his similar discussion of the problem, as we have shown in the body of the chapter. But the correct solution is implied in the remaining part of the selection (passage 2) where he says (under the influence of Buridan) that the longer the movement takes *(diutius et diutius movetur)* the more impetus is acquired and thus the more speed is acquired. The correct solution (not without some confusion) is also mentioned by Leonardo da Vinci (see Doc. 9.4), and by the Dominican Domingo de Soto, and finally by Galileo (see Doc. 9.5).

The reader should be again reminded that the mathematics of the description of uniform acceleration has already been treated in Chapters 5 and 6.

Questiones [subtilissime] Alberti de Saxonia in libros de celo et mundo Aristotelis*

Liber secundus. Quest. 14.... Utrum omnis naturalis sit velocior in fine quam in principio.... Quarta distinctio. Motum intendi potest intelligi dupliciter, uno modo per duplum, triplum, quadruplum, et sic ultra; alio modo quod primo sit aliquantus et secundo addatur sibi aliquis gradus velocitatis, et tertio medietas illius gradus velocitatis, et quarto quarta pars illius gradus velocitatis, et sic ultra.

Tunc sit prima conclusio: Si motus intenderetur primo modo fieret infinitus. Probatur ex quo primo esset aliquantus, et deinde in duplo velocior, et deinde in triplo, et deinde in quadruplo, et sic ultra.

Secunda conclusio: Si intenderetur secundo modo non oporteret ipsum fieri infinitum. Immo in casu posset in infinitum intendi et adhuc numquam fieri triplus ad se ipsum, recte in simili sicut si esset aliquod lignum pedale et adderetur sibi primo medietas semipedalis, et deinde medietas residui de semipedali, et deinde iterum medietas residui, et sic in infinitum, illud lignum quod primo erat pedale in infinitum augeretur secundum infinitas partes proportionales ligni semipedalis sibi addendas, et tamen non propter hoc illud lignum pedale, sic auctum, fieret duplum vel triplum ad se. Immo si totale lignum semipedale esset sibi additum, aggregatum ex preexistente et addito non esset duplum ad preexistens; immo solum sexquialterum; sic in proposito esset dicendum de intensione motus secundo modo dicta.

Tertia conclusio: Motus naturalis in fine, hoc est, eundo versus finem, non stanti quando incipit et vadit versus finem intenditur secundo *(sic, but should be* primo*)* modo. Patet hoc, nam aliter grave non moveretur velocitate infinita antequam veniret ad centrum si distaret distantia infinita, cuius oppositum dicit Aristoteles.

Sed sciendum est quod motus naturalis non intendit per duplum, triplum, et cetera, sic quod in prima parte proportionali hore sit aliquantus, et quod in secunda parte proportionali hore sit in duplo velocior, et sic ultra; nec etiam sic quod quando est pertransita prima pars proportionalis spacii, puta prima medietas, quod tunc sit aliquantus, et quando secunda pars proportionalis spacii sit pertransita, quod tunc in duplo sit velocior, et sic ultra. Nam tunc sequeretur

* Edition of Venice, 1492, sign. f. 2v—f. 3v (ff. 32r–33v). Punctuation and capitalization have been altered.

quod quilibet motus naturalis, qui per quantumcunque tempus parvum duraret vel quo quantuncunque parvum spacium pertransiretur, ad quemcunque gradum velocitatis pertingeret ante finem. Modo hoc est falsum. Consequentia tenet ex hoc, quod talis motus fieret per aliquod tempus in cuius prima parte proportionali esset aliquantus et in secunda duplus et sic ultra, et ita arguitur de spacio. Et ideo tertia conclusio intelligitur, quod intenditur per duplum, triplum, et cetera ad istum intellectum, quod, quando ipso pertransitum est aliquod spacium, est aliquantus; et quando ipso est pertransitum duplum spacium, est in duplo velocior; et quando ipso pertransitum est triplum spacium, est in triplo velocior, et sic ultra.... (32v) Quantum ad tertium articulum est alia opinio de causa velocitationis motus naturalis in fine et illam approbo. Unde secundum istam opinionem imaginandum est quod ipsum grave in suo motu ad suam gravitatem naturalem acquirit sibi quendam impetum seu gravitatem accidentalem, que iuvat gravitatem naturalem ad movendum ipsum grave velocius.... et secundum quod ipsum corpus naturale movetur diutius et diutius, secundum hoc sibi acquiritur maior et maior impetus, et secundum hoc continue movetur velocius et velocius; nisi tamen hoc impediretur per maiorem crescentiam ipsius resistentie quam esset impetus sic acquisitus. Et imaginandum est quod sicut ille impetus acquiritur consequenter ad motum, ita consequenter minoratur vel deficit ad minorationem vel defectum motus....

(33r) Nam grave in descensu acquirit sibi talem impetum qui non subito corrumpitur; et ideo cum invenit obstaculum ille impetus nondum corruptus inclinans ad motum, non potens ipsum movere ulterius deorsum, movet ipsum econverso sursum, tam diu quod ille impetus corrumpitur.

Iuxta illud etiam diceretur quod si terra esset perforata totaliter, et per illud foramen descenderet unum grave bene velociter versus centrum, tunc quando medium gravitatis illius corporis descendentis esset medium mundi, illud corpus adhuc ulterius moveretur versus aliam partem celi propter impetum nondum in ipso corruptum; et in sic ascendendo quando ille impetus deficeret, illud grave econverso descenderet; et in tali descensu iterum acquireret sibi quemdam parvum impetum quo iterum moveretur ultra centrum; et illo impetu corrupto iterum descenderet; et sic moveretur circa centrum, titubando tamdiu quod non amplius in eo esset talis impetus, et tunc quiesceret.

Document 9.3

Nicole Oresme, *On the Book of the Heavens and the World of Aristotle**

BOOK I, CHAPTER 17 ... And when he says that the weight *(pesanteur)* is greater just as the velocity *(l'isneleté)* is greater, we are not to understand by "weight" a natural quality which inclines downward. For if a stone of one pound should descend from a high place so that the movement was swifter in the end than at the beginning, the stone still would have no more natural weight at the one time than at the other. But we ought understand by this "weight" which increases in descent an accidental quality which is caused by the compulsion *(l'enforcement)* of the increase *(l'accressement)* in the velocity, as I have said on another occasion in the seventh [book] of the *Physics*. And this quality can be called "impetuosity" *(impetuosité)*. And it is not weight properly [speaking] because if a passage *(pertuis)* were pierced from here to the center of the earth or still further, and something heavy were to descend in this passage or hole *(treu)*, when it arrived at the center it would pass on further and ascend by means of this accidental and acquired quality, and then it would descend again, going and coming several times in the way that a weight which hangs from a beam *(tref)* by a long cord [swings back and forth]. And so [this impetuosity] is not properly weight, since it causes ascent. And such a quality exists in every movement—natural and violent—as long as the velocity increases, the movement of the heavens being excepted. And such a quality is the cause of the [continued] movement of projected things when they are no longer in contact with the hand or the instrument [of projection] ....

COMMENTARY

I have included this passage for two reasons. It illustrates Oresme's

---

* Translated, from the Old French edition of A. D. Menut and A. J. Denomy, in *Mediaeval Studies*, Vol. *3* (1941), 230–31 concept of impetus as arising from acceleration, and it presents the well-known example of a body oscillating back and forth beyond the center of the earth because of the acquired impetus. What is more, it presents an interesting factual analogy to this hypothetical case of the weight dropping through the earth. The analogy is that of the pendulum. It is the impetus acquired in the downward swing that causes the pendulum weight to swing upward.

Document 9.4

The Manuscripts of Leonardo da Vinci*

1. (*M* 44r) [The] proof of the proportion of the time and movement *(moto)* together with the speed made in the descent of heavy bodies [is found] in the pyramidal figure, because the aforesaid powers *(potentie)* are all pyramidal since they commence in nothing and proceed to increase in degrees of arithmetical proportion *(gradi di proportio ne aritmetricha)*.

If you cut the pyramid at any degree of its height by a line parallel to its base (see Fig. 9.1), you will find that whatever proportion there is

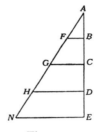

Fig. 9.1

between the height of this section from its base and the whole height of the pyramid, there will be the same proportion between the breadth *(largeza)* of this section and the breadth of the whole base. You see that if *AB* is 1/4 of *AE*, so *FB* is 1/4 of base *NE*.

2. (*M* 44v) This happens *(acade)* in the air of uniform thickness *(groseza)*.

The heavy body *(gravità)* which descends at each degree of time acquires a degree *(grado)* of movement *(moto)* more than in the degree of time preceding, and similarly a degree of swiftness *(velocità)* greater than the degree of the preceding movement. Therefore at each doubled quantity of

---

* I have used the translation by Edward MacCurdy, *The Notebooks of Leonardo da Vinci*, Vol. *1* (New York, 1938), 586–89, but have altered it freely to make it more literal. In so doing I have employed the edition of M. Charles Ravaisson-Mollien, *Les Manuscrits de Léonard de Vinci*, Manuscrits *G*, *L*, and *M* de la Bibliothèque de l'Institut (Paris, 1890). In quoting the Italian of Leonardo I have maintained the old spelling and spacing given by Leonardo.

time the length of the descent *(la lungeza del disscienso)* is doubled and also the swiftness of the movement.

It is here shown that whatever the proportion that one quantity of time has with another (see Fig. 9.2), the one quantity of movement will

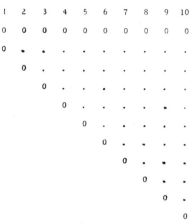

Fig. 9.2

have the same with the other and similarly one quantity of swiftness with the other.

3. (*M* 45r) The heavy body which descends freely *(che libera disciende)* with every degree of time acquires a degree of movement, and with every degree of movement it acquires a degree of speed.

Although the equal division of the movement of time cannot be indicated by degrees as is the movement made by the bodies, nevertheless the necessity of the case constrains me to make degrees after the manner in which they are made among musicians.

Let us say that in the first degree of time it (i.e., the heavy body) acquires a degree of movement and a degree of speed, in the second degree of time it will acquire two degrees of movement and two of speed and so it continues in succession *(successiva)* as has been said above....

4. (*M* 47r–v) Now we have found that the discontinuous quantity when moving acquires in each degree of its movement a degree of speed; and so in each harmonic time they acquire a length of space from each other, and this acquisition is in arithmetical proportion.

How then are we to account for the continuous quantity of liquid bodies in their descent, since in each interval of harmonic time it pours out the same weight, and in each degree of movement it becomes longer and thinner, so that in a long course it shows itself ending in a point as does the pyramid; consequently such a liquid body would not fall to earth, but it would rather be that each great mass of this body would remain in the air even though it should be a very great river which was continually rolling away; and experience shows the contrary; for as much as departs above strikes at the same time below. And if the same weight of this liquid body makes itself thinner, it meets with less resistance from the air and consequently acquires speed; and if by being thinner it has acquired speed, this same weight would also for this second reason come to make itself longer and in consequence still thinner and so would descend more rapidly; and this would go on in succession to infinity. Therefore either nature or necessity has brought it about that in whatever manner the descent comes to assume the form of a pyramid, it makes intersection by changing its extremities from right to left and commences to divide itself; and the more it descends, the more it divides; and thus with many ramifications it comes to lighten itself and to check its irregular movement....

5. (*M* 48r) If two bodies of equal weight and the same shape fall one after the other from the same height, in each degree of time the one will be a degree more distant *(un grado piu disstanti)* than the other....

6. (*M* 49r) The heavy thing descending freely gains a degree of speed with every degree *(grado)* of movement.

And the part of the movement which is made in each degree of time is always longer successively, the new one than that which preceded it.

It may be clearly shown that what is set forth above is true, for during the same time that the weight *a* descends at *c*, *b* which finds itself fifteen times swifter than *a* has covered fifteen times as much space in its descent....

COMMENTARY

Leonardo's treatment in passage 1 of the proportion of speed, movement, and time in the descent of bodies by means of the pyramid seems to be a continuation of the graphic representation of uniform acceleration begun by Oresme and Casali. However, it is significant that it is applied by Leonardo specifically to the fall of bodies. This is the earliest example of such an application known to me. One difficult question raised by this and the succeeding passages concerns Leonardo's concept of quantity of motion as distinct from speed. Duhem in his *Études* (Vol. *3*, 514) believes that Leonardo's quantity of motion is equivalent to Buridan's *impetus* and thus is measured jointly by quantity of matter and speed. But in passages 2 and 4 and elsewhere Leonardo seems to be identifying it with space traversed, so that for each unit of time the descending body acquires an additional unit of space as well as an additional unit of speed, thus making time directly proportional to both speed and distance, an impossibility.

It is worth noting that only in passage 1 do we truly find the infinitesimal aspect of increase in speed represented, although it is presumed that for Leonardo as well as the fourteenth-century philosophers the use of the term "*grado*" gave some connotation of an infinitesimal.

In passage 1 Leonardo is really characterizing the uniform acceleration in about the same manner as Oresme, namely, by uniformity of slope. He does this by saying that the ratio of the whole height to the whole base is as the ratio of the height of any section is to the base of that section.

In passage 4 Leonardo repeats the example of the thinning out of a falling liquid which Strato had originally advanced in a somewhat different form and which Jordanus had repeated in the Middle Ages (see Chapter 5), but this remains a highly confused passage. In passage 5 Leonardo repeats his error to the effect that the time of fall is directly proportional to the distance of fall. We have said elsewhere that this confusion and identification of time and distance was a common one from the time of Buridan through the early period of Galileo's researches.

Finally we should notice that Leonardo, in a short passage just preceding our selection, also says that even when the motion is oblique, i.e., on an inclined plane, "for each degree of it the increase of the movement and the speed is in arithmetical proportion" (*M* 42v: *anchora chel moto sia obliquo esse osserua in ogni suo grado lo accresscimento del moto e della velocita improportione arimetricha*). Galileo, of course, arrived at a similar conclusion, namely, that the balls move on inclined planes such that $S \propto t^2$, just as is the case of free fall.

Document 9.5

Galileo Galilei, *The Two New Sciences**

THE THIRD DAY.—SALV.: .... Place a heavy body upon a yielding material, and leave it there without any pressure except that owing to its own weight; it is clear that if one lifts this body a cubit or two and allows it to fall upon the same material, it will, with this impulse *(percossa)*, exert a new and greater pressure *(pressione)* than that caused by its mere weight; and this effect is brought about by the [weight of the] falling body together with the velocity acquired during the fall, an effect which will be greater and greater according to the height of the fall, that is according as the velocity of the falling body becomes greater. From the quality and quantity of the blow *(percossa)* we are thus enabled to accurately estimate the speed of a falling body....

2. .... SAGR.: From these considerations it appears to me that we may obtain a proper solution of the problem discussed by philosophers, namely, what causes the acceleration in the natural motion of heavy bodies? Since, as it seems to me, the force *(virtù)* impressed by the agent projecting the body upwards diminishes continuously, this force, so long as it was greater than the contrary force of gravitation, impelled the body upwards; when the two are in equilibrium, the body ceases to rise and passes through the state of rest in which the impressed impetus *(impeto)* is not destroyed, but only its excess over the weight of the body has been consumed—the excess which caused the body to rise. Then as the diminution of the outside impetus *(impeto)* continues, and gravitation gains the upper hand, the fall begins, but slowly at first on account of the opposing impetus *(virtù impressa)*, a large portion of which still remains in the body; but as this continues to diminish it also continues to be more and more overcome by gravity, hence the continuous acceleration of motion....

---

* *Le Opere*, Ed. Naz. Vol. *8* (Florence, 1898), 199, 201–4, 212–13. We are not particularly interested in recapturing Galileo's vocabulary in this passage. Hence I have used the Crew–De Salvio translation (Evanston and Chicago, 1946), pp. 157, 158–59, 160–61, 171–72, although I have altered it slightly.

[Galileo, *The Two New Sciences*]

3. ... SALV.: The present does not seem to be the proper time to investigate the cause of the acceleration of natural motion concerning which various opinions have been expressed by various philosophers, some explaining it by attraction to the center, others [attributing it] to repulsion between the very small parts of the body, while still others attribute it to a certain stress in the surrounding medium which closes in behind the falling body and drives it from one of its positions to another. Now, all these fantasies, and others too, ought to be examined; but it is not really worthwhile. At present it is the purpose of our Author merely to investigate and to demonstrate some of the properties of accelerated motion (whatever the cause of this acceleration may be)—meaning thereby a motion, such that the moments of its velocity *(i momenti della sua velocità)* go on increasing after departure from rest, in simple proportionality to the time, which is the same as saying that in equal time intervals the body receives equal increments of velocity; and if we find the properties [of accelerated motion] which will be demonstrated later are realized in freely falling and accelerated bodies, we may conclude that the assumed definition includes such a motion of falling bodies and that their speed *(accelerazione)* goes on increasing as the time and the duration of the motion.

SAGR.: So far as I see at present, the definition might have been put a little more clearly perhaps without changing the fundamental idea, namely, uniformly accelerated motion is such that its speed increases in proportion to the space traversed; so that, for example, the speed acquired by a body in falling four cubits would be double that acquired in falling two cubits and this latter speed would be double that acquired in the first cubit. Because there is no doubt but that a heavy body falling from the height of six cubits has, and strikes with, a momentum *(impeto)* double that it had at the end of three cubits, triple that which it would have if it had fallen from two, and sextuple that which it would have had at the end of one.

SALV.: It is very comforting to me to have had such a companion in error; and moreover let me tell you that your proposition seems so highly probable that our Author himself admitted, when I advanced this opinion to him, that he had for some time shared the same fallacy. But what most surprised me was to see two propositions so inherently probable that they commanded the assent of everyone to whom they were presented, proven in a few simple words to be not only false but impossible.

SIMP.: I am one of those who accept the proposition, and believe that a falling body acquires force *(vires)* in its descent, its velocity increasing in proportion to the space, and that the moment *(momento)* of the falling body is doubled when it falls from a doubled height; these propositions, it appears to me, ought to be conceded without hesitation or controversy.

SALV.: And yet they are as false and impossible as that motion should be completed instantaneously; and here is a very clear demonstration of it. If the velocities are in proportion to the spaces traversed, or to be traversed, then these spaces are traversed in equal intervals of time; if, therefore, the velocity with which the falling body traverses a space of eight feet were double that with which it covered the first four feet (just as the one distance is double the other) then the time intervals required for these passages would be equal. But for one and the same body to fall eight feet and four feet in the same time is possible only in the case of instantaneous [discontinuous] motion; but observation shows us that the motion of a falling body occupies time, and less of it in covering a distance of four feet than of eight feet; therefore it is not true that its velocity increases in proportion to the space....

4. ... But as to whether this [uniform] acceleration [where the space traversed is directly proportional to the square of the time] is that which one meets in nature in the case of falling bodies, I am still doubtful; and it seems to me, not only for my own sake but also for all those who think as I do, that this would be the proper moment to introduce one of those experiments—and there are many of them, I understand—which illustrate in several ways the conclusions reached.

SALV.: The request which you, as a man of science, make, is a very reasonable one; for this is the custom—and properly so—in those sciences where mathematical demonstrations are applied to natural phenomena, as is seen in the case of perspective, astronomy, mechanics, music, and others where the principles, once established by well-chosen experiments, become the foundations of the entire superstructure. I hope, therefore, it will not appear to be a waste of time if we discuss at considerable length this first and most fundamental question upon which hinge numerous consequences of which we have in this book only a small number, placed there by the Author, who has done so much to open a pathway hitherto closed to minds of speculative turn. So far as experiments go, they have not been neglected by the Author; and often, in his company, I have attempted in the following manner to assure myself that the acceleration actually experienced by falling bodies is that above described.

A piece of wooden moulding or scantling, about 12 cubits long, half a cubit wide, and three fingerbreadths thick, was taken; on its edge was cut a channel a little more than one finger in breadth; having made this groove very straight, smooth, and polished, and having lined it with parchment, also as smooth and polished as possible, we rolled along it a hard, smooth, and very round bronze ball. Having placed this board in a sloping position, by lifting one end some one or two cubits above the other, we rolled the ball, as I was just saying, along the channel, noting, in a manner presently to be described, the time required to make the descent. We repeated this experiment more than once in order to measure the time with an accuracy such that the deviation between two observations never exceeded one-tenth of a pulse beat. Having performed this operation and having assured ourselves of its reliability, we now rolled the ball only one-quarter the length of the channel; and having measured the time of its descent, we found it precisely one-half of the former. Next we tried other distances, comparing the time for the whole length with that for the half, or with that for two-thirds, or three-fourths, or indeed for any fraction; in such experiments, repeated a full hundred times, we always found that the spaces traversed were to each other as the squares of the times, and this was true for all inclinations of the plane, i.e., of the channel, along which we rolled the ball. We also observed that the times of descent, for various inclinations of the plane, bore to one another precisely that ratio which, as we shall see later, the Author had predicted and demonstrated for them.

For the measurement of time, we employed a large vessel of water placed in an elevated position; to the bottom of this vessel was soldered a pipe of small diameter giving a thin jet of water, which we collected in a small glass during the time of each descent, whether for the whole length of the channel or for a part of its length; the water thus collected was weighed, after each descent, on a very accurate balance; the differences and ratios of these weights gave us the differences and ratios of the times, and this with such accuracy that although the operation was repeated many, many times, there was no appreaciable discrepancy in the results....

COMMENTARY

1. The first passage I have drawn from Galileo's "Third Day" relates velocity and force of impact in much the same way as did Buridan, and even Simplicius, quoting Strato. This is pointed up in the sentence reading: "From the quality and quantity of the blow we are thus enabled to accurately estimate the speed of a falling body."

2. In the second passage Sagredo is describing the theory of accelera-

tion reported by Simplicius as originating with Hipparchus. We have already discussed it in the beginning of this chapter. It was also repeated by Abū 'l-Barakāt.

3. In passage 3 Salviati sets aside, at least temporarily, the investigation of the cause of acceleration. He indicates that the author (Galileo) wants only in this work to investigate and demonstrate the properties of acceleration, and, particularly, that in uniform acceleration velocity varies directly as time, and then to demonstrate that this acceleration is realized in the movement of falling bodies. Salviati also confesses in this passage that the author had once thought that the velocity increases directly with the distance of fall, a theory we have seen that perhaps Buridan but certainly Albert of Saxony accepted. In connection with Galileo's early belief in this incorrect theory, we can quote from a letter that Galileo wrote to Paolo Sarpi in 1604 (*Opere*, Vol. *10*, 115). The curious point is that Galileo at one and the same time believed that the distance is directly proportional to the time squared and that the velocity is directly proportional to the distance. The former is of course correct, while the latter is erroneous. He says:

...I have arrived at a proposition which is most natural and evident, and with it being assumed I can demonstrate the rest, namely, that spaces traversed in natural motion *(moto naturale)* are in the squared proportion *(proporzione doppia)* of the times, and consequently the spaces traversed in equal times are as the odd numbers *ab unitate*, etc. And the principle is this, that the natural moving body increases its velocity in the proportion that it is distant from the beginning of its motion, as for example, in assuming that a heavy body is falling from terminus *a* through the line *abcd*, I suppose that the degree *(grado)* of velocity which it has in *c* to the degree of velocity it has in *b* is as the distance *ca* is to the distance *ba*, and so consequently in *d* it will have a degree of velocity more than in *c* according as the distance *da* is more than *ca*.

It will be noticed that his statement of the principle is very much like that of Albert of Saxony (Doc. 9.2). It is obvious that Galileo could never get to the formulation $S \propto t^2$ from the formulation $V \propto S$ as he proposes. Galileo's refutation in the *Two New Sciences* of the erroneous view of $V \propto S$, on logical grounds, has been the object of much discussion by Mach and others. Recently, I. B. Cohen has interpreted this refutation in the light of the Merton mean speed theorem.[25] If Galileo did assume the

[25] I. B. Cohen, "Galileo's Rejection of the Possibility of Velocity Changing Uniformly with Respect to Distance," *Isis*, Vol *47* (1956), 231–38. Cf. the earlier and similar suggestion of A. R. Hall that in this passage Galileo "used Oresme's cal-

Merton theorem in this argument, he was obviously guilty of inconsistency, since the Merton theorem implies the correct description of $V \propto t$ and thus it could hardly be used to express the average velocity in the case of $V \propto S$.[26]

It should be recalled by reference to Chapter 6 above that Galileo's statement that the spaces traversed in successive periods of time are related as the odd numbers from unity was foreshadowed by a similar statement made by Oresme for uniformly difform qualities in his *Questiones* on Euclid's *Elements*.

4. The fourth passage includes Galileo's famous inclined plane experiment to show that the acceleration one meets in nature is the simple acceleration he has already described, namely that in which velocity grows directly as the time of fall. He has already reasoned that the character of the motion on an inclined plane is that of free fall. The advantage, of course, is that with an inclined plane he can make measurements with regard to time more easily than with free fall. Notice that his experimental method for measuring time falls back on one of the oldest clocks, the water clock. Mechanical clocks, as then developed, would have been inadequate for the job. This fourth passage appears—at least on the surface—to distinguish Galileo's method from that of his predecessors; he feels that he is confirming that his deductions apply to nature by the use of a *quantitative experiment*. Soto had made a similar deduction but had not related that deduction to nature by an experiment.

It is true that the data of Galileo's experiments are not given in this account from the *Discorsi*, and as a result there has been considerable discussion of the character of Galileo's experiment. It is, further, probable that Galileo was convinced of the veracity of his "law" before making the experiment. But it is not my purpose here to treat Galileo's use of experiment, however interesting that question might be, since my objective in including this passage is merely to contrast it with the medieval treatments of acceleration. We should, however, observe that the form of the statement in which Galileo expresses the results of his experiment is still that

culus of varying qualities." *The Scientific Revolution, 1500–1800* (London, 1954), p. 88, n. 1.

[26] On reading this part of my manuscript, Dr. Curtis Wilson suggested the following: "What about the use of the Merton rule to apply to 'total velocity,' understood as the sum of all the instantaneous velocities, and represented as an area? The base could be space as easily as time; without an explicit understanding in terms of the calculus as to what 'total velocity' might mean, the Merton rule would, it seems to me, be applied without hesitation."

of a simple proportionality statement. Thus Galileo says, "we always found that the spaces traversed were to each other as the squares of the times." However, elsewhere Galileo gives approximate figures for the acceleration of free fall.[27]

---

[27] Galileo mentions the figure of "100 braccia in 5 seconds," *Dialogo, Opere*, Ed. Naz., Vol. *7*, 250, and again in discussing Antonio Rocco's *Esercitazioni philosophiche*, *ibid.*, Vol. *7*, 729; cf. a letter to Baliani of 1639, *ibid.*, Vol. *18*, 76–77. This figure would lead to a value for *g* equal to about one-half of that of the actual value. A second figure—very imprecisely given—is stated by Galileo in an addition he made in his own copy of the *Dialogo* and is put in the mouth of Simplicius, who speaks of a body falling "more than a hundred braccia in four pulse beats" (*ibid.*, Vol. *7*, 54). Even if Galileo meant "four pulse beats" (imprecise enough) rather than "a few pulse beats" (which would also be an acceptable rendering of "quattro battute"), the vague expression "more than 100 braccia" (più di 100 braccia) renders this figure even more imprecise. Another imprecise figure is given in the *Dialogo*, namely "200 braccia in less than 10 pulse beats" (*ibid.*, Vol. *7*, 46). For Mersenne's criticism of Galileo's figure consult A. Koyré, "An Experiment in Measurement," *Proceedings of the American Philosophical Society*, Vol. *97* (1953), 222–37 (and particularly pp. 225–28), and A. Koyré, "A Documentary History of the Problem of Fall from Kepler to Newton," *Transactions of the American Philosophical Society*, New Series, Vol. *45* (1955), 339–41. On Galileo's figures, see Stillman Drake's translation of the *Dialogo*, *Dialogue Concerning the Two Chief World Systems* (Berkeley, Calif., 1953), p. 480.

Chapter 10

Mechanics and Cosmology

TURNING from basic kinematic and dynamic principles, I should like to single out for brief discussion a few questions raised by the schoolmen in the fourteenth century as they applied their mechanical concepts to the larger world. Obviously I have not attempted to give a complete historical treatment of each of the problems discussed, nor have I ranged over the large spectrum of cosmological problems revealed in Pierre Duhem's *Le Système du Monde*. The focus is on certain questions of importance to *mechanics* in the period of the fourteenth century.

MOTION OF THE EARTH AND RELATIVE PERCEPTION OF MOTION

One of the most interesting questions concerned itself with the possibility of the earth's rotation. There is no evidence that medieval philosophers accepted or posited the *complete* heliocentric system of Aristarchus, which apparently assumed both a diurnal rotation of the earth on its axis and an annual revolution of the earth about the sun. For the medieval schoolmen universally believed that the earth was at, or striving for, the center of the universe. But a number of medieval authors knew at least part of the system of Heraclides of Pontus wherein the earth was thought to be rotating on its axis at the center of the universe, while the planets Mercury and Venus were revolving about the sun as satellites, and the remaining planets together with the sun and its satellites were revolving about the earth.[1] One author, Johannes Scotus Eriugena, as early as the ninth century may have extended the Heraclidian system to include Jupiter and Mars, as well as Mercury and Venus, revolving about the sun.[2]

[1] P. Duhem, *Le Système du monde*, Vol. *3* (Paris, 1910), 43–162, traces the history of the doctrine of Heraclides in the Middle Ages.

[2] *Ibid.*, p. 61. Cf. C. A. Lutz, *Iohannis Scotti Annotationes in Marcianum* (Cambridge, Mass., 1939), pp. xviii–xix. The attempt by Erika von Erhardt-Siebold and Rudolph von Erhardt in two small volumes, *The Astronomy of Johannes Scotus Erigena* (Baltimore, 1940), and *The Cosmology in the "Annotationes in Marcianum"*;

Now it was the first part of the Heraclidian system which was discussed at Paris in the fourteenth century, namely, the rotation of the earth on its axis. It had been treated in antiquity by a number of people,[3] including Simplicius in his commentary on the *De caelo*,[4] which no doubt was known by the schoolmen of the fourteenth century. Similarly the rotation of the earth was discussed by Islamic and Indian authors.[5] Rejected by Thomas Aquinas,[6] it was mentioned by Franciscus de Mayronis in the early fourteenth century.[7] Not long after Franciscus, John

More Light on Erigena's Astronomy (Baltimore, 1940), to rehabilitate the whole doctrine and to deny that it is the Heraclidean doctrine that John is describing fails in my opinion.

[3] G. McColley, "The Theory of the Diurnal Rotation of the Earth," *Isis* Vol. 26 (1936–37), 392–402, and particularly pp. 392–95 for antique references to the theory of the earth's rotation.

[4] Simplicius, *In Aristotelis de caelo commentaria*, edition of J. L. Heiberg in *Commentaria in Aristotelem graeca*, Vol. 7 (Berlin, 1894), 444–45 (on II, 7); 517–19 (on II, 13); 541 (on II, 14).

[5] G. Sarton, *Introduction to the History of Science*, Vol. *1* (Baltimore, 1927), 489, 709; Vol. *2* (1931), 764, 868; Vol. *3* (1947), 146. Cf. S. Pines, *Journal asiatique*, Vol. 244 (1956), 301–6. See also for example al-Bīrūnī's discussion of whether the Indians believed in diurnal rotation (E. C. Sachau, *Alberuni's India*, Vol. *1* [London, 1888], 276–77): "As regards the resting of the earth, one of the elementary problems of astronomy, which offers many and great difficulties, this, too, is a dogma with the Hindu astronomers. Brahmagupta says in the *Brahmasiddhânta*: 'Some people maintain that the *first* motion (from east to west) does not lie in the meridian, but belongs to the earth. But Varahamihira refutes them by saying "If that were the case, a bird would not return to its nest as soon as it had flown away from it towards the west." And, in fact, it is precisely as Varahamihira says.' Brahmagupta says in another place in the same book: 'The followers of Âryabhata maintain that the earth is moving and heaven resting. People have tried to refute them by saying that, if such were the case, stones and trees would fall from the earth.' But Brahmagupta does not agree with them, and says that that would not necessarily follow from their theory, apparently because he thought that all heavy things are attracted towards the center of the earth. He says: 'On the contrary, if that were the case, the earth would not vie in keeping an even and uniform pace with the minutes of heavens....' Besides, the rotation of the earth does in no way impair the value of astronomy, as all appearances of an astronomic character can quite as well be explained according to his theory as to the other. There are, however, other reasons which make it impossible. This question is most difficult to solve. The most prominant of both modern and ancient astronomers have deeply studied the question of the moving of the earth, and tried to refute it. We, too, have composed a book on the subject...."

[6] Thomas Aquinas, *In libros Aristotelis de caelo et mundo*, Bk. II. 14, lect. 26, in *Opera omnia*, Vol. *3* (Rome, 1886), 219; cf. recent printing (Rome, 1952), p. 264, c. 1. "Secunda conclusio est quod terra sit immobilis. Quod quidem concluditur ex praemissis sic. Nihil movetur in loco ad quem naturaliter movetur, quia ibi naturaliter quiescit; sed terra naturaliter movetur ad medium mundi; ergo non movetur in medio." Cf. also Lect. 11, n. 2, where is noted the opinion of Heraclitus Ponticus and Aristarchus.

[7] P. Duhem, "François de Meyronnes O.F.M. et la question de la rotation de la

[Miscellaneous Problems

Buridan, whose impetus theory we have discussed in Chapters 8 and 9, presented one of the best discussions of the question of rotation (see Doc. 10.1). Buridan suggested certain important arguments which were taken up and elaborated upon by Nicole Oresme (see Doc. 10.2).

Thus both of these schoolmen stressed the doctrine of the relativity to the observer of the perception of motion. For example, Oresme tells us (Doc. 10.2, passage 3): "Again, I make the supposition that local motion can be sensibly perceived only in so far as one may perceive one body to be differently disposed with respect to another." Both schoolmen illustrated the doctrine by positing observers on ships which are moving relatively to each other, noting that the observer would not detect his own motion (assuming a calm sea) but would describe the motion of another ship in terms of his own ship being at rest (see Docs. 10.1, passage 2, and 10.2, passage 3). Furthermore, both Buridan and Oresme held that because of the relativity of the detection of motion, all astronomical phenomena *(apparentia)* could just as well be accounted for or "saved" *(salvari)* by assuming the earth's diurnal rotation as by assuming the diurnal movement of the heavens. However, Albert of Saxony and another fourteenth century author do not agree that all of the astronomical phenomena are saved.[8]

---

terre," *Archivum Franciscanum Historicum*, Annus VI, Tomus *4* (Quaracchi, 1913), 23–25. Mayronis briefly mentions the rotation doctrine as being held by a certain doctor, presumably at his time at Paris, i.e., about 1320. We read in Mayronis' commentary on the *Sentences*, *Praeclarissima ac multum subtilia egregiaque scripta...in quatuor libros sententiarum* (Venice, 1520), Bk. II, Dist. 14, quaest. 5, f. 150, col. a: "14ª difficultas: De Terra, quae est immobilis, cum ignis et alia sint mobilia. Dicit tamen quidam doctor quod si Terra moveretur et Coelum quiesceret, quod hic esset melior dispositio. Sed hoc impugnatur, propter diversitatem motuum in Coelo, quae non possent salvari." (Quoted through Duhem, "François de Meyronnes," p. 25).

[8] L. Thorndike, *A History of Magic and Experimental Science*, Vol. *3* (New York, 1934), 581, mentions that an anonymous tract on natural philosophy in Paris, BN lat. 6752, 14c, f. 160r, brands the theory of the rotation of the earth as "most false." According to Thorndike, the author argues "that such a velocity of the earth would bring buildings down in ruin and would not serve to explain why such planets as the sun and the moon are nearer at some times than others, or to explain phenomena like eclipses, conjunctions, and oppositions." Albert of Saxony takes up the question very much in the tradition of John Buridan and Nicole Oresme. He does so in his *Questiones subtilissime in libros de celo et mundo Aristotelis* (Venice, 1492), Bk. II, quest. 26 (no pag.): "Quantum ad primum sit prima distinctio, quod terram moveri potest intelligi vel secundum qualitatem vel secundum quantitatem vel ubi vel secundum mutationem eius in substantia, accipiendo motum largo modo. In proposito non intelligo nisi de motu secundum ubi. Secunda distinctio, terram moveri secundum ubi potest imaginari dupliciter, scilicet motu recto vel motu circulari; et intelligo propositum de tota litera (! terra?) cuius medium gravitatis est medium mundi. Tertia distinctio terram moveri circulariter potest intelligi uno modo circa centrum sibi extrinsecum, modo quo stelle moventur circa centrum

Oresme, even more specifically than Buridan, stated that the astronomical tables of celestial movements would apply equally well to each system (see Doc. 10.2, passage 7):

> ...all aspects, conjunctions, oppositions, constellations, figures, and influences would be completely just as they are.... The tables of the movements and all other books would be just as true as they are, except that in regard to the daily movement one would say of it that it is in the heavens "apparently" but in the earth "actually." There is no effect which follows from the one [assumption] more than from the other.

---

mundi, secundo modo circa centrum proprium et super polos proprios; et hoc potest imaginari vel ab oriente ad occidentem vel econverso, vel a meridie ad septentrionem vel econverso. Tunc sit prima conclusio: terra non movetur circulariter circa centrum sibi extrinsecum. Patet hoc, nam tunc centrum gravitatis terre non esset in medio mundi, cuius oppositum dictum est in alia questione. Et etiam secundo, nam tunc alique stelle deberent nobis apparere aliquando maiores aliquando minores. Hoc est falsum. Consequentia tenet ex eo quod per talem motum terre nos aliquando essemus aliquibus stellis propinquiores et aliquando ab eisdem remotiores. Secunda conclusio: nec terra movetur circulariter a meridie ad septentrionem vel econverso circa centrum proprium et circa polos proprios probatur, quia tunc non semper polus apparet nobis equaliter elevatus, cuius oppositum docet experientia. Tertia conclusio: nec terra movetur circulariter ab oriente ad occidentem nec econverso, saltem motu diurno sicut quidam antiqui voluerunt; dixerunt enim celum quiescere et terram moveri. Huius conclusionis probatio et horum antiquorum improbatio fiebat prius in illa questione, utrum motus celi ab oriente in occidentem sit regularis. Circa tamen istam questionem vel conclusionem est advertendum quod unus de magistris meis videtur velle quod non sit demonstrabile quin possit salvari terram moveri et celum quiescere. Sed apparet mihi, sua reverentia salva, quod immo et hoc per talem rationem, nam nullo modo per motum terre et quietem celi possemus salvare oppositiones et coniunctiones planetarum, nec eclipses solis et lune. Verum est quod istam rationem non ponit nec solvit, licet plures alias persuasiones quibus persuaderetur terram quiescere et celum moveri ponat et solvat. Quarta conclusio: bene verisimile est quod semper quelibet pars terre totalis moveatur motu recto, quod persuadetur sic. Nam continue de ista terra elementari discooperta aquis cum fluviis fluunt multe partes terre ad profundum maris et sic augetur terra in parte cooperta aquis et in parte discooperta aquis diminuitur, et per consequens non remanet idem medium gravitatis sicut ante. Medio autem gravitatis mutato illud quod de novo factum est medium gravitatis movetur ut sit medium mundi, et illud quod ante erat medium gravitatis ascendit versus partem discoopertam aquis; et tandem per talem continuum fluxum et motum illa terra que aliquando erat in medio erit in circumferentia et econverso. Et iuxta illud potest apparere quomodo generati sunt magni montes. Nam non est dubium quin alique partes terre magis tenent se simul quam alie, et ideo quando ille partes que non tenent se simul fluunt cum fluviis ad marem, relique tenentes se simul manent et faciunt eminentiam super terram. Sed verum est quod tandem per motum terre vel alio modo evertuntur et cadunt et destruuntur." I have somewhat altered the punctuation and capitalization of the edition. Cf. also Bk. II, quest. 13, where Albert examined the question of "Whether the motion of the heavens from

It will be further noted that the two schoolmen advanced the principle of economy or simplicity in nature as a "persuasion" in favor of the diurnal movement of the earth (see Doc. 10.1, passage 4, and 10.2, passage 9). Since God, the argument ran, does not do things in vain or for nought, He would not arrange the celestial movements in the more complex way demanded by the rotation of the heavens when He could produce the same appearances by the simpler system of the earth's rotation. The modest velocity required for the earth's diurnal rotation is in a sense simpler than that which would be required by the fixed stars if they had to complete a revolution through an almost unbelievably large great circle every twenty-four hours.

Finally, and most important, Buridan hinted at, and Oresme rather specifically outlined, the concept of a closed mechanical system, wherein, due to the relativity of the perception of motion, the observer describes all movements as if they were a part of his system only. This concept is used by Oresme to explain what he believes to be the apparent rectilinear movement of objects falling to the earth. The analogy is again to the man on the ship. As Oresme showed, it would appear to the observer on the ship that his hand was descending in rectilinear motion if he made it slide down the mast (see Doc. 10.2, passage 5):

...the arrow is trajected upwards and [simultaneously] with this trajection it is moved eastward very swiftly with the air through which it passes and with all the mass of the lower part of the universe..., it all being moved with diurnal movement. For this reason, the arrow returns to the place on earth from which it left. This appears possible by analogy: If a person were on a ship moving toward the east very swiftly without his being aware of the movement and he drew his hand downward, describing a straight line against the mast of the ship, it would seem to him that his hand moved with rectilinear movement only.

Oresme also seems to suggest that the basic reason for the phenomena of the closed mechanical system is that all objects therein share the same horizontal velocity (as, per example, all objects on board a ship). Bruno and Galileo further developed this doctrine,[9] and for Galileo the horizontal impetus was still circular rather than rectilinear, at least in his *Dialogo*. It was not until Newton's generation that the correct view was advanced

---

east to west is regular." He adduces several experiences (sign. f. 2r, c.2) against the motion of the earth, all of which Buridan and Oresme discussed.

[9] P. Duhem, *Études sur Léonard de Vinci*, Vol. *3* (Paris, 1913), 257–58. Galileo Galilei, *Dialogo sopra i due massimi sistemi etc.*, in *Le Opere*, Ed. Naz. Vol. 7 (Florence, 1897), 180. G. Bruno, *La cena de la ceneri*, in *Opere italiane*, Vol. *1* (Göttingen, 1888), 167–69. Cf. al-Bīrūnī's discussion as noted by Pines (*op. cit.* in note 5).

on the path of fall (i.e., fall to the east rather than to the west or in a vertical line).[10]

Less important than these mechanical arguments, but still interesting, was the argument used by Oresme to answer the apparent scriptural statements against the diurnal rotation of the earth (see Doc. 10.2, passages 6 and 8). Oresme's answer suggests that the Scriptures were written in the common speech of their time and thus are not to be taken too literally. This was repeated by Galileo later.

While Buridan clearly rejects the idea of the earth's rotation, even after arguing quite persuasively for it, he does believe that the earth has short movements of translation (see Doc. 10.1, passage 13). These rectilinear displacements result from the fact, he believes, that the center of gravity of the earth is continually being altered by erosion and the building up of mountains. Now the center of gravity of the earth continually strives to be at the center of the universe. The result is that the whole mass of the earth moves to bring the altered center of gravity in coincidence with the center of the universe.

We might conclude our remarks on the medieval discussions of the earth's rotation by noting that neither Buridan nor Oresme actually believed that the earth made such a rotation. Buridan definitely appears to reject the idea for empirical reasons (see Doc. 10.1, passage 9). Oresme, on the other hand, rejects it on scriptural grounds (see Doc. 10.2, passage 10), although he has already explained that such an argument is not conclusive. We must conclude, however, that Buridan and Oresme certainly argued most persuasively for the rotation before finally rejecting it.

For the sake of comparison of the medieval discussions with a later treatment of the earth's rotation, I have included excerpts from Copernicus' *De revolutionibus* (Doc. 10.3). It will be noticed that some of the arguments used by Copernicus (and other later writers[11]) reflect those of the medieval authors. That the Parisian mechanics spread to Cracow and throughout Eastern Europe, as well as to Italy is abundantly clear from

---

[10] A. Koyré, "A Documentary History of the Problem of Fall from Kepler to Newton," *Transactions of the American Philosophical Society*, New Series, Vol. 45 (1955), 329–95.

[11] The reader should also compare the medieval discussions of the rotation of the earth with that of Galileo. I have refrained from publishing the latter discussion, since recently two English translations of the *Dialogo* have appeared and are readily available. See the T. Salusbury translation as revised by Giorgio de Santillana, *Dialogue on the Great World Systems* (Chicago, 1953), particularly "The Second Day," pp. 127 *et seq.*, or Stillman Drake's translation, *Dialogue Concerning the Two Chief World Systems* (Berkeley, Calif., 1953), pp. 114 *et seq.* For the Italian text, see *Le Opere*, Ed. Naz., Vol. 7, 139 *et seq.*

[Miscellaneous Problems

manuscript evidence,[12] and it is certainly possible, if not probable, that Copernicus knew of this Parisian question of the earth's rotation during his student days at Cracow.

ONTOLOGY OF MOTION

Now the doctrine of the relativity of the detection of motion—a doctrine so important for the problem of the earth's rotation—can be juxtaposed with the view of William Ockham and his nominalist successors that movement *(motus)* is not an entity having existence independent of the moving body, but is simply a shorthand way of saying that a body occupies first this position, and then this position, and then this position, and so on. I quote from Crombie three passages from works of Ockham in which he develops this approach to movement:[13]

Motion is not such a thing wholly distinct in itself from the permanent body, because it is futile to use more entities when it is possible to use fewer. But without any such thing we can save the motion and everything that is said about motion. Therefore it is futile to postulate such other things. That without such an additional thing we can save motion and everything that is said about it is made clear by considering the separate parts of motion. For it is clear that local motion is to be conceived of as follows: positing that the body is in one place and later in another place, thus proceeding without any rest or any intermediate thing other than the body itself and the agent itself which moves, we have local motion truly. Therefore it is futile to postulate such other things....

I say therefore that the moving thing in such a motion [i.e., projectile motion], after the separation of the moving body from the first projector, is the moved thing itself, not by reason of any power in it; for this moving thing and the moved thing cannot be distinguished. If you say that a new effect has some cause and local motion is a new effect, I say that local motion is not a new effect... because it is nothing else but the fact that the moving body is in different parts of space in such a manner that it is not in any one part, since two contradictories are not both true....

It is clear how "now before" and "now after" are to be assigned, treating

[12] K. Michalski, "Les Courants philosophiques à Oxford et à Paris pendant le XIVe siècle," *Bulletin international de L'Académie polonaise des sciences et des lettres, Classe d'histoire et de philosophie, et de philologie*, Les Années, 1919, 1920 (Cracovie, 1922), pp. 86–88.

[13] A. C. Crombie, *Robert Grosseteste and the Origins of Experimental Science, 1100–1700* (Oxford, 1953), pp. 176–77. In the quotations here given Ockham's rejection of motion as an entity distinct from the moving body is based on the principle of economy. But Ockham also finds other incoherencies in the doctrine that motion is a separate entity. See H. Shapiro, "Motion, Time, and Place according to William Ockham," *Franciscan Studies*, Vol. 16 (1956), 238–39, 248–50.

"now" first: this part of the moving body is now in this position, and later it is true to say that now it is in another position, and so on. And so it is clear that "now" does not signify anything distinct but always signifies the moving body itself which remains the same in itself, so that it neither acquires anything new nor loses anything existing in it. But the moving body does not remain always the same with respect to its surroundings, and so it is possible to assign "before and after," that is, to say: "this body is now at A and not at B," and later it will be true to say: "this body is now at B and not at A," so that contradictories are successively made true....

So far as an observer can tell, then, motion is simply the series of successively different position of one body relative to another. This ontological reduction of motion to its kinematic aspects, relatively detected, perhaps had some influence on the increasing attention paid by the early modern mechanicians to the kinematic aspects of motion and on their evident reluctance to treat of causes in a fashion other than kinematically.

As an example of a nominalist denial of movement as an entity, I have appended an extract from a *Questiones* on the *Physics*, erroneously ascribed to Duns Scotus and apparently composed by Marsilius of Inghen in the second half of the century (see Doc. 10.4). It will also be evident in this selection that in fact the concept of the relative detection of motion played an important part in the arguments on both sides of this question of whether motion is an entity separate from the space traversed and the moving body (see passages 3 and 5). Some of the most interesting arguments are of a cosmological nature, such as the possibility that God could move the whole universe in a rectilinear or circular motion if he wished (passage 3). The author also makes an interesting reference to certain people who "posit a place outside of the heavens, or an infinite space" (see passage 5). This infinite space is suggested as a kind of referent for a possible rectilinear or circular movement of the whole world. It is mainly because of the relationship of mechanical considerations and cosmological arguments that I have included this brief treatment of the ontology of motion in this chapter.

CENTER OF GRAVITY

It has been noticed that when Buridan discussed the problem of whether the earth is at rest at the center of the world, he concluded that although it does not rotate yet it moves in an oscillating movement of translation *in order that the center of gravity which is constantly changing will coincide with the center of the world*. In short, he considered the earth as a whole acting as if its weight were concentrated at a point. The doctrine of centers of gravities that plays such an important part in Archimedes' geometry and mechanics began in the fourteenth century, then, to play a role in the mechanics of large bodies. It is a moot question whether Archimedes directly influenced the Parisian scholastics in the use of the concept of center of gravity in their exposition of celestial mechanics. I would conclude that there is no direct influence. But it will be remembered that the works of Archimedes were translated virtually as a whole by William Moerbeke in 1269, and that there is evidence that parts of this translation were known at Paris in the fourteenth century. It is not specifically known, however, whether Moerbeke's translation of the *De centris graviuum vel de planis eque repentibus* was known. But it is one of the few works of the Moerbeke translation extant today in more than one manuscript copy.[14] Some indirect influence on medieval statics of the doctrine of centers of gravities has already been shown in Chapters 1 and 2.

In connection with Buridan's discussion of center of gravity, it will be noticed (Doc. 10.1, passage 11) that he is concerned with the difficulty as to whether the center of gravity of the earth is its center of magnitude,[15] for he obviously would agree with Marcus Trivisano—his junior contemporary—that in regular (homogeneous) figures, the center of gravity is at the center of magnitude (see Doc. 2.5). Buridan concludes in the case of the earth that they are not identical, since some parts of the earth are of greater density than others: "Now if one body in one part is lighter and in an opposite part is heavier, the center of gravity will not be the center of magnitude. Rather with the center of gravity given, the greater magnitude will be in the lighter part, just as is the case of balances if on one side a stone is placed and on the other side wool [and they balance], the wool will be of greater magnitude."

The interest in and concern with centers of gravities in celestial mechanics further impinged upon a question already hoary by the fourteenth century, namely that of the possible existence of a plurality or infinitude of worlds like the visible cosmos. This problem had a long history from the time of its formulation by pre-Socratic philosophers, where the two principal possibilities were presented: either that there has been a plurality or infinitude of worlds existing successively in time, or that there is a

---

[14] M. Clagett, "The Use of the Moerbeke Translations of Archimedes in the Works of Johannes de Muris," *Isis*, Vol. 43 (1952), p. 240.

[15] Duhem, *Études*, Vol. *1*, pp. 11–19, discusses the similar doctrine of Albert of Saxony, and see footnote 8 above.

plurality or infinitude of worlds coexisting simultaneously. Aristotle's rejection of at least the second doctrine is explicit and is argued from his view of a single finite cosmos of which the earth occupies the center: "There is therefore no infinite body beyond the heaven. Nor again is there anything of limited extent beyond it, and so beyond the heaven there is no body at all."[16] It was customary for most thirteenth-century authors to follow Aristotle in this rejection of the plurality concept (e.g., Michael Scot, Roger Bacon, Albertus Magnus, and Thomas Aquinas).[17] However, in 1277 it was made clear in the doctrines condemned by Étienne Tempier, bishop of Paris, that however reasonable in philosophy the doctrine of a single cosmos might be, one could not, without falling into heresy, say that it is impossible for God to make a plurality of worlds *(quod prima causa non possit plures mundos facere)*.[18] This would be placing a limitation on the plenitude of God's power. Following the spirit of this condemnation, many schoolmen admitted God's power to produce a plurality of worlds, even when they rejected the doctrine on philosophical grounds. In the course of their discussions, they admitted the possibility of the following alteration of the doctrine of the finite cosmos with the earth at the center: If some part of the elements forming a world was detached from it and driven away, then the detached element would tend to move toward another world closer to it. The elements of each world would be inclined to arrange themselves concentrically about the center of gravity of their own world. Now the doctrine of center of gravity bears on this question most clearly in the treatment of Oresme. After saying that there are three ways in which a plurality of worlds can be conceived, he discusses the common way, i.e., the coexistence of worlds, each with its own center. In the course of this discussion he tells us:[19]

... that all the heavy things of this world tend to be conjoined in one mass *(masse)* such that the center of gravity *(centre de pesanteur)* of this mass is the place or center of this world, and the whole constitutes a single body in number. And consequently they all have one [natural] place according to number. And if a part of the [element] earth of the other world was in this world, it would tend towards

---

[16] *De caelo*, (edition of D. J. Allan, Oxford, 1936), I.7.275b, lines 7–9. See also 276a.18 and following where Aristotle explains "why there cannot even be more than one world."

[17] Duhem, *Études*, Vol. 2, 57–96, 408–423, gives a review of the medieval history of the plurality problem through the time of Leonardo da Vinci.

[18] *Ibid.*, Vol. 2, 75–76.

[19] Nicole Oresme, *Le Livre du ciel et du monde*, edition of A. D. Menut and A. J. Denomy in *Mediaeval Studies*, Vol. 3 (1941), 239–44, gives a rather complete discussion of the problem. The particular passage here quoted is on pages 243–44.

the center of this world and be conjoined to its mass, and vice versa. But it does not accordingly follow that the parts of the [element] earth or heavy things of the other world (whatever it is) tend to the center of this world, for in their world they would make a mass *(masse)* which would be a single body according to number and which would have a single place according to number, and which would be ordered according to high and low [in respect to its own center] just as is the mass of heavy things in this world.... I conclude, then, that God can and would be able by his omnipotence *(par toute sa puissance)* to make another world other than this one, or several of them either similar or dissimilar, and Aristotle offers no sufficient proof to the contrary. But as it was said before, *in fact (de fait)* there never was nor will there be any but a single corporeal world....

With this brief glimpse of some of the problems of applying mechanical ideas to cosmology, we can terminate our treatment of the more important mechanical questions that stimulated discussion in the fourteenth century. We are now prepared in the fourth part of our volume to treat briefly the fate of the most original of the mechanical ideas in the late medieval and early modern period.

Document 10.1

John Buridan, *Questions on the Four Books on the Heavens and the World of Aristotle**

1. BOOK II, QUESTION 22. It is sought consequently whether the earth always is at rest in the center of the universe....

This question is difficult. For in the first place there is a significant doubt as to whether the earth is directly in the middle of the universe so that its center coincides with the center of the universe. Furthermore, there is a strong doubt as to whether it is not sometimes moved rectilinearly as a whole, since we do not doubt that often many of its parts are moved, for this is apparent to us through our senses. There is also another difficult doubt as to whether the following conclusion of Aristotle is sound, namely, if the heaven is by necessity to be moved circularly forever, then it is necessary that the earth be at rest forever in the middle. There is also a fourth doubt whether, in positing that the earth is moved circularly around its own center and about its own poles, all the phenomena that are apparent to us can be saved *(possent salvari omnia nobis apparentia)*. Concerning this last doubt let us now speak.

2. It should be known that many people have held as probable that it is not contradictory to appearances for the earth to be moved circularly in the aforesaid manner, and that on any given natural day it makes a complete rotation from west to east by returning again to the west—that is, if some part of the earth were designated [as the part to observe]. Then it is necessary to posit that the stellar sphere would be at rest, and then night and day would take place through such a motion of the earth, so that that motion of the earth would be a diurnal motion *(motus diurnus)*. The following is an example of this [kind of thing]: If anyone is moved in a ship and he imagines that he is at rest, then, should he see another ship which is truly at rest, it will appear to him that the other ship is

* Translated from the Latin edition of E. A. Moody (Cambridge, Mass., 1942), pp. 226–32.

moved. This is so because his eye would be completely in the same relationship to the other ship regardless of whether his own ship is at rest and the other moved, or the contrary situation prevailed. And so we also posit that the sphere of the sun is everywhere at rest and the earth in carrying us would be rotated. Since, however, we imagine that we are at rest, just as the man located on the ship which is moving swiftly does not perceive his own motion nor the motion of the ship, then it is certain that the sun would appear to us to rise and then to set, just as it does when it is moved and we are at rest.

3. It is true, however, that if the stellar sphere is at rest, it is necessary to concede generally that the spheres of the planets are moving, since otherwise the planets would not change their positions relative to each other and to the fixed stars. And, therefore, this opinion imagines that any of the spheres of the planets moved evidently like the earth from west to east, but since the earth has a lesser circle, hence it makes its rotation *(circulatio)* in less time. Consequently, the moon makes its rotation in less time than the sun. And this is universally true, so that the earth completes its rotation in a natural day, the moon in a month, and the sun in a year, etc.

4. It is undoubtedly true that, if the situation were just as this position posits, all the celestial phenomena would appear to us just as they now appear. We should know likewise that those persons wishing to sustain this opinion, perhaps for reason of disputation, posit for it certain persuasions.... The third persuasion is this: To celestial bodies ought to be attributed the nobler conditions, and to the highest sphere, the noblest. But it is nobler and more perfect to be at rest than to be moved. Therefore, the highest sphere ought to be at rest....

The last persuasion is this: Just as it is better to save the appearances through fewer causes then through many, if this is possible, so it is better to save [them] by an easier way than by one more difficult. Now it is easier to move a small thing than a large one. Hence it is better to say that the earth, which is very small, is moved most swiftly and the highest sphere is at rest than to say the opposite.

5. But still this opinion is not to be followed. In the first place because it is against the authority of Aristotle and of all the astronomers *(astrologi)*. But these people respond that *authority does not demonstrate*, and that it suffices astronomers that they posit a method by which appearances are saved, whether or not it is so in actuality. Appearances can be saved in either way; hence they posit the method which is more pleasing to them.

6. Others argue [against the theory of the earth's diurnal rotation] by

many appearances *(apparentiis)*. One of these is that the stars sensibly appear to us to be moved from the east to the west. But they solve this [by saying] that it would appear the same if the stars were at rest and the earth were moved from west to east.

7. Another appearance is this: If anyone were moving very swiftly on horseback, he would feel the air resisting him. Therefore, similarly, with the very swift motion of the earth in motion, we ought to feel the air noticeably resisting us. But these [supporters of the opinion] respond that the earth, the water, and the air in the lower region are moved simultaneously with diurnal motion. Consequently there is no air resisting us.

8. Another appearance is this: Since local motion heats, and therefore since we and the earth are moved so swiftly, we should be made hot. But these [supporters] respond that motion does not produce heat except by the friction *(confricatio)*, rubbing, or separation of bodies. These [causes] would not be applicable there, since the air, water, and earth would be moved together.

9. But the last appearance which Aristotle notes is more demonstrative in the question at hand. This is that an arrow projected from a bow directly upward falls again in the same spot of the earth from which it was projected. This would not be so if the earth were moved with such velocity. Rather before the arrow falls, the part of the earth from which the arrow was projected would be a league's distance away. But still the supporters would respond that it happens so because the air, moved with the earth, carries the arrow, although the arrow appears to us to be moved simply in a straight line motion because it is being carried along with us. Therefore, we do not perceive that motion by which it is carried with the air. But this evasion is not sufficient because the violent impetus of the arrow in ascending would resist the lateral motion of the air so that it would not be moved as much as the air. This is similar to the occasion when the air is moved by a high wind. For then an arrow projected upward is not moved as much laterally as the wind is moved, although it would be moved somewhat....

10. Then I come to the other doubts. One would be whether the earth is situated directly in the middle of the universe. It should be answered in the affirmative. For we suppose that the place [designated] absolutely *(simpliciter)* as "upward," insofar as one looks at this lower world, is the concave [surface] of the orb of the moon. This is so because something absolutely light, i.e., fire, is moved toward it. For since fire appears to ascend in the air, it follows that fire naturally seeks a place above the air,

and this place above the air is at the concave [surface] of the orb of the moon; because no other element appears to be so swiftly moved upward as fire. Now the place downward ought to be the maximum distance from the place upward, since they are contrary places. Now that which is the maximum distance from the heaven is the middle of the universe. Therefore the middle of the universe is absolutely downward. But that which is absolutely heavy—and earth is of this sort—ought to be situated absolutely downward. Therefore, the earth naturally ought to be *in* the middle of the universe or *be* the middle of the universe.

11. But it is a significant difficulty as to whether the center of magnitude in the earth is the same as the center of gravity *(medium gravitatis)*. It seems according to some statements that it is not. This is because if a large region of the earth is not covered with waters due to the habitation of animals and plants, and the opposite part is covered with waters, it is clear that the air which is naturally hot, and the sun, make hot the non-covered part, and thus they make it to some degree more subtle, rarer, and lighter. The covered part remains more compact and heavier. Now if one body in one part is lighter and in an opposite part heavier, the center of gravity will not be the center of magnitude. Rather with the center of gravity given, the greater magnitude will be in the lighter part, just as in the case of balances if on one side a stone is placed and on the other side wool [and they balance], the wool will be of a much greater magnitude.

12. With this understood, it ought to be seen which of those centers is the center of the universe. It should be answered immediately that the center of the universe is the center of gravity of the earth. This is because, as Aristotle says, all parts tend toward the center of the universe through their gravity, and a part which is heavier would displace another, and thus finally it is necessary that the center of the universe coincide with the center of gravity. From these arguments it follows that the earth is nearer to the heaven in the part not covered with waters than in the covered part, and thus at the covered part there is greater declivity, and so the waters flow to that part. So, therefore, the earth, with respect to its magnitude, is not directly in the center of the universe. We commonly say, however, that it is in the center of the universe, because its center of gravity is the center of the universe.

13. By this another doubt is solved, evidently, whether the earth is sometimes moved according to its whole in a straight line. We can answer in the affirmative because from this higher [part of] the earth many parts of the earth (i.e., debris) continually flow along with the rivers to the bot-

tom of the sea, and thus the earth is augmented in the covered part and is diminished in the uncovered part. Consequently, the center of gravity does not remain the same as it was before. Now, therefore, with the center of gravity changed, that which has newly become the center of gravity is moved so that it will coincide with the center of the universe, and that point which was the center of gravity before ascends and recedes, and thus the whole earth is elevated toward the uncovered part so that the center of gravity might always become the center of the universe. And just as I have said elsewhere, it is not apparent how it could be saved unless the mountains were consumed and destroyed sometimes, nay infinite times, if time were eternal. Nor is any other way apparent by which such mountains could be generated. This was spoken of elsewhere, so I shall now desist....

COMMENTARY

The first passage outlines the various "doubts." In passage 2 the relative nature of the perception of motion is introduced: "If anyone is moved in a ship and he imagines that he is at rest, then, should he see another ship which truly is at rest, it will appear to him that the other ship is moved." In the same way, an observer on a moving earth would think he was at rest and the heaven, truly at rest, he would believe to be in motion. On the other hand (passage 3) the planets change their positions relative to each other and the fixed stars. Hence they are in movement. But all celestial phenomena (passage 4) would appear to us as they do regardless of whether the heavens or the earth rotate.

Furthermore, according to Buridan, certain persuasions can be introduced to support the diurnal rotation of the earth. After presenting some that involve the natural superiority of the heavens over the earth, Buridan in the last persuasion gives the principle of economy or simplicity; saying that it is better to save appearances by the simpler and easier movement, i.e., the rotation of the earth.

In passage 5 Buridan notes that the authority of Aristotle is against the diurnal rotation of the earth. But the holders of the theory of the earth's rotation would however answer rightfully, "Authority does not demonstrate." Furthermore, astronomers are only interested in finding the simplest way to save appearances. They are not concerned with whether the theory is so in actuality. Here Buridan is assigning the same sort of role to astronomers as did the Greek astronomer Geminus, whom Simplicius quotes as saying that the astronomer merely seeks hypotheses by the assumption of which the phenomena will be saved; it is not his business, but rather that of the physicist, to determine which hypothesis is actually more suitable to nature.[20]

In passage 7 Buridan presents the argument that, as the result of the diurnal rotation of the earth, one would expect a continually resisting wind. But the answer given is that the earth, water, and air are moved together. This would also explain why there is no heat generated by the friction between the air and the moving earth (passage 8). Finally, Buridan presents as an answer to the apparent rectilinear ascent and descent of arrows or projectiles by saying that the arrows are carried along with the air in the same diurnal motion as the earth. Buridan says this is not true, for the vertical motion of the arrow would resist the air. Perhaps Oresme, later, realized that it is not the air that causes the arrow to move laterally but rather the fact it is a part of the mechanical system of the rotating earth, just as a man moving his hand vertically up and down a mast is part of the mechanical system of the boat.

In passage 11–13, Buridan argues that geological processes are always causing a redistribution of the matter of the earth; hence they are continually changing its center of gravity. But the center of gravity always strives to be at the center of the universe. Hence the earth is constantly shifting about near the center of the universe.

[20] M. R. Cohen and I. E. Drabkin, *A Source Book in Greek Science* (New York, 1948), pp. 90–91.

Document 10.2

Nicole Oresme, *On the Book of the Heavens and the World of Aristotle**

1. BOOK II, CHAPTER 25. Afterwards he (Aristotle) sets forth another opinion.

TEXT: And some say that the earth is at the center of the universe and that it revolves and moves circularly around the pole established for this, just as is written in the book of Plato called the *Timaeus*.

GLOSS: This was the opinion of one called Heraclitus Ponticus who proposed that the earth is moved circularly and that the heavens are at rest. Aristotle does not here refute these opinions, perhaps because it seemed to him that they have little [root in] appearance and are pretty well refuted elsewhere in philosophy and astronomy.

But it seems to me, subject to correction, that one could support well and give luster to the last opinion, namely, that the earth, and not the heavens, is moved with a daily movement. Firstly, I wish to state that one could not demonstrate the contrary by any experience *(experience)*. Secondly [I will show that the contrary cannot be demonstrated] by reasoning. And thirdly, I will put forth reasons in support of it (i.e., the diurnal rotation of the earth).

2. As for the first point, one experience [commonly cited in support of the daily motion of the heaven is the following]: We see with our senses the sun and moon and many stars rise and set from day to day, and some stars turn around the arctic pole. This could not be except by the movement of the heavens, as was demonstrated in chapter 26. Thus [the argument runs], the heaven is moved with a diurnal movement. Another experience [cited] is [this]: If the earth is so moved, it makes a complete turn in a single natural day. Therefore, we and the trees and the houses

---

* Translated from the Old French edition of A. D. Menut and A. J. Denomy, in *Mediaeval Studies*, Vol. *4* (1942), 270–79.

are moved toward the east very swiftly, and so it should seem to us that the air and the wind blow continuously and very strongly from the east, [much] as it does against a quarrel shot, [only] very much more strongly. But the contrary appears by experience. The third [experience] is that which Ptolemy advances: If a person were on a ship moved rapidly eastward and an arrow were shot directly upward, it ought not to fall on the ship but a good distance westward away from the ship. Similarly, if the earth is moved so very swiftly in turning from west to east, and it has been posited that one throws a stone directly above, then it ought to fall, not on the place from which it left, but rather a good distance to the west. But in fact the contrary is apparent.

It seems to me that by [using] what I shall say regarding these experiences, one could respond to all the other [experiences] which might be adduced in this matter....

3. ... Again, I make the supposition that local motion can be sensibly perceived only in so far as one may perceive one body to be differently disposed with respect to another. In support of this [I give the following illustration]: If a person is in one ship called *a* which is moved very carefully [i.e., without pitching or rolling]—either rapidly or slowly—and this person sees nothing except another ship called *b*, which is moved in every respect in the same manner as *a* in which he is situated, I say that it will seem to this person that neither ship is moving. And if *a* is at rest and *b* is moved, it will appear and seem to him that *b* is moved. [On the other hand], if *a* is moved and *b* is at rest, it will appear to him as before that *a* is at rest and that *b* is moved. And thus, if *a* were at rest for an hour and *b* were moved, and then immediately in the following hour the situation were reversed, namely, that *a* were moved and *b* were at rest, this person [on *a*] could not perceive this mutation or change. Rather it would continually seem to him that *b* was moved; and this is apparent by experience. The reason for this is because these two bodies, *a* and *b*, are continually changing their dispositions with respect to each other in the same manner throughout when *a* is moved and *b* is at rest as they were conversely when *b* is moved and *a* is at rest. This is apparent in the fourth book of *The Perspective* of Witelo,[21] [who says] that one can perceive movement only in such

---

[21] Witelo, *Opticae libri decem* (Basel, 1572), Bk. IV, prop. 110, p. 167: "Quoniam enim moveri est aliter se habere quam prius, palam quod facilitas huius comprehensionis motus fit ex comparatione rei motae ad aliud visibile quiescens non motum. Quando enim comprehenditur situs unius rei mobilis respectu alterius rei visibilis, tunc etiam comprehenditur diversitas situs eius respectu illius visibilis, et tunc comprehenditur motus." Cf. Aquinas, *op. cit.* in note 6, Bk. II, Lect. 12, n. 4:

a way as one perceives one body to be differently disposed in comparison with another. I say, then, that if the upper of the two parts of the universe mentioned above should today move with a diurnal movement, just as it is, and the lower part should not, and tomorrow the contrary should prevail, [namely] that the lower should be moved with a diurnal movement while the upper (i.e., the heavens) should not, we could not perceive this change in any way, but everything would seem to be the same today and tomorrow. It would seem to us continually that the part where we are situated was at rest and that the other part was always moved, just as it seems to a person who is in a moving ship that the trees outside are moved. Similarly, if a person were in the heavens and it were posited that they were moved with a diurnal movement, and [furthermore] that this man who is transported with the heaven could see the earth clearly and distinctly and its mountains, valleys, rivers, towns, and chateaux, it would seem to him that the earth was moved with a diurnal movement, just as it seems to us who are on the earth that the heavens move. Similarly, if the earth and not the heavens were moved with a diurnal movement, it would seem to us that the earth was at rest and the heavens were moved. This can be imagined easily by anyone with good intelligence. From this [reasoning] is evident the response to the first experience, since one could say that the sun and the stars appear thus to set and rise and the heavens to turn as the result of the movement of the earth and its elements where we are situated.

4. To the second experience, according to this opinion, the response is [this]: Not only is the earth so moved [diurnally], but with it the water and the air, as was said, in such a way that the water and the lower air are moved differently than they are by winds and other causes. It is like this situation: If air were enclosed in a moving ship, it would seem to the person situated in this air that it was not moved.

5. To the third experience, which seems more effective, i.e., the experience concerning the arrow or stone projected upward etc., one would say that the arrow is trajected upwards and [simultaneously] with this trajection it is moved eastward very swiftly with the air through which it passes and with all the mass of the lower part of the universe mentioned above, it all being moved with a diurnal movement. For this reason the arrow returns to the place on the earth from which it left. This appears possible by analogy: If a person were on a ship moving toward the east

---

"... eo quod nihil differt quantum ad hoc quod aliquid videatur moveri, utrum moveatur visus vel res quae videtur; sicut patet de illis qui navigant circa littora, quod, quia ipsi sunt in motu videtur eis quod montes et terra moveantur."

[Oresme, *On the Heavens and the World* 603]

very swiftly without his being aware of the movement, and he drew his hand downward, describing a straight line against the mast of the ship, it would seem to him that his hand was moved with rectilinear movement only. According to this opinion [of the diurnal rotation of the earth], it seems to us in the same way that the arrow descends or ascends in a straight line....

...In support of this [position, consider the following]: If a man in that ship were going westward less swiftly than the ship was going eastward, it would seem to him that he was approaching the east, when actually he would be moving toward the west. Similarly, in the case put forth above, all the movements would seem to be as if the earth were at rest.

Also, in order to make clear the response to the third experience, I wish to add a natural example verified by Aristotle to the artificial example already given. It posits in the upper region of the air a portion of pure fire called a. This latter is of such a degree of lightness that it mounts to its highest possible point b near the concave surface of the heavens. I say that just as with the arrow in the case posited above, there would result in this case [of the fire] that the movement of a is composed of rectilinear movement, and, in part, of circular movement, because the region of the air and the sphere of fire through which a passes are moved, according to Aristotle, with circular movement. Thus if it were not so moved, a would ascend rectilinearly in the path ab, but because b is meanwhile moved to point c by the circular daily movement, it is apparent that a in ascending describes the line ac and the movement of a is composed of a rectilinear and a circular movement. So also would be the movement of the arrow, as was said. Such composition or mixture of movements was spoken of in the third chapter of the first book [of the *De caelo*].... I conclude then that one could not by any experience whatsoever demonstrate that the heavens and not the earth are moved with diurnal movement.

6. As to the second point relative to the rational demonstration [of the diurnal movement of the heavens, I first note the following]: It seems to me that this [rational demonstration] proceeds from these arguments which follow and to which I shall respond in such a fashion that, using the same reasoning, one could respond to all other arguments pertaining to it....

Again, if the heavens were not moved with diurnal movement, all astronomy *(astrologie)* would be false and a great part of natural philosophy where one supposes throughout this movement of the heavens.

Also, this seems to be against the Holy Scripture which says [in Eccles.

1 : 5–6]: "The sun riseth, and goeth down, and returneth to his place: and there rising again, maketh his round by the south, and turneth again to the north: the spirit goeth forward surveying all places round about, and returneth to his circuits" (Douay translation of Vulgate). And so it is written of the earth that God made it immobile: "For [God] created the orb of the earth, which will not be moved."

Also, the Scriptures say that the sun was halted in the time of Joshua (see Josh. 10 : 12–14) and that it returned (i.e., turned back) in the time of King Ezechias (Hezekiah; see Isa. 38 : 8; II Kings 20 : 11 [Vulgate, IV Kings 20 : 11]). If the earth were moved and not the heavens, as was said, such an arrestment would have been a returning and the returning of which it speaks would rather have been an arrestment. And this is against what the Scriptures state....

7. To the fifth argument, where it is said that if the heavens would not make a rotation from day to day, all astronomy would be false, etc., I answer this is not so because all aspects, conjunctions, oppositions, constellations, figures, and influences of the heavens would be completely just as they are, as is evident clearly from what was said in response to the first experience. The tables of the movements and all other books would be just as true as they are, except that in regard to the daily movement one would say of it that it is in the heavens "apparently" *(selon apparence)* but in the earth "actually" *(selon verité)*. There is no effect which follows from the one [assumption] more than from the other. Apropos of this is the statement of Aristotle in the sixteenth* chapter [of the second book of the *De caelo*], namely, that the sun appears to us to turn and the stars to sparkle and twinkle because, he says, it makes no difference whether the thing one sees is moved or the sight is moved. Also one would say apropos of this matter [of diurnal rotation] that our sight is moved with diurnal rotation.

8. To the sixth argument concerning the Holy Scripture which says that the sun revolves, etc., one would say of it that it is in this part [simply] conforming to the manner of common human speech, just as is done in several places, e.g., where it is written that God is "repentent" and that he is "angry" and "pacified" and other such things which are not just as they sound. Also appropriate to our question, we read that God covers the heavens with clouds—"who covereth the heavens with clouds" (Ps. 146 : 8)—and yet in reality the heavens cover the clouds. Thus one would say that according to appearances the heavens and not the earth are moved

---

* Actually the eighth chapter.

with a diurnal motion, while in actuality the contrary is true. Concerning the earth, one would say it is not moved *from* its place in actuality, nor *in* its place apparently, but that it is moved *in* its place actually. To the seventh argument, one would answer in just about the same way, that according to appearances in the time of Joshua the sun was arrested and in the time of Ezechias it returned, but actually the earth was arrested in the time of Joshua and advanced or speeded up its movement in the time of Ezechias. It would make no difference as to effect whichever opinion was followed. This latter opinion [supporting the diurnal rotation of the earth] seems to be more reasonable than the former, as we shall make clear later.

As to the third [main] point [of this gloss], I wish to put forth persuasions or reasons by which it would appear that the earth is moved as was indicated....

9. Again, all philosophers say that something done by several or large-scale operations which can be done by less or smaller operations is done for nought. And Aristotle says in the eighth* chapter of the first book that God and Nature do not do anything for nought. But if it is so that the heavens are moved with a diurnal movement, it becomes necessary to posit in the principal bodies of the world and in the heavens two contrary kinds of movement, one of an east-to-west kind and others of the opposite kind, as has been said often. With this [theory of the diurnal movement of the heavens] it becomes necessary to posit an excessively great speed. This will become clear to one who considers thoughtfully the height or distance of the heaven, its magnitude, and that of its circuit; for if such a circuit is completed in one day, one could not imagine nor conceive of how the swiftness of the heaven is so marvelously and excessively great. It is so unthinkable and inestimable. Since all the effects which we see can be accomplished, and all the appearances saved, by substituting for this [diurnal movement of the heavens] a small operation, i.e., the diurnal movement of the earth, which is very small in comparison with the heavens, and [since this can be done] without making the [number of necessary] operations so diverse and outrageously great, it follows that [if the heaven rather than the earth is moved] then God and Nature would have made and ordained things for nought. But this is not fitting, as was said.

Again, when it has been posited that the whole heavens are moved with daily movement and in addition that the eighth sphere is moved with another movement, as the astronomers posit, it becomes necessary, ac-

* Actually the fourth chapter.

cording to them, to assume a ninth sphere which is moved with a daily movement only. But when it has been posited that the earth is moved as was said, the eighth sphere is moved with a single slow movement and thus it is not necessary with this theory to dream up or imagine a ninth natural sphere, invisible and without stars; for God and Nature would not have made this sphere for nought, since all things can be as they are by using another method....

10. It is apparent, then, how one cannot demonstrate by any experience whatever that the heavens are moved with daily movement, because, regardless of whether it has been posited that the heavens and not the earth are so moved or that the earth and not the heavens is moved, if an observer *(ouyl)* is in the heavens and he sees the earth clearly, it (the earth) would seem to be moved; and if the observer were on the earth, the heavens would seem to be moved. The sight is not deceived in this, because it senses or sees nothing except that there is movement. But if it is relative to any such body, this judgment is made by the senses from inside that body, just as he [Witelo] stated in *The Perspective;* and such senses are often deceived in such cases, just as was said before concerning the person who is in the moving ship. Afterwards it was demonstrated how it cannot be concluded by reasoning that the heavens are so moved. Thirdly, reasons have been put forth in support of the contrary position, namely that [the heavens] are not so moved. Yet, nevertheless, everyone holds, and I believe, that they (the heavens), and not the earth, are so moved, for "God created the orb of the earth, which will not be moved" (Ps. 92 : 1), notwithstanding the arguments to the contrary. [This is] because they are "persuasions" which do not make the conclusions evident. But having considered everything which has been said, one could by this believe that the earth and not the heavens is so moved, and there is no evidence to the contrary. Nevertheless, this seems prima facie as much, or more, against natural reason as are all or several articles of our faith. Thus, that which I have said by way of diversion *(esbatement)* in this manner can be valuable to refute and check those who would impugn our faith by argument.

COMMENTARY

1. In passage 1, we see that Oresme raises the question of the earth's rotation in connection with the celebrated and much disputed passage in Aristotle's *De caelo* where he refers to the *Timaeus*. Oresme joins the few critics who have interpreted this statement to mean that Plato assumed a

rotating earth.[22] Oresme's argument, as he outlines it here, is to be threefold: (1) One cannot argue against the earth's rotation by an appeal to observation or experience; (2) one cannot argue against it by any test of reason; and (3) one can advance persuasions or reasons for its support.

2. Turning to the observations commonly adduced to refute the earth's rotation, Oresme in passage 2 lists them tersely: (1) Our senses show that the heavens are rotating; (2) there ought to be a prevailing wind from the east caused by the earth's rotation; and (3) a stone or arrow thrown vertically into the air ought to fall to the ground some distance to the west of the spot from which it was thrown.

3. In passage 3, Oresme shows, by advancing the doctrine of the complete subjective relativity of the detection of motion, that there would be no way to tell with the senses whether it was the heavens or the earth which was rotating. The analogy he gives is the familiar one of an observer on a moving ship who considers himself at rest and the other ship in motion so long as there is a perceptible change in the distance between the ships.

4. Oresme then proceeds in passage 4 to answer the second "experience" adduced against the diurnal rotation of the earth, namely the argument that rotation should produce prevailing winds. His answer is that the earth, air, and water all *share* the same movement of rotation. The situation is comparable to a person in a cabin in a ship. The air in the cabin shares the movement of the ship, and thus the observer in the cabin believes it to be at rest.

5. Oresme answers in passage 5 the third "experience" against the theory of diurnal rotation of the earth, i.e., the experience of the projectile or rock shot into the air vertically. Its movement appears rectilinear because it shares the same diurnal rotation as the earth and the air. The analogy advanced to illustrate Oresme's answer is the observer on the moving ship. He moves his hand (which of course shares his motion) up and down the mast of the ship. And it thus *appears to him* that his hand is moved rectilinearly. In relation, then, to his mechanical system, his hand is moving rectilinearly.

6. Oresme is now ready to consider the rational arguments generally used against the diurnal rotation of the earth and for that of the heavens. Among those arguments we have included, in passage 6, the following: (1) All astronomy would be false unless the diurnal movement of the

---

[22] For the literature on this question see H. F. Cherniss, *Aristotle's Criticism of Plato and the Academy* (Baltimore, 1944), pp. 545–65.

heavens was assumed; (2) the Scriptures expressly state (Eccles. 1 : 5–6) that the sun is in daily motion, while the earth is said to be immobile; and (3) the Scriptures further state that Joshua stopped the sun.

7–8. In passage 7, Oresme answers that astronomy would just as well be saved with the earth's diurnal rotation as with that of the heavens. All aspects, conjunctions, oppositions, constellations would remain unchanged. Furthermore, we can answer (see passage 8) the arguments based on scriptural statements by saying that the Holy Scripture in these statements is merely "conforming to the manner of common human speech."

9. In turning (passage 9) to the third main part of his treatment, Oresme advances further these positive reasons, among others, in support of the earth's rotation. (1) It would be simpler to posit a modest speed for the earth than the excessively great speed demanded of the diurnal rotation of the heavens. The principle of economy to the effect that God does nothing in vain has given us the basic reason for rejecting the enormous velocity of the heavens. (2) We can reduce the number of spheres from nine to eight, and the principle of economy would again be saved.

10. In the final passage (10) Oresme notes the completeness of his argument but admits that the arguments on the earth's rotation are not conclusive but only as persuasive as those against it. Then he goes on to say "nevertheless, everyone holds, and I believe, that they (the heavens), and not the earth, are so moved." The very kind of scriptural quotation which he has already answered he uses as his reason. This whole selection shows Oresme at his best and illustrates once more that the criticism of some of the basic tenets of the ancient natural philosophy and determined.

It should be observed finally that Oresme had treated the rotation of the earth in what appears to be one of his very early works, his *Questiones de spera*.[23] He perhaps also treated it in his Latin *Questiones super libros de celo*, but the text of that work was not available to me.

---

[23] *Questiones de spera*, MS Florence, Bibl. Riccard. 117, f. 130r: "... ipsa autem terra et alia elementa usque ad ignem qualibet die revolvuntur et hoc de oriente ad occidens, ita quod ista moventur motu diurno.... Nunc vero circa tertium principale pono ymaginationes. Primo arguitur per rationes sumptas ab experientiis. Prima est ex visu, quia videtur quod apareat cuilibet ad visum quod cellum (!) movetur, quia nos videmus stellas in oriente et postea in occidente. Secunda experientia sumitur ex aere et vento; quia si terra sic moveretur, tunc domus aliqua valde cito iret versus orientem et tamen cum aer impediret apareret quod esset ventus sicut quando hoc esset. Tertia experientia est de sagita, quia tunc sagita proiecta versus occidentem non moveretur, ymo posset moveri versus orientem quia terra versus

eam accureret. Quarta experientia est de lapide qui si dimittatur cadere iam non caderet in loco qui esset sub ipso et causa est quia iste locus iam movisset (?) motu terre versus orientem. Quinto quod lapis descendens velocius moveretur sursum quam deorsum et in principio quam in fine, cuius oppositum videtur. Consequentia patet, quia lapis descendendo cum hoc motu circulari iret versus oriens et sic describeret maiorem lineam sursum quam deorsum. Deinde arguitur aliis rationibus: primo per rationem Aristotelis secundo celli, quia nisi cellum moveretur iam sol et luna non fierent quandoque propinquiores et quandoque remotiores, cuius oppositum videtur; secundo omne grave naturale quiescit in loco suo, igitur terra non movetur in loco suo; tertio aut hoc esset naturaliter et hoc non quia unus motus simplex est unius corporis simplicis primo celli, aut violente et hoc non quia nullum violentum est perpetuum primo metheorologicarum et secundo celli; quarto quare a quo movetur; quinto videtur quod hec positio sit destructio totius astrologie. Ad hoc ad primum dico quod per visum non videmus plus nec habemus plus nisi quod aliquid movetur circulariter, nec est in consequens visum talem decipi sicut aparet de homine et navi, et istudmet confitetur commentator secundo de anima ubi dicit quod visus decipitur quandoque de suo loco visibili ubi est secundum solutum est quia positum est quod ignis et aer et aqua moventur sicut terra, nisi quod quanto impeditur vel iuvatur a vento. Et ad tertium concedo quod sagita non potest moveri versus orientem nisi proiiceretur [amplius] quam terra moveatur et ad salvandum aparentiam suficit quod ipsa minus velociter movetur ad occidentem expectando illud contra quod proiicitur. Ad quartum de lapide dico quod lapis caderet in loco qui est sub ipso, quia dum caderet ipse cum hoc moveretur versus orientem ita quod movetur motu quodam composito ex circulari et recto. Ad quintum concedo quod est possibile quod lapis velocius movetur sursum quam deorsum, tamen non velocius descendit, quia ex alia questione patet quod aliquid potest moveri velocius quam descendat. Ad rationem Aristotelis ipsa est soluta in positione quia ponitur quod sol in anno faciat unam revoluptionem et luna in mense et ex hoc sic variantur. Ad aliam concedo quod corpus quiescit in loco suo naturali, sic quod non movetur, id est, non recedit ab illo, sed non sic quin moveatur in illo sicut patet de igne qui movetur in spera sua. Ad tertiam potest dici quod movetur ab una intelligentia vel a quadem alia natura. Ad quartam dico quod ita bene astrologia quo ad tabulas et quo ad omnia alia salvaretur sicut nunc, quia omnes aspectus provenirent sicut proveniunt equinocialiter. Nunc vero dico quantum ad quartum principale quod veritas est quod terra non sic movetur sed cellum. Dico tamen quod conclusio non potest demonstrari sed persuaderi, sicut patuit ex positione oposita et ideo est credita." It is evident some of the same arguments found developed in Document 10.2 are suggested here. As in the later document Oresme rejects the rotation of the earth, but notes that the diurnal rotation of the heavens "cannot be demonstrated but only argued by persuasion."

Document 10.3

Nicolas Copernicus, *On the Revolutions of the Celestial Orbs**

1. BOOK I, CHAPTER 7: *Why the ancients believed the earth to be at rest in the middle of the universe as its center.*—For this reason the ancient philosophers tried by certain other arguments to prove that the earth is fixed in the middle of the universe. The most powerful cause they allege [for it] is from [the doctrine of] the heavy and the light. For the element earth is the heaviest, and all things of weight move toward it, tending to its center. Hence, since the earth is spherical, and heavy things by nature move vertically to its surface from all sides, they would all rush together to its center if not stopped at the surface, since indeed a straight line perpendicular to a plane tangential to the sphere leads to the center. Now those things which move toward the center must, it seems, on reaching it, remain at rest. Much more, then, will the whole earth remain at rest at the center [of the universe]; and, receiving to itself everything which falls, it will remain immobile by its own weight.

To prove the same thing, they also employ a reasoning based on motion and its nature. Indeed, Aristotle says that the motion of a single and simple body is simple. A simple motion may be either rectilinear or circular. Again, a rectilinear motion may be either up or down. So every simple motion must be either toward the center, namely downward, or away from the center, namely upward, or around the center, namely circular. Now it is a property only of the heavy elements, earth and water, to move downward, that is, to seek the center. But [it is the property] of the light elements, air and fire, to move upward and away from the center. It seems fitting to ascribe rectilinear motion to these four elements.

---

* Translated from the critical edition of F. and C. Zeller, Vol. 2 (Munich, 1949), 17–21. The translation of J. F. Dobson, *Occasional Notes, Royal Astronomical Society*, No. *10* (May, 1947), 12–15, is quite free; and so, while I made use of it, often my translation is new and literal, more in the manner of the French translation of A. Koyré (Paris, 1934), pp. 85–99.

To the celestial bodies, however, [is ascribed a motion] of turning in a circle about the center. So far Aristotle.

If then, says Ptolemy of Alexandria, the earth would turn at least with a diurnal rotation, the result must be the reverse of that described above. For the motion must be most violent and of unsurpassable rapidity, since in twenty-four hours it must impart a complete rotation to the earth. Now things rotating very rapidly resist cohesion, or, if united, are apt to disperse, unless firmly held together. He (Ptolemy) says that the earth [if turning] would have been dissipated long ago and would have destroyed the heavens themselves—which is exceedingly ridiculous—; and [Ptolemy continues] certainly all living creatures and any other heavy bodies free to move would not at all remain stable [but would be shaken off the earth's surface]. Neither would objects falling directly approach perpendicularly the place destined for them, since it (the place) would have moved with some swiftness [from under them]. Moreover clouds and everything else floating in the air we would see continually carried westward.

2. Chapter 8: *The solution of the said arguments and their insufficiency.*—For these reasons certainly, and for similar reasons, they say that the earth is at rest at the center of the universe, and that such is the case without doubt. But if one thinks that the earth turns, he would undoubtedly say that the motion is natural, not violent. And things which happen according to nature produce effects contrary to those due to violence. Things subjected to any force or impulsion *(vis vel impetus)* must be disintegrated and cannot long exist. But those things which arise naturally have the right disposition and are maintained in their optimum condition.

Idle, therefore, is the fear of Ptolemy that the earth and all terrestrial things would be disintegrated through a rotation produced by nature—a thing far different from an artificial act or from that which can be done by human capabilities. But why should he not fear even more for the universe, whose motion must be as much more rapid as the heavens are greater than the earth? Have the heavens become so vast because of the unutterable [centrifugal] force of their motion proceeding from the center, and would they collapse if they stood still? Surely if this reason were so, the size of the heavens will go to infinity. For the more they are forced above by the impulsion *(impetus)* of their motion, the more rapid will become the motion because of the ever-increasing circumference to be traversed in the space of twenty-four hours. And in turn, with the increase of motion, the immensity of the heavens increases. And so they will mutually stimulate each other to infinity—the velocity, the size; the size, the velocity.

And, following the axiom of the *Physics* (III. 4. 204a) [that] "the infinite cannot be traversed nor moved in any way," [one would hold that] the heavens must stand still.

But they say that outside of the heavens there is no body, no space, no void, in fact absolutely nothing, and therefore no place where the heavens could expand. Then surely it is astonishing if something can be contained by nothing. And if the heavens are infinite and bounded only by an interior concavity, then perhaps [the proposition] that nothing is outside the heavens is better verified, since everything, however great, is contained in them, but the heavens will remain immobile. However, the most powerful [argument] which men use to infer that the universe is finite is [that it is in] motion.

Let us then leave to the philosophers the question of whether the universe is finite or not, holding this as certain, namely, that the earth between its poles is bounded by a spherical surface. Why then do we still hesitate to concede to it a mobility naturally appropriate to its form, rather than supposing the revolution of the whole universe, whose limits are unknown and unknowable? And why do we not grant that the diurnal rotation is [only] apparent in the heavens but real in the earth? It is like the saying of Vergil's Aeneas (Aeneid, III. 72), "We sail forth from the harbor, and lands and cities retire." When a ship floats along in a calm sea, all external things seem to the sailors to have the motion that is really that of the ship, while they think that they are at rest along with everything which is with them [in the ship]. No doubt in regard to the motion of the earth, it can happen in the same way that the whole universe is thought to rotate.

We would ask what of the clouds and other objects suspended in the air or sinking and, on the contrary, rising in it? Surely not only the earth, with the watery element joined to it, is moved thus, but also a quantity of air and all things so associated with the earth. Perhaps the contiguous air contains an admixture of earthy or watery matter and so follows the same nature as the earth, or perhaps the motion of the air is a motion which it derives from the perpetually rotating earth by contiguity and absence of resistance. On the other hand, they say—and this is astonishing —that the highest region of the air follows the celestial motion, as suggested by those swiftly moving stars which are called by the Greeks "comets" or "*pogoniae*"—for whose origin they assign this region—stars which, like the other stars, rise and set. We can say that, because of the great distance from the earth, that part of the air is deprived of terrestrial motion. Accordingly, the air nearest the earth, with the objects suspended in it, will appear stationary, unless, as it happens, they are agitated in this way or that by wind or some other impulse. For is wind in the air any different from a current in the sea?

Of objects which fall and rise, we must admit that their movement relative to the universe is a double one, it being generally the resultant of rectilinear and circular [motion]. When, indeed, things by their weight are borne downward, because of their being maximally earthy, the parts doubtless retain the same nature as their whole. For no different reason, things which are fiery are carried up into the higher regions. In effect, this terrestrial fire is nourished particularly by earthy matter, and they say that flame is simply burning smoke. Now it is a property of fire to expand that which it attacks, and this with such force that it cannot in any way or with any machines be restrained from completing its work after it has broken from its prison. The motion is one of extension from the center toward the circumference; and consequently, if something consisting of earthy parts is burned, it is carried from the center to the upward region.

Therefore, they say that the motion of a simple body is simple—this is primarily of circular motion—only so long as the simple body remains in its own natural place as a unit. In that place, indeed, the motion is nothing else but circular, for such motion is wholly self-contained and is similar to being at rest. But if objects wander or are extruded from their natural place, or are outside of it in any way, rectilinear motion supervenes. Nothing is so repugnant to the whole order and form of the universe as that something should be outside of its own place. Therefore, there is no rectilinear motion save of objects not properly disposed [as regards their natural place]; nor is there such motion in naturally perfect objects, since [if so] they would be separated from the whole to which they belong and thus destroy its unity. Moreover, even apart from circular motion, things moving up and down do not produce simple uniform and equal motion. For they are unable to be regulated by their lightness or by the impulsion (*impetus*) of their weight. Thus, all things which fall begin by moving slowly and then augment their speed while they are falling. On the other hand, earthy fire (the only kind we can observe) when carried aloft immediately grows weak, owing to the force and violence of the earthy matter.

A circular motion is always uniform, for it has a never-failing cause of motion, but those things [which have rectilinear motion] terminate the acceleration by which they have reached their natural place [and] cease to be either heavy or light, and their motion ceases. Since circular motion, then, is of things as a whole, while parts may posses rectilinear motion as well, we can say that circular motion may be combined with rectilinear—just as an animal, with sickness. And certainly the fact that Aristotle has divided simple motion into three classes—from the center, to the center, and around the center—will be thought of only as an act of reasoning, just as we distinguish [rationally] a line, a point, and a surface, although one cannot exist without the other, and none can exist without a body.

To these [arguments] one also adds that the state of immobility is thought to be nobler and more divine than that of change and instability, which latter is thus more appropriate to the earth than to the universe. I add also that it would appear to be very absurd to ascribe motion to that which contains or locates, and not rather to that which is contained and located, namely the earth.

Lastly, since it is manifest that the planets approach and recede from the earth, the motion of a single body around the center—which they wish to be the center of the earth—will also be at the same time motion from the center and motion towards it. Therefore, it is necessary to accept motion around the center in a more general sense and to be satisfied so long as every motion is applied to its own center. You see, therefore, that from all these considerations the mobility of the earth is more probable than its rest. This is especially the case with the diurnal rotation, as being particularly proper to the earth.

COMMENTARY

Notice how closely the negative arguments in passage 1 parallel those of Oresme. This is mainly because both of them are rendering the opinions of Aristotle and Ptolemy. But the refutation of these arguments in passage 2 also reflects at some points very closely the reasoning of Buridan and Oresme. Particularly note that Copernicus argues against the tremendous or infinite velocity of rotation which would be required of a heaven rotating daily and that he believes that the air rotates in some way with the earth, that the motion of the earth is comparable to that of a ship moving along in the calm sea, that objects falling combine circular and rectilinear movements in relation to the universe, and that the nobler and more divine heavens should be at rest—all of which were arguments found in either Buridan's or Oresme's treatment.

Document 10.4

*Questions on the Eight Books of the Physics [in the Nominalist Manner]**
Attributed to Marsilius of Inghen

1. BOOK III, QUESTION 7: *Whether local motion is a thing distinct from the moving body.*—[The principal affirmative reasons:] It is argued in the affirmative first by [the authority of] Aristotle in the third [book] of this [work, the *Physics*], text number 70, in the tract "On the infinite," where he says that infinity is found in different ways in motion and magnitude. It is true that in the case of each of them an infinite number of parts is measured to reach the end. But in the case of motion, the parts taken earlier cannot remain with the parts taken later, as they can in the case of magnitude. This situation would not prevail if motion were identical with the magnitude of the moving body.

It is argued in the second place [affirmatively] by Aristotle in the fourth [book] of this [work, the *Physics*], text number 99, where he posits that the *esse* of motion consists in always becoming another thing and yet another thing *(semper fieri aliud et aliud)*. Therefore, motion is not the moving body.

Third [affirmative argument]: If the distinction of movement from the moving body were not accepted, it would follow that when Aristotle, in the sixth book of this [work, the *Physics*], proved that motion is divided in the manner of the division of magnitude, he proved nothing. The consequent is false. The consequence is proved, for if you hold that motion is identical with the magnitude of the moving body, then he has proved nothing else but that the magnitude of a moving body is divisible like the divisibility of magnitude. Now this would have been proving nothing because it was already manifest.

* This fourteenth-century piece is translated from a text believed by some to be by Marsilius of Inghen. It was published with the works of Duns Scotus in the *Opera omnia* Vol. 2 (Lyon, 1639), 188–94.

Fourth [affirmative argument: With the acceptance of the identity of motion and the moving body], it would follow that motion would be moved and that there would be motion with respect to motion *(ad motum esset motus)*, which is impossible because then the arguments of Aristotle in the fifth book of this [work, the *Physics*], texts numbers 1, 2, and 3, would not be valid—arguments by which he proves that motion is not with respect to motion. But such is the obvious consequence of accepting the identity of motion and the moving body.

Fifth [affirmative argument: Movement is distinct from the moving body], by the Commentator in the twelfth [book] of the *Metaphysics*, comment number 41, where he says that nothing corruptible can be perpetuated except the motion of the heavens. Then I ask, what does he understand by the motion of the heavens? Not the heavens themselves, for the heavens are not corruptible. If it is a flux superadded to the heavens, then I have the position we have proposed.

Sixth [affirmative argument]: If the motion of Socrates were Socrates, it would follow that whoever should make Socrates move would make Socrates, and whoever should stop the movement of Socrates would make Socrates be destroyed *(corrumpi)*, which is false. But this consequence is obvious from the identity of the motion of Socrates and Socrates.

Seventh [affirmative argument]: Because motion is a common sensible and the moving body is not, hence motion is not the moving body. The antecedent is apparent from the second [book] of the *De anima*, text number 64. Nor is it valid to answer that motion is a common sensible because magnitude is a common sensible, since then motion and magnitude would not be distinct common sensibles. This is obviously false [as is apparent] in the same second [book of the *De anima*].

Eighth [affirmative argument: The distinction of motion and moving body is affirmed] because when two or three things are not sufficient for the truth of some proposition, it is necessary to add a fourth for it to be true. But this [proposition] is true: "Motion is." And "motor," "moving body," and "space" are not sufficient for its truth. For if they were, then whenever these three things exist then there would be motion. This is false. Hence, in addition to these [three] things it is necessary to add another thing, evidently a flux *(fluxum)* which I call local motion.

2. [Preliminary arguments of the negative side:] The opposite [side of the question] is argued by Aristotle in the third [book of the *Physics*], text number 4, where he says motion is nothing but the things with respect to which there is motion. By this he proves in how many *genera*

motion is to be found. Secondly, [the opposite side is argued] by Aristotle in the eighth [book] of this work [the *Physics*], text number 55, where he proves that in local motion nothing is acquired intrinsically by the moving body. Therefore, only local motion can be perpetual.

3. [Various ways of answering the question:] There is a certain [initial] way of answering this question. [It holds] that in local motion there is acquired a certain flux or disposition inherent in the moving body. It is according to this disposition that the moving body is said to be moved locally, just as whiteness is the disposition inherent in the moving body according to which the moving body is said to be white. They prove this conclusion by an argument *(ratio)* which appears demonstrative to them.

For this [argument] it is supposed in the first place that every "motion" is "change" *(mutatio)*, every "being moved" *(moveri)* is "being changed" *(mutari)*, and that "being changed," is "being disposed in a different way according to before and after" *(aliter se habere secundum prius et posterius)*, or at least "being disposed in some way other than the previous disposition" *(aliqualiter se habere et non taliter se habere prius)*.

Second supposition: One body can be made out of all the bodies of the world. This is proved [1] by reference to the case which Aristotle posits in the second book of this [work, the *Physics*], text number 38, and [2] because this is not contradictory to the divine power, and [3] because even if it is given as not possible, yet it is not contradictory on the basis of the general nature of bodies and moving bodies, as should be observed in the seventh book of this [work, the *Physics*].

The third supposition is that God can move the continuous body made up of all the bodies of the world in a straight line, or circularly, or with any motion pleasing to Him. Even if it were given that He would not, still it is imaginable and it is not contradictory for such a body to be so moved.

With these suppositions the argument is fashioned as follows: [First] unless local motion were a thing *(res)* distinct from the moving body and from space, it would follow that something could be moved locally and still would [not] be disposed in another way, according to before and after, which is impossible and against the first supposition. But the consequence is proved: It was posited in the second supposition that He made one continuum of all of the bodies of the world so that there would not be any moving body other than the world and its parts. Then by the third supposition, let God move that moving body, i.e., the whole world, and let Him move it rectilinearly. Then the whole world is moved locally:

Hence by the first supposition it is disposed differently than before. But this would not be with respect to something else extrinsic because it was posited in the case that nothing extrinsic to it exists. Therefore, it must be differently disposed *intrinsically* by the reception of something in it which is the disposition according to which the moving body is said to be moved locally. And so follows the proposed view.

Second [argument supporting first view identifying motion as a flux distinct from moving body]: It is affirmed that there are only two bodies, A and B. Then during one hour A is moved about B. During the second hour B is moved about A with the same velocity. Then, unless a thing moved locally has some different [intrinsic] disposition than it had earlier and not merely a different extrinsic disposition, it would follow that A and B would be similarly disposed throughout both the first and second hours and thus A would be moved in the same way in the second hour as in the first. But this is against the case posited.

Third [argument for the first way of solving the question follows]: As before, let there be only two bodies, A and B. Let A be moved and B be at rest. Then in the middle of the movement it is posited that God annihilates B. But A is still moved, and yet it is not differently disposed with respect to anything extrinsic because the only extrinsic thing has been destroyed. Therefore, it is differently disposed *intrinsically** by some disposition added to it; according to this disposition it is said to be moved locally. I call this disposition local motion. That, moreover, A is moved with B destroyed, I prove because the motive power of A is equally powerful as before and no resistance to it has been created. Therefore, if it was moved as before, it still moves because no source of its coming to rest is apparent.

The second way [of solving the proposed question] holds that local motion is the space over which the moving body is moved. This is proved by Aristotle in the third book of this [work, the *Physics*], text number 4, where he says that motion is nothing but the thing with respect to which there is motion. But local motion is with respect to space. Hence it is no other thing than space.

Second [argument in support of this view]: Local motion seems to be just like motion of alteration. But alteration is not a thing *(res)* distinct from the form which the altered body acquires. Therefore, local motion is not a thing distinct from the space which is acquired by the body moved.

A third way [of solving the question] holds that local motion is the

---

* Reading *intrinsece* instead of *extrinsece*.

very thing which is moved locally *(ipsum quod movetur localiter)*, so that the locally moving body which is actually moved is not disposed intrinsically in any different way than before unless there will have been some addition or subtraction of parts or some alteration other than the local motion.

In support of this opinion there are some arguments *(rationes)*. The first is that, if local motion were some flux or disposition superadded, then it would follow that the same thing would simultaneously "be" *(esset)* and "become" *(fieret)*. The consequent is impossible because "to become" is to be changed from nonbeing *(non esse)* to being *(esse)*. Nor is it valid to answer that it is not illogical in successive things [for something "to be" and "to become" simultaneously], because Aristotle in the fifth book of this [work, the *Physics*], text number 13, wishing to prove that generation is not generated concludes that then generation would be "generated" when it "existed." And thus he believes it to be impossible for generation, which according to his adversaries is of the genus of successive things, both "to be" and "to be generated." This is obvious by the Commentator in the same work, same place, comment number 13. The consequence is proved because motion exists while a moving body is moved and also because, if something had been produced earlier, then it would be of a permanent nature since it would remain the same before and after.

Second [argument for the theory of the identification of local motion and the moving body]: [If local motion were something distinct,] then it would follow that something would immediately *(primo)* be generated and corrupted in the same time. The consequent is impossible, because if some disposition—for example, whiteness—is acquired by a subject, and another disposition—for example, sweetness—is corrupted, and each immediately in the same time, then whiteness "is" and sweetness "is not." Therefore, if the same thing were immediately generated and corrupted in the same time, that thing would simultaneously "be" and "not be," which is a contradiction. This is obvious through Aristotle in the sixth book of this [work, the *Physics*], text number 74, where he holds that it is impossible for something to begin to be simultaneously with its ceasing to be; and that rather, there must be an intervening period of time between the time of generation and the time of corruption.

The principal conclusion [that from the separate existence of a disposition follows the immediate simultaneous existence of the generation and corruption of something] is proved [as follows]: Let something be moved in a period of time AB. Then, since the moving body is moved in the second part [of the time], [1] the motion by which the body is moved in the second part would last through the [whole] time after it had been generated, and consequently it would not be a thing distinct from every permanent thing; or [2] it would be produced *(fieri)* after AB time or in the second half of the time; and this is not so because then the moving body would be moved before the motion through which it would be moved is produced; which is impossible. In the same way it will be proved that in time AB it will be corrupted immediately, because, if it remained afterwards, it would now be of a permanent nature; [or] if it were corrupted before, then the moving body would be moved and [yet] it would be moved in no motion. Hence, since it is corrupted neither before time A, nor afterwards, nor in the first part, nor in the second, it follows that in AB time it is corrupted immediately.

Third [argument for the view identifying motion and the moving body]: If the motion were such a thing distinct from every permanent thing, it would follow that one thing would have in itself simultaneously an infinite number of motions equal to one motion according to duration, or that it would have successively an infinite number of motions equal according to duration. Either consequent is impossible. The consequence is proved by supposing that that which has in act some entity which it had before in potentiality is changed with respect to that entity. This is obvious from the nominal essence *(quid nominis)* of mutation, because since something now has an entity which it did not have before, it is now differently disposed according to that entity than it was before. Consequently, it has been changed with respect to that thing. Then it is argued: A thing which is moved has motion which it did not have before. Hence by the supposition it has been changed with respect to motion. Then I ask regarding that change whether it is simultaneous with the first given movement or whether it existed before, since it is impossible that it exists afterwards. If it is answered that it is simultaneous, then the thing which is moved has a second given change which it did not have before. Hence by the supposition it has been changed with respect to that change. Consequently, the change would be simultaneous with the first given motion and then it would be changed with respect to that motion. Similarly it is argued thus concerning a third change and so on to infinity and the first consequent results.

If [it is answered that the change is] prior [to the first given movement], then it is argued similarly that the change with respect to the second movement was before that movement and thus to infinity and the second consequent results. The falsity of the [first] consequent is proved because it does not appear to be possible naturally for an infinite number of things equal to a third thing to be simultaneously in the same thing in a simple way. The second [consequent] is impossible because then nothing could be moved unless it has been changed through an everlasting *(aeternum)* time before.

Fourth [argument for the third position: If motion were distinct from the moving body], it would follow that something would be composed out of nonexistent parts, or rather from parts whose existence is impossible. It implies this consequent because, just as it would be said that something would be a composite of a *chimera* and a *tragelaphus* which are mutually continuous, so you say that the parts of a movement—past and future—are continuous in turn through an intermediate "having been changed."

Fifth: If movement were such a flux distinct from every permanent thing, i.e., from space and from the moving body, then it would follow that God could separate it from a moving body and conversely that He could separate a moving body from such a movement with both of them still existing. And even if this is not conceded to be possible, yet it is imaginable from the fact that they are distinct things. But from such a separation follows something unimaginable, evidently, a moving body moving without movement, as an ascent without local change and a descent without a downward place being acquired.

Sixth: If motion were distinct from the moving body, it would follow that motion is not the actualizing* of the being which is in potentiality according as it is in potentiality. The consequent is impossible, for then motion would not be motion, since in a definition things are predicated of the thing defined. The consequence is proved, because the preterite part of the motion is not the actualizing of the moving body, nor also is the future part that actualizing, and furthermore the present part is not the actualizing. For if the actualizing were some present part, it would be now "having been changed" *(mutatum esse)* which is not motion. From these statements it is obvious that motion is not some successive thing inherent in a moving body, distinct from every permanent thing.

4. [Conclusions:] Now certain conclusions are posited. The first conclusion is that local motion is not the space which is acquired in the course of the local motion. Proof: If it were the space, it would follow that something at rest would be moved and that rest would be motion. This is impossible, for these terms "motion" and "rest" are privatively opposite; hence they cannot be spoken of together affirmatively by means of the word "is."

* Or "actuality"; see above, p. 423, note 2.

The consequence is proved, because the space over which there is movement is at rest.

The second conclusion: Local motion is not a flux, or disposition, or successive thing inherent in the moving body [and] distinct from every permanent thing. Proof: If it were so, then it would follow that everything moved locally would be moved according to some *esse*. The consequent is false and impossible as is obvious in the eighth book of this [work, the *Physics*], text number 55, where Aristotle wishes that in local motion no thing *(res)* is acquired by the moving body; and from this he proves that only local motion could be perpetual. The consequence is proved, because every thing moved locally would have that accident in act which before it had in potentiality in a certain amount and which was changed with respect to its *esse*. The consequence is proved in the second place by the six arguments posited before [in support of the third opinion].

The third conclusion: Local motion is the moving body which is moved locally. Proof: Movement exists, as was made clear in another question. Hence it is either the moving body, or the space, or a flux superadded to the moving body. For it does not appear that motion is anything else beyond these three things. But it is neither of the latter two. Therefore it is the moving body.

5. [Answers to the previous arguments supporting the first two opinions:] In answer to the arguments supporting the other opinion [i.e., that motion is flux], the second and third suppositions can be denied on the grounds of reason *(rationabiliter)*. But with the suppositions admitted, I concede the consequent, namely that with the case posited, that which is moved locally is not differently disposed than before. But for something to be moved locally, it is sufficient that it be differently disposed than before with respect to something at rest, if something at rest exists.

But against this whole argument are two objections: First, Aristotle had defined change insufficiently because it is not enough to define "to be changed" as "to be differently disposed before and after." But it is necessary to add conditionally that something is changed which is differently disposed with respect to something at rest, if something were at rest.

Second [objection]: These terms "motion" and "rest" are privatively opposite terms. Now of terms which are privatively opposite, the term which signifies the form is more knowable *(notior)* than the term signifying privation. Hence the term "motion" is more knowable than the term "rest." Therefore, motion ought not be defined by rest; but rather contrariwise, rest ought to be defined by motion.

[Responses to the two objections:] To the first I say that the definition of Aristotle is sufficient in cases where it is possible for natural things to exist by nature. It was with this supposed that he gave his definitions. Now [motion] is impossible naturally unless there is something at rest or something moved in a different way—a something with respect to which the body moved locally is differently disposed than before. And therefore because you posit a case impossible according to nature, it is necessary to correct the definition....

In response to the second objection, I concede that sometimes a positive term can be defined by a privative term, particularly where the privative term is more knowable in the same way as a thing is more knowable under the concept of a positive term. Now it is thus in the proposed question, because it is more knowable that something is at rest than that something is moved. This is obvious by the authority of the ancients (i.e., the Eleatics) who denied that motion exists and yet did not deny that rest exists.

In response to the second argument of the [first] opinion [holding that motion is a flux], let A, B, and C exist. I concede the case and the consequence is denied. For although A and B are similarly disposed in the one hour as in the other, yet in the one hour A is disposed differently than before with respect to B, which is at rest, and B in the other hour is disposed differently than before with respect to A at rest. Therefore, in the one hour A is moved while B is not, and in the other hour the converse is true.

Responding to the third argument, with the case admitted, I say that, assuming B has been destroyed, A will be moved as before. It does not follow that A is differently disposed than before, but only that it would be differently disposed if there were some body at rest. Certain others respond differently, who on account of these arguments posit that place is separate space. Therefore, they posit that there is place outside of the heavens, or an infinite space. Therefore, if God were to move the whole world rectilinearly or circularly, the world would be differently disposed with respect to the place or separate space in which it would be. Therefore, they concede that the world moved locally [in the fashion posited in the case] is disposed differently than before with respect to that which is nothing [i.e., separate space]. But the first solution is better.

6. Answers to the principal reasons [affirming motion as a thing distinct

from the moving body:] In response to the first, I say that [the observation of Aristotle distinguishing infinites in motion and magnitude] ought to be understood as follows. We speak in terms of the different natures *(rationes)* of a moving body infinite in magnitude and a body of infinite motion. For [on the one hand] a body is called "infinite in magnitude" when it is quantitatively extended without limit *(terminum)*. On the other hand, it is said to be of "infinite motion" when a given finite body is moved in infinite time.

In response to the second, [I say that the statement that] the *esse* of motion consists in the becoming of one different thing after another is a locution meaning a moving body which is moved is always in one place after another *(semper est in alio et alio loco)*.

COMMENTARY

On the question of authorship there has been some discussion. We are tentatively settling on Marsilius of Inghen as the author, even though there are important differences between the views expressed in this work and the admittedly genuine *Abbreviationes libri physicorum* of Marsilius. The question of authorship is not important for our purpose, which is to give an example of the Nominalist view of motion. That the author was a Nominalist is not only indicated by the title as it appears in some versions of the text, i.e., "Questions .... according to the Nominalist Way," but is everywhere evident in the opinions held in the various questions treated. The question here reproduced seemed important for the reason that it reviewed the Nominalist and previous views of the nature of movement. The author, it will be noticed, identifies movement with the moving body, as Ockham had done, thus rejecting the view that motion was a permanent thing *(res)* distinct from the moving body.

1. The author in the first passage marshals the "principal reasons" or common arguments for holding motion as a distinct entity. These arguments are largely taken from passages in the works of Aristotle or Averroës. In these arguments motion is characterized as a flux *(fluxus)*.

2. In the second passage he presents two passages of Aristotle that appear to support the opposite view, that motion is not a distinct entity.

3. With the preliminary arguments completed, the author in the long third passage presents three possible positions: (1) Motion is a thing *(res)*, actually a flux or a disposition, which is distinct from both the moving body and the space traversed. (2) Local motion is the space traversed by the moving body. (3) Local motion is the very thing which

[Marsilius of Inghen, *On the Physics*   625]

is moved locally, so that the moving body is not disposed intrinsically in any way differently than it was before. In the course of presenting these different opinions, we see emphasis laid on the Aristotelian definition of motion as "being disposed in a different way according to before and after." In presenting the first view, the author gives the argument that motion must be some intrinsic disposition, for cases can be constructed where one motion is different from another and yet their extrinsic relationships appear to be the same. One such case involves the possibility of God's moving the whole universe. There would be no extrinsic referent, but still there would be motion. Hence motion must consist in something intrinsic. The second case involves two bodies A and B. During one hour, A is moved about B, while during the second hour B is moved about A. There would be no observable extrinsic differences between the motions in the first and second hours (because, we might add, of the relativity of the detection of motion); hence, there must be some intrinsic difference that constitutes motion. The presentation of the third view, which is of course that of the author's, is presented in the third through the sixth passages by controverting the opinion identifying motion as a distinct thing.

4. In the fourth passage are presented conclusions leading to the support of the Nominalist view. This is done by a method of exclusion: Since motion is not the space traversed (the first conclusion), nor is it a distinct flux (the second conclusion), hence the only thing left is that it be identifiable with the moving body locally moved.

5. I have already pointed out in the body of the chapter that the author in the fifth passage mentions those who posit an infinite space beyond the heavens in order that the world itself might have a frame in which to move.

6. In the last passage here given, the author repeats Ockham's view that the "motion" with its definition as the "becoming of one different thing after another" merely is a locution for saying that a moving body is now in this place and now in another place, and so on.

Part IV

THE FATE AND SCOPE
of
MEDIEVAL MECHANICS

Chapter 11

The Reception and Spread of the English and French Physics, 1350-1600

WE are now prepared to trace briefly the fate of medieval mechanics after its initial formation in the thirteenth and fourteenth centuries. To some extent we have already contrasted the medieval views with those of the early modern scientists, particularly with those of Galileo. But we have nowhere indicated properly how the medieval doctrines fared in the course of the fifteenth and sixteenth centuries. To trace the spread of the doctrines in a complete manner would demand another volume of at least the size of this one; and, as a matter of fact, such an account is to be desired. Here, however, we shall undertake only to sketch the course of medieval mechanics in skeleton form, feeling that such a sketch might serve as a necessary historical complement to our analysis of the documents. It will be convenient to make our sketch both geographical and chronological, starting with England in the fourteenth century, then proceeding to France in the fourteenth and fifteenth centuries, Germany and Eastern Europe in the fourteenth and fifteenth centuries, Italy in the fourteenth and fifteenth centuries, and finally returning to France and Italy in the sixteenth century.

THE FOURTEENTH AND FIFTEENTH CENTURIES

England. The *original* contributions of the logicians of Merton College to kinematics and dynamics—contributions we have illustrated and analyzed in Chapters 4, 5, and 7—seem to have ended by about 1350. But we do have the names and works of a few continuators of the great tradition of

Bradwardine, Heytesbury, Swineshead, and Dumbleton. For example, not long after the middle of the century, the techniques and even the substance of the kinematic proofs originally developed by Heytesbury and Swineshead are found in a treatise entitled *De motu* and composed by the logician Richard Feribrigge (Ferebrich Anglicus). And so, for example, Feribrigge defines uniform acceleration in this way: "We have a case of uniform increase of speed *(uniformiter intendere motum)* when in the final instant of any part at all of the time, the speed *(motus)* of a moving body exceeds the speed with which it will be moved in the middle instant of that same part of the time, just as the speed in the same middle instant exceeds the initial speed of the increment [acquired during that part of the time period], i.e., the speed from which it begins to accelerate its movement [during that part of the time chosen]."[1] Feribrigge's account contains the usual distinctions of uniform and nonuniform movements, both as to subject and to time. The proof given for the Merton mean speed theorem is very similar to the proofs of Heytesbury and Swineshead given in Documents 5.2 and 5.3 (passage 1).[2]

[1] *De motu*, MS Rome, Bibl. Angelica, 1017, f. 37r: "Quia uniformiter vel difformiter omne quod movetur, quidam igitur est uniformiter moveri et quidam difformiter. Primitus exponendum quorum tamen expositiones duas suppositiones exiguunt presuppositioni. [1] Prima est, uniformiter velociter motus totalis attenditur penes motum velocissimum in aliqua parte mobilis movente. [2] Secunda est, quelibet intensio motus vel remissio attenditur penes latitudinem acquisitam vel deperditam in tanto tempore vel in tanto. Post hoc ulterius sciendum est quod aliquid uniformiter moveri est dupliciter: vel uniformiter secundum subiectum, vel uniformiter quo ad tempus. Uniformiter moveri secundum ‹subiectum› est quamlibet partem mobilis equevelociter movere sicud aliam eiusdem. Uniformiter movere quo ad tempus est in omni parte temporis in quo est motus equevelociter moveri sicud in alia eiusdem.... Uniformiter difformiter moveri quo ad tempus est uniformiter intendere motum quo ad totius aliquid vel partis. Et uniformiter intendere motum quando in cuiuscunque partis (*MS*, parte) temporis ultimo instanti equaliter excedit motus mobilis motum quo movebitur in medio instanti eiusdem sicud motus in eodem instanti medio excedit extremum latitudinis a qua incipit intendere suum motum...." Feribrigge was one of those logicians singled out by Leonardo Aretino for scorn: "Their very names fill me with horror: Ferabrich, Busser (Heytesbury ?), Ockham, Suisset (Swineshead)...." (See P. Duhem, *Études sur Léonard de Vinci*, Vol. *3* [Paris, 1913], 450–51). Earlier a *Recollecte* on him had been composed by Gaetanus de Thienis. See MS Venice, Bibl. Naz. San Marco, Lat., VI, 160, 118r–122r. Mr. A. B. Emden of Oxford writes that according to the Reg. Thoresby, York, f. 249, Feribrigge became rector of Shelton, Notts., in 1361 and the rector of the mediety of Cotgrave, Notts., in 1367 (f. 238).

[2] *De motu, MS cit.*, f. 37v: "Pro quo sciendum quod cuiuscunque uniformiter intendere dicendo vel remittere motum suum, motus suo medio gradui correspondet, quod ostenditur per hunc modum. Ponatur quod *A* a *C* gradu uniformiter remittat motum suum in medietate illius hore ad non gradum et quod *B* ab eodem

Contemporary with, but probably slightly junior to, Feribrigge was the fellow of Merton College, John Chilmark, who can be placed at Merton College in 1384, who vacated the fellowship in 1393, and who died in 1396. He composed several works in the Merton tradition, works which still remain to be studied in detail, but which on first glance seem to be

C gradu uniformiter intendat motum suum in eadem prima medietate precise ad suum duplum, ita quod nec B in minori tempore intendat motum a C usque ad gradum duplum ad C quam A remittat a C usque ad non gradum nec e converso. Quo posito sequitur quod A et B tantum pertransibunt de spatio quantum pertransirent si continue C gradu moverentur. Quod sic arguitur, nam quantum vero acquiret B de latitudine motus per intensionem sui motus tantum precise deperdet A per remissionem sui motus. Igitur per quantum pertransibit B maius per intensionem sui motus tantum pertransibit A minus. Sequitur quod precise tantum erit pertransitum ab istis acsi continue moverentur C gradu. Arguitur sic, A et B tantum pertransibunt de spatio quantum pertransirent si continue C gradu moverentur. Et A tantum pertransibit in alia medietate hore sicud B, si in prima medietate hore continue et ab eodem gradu intenderet motum suum sicud B intendet motum suum in prima medietate. Igitur tantum pertransiet A in tota sicud [A, B] in alia medietate totius hore. Consimiliter intendet motum sicud B in prima medietate intendet, cum hoc quod in prima medietate remittetur alius motus, sicud positum est. Et precise tantum pertransibit A in prima medietate per remissionem sui motus a C usque ad non gradum sicud per A intensionem motus a non gradu usque ad C gradum. Ergo in tota hora tantum pertransietur et de spatio sicut in prima medietate uniformiter intendet motum suum a non gradu usque ad C et in alia medietate uniformiter a C gradu ad gradum duplum ad C sicud omnes pertransibunt et A, B sicud A per talem remissionem motus et B per talem intensionem (MS motum totum!) motus; tantum pertransibunt de spatio acsi per idem tempus idemque C gradu moverentur gradu (!). A in prima medietate uniformiter intendens motuum suum a non gradu ad C, in alia medietate hore uniformiter intendens motum a C usque ad gradum duplum ad C tantum pertransient de spatio sicud A et B moventes per primam medietatem eiusdem. Et igitur A uniformiter in prima medietate intendens motum suum usque ad gradum duplum ad C tantum pertransibit de spatio acsi continue C gradu moverentur. Et si B (MS A!) similiter intendat (MS intensum) uniformiter motum suum a non gradu usque ad gradum duplum ad C, igitur talis latitudo motus a non gradu usque ad gradum duplum ad C uniformiter acquisita, gradui suo medio correspondet quantum ad pertransitionem spatii, eo quod C gradus est gradus medius inter gradum duplum ad C et non gradum. Et illo modo est arguendum de quacunque latitudine uniformiter acquisita in quocunque tempore parvo sive magno. Et consimiliter est arguendum: si aliquid motum per unam medietatem temporis movetur uniformiter uno gradu et per aliam medietatem uniformiter alio gradu, quidem totus eius motus correspondebit gradui medio inter illos. Ex penultima enim conclusione sequuntur 4 conclusiones, quarum prima est hec: Latitudo motus velocius et velocius acquisita gradui remissiori quam suo medio correspondet. Secunda (est) hec: Latitudo motus tardius et tardius acquisita gradui intensiori quam est suus medius correspondet...." (I have capitalized the letters designating the movements.) The reader is urged to compare this passage with the proofs given by the author of the *Probationes* of Heytesbury (Doc. 5.2) and by Swineshead (Doc. 5.3).

little more than paraphrases of the earlier Merton scholars (and particularly of Dumbleton).³ The substance of the earlier Merton work, it should be observed, about this same time had been simplified and codified in elementary outlines and manuals, which exist in the same manuscripts as the works of Chilmark.⁴ Presumably, most arts students at Oxford in the

---

³ See the article by R. L. Poole on Chilmark in the *Dictionary of National Biography*. Cf. Duhem, *Études*, Vol. *3*, 410, 411; A. B. Emden, *A Biographical Register of the University of Oxford to A.D. 1500*, Vol. *1* (Oxford, 1957), 416. It is Mr. A. B. Emden who writes that Chilmark is first found in the Merton bursar accounts for 1384. He also notices that Chilmark vacates the fellowship in 1393, in which year he became rector of Offord d'arcy, Hunts., and that he died in 1396. Chilmark appears to have been a fellow of Merton College and connected with Exeter in some fashion. Apparently the popular work of Chilmark was a summary compiled from the fourth part of Dumbleton's *Summa de logicis et naturalibus*. It is entitled *De actione elementorum*. In MS Oxford, Bodl., Digby 77, ff. 153–65, it has the following incipit: "Cum materia de alteratione que est una species motus sit quampluribus incognita...." The colophon (f. 165v) reads: "Explicit quoddam compendium de actione elementorum abstractum de quarta parte Dumbletonis secundum Magistrum John Chylmark." The same work (without introduction) exists in MS Bodl., Laud. Misc. 706, ff. 176r–183v. Another work, *De alteratione*, is contained in Bodl., Bodley 676, ff. 76r–101r with the incipit: "Pro materia alterationis in qualitate promulganda primitus...," and with the colophon: "Explicit materia de alteratione secundum Chilmark q. S." It is contained also in MS London Brit. Mus., Royal 12B 19, ff. 2r–14r. Another work in Bodley 676, ff. 52v–75v, begins "Pro materia propositionis...." It is called *De quantitate* by Tanner. Still another work *De motu* appears in MS Brit. Mus., Harl. 2178, ff. 1r–7r. The beginning is missing, but apparently not much is missing, for it has three parts, the first on motion of augmentation (two complete folios left), second on local motion (three folios in all), and the third on motion of alteration (a little more than one folio). My guess is that only one folio is missing. A considerably longer *De motu* is found is MS Bodley 676, ff. 11r–38r with the incipit: "Ad materiam motus intellectui...." and in MS Oxford, New College 289, ff. 138r–162r. This latter manuscript contains, ff. 163r–166v, a *De augmentatione* of Chilmark which may simply be a part of a longer work on motion. It begins, "Pro materia augmentationis...." This manuscript also contains several tracts probably by Chilmark: ff. 171r–173r, *Dubium*, beginning, "Dubitatur an aliqua sic..."; ff. 174v–175v, *Materia de generatione*, beginning, "In materia de generatione notandum..."; ff. 177r–180r, *De aggregatis*, beginning, "Dubium est utrum aliquod est aggregatum..."; ff. 180r–182v, *De successivis*, beginning, "Pro materia de successivis habenda est primo..."; ff. 183r–191v, *De alteratione*, beginning missing (?), it reads, "...in sua spera sunt due qualitates summe et dicitur...." Finally on f. 191v we read: "Explicit tractatus summa (?) Chylmark doctoris philosophie, de alteratione." There is perhaps also the possibility that the *Questions on the Physics* in Erfurt, Stadtbibliothek, Amplon. Q 304, ff. 1–148, attributed to one Johannes de Chymiacho belongs to John.

⁴ For example Oxford, Bodl., Bodley 676, ff. 1r–6v, contains an anonymous *Linea naturalium*, which in outline form distinguishes and defines the various terms of physics. Much the same things are included also on ff. 149r–160v, but not in outline form. The outline (f. 1r) begins:

"Natura⟨Naturans, ut Deus
Naturata, ut materia prima...."

last half of the century were exposed to the Merton studies in one form or another. Hence it is not surprising to find strong evidences of the influence of the Merton studies in the *Tractatus de logica* of John Wyclif (who perhaps was a fellow of Merton). A chapter on motion makes the usual Merton distinctions of uniform and nonuniform movements, including uniformly nonuniform.[5] It repeats the universally accepted opinion

On the other hand, the rhetorical account begins (f. 149r): "Natura est duplex, scilicet natura naturans et natura naturata. Natura naturans est prima causa, i.e., Deus...." To show the interest of the *Linea naturalium* in the Merton kinematics, I quote the following (f. 4r): "*Gradus summus* est qui non est admixtus cum suo contrario vel quo nullus est intensior. *(Gradus) intensus* est qui modicum participat de suo contrario, vel per modicam latitudinem distat a summo gradu. *(Gradus) remissus* est qui modicum participat de suo contrario secundum intensius et remissius. *Latitudo* est natura continens in se difformitatem gradus secundum intensius et remissius. *Intendere inclusive* est per latitudinem modicam gradum adquirere, illum adquirendo. *(Intendere)* est per aliquam latitudinem gradum perfectiorem adquirere. *(Intendere)* est per latitudinem modicam terminatam ad aliquem gradum perfectiorem, non adquirendo illum, hoc est, *exclusive*. *Remittere* per latitudinem modicam gradum perfectiorem deperdere, hoc est, *inclusive*. *(Remittere)* est per latitudinem gradum perfectiorem deperdere. *(Remittere)* est totam latitudinem citra certum gradum deperdere, non deperdendo illum, hoc est, *exclusive*. *Intendere uniformiter* est tantam latitudinem graduum adquirere in una parte temporis quam sicut in alia sibi equali. *(Intendere) difformiter* est intensiorem latitudinem adquirere in una parte temporis quam in alia parte temporis sibi equali. (4v) *Remittere uniformiter* est tantam latitudinem deperdere in una parte temporis quam in alia sibi equali. *(Remittere) difformiter* est intensiorem latitudinem deperdere in una parte temporis quam in alia sibi equali.... (5v) *Qualitas uniformis* est cuius quelibet pars est eque intensa cum suo toto. *(Qualitas) difformis* est cuius aliqua pars est intensior alia secundum intensionem. *(Qualitas) difformiter difformis* est cuius immediate secundum extensionem multum distant. *(Qualitas) uniformiter difformis* est cuius omnes partes que sunt immediate secundum extensionem sunt immediate secundum intensionem. *Motus uniformis* est ille mediante quo in tempore equali spatium pertransitum sit equale. *(Motus) uniformiter difformis* est ille mediante quo uniformiter adquiritur vel deperditur latitudo motus. *(Motus) difformis* est mediante quo in tempore equali spatium pertransitum [est] inequale. *(Motus) difformiter difformis* est ille quo difformiter adquiritur vel deperditur latitudo motus. *Movere uniformiter* est tantum spatium pertransire in una parte (temporis) sicut in alia sibi equali. *(Movere) uniformiter difformiter* est uniformiter adquirere vel latitudinem motus deperdere. *(Movere) difformiter* (6r) est maius spatium vel minus pertransire in una parte temporis quam in alia sibi equali. *(Movere) difformiter difformiter* est difformiter adquirere vel deperdere latitudinem motus. Aliqua dicuntur *immediata secundum intensionem* que terminantur ad eundem gradum, secundum extensionem que terminantur ad eundem punctum...."

[5] Johannis Wyclif *Tractatus de Logica*, edition of M. H. Dziewicki, Vol. *3* (London, 1899), chap. 9, p. 87. "...aliqui dicunt quod [velocitas motus localis] attenditur penes lineam descriptam a puncto velocissime moto (*ed. wrongly has* velocitate motus?).... (p. 94) Istis tribus premissis, dicitur generaliter quod quilibet motus velox est ita velox sicud aliqua eius pars, et per consequens cuiuslibet talis motus velocitas attenditur penes partem eius

of Bradwardine to the effect that the speed of local motion *(velocitas motus localis)* is measured *(attenditur)* by the line described by the fastest moving point in the body, the measure being taken with respect to time *(in comparacione ad tempus motus)*. Thus Wyclif has approached closely to the concept of speed being measured by a ratio. It is presumed that this work of Wyclif dates from the time he was at Oxford sometime close to 1360, although some scholars would date it much later. Thus Wyclif was probably active about the same time as Radulphus Strode, the Merton College Fellow, who composed logical works firmly in the older Merton tradition.[6] In the same general period, a little-known scholar, Edward Upton, composed a work on *Conclusiones de proportione* [*motuum*], which clearly goes back to Bradwardine.[7]

We know very little about the fate of the Merton kinematics in England during the fifteenth century. That it was studied to some extent (or at least read about) seems probable from the fact that fifteenth-century library catalogues of books at colleges of both Oxford and Cambridge indicate some copies of the various early Merton College works.[8] But this problem

velocissimam moventem (? *I have corrected the editor, who has* mensurantem) ... et patet quod motus localis velocitas attenditur penes longitudinem situs linealis descripti, a punctali per se mobili velocissime moto in comparacione ad tempus motus (? *ed. has* mensurans).... (p. 33) Similiter, ex modo loquendi de difformitate motuum quo ad tempus, patet idem. Nam motus uniformiter difformis, in quantum talis, terminatur ad gradum inclusive; quia aliter non esset dare gradum eis (eius?) intensissimum vel medium, et per consequens non esset descriptibilis aliqua descripcione; quia dicere quod gradus ad quem terminatur exclusive, qui non est in mobili, mensurat eius velocitatem vel uniformitatem, est plana contradiccio, cum medium requirat extrema illius cuius est medium. Et dicere quod gradus medius, qui non est extremum vel terminus illius motus, est plane demencie. Gradus ergo medius motus uniformiter difformis quo ad tempus, erit gradus utriusque medietatis successive...."

[6] Thus Strode like the earlier authors composed a *Consequentie*, in MS Copenhagen, Bibl. Reg., Thott. 581, ff. 1–12,

with the incipit: "Consequentia est illatio consequentis...." See Duhem, *Études*, Vol. *3*, 409, 444, the DNB, and see *Recollecte super consequentias Strodi* by Gaetano of Thienis, Venice, Bibl. Naz. San Marco, Lat. VI, 160, ff. 109r–118r. Father J. A. Weisheipl, O.P., tells me that the *Consequentie* is only Tract IV of Strode's complete *Logica* (MS Oxford, Bodl., Canon. Misc. 219, ff. 13–52).

[7] Upton's treatise appears in the Bodleian, Bodley MS 676, ff. 83r–49v. It begins: "Ponatur quod *a* sit una potentia ut 6, *b* resistentia ut 2, *c* alia resistentia ut unum...." The colophon (f. 49v) reads: "Expliciunt septem conclusiones proportionum secundum magistrum Edwardum Uptonum."

[8] For example, M. R. James in his *Descriptive Catalogue of the Manuscripts in the Library of Peterhouse* (Cambridge, 1899), prints the old catalogue of Peterhouse books made in 1418. Among the items are the works of Aristotle, commentaries of Averroës, numerous scholastic commentaries, Roger Bacon's *Communia naturalis philosophie*, Bacon's *Perspectiva*, Jordanus' *De ponderibus*, Euclid (?) *De pon-*

of the continuation of mathematical and mechanical studies in England in the fifteenth and sixteenth century demands further study.

France. The mechanical studies of the scholars at Paris in the second half of the fourteenth century seem somewhat more original than those of Feribrigge, Chilmark, and Upton in England in the same period. Not only were the successors of Buridan, like Nicole Oresme and Albert of Saxony, making significant additions to both the Parisian dynamics and the Oxford kinematics—contributions we have already detailed in Chapters 6—10, but a number of their lesser-known contemporaries were taking up the new mechanics in a more modest fashion.

The mathematician Dominicus de Clavasio did not confine himself to the geometrical studies of his *Practica geometrie* but also composed a commentary on the *De caelo*, a manuscript of which is dated 1357.[9] Dominicus employed the impetus theory both to explain the continuance of projectile motion and to account for the continuing acceleration of the speed of falling bodies. It was perhaps Dominicus who unwittingly helped to prepare Oresme's theory of self-expending impetus produced by and productive of acceleration. Dominicus distinguished *impetus* from the "actual force": "When something moves a stone by violence, in addition to imposing on it an actual force *(virtutem actualem)*, it impresses in it a certain impetus. In the same way gravity *(gravitas)* not only gives motion itself to a moving body, but it also gives it a motive power and an impetus, and the more this power is stronger, the greater is the impetus; and therefore there would be greater velocity *(velocitas)* with the impetus."[10]

deribus, Theodosius' *De speris, auctor de visu, auctor de speculis,* Archimedes' *De mensura circulorum,* John of Dumbleton's *Summa* (which of course summarizes the Merton logic and physics), the *Sophismata* of Heytesbury, numerous astronomical works including Ptolemy's *Almagest,* the *Elements* of Euclid, etc. All in all, the student of Peterhouse could get a good education in both the old and the new physics of the Middle Ages. However, the earlier Merton works continued in scattered fashion only among the collections of books at Merton College in the fifteenth century. See F. M. Powicke, *The Medieval Books of Merton College* (Oxford, 1931), pp. 181, 215. For Clare College, Cambridge, see R. N. Hunt, "Medieval Inventories of Clare College Library," *Transactions of the Cambridge Bibliographical Society,* Vol. 2 (1950) 105–25.

[9] MS Vat. lat. 2185, f. 20v: "Expliciunt questiones super 1º et 2º de celo disputate parisius per magistrum Dominicum de Clavisio . . . quas scripsi et complevi anno domini 1357. Deo gratias. Amen." (See A. Maier, *Zwei Grundprobleme der scholastischen Naturphilosophie,* 2d ed. [Rome, 1951], p. 241, n. 10.)

[10] MS Vat. lat. 2185, f. 17r: "Alia opinio, que ponit quod quando aliquid movet lapidem per violentiam, cum hoc, quod imponit sibi virtutem actualem, imponit sibi quendam impetum. Modo eadem gravitas non solum dat mobili motum eundem actualiter sed etiam dat sibi vir-

We must also make passing reference to other contemporaries of Albert of Saxony and Nicole Oresme at Paris: Marsilius of Inghen and Henry of Hesse. Marsilius was a master at Paris in 1362, rector in 1367 and 1371, and first rector of Heidelberg University in 1386.[11] Marsilius spoke of impetus as a quality of the first species (i.e., a *habitus* or *dispositio*) and also one of the third species (an *actio* or *passio*).[12] Rotary impetus, as in the case of a smith's wheel, he distinguished *in specie* from rectilinear impetus.[13] He also described, without however accepting, Franciscus de Marchia's compromise in giving the medium a supplementary role in the continuance of the projectile's motion. The air as well as the projectile receives

tutem motivam et impetum et quantum illa virtus est fortior, tanto est maior impetus et ideo esset maior velocitas cum impetu...." See K. Michalski, "La physique nouvelle et les différents courants philosophiques au XIVe siècle," *Bulletin international de l'Académie polonaise des sciences et des lettres. Classe d'histoire et de philosophie, et de philologie*, L'Année 1927 (Cracovie, 1928), p. 150. In the same place Dominicus suggests that the impetus is a quality: "Quid autem sit ipse impetus... forte est qualitas et hoc oportet poni, quia aliter non potest esse motus violentus." It is Maier, *Zwei Grundprobleme*, pp. 241–43, who quotes from and analyzes a further question of Dominicus: "Utrum proiecta moveantur velocius in medio quam in fine vel in principio." We see something of the connection between impetus and acceleration later advocated by Oresme in the following statement which we quote from Dominicus through Maier (p. 242): "Et nota quod causa, quare aliquis velocius movetur in medio quam in principio, est, quia in principio movetur ab anima solum, postea acquiritur impetus in motu intrans ad talem motum. Similiter potest dici de corda arcus et sagitta emissa ab arcu, acquirit quendam impetus in motu, quare magis laedit a distantia 6 pedum quam duorum et tamen non sequitur quod sagitta emissa debeat post continue velocitari, quia ille motus retardatur per restentiam inclinationis contra naturam."

[11] For a general account of Marsilius' career, see Gerhard Ritter, "Studien zur Spätscholastik I, Marsilius von Inghen und die okkamistische Schule in Deutschland," *Sitzberungsberichte der Heidelberger Akademie der Wissenschaften*, Philosophisch-historische Klasse (1921), Abh. *4*. Cf. Duhem, *Études*, Vols. *2* and *3*, *passim*, and Maier, *Zwei Grundprobleme*, pp. 274–90. Maier has printed the text of Book VIII, Not. 4, quest. 4 of Marsilius' *Abbreviationes libri physicorum* which takes up the impetus theory; see pp. 279–84.

[12] Maier, *Zwei Grundprobleme*, p. 281: "...impetus ille est qualitas impressa mobili faciens in eo motum, et est de prima specie qualitatis, quia acquisita post productionem mobilis et dispositionem ipsum ad melius vel peius.... Ad tertiam patet ex dictis quod est de prima specie qualitatis vel de tertia. Si enim tali impetui imponeretur nomen ipsum absolute significans, illud nomen esset sub tertia specie qualitatis. Nomen vero ipsum significans prout est acquisitus post productionem rei et dispositionem subiectum aliquod ad melius ad peius esset sub prima specie qualitatis. Verum est quod, si ille talis impetus diceretur ab impetendo sic quod illa res impetus vocaretur ex hoc quia active inclinaret suum subiectum ad motum, tunc esset de praedicamento actionis, sicut sectio, dectio et similia."

[13] *Ibid.*, p. 282: "Secunda propositio: nec omnes impetus violenti sunt unius speciei. Patet [per] proportionales rationes quia aliquis talium est natus de per se movere recte, ut sagittae, aut circulariter, ut molae fabri."

an impetus, and it is this added motive force that might produce the initial acceleration which he (as well as Oresme) believed the projectile to have.[14] We shall see that the assignment of an additional, necessary role to the air was a compromise that particularly appealed to Italian Averroïsts later. Such a theory was more patently "saving" Aristotle. Finally, we may note that Marsilius was at least familiar with Oresme's system of graphing qualities.[15]

Henry of Hesse, also at Paris in the 1360's and 1370's, made but brief mention of the impetus mechanics. Most interesting was his distinction of an impetus of circular motion *(motio circularis)* from an impetus of rectilinear motion *(motio recta)*.[16] One can also mention in passing that at Paris in 1369 one Johannes de Wasia made a compendium of the first three chapters of the *Tractatus de proportionibus velocitatum in motibus* of Bradwardine (MS Erfurt, Stadtbibliothek, Amplon. Q. 325, ff. 47r–51v). And perhaps the same author was responsible for a treatment of the doctrine of the configuration of qualities made so popular by Oresme (*ibid.*, ff. 43r–45r).

Even while ignoring other occasional notices of the impetus mechanics among Parisian masters, we should however not overlook Lawrence of Scotland, a master at Paris in 1393. Like Marsilius of Inghen and Albert of Saxony, Lawrence through his works was particularly influential in carrying the doctrine of impetus into Germany and Eastern Europe, although he himself returned to Scotland to be first rector of St. Andrews. His questions on the *Physics* were used at Prague, Cracow, Erfurt, and Leipzig in the fifteenth century.[17] Lawrence accepted in addition to the

---

[14] *Ibid.*, p. 283: "Sed dices: ille impetus est fortior apud proiciens et per consequens arcus deberet fortius laedere prope se quam ad aliquantulam distantiam, quod est contra experientiam. Ista ratio est bene difficilis et ideo respondetur ad eam evasive probabiliter solum. Uno modo quod proiciens incipit inducere impetum a non gradu et quod in emissione virtutem impressit aeri circumexistenti simul moto cum proiecto, qui aer impetum proiecti a proiciente introductum ad aliquam distantiam intendit et fortificat."

[15] According to Pierre Duhem, *Études*, Vol. *3*, 403, Marsilius at least twice uses the graphing system in his *De generatione* (Bk. I, quest. 18; Bk. II, quest. 6).

[16] *De reductione effectuum*, MS Paris, BN lat. 2831, f. 110r: "Consurgunt diverse species motivarum qualitatum, quas vocant impetus motionis, quorum quidam est motionis circularis, ut apparet in mola fabri, et quidam recte." Cf. Michalski, "La physique nouvelle," p. 156; Maier, *Zwei Grundprobleme*, p. 288.

[17] Michalski, "La physique nouvelle," p. 156. Michalski notes from a manuscript that Lawrence's *Questiones* were used at Prague as early as 1406. Michalski quotes from Book VIII, quest. 9: "Secunda conclusio: Proiecta moventur a quadam qualitate, quae vocatur impetus, quam proi(i)ciens imprimit in proiectum patet, quia non videtur, a quo alio moveretur.... Ponendo istum impetum salvamus omnia." In addition to the manuscripts cited by Michalski, see Vienna, Dominik. 114/81, 115/82, and 116/83, which together con-

impetus as the principal cause of the continuing projectile motion, the motion of the air as a necessary accompanying cause *(causa sine qua non)*.[18] About the same time, Peter of Candia, a master in theology at Paris in 1381 (the later Pope Alexander V), accepted the impetus theory and was also acquainted with Oresme's system of graphing qualities.[19]

After the great beginnings in the fourteenth century, one might expect a further development at Paris in the fifteenth century. Such does not appear to have been the case. We know of only four of the Parisians to profess or describe the Buridan dynamics, excepting John Dullaert at the turn of the sixteenth century (who belongs rightfully to the next century): George of Brussels, John Hennon, Thomas Bricot, and Peter Tartaret.

Although the ideas of George of Brussels have received only brief notice,[20] we know a little more about John Hennon. A bachelor of theology, he composed a *Liber philosophie Aristotelis* at Paris which was copied in 1463 (Paris, BN lat. 6529, f. 327r; cf. f. 281v). He describes (f. 146r) the impetus theory of projectile motion, declaring that it is probable but against Aristotle. Similarly he notes (f. 164r) the use of the impetus theory to explain the acceleration of falling bodies. For him this impetus that continually grows as the body falls is a quality of the second species which is engendered by the mediation of movement from the substantial form, which of course is gravity.

Thomas Bricot developed the theory along the traditional lines in his *Textus abbreviatus* of the *Physics*. As objections to the Aristotelian theory, he posed: a hoop or a top moving with a movement of circumgyration; the fact that you cannot move the air a distance of twenty or thirty feet by stirring it up, since you cannot blow a candle out at that distance; that you cannot move a bean as far as a stone of one pound; and some other experiences that had been employed before.[21] The impetus was

tain his *Questions on the Physics*.

[18] This doctrine is evident at Erfurt in the *Commentaries on the Physics* by a so-called Magister de Stadis, who speaks of Lawrence as one of its supporters. See Michalski, "La physique nouvelle," pp. 157–58.

[19] Michalski, "Les courants philosophiques à Oxford et à Paris pendent le XIVe siècle," *Bulletin international de L'Académie polonaise des sciences et des lettres, Classe d'histoire et de philosophie, et de philologie*, Les Années 1919, 1920 (Cracovie, 1922), p. 81, quoting Paris, BN Nouv. Acquis.

1467, f. 179r (for the impetus theory).

[20] For George of Brussels, see Michalski, "Les courants philosophiques," p. 86; for Hennon, see Duhem, *Études*, Vol. 3, 520–22.

[21] Thomas Bricot, *Textus abbreviatus Aristotelis super octo libris phisicorum, et tota naturali philosophia* (Paris, 1494), f. 101v. c.2: "...trocus movetur motu circumgirationis, et tamen hoc non est ab aere.... per aerem nullus potest commovere flammam candele distantem viginti vel triginta pedes.... aliquis deberet ad maiorem distantiam movere fabam quam

spoken of by Bricot as a second quality *(qualitas secunda)* and as an instrument which begins motion under the influence of a principal particular agent but which continues it alone.[22] Even though the projectile is moved by an intrinsic source *(principium)*, it does not follow that the motion is a natural one, since this *principium* is violent, and against the nature of the projectile.[23] Bricot, almost repeating the words of Buridan, explains that the impetus is a motive power by which the projectile can be moved in the direction intended by the projector.[24] "Moreover, this impetus is a certain quality distinct from the moving body itself; but it is a 'passion,' a passible (nonactive) quality, or a disposition."[25] All impetuses are essentially and specifically different, arising from the diverse methods of impressing them and the diverse movements of the projecting hand. The impetus moving a heavy body up and one moving the same body down are not alike specifically, since the former is corrupted by the natural form of the body (the gravity) and the latter is aided by it. If they were of the same essential species, they should be naturally conserved by the same thing and have similar effects according to species.[26]

Finally, we can remark concerning Bricot that he followed Buridan in accepting impetus as of permanent nature, and Oresme and Marsilius of Inghen in discussing an initial period of acceleration of a projectile.[27] Although somewhat more extended than the other fifteenth-century ac-

pilam vel lapidem unius libre. Consequens est falsum."

[22] *Ibid.*, f. 101r, c. 2: "... licet instrumentum non incipiat motum sine agente principali particulari, bene tamen potest continu[a]re motum inceptum sine agente particulari principali...."

[23] *Ibid.*: "Ad nonam dicitur quod licet proiectum moveatur a principio intrinseco ei violento, non tamen a principio intrinseco sibi naturali. Iste autem impetus inest proiecto contra naturam eius; et ideo movet ipsum proiectum contra naturam."

[24] *Ibid.* f. 101r, c. 1: "... cum proiiciens movet ipsum proiectum imprimit sibi quendam impetum sive quandam virtutem motivam per quam ipsum proiectum potest moveri ad illam partem ad quam proiiciens intendit."

[25] *Ibid.*: "Iste autem impetus est qualitas quedam distincta ab ipso mobili, que est passio sive passibilis qualitas vel dispositio."

[26] *Ibid.* f. 101r, c. 1–c. 2, "Ad dubium dicitur non, quia impetus movens grave sursum corrumpitur a forma gravis et a gravitate, et impetus movens grave deorsum magis conservatur a forma gravis, modo illa que sunt eiusdem speciei essentialis habent ab eodem naturaliter conservari.... Sed quereret aliquis unde diversificantur isti impetus essentialiter et specifice, dicitur quod hoc est propter diversum modum imprimendi et ex diversis motibus manus proiicientis."

[27] *Ibid.*: "Dicitur quod hoc est ideo, quia licet ille impetus in primo impressus in partem quam tangit proiiciens sit fortissimus, tamen in partes distantes impressus est parvus et remissus, et sic in principio impetus non est per totum proiectum eque fortis.... Iste autem impetus dicitur esse nature permanentis."

counts, Bricot's discussion surely suffers in comparison with the pristine account of Buridan.

More brief is the exposition of Peter Tartaret, rector at Paris in 1490. He accepts impetus as a permanent quality; it explains all the phenomena.[28] Further, he repeats Bricot's account of the supposed acceleration of the projectile. It is worth noting that none of these fifteenth-century schoolmen seems to have been interested in the English kinematics or in Oresme's development of it. However, as we shall see shortly, just a few years after Tartaret there was a revival of the English logic and kinematics, as well as a return to the purer form of the fourteenth-century accounts of the impetus mechanics.

Germany and Eastern Europe. The immediate passage of both the English and French mechanics towards the east and towards the south is apparent from abundant evidence. We have in Chapter 4 given some extracts from the *De motu* of John of Holland. We have seen that the *De motu* was primarily a summary of the kinematic ideas of Heytesbury and Swineshead. Since we can place John in Prague in 1369 as a member of the "German" nation,[29] we can infer that the English kinematics was already studied at the University of Prague by that date.

---

[28] *Commentarii Petri Tartareti in libros philosophie naturalis et metaphysice Aristotelis* (Basel, 1514), f. 55v: "...proiecta moventur a virtute ipsis impressa a proiiciente, quam aliqui communiter vocant impetum...et in isto modo concordant fere omnes philosophi. Sed quereres quid est ille impetus, respondeo quod est quedam qualitas impressa mobili faciens in eo motum, et potest poni de prima specie qualitatis vel tertia que acquiritur per actum proiicientis....Respondetur quod est qualitas permanens, nam manet pila descendente...non omnes impetus sunt eiusdem speciei specialissime. Quod aparet, quia forma substantialis gravis cum ipsa gravitate corrumpunt impetum lapidis sursum et tamen eadem forma cum gravitate conservat impetum lapidis deorsum et producit; ...per istum impetum possunt servari omnes apparentie: propter eum pila resilit cadens supra terram; etiam per eum movetur mola fabri postquam faber dimisit molam, etiam trochus pueri postquam est extra manum pueri... et propter defectum istius impetus faba non ita longe proiicitur sicut plumbum, quia sibi non potest imprimi ita magnus impetus. Et si quis querat quare illud quod sic movetur ab impetu velocius movetur aliquando in medio vel in fine quam in principio, ut patet de sagitta, respondetur quod ratio est quia ille impetus in principio non imprimitur omnibus partibus mobilis, sed solum partibus propinquis, et mediantibus illis partibus imprimitur partibus remotis donec impetus est per totum mobile, et tunc velociori motu movetur." Cf. Duhem, *Études, Vol. 3*, 96–98.

[29] MS Oxford, Bodl., Canon. Misc. 177: f. 48v, "Johannes de Ollandria Alemanus," f. 61v, "Explicit tractatus de instanti Magistri Johannis de Holandya in Universitate Pragii sub anno domini millesimo CCC° 69, compilatum et scriptum per Donatum de monte anno domini 1391 die tertia Septembris Padue...."

At the same time, Buridan's impetus mechanics is evident at Prague, for there exists a Vienna manuscript (Nat.-bibl. lat. 5481) of Buridan's *Questions on the Physics* set down at Prague in 1366 (f. 96r, where the date is obscure; but cf. f. 116r where we clearly read M.CCC.LXVII for a later-copied work). There was also copied at Prague in 1385 a commentary of one Hermannus de Curis on Buridan's *Physics* (Erfurt, Stadtbibliothek, Amplon. F. 300, ff. 166–285v). Also, Albert of Saxony probably brought the impetus mechanics to Vienna when he left Paris to become first rector of the University of Vienna in 1366.[30] Whether he completed his *Questions on the Book of the Heavens* while he was still at Paris or after he had moved to Vienna, he no doubt introduced the impetus mechanics into the new university. At any rate, the work was often copied and used at Vienna (e.g., cf. Vienna, Nat.-bibl. lat. 5446, which bears a date of composition of M.CCC.LXIIII, although the last strokes might be LXVII). Similarly, the *Questions on the Physics* of Buridan were popular at Vienna. How early they appeared at Vienna, I do not know, but there is a manuscript of the *Questiones* copied at Vienna in 1390 (Nat.-bibl. lat. 5424, 163r). There is also a number of other Vienna copies of Buridan's work, including one copied in 1413 (Nat.-bibl. lat. 5332, 1r–71v). The Viennese studies were not limited to the impetus mechanics. The Aristotelian commentator Nicholas of Dinkelsbühl lectured on "the latitudes of forms" in 1391–92,[31] possibly dealing with the Oresme graphing system in one form or another. It was still the object of interest at Vienna as late as 1466, when Jacobus de Sancto Martino's shorter treatise *De latitudinibus formarum* (also attributed to Oresme) was again copied (Nat.-bibl. lat. 4953, 1r–17v); and in 1505 and 1515 the short tract was published at Vienna. Furthermore, one Fredericus Stoezlin composed a commentary entitled *Questiones in librum proportionum* which was based on the Bradwardine treatise. And, as a matter of fact, the abbreviated text of Bradwardine's *Proportions*, which we have reproduced in part in Chapter 7, was certainly popular at Vienna

---

[30] George Sarton, *Introduction to the History of Science*, Vol. *3* (Baltimore, 1948), 1429.

[31] *Ibid.*, p. 1433. J. Aschbach, *Geschichte der Wiener Universität* (Vienna, 1865), p. 139, notes this and gives the spelling as Dinkelspuhel. He also notes that *Latitudines formarum* was lectured on by Johann Fluck von Pfullendorf in 1392–93 (p. 143), by Peter von Pulka in 1394–95 (p. 147), by Michael Suchenschatz in 1396–97 (p. 155), by Friedrich von Passau in 1399–1400 (p. 168). Quoting the *Acta* of the Faculty of Arts, Aschbach also notes other works being lectured on in this period of about 1391–1400, such as the *Proportiones* (presumably that of Bradwardine or possibly that of Albert of Saxony), the *Proportiones breves* (presumably the shortened version of Bradwardine's treatise), and logical tracts like *Obligatoria et insolubilia*.

(cf. Vienna, Nat.-bibl. lat. 4951, which, although itself dated 1501, gives a date of 1451 for the *Abbreviatio*, f. 271v; cf. also lat. 4953, 19r–35v, copied in 1466).

A situation similar to that at Vienna existed in some of the other eastern and German universities. Marsilius of Inghen left Paris to become first rector of the University of Heidelberg in 1386.[32] Even granting that he compiled his principal physical tract at Paris, it is reasonable to expect that he would encourage the teaching of the Parisian ideas in the new university. At any rate, his *Abbreviata phisicorum* became quite popular in the eastern universities. For example, in 1413 a Viennese student copied the work in a manuscript now extant at the Nationalbibliothek (lat. 5332, 1r–71v, colophon, 71r).

In the second of the German universities, Cologne, founded in 1388 and thus not long after Heidelberg, in 1398 the study of the "latitude of forms" was obligatory,[33] and about the turn of the century Johannes Dorp of Leiden composed—presumably at Cologne—a commentary on the works of Buridan (see Bibliography below for manuscript). Similarly the new physics spread to Erfurt. Lawrence of Scotland's works appeared at Erfurt. And Lawrence is mentioned by an unknown Magister de Stadis of Erfurt as the author of the theory of the supplementary role of the air in projection.[34] Erfurt was to become a great center of manuscripts in medieval physics, and the catalogue of the Amplonian collection of 1412 has all of the standard Merton, Parisian, and German authors.[35]

We do not intend to list all of the eastern authors who in the course of the fifteenth century mentioned the new mechanics. But we can note an *Exercitium* on the *Physics* based on the *Questiones* of Buridan which appeared quite early in the fifteenth century at Cracow. This work was elaborated by Master Serpens, in the manner it was taught at Cracow (*iuxta cursum Alme Universitatis studii Cracoviensis*). It tells us that the air sometimes concurs in the motion.[36]

In addition to the Polish and German commentaries already mentioned,

---

[32] Sarton, *Introduction*, Vol. *3*, 1435. It is of interest ot note that, just as at Vienna and Cologne, the *Latitudines formarum* constituted one of the arts studied; for from the early acts of the Arts Faculty, the aspirant for the *licentia* has to swear "se audivisse latitudines formarum." See E. Winkelmann, *Urkundenbuch der Universität Heidelberg*, Vol. *1* (Heidelberg, 1886), 38.

[33] S. Gunther, "Le origini ed i gradi di sviluppo del principio delle coordinate," *Bulletino di bibliografia e storia delle scienze matematiche e fisiche*, Vol. *10* (1887), 375.

[34] See note 18.

[35] W. Schum, *Beschreibendes Verzeichniss der Amplonianischen Handschriften-Sammlung zu Erfurt* (Berlin, 1887), pp. 793–96, 798–817.

[36] Michalski, "Les courants philosophiques," p. 87.

the Polish historian Michalski listed without any elaboration the following works which discuss the impetus theory: Benedict of Hesse, *Commentary on the Physics* (MS. Bibl. Jag. 1367); an *Exercitium contra conclusiones Buridani* (MS Bibl. Jag. 1905) dated 1449, which, strangely enough, in contradiction to its title, resolves all the questions in favor of Buridan's opinion *(ad intentionem Buridani)*; Johannes de Thost, *Questiones* (on the *Physics*) (MS Bibl. Jag. 2097) dated 1451; an anonymous *Questiones* (Bibl. Jag. 1946) dated 1458; and, finally, a later group of *Questiones* on the *Physics* (MSS Bibl. Jag. 2024, 2087, and 2088) which explain, but alter, the impetus theory considerably. As a result of this manuscript study, Michalski believed that Copernicus must have certainly been acquainted with this theory when he was at the University of Cracow in 1491 and thus before he went to Italy. [37] Copernicus, however, employs the impetus theory only briefly, if at all. [38]

The impetus theory also received the treatment of Nicholas of Cusa, the German cardinal (d. 1464). In an exposé of some philosophical ideas in a dialogue *De ludo globi*, we find the motion of the last sphere attributed to an *impetus*. [39] The Duke of Bavaria asks how God has created the movement of the last sphere. This sphere, in effect, is not moved directly by God, the Creator, or by the Spirit of God; no more than it is you or your spirit which moves the globe which you see moving before you. It is you, however, who have put it into movement; because the impulsion of your

[37] *Ibid.*, pp. 87–88.
[38] Duhem, *Études*, Vol. *3*, 196.
[39] *De ludo globi*, Book I, in *Opera* (Basel, 1565), p. 210: "CARDINALIS. Recte, sed oportet etiam considerare, lineas, descriptionis motus unius eiusdem globi variari, et nunquam eandem describi, sive per eundem, sive per alium impellatur, quia semper varie impellitur, et in maiori impulsu, descripta linea videtur rectior, et secundum minorem curvior. Quare in principio motus, quando impulsus est recentior, lineae motus sunt rectiores, quam quando modo tepescit. Non enim impellitur globus nisi ad rectum motum. Unde in maiori impulsu, globus a sua natura magis violentatur, ut contra naturam, etiam quantum fieri potest, recte moveatur. In minori vere impulsu, ad motum naturalem minus violentatur, sed aptitudinem naturalem formae suae, motus sequitur.... (pp. 213–14) IOANNES. Quomodo concreavit Deus, motum ultimae sphaerae? CARDINALIS. In similitudine, quomodo tu creas motum globi. Non enim movetur sphaera illa, per Deum creatorem, aut spiritum Dei, sicut ne globus movetur per te, quando ipsum vides discurrere, nec per spiritum tuum, licet posueris ipsum in motu, exequendo per actum manus, voluntate, impetum in ipsum faciendo: quo durante movetur.... CARDINALIS. Attende, motum globi deficere et cessare, manente globosano et integro, quia non est motus, qui globo inest, naturalis, sed accidentalis et violentus. Cessat igitur, impetu, qui impressus est ei, deficienti. Sed si globus ille esset perfecte rotundus (ut praedictum est) quia illi globo rotundo motus esset naturalis, ac nequaquam violentus, nunquam cessaret." Cf. Duhem, *Études*, Vol. 2, 187, and E. Wohlwill, "Die Entdeckung des Beharrungsgesetzes," *Zeitschrift für Volkerpsychologie und Sprachwissenschaft*, Vol. *14* (1883) pp. 376–77.

hand, which follows your will, has produced there an impetus, and as long as this impetus lasts, the globe continues to be moved. The movement ceases, then, when the impetus which has been impressed in this globe is deficient. But as we have said above, if the globe were perfectly round (as in that sphere), circular movement would be natural to it and not violent; then this movement would not cease. Cusa is clearly accepting here a circular impetus which has the property attributed generally to impetus by Buridan, namely, the property of continuing indefinitely without resistance, at least in a body where it is natural—and yet Cusa in a passage just preceding the one noted holds that the greater the impulse given to a ball thrown, the longer it tends to go in a straight line. This is, of course, for bodies subject to weight. As we shall see later, it was just this property of a rotary impetus to last indefinitely which Benedetti was to oppose vigorously when he accepted the tendency of impetus to continue indefinitely in a straight line.

We know the names of a number of other minor figures in Germany who described or mentioned the impetus theory, but the exposition of their brief remarks would not add materially to our account, and it will be much more profitable to turn back to the fate of the new physics in Italy.

Italy. In the second half of the fourteenth century, the English and French physics took firm root in Italy. We have already mentioned the works of Giovanni di Casali (1346) and Francischus de Ferraria (1352) which demonstrate a knowledge of the Merton kinematics as well as dynamics (see Chapters 6 and 7). Casali, we suggested, may have preceded Oresme in the application of a coördinate geometry to the English kinematics. The work of Francischus is of particular interest, for it shows that Bradwardine's treatment of the proportions of motions was not only understood but taught and disputed at Padua at this early date. We know little about the later careers of these two initiators of the new mechanics in Italy. Casali appears to have been a nuncio sent by Pope Gregory XI to Sicily in 1375, and he may still have been living in 1390.[40] He may have been present in Padua in 1376 *(pres.—mag. Johanne physico q. d. Johannis de Caselis)*,[41] while it was perhaps our Francischus who is mentioned at Padua in 1378.[42]

The appearance of the Merton mechanics in Italy antedates the intro-

---

[40] Jo. Sbaralea, *Supplementum et castigatio ad scriptores trium ordinum S. Francisci etc.*, Part II (Rome, 1921), p. 52.
[41] A. Gloria, *Monumenti della Università di Padova (1318–1405)*, Vol. 2 (Padua, 1888), 116.
[42] *Ibid.*, Vol. *1* (Padua, 1888), 508.

duction of Buridan's dynamics. Although the Paduan university records note the presence in 1351 at an arts examination of three Parisian doctors of arts,[43] there is no indication that the impetus theory was discussed at such an early date. Actually, the earliest evidence of the impetus theory at Padua (or in fact in Italy) comes from a manuscript of Buridan's *Questions on the Physics* copied at Padua in 1377 and now in the Vatican Library (Vat. lat. 2163, f. 157v).

Our evidence for the continuing interest in the new physics and its problems in the 1380's is scattered. We know something of the interest that the question of the intension and remission of forms, which lay behind the kinematic development, had at Padua from a copy of Gregory of Rimini's views on the subject made in 1384 by Bartholomew of Mantua, then an arts student and later a doctor of arts of considerable reputation (see MS Venice, Bibl. Naz. San Marco Lat. VI, 160, f. 108r). There is also a copy of Casali's *De velocitate* dating from 1386 at Padua to remind us of the continued concern with the English kinematics (MS Oxford, Bodl. Canon. Misc. 177, f. 228).

The expanding interest in the new physics shown in the late 1380's and early 1390's is well illustrated in the activity of Biagio Pelacani di Parma, or Blasius of Parma.[44] There is a strong tradition that Blasius took his arts doctorate at Pavia in 1374 and that during the 1370's and 1380's he taught successively at Pavia, Bologna, and Padua, and then went back to Pavia during the 1390's. An examination of the manuscripts of his works reveals that practically all of them, and cetainly the major works in which he discusses the new physics, were written before 1400. We can cite as his earliest datable work some lectures on the various works of Aristotle, including the *Physics*, and dated 1385.[45] We have from 1390 a manuscript of some questions on *The Perspective* of Witelo, a work widely used and discussed by the schoolmen, and which has some oft-quoted passages bearing on the nature of movement.[46] At the same time, his interest in the new mechanics is evident. We know that his *Questions on the Proportions of Bradwardine* was copied in 1391 (Venice, Bibl. Naz. San Marco Lat. VII, 38, ff. 8–37). Simultaneously, he seems to have been interested in the problems of the intension and remission of forms, for he composed two short treatises on such problems. The first is entitled *De intensione et remis-*

---

[43] *Ibid.*, p. 42.
[44] E. A. Moody and M. Clagett, *Medieval Science of Weights* (Madison, 1952), pp. 236–37.
[45] L. Thorndike, *A History of Magic and Experimental Science*, Vol. 4 (New York, 1934), 67.
[46] *Ibid.*

sione formarum and dealt with the basic question as to how intension and remission take place. It was copied in 1391.[47] The other work, *Super tractatu de latitudinibus formarum*, is the treatise from which we have already given a selection (see Doc. 6.4). A copy of the second treatise made at Padua and now in the Bodleian Library is dated 1392 (Bodl. Canon. Misc. 177, 97v–100v). This copy of 1392 includes the familiar diagrams and figures illustrating the Oresme graphic method treated in Chapter 6. Thus this work is one of the earliest examples of the spread of the Oresme graphic treatment in Italy in contradistinction to that of Casali. This work is to be compared with an anonymous tract of about the same time also treating the Oresme configuration system and bearing the title *Questio utrum omnis forma habeat latitudinem nobis presentibilem per figuras geometricas* (MS Venice, Bibl. Naz. San Marco, Lat. VI, 62, ff. 63r–68r).

The only other point of interest concerning Blasius' works, is that he was also familiar with the Buridan physics, since he made an arrangement of Buridan's *Questions on the Physics* in about 1396, now extant in a Venetian manuscript.[48] We know, too, that he visited Paris, for in one of his works, *De ponderibus*, he mentions that he has been to Paris.[49]

The works of Blasius of Parma are but one aspect of the expanding influence of the new physics in Italy about 1390. In 1391 we see copies of Heytesbury's *Three Categories of Movement* and of the *Probationes* which were composed on the same subject (MS Venice, Bibl. Naz. San Marco Lat. VIII, 38, ff. 40r–54v, 66v–72r). And in the next year we notice the copying at Padua of the *De motu* of John of Holland, who was, as we have seen, under the strong influence of Heytesbury and Swineshead (MS Oxford, Bodl. Canon. Misc. 177, 109r).

Probably about the same time as the reception of Heytesbury and John of Holland, a little-known Italian schoolman whose only appellation in the manuscripts is Messinus occupied himself with the kinematic problems raised by Heytesbury and Casali. His longest work was some *Questions on the Question of John Casali*. This treatise was not published, but there are manuscripts at Bologna, Venice, and the Escorial.[50] He is designated

---

[47] MS Oxford, Bodl. Canon. Misc. 177, ff. 24r–39r. The date of 1391 is a conjecture from the fact that succeeding works in the same codex and in the same hand were copied during that year.

[48] Thorndike, *Magic and Experimental Science*, Vol. 4, 68.

[49] Moody and Clagett, *Medieval Science of Weights*, pp. 237, 413–14 (n. 21). Cf. Thorndike, *Magic and Experimental Science*, Vol. 4, 71.

[50] MS Bologna, Bibl. Univ., 1227 (2410), ff. 101–63; Venice, Bibl. Naz. San Marco, Lat. VI, 25, ff. 1r–76r; and Escorial, La Real Bibl., f.II.8, ff. 1r–49r. Among the many questions the following are of interest to us (Venice MS, 1r): "Queritur utrum bene describatur hic terminus lati-

in the manuscripts as "Master Messinus, famous doctor of arts and medicine." In the introduction we are told that in reading Casali, and he obviously means lecturing on Casali, certain doubts occurred, and he now intends to discuss these doubts seriously. His discussion ranges over a wide variety of topics, starting with the question as to whether the definition of a uniformly difform latitude presented by Casali is sound. Similarly, he discusses whether every quality is infinitely divisible in intensity, whether the theorem which represents a uniformly difform latitude by its mean degree is correct, whether the exposition of the peripatetic law of movement made popular by Bradwardine is sound, and a number of other similar questions.

The second extant work of Messinus is a *Sententia super de tribus predicamentis motus Hentisberi*, i.e., a commentary on Heytesbury's *Three Categories of Movement*. The copies of this work [51] tell us little except that he left the work incomplete and that it was completed later by the famous Paduan schoolman Gaetano de Thienis, of whom I shall speak briefly later. I have tentatively placed Messinus' activity about 1390–91 simply because I have found mentioned in the records of the University of Pavia for that year a Bolognese named Masino Codronchi, a lecturer in natural philosophy and astrology. [52] If this is our Messinus or Messino, his career was cut

tudo uniformiter difformis, sic, latitudo uniformiter difformis est latitudo cuius quarumlibet partium immediatarum intensissimus gradus qui non est in parte intensiori est remississimus qui non est in parte remissiori.... (3v) Queritur secundo utrum sit possibile quod uniformiter difforme calidum alteretur ad caliditatem per certum tempus ipso manente continue per idem tempus uniformiter difforme et consimilter de aliis.... (6r) Queritur tertio utrum omnis qualitas sit in infinitum indivisibilis intensive.... (7r) Queritur quarto utrum cuiuslibet latitudinis uniformiter difformis incipientis a non gradu et terminantis ad certum gradum gradus medius est precise subduplus ad gradum ad quem terminetur predicta latitudo.... (10r) Queritur quinto utrum omnis latitudo uniformiter difformis correspondeat gradui suo medio.... (70r) Utrum cuiuslibet motus velocitas tanquam penes tantam proportionem proportionum geometricam moventis ad motum sit attenda et proportio proportionum sit sicut proportio motuum ab illis proportionibus productibilium.... (73r) Queritur duodecimo utrum quodlibet quod movetur ita velociter precise moveatur sic eius pars velocissisme mota.... (76r) Expliciunt questiones edite per magistrum Messinum artium et medicine doctorem eximium super questione magistri Johannis de Casali, et cetera, et sic est finis huius operis."

[51] Edition of Venice, Octavianus Scotus per Bonetum Locatellum, 1494, ff. 52v–64v; Venice, Bibl. Naz. San Marco, Lat. VI, 105, ff. 47r–65r. On f. 65r of MS: "Et sic est finis scripti super de tribus predicamentis Hentisberi editi initiative a Messino et completive a Cayetano de Tyenis"; on f. 62v of edition: "Explicit tractatus de tribus predicamentis compositus per famosissimum doctorem Messinum expositorem optimum Tisberi, qui non fuit completus per Messinum, sed completus fuit per famosissimum doctorem Gaetanum de Tenis, ut in Sequentibus."

[52] R. Maiocchi, *Codice diplomatico dell'*

short at Pavia, for this "Masino" vacated his chair in 1391.[53] It seems unlikely that our Messinus could be the earlier Misino di M. Bonfantino dalle Pecore, dated by Alidosi at Bologna in 1372.[54] We do know at least that some of his activity preceded 1401, for in a Paduan Library list of that year is included a third work of his *Questiones super libro Dei Gratias* (i.e., Questions upon the *Questiones in libros physicorum disputate Padue* by Johannes Baptista Gratia Dei, called Esculanus, a schoolman of the first half of the fourteenth century).[55]

Nearly contemporary with Messinus and Blasius was the activity of two Paduan doctors whose work extends over a bit into the fifteenth century: Angelus de Fossambruno and Jacopo da Forlì. We can catch but brief glimpses of Angelus at about the turn of the century. We are told that he received the doctorate of arts at Bologna in 1395 and that he taught there, first logic and then natural philosophy, until 1400, when he came to Padua.[56] We know that one of his relatives (perhaps his father?), Pietro da Forosempronio, was a physician and settled at Padua as early as 1368.[57] Now Angelus is present at examinations given at Padua from 1400 to 1402,[58] in company with Jacopo da Forlì and Bartholomew of Mantua, and there can be little doubt that he was occupied with the new physics at this time. We are told in the colophon of a manuscript of one of his works that he was teaching natural philosophy at Padua in 1402.[59] We have no reliable data on his career after this date, his name no longer appearing in the Paduan records. He was the author of at least four works, of which we can single out two physical works: (1) *Recollecte super Hentisberi de tribus predicamentis*,[60] in which he goes over again the standard English kinematic problems. This treatise makes reference to the opinions of Bradwardine, Heytesbury and Dumbleton. (2) *De reactione*,[61] which takes up a heat problem that gave the schoolmen great concern.[62] In this

Università di Pavia, Vol. *1* (Pavia, 1905), 186.

[53] *Ibid.*, p. 193.

[54] N. P. Alidosi, *I Dottori Bolognesi di teologia, filosofia, medicina, e d'arti liberali, etc.* (Bologna, 1623), p. 140.

[55] Gloria, *Monumenti*, Vol. *1*, 385.

[56] *Ibid.*, pp. 495–96.

[57] *Ibid.*, Vol. *2*, 74.

[58] *Ibid.*, Vol. *1*, 495; cf. Vol. *2*, 398.

[59] *De inductione formarum*, MS Vat. lat. 2130, f. 194v: "Explicit questio de inductione formarum determinata per insigne et famosum artium doctorem magistrum Angelum de Fossambruno, dum Padue philosophiam legeret, anno Mº CCCº 2º."

[60] See the various manuscripts from Venice, Bibl. Naz. San Marco, Lat. VI, 105, ff. 65r–79r; VI, 30, ff. 1r–13r; VI, 160, ff. 224r–239r; VI, 71, ff. 113r–128v; VII, 7, ff. 55v–80r; VI, 155, ff. 159r–178r; and the edition of Oct. Scotus per Bonetum Locatellum (Venice, 1494), ff. 64–73.

[61] MS Venice, Bibl. Naz. San Marco, Lat. VI, 160, ff. 248r–252r.

[62] For a discussion of heat reaction, see

work Angelus cites Swineshead, calling him the Calculator,[63] a name by which he came to be known to Scaliger, Leibniz, and others. Thus Angelus appears to be the first Italian to so designate him.

Let me return for a moment to the first of Angelus' works, the commentary on Heytesbury, since it is this one that particularly reveals the flourishing of the new physics.[64] It covers a wide range of subjects made popular by the English school. Thus there is one part on kinematics as such, entitled "concerning the velocity of movement with respect to effect." Another part doubts Heytesbury's initial conclusion that we can measure the velocity of any body by the line described by its most rapidly moving point. Also, the whole question of how to measure uniform acceleration is gone into again. Furthermore, the dynamical question, raised by Bradwardine, of the role of power and resistance in velocity is examined. One copy of the work, dated 1412, is particularly interesting, for it utilizes the Oresme method of configuration throughout; and, in so representing the various geometric figures, it follows the custom of reproducing underneath the base line a replica of the base line and then clearly labeling it as the measure of time with the left end of the line representing the zero point of time;[65] so when the various kinds of nonuniform movements are compared, it is emphasized that the measure they have in common is time.

Of greater fame than Angelus was his contemporary and colleague at Padua, Jacopo da Forli, an eminent physician and philosopher. Just as clearly as that of Angelus, his work reveals the domination of Italian philosophy by the new physics. We know that Jacopo was the son of Peter de Forlivio, who taught astrology and grammer at Bologna in 1384–85, in which year young Jacopo was also called upon to fill in for the professor of natural philosophy at the same university.[66] Just when Jacopo came to

my *Giovanni Marliani and Late Medieval Physics* (New York, 1941), Chapter 2.

[63] MS *cit.* in note 61, f. 248r. "Utrum omne agens in agendo repatiatur.... In hoc sunt tres opiniones Marsilii, Hentisberi, Dulmentonis, et Calculatoris."

[64] Venice, Bibl. Naz. San Marco Lat. VI, 105 (f. 65v): "De velocitate motus penes effectum....(66v) Motuum localium, ut dicebatur, aliquis uniformis, aliquis difformis. Dubitatur ergo utrum velocitas in motu locali uniformi attenditur penes lineam descriptam a puncto mobilis velocissime moto in tanto vel in tanto tempore....(69r) Sequitur nunc dicere de motu difformi penes quid attenditur velocius moveri vel tardius....Circa materiam de motu difformi dubitatur utrum possit aliquid per certum tempus difformiter moveri....(70r) Dicto de motu locali tam uniformi quam difformi restat dicendum de intensione eiusdem motus localis....(71v) Regule de pertransitione spatii....(72v) Utrum motus uniformiter possit intendi et remitti....(74r) Regule de intensione motus...."

[65] Venice, Bibl. Naz. San Marco Lat. VI, 30, note figures on ff. 6v and 7r.

[66] Gloria, *Monumenti*, Vol. *1*, 437; D. Umberto Dallari, *I Rotuli dei lettori, legisti,*

the University of Padua from Bologna is not clear. By 1400 he was already sponsor at Padua for several doctoral candidates.[67] Presumably he would have been teaching there long enough before 1400 to have acquired students who now in 1400 were at the end of their arts course. Jacopo remains in the records of the University until 1404, at which time he seems to have retired temporarily from Padua as a result of the war between the Carrara family and the Venetian Republic for the control of the city. But his reputation was great, and so he was invited back to the University in 1407 to teach theoretical medicine,[68] the city and the University being safely in Venetian hands. From then until just before his death in 1414, he appears as the real work horse of the arts and medical faculties. He sponsored and sat in on a far greater number of examinations than any of his colleagues.[69] In 1410, for example, he participated in twenty-nine examinations. His most frequent companions at these examinations were (first, in the early years) Angelus de Fossambruno, then after 1407 the famous Blasius of Parma, and the soon-to-be-distinguished Paul of Venice—all of course exponents and commentators of the new physics. There can be little doubt then as we look at the University of Padua that it was the real heart of the study of English and French natural philosophy in Italy, followed closely by Bologna and later Pavia. I might add in passing that in contrast to the great popularity of Jacopo, Blasius seems by this late date to have developed into a disagreeable, greedy pedant, and while there are a number of different stories to account for his unpopularity with the students, it is a fact that he was dismissed by the University in 1411 because his classes were empty.[70] The story is that he exacted high fees from his students and they thus boycotted his classes. The question of his orthodoxy may have had some influence on his declining popularity.[71]

But to return to Jacopo. I am not interested here in his medical works.[72] His wide understanding of the new physics is illustrated principally in one lengthy work entitled *De intensione et remissione formarum*.[73] Like the works

e artisti dello Studio Bolognese dal 1384 al 1799 (Bologna, 1888), pp. 4–5.

[67] Gloria, *Monumenti*, Vol. *1*, 437.

[68] A. Favaro, "Intorno alla vita ed alle opere di Prosdocimo de' Beldomandi," in B. Boncompagni, *Bulletino di bibliografia e di storia delle scienze matimatiche e fisiche*, Vol. *12* (1879) 27–28; cf. F. M. Colle, *Storia scientifico-letteraria dello studio di Padova*, Vol. *3* (Padua, 1825), 235–40.

[69] G. Zonta and I. Brotto, *Acta graduum academicorum gymnasii Patavini ab anno 1406 ad 1450* (Pavia, 1922). See index for the numerous references to his participation in examinations.

[70] Thorndike, *Magic and Experimental Science*, Vol. *4*, 70.

[71] Maier, *Die Vorläufer Galileis im 14. Jahrhundert* (Rome, 1949), pp. 279–99.

[72] Gloria, *Monumenti*, Vol. *1*, 438–39. Gloria gives a listing of Jacopo's works.

[73] MS Venice, Bibl. Naz. San Marco

of Angelus, this work shows a familiarity with the Merton group. A copy of this work illustrates particularly the popularity of Oresme, for statements on motion in the text which are purely of rhetorical nature are interpreted in the margins by the Oresme graphic system.[74]

We note finally in speaking of Jacopo, that he represents an interesting Italian phenomenon: the juncture of medical work and natural philosophy. From the time of Peter of Abano at the beginning of the fourteenth century until the time of Jacopo, that juncture had been close. And, as Randall has shown, that joining of the two was important in the increasingly mature discussion of scientific method that occurs at Padua.[75]

These last few remarks on Italian schoolmen have shown that by about 1400 the reception was complete and extensive. An inventory of a library at Padua of 1400 confirms this completeness.[76] It included works of Heytesbury, Swineshead, Dumbleton, Buridan, Albert of Saxony, Marsilius of Inghen, Messinus, etc.

Now I have carried the detailed story of the spread of the new physics to a little past 1400. It had a vital life in the course of the fifteenth century, particularly at Padua and Pavia. At Padua it was continued in the hands of Jacopo's junior contemporary Paul of Venice,[77] Paul's students Gaetano de Thienis[78] and Giovanni da Fontana,[79] and their contemporary Johannes de Marchanova[80] (at Padua from about 1438 to 1450) who built a fine

Lat. VII, 7, ff. 1r–55r; VI, 155, ff. 134r–159r; edition of Venice, 1496, ff. 16r–42v.

[74] See Venice, Bibl. Naz. San Marco, Lat. VII, 7, particularly f. 1v for use of the Oresme configuration system.

[75] J. H. Randall, "The Development of Scientific Method in the University of Padua," *Journal of the History of Ideas*, Vol. I (1940), 177–206.

[76] For the inventory, see Gloria, *Monumenti*, Vol. 2, 385.

[77] For Paul of Venice, see Duhem, *Le Système du monde*, Vol. 4 (Paris, 1916), 280–83; Gloria, *Monumenti*, Vol. 1, 553 (erroneous death date), Vol. 2, 202; Zonta and Brotto, *Acta graduum*, index; F. Momigliano, *Paolo Veneto e le correnti del pensiero religioso e filosofico nel suo tempo* (Turin, 1907); Maier, *Zwei Grundprobleme*, p. 273, n. 18. Father Weisheipl tells me that Mr. A. B. Emden has evidence of Paul's having studied at Oxford.

[78] Duhem, *Système*, Vol. 4, 301; Zonta and Brotto, *Acta graduum*, index; Clagett, *Giovanni Marliani*, index. Cf. P. S. da Valsanzibio, *Vita e dottrina di Gaetano di Thiene*, 2d ed. (Padua, 1949). Gaetano's commentary on Heytesbury's *Regule solvendi sophismata* was published as interlinear commentary in the Heytesbury edition of 1494. In that publication the familiar Oresme triangle and rectangle were used to illustrate the mean speed theorem (see Plate 5).

[79] For his life see Thorndike, *Magic and Experimental Science*, Vol. 4, chap. 45. See particularly his attention to the doctrine of the difference of center of gravity and center of magnitude of the earth (*ibid.*, p. 176). He has basic kinematic rules including the distinction between curvilinear and angular velocity (*Tractatus de trigono balistario*, MS Oxford, Bodl. Canon. Misc. 47, ff. 217v–218v).

[80] Zonta and Brotto, *Acta graduum*, index; and see the numerous Marchanova

library in the new physics, a library which he left to the Augustinian canons of the monastery of S. Johannes in Viridario and which later became the kernel of the present collection at the Biblioteca Nazionale di San Marco at Venice. It is noteworthy that Nicholas of Cusa also studied at Padua, and, as we have already mentioned, his works show knowledge of the impetus mechanics. After Gaetano at Padua there was Christopher de Recaneto, who edited lectures on Swineshead's *Liber calculationum*.[81] At the same time, at Pavia Giovanni Marliani and his sons concerned themselves with the English and French authors.[82] Toward the end of the century, authors like Bernard Tornius and Benedetto Vittorio occupied themselves with the medieval physics, the first having a particular predilection for Heytesbury's *Tria predicamenta de motu*, which he commented upon (Venice, 1494, 73v–77v), and the latter for Albert of Saxony, whose *Tractatus proportionum* he submitted to commentary (Bologna, 1506).

By the end of the fifteenth century and the beginning of the sixteenth century, most of the important works of the English and French schoolmen were published either in Italy or France, the major tract of Dumbleton being a notable exception. Probably the most important team publishing the medieval physical works in Italy at the turn of the sixteenth century was that of Octavianus Scotus and Bonetus Locatellus at Venice. Practically the whole range of the English mechanics and at least part of the French mechanics with their Italian commentators and continuators appeared in works bearing their imprint, *Octavianus Scotus per Bonetum Locatellum*. Some of the authors whose works they published were Thomas Bradwardine, William Heytesbury, Richard Swineshead, Albert of Saxony, Giovanni di Casali, Blasius of Parma, Nicole Oresme, Angelus de Fossambruno, Messinus, Jacopo da Forlì, Gaetano de Thienis, and Bernard Tornius. In addition, the ever-popular *Liber calculationum* was twice more published by other Italian publishers.[83] And, as we shall see shortly, the best products of the Parisian impetus dynamics such as those of Buridan and Albert of Saxony were almost simultaneously being published in Paris.

codices mentioned by J. Valentinelli, *Bibliotheca manuscripta ad S. Marci Venetiarum*, Vols. *1–6* (Venice, 1868–76). For example, see Lat. VI, 62; Lat. VI, 72; Lat. VI, 79; Lat. VI, 155; Lat. VI, 226; Lat. VIII, 19; Lat. VII, 7, etc.

[81] Venice, Bibl. Naz. San Marco Lat. VI, 149, ff. 31r–49v.

[82] Clagett, *Giovanni Marliani, passim*.

[83] Thorndike, *Magic and Experimental Science*, Vol. *3*, 371–72; and see Chapter 4 above, note 13.

THE SIXTEENTH CENTURY

France, Spain, and Portugal. Concurrent with the publication activity in Italy, the various currents of medieval natural philosophy had a marked revival at the University of Paris in the beginning of the sixteenth century. Pierre Duhem has labeled this activity eclectic,[84] and with reason; but for our purpose we shall do well to single out the revival of French dynamics and of the English kinematics with its Italian commentary. The principal center of the new interest in the older mechanics was the Collège de Montaigu. For example, the masters of Montaigu were responsible for the publication of the crucial works of John Buridan and Albert of Saxony. The Scotch master George Lockert published a collection of works in 1516, which included Albert of Saxony's *Questions* on the *Physics*, the *De caelo*, and the *De generatione*, as well as some less important works of Buridan and another Parisian master Themon.[85] Similarly, two other foreign masters of Montaigu, John Major of Scotland and John Dullaert of Ghent, were responsible for the publication of other works of Buridan, unquestionably the most important being the publication of the *Questions on the Physics* in 1509 under the editorship of Dullaert.[86]

The masters of Montaigu did not limit their interest in the medieval physics to editing and publishing the principal works of that physics but in addition discussed and wrote on the conclusions of the earlier mechanics. Thus John Major (*ca.* 1478—*ca.* 1540), in analyzing the motion of the heavens, drew on Buridan's impetus treatment and compared the substantial or accidental form which God naturally impressed in the heavens for their everlasting motion with the impetus of a smith's wheel.[87] Hence he seems to be perfectly aware that Buridan's impetus would be everlasting were it not for resistance; and so he is parroting the concept of the "inertial" impetus in contrast to the "self-expending" impetus. Major is equally acquainted with the Merton authors—and he notes the Merton law describing a uniformly difform quality in terms of its mean degree.[88]

Major's contemporary at Montaigu, John Dullaert, likewise concerned himself with the impetus theory, reporting on the authority of "certain people," that the impetus engendered by violence is corrupted bit by bit because of the absence of its cause, just as intuitive understanding is

---

[84] Duhem, *Études*, Vol. *3*, 272.
[85] *Ibid.*, pp. 133–34.
[86] *Ibid.*, p. 134.
[87] *Ibid.*, p. 142.
[88] *Ibid.*, pp. 525–26.

destroyed by the absence of its object.[89] This, then, is the self-expending impetus which we have contrasted with the inertial impetus of Buridan. But Dullaert rejected this self-expending impetus: "Some say that the impetus caused by violence is corrupted by the absence of its cause.... But it would be better, I believe, to say that the impetus caused by violence is corrupted by the form itself (i.e., the weight) of the projectile which inclines the body toward a movement contrary to that which the impetus produces...."[90] For Dullaert the quantity of the impetus impressed is dependent on the matter of the projectile, and so an arrow receives more impetus than a feather and can consequently be thrown farther.[91]

It should also be noted that Dullaert was familiar with, and influenced by, the kinematic studies of Heytesbury and Swineshead. He, too, is interested in the question of how to measure speed in local motion, in the motion of augmentation, and in the motion of alteration. In the course of this discussion he gives a Heytesbury-Swineshead type of symmetrical proof of the mean speed theorem,[92] although his general discussion is less

---

[89] John Dullaert, *Questiones super octo libros phisicorum Aristotelis* (Paris, 1506), Bk. VIII, quest. 2 (no pag.): "...aliquis impetus causatur naturaliter et aliquis violenter, circa quod notandum est quod quando aliquod grave detinetur sursum et removetur illud impedimentum illud grave velocius movetur in fine quam in principio, dato quod resistentia sit omnino uniformis, et causa huius est quia in motu illius gravis incipit intendi a non gradu intensionis in illo mobili et semper continue postea intenditur usque ad finem motus (impetus naturalis).... et impetus naturalis nunquam remittitur uniformiter sed semper intenditur usquequo destruitur per detentionem mobilis; opposito modo est in impetu causato violenter, nam post causationem sui semper remittitur usque ad non gradum intensionis impetus... sed impetus violente causatus corrumpitur, ut aliqui dicunt, ab absentia sue cause sicut notitia intuitiva corrumptiur per absentiam sui obiecti."

[90] *Ibid.*: "Sed melius credo esse dicendum quod quilibet talis impetus violente causatus corrumpitur a forma proiecti inclinante mobile ad motum oppositum illi motui quem causat ille impetus."

[91] *Ibid.*: "...motus proiectorum quiescente primo proiiciente fiunt a virtute impressa in ipsis proiectis...et quia sagitta plus recipit de hac virtute quam pluma, ideo ad maiorem distantiam moveri potest."

[92] *Ibid.*, Bk. III, quest. 1 (no pag.; sig. h.2.v): "...omnis latitudo uniformiter difformis, sive incipiat ab aliquo certo gradu sive incipiat a non gradu et terminetur ad aliquem certum gradum, correspondet suo gradui medio. Volo dicere, si essent duo mobilia *a* videlicet et *b* et *a* per totam horam sequentem moveretur uniformiter ut 4, *b* vero moveretur uniformi-

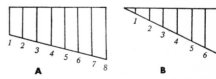

ter difformiter intendendo motum suum a o usque ad 8, dico quod ista duo mobilia equalia spacia pertransibunt, licet *b* per totam secundam medietatem hore movebitur velocius quam *a*, et ratio est quia tantum quantum movetur velocius in illa medietate tantum *a* movebitur velocius in prima medietate...." Figure A: Latitudo

lucid than those of his English predecessors. As in the contemporary Italian works, his discussion of uniformly difform movement is illustrated by the conventional right triangle.

Among the many foreigners at Paris at the turn of the sixteenth century, no group is more interesting than that of the Spaniards and the Portugese. The Spaniard Luiz Coronel, whose principal physical tract was published in 1511, worked side by side with the other Montaigu masters in expounding both the impetus mechanics and Merton kinematics.[93] He reported the impetus mechanics faithfully, although he does follow Marsilius of Inghen in his acceptance of the possibility of a moment of rest in the flight of the projectile when the force of the impetus and that of gravity are equally balanced.[94] Somewhat more fully developed was the exposition of Buridan's theory given by Coronel's countryman Juan Celaya, whose treatment of the *Physics* appeared in 1517.[95] Juan taught at the Parisian Collège de Sainte-Barbe. In the manner of Buridan, he taught that the more matter there is in the projectile, the greater the impetus. For Celaya the quantity of impetus was to be distinguished as "intensive" or "extensive."[96] This is another instance of the interest that began to

uniformiter difformis incipiens a gradu. Figure B: Latitudo uniformiter difformis incipiens a non gradu.

[93] *Physicae perscrutationes*, (Paris, 1511, and edition of Lyon, 1530). See Bk. III, part 1, for his discussion of the impetus mechanics. Like Dullaert and the other Paris masters, Coronel also (Bk. III, part 4) reports the Merton kinematics. In doing so he has used Heytesbury's *Tria predicamenta de motu*, Swineshead's *Liber calculationum*, Albert of Saxony's *Tractatus de proportionibus*, and the works of several Italian masters (cf. Duhem, *Études*, Vol. *3*, 553–554). Needless to say the mean speed theorem is given in the course of the kinematic discussion, although only as applied to qualities. I used the edition of Lyon, 1530, from which I give a few passages (f. 57v): "...proiectum grave quod in sursum vel lateraliter movetur moveri naturaliter non posset post expulsionem nisi qualitas motiva in eo a proiiciente producta quam impetum vocamus assignaretur, illa enim non posita nescirent physici aliud movens dare...; grave enim per medium uniforme descendens prope finem motus velocius descendit, eo quod tempore descensus in ipso vel a gravitate vel a propria forma substantiali vel ab utroque producebatur impetus qualitas motiva deorsum et quia prope talem terminum ad quem intensior est talis impetus quam prope initium motus velocius versus finem grave descendit. Sed mobile violenter motum opposito modo movetur, eo quod tempore motus impetus quo movetur continualiter remittitur et quia prope initium intensior est, prope finem vero remissior tardius et tardius tale mobile fertur.... (75v) qualitatem videlicet uniformiter difformem suo gradui medio correspondere, i.e., esse ita intensam sicut est gradus medius eius inter intensissimum gradum et remississimum ad quem terminatur."

[94] Duhem, *Études*, Vol. *3*, 144.

[95] Juan Celaya, *Expositio in octo libros phisicorum Aristotelis: cum questionibus eiusdem*, (Paris, 1517).

[96] *Ibid.*, f. 201r: "Maior impetus imprimitur maiori proiecto vel intensive... vel extensive... unde non inconvenit quod minor impetus extensive velocius moveatur quam maior impetus extensive,

grow from the fourteenth century in distinguishing "extensive" from "intensive" factors in the quantification of physical phenomena. We have already remarked in Chapter 4 a similar distinction of "extension" from "intension" was made in the analysis of force, of speed, of heat, and of weight. Furthermore, Oresme's geometric method of representing the quantities of qualities and movements was based precisely on considering simultaneously these two factors. And much of the quantification of early modern physics in the seventeenth century turned on the quantitative description and distinction of extensive and intensive factors.

For Celaya, as for Buridan and Dullaert, impetus was not self-expending (i.e., destroyed simply by the absence if its cause), but rather was inertial, being destroyed by the resisting medium and the contrary weight of the projectile.[97] With Buridan, Celaya says that the motion of the heavens could be explained by the use of an everlasting impetus.

Among the various scholars at Paris at this time, there was considerable discussion of the nature of impetus. For Celaya it was a "second quality" comparable to dispositions of the soul.[98] Coronel, on the other hand, identified it as "local motion"—a most intense motion. The weight which moves upward has no other movement than the impetus; in a falling body, the substantial form and weight produce a movement which can be called impetus when it is sufficiently intense.[99] Coronel certainly seems to have

quoniam est intensior altero." Duhem, *Études*, Vol. *3*, 140, with excessive freedom of interpretation says that this doctrine amounts to the following: "The total impetus of a body results from the impetus attributed to each element of the body; other things being equal, however, the elementary impetus is more intensive in proportion as the speed of the element is greater."

[97] Celaya, *op. cit.*, ff. 200v–201r: "Quarta est opinio asserentium quod proiiciens imprimit proiecto quandam virtutem motivam que nata est movere nisi fit aliunde impedimentum ad eandem differentiam positionis ad quam proiiciens proiicit et talis qualitas solet vocari impetus ... (201r) aliquando talis impetus destruitur a medio resistente, aliquando a forma vel virtute proiecte active resistentis, et aliquando ab obstaculo.... unde quando proiicitur aliquod grave sursum forma talis gravis non cooperat illi assensui, immo resistit et diminuit impetum impressum in ipso mobili.... (201v) Et secundum istam opinionem non esset necessarium ponere tot intelligentias quod sunt orbes celestes, diceretur enim quod in quolibet orbe est unus impetus impressus a prima causa qui illum orbem movet; nec corrumpitur talis impetus, quoniam talis orbis celestis nullam habet ad motum oppositum naturalem inclinationem."

[98] *Ibid.*, f. 201r: "(Impetus) est qualitas secunda large capiendo qualitatem secundam pro omni qualitate que non est prima; nec oportet quod omnis talis qualitas causetur a primis, ut patet de noticiis et habitibus inherentibus anime...."

[99] Edition of 1530, f. 58r: "... inter impetum et motum localem nullam aliam differentiam assignarem quam inter magis et minus commune, ita quod quilibet impetus est motus localis.... Impetus enim est motus valde intensus." Cf. Duhem, *Études*, Vol. *3*, 150–51.

moved away from considering impetus simply as a cause of motion to considering it as an effect; and it is an effect that a force such as gravity produces. We seem to be closer to Descartes' concept of quantity of motion with Coronel than with the other sixteenth-century exponents of the impetus theory.

The Spanish and Portugese masters at Paris were, of course, not only acquainted with the dynamics of Buridan, but also with the kinematics of Merton College, particularly as summarized in the works of Heytesbury and Swineshead. We have already mentioned Coronel's knowledge of the uniform acceleration theorem. Celaya is also acquainted with it.[100] Furthermore, Master Alvarus Thomas of Lisbon wrote in 1509 what amounted to a commentary on (and paraphrase of) Swineshead's *Calculationes* and Heytesbury's *Tria praedicamenta de motu*. This work, borrowing its basic structure from Heytesbury's work, was called *Liber de triplici motu*. Thomas not only used the English works, but he was conversant with the Italian commentaries and paraphrases.[101] He, then, has at his command the whole medieval mechanical tradition. In addition to giving the Merton acceleration theorem, Thomas discussed numerous species of infinite series of the kinds used by Heytesbury and Swineshead;[102] and Thomas, like his

[100] Celaya, *op. cit.*, f. 83v: "Alia est opinio Guillelmi Hentisberi et Calculatoris et ferre omnium aliorum philosophorum tenentium quod in tali motu difformi quo ad tempus oportet reducere difformitates ad uniformitatem et penes talem gradum ad quem reducentur velocitas debet mensurari. Ex ista opinione sequuntur aliqua correlaria: primum quod quilibet motus uniformiter difformis a non gradu usque ad aliquem gradum vel a gradu ad aliquem gradum correspondet gradui medio inter gradum et illum non gradum vel gradum. Et ex isto sequitur quod si aliquod mobile intendat motum suum uniformiter a non gradu usque ad octo tantum spacium pertransibit acsi moveretur velocitate ut quattuor. Ratio est quia illi gradui correspondet velocitas illius motus."

[101] Duhem, *Études*, Vol. *3*, 531–43.

[102] Alvarus Thomas, *Liber de triplici motu proportionibus annexis...philosophicas Suiseth calculationes ex parte declarans* (Paris, 1509),sig. n. 1v: "Motus localis uniformis est quo in equalibus temporibus equalia spacia pertranseuntur.... motus vero uniformiter difformiter quo ad tempus est quando cuiuscunque partis accepte secundum tempus, i.e., que adequate est in aliqua parte temporis, gradus medius qui est in medio talis partis tanto excedit extremum remissius quanto exceditur ab intensiori. Exemplum ut si aliquod mobile incipiat moveri a non gradu continuo intendendo uniformiter motum suum per aliquod tempus, tunc talis motus est uniformiter difformis quo ad tempus.... (p. 2v) Prima propositio. Si aliquis motus uniformiter continuo intendatur vel remittatur a certo gradu usque ad certum gradum vel ad non gradum eius velocitas gradui medio correspondet...." Earlier Thomas has told us what it means to have a uniform velocity to which a difform velocity corresponds (sig. o. 6r): "Preterea per quemlibet talem motum difformem in totali tempore adequate pertransitur aliquod spacium adequate, et tale spacium in tali tempore ab aliqua velocitate uniformi natum est pertransiri, igitur illa velocitas uniformis est tanta quanta est velocitas illius motus difformis quo illud spacium in

contemporary John Major, who discussed the infinite in the context of medieval theories of the infinite, must not be left out of any account of the growing application of concepts of the infinitesimal to problems of motion.

Before concluding even a brief history of the Iberian masters at Paris, we must call the reader's attention once more to Domingo de Soto, who returned from study in Paris to publish his *Questions on the Physics* in Spain in 1545. We have already extracted his views on acceleration in Chapter 9. We saw there that he specifically applied the Merton rule for determining distance in a uniformly accelerated motion to the motion of freely falling bodies. While he probably was not the first to make this application, his work remains the first reference to such an application that we now know of, and of course precedes the use of the medieval formula by Galileo by more than a half a century, as we have already noted in Chapter 9 (see particularly note 21). Domingo demonstrates his thorough knowledge of Merton kinematics in the third question of the seventh book of the *Physics*, entitled, "Whether velocity of motion in regard to effect is measured by the quantity of space which is traversed."[103] He separates circular from rectilinear motion, after having declared that he will treat the velocity of motion with respect to effect as well as with respect to cause. It is natural to begin with the investigation according to effect, since our understanding begins by means of the senses and thus through effect. It is in question 4 that he takes up the causal question, and there he examines Bradwardine's law of the proportion of proportions. He follows Bradwardine closely through the later authors: William Heytesbury, Albert of Saxony, and Paul of Venice. He accepts in a further question (quest. 3 on Bk. VIII) the air and impetus as joint movers of the projectile, saying that Aristotle accepted this.

The Parisian revival of Nominalism with its emphasis on the English logical subtleties evoked sharp and uninformed criticism from humanists, just as it had earlier in Italy.[104] But we forbear at this point the discussion

eodem tempore pertransitur adequate." For Thomas' discussion of infinite series—much too long to be treated here—see H. Wieleitner, "Zur Geschichte der unendlichen Reihen im christlichen Mittelalter," *Bibliotheca Mathematica*, Dritte Folge, Vol. *14* (1913–14), 150–68.

[103] Edition of Venice, 1582, pp. 336–44: "Utrum velocitas motus ab effectu attendatur penes quantitatem spatii, quod pertransitur;... tractabimus de velocitate motus: alteram penes effectum et alteram penes causam... cognitio nostra incipit a sensu per illam.... (pp. 344–50) Utrum velocitas motus attenditur ex parte cause penes proportionem proportionum quae sunt velocitatum ad suas ipsarum resistentias.... (p. 369) non solus aer est causa, quae movet proiectum, verum et proiiciens ipse per impetum impressum. Neque Arist. credendus est hoc dubitasse."

[104] Duhem, *Études*, Vol. *3*, 160–81.

of the often antiscientific and antischolastic attitude of the humanists, and we remark only that the Parisian development through its Spanish and Portugese adherents influenced the authors in the new Jesuit movement. We shall limit ourselves here to noting the so-called *Commentarii Collegii Conimbricensis* on both the *Physics* and the *De caelo*, both published in 1592. These commentaries, like the writings of the Italian Averroïsts, attempt to save Aristotle by adopting the view that the medium, while not the exclusive or even principal continuator of the projectile motion, nevertheless assists in it. The principal mover is the impetus impressed in the body by the projector.[105] Other less important authors tended to see the impetus theory as an Aristotelian theory—and there appears to have been a fairly widespread tendency to merge the impetus theory with the older Aristotelian dynamics.[106]

Italy. As in France, so in Italy, the medieval mechanics was quite popular in the sixteenth century, although Averroïsts at the turn of the century, like Nicoletto Vernias, opposed the impetus dynamics,[107] preferring something closer to Aristotelian thought. On the other hand, Vernias' student Agostino Nifo appears, at least in 1506, to have accepted the compromise which allowed an auxiliary role to the air, leaving the principal mover as the impetus.[108] The doctrine of impetus also attracted a little-known Venetian cleric, one Hieronymus Picus, who wrote "in the Church of St. Mark" a *Questio de motu gravium et levium*, which he dedicated to the Venetian Prince Leonardo Laurendando and which he supplemented with a discussion of proportions and other materials he hoped would serve as an introduction to the "Calculationes Suiseth anglici calculatoris subtilissimi."[109] In his treatment of the impetus theory, Pico follows Buridan but

---

[105] Maier, *Zwei Grundprobleme*, p. 302.
[106] *Ibid.*, p. 304.
[107] *Ibid.*, p. 295.
[108] *Ibid.*, pp. 296–97. Cf. Duhem, *Études*, Vol. 3, 116–19.
[109] *Questio de motu gravium et levium*, Venice, Bibl. Naz. San Marco, Lat. VIII, 83, 4r–15v. "(14v) Ideo ponitur altera opinio quam veriorem reputo, dicens proiectum post exitum eius a proiciente per impetum in eo impressum moveri. Qua opinione stante, sustineri posset non esse tot intelligentias moventes, sed celum moveri per impetum et motivam qualitatem impressam in eo per gloriosum Deum, qui non corrumpetur tum quia celum ad oppositum non habet inclinationem tum quia ordine divino non corrumpitur.... Ad secundam dico motum naturalem in fine intendi quia in continuatione motus ibi acquiritur plus de impetu, qui plus facit ad intensionem eius; quanto enim magis grave appropinquatur loco suo tanto plus gravitas eius confortatur.... iste impetus est qualitas de secunda specie qualitatis, id est, habilitas quedam et facilitas ad motum generata a forma substantiali mobilis mediante motu, corrupta vero per absentiam conservantis, i.e. motus, et inclinationem mobilis ad quietem. Et scien-

falls short of his account on at least two counts: (1) Pico does not give an adequate account of the measure of the impetus in terms of velocity and prime matter, although he does mention the quantity of matter as important; (2) he does not stress the potentially permanent nature of the impetus. Like Buridan, he employs the impetus to explain: (1) the continuation of projectile motion; (2) the continuous movement of the heavenly bodies; (3) the acceleration of falling bodies; (4) why one can throw a stone farther than a feather; (5) why a long lance penetrates further than a short lance; (6) why a smith's wheel continues rotation when the principal turning agent is no longer turning the wheel; (7) why a top continues to spin; (8) why a man running fast cannot stop himself; (9) the phenomenon of rebound; and (10) why some parts of flax, when thrown, move back toward the hurler. He has already explained the impetus as a quality of the second species, a facility to movement. However, the air plays a secondary but necessary role in offering the resistance which is necessary for movement.

Much more interesting than the Averroïsts' accounts were the fragmentary statements of Leonardo da Vinci, made very close to the time when Nifo was writing. We should realize immediately that the role of Da

---

dum quod secundum istam opinionem possunt salvari omnia apparentia iuxta motum proiectionis. Primo quidem quare proiciens remotius proicit lapidem quam plumam. Huius quidem ratio est quia lapis plus habet de materia, et in eo materia est magis compacta. Ideo virtus illa motiva fortius et magis imprimitur lapidi quam plume et etiam ab eo diutius retinetur. Item redditur causa quare lancea longa magis percutit quam brevis. Huius causa est quia in lancea longa est plus de materia. Ideo in ea (15r) plus acquiritur de impetu.... Item dicitur molam fabri post amotionem principalis agentis moveri per impetum sibi ab extrinseco acquisitum. Idem patet de trocho velociter circulariter moto a tali impetu impresso, et homine currente qui postquam incepit currere et cucurrit (! concurrit?) bene velociter non potest se bene retinere quando vult, quod non videtur esse nisi propter talem impetum in ipso acquisitum qui adhuc ulterius inclinat ad motum. Preterea potest reddi causa quare grave quum fortiter descendit et invenit resistentiam reflectitur versus partem a qua venit. Unde huius causa est quia grave in descensu acquirit talem impetum qui non subito corrumpitur et ideo quum invenit resistentiam ille impetus nondum corruptus inclinans ad motum, non potens ulterius movere ipsum grave deorsum movet ipsum ad oppositum tandiu quod impetus ille corrumpitur. Item potest reddi causa quare alique partes stuppe versus proiciens moveantur, quia scilicet stuppa habet parum de materia que non est bene compacta. Ideo parum habet de tali impetu et sic ex tali opinione salvantur omnia que apparent in motu proiectionis. Et si quis contra eam vellet sic argumentari quia ex ipsa sequitur aerem in tali motu esse superfluum, huic respondeo quod aer in motu proiectionis propter resistentiam et successionem est necessarius...." Earlier, on ff. 2v–3r preceding the question we read: "Icciro diebus idistis difficultatem quandam de motu gravium et levium cum quibusdam notabilibus ad calculationes Suiseth anglici calculatoris subtilissimi introductoriis composui."

Vinci in the progress and development of physical ideas is obscure, for we know very little about the books he read and the people who read his statements. Furthermore, his references are often so abbreviated as to hide successfully their true meaning. However, with all this granted, his ideas still demand some treatment in an account of the fate of medieval mechanics, although I have made no attempt to give a full picture of his mechanics.

We have already mentioned in Chapter 9 (see Doc. 9.4) that Leonardo more than once said that the velocity of freely falling bodies grows in arithmetic proportion to the time, although he seems at the same time to confuse this correct statement with the erroneous proposition that the velocity increases directly as the distance of fall. He did *not* apply the Merton uniform acceleration theorem to give a correct way of determining the distance of fall. On the other hand, he may have had some familiarity with the impetus mechanics, perhaps through the works of Albert of Saxony.[110] This does not prevent him from assigning in some places a major role to the air, both in continuing the motion of projectile and in accelerating the fall of bodies.[111] For example, he says, "Impetus is the

[110] Duhem, *Études*, Vol. *1*, 108–16, attempts to make a case for the influence of Albert of Saxony on Leonardo. It is true that in several passages which we quote later Leonardo speaks of impetus as impressed power. On one occasion (MS *G*, f. 85v, edition of M. Charles Ravaisson-Mollien [Paris, 1890]) we read that impetus is "derived motion, which arises out of primary movement, that is, when the moveable body is joined to the mover" *(inpeto ecquel cheperaltro nome eddecto moto derivativo il quale nasscie dal moto primitivo cioe quando esso mobile era chongiunto cholsuo motore)*. Now if by *moto* Leonardo meant quantity of motion measurable by quantity of matter and velocity, Leonardo is near the modern momentum concept. General studies of the mechanics of Leonardo are R. Marcolongo, *La meccanica di Leonardo da Vinci* (Naples, 1932), and I. B. Hart, *The Mechanical Investigations of Leonardo da Vinci* (London, 1925).

[111] *Il codice atlantico di Leonardo da Vinci*, edition of Regia Accadèmia dei Lincei, Fasc. II, V. Hoepli (Milan, 1894), p. 569, f. 168v (cf. E. MacCurdy, *The Notebooks of Leonardo da Vinci*, Vol. *1* [New York, 1938], p. 526): "Linpeto einpressione di moto lochale *(fa)* trassmutato dal motore almobile e mantenuto dallari a oddallacqua col moversi all proibitione del uachuo." Cf. also (C. A. *edit. cit*, Fasc. III, p. 778), f. 219v.a.: "Linpeto e una virtu trasmutata dal motore al mobile e mantenuta dallonda chettal motore gienera dellaria in fra laria ecquessta nasscie dal uachuo che chontro all naturale leggie sigienererebbe sellaria anteciedente non rienpi essi iluachuo donde sifuggie laria scacciata *(da)* del suo sito dal pedetto motore ettale aria ante cie dente non rienpierebbe il sito don de sidujse essa anteciedente se unaltra quantita daria non riempiessi ilsito donde tale aria sidujse e chosi *(co)* eneciessario che ssucciessius *(ve)* mente seguiti ex questo moto seguirebee infinjto sellaria nuffusi chondesnsabile injnfinito." (MacCurdy, *Notebooks*, Vol. *1*, p. 529.) Other passages illustrating the role of air in projectile motion are MS *F*, f. 74r–v, in edition of M. Charles Ravaisson-Mollien, MS *M*. ff. 45v–46r *(ibid.).* On the other hand we read in *F* f. 52 r *(ibid.)* that the air is resistant

impression of local motion transmitted from the mover to the movable thing and maintained by the air or by the water as they move in order to prevent the vacuum." "Impetus is a power transmitted from the mover to the movable thing, and maintained by the wave of air within the air which this mover produces." Da Vinci holds in at least one place that the complete projectile motion is the result of the action of *compound* impetus.[112] He appears to believe that impetus from the mover orginally dominates the first part of the movement, thus producing initially rectilinear motion; in the middle part, the path is curved because the impetus of the projectile is having more effect; and finally in the last part of the motion the impetus of the projectile is dominant, producing a downward rectilinear trajectory. Like Oresme, Da Vinci believed that an arrow or projectile shot into the air would fall toward the ground with rectilinear motion with respect to the earth because it shares the circular movement of the earth. The actual path of a heavy substance as it falls he described as being "in the form of a spiral descent."[113]

Sometimes impetus appears with Leonardo to be not so much a continuing power producing motion as rather the effective power something has because it is in motion, as the two quotations on impetus above seem to indicate. In still another note, he claims that impetus is a power impressed by the mover in the moved and furthermore, "every impression tends towards permanence."[114] But this idea is not sufficiently elaborated for us to form any definite conclusions; and the fact that Da Vinci feels that it is necessary to have movement of the air to continue impetus shows that he is still thinking basically in terms of force being necessary for the continuation of movement. It is worth noticing finally that Da Vinci reveals extensive influence from the medieval statical corpus,[115] while showing virtually no knowledge of English kinematics.

Toward the middle of the century, a number of different Italian authors discuss and accept the impetus theory in one form or another. These

---

to the flight of the arrow *(le faccia resistentia)*. Cf. also *G*. f. 72v where he says that the compound impetus which moves a stone projected from a ship does not last long because of the resistance of the air *(...cimanifesta esso mobile trovare resistentia nellaria dallu penetrata)*.

[112] For example, MS *E* f. 35r (cf. MacCurdy, *Notebooks*, p. 545): "Il moto chonposto eddetto quello che participa dell inpet(o) del motore edel inpeto del mobile...."

[113] MS *G*, f. 55r (edition of Ravaisson-Mollien; MacCurdy, *Notebooks*, pp. 565–66) "...discienso churvo a modo dil linea elica."

[114] MS *G*. f. 73r: "Inpeto he inpressione di moto trasmutato dal motore al mobile. Inpeto evna potentia inpressa dal motore nel mobile—ogni impressione attende alle premanentia (!) over desidera permanentia."

[115] See Chapter 2 above, note 51.

include Piccolomini, Cardano, Scaliger, Tartaglia, and Telesio, all of whom have been treated biefly by Duhem in his *Études* and /or by Maier in her *Zwei Grundprobleme*. But I should like to hasten beyond all of these authors to discuss the more significant treatments of G. Battista Benedetti and Giordano Bruno.

Benedetti demands our particular attention, for it is ordinarily agreed that, in his directional analysis of impetus and the movement of bodies, he foreshadowed the inertial concept more closely than any of his predecessors. This he did in his *Diversarum speculationum mathematicarum et physicarum liber* (1585), a work that stands as one of the classics of early modern mathematical mechanics. In his description of impetus we are told that "every body moved naturally or violently receives in itself an impression *(impressio)* and impetus of movement, so that separated from the motive power, it would be moved of itself through a space in some time. For if a body is moved naturally, it will always increase its velocity, since the impression and impetus in it would be always increased. This is so because it has contact with the motive power perpetually."[116] It is clear that Benedetti has sharpened Buridan's application of the impetus theory to the fall of bodies. Since the gravity as a motive force continually acts, the impetus of motion, and hence the velocity of motion, increases continually.

On the other hand, let us analyze a form of violent motion, namely projection from a sling. "The true reason why a heavy body is impelled further by a sling than by a hand arises from the fact that in revolving the sling a greater impression of the impetus of motion *(impressio impetus motus)* takes place in the heavy body than takes place by using the hand. For when the body is freed from the sling, directed by nature, it assumes a trajectory from the point of its liberation, which trajectory is a line tangent to the final circle of revolution which it made. And it is not to be

---

[116] G. B. Benedetti, *Diversarum speculationum mathematicarum et physicarum liber*, (Turin, 1585), p. 286. Cf. A. Koyré, *Études Galiléennes*, Vol. *1* (Paris, 1939), 42. Duhem was the first to analyze the scholastic elements in Benedetti's exposition of impetus (Duhem, *Études*, Vol. *3*, 217). We should note also the Benedetti passage (p. 184): "Huiusmodi igitur corporis separatim a primo movente velocitas oritur a quadam naturali impressione, ex impetuositate recepta a dicto mobili, quae impressio et impetuositas, in motibus rectis naturalibus continuo crescit, cum perpetuo in se causam moventem, id est propensionem eundi ad locum ei a natura assignatum habeat... quanto longius distat a termino a quo tanto velocius existit, quia tanto maior fit semper impressio, quanto magis movetur naturaliter corpus, et continuo novum impetum recipit, cum in se motus causam contineat, quae est inclinatio ad locum suum eundi, extra quem per vim consistit." I have changed the forms of *impraessio* in the edition to *impressio*.

doubted that the sling can impress greater impetus in the body, since out of its many rotations a continually greater impetus accrues to the body.... Indeed the circular motion, as the body rotates, makes it so that the body by natural inclination *(naturalis inclinatio)* and under the force of impetus already initiated pursues a rectilinear path *(recta iter peragere).*"[117] In short, by its rotation the body has received an impetus, and when freed from the constraint of the sling, the impressed impetus carries it in a straight line, for rectilinear motion is natural to freely moving bodies. Clearly Buridan's vaguely expressed inertial impetus has become restricted, so that we are left with a naturally rectilinear inertial impetus. For Benedetti the impetus continually decreases under the influence of the gravity of the body, and the gravity is "compounded" *(missens)* with the impressed impetus received by the body. For this reason the path does not remain a straight line for very long. Actually we receive the impression from Benedetti that even at the beginning, when the impetus is still strong, the gravity is acting to some extent to deflect it from rectilinear movement. But the greater the velocity impressed in the projectile, the greater is the propensity to go in a straight line.[118] Here again Benedetti has given precision to the compound impetus of the scholastics and Leonardo da Vinci. With Benedetti we are a long step closer to Galileo's analysis of the path of the projectile as a compound of a uniform motion with a uniformly accelerated motion.

To Benedetti, working from his idea that a particle completely free to move under the influence of an impressed impetus moves in a straight line and that only such movement would tend to be permanent, it appeared impossible that a smith's wheel, even under perfect conditions, could rotate indefinitely once an impetus had been imparted. Scholastics like John Buridan, Albert of Saxony, and later Nicholas of Cusa, held that an indefinitely rotating wheel was possible. Benedetti opposed this on the ground that, in being forced into rotary movement, each of the various parts is given an impetus which tends to move it in a straight line, but, being constrained into circular movement, each of the partial impetuses is being violated and thus a certain amount destroyed.[119] The natural straight-line move-

---

[117] Benedetti, *Diversarum speculationum... liber*, p. 160.

[118] *Ibid.*, p. 285, and p. 161: "...quanto maior impetus motus ipsi *a* est impressus, tanto magis dictum corpus *a* ad rectum iter peragendum inclinatur, unde ut recta incedat, tanto maiore quoque vi trahit."

[119] *Ibid.*, p. 159: "...quelibet pars corporea, quae a se movetur, impetu eidem a qualibet extrinseca virtute movente impraesso, habet naturalem inclinationem ad rectum iter, non autem curvum, unde si a dicta rota particula aliqua suae circumferentiae disiungeretur...."

ment toward which all points in a rotating body tend also explains for Benedetti the stability of a rotating top. "From the inclination toward rectilinear movement of the parts of the round body, it results that a top *(trochus)* rotating with great violence remains for some space of time completely upright on its iron point and so does not topple over toward the center of the world any more in one direction than in another. This is because none of its parts inclines at all toward the center of the world but rather tends to move out at right angles to the line of direction, i.e., to the vertical or to the axis of the horizontal."[120]

Finally, we should observe in connection with Benedetti that he opposed the Peripatetic law of movement (see Chapter 7). The "unshakable foundation of mathematical philosophy" (and in fact Benedetti's knowledge of Archimedes' hydrostatics) led him to two important modifications of Aristotle's dynamics: (1) The first was that the speed of fall in the same medium is not directly proportional to gross weight but varies rather with the surpluses of specific weight. In fact, two bodies of the same specific weight but of different gross weights would, according to Benedetti, have the same velocity in the same medium or in a vacuum.[121] (2) In the second place, the speed of fall is not inversely proportional to the resistance, as the Aristotelian law appears to hold. For Benedetti the resistance is to be *subtracted* from rather than *divided* into the weight, leaving a certain residual specific weight which determines the speed of fall.[122] It should be noted that Galileo in his early *De motu* was to follow Benedetti in both his "corrections" (see note 130). Benedetti knew of and opposed some of the conclusions of Jordanus' *De ratione* as presented by Tartaglia.[123]

Like Benedetti, Giordano Bruno was familiar with, but added to, the medieval mechanics. He advanced and sharpened the concept already

---

[120] *Ibid.*, p. 285.

[121] Benedetti, *Demonstratio proportionum motuum localium contra Aristotelem et omnes philosophos* (Venice, 1554), without pag., but see p. 7: "Nunc autem demonstrabo, quo pacto corpora unius et eiusdem speciei, itidem et figurae, aequalia invicem, vel inaequalia, in eodem medio, per aequale spatium, in eodem tempore ferentur." Cf. *Diversarum speculationum...liber*, p. 174: "Quod in vacuo corpora eiusdem materiae aequali velocitate moverentur."

[122] *Demonstratio proportionum*, p. 10: "Praeterea si fuerint duo corpora eiusdem figurae, sed diversae homogeneitatis, inaequalis etiam corporeitatis, et utrunque eorum (exempli gratia) gravius medio, per quod feruntur, sit etiam minus eorum, gravioris speciei quam maius, sed maius, plus ponderet minori, tunc dico quod minus, velocius erit in motu, eaque proportione erit temporis in quo minus, ad tempus in quo maius, quae est gravitatis speciei maioris ad speciem minoris, sublata tanta gravitate ab utroque, quanta est medii in unoquoque ipsorum."

[123] See Chapter 2 above, note 52.

familiar to Oresme, that bodies on the earth participate in its movement in the same way that bodies in a ship participate in the motion of the ship.[124] They share a certain horizontal impetus impressed by the movement of the earth on which they rest. With Bruno the concept of the closed mechanical system had emerged more clearly than with Oresme and Buridan, but that we are dealing with basically the same argument can scarcely be doubted. Bruno's role in the substitution of a geometrical infinite space for the physical space of Aristotelian physics is of great interest in the rise of modern physics, but we must refrain from discussing it in this context. Let us suffice ourselves here with A. Koyré's summary of Bruno's importance for modern mechanics:[125] "We remain astonished before the boldness and radicalism of Bruno's thought, which works a transformation of—a veritable revolution in—the traditional image of the world and of physical reality. The infinity of the universe, the unity of nature, the geometrization of space, the denial of place, the relativity of movement— we are very near to Newton." If this is a just estimate of Bruno, we can agree that, in some respects, in his concept of the relativity of movement as well as that of a closed mechanical system, and even in the unity of nature, he was foreshadowed by the medieval critics of Aristotle.

We must finally note most briefly something of the knowledge of medieval mechanics possessed and used by Galileo. We have already seen in Chapters 5 and 6 that in his mature work Galileo was clearly the heir of the medieval kinematicists. For he had studied the intension and remission vocabulary, and he used the Merton form of the uniform acceleration theorem before passing on to his own form of the law (see Docs. 4.7 and 6.5). He also used the traditional geometric proof of the medieval form of the law which originated with Oresme or Casali, and he did not improve it materially. It was Beeckman who went beyond Oresme and Galileo by introducing sound infinitesimal considerations into Oresme's "triangular" proof (see Doc. 6.6). It is worth noting incidentally that Descartes, like Galileo in the earlier period, had attempted to use the "triangular" proof to demonstrate the erroneous law that velocity is directly proportional to the distance of fall in the case of a uniformly accelerating body.[126]

It is also clear that Galileo took up the impetus dynamics in his youthful work *De motu*, perhaps having been instructed in it by his teacher Bonamico. Like Bonamico, it was the self-expending rather than inertial

---

[124] G. Bruno, *La Cena de la ceneri*, in *Opere italiane*, Vol. *1* (Göttingen, 1888), 167–69.

[125] Koyré, *Études Galiléennes*, Vol. *3*, 21.

[126] *Ibid.*, Vol. 2, 40–42.

impetus which Galileo professed at that time,[127] and, like Bonamico, he accepted then the pretended initial acceleration of projectiles.[128] Furthermore, he was attracted to the Hipparchian explanation of the acceleration of falling bodies in terms of decreasing residual resistance[129] (see Chapter 9) and to the Philoponus–Avempace–Benedetti analysis and criticism of the Peripatetic law of movement.[130]

Now it is perfectly clear that Galileo abandoned much of the impetus mechanics in his *Two New Sciences*. Impetus often appears to be for Galileo an effect rather than a cause,[131] although it is significant that its "self-expending impetus." By the time he wrote his *Dialogo sopra i due massimi sistemi del mondo*, his *impeto* (when identifiable with *virtù impressa*) had become inertial. See *Dialogo* in *Le Opere*, Ed. Naz. Vol. 7, 175–83, and particularly p. 180 where he talks of the horizontal motion of a ship impressing an impetus "indelibly" on a stone falling from the mast of the ship: "... l'impeto col quale si muove la nave resti impresso indelebilmente nella pietra, dopo che s' è separate dall' albero...." And equally important, he considers this impressed *impeto* as an effect identifiable with the movement itself rather than as a separate cause of that movement (*ibid.*, Vol. 7, 177–78): "Salv. Talchè quell' impeto e quella mobilità, qualunque se ne sia la causa, più lungamente si conserva nelle materie gravi che nelle leggieri.... (p. 182) Salv. Così voglio io che segua. Ma quando voi la tirate col braccio, che altro rimane alla palla, uscita che ella vi è di mano, che il moto concepito dal vostro braccio, il quale, in lei conservato, continua di condurla innanzi? ora, che importa che quell' impeto sia conferito alla palla più dal vostro braccio che dal cavallo....?" Galileo continues to equate *impeto* with *virtù impressa* (and no doubt conceived as an effect) in his *Discorsi* (see Doc. 9.5, passage 2). In line with the change of impetus from cause to effect, he often loosely uses *impeto* interchangeably with *velocità* or *grado di velocità*, e.g., see the *Dialogo* (in *Le Opere*, Ed. Naz. Vol. 7, 46): "Salv.... Però ditemi se voi avete difficultà nessuna in concedere che quella palla, nello scendere, vadia sempre aquis-

---

[127] Galileo Galilei, *De motu*, in *Le Opere*, Ed. Naz., Vol. *1* (Florence, 1891), p. 310: "virtus impressa successiva remittitur in proiecto a proiciente absente." For impressed force, see pp. 307–14.

[128] *De motu*, edit. cit., p. 308; Cf. Koyré, *Études*, Vol. *1*, 24–25.

[129] *De motu*, edit. cit., p. 319.

[130] *Iuvenilia*, in *Le Opere*, Ed. Naz., Vol. *1*, 263: "Dicimus ergo, mobilia eiusdem speciei (eiusdem autem speciei vocentur quae ex eadem materia, ut plumbo vel ligno etc., conflantur), quamvis mole differant, tamen eadem cum celeritate moveri, nec citius descendere maior lapis quam minor.... (p. 267, n. 1) Dicimus mobilia eam inter se in motibus proportionem servare, quam habent eorum gravitates (i.e., specific gravities, as the context shows) dum in eo medio ponderentur, in quo fieri debet motus.... (p. 272) Constat enim idem mobile in diversis mediis descendens eam, in suorum motuum celeritate, servare proportionem, quam habent inter se excessus quibus gravitas sua mediorum gravitates excedit." (Note that in all the examples the excesses are merely arithmetical.) Galileo notices however that experience does not confirm his theory—i.e., the case of actually dropping two bodies of different weight and material; but the nonuniformity he believes due to accidental causes (p. 273). Cf. R. Giacomelli, *Galileo Galilei giovane e il suo "De motu"* (Pisa, 1949), 36–50.

[131] Koyré, *Études*, Vol. *3*, 70–71. We have already noted that Galileo accepted in his early *De motu* the medieval impetus theory in the form which I have called

measure remained similar to that given by Buridan, except that for Galileo (who confuses mass and weight) the measure of impetus is the product of *weight* and velocity, rather than of *quantity of matter* and velocity as it is for Buridan. Furthermore Galileo was still somewhat within the framework of the impetus mechanics in those places where he describes a circular inertia. Incidentally, Galileo's views of inertia are much mooted. It appears that when Galileo conceived of the motion of a body whose trajectory was of significant magnitude in relationship to the radius of the earth, he employed a circular inertia, but that when he analyzed physical problems such as the force necessary to sustain or move bodies on inclined and horizontal planes, the path of projectile motion, the conservation of impetus—all problems where the horizontal trajectory can be considered as insignificant in comparison with the radius of the earth—he conceived of the horizontal trajectory over which inertial motion takes place as rectlinear.[132]

tando maggior impeto e velocità.... (p. 52) Bisogna ora che voi sappiate, che l'impeto, cioè il grado di velocità, che la palla si trova avere aquistato quando arriva al punto A è tale, che quando ella continuasse di muoversi con questo medesimo grado uniformemente, cioè senza accelerarsi o ritardarsi...." Cf. also in the *Discorsi*, Document 4.7. In a large number of passages Galileo identifies or relates *impeto* with *momento*, both primarily dependent on weight and velocity. Thus in *Le Mecaniche* (*Le Opere* Ed. Naz. Vol. *2*, 159) he tells us that "moment is the impetus of descent, compounded of gravity, position, and anything else by means of which such a propensity might be caused (È dunque il momento quell' impeto di andare al basso, composto di gravità, posizione e di altro, dal che possa essere tal propensione cagionata)." The dependence of moment on weight and velocity is shown particularly in a passage from his *Discorso intorno alle cose che stanno in su l'acqua* (*Le Opere* Ed. Naz. Vol. *4*, 68): "Momento, appresso i meccanici, significa quella virtù, quella forza, quella efficacia, con la quale il motor muove e'l mobile resiste; la qual virtù depende non solo dalla semplice gravità, ma dalla velocità del moto, dalle diverse inclinazioni degli spazi sopra i quali si fa il moto.... Il secondo principio è, che il momento e la forza della gravità venga accresciuto dalla velocità del moto; sì che pesi assolutamente equali, ma congiunti con velocità diseguali, sieno di forza, momento e virtù diseguale, e più potente il più veloce, secondo la proporzione della velocità sua alla velocità dell' altro." It is well known that in lever problems Galileo used *momento* to express the concept now known as static moment and measurable by weight and the horizontal distance of that weight from the center of rotation, see *Le Mecaniche* (*Le Opere*, Ed. Naz. Vol. *2*, 159): "... nella stadera si vede un picciolo contrapeso alzare un altro peso gradissimo, non per eccesso di gravità, ma sì bene per la lontananza dal punto donde viene sostenuta la stadera; la quale, congiunta con la gravità del minor peso, gli accresce momento ed impeto di andare al basso, col quale può eccedere il momento dell' altro maggior grave." Often, however, he uses *momento* to stand for the fuller expression *momento della potenza*, which simply means the quantity or magnitude of force (cf. *Discorsi, Le Opere* Ed. Naz., Vol. *8*, 155–56).

[132] A. Koyré in his *Études*, Vol. *3*, 113–16, has argued that even in those passages where the inertial motion is con-

We have already commented on his sagacious shyness of any causal analysis of acceleration in his mature work. However, if he had followed through the implications of his changed concept of impetus and carried further his first efforts toward an inertial principle, he might have arrived at the Newtonian concept of force.

---

ceived as taking place on a Euclidean plane instead of the surface of a sphere, Galileo has not freed himself from considering weight as an essential property of bodies. He has, in short, confused mass and weight. The reader can appreciate the distinction between Galileo's circular and rectilinear inertial motions by examining some of the more significant of the various passages in which he uses the inertial concept. (1) Consider first those early statements in his *Le Mecaniche* where Gaileo indicates that no force is necessary to keep a body in motion on a horizontal (and, quite obviously, rectilinear) plane (see Doc. 2.6, passages 3 and 4; *Le Mecaniche*, in *Le Opere*, Ed. Naz. Vol. 2, 181: "...non si ricercando forza sensibile—rimossi l'impedimenti accidentarii, che dal teorico non si considerano—per muovere il dato peso nell' orizonte....(p. 186) i corpi gravi non fanno resistenza a i moti transversali"; and compare similar similar remarks in his *Discorsi e dimonstrazioni matematiche intorno a due nuove scienze*, in *Le Opere*, Ed. Naz. Vol. 8, 215, 216–17). (2) Then consult Galileo's earliest exposition of the circular inertia in his *Seconda lettera...delle macchie solari*, in *Le Opere*, Ed. Naz. Vol. 5, 134–35, where Galileo holds that heavy bodies by an intrinsic principle tend to move downward and are repugnant to upward motion, but are indifferent to rest or motion on a spherical surface concentric with the earth, so that if all external impedimenta are removed, they will remain perpetually in whatever motion they have been placed, or if they have been put at rest they will remain at rest so long as no external force is added; and that, if all external impediments could be removed, a ship having once received an impetus would on a calm sea move around the earth pertually without ever stopping, or if placed at rest would perpetually remain at rest so long as no external motive force were added to it. ("Imperò che mi par di osservare che i corpi naturali abbino naturale inclinazione a qualche moto, come i gravi al basso, il qual movimento vien da loro, per intrinseco principio e senza bisogno di particolar motore esterno, esercitato, qual volta non restino da qualche ostacolo impediti; a qualche altro movimento hanno repugnanza, come i medesimi gravi al moto in su, e però già mai non si moveranno in cotal guisa, se non cacciati violentemente da motore esterno; finalmente, ad alcuni movimenti si trovano indifferenti, come pur gl' istessi gravi al movimento orizontale, al quale non hanno inclinazione, poi che ei non è verso il centro della Terra, nè repugnanza, noi si allontanando dal medesimo centro: e però, rimossi tutti gl' impedimenti esterni, un grave nella superficie sferica e concentrica alla Terra sarà indifferente alla quiete ed a i movimenti verso qualunque parte dell' orizonte, ed in quello stato si conserverà nel qual una volta sarà stato posto; cioè se sarà messo in stato di quiete, quello conserverà, e se sarà posto in movimento, v.g. verso occidente, nell' istesso si manterrà: e così una nave, per essempio, avendo una sol volta ricevuto qualche impeto per il mar tranquillo, si moverebbe continuamente intorno al nostro globo senza cessar mai, e postavi con quiete, perpetuamente quieterebbe, se nel primo caso si potessero rimuovere tutti gl' impedimenti estrinseci, e nel secondo qualche causa motrice esterna non gli sopraggiugnesse.") The same doctrine is elaborated in his *Dialogo* in *Le Opere*, Ed. Naz. Vol. 7, 173–74, with the additional idea that not only would a ship move indefinitely in circular motion on a calm sea, if the external accidents could be removed, but also that an object

So much, then, for the highlights in the passage of the medieval mechanics to the early modern times. We can, I believe, conclude that the medieval mechanics occupied an important middle position between the terms of Aristotelian and Newtonian mechanics—even when it was almost

dropped from the mast of the ship would share in that circular horizontal motion, as it fell. (P. 174: "Salv. Adunque una nave che vadia movendosi per la bonnacia del mare, è un di quei mobili che scorrono per una di quelle superficie che non sono nè declici nè acclivi, e però disposta, quando le fusser rimossi tutti gli ostacoli accidentarii ed esterni, a muoversi, con l'impulso concepito una volta, incessabilmente e uniformemente.... Simpl. Voi volete dir per ultima conclusione, che movendosi quella pietra d' un moto indelebilmente impressole, non l' è per lasciare, anzi è per seguire la nave, ed in ultimo per cadere nel medesimo luogo dove cade quando la nave sta ferma; e così dico io ancora che seguirebbe quando non ci fussero impedimenti esterni, che sturbassero il movimento della pietra dopo esser posta in libertà.") The circular inertial movement is also assumed in his curious and erroneous analysis of the path of fall as a semicircle (*ibid.* pp. 190–92). (3) On the other hand, Galileo in discussing gyratory motion of projection exposes the idea of Benedetti that such a motion is compounded of a horizontal tangential motion and motion toward the center, the former however never completely predominating. (*Ibid.*, p. 219: "Simpl. ... Dalle cose dette si reccoglie che il proietto, mosso velocemente in giro dal proiciente, nel separarsi da quello ritiene impeto di continuare il suo moto per la linea retta che tocca il cerchio descritto dal moto del proiciente nel punto della separazione.... (p. 222). Salv. Talchè qui cascano in considerazione due moti: uno della proiezione, che comincia dal punto del contatto e segue per la tangente; e l' altro dell' inclinazione all' ingiù, che comincia dal proietto e va per la segante verso il centro: ed a voler che la proiezione segua, bisogna che l'impeto per la tangente prevaglia all'inclinazione per la segante....") (4) The treatment by use of the idea of rectilinear horizontal inertial motion of the problem of projection is of course found in the famous passages from the fourth day of his *Discorsi*, where projectile motion is described as compounded of uniform horizontal rectilinear motion and uniformly accelerated vertical motion; in his abstractive technique he follows Archimedes in thinking of the radius as effectively infinite and thus the weight as acting at right angles to the horizontal inertial plane. (See *Le Opere*, Vol. *8*, p. 268: "Mobile quoddam super planum horizontale proiectum mente concipio, omni secluso impedimento: iam constat, ex his quae fusius alibi dicta sunt, illius motum aequabilem et perpetuum super ipso plano futurum esse, si planum in infinitum extendatur; si vero terminatum et in sublimi positum intelligamus, mobile, quod gravitate praeditum concipio, ad plani terminum delatum, ulterius progrediens, aequabili atque indelebi priori lationi superaddet illam quam a propria gravitate habet deorsum propensionem, indeque motus quidam emerget compositus ex aequabili horizontali et ex deorsum naturaliter accelerato, quem proiectionem voco.") (5) Finally, we can observe that in the *Discorsi* Galileo employs the horizontal rectilinear inertial motion in his treatment of the conserved impetus that raises a body through the same vertical distance it falls. (*Ibid.*, pp. 243–44: "Attendere insuper licet, quod velocitatis gradus, quicunque in mobili reperiatur, est in illo suapte natura indelebiliter impressus, dum externae causae accelerationis aut retardationis tollantur, quod in solo horizontali plano contingit; nam in planis declivibus adest iam causa accelerationis maioris, in acclivibus vero retardationis: ex quo pariter sequitur, motum in hori-

completely abandoned or altered, as in the case of the impetus theory. And regardless of the "modernity" of medieval mechanics, it remains an important link in man's efforts to represent the laws that concern bodies at rest and in movement.

---

zontali esse quoque aeternum; si enim est aequabilis, non debilitatur aut remittitur, et multo minus tollitur. Amplius, existente gradu celeritatis per naturalem descensum a mobili acquisito, suapte natura indelebili atque aeterno, considerandum occurrit, quod si post descensum per planum declive fiat reflexio per aliud planum acclive, iam in isto occurrit causa retardationis: in tali enim plano idem mobile naturaliter descendit; quare mixtio quaedam contrariarum affectionum exurgit, nempe gradus illius celeritatis acquisitae in praecedenti descensu, qui per se uniformiter mobile in infinitum abduceret, et naturalis propensionis ad motum deorsum iuxta illam eandem proportionem accelerationis iuxta quam semper movetur. Quare admodum rationabile videbitur si, inquirentes quaenam contingant accidentia dum mobile post descensum per aliquod planum inclinatum reflectatur per planum aliquod acclive, accipiamus, gradum illum maximum in descensu acquisitum, idem per se perpetuo in ascendente plano servari; attamen in ascensu ei supervenire naturalem inclinationem deorsum, motum nempe ex quiete acceleratum iuxta semper acceptam proportionem.")

Chapter 12

Medieval Mechanics in Retrospect

AS we close this volume one final task is left for us, to summarize in a single and coherent whole some of the principal medieval mechanical concepts which still circulated in the sixteenth century, either in printed editions of the works of medieval authors or in the works of sixteenth-century authors who still parroted the views and words of their predecessors. We shall not attempt in this summary to pay much attention to the order in which we have already presented these concepts in the previous chapters. Our hope here is more for coherence than for strict recapitulation.

In singling out and numbering the principal medieval concepts, we have sometimes quoted directly from the documents given in earlier chapters and at other times we have for the sake of economy contented ourselves with an accurate paraphrase. But in either case reference has been made to the appropriate documentary source and discussions already presented. Finally, it should be noticed that we make no attempt here to summarize the general Aristotelian framework assumed in medieval mechanics but only the distinctive and influential concepts taken up or invented by the medieval schoolmen.

At the heart of both ancient and modern systems of mechanics lie the nature, definition, and role of weight. All of the principal systems of mechanics from Aristotle to Newton accepted weight as a force. But the concept and definition of force changed from a *cause of motion* in ancient and medieval mechanics to a *cause of change of motion* in Newtonian mechanics. Among the most important distinctions with regard to weight that began to develop from the time of Aristotle was the distinction between gross weight and specific weight. This distinction, present in germinal form in the *De caelo* of Aristotle, was made more quantitatively precise in Archimedes' genuine work *On Floating Bodies*, and it proved to be of some importance for both the statics and dynamics of the Middle Ages. In Arabic treatises concerned with specific gravities, the distinction was sharpened beyond the statements found in Archimedes' work. Whether original with them or not, the Arabic precision was duplicated in at least one medieval Latin work which stemmed from the Arabic tradition, and we should like to single out as our first concept the definitions of gross and specific weight given in the pseudo-Archimedean *De insidentibus in humidum* of the thirteenth century:

1. *"One body is said to be heavier than another numerically if, when these bodies are suspended at the ends of the balance beam, its arm of the balance inclines downward; or, if its weight is equal to a greater number of calculi....Of two bodies equal in volume, the one whose weight is equal to a greater number of calculi is of greater specific weight. Of two bodies of the same kind* (i.e., of the same specific gravity), *the proportion of volumes to weights is the same."* (See Chapter 2, p. 94.)

Starting, then, with a distinction between gross weight and specific weight, many mechanicians accepted the following simple extension of the basic Aristotelian law relating force (and thus weight) to velocity:

2. *The force exerted by a heavy body is proportional to its weight, and thus for a given volume, force is proportional to density; the speed of a heavy body is proportional to the force it exerts, and so with a given volume assumed, the speed of fall is proportional to the density of the heavy body.* (See Chapter 7, pp. 429–30).

This extension of the Peripatetic law was actually found in the *De caelo* of Aristotle and elaborated in the popular *Liber de ponderoso et levi* attributed to Euclid and composed in late antiquity. This work was translated from the Arabic and circulated widely in the thirteenth and fourteenth centuries. Its basic ideas were expressed by Johannes de Muris, Albert of Saxony, Blasius of Parma, and no doubt many others in the course of the fourteenth and fifteenth centuries. As we have seen in Chapter 11, Benedetti and Galileo (in his youth), both under the influence of Archimedes' hydrostatics, went one step further and singled out specific gravity, without regard for the volume or absolute weight involved, as the principal determiner of the velocity of fall—a uniform medium assumed. Now granting the importance of the motivating force—whether gross or specific weight—one still had to assess the role of the resisting medium. We find some tendency (starting in late antiquity) to refute the role assigned to the resisting medium by Aristotle. The tenor of this criticism we can paraphrase as follws:

3. *The effect of the resistance of a medium is an effect to be subtracted from the*

original force of movement rather than one to be divided into the original force of movement. (See Chapter 7, pp. 433–35, 439.)

This view, also adopted by Benedetti and Galileo (in his early *De motu*), arose historically out of John Philoponus' criticism of Aristotle's "law" of dynamics in the sixth century A.D., although it was not precisely stated in this manner by Philoponus. From Philoponus the germ of this view was taken up by Avempace, and his views were known to the Latin West as the result of their criticism by Averroës. The first precise mathematica statement of this view was made in the fourteenth century by Bradwardine, who rejected it and attempted to save Aristotle with a new and quite different form of Aristotle's law, expressible as follows:

4. *Velocity follows the "geometric proportionality"* (i.e., is an exponential function) *of the ratio of the motive force to the resistance* (expressed in modern terminology as $V \propto \log F/R$). (See Chapter 7, pp. 438–39.)

This law of Bradwardine was very widely accepted by the schoolmen of the fourteenth and fifteenth centuries. It was particularly important for introducing exponential functions to represent physical occurrences. But even more important, it was a law which attempted to describe instantaneous changes in velocity rather than average velocity. Thus it foreshadowed the differential equation used so universally in modern mechanics. Bradwardine's law was, however, without adequate physical verification; and it passed out of vogue by the sixteenth century. Its great significance was that, being like a "differential" equation, it not only brought to a head the consideration of instantaneous velocity, but it had the most significant kinematic consequences at Merton College—consequences we shall summarize shortly.

Concerned so paramountly with weights, forces, and resistances, medieval mechanics from the thirteenth century occupied itself with problems of the equilibrium of forces and weights as represented by balances, levers and other statical situations. Heir of Hellenistic mechanics, the medieval mechanician sharpened the concept of virtual displacements, which had had its origin in the Pseudo-Aristotelian treatise *Mechanica*. The form of the principle of virtual displacements used by the thirteenth-century mathematician Jordanus and expressed by one of his thirteenth- or fourteenth-century commentators (in the so-called *Aliud commentum* published in 1533 by Peter Apianus) was essentially of this form:

5. *What suffices to lift a weight* W *through a vertical distance* H *will lift a weight* kW *through a vertical distance* H/k *or a weight* W/k *through a vertical distance* kH. (See Chapter 2, p. 78.)

This principle was employed by Jordanus to prove the law of the lever as applied to both a straight and a bent lever. It was also used in the elegant proof of the proposition concerning the equilibrium of interconnected weights on oppositely inclined planes whose lengths vary directly as the weights. Furthermore, in the proposition concerning the bent lever, what is essentially the concept of static moment was conceived:

6. *The effective force of a weight on any lever arm, straight or bent, depends on both its weight and its horizontal distance from the vertical line passing through the fulcrum.* (See Chapter 2, pp. 82–83.)

Still another fertile statical concept was used by the author of the *De ratione ponderis*. This was the concept of positional gravity used to determine the component of natural gravity acting along an inclined plane. This concept can be expressed as follows:

7. *The ratio of the effective weight (positional gravity) of a body in the direction of the inclined plane on which it rests to the free natural gravity is equal to the ratio of the vertical component of any given potential trajectory along that plane to that trajectory.* (See Chapter 2, pp. 74–75.)

This concept is equivalent to the modern formulation $F = W \sin a$, where F is the force along the plane, W is the free weight and a is the angle of inclination of the plane. It was employed by the medieval author to establish the law of equilibrium of interconnected weights on oppositely inclined planes. This concept and its use marked an important step in the rise of a vectorial analysis of forces.

Fortunately the medieval mechanicians did not limit themselves to propositions concerning statics and dynamics but also pursued in an original manner some of the kinematic possibilities made pregnant by Bradwardine's discussions of dynamics. In this kinematic investigation they outstripped antique efforts. In the first place, the followers of Bradwardine at Merton College, Oxford, made more explicit the distinction between dynamics and kinematics already clear to William Ockham and Bradwardine himself. As Swineshead puts the distinction, it takes this form:

8. *In the first place the velocity in every successive movement is measured causally by the proportion of the motive force to the resisting force; while in the second place it is to be measured by its effects, i.e., by the distance traversed in comparison to the time.* (See Chapter 4, p. 208.)

In pursuing the kinematic aspects of movement, the junior contemporaries of Bradwardine inherited from their thirteenth-century predecessor, Gerard of Brussels, this general kinematic statement:

9. "*The proportion of the movements* (i.e., speeds) *of points is that of the lines described in the same time.*" (See Chapter 3, p. 186.)

Furthermore, as the result of past kinematic statements, the Merton College kinematicists agreed on this precise definition of uniform speed:

10. "*Uniform local motion is one in which in every equal part of the time an equal distance is described.*" (From Swineshead; see Chapter 4, p. 243.)

This Merton definition was precisely the definition adopted by Galileo later. Similarly, the masters of Merton were also agreed upon this general definition of acceleration later concurred in by Galileo:

11. "*Any motion whatever is uniformly accelerated if, in each of any equal parts of the time whatsoever, it acquires an equal increment of velocity.*" (From Heytesbury's *Regule*; see Chapter 4, p. 237.)

Now in order to treat properly problems of acceleration, the Merton College kinematicists arrived at a concept of "qualitative" or "instantaneous" velocity. Such velocity had its measure in "degrees," as did any qualitative measure. For the practical determination of instantaneous velocity, the schoolmen were to anticipate almost exactly the words of Galileo. The Merton definition of the measure of instantaneous velocity was expressed in this way:

12. "*Instantaneous velocity is not measured by the distance traversed, but by the distance which would be traversed by such a point, if it were moved uniformly over such or such a period of time at that degree of velocity with which it is moved in that assigned instant.*" (Heytesbury; see Chapter 4, p. 236.)

In their analysis of the many problems involved in acceleration and deceleration, the Merton schoolmen discovered the following form of the theorem describing uniform acceleration:

13. "*A moving body uniformly acquiring or losing that increment* (of speed) *will traverse in some given time a magnitude completely equal to that which it would traverse if it were moving continuously through the same time with the mean degree [of velocity].*" (Heytesbury (?); see Chapter 5, p. 284.)

This form of the acceleration theorem was proved in a number of ways in the course of the fourteenth and fifteenth centuries, the most important of which was employed by Oresme at Paris just after 1350. Using a graphing system, Oresme showed that a right triangle representing uniform acceleration (i.e., the distance traversed in some movement of uniform acceleration) is equal to the rectangle representing the distance traversed in a uniform movement, where the altitude of the rectangle (the altitude representing the uniform speed) is half the altitude of the triangle (this altitude representing the final speed in the movement of uniform acceleration).

Both the medieval acceleration theorem (in slightly modified form) and Oresme's (or Casali's) geometric type of proof were given by Galileo before he passed on to his well-known form: $S \propto t^2$. But note well that the Merton formula does not appear to have been applied in the fourteenth century to the problem of falling bodies, as of course it was later applied in the sixteenth century by Domingo de Soto and in the seventeenth by Galileo and Beeckman.

The schoolmen of the thirteenth and fourteenth century were actually much confused as to the measure of the speed of falling bodies. Jordanus in the thirteenth century seemed to accept the proportionality of speed and time in a germinal fashion (see pp. 548–49), while Oresme accepted it somewhat more clearly (pp. 553–54). But his teacher, Buridan, at Paris appears to have held these two contradictory conclusions without realizing it:

14. *The speed of falling bodies increases directly with the distance of fall. The same speed increases directly with the time of fall.* (See Chapter 9, pp. 554–55, 563.)

The same confusion is evident in the treatment of the problem by Albert of Saxony, who, when specifically talking about the measure of the speed of fall, says that it grows as the distance of fall. "When some space has been traversed by this [motion], it has a certain velocity, and when a double space has been traversed by it, it is twice as fast, and when a triple space ... it is three times as fast, and so on." (See Chapter 9, p. 566.) And yet, in the same question a little bit farther on, when Albert is talking about the cause of acceleration he seems to accept the direct proportionality of speed and time. A similar confusion plagued Leonardo da Vinci, Galileo in his earlier fragments, and Descartes. (See pp. 555, 574–75, 580.)

Except for Oresme and Albert of Saxony, the authors at the University of Paris were somewhat less interested in kinematics than the English. Their glory lay in a new form of an earlier dynamics critical of Aristotle. This was the so-called impetus dynamics. The most original exponent of the impetus mechanics at Paris was John Buridan. In Buridan's scheme the continuance of projectile motion was no longer considered as resulting from the action of the surrounding air, as in the Aristotelian explanation; rather the following explanation was given:

15. "*The motor in moving a moving body impresses in it a certain impetus or a certain motive force of the moving body, [which impetus acts] in the direction toward which the mover was moving the moving body, either up or down, or laterally or circularly.*" (Buridan; see Chapter 8, p. 523.)

The impetus then appears to be some kind of intrinsic mover impressed in the projectile by the projector. So far, this is the same explanation as that offered by John Philoponus in the sixth century A.D. But the originality of Buridan's discussion of the impetus lies in the measure he assigns the impetus:

16. "*By the amount the motor moves that moving body more swiftly, by the same amount it will impress in it a stronger impetus.*" (See Chapter 8, p. 522.)

Hence for Buridan the impetus varies directly as the speed impressed in the moving body by the original mover. But it is also measured by the amount of prime matter in the body:

17. "*By the amount more there is of matter, by that amount can the body receive more of that impetus and more intensely. Now in a dense and heavy body, other things being equal, there is more of prime matter than in a rare and light one. Hence a dense and heavy body receives more of that impetus and more intensely.... And so also if light wood and heavy iron of the same volume and of the same shape are moved equally fast by a projector, the iron will be moved farther because there is impressed in it a more intense impetus, which is not so quickly corrupted as the lesser impetus.*" (See Chapter 8, p. 535.)

The two factors taken conjointly of which impetus is a function, for Buridan, are clearly velocity and prime matter, and so using medieval "proportional" expressions, we can write the following relationships between the impetus (I), the velocity (V) and the quantity of matter (m):

$$I_1 : I_2 :: V_1 : V_2 \text{ when } m_1 = m_2,$$
$$I_1 : I_2 :: m_1 : m_2 \text{ when } V_1 = V_2.$$

Taken conjointly, these expressions are equivalent to this modern metric definition: $I = m V$. The resemblance to the modern definition of momentum is evident.

One other extremely important feature of Buridan's impetus was that it was considered as something of permanent nature, not self-corrupting. Its destruction was accomplished by the resistance of the medium and the weight or contrary inclination of the body:

18. The "*impetus is a thing of permanent nature, distinct from the local motion in which the projectile is moved....[It] is a quality naturally present and predisposed for moving a body in which it is impressed....[It] is impressed in the moving body along with the motion by the motor; so with the motion it is remitted, corrupted, or impeded by resistance or a contrary inclination.*" (See Chapter 8, p. 537.) *The impetus would last indefinitely if it were not diminished by a contrary resistance or by an inclination to a contrary motion.*" (See p. 524.)

With impetus considered as tending toward permanence, it is not surprising that Buridan should suggest that the everlasting movement of the heavens might be explained by the impression of impetus by God; the impetus being thought of as lasting indefinitely, since in the heavenly regions it would be without resistance. Thus Buridan thought it possible to impress circular motion and circularly acting impetus in something of the same way as rectilinear motion and impetus. It was not until Benedetti that a freely moving body under the influence of impetus was thought to move only in a straight line. (See Chapter 11, pp. 633–64.)

Of the numerous applications of the impetus theory to phenomena, none was more fertile than Buridan's use of the theory to explain the acceleration of falling bodies, which he stated as follows:

19. "*A heavy body not only acquires motion unto itself from its principal mover i.e., its gravity, but it also acquires unto itself a certain impetus with that motion. This impetus has the power of moving the heavy body in conjunction with the permanent natural gravity....from the beginning the heavy body is moved by its natural gravity only; hence it is moved slowly. Afterwards it is moved by that same gravity and by the impetus acquired at the same time; consequently, it is moved more swiftly. And because the movement becomes swifter, therefore the impetus also becomes greater and stronger, and thus the heavy body is moved by its natural gravity and by that greater impetus simultaneously, and so it again will be moved faster; and thus it will always and continually be accelerated to the end.*" (See Chapter 9, pp. 560–61.)

Buridan's application of the impetus theory to the acceleration of falling bodies was essentially the same as that of Abū 'l-Barakāt earlier. Buridan's somewhat clumsy expression of the impetus explanation of acceleration was sharpened by later authors, for example by Benedetti, who said that "if a body is moved naturally, it will always increase its velocity, since the impression and impetus in it would always be increased. This is so because it has contact with the motive power perpetually." (See Chapter 11, p. 663.)

The Buridan doctrine, then, by the sixteenth century had clearly become the following: Continually acting gravity (the primary force) impresses continually increasing impetus (the intermediary force) in the falling body which results in continually increasing velocity. Now, when in modern mechanics impetus is conceived as an *effect* rather than a *cause*, impetus loses its position as an intermediary force, and we are left with the familiar Newtonian doctrine, a continually acting force results in acceleration. And furthermore, there is evidence that some of the adherents to the impetus mechanics were tending toward the conversion of impetus into an effect in the sixteenth and early seventeenth centuries. Finally, the impetus analysis of fall was of importance in stressing the continuous action of gravity and impetus *in time*, thus bringing to the fore time of fall rather than distance of fall as the direct measure of the speed of fall.

Not all of the medieval schoolmen who accepted the impetus mechanics accepted Buridan's "inertial" impetus. Many of his successors returned to an old-fashioned "self-expending" impetus. Furthermore, Buridan's principal successor at Paris, Oresme, conceived of impetus as being produced by the *acceleration* of a body and in turn producing further acceleration. With this interpretation of impetus he also returned to a peculiar antique theory that projectiles "accelerated" at the beginning of their flight.

The same Parisian masters who developed the theory of the impetus, and particularly John Buridan and Nicole Oresme, also considered seriously the possibility of the earth's rotation on its axis. In so arguing this, these men advanced and developed important mechanical concepts, not the least of which was their use of an older idea of the complete relativity of the detection of motion. In the words of Oresme we are told:

20. "*Local motion can be sensibly perceived only in so far as one may perceive one body differently disposed with respect to another.*" (See Chapter 10, p. 585.)

The example of this doctrine used by the medieval and the early modern authors was that of observers on ships who are able only to detect their movement relative to each other (so that if both were moving at the same speed in the same direction it would seem "that neither ship is moving"). The observer on the earth is in precisely the same boat, so far as detecting the movement of the heavenly bodies, as the observer on one ship trying to decide, as the distance to another ship varies, which of the ships is moving. He cannot tell whether in fact the heavens move in diurnal rotation and the earth is at rest or vice versa.

Furthermore, Oresme came particularly close in the discussion of this problem to the concept of a closed mechanical system. An arrow shot from a moving earth would share the same horizontal velocity as the earth, and hence only *its motion relative to the earth*, i.e., its vertical rectilinear path, could be detected by an observer on the earth, and he would be unable to say anything about its absolute movement. The analogy given by Oresme to illustrate this idea is the observer on a ship which is moving eastward rapidly. He draws his hand down the mast of the ship, and "it would seem to him that his hand was moved with rectilinear movement only. According to this opinion, it seems to us in the same way that an arrow descends or ascends in a straight line." (See Chapter 10, p. 581.)

We have, then, come to the end of our review of some significant medieval mechanical doctrines. Admittedly, most of these concepts we have singled out as important were formed within the Aristotelian framework of mechanics. But these medieval doctrines contained within them the seeds of a critical refutation of that mechanics. The medieval mechanicians, as they propagated these concepts, were attempting to amend the system at the very points it was weakest, and in so doing they focused attention on those weaknesses, at the same time making preliminary and not completely unsuccessful efforts to solve the crucial problems—the problems arising from considering the balance and the lever in operation, the stone in its fall, and the arrow in its flight.

Bibliography

In general only mechanical and mathematical works important for mechanics are given in this bibliography. In listing manuscripts and early editions, I have given only those which I consulted. There are in most instances other manuscripts and editions. At times I did not use the first edition, either because it was not available to me or because it was defective. In many cases a *full* bibliographical title has not been given. However, I trust that in every case sufficient information has been given for proper identification. The abbreviations used for manuscripts are those followed in my list of microfilm reproductions in *Isis*, Vol. 44 (1953), 372. They should, however, be immediately recognizable to anyone familiar with the principal manuscript collections of Europe.

Achillini, A. *De proportione motuum*, in *Opera omnia*. Venice, 1545.

Albert of Saxony. *Questiones subtilissime in libros de celo et mundo Aristotelis*. Venice, 1492. (MS Paris, BN lat. 14723, ff. 113r–162r.) Cf. Latin texts of Documents 2.4, and 9.2.

———. *Questiones... in octo libros physicorum Aristotelis*. Paris, 1516, 1518. (MSS Bruges, Stadsbibliotheek 477, ff. 60v–164v; Paris, BN lat. 6526, ff. 1r–165v.)

———. *Tractatus de proportionibus*. Venice, ca. 1480, 1496. (MSS Paris, BN lat. 2831, ff. 116r–122v; 7368, ff. 14r–26v; Venice, Bibl. Naz. San Marco Lat. VI, 62,ff. 111v–117r; VI, 71, ff. 42r–46r; VI, 149, ff. 20r–24r.)

Allan, D.J. "Mediaeval Versions of Aristotle, *De Caelo*, and of the Commentary of Simplicius," *Mediaeval and Renaissance Studies*, Vol. 2 (1950), 82–120.

Ametus filius Josephi. *Epistola de proportione et de proportionalitate*. MS Paris, BN lat. 9335, ff. 64r–75r.

Amodeo, F. "Appunti su Biagio Pelicani da Parma," *Atti del IV Congresso Internazionale dei Matematici*, Vol. *3* (1908), 549–53.

Anastos, M.V. "Aristotle and Cosmas Indicopleustes on the Void," *Prosphora eis Stilpōna P. Kuriakidēn*. Thessalonica, 1953, pp. 35–50.

Angelus de Fossambruno. *De inductione formarum*. MSS Vat. lat. 2130, ff.188r–194v; Vat. lat. 3026, ff. 115r–121r.

———. *De reactione*. MS Venice, Bibl. Naz. San Marco Lat. VI, 160, ff. 248r–252r.

———. *De tribus predicamentis Hentisberi*. Venice, 1494, ff. 64–73r. (MSS consulted are listed in Chapter 11, note 60.)

Anonymous. *A est unum calidum*. MSS Paris, BN lat. 16134, ff. 73r–80v; Venice, Bibl. Naz. San Marco Lat. VI, 30, ff. 113r–133r, and VI, 155, ff. 65r–82v.

———. *"Aliud Commentum" in librum de ponderibus*. Nuremberg, 1533.(MSS Erfurt, Stadtbibliothek, Amplon. Q. 385, pp. 641–44, and Q. 387, ff. 52v–57r; Florence Bibl. Naz. J.VI. 36, ff. 2r–8r; Vienna, Nat.-bibl. 5203, ff. 174r–180v.)

———. *De canonio*. For Latin text and English translation, see E. Moody and M. Clagett, *Medieval Science of Weights*.

———."*Corpus Christi" Commentum in Elementa Jordani de ponderibus*. MSS Oxford, Corpus Christi College 251, ff. 10r–12r; Cambridge Univ. Library Mm.3.11, ff. 151r–154v. Cf. Latin text of Document 2.2.

———. *Linea naturalium*. MS Oxford, Bodl., Bodley 676, ff. 1r–6v; cf. ff. 149r–160v.

———. *"Pseudo-Euclid" Versio commenti in Elementa Jordani de ponderibus*. MSS Cambridge, Gonv. and Caius 504/271, ff. 97v–100v; Florence, Bibl. Naz. J.IV. 29, ff. 61v–67r; Milan, Bibl. Ambros. T. 100 Sup., ff. 149v–154v; Vat. lat. 2975, ff. 164r–171v; Paris, BN lat. 7310, ff. 110r–121r; Vienna, Nat.-bibl. 5304,ff. 128r–134.

———. *Questio utrum omnis forma habeat latitudinem nobis presentabilem per figuras geometricas*. MS Venice, Bibl. Naz. San Marco Lat. VI, 62, ff. 63r–68r.

———. *Summulus de motu*. With *Proportiones* of Thomas Bradwardine, Venice, 1505, ff. 9r–10r. Cf. Latin text of Document 7.1.

———. *Tractatus de motu*. MS Venice, Bibl. Naz. San Marco Lat. VIII, 19, ff. 193r–243r. Cf. Document 5.5.

———. *Tractatus de sex inconvenientibus*. Venice, 1505.

———. *De uniformi et difformi secundum figuras*. MS Erfurt Stadtbibliothek, Amplon. Q. 325, ff. 43r–45r.

Apianus, Petrus, ed. *Liber Iordani Nemorarii...de ponderibus propositiones xiii et earundem demonstrationes*. Nuremberg, 1533. An edition of Version *P* and the "Aliud Commentum."

Aquinas, Thomas. *In libros Aristotelis de caelo et mundo expositio*, *Opera omnia*. Vol. *3*, Rome, 1886; and the recent printing, Rome, 1952.

Archimedes. *Opera omnia*. In medieval Latin translation of William of Moerbeke, MS Rome, Vat. Ottob. lat. 1850, ff. 11r–60r.

———. *Opera omnia*. In modern edition of J. L. Heiberg, 3 vols., Leipzig, 1910–15.

———. *The Works of Archimedes*. Translated and paraphrased by T. L. Heath, Cambridge, 1897 (reissued by Dover Publications, New York, 1950). See also the important new English version of E. J. Dijksterhuis, Copenhagen, 1956.

———. *Floating Bodies*. See Zotenberg.

Archimedes (Pseudo-). *De ponderibus* (or *De insidentibus in humidum*). For Latin text and English translation, see Moody and Clagett, *Medieval Science of Weights*. See

also L. Thorndike, *Isis*, Vol. *45* (1954), 98, note.
Aristotle. *De caelo libri quattuor*. Edition of D. J. Allan, Oxford, 1936. English translation of W. K. C. Guthrie, London, Cambridge, Mass., 1945. Latin translations in edition of Averroës' Commentary, Venice, 1574, and with the *Expositio* of Aquinas, which see.

———. *Physica*. Medieval translations in MSS Paris, BN lat. 16141, 16142, 6320, 6505, and 16672. Edition of Venice, 1495; modern edition of W. D. Ross, Oxford, 1936, 1950; English translation of P. H. Wicksteed and F. M. Cornford, 2 vols., London, New York, Cambridge, Mass., 1929–34.

———. *Meteorologicorum libri quattuor*. Edition of F. H. Fobes, Cambridge, Mass., 1919.

Aristotle (Pseudo-). *Mechanica*. Edition of O. Apelt, Leipzig, 1888; English translation of E. S. Forster, Oxford, 1913.

Averroës. *Commentaria in de caelo libros*. With Aristotle's *De caelo*, Venice, 1574.

———. *Commentarium in physicam*. With Aristotle's *Physica*, Venice, 1495.

Avicenna. See Ibn Sīnā.

Bacon, Roger, (Pseudo?). *De graduatione medicinarum compositarum*, in *Opera hactenus inedita Rogeri Baconi*. Edition of A. G. Little and E. Withington, fasc. IX, Oxford, 1928, pp. 144–49.

———. *Questiones supra libros octo physicorum Aristotelis*, in *Opera hactenus inedita Rogerii Baconi*. Fasc. XIII, edition of F. M. Delorme, O.F.M., with the collaboration of R. Steele, Oxford, 1935.

Bardi, Giovanni. *Eorum quae vehuntur in aquis experimenta etc.* Rome, 1614.

Bauerreis, H. *Zur Geschichte des spezifischen Gewichtes in Altertum und Mittelalter*. Erlangen, 1914.

Beeckmann, Isaac. *Journal tenu par Isaac Beeckman de 1604 à 1634*. Edition of C. de Waard, Vol. *1*, The Hague, 1939.

Benedetti, Giovanni Battista. *Diversarum speculationum mathematicarum et physicarum liber*. Turin, 1585.

———. *Demonstratio proportionum motuum localium contra Aristotelem et omnes philosophos*. Venice, 1554. Cf. G. Libri, *Histoire des sciences mathématiques en Italie*, Vol. *3*, Paris, 1840, 258–64.

Bernoulli, J. *Discours sur les loix de la communication du mouvement etc.*, in *Opera omnia*. Vol. *3*. Lausanne, 1742.

Berthelot, P. E. M. *La Chimie au moyen âge*. Vol. *1*. Paris, 1893.

al-Bīrūnī. See Sachau.

al-Bitrūjī. *De motibus celorum*. Edition of F. J. Carmody, Berkeley, Calif., 1952.

Blasius of Parma. *De intensione et remissione formarum*. MSS Oxford, Bodl., Canon. Misc. 177, ff. 24r–39r; Venice, Bibl. Naz. San Marco Lat. VI, 62, ff. 1r–18r.

———. *De ponderibus*. For Latin text and English translation, see Moody and Clagett, *Medieval Science of Weights*.

———. *Questiones super tractatu de latitudinibus formarum*. Venice, 1505, ff. 30r–32r.

(MSS Oxford, Bodl., Canon. Misc. 177, ff. 97v–100v; Venice, Bibl. Naz. San Marco Lat. VI, 155, ff. 88r–92r; Milan, Bibl. Ambros. F. 145 Sup., ff. 1r–5r.) Cf. Latin text of Document 6.4.

——. *Questiones super tractatu de proportionibus.* MSS Venice, Bibl. Naz. San Marco Lat. VIII, 38, ff. 8v–37r; Oxford, Bodl., Canon. Misc. 177, ff. 69r–97v; Vat. lat. 3012, ff. 137r–163v; Milan, Bibl. Ambros. F. 145 Sup., ff. 5r–18r.

Boas, G. "A Fourteenth-Century Cosmology," *Proceedings of the American Philosophical Society*, Vol. *98* (1954), 50–59.

Boehner, P. *Ockham, Philosophical Writings.* Edinburgh, 1957.

Borchert, E. *Die Lehre von der Bewegung bei Nicolaus Oresme. (Beiträge zur Geschichte der Philosophie und Theologie des Mittelalters*, Bd. *31*, Heft 3). Münster, 1934.

Boutroux, P. "L'Histoire des principes de la dynamique avant Newton," *Revue de métaphysique at de morale*, Vol. *28* (1921), 657–88.

Bradwardine, Thomas. *De continuo.* MSS Thorn, R. 4° 2, pp. 153–92; Erfurt, Stadtbibliothek, Amplon. Q. 385, ff. 17r–48r; Paris, BN, Novelles Acquis. lat. 625, f. 71v. Cf. Latin text of Document 4.3.

——. *Tractatus brevis proportionum: abbreviatus ex libro de proportionibus.* Vienna, 1515. (MSS Cambridge, Gonv. and Caius 182/215, pp. 119–31; London, Brit. Mus. Royal 8.A.XVIII, ff. 75r–81v; Vat. lat. 1108, ff. 140r–143v. Vienna, Nat.-bibl. 4951 ff. 260r–271v; 4953, ff. 19r–35v; 4698, ff. 123r–129r. Note these are not all of the same "abbreviations," although there are similarities.) Cf. Latin text of Document 7.3.

——. *Tractatus de proportionibus.* Edition of H. Lamar Crosby, Jr., Madison, 1955.

Bricot, Thomas. *Textus abbreviatus Aristotelis super octo libris phisicorum et tota naturali philosophia.* Paris, 1494.

Brodrick, G. C. *Memorials of Merton College.* Oxford, 1885.

Bruno, G. *Le Opere italiane.* 2 vols. Göttingen, 1888.

Buchner, F. "Die Schrift über den Qarastun von Thabit ben Qurra," *Sitzungsberichte der Physikalisch-medizinischen Sozietät in Erlangen*, Vols. *52–53* (1920–21), 141–88.

Bulliot, J. "Jean Buridan et le mouvement de la terre, etc.," *Revue de philosophie*, Vol. *25* (1914), 5–24.

Buridan, John. *Quaestiones super libris quattuor de caelo et mundo.* Edition of E. A. Moody, Cambridge, Mass., 1942.

——. *Expositio physicorum.* MS Paris, BN lat. 16130, ff 36r–59r.

——. *Questiones super octo phisicorum libros Aristotelis.* Paris, 1509. (MSS Paris, BN lat. 14723, ff. 2r–107v; Carpentras 293; Vat. lat. 2163, ff. 1r–157v, and 2164, ff. 1r–120r.) A different version in MS Erfurt, Stadtbibliothek, Amplon. F. 298, and Haute-Garonne MS 6.

Burley, Walter of. *Super libros Aristotelis de physica auscultatione...commentaria.* Venice, 1589. (MS Florence, Bibl. Naz. conv. soppr. A 1; for MSS, see Michal-

ski, "La physique nouvelle," pp. 98–99.)
Cajori, F. "Falling Bodies in Ancient and Modern Times," *Science and Mathematics*, Vol. *21* (1921), 638–48.
Carton, R. *L'Expérience physique chez Roger Bacon etc.* Paris, 1924.
Casali, Giovanni di. *De velocitate motus alterationis.* Venice, 1505, ff. 57r–70v. (MSS Oxford, Bodl., Canon. Misc. 376, ff. 32r–47r; Florence, Bibl. Riccard. 117, ff. 135r–144v; Vienna, Nat.-bibl. lat. 4217, ff. 154r–172r.) Cf. Latin text of Document 6.2.
Celaya, Juan. *Expositio in octo libros phisicorum Aristotelis cum questionibus eiusdem.* Paris, 1517.
Child, J. M. "Archimedes' Principle of the Balance and Some Criticisms Upon It," in C. Singer, *Studies in the History and Method of Science.* Vol. 2. Oxford, 1921, 490–520.
Chilmark, Johannes. *De actione elementorum.* MS Oxford, Bodl., Digby 77, ff. 153v–165v.
———. *De alteratione.* MS Oxford, Bodl., Bodley 676, ff. 76r–101r.
———. *De motu* (shorter version). MS London, Brit. Mus., Harl. 2178, ff. 1r–7r.
———. *De motu* (longer version). MS Oxford, Bodl., Bodley 676, ff. 11r–38r.
Christopher de Recaneto. *Super librum calculationum.* MS Venice, Bibl. Naz. San Marco. Lat. VI, 149, ff. 31r–49v.
Clagett, M. "Archimedes in the Middle Ages: The *De mensura circuli*," *Osiris*, Vol. *10* (1952), 587–618.
———. *Giovanni Marliani and Late Medieval Physics.* New York, 1941.
———. *Greek Science in Antiquity.* New York, 1955.
———. "The *Liber de Motu* of Gerard of Brussels and the Origins of Kinematics in the West," *Osiris*, Vol. *12* (1956), 73–175.
———. "Medieval Mathematics and Physics: A Checklist of Microfilm Reproductions," *Isis*, Vol. *44* (1953), 371–81.
——— (with Ernest Moody). *The Medieval Science of Weights.* Madison, 1952.
———. Review of A. C. Crombie, *Augustine to Galileo etc.*, in *Isis*, Vol. *44* (1953), 398–403.
———. "Richard Swineshead and Late Medieval Physics," *Osiris*, Vol. *9* (1950), 131–61.
———. "Some General Aspects of Medieval Physics," *Isis*, Vol. *39* (1948), 29–44.
———. "The *De curvis superficiebus Archimenidis*: A Medieval Commentary of Johannes de Tinemue on Book I of the *De sphaera et cylindro* of Archimedes," *Osiris*, Vol. *11* (1954), 295–358.
———. "The *Quadratura per lunulas*, A Thirteenth Century Fragment of Simplicius' Commentary on the *Physics* of Aristotle," in *Essays in Medieval Life and Thought Presented in Honor of Austin Patterson Evans*, New York, 1955, pp. 99–108.
———. "The Use of the Moerbeke Translations of Archimedes in the Works of Johannes de Muris," *Isis*, Vol. *43* (1952), 236–42.

Claius, Magister. *Conclusiones que sunt necessaria ad dicta anglicarum intelligenda*. MS Paris, BN lat. 16621, ff. 212v–213v.

Cohen, I. B. "Galileo's Rejection of the Possibility of Velocity Changing Uniformly with Respect to Distance," *Isis*. Vol. *47* (1956), 231–38.

Cohen, M. R. (with I. E. Drabkin). *A Source Book in Greek Science*. New York, 1948.

Copernicus, N. *De revolutionibus orbium*. Partial translation of J. F. Dobson in *Occasional Notes, Royal Astronomical Society*, No. 10 (May, 1947). New critical edition F. and C. Zeller, *Nikolaus Kopernikus Gesamtausgabe*, Vol. *2* (Munich, 1949). See the French translation of Bk. I by A. Koyré, Paris, 1934.

Cornford, F. M. *The Laws of Motion in the Ancient World*. Cambridge, 1931.

Coronel, L. *Physicae perscrutationes etc.* Paris, 1511; Lyon, 1530.

Crombie, A. C. *Robert Grosseteste and the Origins of Experimental Science 1100–1700*. Oxford, 1953.

Curtze, M. "Ein Beitrag zur Geschichte der Physik im 14. Jahrhundert," *Bibliotheca Mathematica*, Neue Folge, Vol. *10* (1896), 43–49.

———. "Eine Studienreise," *Centralblatt für Bibliothekswesen*, XVI. Jahrg., 6. u. 7. Heft (1899), 257–306.

———. "Über die Handschrift R. 4^0 2: Problematum Euclidis Explicatio, des Königl. Gymnasial Bibliothek zu Thorn," *Zeitschrift für Mathematik und Physik*, Vol. *13* (1868), Suppl. pp. 85–91.

———. "Zwei Beiträge zur Geschichte der Physik," *Bibliotheca Mathematica*, 3. Folge, Vol. *1* (1900), 51–54.

Cusa, Nicholas of. *De ludo globi*, in *Opera*. Basel, 1565.

———. *Idiota de staticis experimentis*, in *Opera omnia iussu et auctoritate Academiae Litterarum Heidelbergensis ad codicum edita*. Edition of L. Bauer, Leipzig, 1937. See also, *Opera*, Basel, 1565. For an English translation see Henry Viets, "De staticis experimentis of Nicolaus Cusanus," *Annals of Medical Science*, Vol. *4* (1922), 115–35.

Delisle, L. *Le Cabinet des manuscrits de la Bibliothèque Nationale*. Vol. 2. Paris, 1874.

Deshayes, M. *Le Découverte de l'inertie. Essai sur lois générales du mouvement de Platon à Galilée*. Paris, 1930.

Dijksterhuis, E. J. *De Mechanisering van het Wereldbeeld*. Amsterdam, 1950. (German translation, 1955.)

———. Review of Moody and Clagett, *Medieval Science of Weights*, in *Archives internationales d'histoire des sciences*, Vol. *6* (1953), 504–7.

———. Review of Crosby's edition of Bradwardine's *Proportionibus*, ibid., Vol. *8* (1955), 390–93.

———. *Val en Worp*. Groningen, 1924.

———. And see Archimedes and Stevin.

Dingler, H. "Uber die Stellung von Nicolaus Oresme in der Geschichte der Wissenschaften," *Philosophische Jahrbuch*, Vol. *45* (1932), 58–64.

Dominicus de Clavasio. *Questiones super libros de celo*. MS Vat. lat. 2185, ff. 15v–

20v.

Dominicus de Soto. See Soto.

Doncœur, P. "Le Nominalisme d'Occam. Théories du mouvement, du temps et du lieu," *Revue de philosophie*, Vol. *26* (1921), 237–49.

Dorp, Johannes. *Commentarii in opera Johannis Buridani*. MS Erfurt, Stadtbibliothek, Amplon. F. 300, ff. 1r–163r.

Drabkin, I. E. (with M. R. Cohen). *A Source Book in Greek Science*. New York, 1948.

———. "Aristotle's Wheel: Notes on the History of a Paradox," *Osiris*, Vol. *9* (1950), 162–98.

———. "Notes on the Laws of Motion in Aristotle," *American Journal of Philology*, Vol. *59* (1938), 60–84.

Drake, S. *Discoveries and Opinions of Galileo*. New York, 1957.

Dugas, P. *Histoire de la mécanique*. Neuchatel, 1950.

———. *La Mécanique au xviie siècle (Des antécedents scolastiques à la pensée classique)*. Neuchatel, 1954.

Duhem, P. "De l'accélération produite par une force constante etc." In *Extrait des comptes rendus du 2eme Congres Inter. de Philosophie*. Genève, 1904.

———. *Études sur Léonard de Vinci*. 3 Vols. Paris, 1906–13.

———. "François de Meyronnes O.F.M. et la question de la rotation de la terre," *Archivum Franciscanum Historicum*, Annus VI, Tomus *4* (Quaracchi, 1913), 23–25.

———. *L'Évolution de la mécanique*. Paris, 1903.

———. "Le Mouvement absolu et le mouvement relatif," *Revue de philosophie*, Vol. *11* (1907), 221–35, 347–62, 548–73; Vol. *12* (1908), 134–50, 246–65, 389–400, 486–98, 607–23; Vol. 13 (1908), 143–65, 275–87, 515–19, 635–65; Vol. *14* (1909), 149–79, 306–17, 436–58, 499–508.

———. *Les Origines de la statique*. 2 Vols. Paris, 1905–6.

———. *Le Système du monde*. 8 Vols. Paris, 1913–16 (Vols. *1–5*); 1954–58 (Vols. *6–8*).

———. "Le Temps et le mouvement selon les scholastiques," *Revue de philosophie*, Vol. *23*, (1913), 453–78; Vol. *24* (1914), 5–15, 136–49, 225–41, 361–80, 470–80; Vol. *25* (1914), 109–52.

———. "Un Précurseur français de Copernic: Nicole Oresme," *Revue générale des sciences pures et appliquées*, Vol. 20 (1909), 866–73.

Dullaert, J. *Questiones super octo libros phisicorum Aristotelis*. Paris, 1506.

Dumbleton, John. *Summa de logicis et naturalibus*. MSS Paris, BN lat. 16146; Cambridge, Peterhouse 272; Gonv. and Caius 499/268; Vat. lat. 954. Cf. Latin text of Document 5.4.

———. *Declaratio super sex conclusiones quarti capituli tractatus proportionum mag. Thome Bradwardin*. MS Paris BN Nouv. Acquis. lat. 625, ff. 70v–71v.

Erhardt-Siebold, Erica von, and Rudolph von Erhardt. *The Astronomy of Johannes Scotus Erigena*. Baltimore, 1940.

———. *The Cosmology in the "Annotationes in Marcianum"; More Light on Erigena's Astronomy*. Baltimore, 1940.

Euclid (Pseudo-). *The Book of Euclid on the Balance*. Arabic text and French translation by F. Woepcke in *Journal asiatique*, Ser. 4, Vol. *18* (1851), 217–32.

———. *De ponderoso et levi*. For Latin text and English translation, see Moody and Clagett, *Medieval Science of Weights*.

Faventini. See Vittorio, B.

Feribrigge (Ferebrich), Richard. *De motu*. MS Rome, Bibl. Angelica 1017, ff. 37r–48v.

Fontana, Giovanni da. *Tractatus de trigono balistario*. MS Oxford, Bodl. Canon. Misc. 47.

Forlì. See Jacobus Forliviensis.

Francischus de Ferraria. *Questio de proportionibus motuum*. MS Oxford, Bodl. Canon. Misc. 226, ff. 58r–63r. Cf. Latin text of Document 7.4

Franciscus de Marchia. *In libros sententiarum*. For MSS see Michalski, "La physique nouvelle," p. 94.

Franciscus de Mayronis. *Expositio in VIII libros physicorum*. Ferrara, 1495.

———. *Praeclarissima ac multum subtilia egregiaque scripta... in quatuor libros sententiarum*. Venice, 1520.

Gaetano de Thienis. *Declaratio super tractatu Hentisberi regularum etc*. Venice, 1494, ff. 7r–52r.

———. *De intensione et remissione formarum*. Venice, 1491.

———. *Recollecte super consequentias*. MS Venice, Bibl. Naz. San Marco Lat. VI, 160, ff. 109r–118r.

———. *Recollecte super octo libros phisicorum Aristotelis*. Venice, 1496.

Galilei, Galileo. *Le Opere*. Ed. Naz. 23 Vols, Florence, 1891–1909. Particularly Vols. *1*, *2*, and *8*. And see S. Drake above and Thomas Salusbury below.

———. *Dialogue on the Great World Systems*. English translation of T. Salusbury as revised by G. de Santillana, Chicago, 1953. Cf. translation of S. Drake, Berkeley, 1953.

———. *Dialogues Concerning Two New Sciences*. English translation of the *Discorsi* by H. Crew and A. de Salvio, Evanston and Chicago, 1946.

Gautier, L. *Antecédents gréco-arabes de la psychophysique*. Beyrouth, 1939.

Gerard of Brussels. *Liber de motu*. For Latin text, see above under Clagett.

Giacomelli, R. *Galileo Galilei giovane e il suo "De motu."* Pisa, 1949.

Ginzburg, B. "Duhem and Jordanus Nemorarius," *Isis*, Vol. *25* (1936), 341–62.

Grabmann, M. *Guglielmo di Moerbeke, O.P., Il traduttore delle opere di Aristotele*. Rome, 1946.

Gunther, S. "Le origini e i gradi di sviluppo del principio delle coordinate," *Bulletino di bibliografia e storia delle scienze matematiche e fisiche*, Vol. *10* (1887), 363–406.

Haas, A. E. "Die Anfänge der mathematischen Physik," *Archiv für die Geschichte der Naturwissenschaften und Technik*, Vol. *1* (1909), 402–9.

———. "Die Grundlagen der antiken Dynamik," *ibid.*, Vol. *1* (1908), 19–47.

[Bibliography 691]

Hall, A. R. *The Scientific Revolution, 1500–1800.* London, 1954.
Hart, I. B. *The Mechanical Investigations of Leonardo da Vinci.* London, 1925.
Haskins, C. H. *Studies in Medieval Science.* 2d ed. Cambridge, Mass., 1927.
Hay, W. H. "Nicolaus Cusanus: The Structure of His Philosophy," *The Philosophical Review*, Vol. *61* (1952), 14–25.
Heath, T. L. *A History of Greek Mathematics.* Vol. *1.* Oxford, 1921.
———. *Mathematics in Aristotle.* Oxford, 1949.
———. *The Works of Archimedes.* Cambridge, 1897.
Henry of Hesse. *De reductione effectuum.* MSS Paris, BN lat., 2831, ff. 103r–115r, and 16401, ff. 96r–106v; London, Brit. Mus., Sloane 2156, ff. 116v–130v.
Hermanus de Curis. *Commentarius in questiones a Johanne Buridano de libris physicorum Aristotelis.* MS Erfurt, Stadtbibliothek, Amplon. F. 300, ff. 166r–285v.
Hero of Alexandria. *Mechanics.* Arabic text and French translation by Carra de Vaux, as "Les Mécaniques ou L'Élévateur de Héron d'Alexandrie," *Journal asiatique*, Ser. 9, Vol. *1* (1893), 386–472; Vol. *2* (1893), 152–269, 420–514. Also published separately, Paris, 1894. See also the new text based on several manuscripts with a German translation by L. Nix in *Heronis Alexandrini opera quae supersunt omnia*, Vol. *2*, fasc. II (Leipzig, 1900).
Heytesbury, William. *Regule solvendi sophismata.* Venice, 1494, ff. 4v–52r. (MSS Bruges, Stadsbibliotheek 497, ff. 46r–59v; Bruges, Stadsbibliotheek 500, ff. 33r–71v; Vat. lat. 2136, ff. 24v–33r.) Cf. Latin texts of Documents 4.4, 5.1.
——— (?). *Probationes conclusionum tractatus regularum solvendi sophismata.* Venice, 1494, ff. 188v–203v. (MS Venice, Bibl. Naz. San Marco Lat. VI, 71, ff. 128v–136r). Cf. Latin text of Document 5.2.
Hölder, O. *Die mathematische Methode.* Berlin, 1924.
Hugh of Siena. *De equali ad pondus.* MS Venice, Bibl. Naz. San Marco Lat. VI, 96, ff. 12–16.
Hultsch, F. *Metrologicorum scriptorum reliquae.* 2 vols. Leipzig, 1866.
Ibel, Th. *Die Wage in Altertum und Mittelalter.* Erlangen, 1908.
Ibn Sīnā. *Kitāb al-Shifā.* 2 Vols. Teheran, 1885.
Jacobus Forliviensis. *De intensione et remissione formarum.* Venice, 1496, ff. 16r–42v. (MSS Venice, Bibl. Naz. San Marco Lat. VI, 155, ff. 134r–159r; and VII, 7, ff. 1r–55r.)
Jacobus de Sancto Martino. *Tractatus de latitudinibus formarum.* Critical text by Thomas Smith as a thesis on deposit in the University of Wisconsin Library. For MSS and earlier editions, see Latin text of Document 6.3.
Jammer, Max. *Concepts of Force.* Cambridge, Mass., 1957.
———. *Concepts of Space.* Cambridge, Mass., 1954.
Jansen, B. "Olivi der älteste scholastische Vertreter der heutigen Bewegungsbegriffs," *Philosophisches Jahrbuch der Görresgesellschaft*, Vol. *33* (1920), 137–52.
Johannes de Casali. See Casali.
Johannes de Muris. See Muris.

Johannes de Wasia. See Wasia.

John of Holland. *De motu*. MSS Oxford, Bodl. Canon. Misc. 117, ff. 100v–109r; Venice, Bibl. Naz. San Marco Lat. VIII, 19, ff. 1r–26r. Cf. Latin text of Document 4.6.

———. *De instanti*. MSS Oxford, Bodl. Canon. Misc. 177, ff. 48v–61v; Venice, Bibl. Naz. San Marco Lat. VI, 155, ff. 43r–64v.

John the Scot (Eriugena). *Annotationes in Marcianum*. Cambridge, Mass., 1939.

Jordanus de Nemore. *Elementa de ponderibus*. For Latin text and English translation, see Moody and Clagett, *Medieval Science of Weights*. This volume also contains texts of the *Liber de ratione ponderis* and the *Liber de ponderibus* sometimes attributed to Jordanus.

———. *Geometria vel de triangulis libri IV*. Edition of M. Curtze, in *Mitteilungen des Coppernicus-Verein für Wissenschaft und Kunst zu Thorn*, Heft *6* (1887). (MSS London, Brit. Mus., Harl. 625, ff. 123r–130r; Brit. Mus., Sloane 285, ff. 90r–92v; Bruges, Stadsbibliotheek 530, 14 c, ff. 1r–8v.)

al-Khāzinī. *Book of the Balance of Wisdom*. Partial edition and translation into English by N. Khanikoff in the *Journal of the American Oriental Society*, Vol. *6* (1860), 1–128. For a list of other parts translated into German, see E. Wiedemann's summary in the *Sitzungsberichte der Physikalisch-medizinischen Sozietät in Erlangen*, Vol. *40* (1908), 158–59.

al-Kindī. *De investigandis compositarum medicinarum gradibus, Supplementum in secundum librum compendii secretorum medicinae Joannis Mesues*. Venice, 1581.

Koyré, A. "A Documentary History of the Problem of Fall from Kepler to Newton," *Transactions of the American Philosophical Society*, New Series, Vol. *45* (1955), 329–95.

———. *Études Galiléennes*. 3 Vols. (*Actualités Scientifiques et Industrielles, Nos. 852–54*). Paris, 1939.

———. "An Experiment in Measurement," *Proceedings of the American Philosophical Society*, Vol. *97* (1953), 222–37.

———. "Le Vide et l'espace infini au xive siècle," *Archives d'histoire doctrinale et littéraire du moyen âge*, Vol. *17* (1949), 45–91.

Krazer, A. "Zur Geschichte der graphischen Darstellung von Funktionen," *Jahresbericht der deutschen Math-Verein*, Bd. *24* (1915), 340–63.

Lacoin, M. "De la scholastique à la science moderne. Pierre Duhem et Anneliese Maier," *Revue de questions scientifiques*, Vol. *17* (1956), 325–43.

Lacombe, G. *Aristoteles latinus*. 2 parts. Rome, 1939; Cambridge, 1955.

Lagrange, J. L. *Mécanique analytique*, in *Oeuvres de Lagrange*. Vol. *11*. Paris, 1888.

Lange, H. *Geschichte der Grundlagen der Physik*. Vol. *1*. Freiburg and Munich, 1954.

Lawrence of Scotland (Laurentius Londorius). *Questiones super libros physicorum*. MSS Vienna, Dominik. 114/81, 115/82, and 116/83; Erfurt, Stadtbibliothek, Amplon. F. 337^a, ff. 64r–116v, 118r–129r, and F. 343, ff. 1r–179v.

Lenzen, V. "Archimedes' Theory of the Lever," *Isis*, Vol. *17* (1932), 288–89.

———. "Reason in Science," *Reason*, Vol. *21* (1939), 81–83.
Leonardo da Vinci. *Il Codice atlantico*. Edition of Regia Accadèmia dei Lincei, Milano, 1894–1904.
———. *Les Manuscrits . . . de la Bibliothèque de l'Institut*. Edition of Charles Ravaisson-Mollien, Paris, 1881–91.
———. *Notebooks of Leonardo da Vinci*. Translation of E. MacCurdy, Vol. *1*, New York, 1938.
Mach, E. *The Science of Mechanics*. English translation of T. J. McCormack, 5th ed., LaSalle, Ill., 1942.
Maier, A. *An der Grenze von Scholastik und Naturwissenschaft*. 2d ed. Rome, 1952.
———. *Die Mechanisierung des Weltbildes in 17. Jahrhundert*. Leipzig, 1938.
———. *Metaphysische Hintergründe der spätscholastischen Naturphilosophie*. Rome, 1955.
———. "Die naturphilosophische Bedeutung der scholastischen Impetustheorie," *Scholastik*, 30 Jahrg., Heft *3* (1955), 321–43.
———. *Die Vorläufer Galileis im 14. Jahrhundert*. Rome, 1949.
———. *Zwei Grundprobleme der scholastischen Naturphilosophie*. 2d ed. Rome, 1951.
———. *Zwischen Philosophie und Mechanik*. Rome, 1958. (This valuable volume appeared too late to use in my work.)
———. Review of Moody and Clagett, *Medieval Science of Weights*, *Isis*, Vol. *46* (1955), 297–300.
Mansion, A. *Introduction à la physique aristotélicienne (Aristote: traductions et études)*. Louvain-Paris, 1946.
Marchanova, Johannes de. *Expositio commentariorum Averrois in libros octo physicorum Aristotelis*. MS Venice, Bibl. Naz. San Marco Lat. VI, 103–4.
Marchia. See Franciscus de.
Marcolongo, R. *La Meccanica di Leonardo da Vinci*. Naples, 1932.
Marliani, Giovanni. *Opera subtilissima*. 2 Vols. Vol. *1*: *Questio de proportione*, Pavia, 1482; Vol. *2*: *Disputatio . . . cum Johanne de Arculis etc.*, Pavia (s.d.). (For MSS see M. Clagett, *Giovanni Marliani*, pp. 22–33.)
Marsilius of Inghen. *Abbreviationes libri physicorum*. MS Vienna, Nat.-bibl. lat. 5437, ff. 1r–67v.
——— (?). *Quaestiones super octo libros physicorum Aristotelis* [*secundum nominalium viam*]. In *Opera omnia* of Duns Scotus, Vol. 2, Lyon, 1639.
McColley, Grant. "The Theory of the Diurnal Rotation of the Earth," *Isis*, Vol. *26* (1936–37), 392–402.
Messinus. *Questiones super questione Johannis de Casali*. MSS Bologna, Univ. Libr. 1227 (2410), ff. 101–163; Venice, Bibl. Naz. San Marco Lat. VI, 225, ff. 1r–76r; Escorial, La Real Bibl. f. II. 8, ff. 1r–49r.
———. *Sententia (tractatus) de tribus predicamentis motus*. Venice, 1494, ff. 52v–64v. (MS Venice, Bibl. Naz. San Marco Lat. VI, 105, ff. 47r–65r.)
Meyronne. See Franciscus de Mayronis.

Michalski, K. "Les Courants philosophiques à Oxford et à Paris pendant le XIVe siècle," *Bulletin international de L'Académie polonaise des sciences et des lettres, Classe d'histoire et de philosophie, et de philologie*, Les Années, 1919, 1920 (Cracovie, 1922), 59–88.

———. "Le Criticisme et le scepticisme dans la philosophie du XIVe siècle," *ibid.*, L'Année 1925, Part I (1926), 41–122.

———. "Les Courants critiques et sceptiques dans la philosophie du XIVe siècle," *ibid.*, L'Année 1925, Part II (1927), 192–242.

———. La Physique nouvelle et les différents courants philosophiques au XIVe siècle," *ibid.*, L'Année 1927 (1928), 93–164.

Michel, Paul-Henri. *De Pythagore à Euclide*. Paris, 1950.

Minio-Paluello, L. "Iacobus Veneticus Grecus: Canonist and Translator of Aristotle," *Traditio*, Vol. 7 (1952), 265–304.

Momigliano, F. *Paolo Veneto e le correnti del pensiero religioso e filosofico nel suo tempo.* Turin, 1907.

Moody, E. A. "Galileo and Avempace: The Dynamics of the Leaning Tower Experiment," *Journal of the History of Ideas*, Vol. *12* (1951), 163–93, 375–422.

———. "Laws of Motion in Medieval Physics," *The Scientific Monthly*, Vol. *72* (1951), 18–23.

——— (with M. Clagett). *The Medieval Science of Weights*. Madison, 1952.

———. "Ockham and Aegidius of Rome," *Franciscan Studies*. Vol. *9* (1949), 417–42.

———. "The Rise of Mechanism in 14th Century Natural Philosophy." Unpublished, mimeographed, New York, 1950.

Muris, Johannes de. *Quadripartitum numerorum*. MSS Paris, BN lat. 7190, 14 c, ff. 1r–100v; BN lat. 14736, 15 c. ff. 1r–108r; Vienna, Nat.-bibl. 4770, 15c, ff. 188v–324v; and 10954, 16c, ff. 1r–135r. Cf. Latin text of Document 2.3.

Ockham, William. *Expositio librorum physicorum*. MS Oxford, Merton College 293, ff. 1r–149v; Vat. lat. 3062.

———. *Questiones super libros physicorum*. MSS Vat. lat. 956, ff 32v–59v.

———. *Questiones super quattuor libros sententiarum*. Lyons, 1495.

———. *Summule librorum physicorum*. Venice, 1506, ff. 1–32. (MS Vat. Pal. lat. 1202, ff. 325r–372v.)

———. *The Tractatus de successivis etc*. Edition of P. Boehner (Franciscan Institute Publications, No. 1), St. Bonaventure, New York, 1944.

Olivi, Peter John. *Quaestiones in secundum librum sententiarum*. Edition of B. Jansen, Vol. *1*, Quaracchi, 1922.

Oresme, Nicole. *Algorismus proportionum*. Edition of M. Curtze, Berlin, 1868. (MSS Bruges, Stadsbibliotheek 530, 14 c, ff. 25r–30v; Paris, BN lat. 7197, ff. 74r–79v; Basel, Bibl. Univ. F. II. 33, ff. 95v–98v; Florence, Bibl. Naz. J.IX.26, ff. 37r–45r; Vat. lat. 4082, a. 1401, ff. 109r–113v; Paris, Bibl. de l' Arsenal 522, ff. 121r–122v; Venice, Bibl. Naz. San Marco Lat. VI, 133, ff. 66r–70v.)

———. *De configurationibus qualitatum*. For MSS and partial edition see Latin text of Document 6.1.

———. *De proportionibus proportionum*. Venice, 1505, ff. 17r–26v. (MSS Paris, BN lat. 7371, ff. 269r–278v; Venice Bibl. Naz. San Marco Lat. VI, 133, ff. 50v–65v; Cambridge, Peterhouse 277, ff. 93v–110v; Paris, BN lat. 16621, ff. 94r–110v; Erfurt, Stadtbibliothek, Amplon. Q. 385, ff. 67r–82v.)

———. *De proportionibus velocitatum in motibus*. MS Paris, Bibl. de l'Arsenal 522, ff. 126r–168v.

———. *Le Livre du ciel et du monde*. Edition of A. D. Menut and A. J. Denomy, *Mediaeval Studies*, Vols. *3–5* (1941–43).

———. *Questiones de spera*. MS Florence, Bibl. Riccard. 117, ff. 125r–135r.

———. *Questiones in libro Euclidis Elementorum*. MSS Vat. lat. 2225, ff. 90r–98v; Vat. Chis. lat. F.IV.66, ff. 22v–40r.

———. *Questiones super de celo*. Erfurt, Stadtbibliothek, Amplon. Q. 299, ff. 1r–50r.

———. For the *De latitudinibus formarum* attributed to Oresme, see Jacobus de Sancto Martino.

Pappus of Alexandria. *Mathematical Collection*. Edition of F. Hultsch, *Pappi Alexandrini collectionis quae supersunt; e libris manuscriptis edidit, latina interpretatione et commentariis instruxit Fridericus Hultsch*. 3 Vols. Berlin, 1876–78.

Paul of Venice. *Summa naturalium*. Venice, 1503.

Phillipps, Thomas. "The *Mappae clavicula*: A Treatise on the Preparation of Pigments during the Middle Ages," *Archeologia*, Vol. *32* (1847), 183–244.

Philoponus, John. *In Aristotelis physicorum libros commentaria*. Edition of H. Vitelli, *Commentaria in Aristotelem graeca*. Vol. *17*. Berlin, 1888.

Picus, H. *Questio de motu gravium et levium*. MS Venice, Bibl. Naz. San Marco Lat. VIII, 83, ff. 4r–15v.

Pines, S. "Études sur Aḥwad al-Zamân Abu 'l-Barakât al-Baghdâdî," *Revue des études juives*, New Series, Vol. *3* (1938), 3–64; Vol. *4* (1938), 1–33.

———. "La Théorie de la rotation de la terre à l'époque d'al-Bîrûnî," *Journal asiatique*, Vol. *244* (1956), 301–6.

———. "Les Précurseurs musulmans de la théorie de l'impetus," *Archeion*, Vol. *21* (1938), 298–306.

———. "Quelques tendances antipéripatéticiennes de la pensée scientifique islamique," *Thales*, Les Années, 1937–39 (1940), 210–219.

———. "Un précurseur Bagdadien de la théorie de l'impetus," *Isis*, Vol. *44* (1953), 247–51.

Politus, Bassianus. *Tractatus proportionum introductorius ad Calculationes Suisset*. Venice, 1505.

Randall, J. R. "The Development of Scientific Method in the University of Padua," *Journal of the History of Ideas*, Vol. *1* (1940), 177–206.

Recaneto. See Christopher.

Reimann, D. "Historische Studie über Ernst Machs Darstellung der Entwicklung des Hebelsatzes," *Quellen und Studien zur Geschichte der Mathematik, Astronomie, und Physik*, Abt. B, Bd. *3* (1936), 554–92.

Ritter, G. "Studien zur Spätscholastik, I. Marsilius von Inghen und die okkamistische Schule in Deutschland," in *Sitzungsberichte der Heidelberger Akademie der wissenschaften*, Philosophisch-historische Klasse, Jahrg. 1921. 4 Abhandlung.

Sachau, E. C. *Alberuni's India*. Vol. *1*. London, 1888.

Salusbury, Thomas. *Mathematical Collections and Translations*. 2 Vols. London, 1661–65. Includes translations of a number of Galileo's works.

Sarton, George. *Introduction to the History of Science*. 3 Vols. in 5 parts. Baltimore, 1927–1948.

Schum, W. *Beschreibendes Verzeichniss der Amplonianischen Handschriften-Sammlung zu Erfurt*. Berlin, 1887.

Shapiro, H. "Motion, Time, and Place According to William Ockham," *Franciscan Studies*, Vol. *16* (1956), 213–303, 319–372.

Simplicius. *In Aristotelis de caelo libros commentaria*. Edition of J. L. Heiberg, *Commentaria in Aristotelem graeca*. Vol. *7*. Berlin, 1894.

———. *In Aristotelis physicorum libros quattuor posteriores commentaria*. Edition of H. Diels. *Commentaria in Aristotelem graeca*. Vol. *10*. Berlin, 1895.

Soto, Domingo de. *Super octo libros physicorum Aristotelis commentaria*. Salamanca, 1582.

———. *Super octo libros physicorum Aristotelis subtilissime quaestiones*. Salamanca, 1555, 1582; Venice, 1582.

Stamm, E. "Tractatus de continuo von Thomas Bradwardina," *Isis*, Vol. *26* (1936), 13–32.

Stein, W. "Der Begriff des Schwerpunktes bei Archimedes," *Quellen und Studien zur Geschichte der Mathematik, Astronomie, und Physik*, Series B, Studien, Bd. *1* (1931), 221–44.

Stevin, S. *Les Oeuvres mathématiques*. Leyden, 1634.

———. *The Principal Works of Simon Stevin*. Vol. *1*, the *Mechanics* text and English translation of E. J. Dijksterhuis, Amsterdam, 1955.

Stoezlin, F. *Questiones in librum proportionum*. MS Vienna, Nat.-bibl. lat. 5222, ff. 1r–47v.

Strode, R. *Consequentie*. MS Copenhagen, Bibl. Reg. Thott. 581, ff. 1–12. This is Part IV of Strode's *Logica* (MS Oxford, Bodl., Canon. Misc. 219, ff. 13–52).

Swineshead, Richard. *De motibus naturalibus (De primo motore)*. MS Erfurt, Stadtbibliothek, Amplon. F. 135, ff. 24v–47r. This is perhaps by Roger Swineshead.

———. *Liber calculationum*. Padua, ca. 1477; Pavia, 1498; Venice, 1520. (MS Pavia, Univ. Bibl. Aldini 314.) Cf. Latin text of Document 5.3. and Chapter 4, note 13, for a list of manuscripts.

———(?). *De motu* (2 fragments). MS Cambridge, Gonv. and Caius 499/268, ff.

212r–213r, 213r,–215r. Cf. Latin text of Document 4.5.

Tannery, P. "Sur les problèmes mécaniques attribués à Aristote," *Memoires scientifiques*. Vol. *3*. Paris, 1915, pp. 32–36.

Tartaglia, N. *Quesiti et inventioni diversi etc.* Venice, 1546.

———. *Jordani Opusculum de ponderositate* (an edition of the *Liber de ratione ponderis*). Venice, 1565.

Tartaretus, P. *Commentarii in libros philosophie naturalis et metaphysice Aristotelis.* Basel, 1514.

Thomas. See Aquinas.

Thomas, Alvarus. *Liber de triplici motu proportionibus annexis...philosophicas Suiseth calculationes ex parte declarans.* Paris, 1509.

Thorndike, L. *A History of Magic and Experimental Science*. 8 Vols. New York, 1923–58.

———. *Science and Thought in the Fifteenth Century.* New York, 1929.

Thurot, C. "Recherches historiques sur le Principe d'Archimède," *Revue archeologique*, New Series, Vol. *18* (1868), 389–406; Vol. *19* (1869), 42–49, 111–23, 284–99, 345–60; Vol. 20 (1869), 14–33.

Tornius, Bernard. *In capitulum de motu locali Hentisberi quedam annotata.* Venice, 1494, ff. 73v–77v.

Trivisano, Aurelius Marcus. *De macrocosmo, i.e. de maiori mundo.* Johns Hopkins MS; unpublished transcription of George Boas. Cf. Latin text of Document 2.5.

Ucelli, G. *Scritti della meccanica.* Milano, 1940.

Upton, Edward. *Tractatus proportionum.* MS Oxford, Bodl., Bodley 676, ff. 38r–49v.

Vailati, G. *Scritti.* Leipzig, Florence, 1911. (All of his splendid essays on mechanics are republished in this edition. Note also the Spanish translation of mechanical essays by Hugo Incarnato, Buenos Aires, 1947.)

Valsanzibio, P. S. da. *Vita e dottrina di Gaetano di Thiene.* 2d ed. Padua, 1949.

Venturi, J. B. *Essai sur les ouvrages physico-mathématiques de Léonard de Vinci etc.* Milan, 1911.

Vinci. See Leonardo da.

Vitruvius. *De architectura.* Edition of F. Krohn, Leipsig, 1912; edition and English translation by F. Granger, 2 Vols., London, 1931–34.

Vittorio, B. *Commentaria in tractatum proportionum Alberti de Saxonia.* Bologna, 1506.

Wasia, Johannes de. *Compendium de proportionibus Bradwardini.* MS Erfurt, Stadtbibliothek Amplon. Q. 325, 14 c, ff. 47r–51 v.

Weisheipl, J. A., O. P. *De natura et gravitatione.* River Forest, Ill., 1955.

Wiedemann, E. "Beiträge zur Geschichte der Naturwissenschaften VI. Zur Mechanik und Technik bei den Arabern," *Sitzungsberichte der Physikalisch-medizinischen Sozietät in Erlangen*, Vol. *38* (1906), 1–56.

———. "Beiträge...VII. Über arabische Auszüge aus der Schrift des Archimedes über die schwimmenden Körper," *ibid.*, Vol. *38* (1906), 152–62.

———. "Beiträge... VIII. Über Bestimmung der spezifischen Gewichte," *ibid.*, Vol. *38* (1906), 163–80.

———. "Beiträge... XV. Über die Bestimmung der Zusammensetzung von Legierungen," *ibid.*, Vol. *40* (1908), 105–59.

———. "Die Schrift über den Qarastūn," *Bibliotheca Mathematica*, 3. Folge, Vol. *12* (1911–12), 21–39.

———. "Über Thabit ibn Qurra...," *Sitzungsberichte der Physikalisch-medizinischen Sozietät in Erlangen*, Vol. *52* (1920, published 1922), 189–219.

———. Über die Kenntnisse der Muslime auf dem Gebiete der Mechanik und Hydrostatik," *Archiv für die Geschichte der Naturwissenschaften und der Technik*, Vol. *2* (1910), 394–98.

———. "Das Gesetz vom freien Falle in der Scholastik, bei Descartes und Galilei," *Zeitschrift für mathematischen und naturwissenschaften Unterricht aller Schulgattungen*, *45*, Jahrg. (1914), 209–28.

Wieleitner, H. "Der 'Tractatus de latitudinibus formarum' des Oresme," *Bibliotheca Mathematica*, 3. Folge, Vol, *13* (1913), 115–45.

———. "Über den Funktionsbegriff und die graphische Darstellung bei Oresme," *ibid.*, *Vol.* *14* (1914), 193–248.

———. "Zur Geschichte der unendlichen Reihen in christlichen Mittelalter," *ibid.*, Vol. *14* (1914), 150–68.

Wilson, C. *William Heytesbury*: *Medieval Logic and the Rise of Mathematical Physics*. Madison, 1956.

Winters, H. J. J. "Muslim Mechanics and Mechanical Appliances," *Endeavor*, Vol. *15* (1956), 25–28.

Witelo. *Opticae libri decem*. Basel, 1572.

Woepcke, F. "Notice sur des traductions arabes de deux ouvrages perdus d'Euclide," *Journal asiatique*, Ser. 4, Vol. *18* (1851), 217–38.

Wohlwill, E. "Die Entdeckung des Beharrungsgesetzes," *Zeitschrift für Völkerpsychologie und Sprachwissenschaft*, Vol. *14* (1883), 365–410; Vol. *15* (1884), 70–135, 337–87.

———. "Eine Vorgänger Galileis im 6. Jahrhundert," *Physikalische Zeitschrift*, Vol. *7* (1906), 23–32.

Wolfson, H. A. *Crescas' Critique of Aristotle*. Cambridge, Mass., 1929.

Wundt, W. *Logik*. Vol. 2. Stuttgart, 1907.

Wyclif, John. *Tractatus de logica*. Edition of M. H. Dziewicki, Vol. *3*, London, 1899.

Zotenberg, H. "Traduction arabe du traité des corps flottants d'Archimède," *Journal asiatique*, Ser. 7, Vol. *13* (1879), 509–15.

Index

I have not indexed my own works which are often referred to; nor have I indexed the citations to E. A. Moody as the co-author of the *Medieval Science of Weights*, although all other citations to Moody have been included. I have not indexed the Bibliography, which is alphabetically arranged and thus is easily consulted. Only significant place names have been included; thus the present locations of manuscripts have not been indexed. In indexing proper names, the first names of medieval and Renaissance people are written out, while those of modern scholars are abbreviated by the use of initials. Incidentally, I have tended to index medieval schoolmen by their first names, except in cases like Bradwardine, Buridan, and Ockham where the person is most familiarly known by his last name. Fifteenth- and sixteenth-century figures are most often indexed by their last names even if these names appear to be place names. However, the reader is advised to try both first and last names before deciding that a given person is not mentioned. The running page number form (e.g., 124–25) does not necessarily mean that there is a continuous discussion of the indexed item, for it may merely indicate that the item appears on each page in the series. If an item is mentioned in both the text and notes of a given page, only the page number is given. If it is mentioned only in the notes, an "n" is added to the page number (e.g., 52n).

Abū 'l-Barakāt: on *mail*, 513–14, 523; on the acceleration of fall, 525, 545, 547–48, 580, 680; mentioned, 510n

Abū 'l-Ḥusain, 28–29

Abū Manṣūr al-Nairīzī, 65

Abū Sahl, 19n, 50, 58

Acceleration: Aristotle on, 258; changing rates of, 267–68, 274, 292–94, 297, 302–4; of falling bodies, xxviii, 514, 525, 613, Chapter 9, *and see* Acceleration, uniform; rules for distance determination in movements of, 204n, 266, 272–74, 276–77, 280–82, 291, 295–96, 300, 312, 324, 344, 360–61, 380–81, 412–14; *and see Velocitatio*

Acceleration, uniform: defined at Merton, 205, 211, 214; defined anonymously, 326, 451–52, 633n; defined by Feribrigge, 630; defined by Galileo, 251–52; defined by Heytesbury, 236–37, 241, 677; defined by John of Holland, 247–50; defined by Oresme, 342–43, 357, 365, 377; defined by Swineshead, 243, 245; Merton theorem of, in general, xxv–xxvi, 205, 556, Chapter 5; Merton theorem of, noted by Alvarus Thomas, 657, noted by Celaya, 656, noted by Coronel, 655n, noted by Heytesbury, 262–63, 270–71, 277–78, 677, noted by Swineshead, 244–46, 263; Merton theorem of, proved anonymously, 326–29, proved by Beeckman, 417–18, proved by Blasius of Parma, 402–8, proved by Casali, 384–85, proved by Dullaert, 654–55, proved by Dumbleton, 266–67, 305–25, proved by Feribrigge, 630n–631n, proved by Galileo, 409–16, proved by Heytesbury (?), 264–65, 284–89, proved

[700   Index]

by John of Holland, 268n–269n, proved by Messinus, 647, proved by Oresme, 343, 358–59, 379–80, 442n–443n, 677–78, proved by Swineshead, 266, 290–302; *and see* Motion, uniformly difform

Achillini, Alessandro, 443

Aegidius Romanus, 207, 550

Aerometer, 90–91

Albert of Saxony: and hydrostatics, xxiv, 96, 124–25, 136–45; and statics, 99; his *Proportiones* and the Bradwardine dynamics, 442–43n, 652, 655n, 658; on motion of rotation, 217, 223–29, 262n, 664; on the acceleration of natural motion, 258, 551, 553–55, 565–67, 580, 678; *On the De caelo of Aristotle*, 96, 565–67, 585n, 641; on the earth's center of gravity, 585n, 591; on the earth's rotation, 585–87n; *On the Physics of Aristotle*, 223–29, 262n, 443n; the publication and use of his works, 652–53, 655n, 658, 651, 661; mentioned, 522, 563, 635–36, 674

Albertus Magnus, 592

Albiruni, 64, 584n, 587n

Alexander of Aphrodisias, 258–59, 543–45, 555

Alexander the Great, 57, 202n

Alexander V, Pope, 638

Algazel, 425n

Alhazen, 19n, 50, 58

'Alī ibn Riḍwān, 338n

Aliud commentum de ponderibus, 78–80, 100, 675

Alkindi, 53, 439n, 441n

Allan, D. J., 428n, 542n, 545n, 592n

Alloys. *See* Weight, specific

Alpetragius, 514–15n

Ametus filius Josephi, 69n–71n

Amiens, 73

Amplonian collection, 642, *and passim* for particular manuscripts

Analytic geometry, 331, 341–42

Anastos, M., 435n

Angelus de Fossambruno, 648–49, 651–52

Anglegena, Henry, 100

Anstey, H., 200n

Antiperistasis, 505–9, 532

Apelt, O., 4

Apian, Peter, 101, 675

Aquinas, St. Thomas: on projectile motion, 516–17; on relativity of motion, 601n–602n; on the earth's rotation, 584; on the plurality of worlds, 592; mentioned, 207, 425n, 428n–429n, 542n

Archimedes: and the crown problem, 57, 64, 85–89; and the law of the lever, 9–12, 14, 49, 156; cited by Ametus, 69n–70n; cited by Hero, 42–43, 45, 49; his *Dimensio circuli*, 170, 184, 188, 635n; his influence on Benedetti, 665; his *Method*, 186; kinematics of, 171–75; *On Balances*, 19n, 33n, 49–50, 66, 70n; *On the Equilibrium of Planes*, 10–12, 16, 18–19, 31–37, 43, 66, 69, 591; *On Floating Bodies*, xxiv, 21–22, 52–55, 64, 67, 70, 88, 95, 124–25, 141, 673–74; *On the Sphere and the Cylinder*, 170, 184; *On Spiral Lines*, 171–75, 217; the *De insidentibus* of Pseudo-, xxiv, 21, 70, 85, 93–97, 120–21, 123–25, 141, 159, 674; the mathematical approach of, xxiii, 103; the method of exhaustion of, employed, 36, noted, 170; the principle of, xxiv, 64, 67, 88–89, 95, 113–14, 121, 127–28, 141; the spurious *De curvis superficiebus* attributed to, 80, 170–71; mentioned, xxiii, 15–16, 68, 167

Archytas of Taras, 3–4

Aretino, Leonardo, 630n

Aristarchus, 583–84n

Aristotle: dynamics of, xxvi–xxvii, 7n, 19–20, 65–67, 84, 97, 546, 659, 665, 673–74, *and particularly* Chapter 7; his *De anima*, 609n, 616; his *De caelo*, 65, 139–40, 203n, 339, 340, 428–30, 463, 532, 540, 542–43, 558, 561, 592, 600, 603, 609n, 659, 673–74; his definition of motion, 423, 445, 615, 622, 625; his *Meteorology*, 140; his *Physics*, 175–82, 207, 256–58, 361n, 423n, 426–28, 430–33, 441n, 463, 476, 487, 490, 496, 500, 505–8, 515, 517, 527–28, 532, 546, 558, 570, 612, 614–19, 622–24, 638, 645, 654n–655n, 658–59; his *Topics*, 94; kinematics of, 164, 175–82, 256–58; mechanics of, xix, 30, 68, 682, *and particularly* Chapter 7; natural philosophy of, xxix, 103, 139, 421–25; on acceleration of natural motion, 542–44; on plurality of worlds, 592; on projectile motion, 505–7, 517, 519, 527–28, 530, 532–34, 637–38, 659; on the "quicker," 177–79, 256, 258–60; on

[Index 701]

weight and density, 139–40, 597, 610, 673–74; spurious *Mechanica*, attributed to, xxiii, 4–7, 8–9, 16, 18–19, 48, 50, 71–72, 76, 83–84, 101, 156, 175, 182–84, 675; wheel of, 40, 48, 184; mentioned, 202n, 333, 339n, 593, 595
Âryabhata, 584n
Aschbach, J., 641n
Astronomy and astronomers: 164, 222, 585–86, 595, 598, 603–4, 608; *and see* Earth, rotation of
Autolycus of Pitane, 164–67, 174–75, 217
Avempace, 207, 436–37n, 667, 675
Averroës: his *Colliget*, 439; *On the De caelo*, 557, 561; *On the Metaphysics*, 616; *On the Physics*, 175–77n, 181, 207, 426n, 428n, 436–37, 505n, 527, 529–30, 542n, 558, 619; mentioned, 551, 675
Averroists, 637, 659–60
Avicenna, 425n, 510–14, 547
Avignon, 200

Bacon, Roger: on acceleration of fall, 549–50; on calculating intensities, 334–35; on plurality of worlds, 592; on projectile motion, 515–16n; on the multiplication and influence of species, 519; used at Cambridge, 634n
Balance: for the theory of, *see* Lever principle *and* Equilibrium
Balance, hydrostatic, 54, 56–58, 64, 87–89, 97n–98n
Baliani, Giovanni Battista, 582n
Banū Mūsā, 28
Bardi, Giovanni, 159
Bartholomew of Mantua, 645, 648
Baryllium, 91n
Bauer, L., 97n
Beeckman, Isaac, 346, 417–18, 666, 678
Benedetti, G. Battista: on impetus, 644, 663–65, 680; on Aristotle's laws of motion, 435, 443, 665, 674–75; statics of, 101
Benedict of Hesse, 643
Benvenuto da Ferrara, 498
Bernoulli, John, xxiv, 8
Berthelot, M., 90n, 92n
βία, 425n, 509
Biagio Pelacani. *See* Blasius of Parma
Bidrey, E., 337n
al-Bīrūnī, 64, 584n, 587n
al-Bitrujī, 514–15n
Blasius of Parma: and Bradwardine's dynamics, 443, 645; hydrometer of, 96, 141; life and works, 645–46, 650; *On the Tract on the Latitudes of Forms*, 346, 402–8, 646; statics of, 100, 646; mentioned, 414, 652, 674
Boas, G., 97n, 147–48n
Bocheński, I. M., 284n
Boehner, P., 284n, 521n
Bologna, 71, 332, 645, 648–50
Bonamico, Francesco, 666–67
Borchert, E., xxi, 524n
Bradwardine, Thomas: his *Abbreviatus de proportionibus*, 440n, 465–94, 641–42; his *De continuo*, 199, 217, 230–34; his *De proportionibus*, xxvii, 186, 199, 202–3, 205, 207–8, 215, 217–18, 220–22, 231, 425, 437–38n, 453, 637, 644–45; his distinction of dynamics from kinematics, 207–8, 676; laws of motion and, xxvii, 437–40, 523, 647, 649, 675, *and* Chapter 7, *passim*; life, 199–200; on the comparison of velocities, 217–18, 230–31, 233; on the measure of the speed of local motion, 221–22; on the motion of rotation, 216, 226, 231, 233–34; later influence in England, 634, in France, 339, 441–42, 637, in Germany, 641–42, in Italy, 645, 649, 652; mentioned, 99, 206, 522, 630
Brahmagupta, 584n
Bricot, Thomas, 638–40
Brodrick, G. C., 199n–201n, 204n–205n
Brotto, I., 650n–651n
Bruno, Giordano, 587, 665–66
Buridan, John: and Bradwardine's dynamics, 441n–442n; and the impetus theory, xxi, xxvii–xxviii, 585, 639, 641, 653–56, 659–60, 663–64, 668, 678–81, *and particularly* Chapters 8 and 9; and the relativity of motion, 587; his later influence in France, 635–40, 653–59, in Germany, 641–44, in Italy, 645–46, 659–66; life of, 521–22; on *antiperistasis*, 508, 532; on the acceleration of natural motion, 545, 548, 551–52, 554–55, 557–64, 580, 678, 680–81; on the concept of a closed mechanical system, 587, 666; *On the De caelo*, 524n, 557–62, 594–99; on the earth's center of gravity, 590, 597–99;

on the earth's rotation, 584–88, 594–99;
On the Metaphysics, 524n; *On the Physics,*
216, 336n, 532–40, 641–43, 645–46, 653;
on the speed of revolution, 216–17n;
the publication of his works, 652–53;
mentioned, xxix, 337, 614, 635
Burley, Walter, 257n, 426n–427n, 430n, 550–51
Busser (Heytesbury?), 630n

Calculator and the English calculators: 247–48, 266, 268, 649; *and see* Swineshead, Richard
Cambridge, 634–35n
Campanus of Novara, 333, 361–62n
Canonio, Tractatus de, 9, 13–14, 20, 69, 83, 99
Canterbury, 200
Caraston: 69n–71n; *and see* Thabit ibn Qurra
Cardano, Girolamo, 663
Carmen de ponderibus, 54, 85–93, 95–96, 141
Carmody, F. J., 515n
Carra de Vaux, B., 15n, 19n
Carrara, 650
Casali, Giovanni di: his life and dates, 332, 644; *On the Velocity of Motion of Alteration,* 218, 332, 345, 382–91, 404, 499, 645; influence of his work, 414, 645–47, 652, 678; mentioned, 255, 333, 337, 361, 574
Cause karastonis, 14
Causes, in mechanics, 17
Celaya, Juan, 655–57
Charles V, king of France, 337, 340
Cherniss, H. F., 607n
Child, J. M., 12
Chilmark, John, 631–32
Chymiacho, Johannes de, 632n
"Circle, principle of," 5, 16, 45–46, 50, 182–83
Cistercian Order, 204
Clare College, Cambridge, 635n
Clavasio, Dominicus de, 635–36n
Clymiton, Richard, 203n
Cohen, I. B., x, 580n
Cohen, M. R., 15n, 49, 164n–165n, 258n, 429n, 508n, 543n, 599n
Colle, F. M., 650n
Cologne, 642
Comets, 612
Commentator: 527, 529, 557–58, 616, 619; *and see* Averroës
Conimbricensis, Commentarii Collegii, 659
Continuum, 230–34, 617
Coopland, G. W., 338n
Copernicus: and impetus, 611, 613, 643; on the earth's rotation, xxi, 588–89, 610–14
Corialti, da Tossignano, 498
Cornford, F. M., 177n, 180n, 426n
Coronel, Luiz, 655–57
Cosmas Indicopleustes, 434n–435n
Cracow, 588–89, 637, 642–43
Crescas, 423n, 510n
Crew, H., 251n, 576n
Crombie, A. C., x, xxii, 589
Crosby, H. L., Jr., 199n, 208n, 212n, 220n, 437n, 440n–441n
Curtze, M., 72n, 200n
Cusa. *See* Nicholas of

Dallari, D. U., 649n
Da Vinci. *See* Leonardo
Delisle, L., 73n
Delorme, F. M., 515n
Denomy, A. J., 337n, 570n, 592n, 600n
Density: xxiv, 58, 65, 87, 89, 94, 96–97, 430; *and see* Weight, specific
Descartes, xix, 418, 657, 666, 678
De Visch, C., 204n
Diels, H., 258n, 507n, 545n
Dijksterhuis, E. J., x, xxi, 12n, 85n, 101n, 436, 519n
Dinkelsbühl (Dinkelspuhel), Nicholas of, 641
Diogenes Laertius, 4
Diurnal motion. *See* Earth, rotation of
Dobson, J. F., 610n
Domingo de Soto. *See* Soto
Dominicus de Clavasio, 635–36n
Domitian, 57
Donatus de Monte, 640n
Dorp, Johannes, 642
Drabkin, I. E., 15n, 48–49, 164n–165n, 258n, 429n, 431n, 508n–509n, 543n, 545n–546n, 599n
Drake, S., xi, 150n, 582n, 588n
Duhem, P.: his *Études,* xx, 100n, 515n, 549, 574, 591n–592n, 630n, 634n, 636n–638n, 640n, 643n, 653, 655n–659n, 663; his *Le Système du monde,* xxi, 583; his *Les Origines de la statique,* xx, 7, 77, 100; his

[Index   703]

methods of scholarship, xxi–xxii; another work of, 584n
Dullaert, John, 638, 653–56
Dumbleton. *See* John *and* Thomas of
δύναμις, 425n, 509
Duns Scotus, 206, 590, 615n
Dynamics: Aristotelian, xxv–xxvii, Chapter 7, *and see* Aristotle; defined and distinguished from kinematics, 163, 204n, 206–9, 676; Greek and Latin terms used in, 425n; medieval, xxvi–xxviii, 674–75, *and* Part III, *passim*
Dziewicki, M. H., 633n

Earth, rotation of, xxviii, 583–89, 594–614, 666, 681
Economy, principle of, 146–47, 520, 529, 587, 589, 595, 598, 605–6, 608
Elvehjem, C., x
Emden, A. B., 199n–201n, 204n, 630n, 651n
Energy, Greek concept of, 423n, 510
English Mechanics. *See* Merton College
Entelechy, 423n, 510n
Equilibrium: in *Book of the Balance*, 9–10, 24–28; in *De canonio*, 13; in *Elementa de ponderibus*, 76; in *Equilibrium of Planes*, 10–11, 31–36; in Hero's *Mechanics*, 16, 41–43; in *Liber karastonis*, 14, 19–20; in *Mechanica of* Pseudo-Aristotle, 7, 8–9; in *De ratione ponderis*, 50, 77, 81, 83, 104–7, 676; stable and unstable, 8–9, 83–84
Erfurt, 637–38n, 642
Erhardt-Siebold, E. von, and R. von Erhardt, 583n–584n
Eriugena, John Scot, 583–584n
Esculanus, 648
Euclid: *Book on the Balance* attributed to, 9–11, 24–30; *De ponderoso et levi* attributed to, 21–22, 65–66, 69, 96, 141, 430, 432, 435–36, 674; his *Elements*, 35, 66, 174, 331, 338n–339, 344, 361, 403, 469, 473, 484, 487, 635n; theory of proportionality of, 167, 217n; version of *Elementa de ponderibus* attributed to, 99, 634n–635n; mentioned, xxiii, 103, 167–68, 185
Eudoxus, 170
Experience and *experientie*, 533–34, 600–602, 608n–609n
Experiment: Galileo and, 546–47, 556, 578–79, 581–82; lack of, at Merton, 206; thought, 97, 141, 473, 475, 556
Experimentum, 97, 473, 475, 524n

Fakhr al-Dīn al-Rāzī, 54, 514n, 547–48
Falling bodies. *See* Acceleration
Faral, E., 521n
Favaro, A., 650n
Feribrigge (Ferebrich), Richard, 630–31
Fitzgerald, A., 91n
Flanders, 200
Fontana, Giovanni da, 651
Force: impressed, xxvii–xxviii, Chapters 8 and 9, particularly 509–14, 516–31, *and see* Impetus; intensive and extensive, distinguished, 212; of impact, *see* Impact; of motion, 19; of weight, 27, 29, 58, 146–47; related to density, 58–59, 65; related to speed, 58–59, 66, 207–9, 425–41, *and see* Dynamics; resolution and components of, 3, 48, 74–75, 107–8, 152–55, 158, 676; terms for, and definitions of, xxvi–xxvii, 58, 425n, 673
Forza, 154–55, 157, 668n
Forlì, Jacopo da, 648–52
Forlivio, Peter de, 649
Forosempronio, Pietro da, 648
Forster, E. S., 4n
Fossambruno, Angelus de, 648–49, 651–52
Fournival, Richard, 73
France, kinematics and dynamics in, 331, 635–40, 653–59
Franceschini, E., 175n
Franciscus de Ferraria, 99, 332, 443, 495–503, 644
Franciscus de Marchia, 519–21, 523, 526–31, 636
Franciscus de Mayronis, 584–85n
Frederick II, 71
Freiling, L., xi
Functionality, 205, 208n, 341

Gaetano de Thienis, 630n, 634n, 647, 651–52
Galen, 333
Galileo Galilei: and a closed mechanical system, 587; and force of impact, 539n, 576; and force on an inclined plane, 15n, 49, 102, 152–55, 158–59; and instantaneous velocity, 215, 252; and *momento* and *impeto*, 150, 152–54, 576–78, 587, 666–

68; and the law of the lever, 150–52; and virtual velocities, displacements, or work, 102, 151, 154–58; his concept of inertia, 158–59, 668–71n; his *De motu*, 563, 665–67n, 675; his *Mechanics*, 102, 150–59, 668n–669n; his *Two Chief Systems (Dialogo)*, 156–58, 414–16, 587n–588n, 667n–670n; his *Two New Sciences (Discorsi)*, 158–59, 218, 251–52, 409–14, 556, 576–82, 667, 668n–671n; on uniform acceleration and the acceleration of falling bodies, 205, 244, 251–52, 255, 258, 266, 346, 409–16, 542, 546, 555–56, 567, 575–82, 658, 665, 677–78; on qualities uniformly difform and uniformly difformly difform, 212, 252–53, 396; on the earth's rotation, 588n; other works of, 156, 158, 668n–669n; mentioned, xix, xxiii–xxv, 101, 365, 418, 563–64, 629

Gauthier, L., 432n, 439n
Geminus, 165n, 598
George of Brussels, 638
Gerard of Brussels: and the development of kinematics, 199, 206, 676–77, and Chapter 3; his comparison of uniform velocities, 175, 217–18; his definition of the "quicker," 182; his *De motu*, xxv, 163, 184–85, 187–97; on the comparison of rectilinear and curvilinear motions, 182, 185–97, 216, 226, 261–62n; the dating of, 185; mentioned, xxv, 205
Gerard of Cremona, 69–70n, 164, 167, 170, 175–76n, 430n, 439n
Germany and Eastern Europe, mechanics in, 218, 248, 637, 640–44
Gernardus, 185
al-Ghazalī, 425n
Giacomelli, R., 667n
Gilson, E., 515n
Giovanni da Fontana, 651
Gloria, A., 72n, 498, 644n–645n, 648n–651n
Grabmann, M., 175n, 545n
Gradus motus (velocitatis): defined, in general, 210; defined by Brawardine, 217, 230–33; how measured by Oresme, 355–56, 364, 375–76; how represented by Dumbleton, 335; in a uniformly difform motion, 451–52, 461; its measure according to Swineshead, 243–45; related anonymously to uniform and uniformly difform motions, 326; used by Galileo, 251–52, 409–10, 412–14, 671n
Gradus qualitatis (e.g., caliditatis): Bacon's measure of, 334–35; defined by Casali, 382, 385–87; defined by Francischus de Ferraria, 498–99, 503; in a uniformly difform quality, 382–83, 387, 449–50, 458–59, 499, 502–3
Granger, F., 85n
Grant, E., xi 338n, 437n, 481
Gratia Dei, Johannes Baptista, 648
Gravity: accidental, 550–51, 561, 566; "actual" and "habitual," 139, 141, 145, 635; Aristotle on, 140; centers of, 11–14, 16, 19, 22, 31–37, 49, 58–61, 147, 566, 569, 590–93, 597–99, 651n; dynamic definition of, 58–59; positional (*gravitas secundum situm*), 3, 74–77, 101n–102, 107–8, 676; specific, 21, 64–65, 93–95, 113–45, *and see* Weight, specific; Trivisano on, 146–49; *and see* Weight *and* Heaviness
Gregory of Rimini, 645
Gregory XI, Pope, 644
Grosseteste, Robert, 333, 361, 545
Gunther, S., 642n

Hall, A. R., 580n
Hart, I. B., 661n
Haskins, C. H., 71
Hauréau, M., 515n
Heat: intensity of, and quantity of, 212, 333–35; reaction, 648–49n
Heath, T. L., 11n, 31n, 85n, 169n, 173n, 427n, 429n–431n, 505n, 507n, 542n
Heaviness and heavy body: Albert of Saxony on, 139, 141–42, 145; Arabic definitions of, 22, 52, 58–60, 65; Aristotle on, 65, 97, 140–41; in *Book on the Balance*, 24; Trivisano on, 97, 146–47; *and see* Gravity
Heiberg, J. L., 31n, 49n, 54, 172n, 543n, 584n
Heidelberg, 636, 642
Hennon, John, 638
Henry of Hesse, 636–37
Heraclides of Pontus, 583–84n
Heraclitus the Pre-Socratic, 168
Hernannus de Curis, 641
Hero of Alexandria: and the bent lever, 15, 43–44, 100, 107, 158; and the paral-

lelogram of velocities, 5–6, 41, 48–49, 184; and virtual work, 8, 17–18, 47; cited by Ametus, 70n; his *Mechanics*, 14–19, 38–51, 84, 158, 175, 183–84, 433; on kinematics, 183–84; on weight and speed, 433; the five machines of, 9, 16, 45–46; mentioned, xxiii, 12, 30, 429n

Heytesbury, William: and instantaneous velocity, 210, 214, 218, 236, 252, 677; and motion uniformly difformly difform, 211; and the Merton acceleration theorem, 256, 262–63, 265–66, 270, 272–73, 277–78, 284–89, 414, 677; his influence in France, 443n, 654–55n, 657–58, in Italy, 269, 646–49, 651–52; life, 200–201; on changing rates of acceleration, 268, 274, 277; on uniform acceleration, 218, 236–37, 241–42, 677; on uniform motion, 218, 235–37; *Probationes* attributed to, 200–201, 263, 265–66, 274, 284–89, 403, 631n, 646; the *Regule solvendi sophismata* (including *Tria predicamenta de motu*) of, 200–201, 218, 235–42, 262, 270–83, 414, 646–48, 651n–652, 655n, 677; the *Sophismata* of, 635n; mentioned, 171, 199, 244, 441n, 630, 640

Hiero of Syracuse, 57, 64, 85
Hipparchus, 259, 514, 543–44, 580, 667
Hippias of Elis, 169, 184
Hölder, O., 12n
Holy Scriptures, 561, 588, 603–5, 608
Hoskin, M., xi
Hultsch, F., 15n, 85n, 90n, 169n
Hunt, R. N., 635n
Hydrometer, 90–91, 96, 138, 141, 143
Hydrostatics: Albert of Saxony and, 96, 136–45; Arabic, 52–57, 62, 64–65; in the *Carmen de ponderibus*, 85–92; in the *Mappae clavicula*, 92–93; in Pseudo-Archimedes' *De insidentibus*, 93–96; Johannes de Muris and, 95–96, 113–35; medieval, xxiv–xxv, 21, 84–97, 673–74; *and see* Archimedes, *On Floating Bodies*
Hypatia, 91

Ibn al-Haitham, 19n, 50, 58
Ibn Rushd. *See* Averroës
Ibn Sīnā, 425n, 510–14, 547
Ihde, A., xi
Impact, 425n, 539n, 546, 564, 576, 579
Impetuosité, 552–53, 570

Impetus: Benedetti and, 663–65; Bricot on, 638–40; Buridan's theory of, xxvii–xxviii, 522–25, 532–40, 561, 678–81, *and* Chapters, 8, 9 and 11, *passim*; Celaya on, 655–56; Coimbra Jesuits on, 659; Copernicus and, 611, 613, 643; Coronel on, 655–57; Da Vinci on, 661–62; Dominicus de Clavasio on, 635–36n; Dullaert on, 653–54; Galileo, *impeto*, and, 152–54, 576–77, 587, 666–68; Hennon on, 638; Henry of Hesse on, 637; Lawrence of Scotland and, 637–38; Major on, 653; Nicholas of Cusa and, 643–44; Oresme's theory of, 66, 525, 552–53, 635, 681; other treatments of, 643, 663; Peter of Candia and, 638; Picus on, 659–60; Tartaret on, 640; Thierry of Chartres and, 515n; Vernias on, 659

Impulse and impulsion, 425n, 518n–519, 539n, 549, 576, 611, 613, 643

Inclined plane, force on: Galileo and, 18, 49, 102, 152–55, 158–59; Hero and, 15, 41–42, 49; Jordanus and, 15, 49, 74–75, 83, 106–8, 159, 676; Pappus and, 15n, 18, 49, 152; Stevin and 49, 102

Inclined plane, motion on: Galileo and, 556, 575, 578–79, 581–82; Leonardo and, 575

Inertia: Aristotle and, 426n; Benedetti and, 663–65; Galileo and, 152, 158–59, 666–71n; in the *Mechanica* of Pesudo-Aristotle, 48; Newton and, 508, 512, 525; Ockham and, 519–20; mentioned, xxviii

Infinite: definitions of, 230, 233; in relation to motion, 257, 266, 451, 453, 460; series, 266–67, 271–72, 276, 279, 291, 295–96, 300, 344, 366–67, 380–81, 657–58n; space, 590, 623, 625, 666

Ingraham, M., ix
'Inḥirāf (tendency), 42
Intelligences, as movers, 520, 524, 530–31, 536, 540

Intension and remission: of forms and qualities, 206, 212, 448–50, 453, 457–59, 645–46, 650; of motion, 211, 236–37, 241–43, 245, 248–49, 251–52, 257, 451, 460–61; *and see* Chapters 4–6, *passim*

ἰσχύς, 5, 425n

Italy, mechanics in, 331–32, 588, 644–52, 659–71

'I'timād (propension), 511–12

Jacobus de Sancto Martino and his *De latitudinibus formarum*, 253, 345–46, 365, 392–401, 404, 641
Jacopo da Forlì, 648–52
Jadhbu (attraction), 433
James of Venice, 175
James, M. R., 634n
Jansen, B., xxi, 518n
Jöcher, C. G., 204n
Johannes de Casali. *See* Casali
Johannes de Chymiacho (Chilmark?), 632n
Johannes Dorp, 642
Johannes de Muris: his *Quadripartitum*, 93n, 95–96, 113–35, 141; mentioned, xxiv, 90, 674
Johannes de Tinemue, 170–71n
Johannes de Wasia, 637
John Chilmark, 631–32
John of Dumbleton: and Bradwardine's dynamics, 441n; and the Merton acceleration theorem, 256, 263, 266–67, 269, 287, 305–25; his life and works, 204–5; his *Summa de logicis et naturalibus*, 205, 305–25, 335, 362–63n, 396, 441n, 635n; influence in Italy, 648–49n, 651; summaries by Chilmark of, 632; mentioned, 199, 630, 652
John II, king of France, 338n
John of Holland: *De motu* and kinematics of, 218, 247–50, 261, 268, 274, 640, 646; his proof of the acceleration theorem, 268n–269n
Jordanus de Nemore: and acceleration, 258–61, 548–49, 551, 575, 678; and the bent lever, 3, 81–83, 100, 105–7, 158, 676; and the concept of positional gravity, 3, 74–75, 676; and the inclined plane, 3, 74–75, 83, 103, 106–8, 676; and the law of the lever, 3, 77–78, 104–5, 107, 676; his concept of virtual work or velocities, xx, xxiv, 3, 77–83, 103, 107, 675–76; his *Elementa de ponderibus*, 29, 72–78, 80–81, 99, 101, 441n; his *Liber philotegni*, 73n–74; life of, 72; the *Aliud commentum* and, 78–80, 100, 675; the Corpus Christi Version of the *Elementa* of, 99–100, 109–12; the *Liber de ponderibus* attributed to, 29, 73, 78, 84, 99, 101, 498, 503, 634n; the *De ratione ponderis* attributed to, 15, 49–50, 71–77, 80–84, 99–101, 104–8, 159, 258–60, 519, 548–49, 665, 676; the Pseudo-Euclid Version of the *Elementa* of, 99–100, 634n–635n; mentioned, 23
Jowett, B., 507n

Khanikoff, N., 12n, 50, 56n, 91n
al-Khāzinī: and centers of gravity, 12n, 19n, 22, 50, 58–61; his *Book of the Balance of Wisdom*, 12n, 19n, 22, 53-54, 56–58, 91; life of, 63–64; mentioned, xxiii
al-Kindī, 53, 439n, 441n
Kinematics: defined and distinguished from dynamics, 163, 204n, 206–9, 676; Greek, 164–84, 256–59; medieval, summarized, xxv–xxvi, 676–78, *and* Part II, *passim*
Koyré, A., x, xxii, 414, 418, 582n, 588n, 610n, 663n, 666–68n

Lagrange, J. L., xxiv, 8
Latitude: of forms and qualities, 219, 253, 348–50, 362–63, 369–71, 382–408 (*passim*), 449–50, 458–59, 499, 641–42, 647; of motion or velocity, 210, 237, 241–42, 244–45, 248–50, 270–74, 277–83, 305n, 451–52, 461; *and see* Intension and remission
Latitudinibus formarum, Tractatus de. See Jacobus de Sancto Martino
Laukwarī, 514n
Laurendando, Leonardo, 659
Lawrence of Scotland, 637–38, 642
Leibniz, G. W., 649
Leipzig, 637
Lenzen, V., 12n
Leonardo da Vinci: his statics, 50, 100, 662; on acceleration, 555, 564, 567, 572–75, 661, 678; on impetus, 661–62; mentioned, 592n
Lever principle or law: anonymous statement of, 22; Archimedes' proof of, 11–12, 14, 16, 31–37, 45, 156; Euclid's proof of, 9–10, 24–30; Galileo's proofs of, 102, 150–52, 156; Hero and, 15–16, 43–47, 156; in irregular or bent levers, xxiii–xxiv, 15, 43–44, 50, 81–83, 105–7; in the Corpus Christi Version, 109–12; Jordanus' proof of, 3, 74, 77–78, 104–5, 107, 676; al-Khāzinī's statement of, 61, 67; Pseudo-Aristotle and, 5–7, 156;

Thabit's proof of, 14, 10–29; *and see* Chapters 1–2, *passim*
Linea naturalium, 632n–633n
Little, A. G., 334n
Locatellus, Bonetus, 647n, 652
Lockert, George, 653
Logic, medieval, 201n–202, 284, 633–34, 641n
Lutz, C. A., 583n
Lyceum, 4, 140, 182, 258

McColley, G., 584n
MacCurdy, E., 572n, 661n–662n
Mach, E., 7, 11–12
Machines, the basic Greek, 4, 9, 16, 45–46
Magnet and magnetism, 537, 558, 562
Maier, A.: general contributions of, x, xxi–xxii, 424; her *An der Grenze*, 200n, 202n, 332n, 334n, 542, 549–50n, 553n, 562n; her *Die Vorläufer*, xxi, 96n, 439, 443n, 523n, 650n; her *Zwei Grundprobleme*, 206, 514n–515n, 519n–520, 526n, 530, 532n, 552–53n, 635n–637n, 659n, 663; other works cited, 333, 441n, 515, 519
Mail (inclination): Abū 'l-Barakāt on, 513–14, 520, 547–48; Avicenna on, 511–14, 547; in translation of Hero, 42; other authors on, 514n, 547–48
Maiocchi, R., 647n
Major, John, 653, 658
al-Ma'mūn, 57
Mappae clavicula, 92–93
Marchanova, Johannes de, 651–52
Marcolongo, R., 661n
Marcus Trivisano, 97, 140, 146–49, 591
Marliani, Giovanni, 269, 443, 652
Marsilius of Inghen: his *Abbreviata (Abbreviationes) phisicorum*, 636n, 642; life of, 636; on impetus, 636–37, 639; on instantaneous motion, 215n; on the kinematic measure of velocity, 209; on the ontology of motion, 590, 615–25; on reaction, 649n; mentioned, 99, 522, 651
Masino Codronchi (Messinus?), 647
Mass: as bulk, 94; modern concept of, 96, 539, 668–69n; *and see* Matter, quantity of
Mathematics: as applied to kinematics, 165, 167; in astronomy, 164; in mechanics, xxii–xxiii, 12, 103

Matter, quantity of, 94, 96, 522–23, 535, 539, 668, 679
Mayronis, Franciscus de, 584–85n
Menelaus, 54, 57, 64
Menut, A. D., 337n, 570n, 592n, 600n
Mersenne, Marin, 582n
Merton College: discussion of density at, 96–97; mechanics at, xxv–xxvi, 205, 210–12, 252, 261, 331–32, 443–44, 629–35, 644, 655, 676–77, *and see* Chapters 4–5, *passim*; schoolmen at, xxv, 199–206, 215, 248, 335, 361, 522, 631–34; theorem of uniform acceleration, *see* Acceleration, uniform
Merv, 63
Messinus, 646–48, 651–52
Method of exhaustion. *See* Archimedes
Meunier, F., 337n
Michael Scot, 175–76n, 428n, 430n, 542n, 592
Michalski, K., 522, 589n, 636n–637n, 642n–643
Michel, P. H., 164n, 169n
Minio-Paluello, L., 175n
Misino di M. Bonfantino, 648
Mixtures, weights of components in: 95–96, 121–24; *and see* Weight, specific
Moerbeke. *See* William.
Mogenet, J., 165n
Moles, 89, 555n
Momento, 150–54, 410, 577–78, 668n
Momentum and impetus, xxi, xxviii, 66, 523, 661n, 679, *and see* Impetus
Momigliano, F., 651n
Monselice, Giovanni Francesco da, 498
Montaigu, Collège de, 653
Moody, E.: and medieval statics, xxiii, 3, 436, *and* Chapters 1–2, *passim*; as translator of Heytesbury, 235, 271; his edition of Buridan's *De caelo*, xxii, 521n, 557n, 594n; his "Galileo and Avempace," 436n; his "Rise of Mechanism," 441n; on Buridan's life, 521n; on Jordanus and *impulsus*, 519; on medieval logic, 284n; on the distinction of dynamics from kinematics, 207; mentioned, x
Motion: as speed, 194, 210; intension and remission of, *see* Intension and remission; mixed, 5–6, 603, 614, 662; natural, xxvi, 423–24, 429–31, 445, 454, 514, Chapter 9; nature of, 422–23, 425–26, 445, 453–

54, 519–20, 540, 589–90, 615–25; of rotation and revolution, 166–67, 170, 194–97, 216–17, 221, 223–29, 231–34, 343, 355–56, 376, 613–14; quality and quantity of, 210, 212–14, *and see* Qualities; perpetual, 102; projectile, xxvii–xxviii, 427–29, Chapter 8, *and see* Impetus; relativity of its detection, 585, 594–95, 598, 601–2, 607, 612, 625, 681; uniform, 165–66, 173–75, 194, 211, 213–14, 231, 234–35, 237–38, 243, 245, 247–48, 251–52, 326, 342, 355, 357, 364–65, 375, 377, 451, 453–54, 460, 630n, 633, 657n, 677, Chapter 4, *passim*; uniformly difform, 211, 236–37, 241–42, 244–45, 247–49, 263, 342–43, 357, 365, 377, 451–52, 460–61, 497–98, 502–3, 630, 633–34, 647, 657n, 677, *and see* Acceleration, uniform; uniformly difformly difform, 211–12, 252–53, 394–96, 399–400; velocity of, *see* Velocity; *and see also* Dynamics *and* Kinematics

Murdoch, J., x, 200n

Naṣīr al-Dīn al-Ṭūsī, 514n
Navarre, Collège de, 337–39n
Neugebauer, O., 14n
Newton: and inertia, 508, 512, 525; his concept of force, xxvi, xxviii, 163, 548, 673, 680; mentioned, xix, 587
Nicholas of Cusa, xxv, 97–99n, 643–44, 652, 664
Nicholas of Dinkelsbühl, 641
Nifo, Agostino, 659–60
Nix, L., 15n, 50
Nominalism and Nominalists, 206, 519, 589, 615, 624–25, 658

Ockham, William: his distinction of dynamics from kinematics, 206–7, 437, 676; on projectile motion, 519–21, 540; on the nature of motion, 519–20, 589–90; mentioned 630n
Olivi, Peter John, 517–19
Omar Khayyam, 65
Oresme, Nicole: and kinematics, 218, 342–43, 354–57, 374–78, 442n–443n, 575; the concept of a closed mechanical system, 587, 666, 681; his concept of impetus, 66, 525, 552–53, 570–71, 635, 639, 681; his configuration system, xxi, xxvi, 297, 306–7, 310, 574–75, 637–38, 641, 646, 649, 651, 656, 677, *and particularly* Chapter 6; his *De caelo*, 552–54; his *De configrationibus qualitatum*, 171–72n, 347–81, Chapter 6, *passim*; his *Du ciel*, 463–64, 570, 592–93, 600–609; his proof of the acceleration theorem, 255, 343, 358–59, 366, 379–80, 677–78; his *Questiones de spera*, 553n, 608n; his scientific works listed, 338n–339n; his use of *velocitatio*, 211, 343, 356, 376, 552; life of, 331, 337–40; *On Euclid's Elements*, 331, 338n, 344, 581; on proportions, 480–81; on rotary motion and angular velocity, 217, 355–56, 365, 376, 442n; on the acceleration of falling bodies, 553–54, 563, 570–71, 678; on the dynamic laws of motion, 440, 442, 463–64; on the earth's rotation, 585–88, 600–609, 614; on the plurality of worlds, 592–93; mentioned, xxix, 522, 635–37, 652, 662
ὁρμή, 425n
Oxford, xx, xxvi, 199, 205, 262n, 332, 421, 632, 634–35, *and see* Merton College

Padua, 72, 332, 443, 495, 498–99, 644–46, 648–52
Pappus of Alexandria: and force on an inclined plane, 15, 18, 49, 152; and the quadratrix, 169; the hydrometer of, 91; mentioned 12
Paris, mechanics at, xx, xxvii, xxix, 205, 217, 255, 521, 584, 635–40, 653–59, 677–78, *and see* Buridan *and* Oresme
Parmenides, 168
Passau, Friedrich von, 641n
Paul of Venice, 650–51, 658
Pavia, 269, 645, 647–48, 650–52
Pendulum, 570–71
Percossa, 158, 576
Peripatetic law. *See* Aristotle, dynamics of
Peter de Forlivio, 649
Peter John Olivi, 517–19
Peter of Abano, 439n, 651
Peter of Candia, 638
Peter the Lombard and his *Sentences*, 204, 517, 520, 526–31
Peterhouse College, 634n–635n
Pfullendorf, Johann Fluck von, 641n
Phenomena, saving the, 520, 529, 535, 585–87, 589, 594–95, 598, 605, 609n

[Index   709]

Philip of Meaux, 338n
Phillipps, T., 92n
Philoponus, John: on Aristotle's laws of motion, xxvi, 433–36, 439, 675; on falling bodies, 546–47, 667; on projectile motion, 508–10, 512, 679; mentioned, 432
Piccolomini, Alessandro, 663
Picus, Hieronymus, 659–60
Pietro da Forosempronio, 648
Pines, S., 510n, 512–14n, 547n, 584n, 587n
Place, natural: Albert of Saxony and, 139, 142, 145; Aristotelian doctrine of, 140, 142, 423–24, 445–46, 454–55; proximity to, as a cause of acceleration, 542–45, 547, 558–59, 562, as a measure of acceleration, 542, 550, 558
Plato: on *antiperistasis*, 507; on the earth's rotation, 600, 606–7; spurious work on specific weight attributed to, 65
Plato of Tivoli, 170, 338n
Plurality or infinitude of worlds, 591–93
Ponderoso et levi, Liber de. *See* Euclid
Poole, R. L., 632n
Posidonius, 42, 50
Potentia, 425n, 427n, 436n, 438n, 440n–441n, 487–94, 500–501, 506n, 516, 539, 572
Powicke, F. M., 635n
Prague, 218, 247–48, 637, 640–41
Priscian, 85
Proportions, the medieval theory of, 167, 464–72, 480–87, *and see* Bradwardine
Pseudo-Aristotle. *See* Aristotle
Ptolemy, 74, 338n, 601, 611, 635n
Pulka, Peter von, 641n
Pythagoras and the Pythagoreans, 164

Quadratrix, 169
Qualities, their measure and kinematics, 206, 333–36, 340–42, 347–54, 357–64, 366–74, 378–81, 448–50, 458–59, Chapters 4–6, *passim*
Queen's College, 200, 205
Qūwat (force or power), 16, 27, 425n

Radulphus Strode, 634
Randall, J. H., 651
Ravaisson-Mollien, C., 572n, 661–662n
Rebound, 537, 660

Recaneto, Christopher de, 652
Rédeur, 553
Reimann, D., 12n
Relativity of motion. *See* Motion
Richard de Fournival, 73
Richard Swineshead. *See* Swineshead
Ritter, G., 636n
Rocco, Antonio, 582n
Roger Royseth, 204n
Roger Swineshead, 201–2
Roman balance, 13, 20–21, 83, 156, *and see De canonio and* Thabit ibn Qurra
ῥοπή, 425n, 430
Ross, W. D., 177n, 257n, 426n, 505n

Sachau, E. C., 584n
St. Andrews University, 637
Sainte-Barbe, Collège de, 655
Salusbury, Thomas, 150n, 156, 414
Salvio, A. de, 251n, 576n
Santillana, G. de, 156, 414, 588n
Sarpi, Paolo, 580
Sarton, G., 199n–200n, 584n, 641n–642n
Sbaralea, J., 644n
Scaliger, Julius Caesar, 649, 663
Schum, W., 202n–203n, 642n
Scot. *See* Michael
Scotus. *See* Duns
Scotus, Octavianus, 647n, 652
Serpens, Magister, 642
Sex inconvenientibus, Tractatus de: its proof of the acceleration theorem, 263–65n; on motion of rotation, 216, 262n; on the causes of the acceleration of natural motion, 551, 562n; mentioned, 99
Shapiro, H., 589n
Simplicius: on *antiperistasis*, 507–8; on the causes of acceleration, 514, 543–45, 562, 580; on the earth's rotation, 584; quoting Geminus, 598; quoting Strato on acceleration, 258–61, 545–46, 579
"Slowing up," principle of: 17–18, 47, 51; *and see* Virtual velocities *and* Work
Smith, T., xi, 334n–335n, 340n, 392n, 397n–398n
Socrates (Sortes), used as a name in discussions of motion, 249–50, 269n, 307–8, 310–12, 319, 322–24, 616
Soto, Domingo de: his knowledge of Merton kinematics, 658; on the acceleration of natural motion, 255, 257n,

555–56, 567, 678
Speculis, Auctor de, 635n
Speed, 194–97, 210, 213n, *and see* Velocity
Stadis, Magister de, 638n, 642
Staltwicht, 102, *and see* Gravity, positional
Stamm, E., 200n
Static moment, 12, 15, 19, 50, 107, 158, 668n, 676, *and see* Lever principle
Statics: Arabic, 18–23, Chapter 1, *passim*; early modern, 100–102, 150–59, 668n; Greek, xxiii, 3–18, Chapters 1–2, *passim*; medieval, xxiii–xxv, 102–3, 675–76, Chapter 2, *passim*
Steele, R., 515n
Stein, W., 12n
Stevin, Simon: and falling bodies, 546–47; and the dynamic concept of statics, xxiv, 101–2; and the inclined plane problem, xxiii, 49, 102
Stoezlin, Fredericus, 641
Strato of Lampsacus, as possible author of the *Mechanica*, 4, 182; *On Motion*, 182, 258, 545–46; on relative heaviness, 140; on the acceleration of falling bodies, 258–61, 545–46, 555, 564, 575; mentioned, 579
Strode, Radulphus, 634
Suchenschatz, Michael, 641n
Suisset (Suiseth). *See* Swineshead, Richard
Swineshead, John, 201–2
Swineshead, Richard: and Bradwardine's dynamics, 440n–441n; and the Merton acceleration theorem, 256, 263, 266, 287, 290–92, 294–302, 403, 631n; his fragments *On Motion*, 203–4, 208–9, 218, 243–46, 677; his distinction of dynamics from kinematics, 208–9, 676; his influence in Italy, 646, 649, 651–52, in France, 443n, 654–55n, 657, 659; his *Liber calculationum*, 96n, 202–3, 274, 290–304, 440n, 652, 655n, 659; life and works of, 201–4; on changing rates of acceleration, 268, 292–94, 297, 302–4, 443n; on density, 96n–97n; on instantaneous velocity, 215, 243–45, 252; on uniform motion, 243, 245, 677; mentioned, 199, 211, 269, 630, 640
Swineshead, Roger, 201–2
Synesius of Cyrene, 91

Tartaglia, Niccolò, 101, 159, 663, 665
Tartaret, Peter, 638, 640
Telesio, Bernardino, 663
Temperature. *See* Heat
Tempier, Etienne, 592
Thabit ibn Qurra, 14, 19–21, 55, 69–70, 99, 156
Themon Judaeus, 653
Theodosius, 635n
Theophrastus, 140, 258
Thierry of Chartres, 515n
Thomas of Dumbleton, 204
Thomas, Alvarus, 657–58
Thomas, Roger, 99
Thorndike, L., x, 201n, 203n, 337n, 339n, 585n, 645n–646n, 650–652n
Thost, Johannes de, 643
Thurot, C., xx, 74, 92n, 94n, 139
Time: as an element in kinematic equations, 165–66, 173–74, 177–81, 187, 208–9, 213–15, 217–18, 231, 233; as a variable in functions, 247–49, 251, 541, 549, 554–56, 563, 567, 574–75, 580–82, *and see* Acceleration *and* Velocity; definition of, 230, 232
Tisberus (Heytesbury), 647n
Tornius, Bernard, 652
Tossignano, Pietro Corialti da, 498
Toulous, 72
Trivet, Nicholas, 72
Trivisano. *See* Marcus

'Umar al-Khayyāmī, 65
Upton, Edward, 634

Vailati, G., xx, 7, 12, 19n, 50, 78n
Valsanzibio, P. S., da, 651n
Varahamihira, 584n
Velocitatio (acceleration), 211, 343, 356, 376, 551–52, 558
Velocity: as a magnitude, 217–18, 231–34; as a vector, 210, 213n; instantaneous, xxv, 210, 212, 214–15, 236, 240–41, 243–44, 251–52, 261, 342, 357, 365, 377, 677; its measure by the fastest moving point, 216, 221, 225, 228, 235, 238, 243, 245, 442n, 452, 461, 630n, 633n–634, 647n, 649; of rotation and revolution, 5, 38–40, 166–67, 170, 194–97, 216–17, 221–29, 231–34, 343, 355–56, 376, 452, 462, 651n; parallelogram of, 5, 41, 48; quality and quantity of, xxv, 210, 212–